Strategies for Sustainability

Series Editors
Lawrence Susskind
Ravi Jain

For further volumes:
http://www.springer.com/series/8584

Strategies for Sustainability

Aims and Scope

The series, will focus on "implementation strategies and responses" to environmental problems – at the local, national, and global levels. Our objective is to encourage policy proposals and prescriptive thinking on topics such as: the management of sustainability (i.e. environment-development trade-offs), pollution prevention, clean technologies, multilateral treaty-making, harmonization of environmental standards, the role of scientific analysis in decision-making, the implementation of public-private partnerships for resource management, regulatory enforcement, and approaches to meeting inter-generational obligations regarding the management of common resources. We will favour trans-disciplinary perspectives and analyses grounded in careful, comparative studies of practice, demonstrations, or policy reforms. We will not be interested in further documentation of problems, prescriptive pieces that are not grounded in practice, or environmental studies. Philosophically, we will adopt an open-minded pragmatism – "show us what works and why" – rather than a particular bias toward a theory of the liberal state (i.e. "command-and-control") or a theory of markets.

We invite Authors to submit manuscripts that:
Prescribe how to do better at incorporating concerns about sustainability into public policy and private action.

Document what has and has not worked in practice.

Describe what should be tried next to promote greater sustainability in natural resource management, energy production, housing design and development, industrial reorganization, infrastructure planning, land use, and business strategy.

Develop implementation strategies and examine the effectiveness of specific sustainability strategies. Focus on trans-disciplinary analyses grounded in careful, comparative studies of practice or policy reform.

Provide an approach "...to meeting the needs of the present without compromising the ability of future generations to meet their own needs," and do this in a way that balances the goal of economic development with due consideration for environmental protection, social progress, and individual rights.

The Series Editors welcome any comments and suggestions for future volumes

SERIES EDITORS
Lawrence Susskind
susskind@mit.edu
Professor Ravi Jain
rjain@pacific.edu

Abdul Malik • Elisabeth Grohmann
(Editors)

Environmental Protection Strategies for Sustainable Development

 Springer

Editors
Abdul Malik
Department of Agricultural Microbiology
Aligarh Muslim University
Aligarh 202002
India
ab_malik30@yahoo.com

Elisabeth Grohmann
Department of Infectious Diseases
University Hospital Freiburg
Hugstetter Straße 55
79106 Freiburg
Germany
elisabeth.grohmann@uniklinik-freiburg.de

ISBN 978-94-007-1590-5   e-ISBN 978-94-007-1591-2
DOI 10.1007/978-94-007-1591-2
Springer Dordrecht Heidelberg London New York

Library of Congress Control Number: 2011932795

© Springer Science+Business Media B.V. 2012
No part of this work may be reproduced, stored in a retrieval system, or transmitted in any form or by any means, electronic, mechanical, photocopying, microfilming, recording or otherwise, without written permission from the Publisher, with the exception of any material supplied specifically for the purpose of being entered and executed on a computer system, for exclusive use by the purchaser of the work.

Printed on acid-free paper

Springer is part of Springer Science+Business Media (www.springer.com)

Preface

Sustainable development means that the needs of the present generation should be met without compromising the ability of future generations to meet their own needs. Sustainability is the key to preventing or reducing the effect of environmental issues. Environmental sustainability is the process of making sure current processes of interaction with the environment are pursued with the idea of keeping the environment as pristine as naturally possible based on ideal-seeking behavior. Ecosystems are dynamic interactions between plants, animals, and microorganisms and their environment working together as a functional unit. Ecosystems will fail if they do not remain in balance. No community can carry more organisms than its food, water, and shelter can accommodate. Food and territory are often balanced by natural phenomena such as fire, disease, and the number of predators. Each organism has its own niche, or role, to play. The environment of our planet is degrading at an alarming rate because of non-sustainable urbanization, industrialization and agriculture. Our air, water, land and food are polluted. Pollution rate has exceeded the manageable capacity of nature at many places. Almost 50% of the land is eroded and robbed of its fertility. The extent of damage done to the world's biological diversity and ecosystem cannot be assessed. Our renewable and non-renewable resources are being alarmingly exhausted due to increasing population pressure posing difficulty to manage threat to future generation. Environmental issues are receiving utmost attention and have been debated at various international forums e.g. the first Earth Summit held in Stockholm, Sweden in June 1972; the second one in Rio de Janeiro, Brazil in 1992. The European Council in Göteborg (2001) adopted the first EU Sustainable Development Strategy (SDS). This was complemented by an external dimension in 2002 by the European Council in Barcelona in view of the World Summit on Sustainable Development in Johannesburg (2002), the Montreal and Kyoto Protocols etc. The European Council of June 2006 adopted an ambitious and comprehensive renewed Sustainable Development Strategy (SDS) for an enlarged EU. The European Commission adopted in October 2007 the first progress report on the Sustainable Development Strategy and in July 2009 reviews of EU SDS. Declarations of far reaching consequences were made at these summits. But current approaches in environmental protection have shifted from the end of pipe mitigation to zero emission strategies and to 3 R's: reduce, reuse and recycle waste.

Every nation desires economic growth, and at the same time it craves for eco-conservation and sustainable development. Administrative authorities are required to frame plans, programs and policies for a better scientific and technological development of production, distribution and consumption processes with sustainability. Green technology concepts are emerging as the future strategy for environmental management. It has become a challenge for scientists to devise remedial measures to control pollution levels and safeguard the future.

The book has seventeen chapters and attempts to present balanced accounts on various strategies for sustainable development. Chapter 1 gives an overview of different environmental protection strategies for sustainable development, while Chap. 2 describes the potential of rhizospheric microorganisms in the sustainable plant development in anthropogenic polluted soils. Other chapters deal with bioremediation of pesticides from soil and wastewater, remediation of toxic metals from soil, biological treatment of pulp and paper industry wastewater, sustainable solutions for agro processing waste management, impact of solid waste management on climate change and human health, environmental impact of dyes and its remediation. Various methods for genotoxicity testing of environmental pollutants are also discussed. Separate chapters have been devoted to molecular detection of resistance and transfer genes in the environmental samples, and biofilm formation by the environmental bacteria. Chapters on biochemical attributes to assess soil ecosystem sustainability, application of rhizobacteria in biotechnology, role of peroxidases as a tool for the decolorization and removal of dyes are also included. The role of biopesticides in sustainable agriculture is dealt in detail in the last chapter.

The book is not an encyclopedic review. The chapters incorporate both theoretical and practical aspects, and serve as baseline information for future research through which significant developments can be expected. This book will be of great interest to the research scientists, students, teachers, and environmental engineers working in the area of Environmental microbiology, ecotoxicology, soil microbiology, biotechnology and agricultural microbiology and would also serve as a valuable resource for environmental regulatory and protection agencies.

With great pleasure, we extend our sincere thanks to all our well-qualified and internationally renowned contributors from different countries for providing the important, authoritative and cutting edge scientific information/technology to make this book a reality. All chapters are well illustrated with appropriately placed tables and figures and enriched with up to date information. We are also thankful to the reviewers who carefully and timely reviewed the manuscript.

We are extremely thankful to Springer, Dordrecht, the Netherlands for completing the review process expeditiously to grant acceptance for publication. We appreciate the great efforts of the book publishing team especially Tamara Welschot, Senior Publishing Editor Environmental Sciences and Prof. Ravi Jain, Series Editor, Strategies and Sustainability in responding to all queries very promptly.

We express sincere thanks to our family members for all the support they provided, and regret the neglect and loss they suffered during the preparation of this book.

Abdul Malik
Elisabeth Grohmann

About the Editors

Dr. Abdul Malik is Chairman, Department of Agricultural Microbiology, Aligarh Muslim University, Aligarh, India. He received his Ph.D. Degree in Agricultural Microbiology from Aligarh Muslim University, Aligarh, India. Dr. Malik has been awarded several Fellowships and honors including DAAD fellowship (Germany), BOYSCAST overseas Fellowship, Jawahar Lal Nehru Memorial Fund award etc. He has been awarded Young Scientist project of Department of Science & Technology, Govt. of India twice. Dr. Malik has visited Technical University Berlin as Visiting Scientist in 2006, 2008 and 2009 and Universitätsklinikum Freiburg im Breisgau, Germany in 2010 under DAAD research programme. He has collaborative projects with Technical University Berlin, Germany and University of Minho, Braga, Portugal. He has received research grants from several funding agencies of the Govt. of India including DST, CSIR, UGC as Principal Investigator. His major areas of Research are Environmental Microbiology, Ecotoxicology, conjugative plasmids in bacteria, degradation of pesticides. Dr. Malik has published a number of research papers in leading scientific journals with high impact factor.

Dr. Elisabeth Grohmann is Group Leader, Department of Infectious Diseases, University Hospital Freiburg, Germany. She received her Ph.D. degree in Molecular Biology and Biochemistry from Technical University, Graz, Austria. She completed her habilitation (Venia legendi) at the Faculty of Process Sciences/Department of Environmental Engineering of the Technical University, Berlin Germany. She has been awarded an Erwin-Schrödinger post doctoral fellowship. Up to February 2010 she was Molecular Biology Group Leader at the Technische Universität Berlin. She has received research grants from DFG, BMBF, EU, ESA, DLR etc. She has collaborative projects with Karl-Franzens-University, Graz, Austria, with universities and research centers in Belgium, USA, The Netherlands, Italy, Spain, Argentina, Chile, Norway, Mexico, Turkey and India. Her thrust area of research is conjugative plasmids (molecular mechanisms of DNA transfer) in bacteria. Dr. Grohmann has published a number of research papers in leading scientific journals with high impact factor in different research fields, bacteriology, environmental microbiology, biochemistry and molecular biology.

Contents

1 Environmental Protection Strategies: An Overview 1
Abdul Malik, Mashihur Rahman, Mohd Ikram Ansari, Farhana Masood
and Elisabeth Grohmann

**2 The Potential of Rhizosphere Microorganisms to Promote
the Plant Growth in Disturbed Soils** 35
Katarzyna Hrynkiewicz and Christel Baum

**3 Sustainable Solutions for Agro Processing Waste
Management: An Overview** 65
C. M. Ajila, Satinder K. Brar, M. Verma and U. J. S. Prasada Rao

4 Dyes—Environmental Impact and Remediation 111
Luciana Pereira and Madalena Alves

**5 Molecular Detection of Resistance and Transfer Genes
in Environmental Samples** 163
Elisabeth Grohmann and Karsten Arends

**6 Key Biochemical Attributes to Assess Soil
Ecosystem Sustainability** 193
Vito Armando Laudicina, Paul G. Dennis, Eristanna Palazzolo
and Luigi Badalucco

7 Methods for Genotoxicity Testing of Environmental Pollutants 229
Farhana Masood, Reshma Anjum, Masood Ahmad and Abdul Malik

8 Trends in Biological Degradation of Cyanobacteria and Toxins 261
Fatma Gassara, Satinder K. Brar, R. D. Tyagi and R. Y. Surampalli

9 Bioremediation of Pesticides from Soil and Wastewater 295
Reshma Anjum, Mashihur Rahman, Farhana Masood and Abdul Malik

ix

10 Isolation and Characterization of Rhizobacteria Antagonistic to *Macrophomina phaseolina* (Tassi) Goid., Causal Agent of Alfalfa Damping-Off 329
L. B. Guiñazú, J. A. Andrés, M. Rovera and S. B. Rosas

11 Biofilm Formation by Environmental Bacteria 341
Mohd Ikram Ansari, Katarzyna Schiwon, Abdul Malik and Elisabeth Grohmann

12 Biochemical Processes of Rhizobacteria and their Application in Biotechnology 379
M. S. Dardanelli, D. B. Medeot, N. S. Paulucci, M. A. Bueno, J. C. Vicario, M. García, N. H. Bensi and A. M. Niebylski

13 Pulp and Paper Industry—Manufacturing Process, Wastewater Generation and Treatment 397
Saima Badar and Izharul Haq Farooqi

14 A Review of Environmental Contamination and Remediation Strategies for Heavy Metals at Shooting Range Soils 437
Mahtab Ahmad, Sang Soo Lee, Deok Hyun Moon, Jae E. Yang and Yong Sik Ok

15 Peroxidases as a Potential Tool for the Decolorization and Removal of Synthetic Dyes from Polluted Water 453
Qayyum Husain and Maroof Husain

16 Solid Waste Management Options and their Impacts on Climate Change and Human Health 499
Muna Albanna

17 Potential of Biopesticides in Sustainable Agriculture 529
M. Shafiq Ansari, Nadeem Ahmad and Fazil Hasan

Index 597

Contributors

C. M. Ajila INRS-ETE, Université du Québec, 490, Rue de la Couronne, G1K 9A9, Québec, Canada

Muna Albanna Water and Environmental Engineering Department, German Jordanian University, P.O. Box: 35247, Amman 11180, Jordan
e-mail: Muna.Albanna@gju.edu.jo

Madalena Alves IBB-Instituto Biotecnologia e Bioengenharia, Centro Engenharia Biológica, Universidade do Minho, Campus de Gualtar, 4710-057 Braga, Portugal
e-mail: madalena.alves@deb.uminho.pt

J. A. Andrés Laboratorio de Interacción Planta-Microorganismo, Facultad de Ciencias Exactas, Físico-Químicas y Naturales, Universidad Nacional de Río Cuarto, Ruta 36 Km 601, Río Cuarto, Argentina

Reshma Anjum Department of Agricultural Microbiology, Faculty of Agricultural Sciences, Aligarh Muslim University, Aligarh-202002, India
e-mail: reshmabiotec@gmail.com

Mohd Ikram Ansari Department of Agricultural Microbiology, Faculty of Agricultural Sciences, Aligarh Muslim University, Aligarh-202002, India
e-mail: ikram_ansari21@yahoo.com

M. Shafiq Ansari Department of Plant Protection, Faculty of Agricultural Sciences, Aligarh Muslim University, Aligarh-202002, India
e-mail: mohdsansari@yahoo.com

Karsten Arends Department of Environmental Microbiology, Technical University Berlin, Franklinstrasse 28/29, 10587 Berlin, Germany
e-mail: karsten.arends@tu-berlin.de

Luigi Badalucco Dipartimento di Ingegneria e Tecnologie Agroforestali, University of Palermo, Palermo, Italy
e-mail: badalucc@unipa.it

Saima Badar Environmental Engineering Section, Department of Civil Engineering, Aligarh Muslim University, Aligarh-202002, India

Christel Baum Institute of Land Use, University of Rostock, Justus-von-Liebig-Weg 6, 18059 Rostock, Germany
e-mail: christel.baum@uni-rostock.de

N. H. Bensi Departamento de Biología Molecular, Facultad de Ciencias Exactas, Físico-Químicas y Naturales, Universidad Nacional de Río Cuarto, Ruta Nacional N 36, Km. 601, CP X5804BYA Río Cuarto, Córdoba, Argentina

M. A. Bueno Departamento de Biología Molecular, Facultad de Ciencias Exactas, Físico-Químicas y Naturales, Universidad Nacional de Río Cuarto, Ruta Nacional N 36, Km. 601, CP X5804BYA Río Cuarto, Córdoba, Argentina

M. S. Dardanelli Departamento de Biología Molecular, Facultad de Ciencias Exactas, Físico-Químicas y Naturales, Universidad Nacional de Río Cuarto, Ruta Nacional N 36, Km. 601, CP X5804BYA Río Cuarto, Córdoba, Argentina
e-mail: mdardanelli@exa.unrc.edu.ar

M. García Departamento de Biología Molecular, Facultad de Ciencias Exactas, Físico-Químicas y Naturales, Universidad Nacional de Río Cuarto, Ruta Nacional N 36, Km. 601, CP X5804BYA Río Cuarto, Córdoba, Argentina

Fatma Gassara INRS-ETE, Université du Québec, 490, Rue de la Couronne, G1 K 9A9, Québec, Canada
e-mail: mail: fatmagassara@yahoo.fr

Elisabeth Grohmann Department of Infectious Diseases, University Hospital Freiburg, Hugstetter Strasse 55, 79106 Freiburg, Germany
e-mail: elisabeth.grohmann@uniklinik-freiburg.de

L. B. Guiñazú Laboratorio de Interacción Planta-Microorganismo, Facultad de Ciencias Exactas, Físico-Químicas y Naturales, Universidad Nacional de Río Cuarto, Ruta 36 Km 601, Río Cuarto, Argentina
e-mail: lguinazu@exa.unrc.edu.ar

Fazil Hasan Department of Plant Protection, Faculty of Agricultural Sciences, Aligarh Muslim University, Aligarh-202002, India
e-mail: fazilento10@gmail.com

Katarzyna Hrynkiewicz Department of Microbiology, Institute of General and Molecular Biology, Nicolaus Copernicus University, Gagarina 9, 89-100 Torun, Poland
e-mail: hrynk@umk.pl

Maroof Husain Department of Microbiology, School of Medicine, University of Colorado Health Sciences Center, Aurora, Colorado, USA

Qayyum Husain Faculty of Applied Medical Sciences, Jazan University, Jazan, Post Box 2092, Kingdom of Saudi Arabia
e-mail: qayyumbiochem@gmail.com

Contributors xiii

Izharul Haq Farooqi Department of Civil Engineering, Environmental Engineering Aligarh Muslim University, Aligarh-202002, India
e-mail: farooqi_izhar@yahoo.com

Jae E. Yang Department of Biological Environment, Kangwon National University, Chuncheon 200-701, Korea

Vito Armando Laudicina Dipartimento di Ingegneria e Tecnologie Agroforestali, University of Palermo, Palermo, Italy

Mahtab Ahmad Department of Biological Environment, Kangwon National University, Chuncheon 200-701, Korea
e-mail: mah_tabee@yahoo.com

Abdul Malik Department of Agricultural Microbiology, Faculty of Agricultural Sciences, Aligarh Muslim University, Aligarh-202002, India
e-mail: ab_malik30@yahoo.com

Mashihur Rahman Department of Agricultural Microbiology, Faculty of Agricultural Sciences, Aligarh Muslim University, Aligarh-202002, India
e-mail: mashihur@gmail.com

Masood Ahmad Department of Biochemistry, Faculty of Life Science, Aligarh Muslim University, Aligarh-202002, India
e-mail: masoodahmad1952@gmail.com

Farhana Masood Department of Agricultural Microbiology, Faculty of Agricultural Sciences, Aligarh Muslim University, Aligarh-202002, India
e-mail: farhanamasud4@gmail.com

D. B. Medeot Departamento de Biología Molecular, Facultad de Ciencias Exactas, Físico-Químicas y Naturales, Universidad Nacional de Río Cuarto, Ruta Nacional N 36, Km. 601, CP X5804BYA Río Cuarto, Córdoba, Argentina

Deok Hyun Moon Department of Environmental Engineering, Chosun University, Gwangju 501-759, Korea

Nadeem Ahmad Department of Plant Protection, Faculty of Agricultural Sciences, Aligarh Muslim University, Aligarh-202002, India
e-mail: nadeem777in@yahoo.com

A. M. Niebylski Departamento de Biología Molecular, Facultad de Ciencias Exactas, Físico-Químicas y Naturales, Universidad Nacional de Río Cuarto, Ruta Nacional N 36, Km. 601, CP X5804BYA Río Cuarto, Córdoba, Argentina

Eristanna Palazzolo Dipartimento di Ingegneria e Tecnologie Agroforestali, University of Palermo, Palermo, Italy

Paul G. Dennis Dipartimento di Ingegneria e Tecnologie Agroforestali, University of Palermo, Palermo, Italy

N. S. Paulucci Departamento de Biología Molecular, Facultad de Ciencias Exactas, Físico-Químicas y Naturales, Universidad Nacional de Río Cuarto, Ruta Nacional N 36, Km. 601, CP X5804BYA Río Cuarto, Córdoba, Argentina

Luciana Pereira IBB-Instituto Biotecnologia e Bioengenharia, Centro Engenharia Biológica, Universidade do Minho, Campus de Gualtar, 4710-057, Braga, Portugal
e-mail: lucianapereira@deb.uminho.pt

U. J. S. Prasada Rao Department of Biochemistry and Nutrition, Central Food Technological Research Institute, Mysore 570020, India

S. B. Rosas Laboratorio de Interacción Planta-Microorganismo, Facultad de Ciencias Exactas, Físico-Químicas y Naturales, Universidad Nacional de Río Cuarto, Ruta 36 Km 601, Río Cuarto, Argentina

M. Rovera Laboratorio de Interacción Planta-Microorganismo, Facultad de Ciencias Exactas, Físico-Químicas y Naturales, Universidad Nacional de Río Cuarto, Ruta 36 Km 601, Río Cuarto, Argentina

Satinder K. Brar INRS-ETE, Université du Québec, 490, Rue de la Couronne, G1K 9A9 Québec, Canada
e-mail: satinder.brar@ete.inrs.ca

Katarzyna Schiwon, FG Umweltmikrobiologie/Genetik, Technische Universitaet Berlin, 10587 Berlin, Germany
e-mail: Katarzyna.Schiwon@tu-berlin.de

Sang Soo Lee Department of Biological Environment, Kangwon National University, Chuncheon 200-701, Korea

R. Y. Surampalli US Environmental Protection Agency, KS 66117 Kansas City, P.O. Box 17-2141 USA

R. D. Tyagi INRS-ETE, Université du Québec, 490, Rue de la Couronne, Québec, Canada G1 K 9A9

M. Verma Institut de recherche et de développement en agroenvironnement inc. (IRDA), 2700 rue Einstein, G1P 3W8 Québec, Canada

J. C. Vicario Departamento de Biología Molecular, Facultad de Ciencias Exactas, Físico-Químicas y Naturales. Universidad Nacional de Río Cuarto, Ruta Nacional N 36, Km. 601, CP X5804BYA Río Cuarto, Córdoba, Argentina

Yong Sik Ok Department of Biological Environment, Kangwon National University, Chuncheon 200-701, Korea
e-mail: soilok@kangwon.ac.kr

Chapter 1
Environmental Protection Strategies: An Overview

Abdul Malik, Mashihur Rahman, Mohd Ikram Ansari, Farhana Masood and Elisabeth Grohmann

Abstract Environment protection and sustainability are harmonious and sustainability can be achieved by protecting our natural resources. This chapter presents an overview of the different types of problems affecting the environment and recent advances in environmental protection strategies. The role and potential of rhizospheric microorganisms in plant growth in disturbed soils is presented. Agro-industrial wastes and municipal solid wastes management options are discussed; the various sustainable solutions are also highlighted. The health effects of dyes and their different remedial treatment process and also the potential of peroxidases for treatment of dyes are discussed. The resistance and transfer genes in microorganisms and their molecular detection methods are explained along with the ability of environmental bacteria to form biofilms. The biochemical attributes for the assessment of soil ecosystem sustainability and the various methods involved in genotoxicity testing of environmental pollutants are summarized. Pesticides bioremediation strategies from soil and wastewater and the biodegradation of cyanobacteria and their toxins are outlined. The cause of Alfalfa damping off and the characterization of the causal agent are discussed. The significance of biochemical compounds derived from legumes and rhizobacteria (rhizodeposits) with potential in biotechnology are explained. The pulp and paper industry is a big sector and generates large amounts of wastewater; the treatment processes are briefly presented. The contamination of shooting range soils with heavy metals is a matter of concern and the remediation processes are discussed. The chapter ends with the role of biopesticides in sustainable agriculture. Thus, whole work concluded the ill effects of different types of pollution and the waste generated by human activities in the environment. Current trends involved in the remediation and the technologies used for this purpose are presented in detail.

Keywords Agriculture • Bioremediation • Environmental protection • Pollutants • Sustainability

---

A. Malik (✉)
Department of Agricultural Microbiology, Faculty of Agricultural Sciences, Aligarh Muslim University, Aligarh-202002, India
e-mail: ab_malik30@yahoo.com

A. Malik, E. Grohmann (eds.), *Environmental Protection Strategies for Sustainable Development*, Strategies for Sustainability,
DOI 10.1007/978-94-007-1591-2_1, © Springer Science+Business Media B.V. 2012

1.1 Introduction

Sustainable development is now of primary importance as the key to future use and management of finite world resources. It recognizes the need for development opportunities while maintaining a balance between these and the environment. As stated by the UN Bruntland Commission in 1987, sustainable development should "meet the needs of the present generation without compromising the ability of future generations to meet their own needs, encompasses, e.g. keeping population densities below the carrying capacity of a region, facilitating the renewal of renewable resources, conserving and establishing priorities for the use of non-renewable resources, and keeping environmental impact below the level required to allow affected systems to recover and continue to evolve". Environmental sustainability can be viewed as balancing the "three pillars" of economic and social development with environmental protection.

Soil and water are basic resources essential for sustainable agriculture. But unfortunately very few people realize the importance of conserving and judiciously utilizing the soil as a basic resource by successfully managing the soil internal, external, renewable or non renewable inputs. Sustainable agriculture is defined as the successful management of resources for agriculture production to satisfy changing human needs while maintaining or enhancing the quality of the environment and conserving natural resources (CGIAR 1989). The key words found in the above definition are (a) resource management (b) changing human needs (c) environmental quality and natural resources. While keyword (b) is involved in the concept of agricultural productivity, keyword (c) is involved in the protection of the environment concept. Both concepts are combined in keyword (a) which refers to resource management. Therefore, the basic concept of sustainable is to define the combined concept of productivity (or profitability) and environmental soundness in farm management or in agriculture at the national level.

The World Commission on Environment and Development (1987) defined sustainability as "a system which can be considered sustainable if it ensures that today's economic development is not at the expense of tomorrow's development prospects".

Sustainability is not a fixed phenomenon; it changes with time, requirement, and places and so on. A system, which is sustainable today, may not be sustainable tomorrow and vice versa. The Rio declaration of 1992, while enunciating many principles for sustainability of environment and development, emphasized the following principles most relevant to sustainable agriculture.

- Human beings are at the centre of concern for sustainable development and environmental issues are best handled with the participative co-operation of all concerned.
- No development can be considered complete unless it meets the criteria of productivity, equity and environmental safety for the present and future generations.
- To ensure development and sustainability, it is necessary to remove all the negative factors leading to unsustainability.

- Environmental policies must form an integral part of development policies and strategies.

The definition of "sustainable land use" is in consonance with "safe environment" or "environmental protection", because harmonization of all the land uses in a given area guarantees safe environment conditions in the sense of a safe protected habitat for "human and other living organism". Therefore sustainable land use creates and guarantees a safe environment by definition (Blum 1994).

Sustainable Agriculture has several positive points such as lesser use of pesticides and chemical fertilizers because when these are used in excess they not only pollute the environment but also cause health hazards as these pollutants might be reaching to the food grains, vegetables, fruits, fodder and milk. However, sustainable agriculture should be not equated with subsistence farming and perpetual low yields.

An alternative to increase agricultural productivity in a sustainable manner, there is increasing reliance on manipulation of microorganisms that benefit soil and plant health (Kloepper et al. 1989). Plant growth promoting rhizobacteria (PGPR) comprise a diverse group of rhizosphere colonizing bacteria and diazotrophic microorganisms which, when grown in association with a plant, stimulate growth of the host. PGPR can affect plant growth and development indirectly or directly (Glick 1995; Vessey 2003; Banchio et al. 2008). In indirect promotion, the bacteria decrease or eliminate certain deleterious effects of a pathogenic organism through various mechanisms, including induction of host resistance to the pathogen (Van Loon and Glick 2004; Van Loon 2007). In direct promotion, the bacteria may provide the host plant with synthesized compounds; facilitate uptake of nutrients; fix atmospheric nitrogen; solubilize minerals such as phosphorus; produce siderophores, which solubilize and sequester iron; synthesize phytohormones, including auxins, cytokinins, and gibberellins, which enhance various stages of plant growth; or synthesize enzymes that modulate plant growth and development (Lucy et al. 2004; Gray and Smith 2005).The large-scale application of PGPR to crop as inoculants would be attractive as it would substantially reduce the use of chemical fertilizers and pesticides, which often pollute the environment. This has a heavy impact on the natural and human environment, as well as on human health, through the pollution of soils, waters, and the whole food supply chain.

The use of arbuscular mycorrhizal fungi in ecological restoration projects has been shown to enable host plant establishment on degraded soil and improve soil quality and health (Jeffries et al. 2003; Siddiqui et al. 2008). Disturbance of native plant communities in desertification-threatened areas is often followed by degradation of physical and biological soil properties, soil structure, nutrient availability and organic matter. When restoring disturbed land it is essential to not only replace the above ground vegetation but also the biological and physical soil properties (Jeffries et al. 2003). A relatively new approach to restore land and protect against desertification is to inoculate the soil with arbuscular mycorrhizal fungi (AMF) with the reintroduction of vegetation. The benefits observed were an increased plant growth and soil nitrogen content, higher soil organic matter content and soil aggregation. The improvements were attributed to the higher legume nodulation in the presence of AMF, better water infiltration and soil aeration due to soil aggregation. Inoculation with native AM fungi increased plant uptake of phosphorus, improving plant growth and health. AM fungi can contribute to plant growth, particularly in disturbed or heavy metal contaminated sites, by increasing plant access to relatively immobile minerals such as P (Vivas et al. 2003; Yao et al. 2003; Gosling et al. 2006), improving soil texture by binding soil particles into stable aggregates that resist wind and water erosion (Rillig and Steinberg 2002; Steinberg and Rillig 2003; Siddiqui et al. 2008) and by binding heavy metals into roots that restricts their translocation into shoot tissues (Dehn and Schuepp 1989; Kaldorf et al. 1999; Gosling et al. 2006). Furthermore, the fungi can accelerate the revegetation of severely degraded lands such as coal mines or waste sites containing high levels of heavy metals (Marx 1975; Marx and Altman 1979; Gaur and Adholeya 2004).

For successful biotechnological application in the field, the selection of microbial inoculums is very important as maximum on-site benefits will only be obtained from inoculation with efficient fungi and/or bacteria in compatible host/microorganism/site combinations. The basic criteria for subsequent selection and later application of microbial inoculum useful for plant-growth promotion is cultivability and fast multiplication of microorganisms. Information on critical factors influencing plant-microbe-pollutant interactions in soils could lead to an improved selection of microbial inoculum for a microbial-assisted bioremediation. Thus an improved fundamental knowledge of physiological traits of rhizosphere microorganisms and their impact on rhizosphere processes, which are especially relevant for the remediation of disturbed soils will be essential to allow an increased and successful use of microbial inoculum in the field.

Agro-industrial wastes can be generally organized into different categories, such as food processing wastes, energy crops and biofuel production wastes and crop residues. Agricultural wastes comprise almost 15% of total waste generated by each country (Hsing et al. 2001; Arvanitoyannis et al. 2006). It has been estimated that approximately 30% of global agricultural products are becoming residues and refuses. Large volumes of solid and liquid wastes are generated from the agro-processing industries. Waste from agriculture and food processing can become one of the most serious sources of pollution (Di Blasi et al. 1997; Monspart-Sényi 2007). The efficient utilization of agro-waste can lead to improvement in agricultural yield and environmental health by decreasing the pollution caused by the agro wastes. The most commonly used methods by which the agro wastes are managed in different countries are: landfilling, incineration, composting and recycling.

Agro-industrial wastes and by-products (substances that originate during processing) can be further utilized in several other ways. Many by-products of the agro industry can be fed to animals directly as such without any modification or can be used after fermentation of the agro-residues. Recovery of by-products for use as animal feed can help agro industry save money by reducing waste discharges and can cut waste management costs and also can prevent environmental pollution. Microorganisms are grown on food processing by-products and utilized in the production of enzymes, single cell protein, amino acids, lipids, carbohydrates and organic acids. Agro-by-products can also be beneficially used as soil conditioner or fertilizer. Over the last century, energy consumption has increased beyond control as a result of growing world population and industrialization (Sun and Cheng 2002). The limited number of known fossil fuel deposits and the threat to environment due to the use of these fossil fuels has made it essential to look for alternative and renewable sources of fuels. Renewable energy sources, such as ethanol, methane, bio-hydrogen can be produced by fermentation of sugars. Owing to diminishing natural oil and gas resources, interest in the bioconversion of renewable cellulosic biomass into fuel ethanol as an alternate to petroleum is rising around the world (Stevenson and Weimer 2002; Reddy et al. 2010). Biomass is the earth's most attractive alternative among fuel sources and sustainable energy resource. Agro-industrial residues produce ethanol, bioethanol, a product of high potential value containing minor quantities of soluble sugars, pectin, proteins, minerals and vitamins. Bioethanol produced from renewable biomass has received considerable attention in current years. Using ethanol as a gasoline fuel helps to alleviate global warming and environmental pollution. They also have potential to produce biogas under anaerobic fermentation conditions. Biological conversion offers a potential for radical technical advances through application of the powerful tools of modern biotechnology to realize truly low costs. In the last few decades, vermicomposting technology has been arising as a sustainable tool for the efficient utilization of the agro-industrial processing wastes and to convert them into value added products for land restoration practices. The product of the process, i.e., vermicompost is humus like, finely granulated and friable material which can be used as a fertilizer to reintegrate the organic matter to the agricultural soils (Garg and Gupta 2009).

Adsorption process has been proven one of the best water treatment technologies around the world and activated carbon is undoubtedly considered as universal adsorbent for the removal of diverse types of pollutants from wastewater. However, widespread use of commercial activated carbon is sometimes restricted due to its higher costs. Attempts have been made to develop inexpensive adsorbents utilizing numerous agro-industrial and municipal waste materials. Use of waste materials as low-cost adsorbents is attractive due to their contribution in the reduction of costs for waste disposal, therefore contributing to environmental protection. Agricultural materials have cellulose, hemicelluloses, lignin, sugars, proteins, and starch containing various functional groups that facilitate metal complexion which in turn helps in the sequestration of heavy metals (Hashem et al. 2007; Bhatnagar and Sillanpää 2010).

By-products of plant food processing may be used because of their favourable nutraceutical properties. There are large varieties of value-added compounds in the by-products and wastes of biological origin. These products may be used as such or may serve as a starting material for the preparation of novel compounds like antioxidants, carbohydrates, dietary fibers, fat and oils, pigments, proteins and starch. Ensuring environmental safety and sustainable development through waste utilization aims to ensure that the development needs of the present do not compromise the needs of future generations (Ahmad et al. 2010).

A variety of synthetic dyestuffs released by the textile industry pose a threat to environmental safety. Azo dyes account for the majority of all dyestuffs produced, because they are extensively used in the textile, paper, food, leather, cosmetics and pharmaceutical industries. Dye-house effluent typically contains only 0.6–0.8 g L^{-1} dye, but the pollution it causes is mainly due to durability of the dyes in the wastewater (Jadhav et al. 2007). Existing effluent treatment procedures are unable to remove recalcitrant azo dyes completely from effluents because of their color fastness, stability and resistance to degradation. Therefore, it is necessary to search for and develop effective treatments and technologies for the decolorization of dyes in such effluents. Various physical/chemical methods, such as adsorption, chemical precipitation, photolysis, chemical oxidation and reduction, electrochemical treatment, have been used for the removal of dyes from wastewater. Moreover, there are many reports on the use of physicochemical methods for the color removal from dye containing effluents (Vandevivere et al. 1998; dos Santos et al. 2007; Wang et al. 2009a, b, c). Several physicochemical methods have been used for the removal of dyes from wastewater effluent. However, implementation of physical/chemical methods has the inherent drawbacks of being economically unfeasible (as they require more energy and chemicals), being unable to completely remove the recalcitrant azo dyes and/or their organic metabolites, generating a significant amount of sludge that may cause secondary pollution problems, and involving complicated procedures (Forgacs et al. 2004; Zhang et al. 2004). However, microbial or enzymatic decolorization and degradation is an eco-friendly cost-competitive alternative to chemical decomposition process that could help reduce water consumption compared to physicochemical treatment methods (Verma and Madamwar 2003; Rai et al. 2005).

The use of microorganisms for the removal of synthetic dyes from industrial effluents offers considerable advantages. The process is relatively inexpensive, the running costs are low and the end products of complete mineralization are not toxic. The various aspects of the microbiological decomposition of synthetic dyes have been previously reviewed by Stolz (2001). Besides the traditional wastewater cleaning technologies, other methods have been employed in the microbial decolorization of dyes. For instance, an activated sludge process was developed for the removal of Methyl violet and Rhodamine B from dyestuff effluents, using microorganisms that were derived from cattle dung (Kanekar and Sarnaik 1991). Also in biofilms, efficient biodegradation of Acid Orange 7 has been demonstrated (Harmer and Bishop 1992; Zhang et al. 1995). Azo dyes did not inhibit the capacity of biofilms in the removal of organics from wastewater (Fu et al. 1994). A multistage rotating biological contactor was used for the biodegradation of azo dyes, where an azo dye assimilating bacterium was immobilized in the system (Ogawa and Yatome 1990).

The emergence of bacterial antibiotic resistances as a consequence of the wide-scale use of antibiotics by humans has resulted in a rapid evolution of bacterial genomes. Mobile genetic elements such as transferable plasmids, transposons and integrons have played a key role in the dissemination of antibiotic resistance genes amongst bacterial populations and have contributed to the acquisition and assembly of multiple antibiotic resistance in bacterial pathogens (Tschäpe 1994; Salyers and Shoemaker 1994; Mazel 2006; Rahube and Yost 2010). Bacteria resistant to multiple antibiotics are not restricted to clinical environments but can easily be isolated from different environmental samples and food (Perreten et al. 1997; Feuerpfeil et al. 1999; Dröge et al. 2000). There is substantial movement of antibiotic resistance genes and antibiotic resistant bacteria between different environments. In assessing the antibiotic resistance problem, a number of factors can be identified which have contributed to the antibiotic resistance problem: the antibiotic itself and the antibiotic resistance trait (Levy 1997; Andersson and Hughes 2010). The genetic plasticity of bacteria has largely contributed to the efficiency by which antibiotic resistance has emerged. However, horizontal gene transfer events have no *a priori* consequence unless there is antibiotic selective pressure (Levy 1997). Since bacteria circulate between different environments and different geographic areas, the global nature of the problem of bacterial antibiotic resistances requires that data on their prevalence, selection and spread are obtained in a more comprehensive way than before (Shaw et al. 1993; Wright 2010). DNA probes and PCR-based detection systems allow us not only to analyze the dissemination of antibiotic resistance genes in the culturable fraction of bacteria but also to extend our knowledge to the majority of bacteria which are not accessible to traditional cultivation techniques (Smalla and van Elsas 1995; Heuer et al. 2002). Studies on the dissemination of the most widely used marker gene, *npt*II, in bacteria from sewage, manure, river water and soils demonstrated that in a high proportion of kanamycin-resistant enteric bacteria the resistance is encoded by the *npt*II-gene (Leff et al. 1993; Smalla et al. 1993; Lynch et al. 2004).

In recent years, approaches have been implemented to characterize the diversity and prevalence of resistance in soil bacteria-the soil antibiotic resistome-as an important reservoir of resistance (Wright 2007). Riesenfeld et al. (2004) investigated resistance in the soil, concentrating on unculturable organisms, bacteria that have yet to be characterized and thus underappreciated because of challenging culture conditions (Riesenfeld et al. 2004). By creating a functional metagenomic library (Handelsman 2004) in which cloned genomic fragments were expressed from DNA isolated directly from soil and selecting for resistance, traditional challenges associated with studying genes of unknown sequence were circumvented. Specifically, these functional analyses revealed novel antibiotic resistance proteins that were previously of unknown function and unrecognizable by sequence alone. Thus, this work not only allowed for the identification of aminoglycoside *N*-acetyltransferases, the *O*-phosphotransferases, and a putative tetracycline efflux pump but also a construct with a novel resistance determinant to the aminoglycoside butirosin (Riesenfeld et al. 2004). This shows the power of the functional metagenomic approach when applied to a search of activity with a highly selectable phenotype such as antibiotic resistance.

It is important to remark that several antibiotics are produced by environmental microorganisms (Waksman and Woodruff 1940). Conversely, antibiotic resistance genes, acquired by pathogenic bacteria through Horizontal Gene Transfer (HGT) have been originated as well in environmental bacteria (Davies 1997), although they can evolve later on under strong antibiotic selective pressure during the treatment of infections (Martinez and Baquero 2000; Martinez et al. 2007). To understand in full the development of resistance, we will thus need to address the study of antibiotics and their resistance genes, not just in clinics but in natural non-clinical environments also (Martinez 2008).

Understanding heavy metal resistance in natural ecosystems may help as well to understand antibiotic resistance in the environment. The elements involved in the resistance to heavy metals are encoded in the chromosomes of bacteria like *Ralstonia metallidurans* (Mergeay et al. 2003), which are well adapted for surviving in naturally heavy metals-rich habitats (e.g. volcanic soils). However, strong selective pressure due to anthropogenic pollution has made that these chromosomally-encoded determinants are now present in gene-transfer units, so that they can efficiently spread among bacterial populations (Silver and Phung 1996, 2005; Nies 2003). Similarly, antibiotic resistance genes that were naturally present in the chromosomes of environmental bacteria (D'Acosta et al. 2006; Wright 2007; Fajardo and Martinez 2008) are now present in plasmids that can be transferred to human pathogens. It has been highlighted that the contact of bacteria from human-associated microbiota with environmental microorganisms in sewage plants or in natural ecosystems is an important feature to understand the emergence of novel mechanisms of resistance in human pathogens (Baquero et al. 2008). A key issue for this emergence will be the integration of antibiotic resistance genes in gene-transfer elements (e.g. plasmids), a feature that is favoured by the release of antibiotics in natural ecosystems (Cattoir et al. 2008).

It was suggested by Rysz and Alvarez (2004) that ARGs themselves could be considered as environmental "pollutants", since they are widely distributed in various environmental compartments, including wastewater and sewage treatment plants (STPs), surface water, lagoon water of animal production areas, aquaculture water, sediments and soil, groundwater, and drinking water. Pruden et al. (2006) have also pointed out that ARGs may be thought as emerging "contaminants", for the public health problems resulting from the widespread dissemination of ARGs. So far, the methods used for detection, typing, and characterization of ARGs have covered, but not been limited to, specific and multiplex polymerase chain reaction (PCR), real-time PCR, DNA sequencing, and hybridization-based techniques including microarray.

The widespread pollution of soils is an increasingly urgent problem because of its contribution to environmental deterioration on a global basis (Bezdicek et al. 1996; Dick 1997; Lal 1997; van Beelen and Fleuren-Kemilá 1997). Until relatively recently, soil was widely regarded as just an environmental filter ensuring the quality of both water and atmosphere. However, in the context of the pursuit of sustainability, it is now recognised that soil is not only an effective de-contaminant of potential pollutants but that its chemical, physical and biological quality must be maintained (Hornick 1992; Parr et al. 1992). From the point of view of sustainability, a high-quality soil is a soil that is capable of producing healthy and abundant crops; decontaminating the water passing through it; not emitting gases in quantities detrimental to the environment; and behaving as a mature, sustainable ecosystem capable of degrading organic input (Doran and Parkin 1994; Gregorich et al. 1994; Brookes 1995; Pankhurst et al. 1995). This view clearly implies that diagnosis of soil pollution should be carried out on the basis of observed alterations in the soil properties controlling the behaviours described above, ideally in a way that allows any loss of soil quality to be quantified as well as identified qualitatively (Larson and Pierce 1991; Doran and Parkin 1994).

The concept of soil quality gives rise to more controversy than that of water or air quality. However, despite the difficulty in providing a definition, the maintenance of soil quality is critical for ensuring the sustainability of the environment and the biosphere. Literature exhibits a great number of soil quality indices for both agro-ecosystems and natural or contaminated soils. The book has reviewed some of the soil quality indices established up to date as well as of the parameters that make up them, and to offer a reflection on the lack of consensus concerning the use of these indices. We have focused on those indices including biological parameters. The most straightforward index used in the literature is the metabolic quotient (qCO_2) (respiration to microbial biomass ratio), widely used to evaluate ecosystem development, disturbance or system maturity. However, qCO_2 and other indices integrating only two parameters provide insufficient information about soil quality or degradation. For this, lately there has been a wide development of multiparametric indices that clearly establish differences between management systems, soil contamination or density and type of vegetation. These indices integrate different parameters, among which the most important are the biological and chemical ones, such as pH, organic matter, microbial biomass C, respiration or enzyme activities. The major part of multiparametric indices has been established based on either, expert opinion (subjective), or using mathematical statistics methods (objective).

Molecular indicators have not yet been used for soil quality indices establishment. However, the development of genomic, transcriptomic or proteomic methodologies could have importance in the evaluation of soil quality, not only in a diversity sense but also in a functional way. These methods can provide information about what is the role of specific microorganisms and their enzymes in key processes related to soil functionality. Despite of the great diversity of indices, they have never been used on larger scales, nor even in similar climatological or agronomic conditions. The lack of applicability of soil quality indices resides on: (i) poor standardization of some methodologies; (ii) some methods are out of reach in some parts of the world; (iii) spatial scale problems (soil heterogeneity); (iv) poor definition of soil natural conditions (climate and vegetation); and (v) poor definition of soil function to be tested for soil quality.

New chemicals are being added each year to the existing burden of toxic substances in the environment. This has led to increased pollution of ecosystems as well as deterioration of the air, water and soil quality. Excessive agricultural and industrial activities adversely affect biodiversity, threatening the survival of species in a particular habitat as well as posing disease risks to humans. Test systems that help in hazard prediction and risk assessment are important to assess the genotoxic potential of chemicals before their release into the environment or for commercial use.

Currently, standardized prokaryotic genotoxicity procedures include the Ames test, the umu-test and the SOS chromotest which are based on genetically engineered *Salmonella typhimurium* strains. Tests with eukaryotic cells or organisms might be more relevant for human and ecological risk assessment, but generally they are much more time-consuming. Several tests have been developed using the integrity of DNA as an unspecific endpoint of genotoxicity e.g. Comet Assay, Alkaline DNA-eluation assay, DNA alkaline unwinding assay, UDS-assay; the Comet assay probably the most cost-efficient test among these.

To date, numerous *in vivo* tests have been developed that take into account uptake and elimination, internal transport and metabolism of pollutants. *In vivo* methodologies for assessing genotoxicity as part of routine toxicity testing are now available, and it is time to move to a regulatory paradigm that includes an assessment of *in vivo* factors that determines genotoxic outcome, where possible integrating genotoxicity determinations into routine toxicity tests. *In vivo* mutation assays are based on the following principles: (1) selection of mutants based on enzymatic activity of an endogenous enzyme in cells isolated after exposure *in vivo* and then cultured *in vitro* (e.g., the *hprt* or *tk* assays in lymphocytes), (2) recovery of a reporter transgene that is subsequently tested for expression in a recipient cell *in vitro* (e.g., the *lacI* or *lacZ* transgenic rodent systems), or (3) antibody-based methods to identify structural alterations in cell or cell-surface proteins (e.g., the glycophorin A or T-cell receptor assays) (MacGregor 1994). Newer technologies such as transcriptomics, proteomics and metabolomics provide the opportunity to gain insight into genotoxic mechanisms and also to provide new markers *in vitro* and *in vivo*. There is also an increasing number of animal models with relevance to genotoxicity testing. These types of models will undoubtedly have an impact on genotoxicity testing in the future.

Cyanobacteria are a widespread group of organisms colonizing all ecosystems. They are common inhabitants of freshwater bodies throughout the world and several of them form surface scums (blooms). Under favourable conditions several species of cyanobacteria may become dominant in the phytoplankton of water bodies. Cell densities may reach many millions per litre (Chorus and Bartram 1999). Cyanobacteria are known to produce several metabolites significant from the public health perspective of acute exposure: lipopolysaccharides (Stewart et al. 2006), and cytotoxic, tumor promoting and enzyme inhibiting metabolites like cyclic depsipeptides, cyclic peptides (anabaenopeptinsandnostophycins), linear peptides (aeruginosins and microginins) (Bickel et al. 2001; Forchert et al. 2001; Welker and Von Dohren 2006). A neurotoxin non-protein amino acid (*N*-methylamino-L-alanine, BMAA), widely produced among cyanobacteria (Cox et al. 2005), has been associated with neurodegenerative disease such as Alzheimer's disease, amyotrophic lateral sclerosis/parkinsonism-dementia complex (Murch et al. 2004). Many cyanobacterial species can produce several categories of powerful toxins that are unique to this group of organisms, with the exception of saxitoxins. Cyanotoxin poisoning in humans was mainly caused by three toxic groups: microcystins (MCYSTs), cylindrospermopsin and anatoxin-a(ANA-a), and occurred through exposure to contaminated drinking water supplies (Annadotter et al. 2001; Falconer 2005), recre-

1 Environmental Protection Strategies: An Overview

ational waters (Chorus and Bartram 1999; Behm 2003), medical dialysis (Azevedo et al. 2002).

Cyanobacteria toxins have quickly risen in infamy as important water contaminants that threaten human health (Svrcek and Smith 2004). Toxins are introduced to the environment, in general, through the rupture of algal cells which may arise by the effect of certain substance used during water treatment (Carmichael 1992) and/or in different water management processes, e.g. in algaecide treatment of the natural medium, pumping of raw water, conveyance of raw water.

Microcystins are chemically stable in water and cannot be effectively removed by conventional water treatment processes. It is possible that the treatment process may cause cell lysis and release the intracellular metabolites containing toxins (Tsuji et al. 1997; Chow et al. 1999). The water treatment process for removing algae and microcystins from source water urgently needs to be studied.

Microcystins are normally present inside cyanobacterial cells and enter the surrounding water after cell lysis. The major route of detoxification of MC-LR is probably biodegradation (Lahti et al. 1997; Miller and Fallowfield 2001; Ishii et al. 2004). Interestingly, bacteria like *Pseudomonas* spp. isolated from the surface water of lakes, rivers and dams decreased microcystins. Takenaka and Watanabe (1997) isolated four kinds of bacteria classed in the genera of *Pseudomonas*, *Citrobacter*, *Enterobacter* and *Klebsiella* from the surface water of a Japanese lake where a heavy water bloom occurs every year, and tested the bacterial degradation ability of microcystin LR. They found that only a bacterium identified as *Pseudomonas aeruginosa* degraded microcystin LR. In some laboratory studies, dissolved microcystins have been rather resistant to degradation.

Biodegradation after a lag period of a few days or weeks has been the most important means of detoxification in laboratory experiments. Microcystins produced in natural waters can be degraded by indigenous microorganisms, although the process occasionally is slow and may require a period of adaption (Cousins et al. 1996).

A new biological treatment method of purification of source water quality in a eutrophic lake using indigenous enrichment microbes by artificial media was used by Ji et al. (2009). The test of algae and microcystins degradation by enrichment microbes on the artificial media (assembled medium, elastic medium and non-woven fabric medium) revealed that the average removal efficiency of chlorophyll-*a* was above 60%, and the removal effect of microcystins was 40–67%. Enrichment microbes on the artificial media could effectively degrade algae and microcystins in Lake Taihu. PCR based results showed that there were algae-lysing bacteria in the natural source water (Ji et al. 2009).

Pesticides are extensively used to increase agricultural production by preventing losses due to pests. However, some are among the highly persistent, toxic, and bioaccumulative contaminants in the environment generally referred to as persistent organic pollutants (POPs) (WHO 2009). These toxicants get into the human body through the food chain, and can cause serious health problems (Albert and Rendon 1988).

Some pesticides are known to resist biodegradation and therefore, they can be recycled through food chains and produce a significant bioaccumulation at the higher end of the chain (Shukla et al. 2006). For this reason, pesticide residue analysis in environmental samples has received increasing attention in the last few decades, resulting in numerous environmental monitoring programs in various countries for a broad range of pesticides. A common consequence of such persistent pollution is the contamination of surface waters with pesticide residues. This calls for urgent attention in two areas: (a) re-evaluation of environmental persistence and risks of currently registered and applied pesticides, and (b) thorough monitoring of potentially water-contaminating pesticides in surface waters and in natural bodies (Mukherjeez and Gopal 2002; Donald et al. 2007; Maloschik et al. 2007).

Biodegradation is a natural process, where the degradation of a xenobiotic chemical or pesticide by an organism is primarily a strategy for their own survival. Most of these microbes work in natural environment but some modifications can be brought about to encourage the organisms to degrade the pesticide at a faster rate in a limited period. This capability of microbes is sometimes utilized as technology for removal of contaminant from actual site. Knowledge of physiology, biochemistry and genetics of the desired microbe may further enhance the microbial process to achieve bioremediation with precision and with limited or no scope for uncertainty and variability in microbe functioning. Genes encoding enzymes for degradation of several pesticides, have been identified, which will provide new inputs in understanding the microbial capability to degrade a pesticide and develop a super strain to achieve the desired result of bioremediation in a short time (Singh 2008). Genetically modified microbes are used to enhance the capability of degradation. Yet, the use of genetic engineering for the use in environment is still controversial because an adverse genotype can be readily mobilized in the environment. In a development of technology for degradation following points should be taken care of i.e. (1) heterogeneity of contaminant, (2) concentration of contaminant and its effect on bio-degradative microbe, (3) persistence and toxicity of contaminant, (4) behaviour of contaminant in soil environment and (5) conditions favourable for biodegradative microbe or microbial population. The use of technology at the actual site requires (1) the knowledge of the natural bioprocess at the contaminated site, (2) detailed and valid data of microbial biodegradation developed in the laboratory, (3) monitoring of the onsite biodegradation process.

Most of the bioremediation technologies for the field are designed to remove the pollutant once it is generated or released into the environment. Usually, these technologies include, bioaugmentation (addition of organism or enzyme to the contaminant), biostimulation (use of nutrients to stimulate naturally occurring organisms), biofilters (removal of organic gases by passing air through compost or soil containing microorganism), bioreactors (treatment of contaminant in a large tank containing organism or enzyme), bioventing (involves the venting of oxygen through soil to stimulate the growth of natural microorganisms capable of degrading contaminant), composting (involves mixing of contaminant with compost containing bioremediation organisms) and landfarming (use of farming, tilling and soil amendment techniques to encourage the growth of bioremediation organism at contaminated site).

Forage legumes are essential for efficient animal-based agriculture worldwide. Besides providing high quality feed for livestock, they are a key component in the sustainability of crop-pasture rotations. Their value lies essentially in their ability to fix nitrogen (N_2) in symbiosis with root nodule soil bacteria, collectively called rhizobia. Microbial-based strategies that improve forage legume establishment and optimize N_2 fixation have been deployed worldwide through rhizobial inoculant technology (Catroux et al. 2001; Höfte and Alteir 2010). However, the study of rhizospheric bacteria for plant-growth promotion remains a challenge (Handelsman et al. 1990; Jones and Samac 1996; Xiao et al. 2002; Villacieros et al. 2003; Höfte and Alteir 2010).

Alfalfa, *Medicago sativa* L., is among the most prized of forages and is grown worldwide as a feed for all classes of livestock. It is one of man's oldest crops, and its cultivation probably predates recorded history. In addition to its versatility as a feed, alfalfa is well known for its ability to improve soil structure and, as a legume, is an effective source of biological nitrogen. Their symbiotic association with rhizobia makes the atmospheric nitrogen available for themselves and other crops in the rotation. Alfalfa (*Medicago sativa*) is an important forage crop owing to its unique characteristics: high yield of excellent quality forage, hydric-stress tolerance and good persistence. However, rapid seedling emergence and adequate pasture establishment are crucial to maximize its potential. Like most crops, alfalfa is attacked by many disease-causing organisms. Seedlings as well as seeds, stems, leaves, and roots of older plants all serve as food sources for a number of disease-causing organisms.

Seedling diseases caused by soil-borne pathogens, primarily *Macrophomina phaseolina* (Tassi) Goid and other Oomycetes, are a critical factor, which limits alfalfa establishment causing pre- and/or post-emergence seedling damping-off. Damping off is a name given to a condition where seeds are killed before germination (pre emergence) or seedlings (post emergence) are stunted or collapse and die. Seeds destroyed before germination are discolored and soft. After seed germination, symptoms include brown necrotic lesions along any point of the seedling. Lesions that girdle the young root or stem lead to plant death. Partially girdled plants, as well as those subject to continued root tip necrosis, may be stunted and yellowish in color to varying degrees.

The indiscriminate use of chemical fungicides is not recommended for the management of alfalfa diseases because of their collateral adverse effects on the environment, along with negative effects on animal and human health. Moreover, their efficacy has been reduced by the appearance of microbial resistance (Sanders 1984; Cook and Zhang 1985; Quagliotto et al. 2009) and their detrimental effect on the biological nitrogen fixation by rhizobia. A high-density sowing practice is usually the strategy followed by farmers to cope with alfalfa damping-off, but this approach significantly increases pasture establishment costs. In recent years, plant growth promoting rhizobacteria (PGPR) have been extensively examined for their role in biomanagement of pathogens. Although there are other bacterial species that have been recognized as biocontrol agents against soil-borne pathogens of agricultural crops (Glick et al. 1999), fluorescent pseudomonads make up a dominant population in the rhizosphere and possess several properties that have made them potential plant growth promoting biocontrol agents of choice. They have been reported for biological control of different fungal species such as *Rhizoctonia, Fusarium, Sclerotium, Pythium* and *Macrophomina* (Negi et al. 2005; Höfte and Alteir 2010). The broad spectrum of the antagonistic activity of pseudomonads is executed by the secretion of a number of metabolites including antibiotics (Haas and Keel 2003), volatile hydrogen cyanide (HCN) (Bhatia et al. 2003), siderophores (Gupta et al. 2002), lytic enzyme chitinases and β-1,3 glucanases (Lim and Kim 1995). Therefore, these attributes make fluorescent pseudomonads effective biocontrol agents. In addition, the use of *Bacillus* spp. (Handelsman et al. 1990) and *Streptomyces* spp. (Jones and Samac 1996; Xiao et al. 2002; Bakker et al. 2010) has been explored to control alfalfa seedling damping-off.

Bacterial biofilms are complex communities of microorganisms embedded in a self-produced matrix and adhering to inert or living surfaces (Costerton et al. 1999). Biofilms have been observed on a variety of surfaces and in a variety of niches, and are considered to be the prevailing microbial lifestyle in most environments. From a medical perspective, biofilm associated bacteria on implants or catheters are of great concern because they can cause serious infections. For the food industry in particular, the formation of biofilms on food and food processing surfaces, and in potable water distribution systems, constitutes an increased risk for product contamination with spoilage or pathogenic micro-flora (Carpentier and Cerf 1993; Donlan 2002). The development of biofilms can be seen as a five-stage process (Stoodley et al. 2002): (1) initial reversible adsorption of cells to the solid surface, (2) production of extracellular polymeric matrix substances resulting in an irreversible attachment, (3) early development of biofilm architecture, (4) maturation, and (5) dispersion of single cells from the biofilm. The bacterial phenotype in this differentiated, complex, mature biofilm differs profoundly from that in the planktonic population. One difference with major implications is the increased resistance of biofilm bacteria towards antimicrobial agents (Lewis 2001).

Fouling of nanofiltration (NF) and reverse osmosis (RO) membranes is a widespread problem that severely limits membrane performance in many water treatment applications (Ridgway and Flemming 1996). Biofouling is especially critical in wastewater reclamation because municipal wastewater effluents can contain significant concentrations of bacteria, dissolved organics, and nutrients (Ridgway et al. 1983; Bailey et al. 1974). When oligotrophic conditions prevail and/or residual disinfectant is maintained, bacterial growth and replication are minimal. Hence, fouling is almost certainly caused by deposition of cell debris and effluent organic matter onto the membrane surface, which leads to a continuous increase in pressure drop across the membrane (Subramani and Hoek 2008).When nutrients are abundant and/or there is no residual disinfectant bacterial growth and exopolymer production are most likely the dominant causes of fouling. In this case, viable bacterial cell deposition and nutrient concentration polarization in regions of stagnant cross-flow (such as where feed spacers contact the membrane surface) initiate fouling, but rapid biogrowth and exopolymer production increase module pressure drop by clogging feed spacer voids and increasing cross-flow drag (Vrouwenvelder et al. 2009a, b). Now a number of techniques are used to detect biofilm formation. Most commonly used technique are staining with dyes like crystal violet, fluorescein isothiocyanate (FITC) or tetramethyl rhodamine isothiocyanate (TRITC) and Cyanine (CY5) or the GFP labeling of the bacteria used in biofilm formation (Neu et al. 2001). Recently some other techniques have been used to directly visualize the biofilm in the environment like laser scanning microscopy (LSM), magnetic resonance imaging (MRI), scanning transmission X-ray microscopy (STXM), Raman microscopy (RM), surface-enhanced Raman scattering (SERS) and atomic force microscopy (AFM) (Neu et al. 2010; Ivleva et al. 2010; Wright et al. 2010).

The rhizosphere of a plant is a zone of intense microbial activity. Rhizobacteria that exert beneficial effects on plant growth and development are referred to as plant growth promoting rhizobacteria (PGPR) because their application is often associated with increased rates of plant growth, development and yield. PGPR can affect plant growth directly or indirectly. Indirect promotion of plant growth occurs when introduced PGPR lessen or prevent deleterious effects of one or more phytopathogenic organisms in the rhizosphere. The direct promotion of plant growth by PGPR may include the production and release of secondary metabolites such as plant growth regulators (phytohormones) or facilitating the uptake of certain nutrients from the root environment. Kloepper et al. (1991) indicated that different strains of PGPR can increase crop yields, control root pathogens, increase resistance to foliar pathogens, promote legume nodulation, and enhance seedling emergence. Co-inoculation of legumes with rhizobia and PGPR is even more effective for improving nodulation and growth of legumes.

It is well known that the rhizosphere is characterized by biological, chemical and physical interactions between plants, microorganisms and soil, and that these interactions have an effect on nutrition, growth and the general health of the plants and consequently on their productivity. Actually, several authors suggest that the plants control these interactions with the rhizodeposition or root exudation of organic compounds. Indeed, plant roots release a wide range of organic compounds and most of them are involved in the nutrient acquisition mechanisms (Neumann and Römheld 1999). Three broad types of rhizodeposits can be determined (Brady and Weil 1999). First, low-molecular-weight organic compounds are passively exuded by root cells, including organic acids, sugars, amino acids, and phenolic compounds. Second, high-molecular-weight mucilages actively secreted by root-cap cells and epidermal cells near apical zones form a substance called mucigel when mixed with microbial cells and clay particles. Third, cells from the root cap and epidermis continually slough off as the root grows or get digested by bacteria. These lysates enrich the rhizosphere with a wide variety of cell contents. High molecular weight compounds such as carbohydrates, proteins and enzymes and low molecular weight compounds such as organic acids, phenols and amino acids are exuded into the rhizosphere changing the physical, chemical and biological properties of the rhizosphere and enhance adaptation to particular environments (Jones et al. 2004, 2009).

Most of the legumes are used as forages, however during the last few years they have been found to contain medicines, nutraceuticals, and pesticides to promote their use as new bio-functional crops. These legumes are important sources of forage, nutritional supplements, medicines, pesticides, primary and secondary metabolites plus other industrial and agricultural raw material (Graham and Vance 2003). Legumes also provide essential minerals required by humans (Grusak 2002a) and produce health promoting secondary compounds that can protect against human cancers (Grusak 2002b; Madar and Stark 2002) and protect the plant against the onslaught of pathogens and pests (Dixon et al. 2002; Ndakidemi and Dakora 2003). In addition to their blood cholesterol-reducing effect (e.g. Andersen et al. 1984), grain legumes generally also have a hypoglycemic effect, reducing the increase in blood glucose after a meal and, hence, blood insulin. Legumes are, therefore, included in the diet of insulin-dependent diabetics (Jenkins et al. 2003). Genomics approaches, including metabolomics and proteomics, are essential to understanding the metabolic pathways that produce these antinutritional compounds and to eliminating these factors from the plant. The general theme of improvement of food and feed represents a clear vision for the future for legume genomics, as well as an emphatic statement directed primarily toward the public, who will be the ultimate beneficiary of genomic activities. This unified theme combines several areas of research. First, it recognizes the importance of grain legumes (also known as pulses) as essential sources of dietary protein for humans and animals, as well as health-related phytochemicals such as dietary fiber, hormone analogs, and antioxidants. Genomics provide essential tools to fully understand the molecular and metabolic basis of the synthesis of these compounds, to increase their content in seeds and pods, and to better manipulate interactions between the plant's genetic makeup and its environment. A focus on seeds also underscores the importance of the genomics of reproductive biology in the development of higher-yielding, more nutritious legume cultivars. With the advent of biotechnology many plant constituents have found their way in food, chemical and energy industries. Plants now are used as bioreactors for production of bioactive peptides, vaccines, antibodies and a range of enzymes mostly for the pharmaceutical industry. For the chemical industry, plants can be used to produce, e.g., polyhydroxybutyrate for the production of biodegradable thermoplastics, and cyclodextrins, which form inclusion complexes with hydrophobic substances (Altman 1999; van Beilen 2008).

The pulp and paper industry involves three basic areas: paper making, paper converting and printing. The pulp and paper industry converts fibrous raw materials into pulp, paper and paperboard. About 500 different chlorinated organic compounds have been identified including chloroform, chlorate, resin acids, chlorinated hydrocarbons, phenols, catechols, guaiacols, furans, dioxins, syringols, vanillins, etc. (Suntio et al. 1988; Freire et al. 2000). In wastewater, these compounds are estimated collectively as adsorbable organic halides (AOX). These compounds are usually biologically persistent, recalcitrant and highly toxic to the environment (Baig and Liechti 2001; Thompson et al. 2001). The toxic effects of AOX range from carcinogenicity, mutagenicity to acute and chronic toxicity (Savant et al. 2006).

Increased awareness of the harmful effects of these pollutants has resulted in stringent regulations on AOX discharge into the environment (Bajpai and Bajpai 1994; Deshmukh et al. 2009). There are several modifications made to reduce the generation of chlorinated organic compounds from bleach plant effluents using one or more of the following strategies: (1) removing more lignin before starting the chlorination, i.e., reducing the kappa number of unbleached pulp, (2) modifying the conventional bleaching process to elemental chlorine free bleaching (ECF) and total chlorine free bleaching (TCF). These methods may be physicochemical or biochemical in nature or a combination thereof (Bajpai and Bajpai 1994; Bajpai 2001). Though these modifications reduce chlorinated compounds, but the major drawback in most of these technologies is the ultimate disposal of sludge or concentrates which is more difficult and costly than the initial removal and separation. This necessitates consideration of developing economical and eco-friendly methods for removal of AOX compounds. Aerobic treatments are applicable where sufficient molecular oxygen is available. The rate of degradation is proportional to dissolved oxygen and therefore the process demands large inputs of energy, making it expensive. Anaerobic treatment is a technically simple, relatively inexpensive technology and consumes little energy. It also requires less space and produces less amount of sludge. Anaerobic microorganisms can be preserved unfed for long periods of time without any serious deterioration of their activity. The nutrient requirement for anaerobic treatment is low. It is less sensitive to toxic substances. Hence, it is proving to be a viable technology for pulp and paper wastewater treatment. The major treatment methods include anaerobic lagoon, anaerobic contact processes, Up-flow Anaerobic Sludge Blanket (UASB), Sequencing Batch Reactors (SBR), fluidized bed, anaerobic filters and hybrid processes. Now-a-days, high rate advanced anaerobic reactors are being increasingly used. Anaerobic treatment remains the most reliable and economically viable method of AOX removal at present.

Shooting ranges are of increasing environmental concern in many countries (Lin 1996; Mozafar et al. 2002; Sorvari 2007). High amounts of ammunition are often deposited in the soils of these sites due to their use for shooting sports or military activities. Depending on firing activities, considerable amounts of inorganic contaminants accumulate in shooting range soils and backstop materials. Due to the alloys used for bullets and jacket housings lead, antimony, arsenic, bismuth, silver, copper, and nickel may be present (Hardison et al. 2004; Johnson et al. 2005). An environmental risk at shooting ranges has been determined from contamination levels of groundwater and surface water (Sorvari et al. 2006; Heier et al. 2009), soil enzymatic activity (Lee et al. 2002), and accumulation of heavy metals into plant tissues (Labare et al. 2004), human being or other animals (Migliorini et al. 2004). Current technologies for treating lead-contaminated soils mainly include solidification/stabilization, vitrification, capping, secondary smelting, and soil washing.

Stabilisation of inorganic contaminants in soils is based on the modification of pollutant characteristics (e.g. speciation, valence) and soil properties (sorption capacity, buffering potential, etc.) by means of additives (Diels et al. 2002). These amendments induce or enhance physicochemical and/or microbial processes, which render pollutants less mobile and less bioavailable. Due to specific interactions with the constituents of the solid phase, cationic and anionic contaminants require different additives. Cation exchange capacity may be increased by addition of synthetic or natural clay minerals and iron oxides (McBride 1994; Lothenbach et al. 1999; Bigham et al. 2002). An alternative approach involves the addition of soluble salts, which provide anions to react with cationic contaminants forming leaching resistant minerals. A typical example is the addition of phosphate using commercially available phosphate fertilisers to stabilise heavy metals by precipitation of minerals with low solubility like chloropyromorphite (McGowen et al. 2001) and thereby minimise both plant uptake and leaching (Cao et al. 2002). Anion sorption capacity in soils of the temperate zone is primarily controlled by iron(III)- and aluminium(III)-(hydr)oxides like ferrihydrite, goethite, gibbsite, etc. Oxyanions like arsenate, chromate, molybdate, etc. as well as cations like cadmium, copper, lead, and zinc are sorbed specifically by these media (Richard and Bourg 1991; Bowell 1994; Martinez and McBride 1999; Trivedi et al. 2003). The resulting inner sphere complexes are resistant to competing anions/cations at typical levels in soil solutions. In addition, sorption may be accompanied by redox processes (Sun and Doner 1998) leading to less toxic contaminant species. Following the reduction of mobility also bioavailability can be expected to be reduced in stabilised soils.

In situ chemical immobilization is, in particular, a practical remediation technology that is capable of reducing cost and environmental impacts (Saikia et al. 2006). In situ chemical immobilization technologies can be employed in conjunction with a plant application as a form of phytostabilization technology that stabilizes the soil and prevents contaminant migration via wind and hydrological processes (Brown et al. 2003; Mench et al. 2003). Kucharski et al. (2005) used calcium phosphate to immobilize metal contaminants in soil with indigenous plant coverage that increased water retention in the soil and reduced the volume of metal-containing leachate. Use of plants in parallel with chemical immobilization technology is particularly important when the contaminated site is required to recover the vegetation that has been degraded by metal toxicity.

Stabilization/solidification (S/S) is gaining prominence in the treatment and remediation of hazardous wastes and contaminated soils due to its cost-effectiveness, rapid implementation and its use of well-established techniques (Palomo and Palacios 2003; Dermatas and Meng 2003; Terzano et al. 2005). There are various techniques currently used for S/S, including pozzolanic, (cementitious or solidifying) based solidification systems and chemical based stabilization systems. The most commonly applied pozzolanic materials are Portland cement, lime, and/or fly ash (Dermatas and Meng 2003; Palomo and Palacios 2003; Terzano et al. 2005). The pozzolanic-based S/S techniques immobilize contaminants by adsorption, incorporation into the pozzolanic products (e.g., calcium alumina hydrate C-A-H or calcium silicate hydrate C-S-H), or via precipitation as metal hydroxides under alkaline pH associated with cement and lime (Gougar et al. 1996; Moulin et al. 1999). Physical entrapment of heavy metals adsorbed to particle surfaces in a low permeability cementitious matrix is also very likely (Moulin et al. 1999; Badreddine et al. 2004). Chemical-based S/S approach is based on the formation of thermodynamically stable and insoluble precipitate end-products with the contaminants. More effective chemical additives include phosphates and Fe-Mn oxides (Ma et al. 1995; Hettiarachchi et al. 2000; Basta et al. 2001; Seaman et al. 2001; Cao et al. 2002; Scheckel & Ryan 2004). The use of readily available and cost-advantageous materials as immobilizing amendments becomes more significant when the remediation targets vast amounts of contaminated soil such as shooting ranges.

1 Environmental Protection Strategies: An Overview

Dyes are an important class of pollutants, and can even be identified by the human eye. Disposal of dyes in precious water resources must be avoided, however, and for that, various treatment technologies are in use. Treatment of synthetic dyes in wastewater is a matter of great concern. Several physical and chemical methods have been employed for the removal of dyes (Robinson et al. 2001). However, these procedures have not been widely used due to high cost, formation of hazardous by products and intensive energy requirement (Hai et al. 2007).

Among various methods adsorption occupies a prominent place in dye removal. The growing demand for efficient and low-cost treatment methods and the importance of adsorption has given rise to low-cost alternative adsorbents (LCAs). Extensive research has been directed towards developing processes in which enzymes are employed to remove dyes from polluted water (Bhunia et al. 2001; Shaffiqu et al. 2002; Torres et al. 2003; Lopez et al. 2004; Husain 2006). Biological treatment is the most common and widespread technique used in dye wastewater treatment (Zhang et al. 1998; Bromley-Challenor et al. 2000; van der Zee and Villaverde 2005; Frijters et al. 2006; Barragan et al. 2007; dos Santos et al. 2007). A large number of species have been used for decolouration and mineralization of various dyes. The methodology offers considerable advantages like being relatively inexpensive, having low running costs and the end products of complete mineralization not being toxic. The process can be aerobic, anaerobic or combined aerobic–anaerobic.

Bacteria and fungi are the two microorganisms groups that have been most widely studied for their ability to treat dye wastewaters. In aerobic conditions, enzymes secreted by bacteria present in the wastewater break down the organic compounds. The work to identify and isolate aerobic bacteria capable of degrading various dyes has been going on since more than two decades (Rai et al. 2005). A number of triphenylmethane dyes, have been found to be efficiently decolourized (92–100%) by the strain *Kurthia* sp. (Sani and Banerjee 1999a). Nevertheless, it is worthwhile pointing that synthetic dyes are not uniformly susceptible to decomposition by activated sludge in a conventional aerobic process (Husain 2006). Attempts to develop aerobic bacterial strains for dye decolourization often resulted in a specific strain, which showed a strict ability on a specific dye structure (Kulla 1981). Fungal strains capable of decolourizing azo and triphenylmethane dyes have been studied in detail by various workers (Bumpus and Brock 1988; Vasdev et al. 1995; Sani and Banerjee 1999b). Various factors like concentration of pollutants, dyestuff concentration, initial pH and temperature of the effluent, affect the decolourisation process.

In order to get better remediation of coloured compounds from the textile effluents, a combination of aerobic and anaerobic treatment is suggested to give encouraging results. An advantage of such system is the complete mineralization which is often achieved due to the synergistic action of different organisms (Stolz 2001). Also, the reduction of the azo bond can be achieved under the reducing conditions in anaerobic bioreactors (Brown and Laboureur 1983a) and the resulting colourless aromatic amines may be mineralized under aerobic conditions (Brown and Laboureur 1983b), thereby making the combined anaerobic–aerobic azo dye treatment system attractive. Thus, an anaerobic decolourization followed by aerobic post treatment is generally recommended for treating dye wastewaters (Brown and Hamburger 1987).

Oxidoreductive enzymes such as peroxidases and polyphenol oxidases are participating in the degradation/removal of aromatic pollutants (Klibanov et al. 1983; Dec and Bollag 1994). These enzymes can act on a broad range of substrates and can also catalyze the degradation or removal of organic pollutants present in very low concentration at the contaminated sites. In view of the potential of these enzymes in treating the phenolic compounds several microbial and plant peroxidases and polyphenol oxidases have been considered for the treatment of dyes but none of them has been exploited at the large scale due to low enzymatic activity in biological materials and high cost of purification (Bhunia et al. 2001; Shaffiqu et al. 2002; Verma and Madamwar 2002). The major reason that enzymatic treatments have not yet been applied on an industrial scale is the huge volume of polluted wastewater demanding remediation. Soluble enzymes suffer from certain drawbacks such as thermal instability, susceptibility to attack by proteases, activity inhibition, etc. (Husain and Jan 2000). An important disadvantage of using soluble enzymes in the detoxification of hazardous aromatic pollutants is that the free enzyme cannot be used in continuous processes. To overcome all these limitations enzyme immobilization is the best alternative to exploit the enzymes at the industrial level.

The potential advantages of enzymatic treatment as compared to microbial treatment are mainly associated to several factors; shorter treatment period; operation of high and low concentrations of substrates; absence of delays associated with the lag phase of biomass, reduction in sludge volume and ease of controlling the process (Lopez et al. 2002; Akhtar and Husain 2006). However, the use of soluble enzymes has some inherent limitations as compared to immobilized form of enzymes, which has several advantages over the soluble enzymes such as enhanced stability, easier product recovery and purification, protection of enzymes against denaturants, proteolysis and reduced susceptibility to contamination (Husain and Jan 2000; Zille et al. 2003; Matto and Husain 2006).

Numerous methods have been employed for the immobilization of peroxidases from various sources but most of the immobilized enzyme preparations either use commercially available enzyme or expensive supports, which increase the cost of the processes (Norouzian 2003; Husain 2006). Such immobilized enzyme systems cannot fulfill the requirements for the treatment of hazardous compounds coming out of the industrial sites.

Besides, the known techniques used for the immobilization of enzyme, physical adsorption on the basis of bioaffinity is useful as this process can immobilize enzyme directly from crude homogenate and thus avoid the high cost of purification. The ease of immobilization, lack of chemical modification and usually accompanying an enhancement in stability are some of the advantages offered by the adsorption procedures (Akhtar et al. 2005a, b; Kulshrestha and Husain 2006). Besides the mentioned advantages offered by the bioaffinity-based procedures, there is an additional benefit, such as proper orientation of enzyme on the support (Mislovicova et al. 2000; Khan et al. 2005). These supports provide high yield and stable immobilization of glycoenzymes/enzymes.

1 Environmental Protection Strategies: An Overview

Waste is an unavoidable by product of human activities. Rapid population growth, urbanization and industrial growth have led to increase in the quantity and complexity of generated waste and severe waste management problems in most cities of third world countries. The large quantity of waste generated necessitates a system of collection, transportation and disposal. It requires knowledge of what the wastes are comprised of, and how they need to be collected and disposed. Solid waste in general, comprises of municipal solid waste (MSW) which includes household and commercial wastes; agricultural waste; and non-hazardous industrial waste; and construction and demolition waste. When solid waste is disposed off on land in open dumps or in improperly designed landfills (e.g. in low lying areas), it causes the following impact on the environment: ground water contamination by the leachate generated by the waste dump, surface water contamination by the run-off from the waste dump, bad odour, pests, rodents and wind-blown litter in and around the waste dump, generation of inflammable gas (e.g. methane) within the waste dump. Some commonly used methods by which the waste could be managed are: land filling, incineration, composting and recycling.

Municipal Solid Waste Management involves the application of the principle of Integrated Solid Waste Management (ISWM) to municipal waste. ISWM is the application of suitable techniques, technologies and management programs covering all types of solid wastes from all sources to achieve the twin objectives of (a) waste reduction and (b) effective management of waste still produced after waste reduction. An effective system of solid waste management must be both environmentally and economically sustainable.

Solid waste incineration is another method of waste management. Trash is put into large incinerators which convert it into steam, gas, heat and ash. This process can sometimes be time intensive. However, it is effective in disposing hazardous waste. The most significant negative outcome of incineration is the emissions that result from combustion. This air pollution has both a harmful effect on the local area and on the planets climate (Buonanno et al. 2008). Greenhouse gas emissions (GHGs) mainly in the form of CO_2 and N_2O are the main contributors to climate change through incineration (Gutierrez et al. 2005).

Recycling involves (a) the separation and sorting of waste materials; (b) the preparation of these materials for reuse or reprocessing; and (c) the reuse and reprocessing of these materials. Recycling is an important factor which helps to reduce the demand on resources and the amount of waste requiring disposal by landfilling. Recycling of waste proves to be an effective management option because it does not involve the emission of many greenhouse gases and water pollutants. Aside from the traditional methods of waste management, biowaste has been used in the production of clean energy where it replaces coal, oil or natural gases to generate electricity through combustion. This waste-to-energy conversion process has been proved to be safe, environment friendly and reduces the incoming volume by 90%; the remaining ash is used as a roadbed material or as a landfill material.

Landfilling involves the controlled disposal of wastes on or in the earth's mantle. Landfills are used to dispose of solid waste that cannot be recycled and is of no further use, the residual matter remaining after solid wastes have been pre-sorted at a materials recovery facility and the residual matter remaining after the recovery of conversion products or energy. It is by far the most common method of ultimate disposal for waste residuals. Many countries use uninhabited land, quarries, mines and pits as landfill sites. Biological reprocessing methods like composting and anaerobic digestion are natural ways to decompose solid organic waste. Composting is nature's way of recycling organic wastes. Composting is a method of decomposing waste for desposal by microorganisms (mainly bacteria and fungi) to produce a humus-like substance that can be used as a fertilizer. This process converts waste which is organic in nature to inorganic materials that can be returned to the soil as fertilizer i.e. biological stabilization of organic material in such a manner that most of the nutrient and humus that are so necessary for plant growth are returned to the soil.

Health issues are associated with every step of the handling, treatment and disposal of waste, both directly (via recovery and recycling activities or other occupations in the waste management industry, by exposure to hazardous substances in the waste or to emissions from incinerators and landfill sites, vermin, odours and noise) or indirectly (e.g. via ingestion of contaminated water, soil and food). The main pathways of exposure are inhalation (especially due to emissions from incinerators and landfills), consumption of water (in the case of water supplies contaminated with landfill leachate), the foodchain (especially consumption of food contaminated with bacteria and viruses from landspreading of sewage and manure, and food enriched with persistent organic chemicals that may be released from incinerators). It is also important to remember that occupational accidents in the waste management industry can be relatively common, higher than national average for other occupations (HSE 2004), and often higher than the potential cases of adverse effects to the resident population investigated by epidemiological studies. The main cause of global warming is the increasing amount of greenhouse gases (CO_2, CH_4 and N_2O) in the atmosphere, a significant contribution comes from waste management practices (Smith et al. 2001).

Agriculture and forests form an important resource to sustain global economical, environmental and social system. Their protection against pests is a priority and due to the adverse impact of chemical insecticides, use of biopesticides is increasing (Marrone 1999). A number of biopesticides (bacteria, fungi, virus, pheromones, plant extracts) have been already in use to control various types of insects responsible for the destruction of forests and agricultural crops. *Bacillus thuringiensis* (Bt) based biopesticides are of utmost importance and occupy almost 97% of the world biopesticide market (Cannon 1993). A biological pesticide is effective only if it has a potential major impact on the target pest, market size, variability of field performance, cost effectiveness, end-user feedback and a number of technological challenges namely, fermentation, formulation and delivery systems (Jacobsen and Backman 1993; Copping 1998). Development cost, time and ease of registration and potential growing market in contrast to chemical pesticides make biopesticides interesting proponents to investigate. Despite, extensive research in the field of Bt biopesticides, many formulations do not deliver effectively in field owing to variable environmental stress (for example, forestry and agriculture). Another reason could be adoption of integrated approach which can play an important role in biopesticide development, in other words, tailoring fermentation and harvesting processes to produce higher potency efficacious formulations. Biopesticide research has been comprehensively detailed in Burges (1998), but there are recent advances, which have taken place henceforth. Meanwhile, wastewater (WW) and wastewater sludge (WWS) based Bt formulations also need to be elaborated and discussed due to their inherent positive features.

The Environmental Protection Strategy is entirely based on the principles of sustainable development. The uprising population and the environmental deterioration face the challenge of sustainable development. The remedy of environmental problems requires great financial resources, for instance, the remediation in the solid waste and wastewaters sectors requires huge investments. Bioremediation uses relatively low-cost, low-technology techniques, which generally have a high public acceptance. The treatment of aqueous and solid wastes of industrial, agricultural and domestic origin offers a number of opportunities to apply a wide range of biotechnological methods. The most essential resources for food production are water, soil and energy. Biotreatment and bioremediation techniques are useful tools to control water quality, monitor pollution, decontaminate wastewaters and prevent pollution. Bioremediation seems to be a good alternative to conventional clean-up technologies and research in this field is rapidly increasing. When used as a component of Integrated Pest Management (IPM) programs, biopesticides can greatly decrease the use of conventional pesticides, while crop yields remain high and the pollution problems caused by conventional pesticides are avoided. The potential of rhizospheric microorganisms like mycorrhizal fungi and rhizobacteria, which contribute essentially to increase the soil fertility and remediate physically and chemically disturbed soils should be utilised. The major advances in molecular methodologies that have been achieved in the recent past will provide improved *in vivo* models for mutagenecity and genotoxicity testing in the near future. The potential of enzymes in cleaning up wastes and biodegradation of contaminants might be even greater as a result of "directed evolution", which has led to the production of highly efficient enzymes.

The various technologies summarized in this chapter could play a major role in most of these fields but will they, in all situations, be efficient and effective enough to justify the necessary investment? A critical evaluation of current approaches and results is needed in order to determine this.

References

Ahmad LA, Hassan DR, Hemeda HM (2010) Antihyperglycemic effects of *Okara*, corn hull and their cobition in alloxan induced diabetic rats. World Appl Sci J 9:1139–1147

Akhtar S, Husain Q (2006) Potential of immobilized bittergourd (*Momordica charantia*) peroxidase in the removal of phenols from polluted water. Chemosphere 65:1228–1235

Akhtar S, Khan AA, Husain Q (2005a) Simultaneous purification and immobilization of bitter gourd (*Momordica charantia*) peroxidases on bioaffinity support. J Chem Technol Biotechnol 80:198–205

Akhtar S, Khan AA, Husain Q (2005b) Partially purified bittergourd (*Momordica charantia*) peroxidase catalyzed decolorization of textile and other industrially important dyes. BioresTechnol 96:1804–1811

Albert LA, Rendon J (1988) Contamination by organochlorine compounds in some foodstriffs from region of Mexico. Rev Saude Publica 22:500–506

Altman A (1999) Plant biotechnology in the 21st century: the challenges ahead. Electron J Biotechnol 2:51–55

Andersen JW, Story L, Sieling B, Chen W-JL, Petro MS, Story J (1984) Hypocholesterolemic effects of oat-bran or bean intake for hypercholesterolemicmen. Am J Clin Nutr 40:1146–1155

Andersson DI, Hughes D (2010) Antibiotic resistance and its cost: is it possible to reverse resistance? Nat Rev Microbiol 8:260–271

Annadotter H, Cronberg G, Lawton LA, Hansson HB, Gothe U, Skulberg OM (2001) An extensive outbreak of gastroenteritis associated with the toxic cyano bacterium *Planktothrix aghardii* (Oscillatoriales, Cyanophyceae) in Scania, South Sweden. In: Chorus I (ed) Cyanotoxins-occurrence, causes, consequences. Springer, Berlin, pp 200–208

Arvanitoyannis IS, Ladas D, Mavromatis A (2006) Potential uses and applications of treated wine waste: a review. Int J Food Sci Technol 41:475–487

Azevedo SMFO, Carmichael WW, Jochimsen EM, Rinehart KL, Lau S, Shaw GR, Eaglesham GK (2002) Human intoxication by microcystins during renal dialysis treatment in Caruaru-Brazil. Toxicology 181:441–446

Badreddine R, Humez AA, Mingelgrin U, Benchara A, Meducin F, Prost R (2004) Retention of trace metals by solidified/stabilized wastes: assessment of long-term metal release. Environ Sci Technol 38:1383–1398

Baig S, Liechti PA (2001) Ozone treatment for biorefractory COD removal. Water Sci Technol 43:197–204

Bailey DA, Jones K, Mitchell C (1974) The reclamation of water from sewage effluents by reverse osmosis. J Water Pollut Control Fed 73:353–364

Bajpai P (2001) Microbial degradation of pollutants in pulp mill effluents. Adv Appl Microbiol 48:9–134

Bajpai P, Bajpai PK (1994) Biological colour removal of pulp and paper mill wastewaters. J Biotechnol 33:211–220

Bakker MG, Glover JD, Mai JG, Kinkel LL (2010) Plant community effects on the diversity and pathogen suppressive activity of soil streptomycetes. Appl Soil Ecol 46:35–42

Banchio E, Bogino PC, Zygadlo J, Giordano W (2008) Plant growth promoting rhizobacteria improve growth and essential oil yield in *Origanum majorana* L. Biochem Syst Ecol 36:766–771

Baquero F, Martinez JL, Canton R (2008) Antibiotics and antibiotic resistance in water environments. Curr Opin Biotechnol 19:260–265

Barragan BE, Costa C, Carmen Marquez M (2007) Biodegradation of azo dyes by bacteria inoculated on solid media. Dyes Pigments 75:73–81

Basta NT, Gradwohl R, Snethen KL, Schroder FL (2001) Chemical immobilization of lead, zinc, and cadmium in smelter-contaminated soils using biosolids and rock phosphate. J Environ Qual 30:1222–1230

Behm D (2003) Coroner cites algae in teen's death. In: Milwaukee Journal Sentinel, Milwaukee

Bezdicek DF, Papendick RI, Lal R (1996) Importance of soil quality to health and sustainable land management. In: Doran JW, Jones AJ (eds) Methods for assessing soil quality. Soil Science Society of America, Madison, WI, pp 1–8

Bhatia S, Dubey RC, Maheshwari DK (2003) Antagonistic effect of fluorescent pseudomonads against *Macrophomina phaseolina* that causes charcoal rot of ground nut. Indian J Exp Biol 41:1441–1446

Bhatnagar A, Sillanpää M (2010) Utilization of agro-industrial and municipal waste materials as potential adsorbents for water treatment – a review. Chem Eng J 157:277–296

1 Environmental Protection Strategies: An Overview

Bhunia A, Durani S, Wangikar PP (2001) Horseradish peroxidase catalyzed degradation of industrially important dyes. Biotechnol Bioeng 72:562–567

Bickel H, Neumann U, Weckesser J (2001) Peptides and depsipeptides produced by cyanobacteria. In: Chorus I (ed) Cyanotoxins-occurrence, causes, consequences. Springer, Berlin, pp 281–286

Bigham JM, Fitzpatrick RW, Schulze DG (2002) Iron oxides. In: Dixon JB, Schulze DG (eds) Soil mineralogy with environmental applications, vol. 7. Soil Science Society of America Book Series, Madison, WI, pp 322–366

Blum WEH (1994) Sustainable land use and environment. Proc Ind Soc Soil Sci, New Delhi, pp 21–30

Bowell RJ (1994) Sorption of arsenic by iron oxides and oxyhydroxides in soils. Appl Geochem 9:279–286

Brady NC, Weil RR (1999) The nature and properties of soils, 12th edn. Prentice-Hall, Upper Saddle River, NJ

Bromley-Challenor KCA, Knapp JS, Zhang Z, Gray NCC, Hetheridge MJ, Evans MR (2000) Decolorization of an azo dye by unacclimated activated sludge under anaerobic conditions. Water Res 34:4410–4418

Brookes PC (1995) The use of microbial parameters in monitoring soil pollution by heavy metals. Biol Fertil Soils 19:269–275

Brown D, Hamburger B (1987) The degradation of dye stuffs. Part III. Investigations of their ultimate degradability. Chemosphere 16:1539–1553

Brown D, Laboureur P (1983a) The degradation of dyestuffs: part I. Primary biodegradation under anaerobic conditions. Chemosphere 12:397–404

Brown D, Laboureur P (1983b) The aerobic biodegradability of primary aromatic amines. Chemosphere 12:405–414

Brown SL, Henry CL, Chaney R, Compton H, DeVolder PS (2003) Using municipal biosolids in combination with other residuals to restore metal contaminated mining areas. Plant Soil 249:203–215

Bumpus JA, Brock BJ (1988) Biodegradation of crystal violet by the white rot fungus *Phanerochaete chrysosporium*. Appl Environ Microbiol 54:1143–1150

Buonanno G, Ficco G, Stabile L (2008) Size distribution and number of particles at the stack of a municipal waste incinerator. Waste Manag 29:749–755

Burges HD (1998) Formulation of microbial biopesticides: beneficial organisms, nematodes and seed treatments. Kluwer Academic Publishers, Dordrecht

Cannon RJC (1993) Prospects and progress for *Bacillus thuringiensis* based pesticides. Pestic Sci 37:331–335

Cao X, Ma LQ, Chen M, Singh SP, Harris WG (2002) Impacts of phosphate amendments on lead biogeochemistry at a contaminated site. Environ Sci Technol 36:5296–5304

Carmichael WW (1992) A status report on planktonic cyanobacteria (Bluegreen Algae) and their toxins summary report, EPA/600/SR-92-079, Environmental Monitoring Systems Laboratory, Office of Research and Development, US EPA, Cincinnati, OH

Carpentier B, Cerf O (1993) Biofilms and their consequences with particular references to hygiene in the food industry. J Appl Bacteriol 75:499–511

Catroux G, Hartmann A, Revellin C (2001) Trends in rhizobial inoculants production and use. Plant Soil 230:21–30

Cattoir V, Poirel L, Aubert C, Soussy CJ, Nordmann P (2008) Unexpected occurrence of plasmid-mediated quinolone resistance determinants in environmental *Aeromonas* spp. Emerg Infect Dis 14:231–237

Chorus I, Bartram J (1999) Toxic cyanobacteria in water. World Health Organisation, E & FN Spon, London

Chow CWK, Drikas M, House J, Burch MD, Velzeboer RMA (1999) The impact of convertion-alwater treatment process on cells of the cyanobacterium *Microcystis aeruginosa*. Water Res 33:3253–3262

Consultative Group on International Agricultural Research (CGIAR) (1989) Sustainable agricultural production: implications for international agricultural research. FAO, Research and Technology paper no. 4, p 131

Cook RJ, Zhang BX (1985) Degrees of sensitivity to metalaxyl within the *Pythium* spp. pathogenic to wheat in the Pacific Northwest. Plant Disease 69:686–688

Copping L (1998) The biopesticides manual. British Crop Protection Council, UK, p 333

Costerton JW, Stewart PS, Greenberg EP (1999) Bacterial biofilms: a common cause of persistent infections. Science 284:1318–1322

Cousins IT, Bealing DJ, James HA, Sutton A (1996) Biodegradation of microcystin-LR by indigenous mixed bacterial populations. Water Res 30:481–485

Cox PA, Banack SA, Murch S, Rasmussen U, Tien G, Bidigare RR, Metcalf JS, Morrison L, Codd JA, Bergman B (2005) Diverse taxa of cyanobacteria produce (beta)-N-methylamino-L-alanine, aneurotoxic amino acid. PNAS 102(14):5074–5078

D'Acosta VM, McGrann KM, Hughes DW, Wright GD (2006) Sampling the antibiotic resistome. Science 311:374–377

Davies JE (1997) Origins, acquisition and dissemination of antibiotic resistance determinants. Ciba Found Symp 207:15–27

Dec J, Bollag JM (1994) Dehalogenation of chlorinated phenols during oxidative coupling. Environ Sci Technol 28:484–490

Dehn B, Schuepp H (1989) Influence of VA mycorrhizae on the uptake and distribution of heavy metals in plants. Agric Ecosyst Environ 29:79–83

Dermatas D, Meng X (2003) Utilization of fly ash for stabilization/solidification of heavy metal contaminated soils. Eng Geol 70:377–394

Deshmukh NS, Lapsiya KL, Savant DV, Chiplonkar SA, Yeole TY, Dhakephalkar PK, Ranade DR (2009) Upflow anaerobic filter for the degradation of adsorbable organic halides (AOX) from bleach composite wastewater of pulp and paper industry. Chemosphere 75:1179–1185

Di Blasi C, Tanzi V, Lanzetta MA (1997) Study on the production of agricultural residues in Italy. Biomass Bioenerg 12:321–331

Dick RP (1997) Soil enzyme activities as integrative indicators of soil health. In: Pankhurst CE, Doube BM, Gupta VVSR (eds) Biological indicators of soil health. CAB International, Wellingford, pp 121–156

Diels L, Van Der Lelie N, Bastiaens L (2002) New developments in treatment of heavy metal contaminated soils. Environ Sci Biotechnol 1:75–82

Dixon RA, Achnine L, Kota P, Liu C-J, Reddy MSS, Wang L (2002) The phenylpropanoid pathway and plant defence: a genomics perspective. Mol Plant Pathol 3:371–390

Donald DB, Cessna AJ, Sverko E, Glozier NE (2007) Pesticides in surface drinking water supplies of the Northern Great Plains. Environ Health Perspect 115:1183–1191

Donlan RM (2002) Biofilms: microbial life on surfaces. Emerg Infec Dis 8:881–890

Doran JW, Parkin TB (1994) Defining and assessing soil quality. In: Doran JV, Coleman DC, Bezdicek DF, Stewart BA (eds) Defining soil quality for a sustainable environment. Soil Science Society of America, American Society of Agriculture, Madison, WI, pp 3–21.

dos Santos AB, Cervantes FJ, van Lier JB (2007) Review paper on current technologies for decolourisation of textile wastewaters: perspectives for anaerobic biotechnology. Bioresour Technol 98:2369–2385

Dröge M, Pühler A, Selbitschka W (2000) Phenotypic and molecular characterization of conjugative antibiotic resistance plasmids isolated from bacterial communities of activated sludge. Mol Gen Genet 263:471–482

Fajardo A, Martinez JL (2008) Antibiotics as signals that trigger specific bacterial responses. Curr Opin Microbiol 11:161–167

Falconer IR (2005) Cyanobacterial toxins of drinking water supplies: cylindrospermopsins and microcystins. CRC Press, Boca Raton, FL.

Feuerpfeil I, Lopez-Pila J, Schmidt R, Schneider E, Szewzyk R (1999) Antibiotika resistente Bakterien und Antibiotika in der Umwelt. Bundesgesundheitsbl-Gesundheitsforsch-Gesundheitsschutz 42:37–50

Forchert A, Neumann U, Papendorf O (2001) New cyanobacterial substances with bioactive properties. In: Chorus I (ed) cyanotoxins-occurrence, causes, consequences. Springer, Berlin, pp 295–315

Forgacs E, Cserhati T, Oros G (2004) Removal of synthetic dyes from wastewaters: a review. Environ Int 30: 953–971

Freire RS, Kunz A, Duran N (2000) Some chemical and toxicological aspects about paper mill effluent treatment with ozone. Environ Technol 21:717–721

Frijters CTMJ, Vos RH, Scheffer G, Mulder R (2006) Decolorizing and detoxifying textile wastewater, containing both soluble and insoluble dyes, in a full scale combined anaerobic/aerobic system. Water Res 40:1249–1257

Fu YC, Jiang H, Bishop P (1994) An inhibition study of the effect of azo dyes on bioactivity of biofilms. Water Sci Technol 29:365–372

Garg VK, Gupta R (2009) Vermicomposting of agro-industrial processing waste. In: Nigam PS, Pandey A (eds) Biotechnology for agro-industrial residues utilisation. Part V. Springer, Netherlands, pp 431–456

Gaur A, Adholeya A (2004) Prospects of arbuscular mycorrhizal fungi in phytoremediation of heavy metal contaminated soils. Curr Sci 86:528–534

Glick BR (1995) The enhancement of plant growth by free-living bacteria. Can J Microbiol 41:109–117

Glick BR, Patten MCL, Holguin G, Penrose DM (1999) Biochemical and genetic mechanism used by plant growth promoting bacteria. Imperial College Press, London, UK

Gosling P, Hodge A, Goodlass G, Bending GD (2006) Arbuscular mycorrhizal fungi and organic farming. Agric Ecosyst Environ 113:17–35

Gougar MLD, Scheetz BE, Roy DM (1996) Ettringite and C-S-H Portland cement phases for waste ion immobilization: a review. Waste Manag 16:295–303

Graham PH, Vance CP (2003) Legumes. Importance and constraints to greater use. Plant Physiol 131: 872–877

Gray EJ, Smith DL (2005) Intracellular and extracellular PGPR: commonalities and distinctions in the plant-bacterium signalling processes. Soil Biol Biochem 37:395–412

Gregorich EC, Carter MR, Angers DA, Monreal CM, Ellert BH (1994) Towards a minimum data set to assess soil organic matter quality in agricultural soils. Can J Soil Sci 74:367–385

Grusak MA (2002a) Enhancing mineral content in plant food products. J Am Coll Nutr 21:178S–183S

Grusak MA (2002b) Phytochemicals in plants: genomics-assisted plant improvement for nutritional and health benefits. Curr Opin Biotechnol 13:508–511

Gupta CP, Dubey RC, Maheshwari DK (2002) Plant growth enhancement and suppression of Macrophomina phaseolina causing charcoal rot of peanut by fluorescent *Pseudomonas*. Biol Fertil Soils 35:399–405

Gutierrez MJF, Baxter D, Hunter C, Svoboda K (2005) Nitrous oxide (N_2O) emissions from waste and biomass to energy plants. Waste Manag Res 23:133–147

Haas D, Keel D (2003) Regulation of antibiotic production in root colonizating *Pseudomonas* spp. and relevance for biological control of plant diseases. Annu Rev Phytopathol 41:117–153

Hai FI, Yamamoto K, Fukushi K (2007) Hybrid treatment systems for dye wastewaters. Crit Rev Environ Sci Technol 37:315–377

Hall RM (1997) Mobile gene cassettes and integrons: moving antibiotic resistance genes in Gram-negative bacteria. In: Antibiotic resistance: origins, evolution, selection and spread. Ciba Found Symp 207, Wiley, Chichester, S. 192–205

Handelsman J (2004) Metagenomics: application of genomics to uncultured microorganisms. Microbiol Mol Biol Rev 68:669–685

Handelsman J, Raffel S, Mestea EH, Wunderlich L, Grass CR (1990) Biological control of damping off of alfalfa seedlings with *Bacillus cerens* UW85. Appl Environ Microbiol 56:713–718

Hardison DW Jr, Ma LQ, Luongo T, Harris WG (2004) Lead contamination in shooting range soils from abrasion of lead bullets and subsequent weathering. Sci Total Environ 328:175–183

Harmer C, Bishop P (1992) Transformation of azo dye AO-7 by wastewater biofilms. Water Sci Technol 26:627–636

Hashem MA, Abdelmonem RM, Farrag TE (2007) Human hair as a biosorbent to uptake some dyestuffs from aqueous solutions. Alexandria Eng J 1:1–9

Heier LS, Lien IB, Stromseng AE, Ljønes M, Rosseland BO, Tollefsen KE, Salbu B (2009) Speciation of lead, copper, zinc and antimony in water draining a shooting range-time dependant metal accumulation and biomarker responses in brown trout (*Salmo trutta* L.). Sci Total Environ 407:4047–4055

Hettiarachchi GM, Pierzynski GM, Ransom MD (2000) In situ stabilization of soil lead using phosphorus and manganese oxides. Environ Sci Technol 21:4614–4619

Heuer H, Krögerrecklenfort E, Wellington EMH, Egan S, van Elsas JD et al (2002) Gentamicin resistance genes in environmental bacteria: prevalence and transfer. FEMS Microbiol Ecol 42:289–302

Höfte M, Altier N (2010) Fluorescent pseudomonads as biocontrol agents for sustainable agricultural systems. Res Microbiol 161:464–471

Hornick SB (1992) Factors affecting the nutritional quality of crops. Am J Alt Agric 7:63–68

HSE (2004) Mapping health and safety standards in the UK waste industry. Research Report 240. Health & Safety Executive (HSE), HMSO, Norwich, UK

Hsing H, Wang W, Chiang P et al (2001) Hazardous wastes transboundary movement management-case study in Taiwan. Resour Conserv Recy 40:329–342

Husain Q (2006) Potential applications of the oxidoreductive enzymes in the decolorization and detoxification of textile and other synthetic dyes from polluted water: a review. Crit Rev Biotechnol 60:201–221

Husain Q, Jan U (2000) Detoxification of phenols and aromatic amines from polluted wastewater by using phenol oxidases. J Sci Ind Res 59:286–293

Ishii H, Nishijima M, Abe T (2004) Characterization of degradation process of cyanobacterial hepatotoxins by a gram-negative aerobic bacterium. Water Res 38:2667–2676

Ivleva NP, Wagner M, Horn H, Niessner R, Haisch C (2010) Raman microscopy and surface-enhanced Raman scattering (SERS) for in situ analysis of biofilms. J Biophotonics 3(8–9):548–556

Jacobsen BJ, Backman PA (1993) Biological and cultural plant disease controls: alternatives and supplements to chemicals in IPM systems. Plant Disease 77:311–315

Jadhav JP, Parshetti GK, Kalme SD, Govindwar SP (2007) Decolourization of azo dye methyl red by *Saccharomyces cerevisiae* MTCC463. Chemosphere 68:394–400

Jeffries P, Gianinazzi S, Perotto S, Turnau K, Barea J (2003). The contribution of arbuscular mycorrhizal fungi in sustainable maintenance of plant health and soil fertility. Biol Fertil Soils 37:1–16

Jenkins DJA, Kendall CWC, Marchie A, Jenkins AL, Augustin LSA, Ludwig DS, Barnard ND, Anderson JW (2003) Type 2 diabetes and the vegetarian diet. Am J Clin Nutr 78:610S–616S

Ji RP, Lu XW, Li XN, Pu YP (2009) Biological degradation of algae and microcystins by microbial enrichment on artificial media. Ecol Eng 35:1584–1588

Johnson CA, Moench H, Wersin P, Kugler P, Wenger C (2005) Solubility of antimony and other elements in samples taken from shooting ranges. J Environ Qual 34:248–254

Jones CR, Samac DA (1996) Biological control of fungi causing alfalfa seedling damping-off with a disease-suppressive strain of *Streptomyces*. Biol Control 7:196–204

Jones DL, Hodge A, Kuzyakov Y (2004) Plant and mycorrhizal regulation of rhizodeposition. New Phytol 163:459–480

Jones DL, Nguyen C, Finlay RD (2009) Carbon flow in the rhizosphere: carbon trading at the soil-root interface. Plant Soil 321:5–33

Kaldorf M, Kuhn AJ, Schroder WH, Hildebrandt U, Bothe H (1999) Selective element deposits in maize colonized by a heavy metal tolerance conferring arbuscular mycorrhizal fungus. J Plant Physiol 154:718–728

Kanekar P, Sarnaik S (1991) An activated sludge process to reduce the pollution load of a dye-industry waste. Environ Pollut 70:27–33

Khan AA, Akhtar S, Husain Q (2005) Simultaneous purification and immobilization of mushroom tyrosinase on immunoaffinity support. Process Biochem 40:2379–2386

Klibanov AM, Tu TM, Scott KP (1983) Peroxidase catalyzed removal of phenols from coal conversion wastewater. Science 221:259–261

Kloepper JW, Lifshitz R, Zablotowicz RM (1989) Free-living bacterial inocula for enhancing crop productivity. Trends Biotechnol 7:39–44

Kloepper JW, Zablotowicz RM, Tipping EM, Lifshitz R (1991) Plant growth promotion mediated by bacterial rhizosphere colonizers. In: Keister D, Cregan P (eds) The rhizosphere and plant growth. Kluwer Academic Publishers, Dordrecht, pp 315–326

Kucharski R, Sas-Nowosielska A, Malkowski E, Japenga J, Kuperberg JM, Pogrzeba M, Krzyzak J (2005) The use of indigenous plant species and calcium phosphate for the stabilization of highly metal-polluted sites in southern Poland. Plant Soil 273:291–305

Kulla HG (1981) Aerobic bacterial degradation of azo dyes. FEMS Microbiol Lett 12:387–399

Kulshrestha Y, Husain Q (2006) Bioaffinity-based an inexpensive and high yield procedure for the immobilization of turnip (*Brassica rapa*) peroxidase. Biomol Eng 23:291–297

Labare MP, Butkus MA, Reigner D, Schommer N, Atkinson J (2004) Evaluation of lead movement from the abiotic to biotic at a small-arms firing range. Environ Geol 46:750–754

Lahti K, Rapala J, Fardig M, Niemela M, Sivonen K (1997) Persistence of cyanobacterial hepatotoxin, microcystin-LR in particulate material and dissolved in lake water. Water Res 31:1005–1012

Lal R (1997) Soil quality and sustainability. In: Lal R, Blum WH, Valentine C, Stewart, BA (eds) Methods for assessment of soil degradation. CRC Press, New York, pp 17–30

Larson WE, Pierce FJ (1991) Conservation and enhancement of soil quality. In: Evaluation for sustainable land management in the developing world, vol. 2. International Boarding for Soil Research and Management Proceedings, Bangkok.

Lee IS, Kim OK, Chang YY, Bae B, Kim HH, Baek KH (2002) Heavy metals concentrations and enzyme activities in soil from a contaminated Korean shooting range. J Biosci Bioeng 94:406–411

Leff LG, Dana JR, Mcarthur JV, Shimkets LJ (1993) Detection of Tn5-like sequences in kanamycin-resistant stream bacteria and environmental DNA. Appl Environ Microbiol 59:417–421

Levy SB (1997) Antibiotic resistance: an ecological imbalance. In: antibiotic resistance: origins, evolution, selection and spread. Ciba Found Symp 207. Wiley, Chichester, S. 1–14

Lewis K (2001) Riddle of biofilm resistance. Antimicrob Agents Chemother 45:999–1007

Lim HS, Kim SD (1995) The role andcharacterization of b–1, 3-glucanase in biocontrol of *Fusarium solani* by *Pseudomonas stutzeri* YLPI. Curr Microbiol 33: 295–301

Lin Z (1996) Secondary menral phases of metallic lead in soils of shooting ranges from Örebro County, Sweden. Environ Geol 27:370–375

Lopez C, Mielgo I, Moreira MT, Feijoo G, Lema JM (2002) Enzymatic membrane reactors for biodegradation of recalcitrant compounds: application to dye decolourisation. J Biotechnol 29:249–257

Lopez C, Moreira MT, Feijoo G, Lema JM (2004) Dye decolorization by manganese peroxidase in an enzymatic membrane bioreactor. Biotechnol Prog 20:74–81

Lothenbach B, Furrer G, Schärli H, Schulin R (1999) Immobilization of zinc and cadmium by montmorillonite compounds: effects of aging and subsequent acidification. Environ Sci Technol 33:2945–2952

Lucy M, Reed E, Glick BR (2004) Applications of free living plant growth-promoting rhizobacteria. Antonie Leeuwenhoek 86:1–25

Lynch JM, Benedetti A, Insam H, Nuti MP, Smalla K, Torsvik V, Nannipieri P (2004) Microbial diversity in soil: ecological theories, the contribution of molecular techniques and the impact of transgenic plants and transgenic microorganisms. Biol Fertility Soils 40:363–385

Ma LQ, Logan TJ, Traina SJ (1995) Lead immobilization from aqueous solutions and contaminated soils using phosphate rocks. Environ Sci Technol 29:1118–1126

MacGregor JT (1994) Environmental mutagenesis: past and future directions. Mutat Res 23:73–77

Madar Z, Stark AH (2002) New legume sources as therapeutic agents. Br J Nutr 88:S287–S292

Maloschik E, Ernst A, Hegedûs G, Darvas B, Székacs A (2007) Monitoring water-polluting pesticides in Hungary. Microchem J 85:88–97

Marrone PG (1999) Microbial pesticides and natural products as alternatives. Outlook Agric 28:149–154

Martinez CE, McBride MB (1999) Dissolved and labile concentrations of Cd, Cu, Pb, and Zn in aged ferrihydrite-organic matter systems. Environ Sci Technol 33:745–750

Martinez JL (2008) Antibiotics and antibiotic resistance genes in natural environments. Science 321:365–367

Martinez JL, Baquero F (2000) Mutation frequencies and antibiotic resistance. Antimicrob Agents Chemother 44:1771–1777

Martinez JL, Baquero F, Andersson DI (2007) Predicting antibiotic resistance. Nat Rev Microbiol 5:958–965

Marx DH (1975) Mycorrhizae and the establishment of trees on stripmined land. Ohio J Sci 75:88–297

Marx DH, Altman JD (1979) *Pisolithus tinctorius* ectomycorrhiza improve survival and growth of pine seedlings on acid coal spoil in Kentucky and Virginia. In: Mycorrhizal Manual. Springer, Berlin, pp 387–399

Matto M, Husain Q (2006) Entrapment of porous and stable Con A-peroxidase complex into hybrid calcium alginate-pectin gel. J Chem Technol Biotechnol 81:1316–1323

Mazel D (2006) Integrons: agents of bacterial evolution. Nat Rev Microbiol 4:608–620

McBride M (1994) Environmental chemistry of soils. Oxford University Press, Oxford

McGowen SL, Basta NT, Brown GO (2001) Use of diammonium phosphate to reduce heavy metal solubility and transport in smelter-contaminated soil. J Environ Qual 30:493–500

Mench M, Bussiere S, Boisson J, Castaing E, Vangronsveld J, Ruttens A, De Koe T, Bleeker P, Assuncao A, Manceau A (2003) Progress in remediation and revegetation of the barren Jales gold mine spoil after in situ treatments. Plant Soil 249:187–202

Mergeay M, Monchy S, Vallaeys T, Auquier V, Benotmane A, Bertin P, Taghavi S, Dunn J, van der Lelie D, Wattiez R (2003) *Ralstonia metallidurans*, a bacterium specifically adapted to toxic metals: towards a catalogue of metal responsive genes. FEMS Microbiol Rev 27:385–410

Migliorini M, Pigino G, Bianchi N, Bernini F, Leonzio C (2004) The effects of heavy metal contamination on the soil arthropod community of a shooting range. Environ Poll 129:331–340

Miller MJ, Fallowfield HJ (2001) Degradation of cyanobacterial hepatotoxins in batch experiments.Water Sci Technol 43:229–232

Mislovicova D, Gemeiner P, Sandula J, Masarova J, Vikartovska A, Doloomansky P (2000) Examination of bioaffinity immobilization by precipitation of mannan and mannan containing enzymes with legume lectins. Biotechnol Appl Biochem 31:153–159

Monspart-Sényi J (2007) Fruit Processing Waste Management. In: Hui YH (ed) Handbook of Fruits and Fruit Processing. Blackwell Publishing, Ames, IA

Moulin I, Stone WEE, Sanz J, Bottero JY, Haehnel C (1999) Lead and zinc retention during hydration of tri-calcium silicate: A study by sorption isotherms and 29 Si nuclear magnetic resonance spectroscopy. Langmuir 15:2829–2835

Mozafar A, Ruh R, Klingel P, Gamper H, Egli S, Frossard E (2002) Effect of heavy metal contaminated shooting range soils on mycorrhizal colonization of roots and metal uptake by leek. Environ Monit Assess 78:177–191

Mukherjeez I, Gopal M (2002) Organochlorine insecticide residues in drinking and roundwater in and around Delhi. Environ Monit Assess 76:185–193

Murch SJ, Cox PA, Banack SA (2004) A mechanism for slow release of biomagnified cyanobacterial neurotoxins and neurodegeneratuive disease in Guam. PNAS 101(33):12228–12231

Ndakidemi PA, Dakora FD (2003) Review: legume seed flavonoids and nitrogenous metabolites as signals and protectants in early seedling development. Funct Plant Biol 30:729–745

Negi YK, Garg SK, Kumar J (2005) Cold tolerant fluorescent *Pseudomonas* isolates from Garhwal Himalayas as potential plant growth promoting and biocontrol agents in pea. Curr Sci 89:2151–2156

Neu TR, Manz B, Volke F, Dynes JJ, Hitchcock AP, Lawrence JR (2010) Advanced imaging techniques forassessmentof structure, compositionand function in bioçlm systems. FEMS Microbiol Ecol 72:1–21

Neu TR, Swerhone GDW, Lawrence JR (2001) Assessment of lectin-binding analysis for *in situ* detection of glycoconjugates in biofilm systems. Microbiol 147:299–313.

Neumann G, Römheld V (1999) Root excretion of carboxylic acids and protons in phosphorus-deficient plants. Plant Soil 211:121–130

Nies DH (2003) Efflux-mediated heavy metal resistance in prokaryotes. FEMS Microbiol Rev 27:313–339

Norouzian D (2003) Enzyme immobilization: the state of art in biotechnology. Iran J Biotechnol 1:191–205

Ogawa T, Yatome C (1990) Biodegradation of azo dyes in multistage rotating biological contactor immobilized by assimilating bacteria. Bull Environ Contam Toxicol 44:561–566

Palomo A, Palacios M (2003) Alkali-activated cementitious materials: alternative matrices for the immobilization of hazardous wastes part II. stabilization of chromium and lead. Cement Concrete Res 33:289–295

Pankhurst CE, Hawke BG, McDonal HJ, Kirkby CA, Buckerfield JC, Michelsen P, O'Brien KA, Gupta VVSR, Doube BM (1995) Evaluation of soil biological properties as potential bioindicators of soil health. Aust J Exp Agric 35:1015–1028

Parr JF, Papendick RI, Hornick SB, Meyer RE (1992) Soil quality: attributes and relationship to alternative and sustainable agriculture. Am J Alt Agric 7:5–11

Perreten V, Schwarz F, Cresta L, Boeglin M, Dasen G, Teuber M (1997) Antibiotic resistance spread in food. Nature 389:801–802

Pruden A, Pei R, Storteboom H, Carlson K (2006) Antibiotic resistance genes as emerging contaminants: Studies in Northern Colorado. Environ Sci Technol 40:7445–7450

Quagliotto L, Azziz G, Natalia Bajsa N, Vaz P, Pérez C, Ducamp F, Cadenazzi M, Altier N, Arias A (2009) Three native *Pseudomonas fluorescens* strains tested under growth chamber and field conditions as biocontrol agents against damping-off in alfalfa. Biol Cont 51:42–52

Rahube Teddie O, Yost Christopher K (2010) Antibiotic resistance plasmids in wastewater treatment plants and their possible dissemination into the environment. African J Biotechnol 9: 9183–9190

Rai H, Bhattacharya M, Singh J, Bansal TK, Vats P, Banerjee UC (2005) Removal of dyes from the effluent of textile and dyestuff manufacturing industry: a review of emerging techniques with reference to biological treatment. Crit Rev Environ Sci Technol 35:219–238

Reddy YHK, Srijana M, Reddy DM, Reddy G (2010) Coculture fermentation of banana agro-waste to ethanol by cellulolytic thermophilic *Clostridium thermocellum* CT2. Afr J Biotechnol 9:1926–1934

Richard F, Bourg A (1991) Aqueous geochemistry of chromium: a review. Water Res 25:807–816

Ridgway HF, Flemming HC (1996) In: Mallevialle J, Odendaal PE, Wiesner MR (eds) Water treatment membrane processes. McGraw-Hill, New York, pp 6.1–6.62

Ridgway HF, Kelly A, Justice C, Olson BH (1983) Microbial fouling of reverse osmosis membranes used in advanced wastewater treatment technology: chemical, bacteriological, and ultrastructural analysis. Appl Environ Microbiol 45:1066–1084

Riesenfeld CS, Goodman RM, Handelsman J (2004) Uncultured soil bacteria are a reservoir of new antibiotic resistance genes. Environ Microbiol 6:981–989

Rillig MC, Steinberg PD (2002) Glomalin production by an arbuscular mycorrhizal fungus: a mechanism of habitat modification. Soil Biol Biochem 34:1371–1374

Robinson T, McMullan G, Marchant R, Nigam P (2001) Remediation of dyes in textile effluents: a critical review on current treatment technologies with a proposed alternative. Biores Technol 77:247–255

Rysz M, Alvarez PJ (2004) Amplification and attenuation of tetracycline resistance in soil bacteria: aquifer column experiments. Water Res 38:3705–3712

Saikia N, Kato S, Kojima T (2006) Behavior of B, Cr, Se, As, Pb, Cd, and Mo present in waste leachates generated from combustion residues during the formation of ettringite. Environ Toxicol Chem 25:1710–1719

Salyers AA, Shoemaker NB (1994) Broad host range gene transfer: plasmids and conjugative transposons. FEMS Microbiol Ecol 15:15–22

Sanders PL (1984) Failure of metalaxyl to control *Pythium* blight on turfgrass in Pennsylvania. Plant Disease 68:776–777

Sani R, Banerjee U (1999a) Decolorization of triphenylmethane dyes and textile and dye-stuff effluent by *Kurthia* sp. Enzyme Microb Technol 24:433–437

Sani R, Banerjee U (1999b) Decolorization of acid green 20, a textile dye, by the white rot fungus *Phanerochaete chrysosporium*. Adv Environ Res 2:485–490

Savant DV, Abdul-Rahman R, Ranade DR (2006) Anaerobic degradation of adsorbable organic halides (AOX) from pulp and paper industry wastewater. Biores Technol 97:1092–1104

Scheckel KG, Ryan JA (2004) Spectroscopic speciation and quantification of lead in phosphate-amended soils. J Environ Qual 33:1288–1295

Seaman JC, Meehan T, Bertsch PM (2001) Immobilization of cesium-137 and uranium in contaminated sediments using soil amendments. J Environ Qual 30:1206–1213

Shaffiqu TS, Roy JJ, Nair RA, Abraham TE (2002) Degradation of textile dyes mediated by plant peroxidases. Appl Biochem Biotechnol 102–103:315–326

Shaw KJ, Rather PN, Hare RS, Miller GH (1993) Molecular genetics of aminoglycoside resistance genes and familial relationships of the aminoglycoside-modifying enzymes. Microbiol Rev 57:138–163

Shukla G, Kumar A, Bhanti M, Joseph PE, Taneja A (2006) Organochlorine pesticide contamination of ground water in the city of Hyderabad. Environ Int 32:244–247

Siddiqui ZA, Akhtar MS, Futai K (Eds) (2008) Mycorrhizae: sustainable agriculture and forestry. Springer, New York, p 362

Silver S, Phung LT (1996) Bacterial heavy metal resistance: new surprises. Annu Rev Microbiol 50:753–789

Silver S, Phung LT (2005) A bacterial view of the periodic table: genes and proteins for toxic inorganic ions. J Ind Microbiol Biotechnol 32:587–605

Singh DK (2008) Biodegradation and bioremediation of pesticide in soil: concept, method and recent developments. Ind J Microbiol 48:35–40

Smalla K, Van Overbeek LS, Pukall R, Van Elsas JD (1993) Prevalence of nptII and Tn5 in kanamycin resistant bacteria from different environments. FEMS Microbiol Ecol 13:47–58

Smalla K, Van Elsas JD (1995) Application of the PCR for detection of antibiotic resistance genes in environmental samples. In: Trevors JT, van Elsas JD (eds) Nucleic acids in the environment-methods and applications. Springer-Verlag, Berlin, New York

Smith A, Brown K, Ogilvie S, Rushton K, Bates J (2001) Waste management options and climate change final report to the European Commission, DG Environment

Sorvari J (2007) Environmental risks at Finnish shooting ranges – a case study. Human Ecol Risk Assess 13:1111–1146

Sorvari J, Antikainen R, Pyy O (2006) Environmental contamination at Finnish shooting ranges the scope of the problem and management options. Sci Total Environ 366:21–31

Steinberg PD, Rillig MC (2003) Differential decomposition of arbuscular mycorrhizal fungal hyphae and glomalin. Soil Biol Biochem 35:191–194

Stevenson DM, Weimer PJ (2002) Isolation and characterization of a *Trichoderma* strain capable of fermenting cellulose to ethanol. Appl Microbiol Biotechnol 59:721–726

Stewart I, Schluter PS, Shaw GR (2006) Cyanobacterial lipopolysaccharides and human health – a review. Environ Health 5:7–53

Stolz A (2001) Basic and applied aspects in the microbial degradation of azo dyes. Appl Microbiol Biotechnol 56:69–80

Stoodley P, Sauer K, Davies DG, Costerton JW (2002) Biofilms as complex differentiated communities. Annu Rev Microbiol 56:187–209

1 Environmental Protection Strategies: An Overview

Subramani A, Hoek EMV (2008) Direct observation of initial microbial adhesion onto reverse osmosis and nanofiltration membranes. J Membr Sci 319:111–125

Sun X, Doner HE (1998) Adsorption and oxidation of arsenite on goethite. Soil Sci 163:278–287

Sun Y, Cheng J (2002) Hydrolysis of lignocellulosic materials for ethanol production: a review. Biores Technol 83:1–11

Suntio LR, Shiu WY, Mackay D (1988) A review of the nature and properties of chemicals present in pulp mill effluents. Chemosphere 17:1249–1290

Svrcek C, Smith DW (2004) Cyanobacteria toxins and the current state of knowledge on water treatment options: a review. Environ Eng Sci 3:155–185

Takenaka S, Watanabe MF (1997) Microcystin LR degradation by *Pseudomonas aeruginosa* alkaline protease. Chemosphere 34:749–757

Terzano R, Spagnuolo M, Medici L, Vekemans B, Vincze L, Janssens K, Ruggiero P (2005) Copper stabilization by zeolite synthesis in polluted soils treated with coal fly ash. Environ Sci Technol 39:6280–6287

Thompson G, Swain J, Kay M, Forster CF (2001) The treatment of pulp and paper mill effluent: a review. Biores Technol 77:275–286

Torres E, Bustos-Jaimes I, Le Bogne S (2003) Potential use of oxidative enzymes for the detoxification of organic pollutants. Appl Catal B: Environ 46:1–15

Trivedi P, Dyer JA, Sparks DL (2003) Lead sorption onto ferrihydrite. 1. A macroscopic and spectroscopic assessment. Environ Sci Technol 37:908–914

Tschäpe H (1994) The spread of plasmids as a function of bacterial adaptability. FEMS Microbiol Ecol 15:23–32

Tsuji K, Watanuki T, Kondo, F, Watanabe MF, Nakazawa H, Suzuki M, Uchida H, Harada K (1997). Stability of microcystins from cyanobacteria-IV. Effect of chlorination on decomposition. Toxicon 35:1033–1041

van Beelen P, Fleuren-Kemilá K (1997) Influence of pH on the toxic effects of zinc, cadmium, and pentachlorophenol on pure cultures of soil microorganisms. Environ Toxicol Chem 16:146–153

van Beilen JB (2008) Transgenic plant factories for the production of biopolymers and platform chemicals. Biofuels Bioproducts Biorefining 2:215–228

van der Zee FP, Villaverde S (2005) Combined anaerobic-aerobic treatment of azo dyes-a short review of bioreactor studies. Water Res 39:1425–1440

Van Loon LC (2007) Plant response to plant growth-promoting rhizobacteria. Eur J Plant Pathol 119:243–254

Van Loon LC, Glick BR (2004) Increased plant fitness by rhizobacteria. In: Sandermann H (ed) Molecular ecotoxicology of plants. Springer-Verlag, Berlin, pp 178–205

Vandevivere PC, Bianchi R, Verstraete W (1998) Treatment and reuse of wastewater from the textile wet-processing industry: review of emerging technologies. J Chem Technol Biotechnol 72:289–302

Vasdev K, Kuhad RC, Saxena RK (1995) Decolorization of triphenylmethane dyes by a bird's nest fungus *Cyathus bulleri*. Curr Microbiol 30:269–272

Verma P, Madamwar D (2002) Decolorization of synthetic textile dyes by lignin peroxidase of *Phanerochaete chrysosporium*. Folia Microbiol 47:283–286

Verma P, Madamwar D (2003) Decolorization of synthetic dyes by a newly isolated strain of *Serratia maerascens*. World J Microbiol Biotechnol 19:615–618

Vessey KJ (2003) Plant growth promoting rhizobacteria as bioferitilizers. Plant Soil 255: 571–586

Villacieros M, Power B, Sánchez-Contreras M, Lloret J, Oruezabal RI, Martín M, Fernández-Pinãs F, Bonilla I et al (2003) Colonization behaviour of *Pseudomonas fluorescens* and *Sinorhizobium meliloti* in the alfalfa (*Medicago sativa*) rhizosphere. Plant Soil 251:47–54

Vivas A, Marulanda A, Gómez M, Barea JM, Azcón R (2003) Physiological characteristics (SDH and ALP activities) of arbuscular mycorrhizal colonization as affected by *Bacillus thuringiensis* inoculation under two phosphorus levels. Soil Biol Biochem 35:987–996

Vrouwenvelder JS, Hinrichs CWGJ, Van der Meer WG, Van Loosdrecht MC, Kruithof JC (2009a) Pressure drop increase by biofilm accumulation in spiral wound RO and NF membrane systems: role of substrate concentration, flow velocity, substrate load and flow direction. Biofouling 25:543–555

Vrouwenvelder JS, von der Schulenburg DAG, Kruithof JC, Johns ML, van Loosdrecht MCM (2009b) Biofouling of spiral-wound nanofiltration and reverse osmosis membranes: a feed spacer problem. Water Res 43:583–594

Waksman SA, Woodruff HB (1940) The soil as a source of microorganisms antagonistic to disease-producing bacteria. J Bacteriol 40:581–600

Wang RC, Fan KS, Chang JS (2009a) Removal of acid dye by $ZnFe_2O_4/TiO_2$- immobilized granular activated carbon under visible light irradiation in a recycle liquid-solid fluidized bed. J Taiwan Inst Chem Engrs 40:533–540

Wang H, Su JQ, Zheng XW, Tian Y, Xiong XJ, Zheng TL (2009b) Bacterial decolorization and degradation of the reactive dye Reactive Red 180 by *Citrobacter* sp. CK3. Int Biodeter Biodeg 63:395–399

Wang H, Zheng XW, Su JQ, Tian Y, Xiong XJ, Zheng TL (2009c) Biological decolorization of the reactive dyes reactive black 5 by a novel isolated bacterial strain *Enterobacter* sp. EC3. J Hazard Mater 171:654–659

Welker M, Von Dohren H (2006) Cyanobacterial peptide-nature's own combinatorial biosynthesis. FEMS Microbiol Rev 30:530–563

WHO (World Health Organization) (2009) Pesticides in drinking water. WHO Seminar Pack for drinking-water quality.

Wright CJ, Shah MK, Powell LC, Armstrong I (2010) Application of AFM from microbial cell to biofilm. Scanning Vol 31:1–16

Wright GD (2007) The antibiotic resistome: the nexus of chemical and genetic diversity. Nat Rev Microbiol 5:175–186

Wright GD (2010) Antibiotic resistance: where does it come from and what can we do about it? BMC Biol 8:123–128

Xiao K, Kinkel LL, Samac DA (2002) Biological control of *Phytophthora* root rots on alfalfa and soybean with *Streptomyces*. Biol Contr 23:285–295

Yao Q, Li X, Weidang A, Christie P (2003) Bi-directional transfer of phosphorus between red clover and perennial ryegrass via arbuscular mycorrhizal hyphal links. Eur J Soil Biol 39:47–54

Zhang F, Yediler A, Liang X, Kettrup A (2004) Effects of dye additives on the ozonation process and oxidation by-products: a comparative study using hydrolyzed CI reactive red 120. Dyes Pigments 60:1–7

Zhang FM, Knapp JS, Tapley KN (1998) Decolourisation of cotton bleaching effluent in a continuous fluidized-bed bioreactor using wood rotting fungus. Biotechnol Lett 20:717–723

Zhang TC, Fu YC, Bishop PL (1995) Transport and biodegradation of toxic organics in biofilms. J Hazard Mater 41:267–285

Zille A, Tzanov T, Guebitz GM, Cavaco-Paulo A (2003) Immobilized laccase for decolorization of Reactive Black 5 dyeing effluent. Biotechnol Lett 25:1473–1477

Chapter 2
The Potential of Rhizosphere Microorganisms to Promote the Plant Growth in Disturbed Soils

Katarzyna Hrynkiewicz and Christel Baum

Contents

2.1	Introduction	36
2.2	The Rhizosphere—A Hot Spot of Microbial Activities	36
2.3	Role of Rhizosphere Microorganisms in the Improvement of Plant Fitness	38
	2.3.1 Mycorrhizal Symbiosis	38
	2.3.2 Plant Growth Promoting Rhizobacteria (PGPR)	40
	2.3.3 Mycorrhization Helper Bacteria	41
2.4	Effect of Rhizosphere Microorganisms on the Plant Growth in Disturbed Soils	42
	2.4.1 Nutrient- and Water-Deficiency	42
	2.4.2 Extreme Soil pH	44
	2.4.3 Soil-Borne Pathogens	45
	2.4.4 Heavy Metal Contamination	46
	2.4.5 Organic Pollutants	47
2.5	Selection and Use of Microorganisms for the Promotion of Plant Growth and Soil Remediation	48
	2.5.1 Selection Criteria of Microorganisms for Inoculation of Unfavourable Soils	48
	2.5.2 Chances and Risks of In Vitro Selection for Field Applications	52
	2.5.3 Status and Perspectives of Commercialisation of Microbial Inoculum	53
2.6	Concluding Remarks and Outlook	54
References		58

Abstract The significance of rhizosphere microorganisms, especially mycorrhizal fungi and bacteria, in polluted soils can be enormous, since they are able to increase the tolerance of plants against abiotic stress, stimulate plant growth and contribute in this way to an accelerated remediation of disturbed soils. The majority of known higher plant species is associated with mycorrhizal fungi, which can increase the tolerance of plants against abiotic stress, e.g. by an improved nutrient supply or by detoxification of pollutants. Rhizosphere bacteria can strongly promote the growth of plants solely and in interaction with mycorrhizal fungi. They can contribute to

---

K. Hrynkiewicz (✉)
Department of Microbiology, Institute of General and Molecular Biology, Nicolaus Copernicus University, Gagarina 9, 89-100, Torun, Poland
e-mail: hrynk@umk.pl

A. Malik, E. Grohmann (eds.), *Environmental Protection Strategies for Sustainable Development*, Strategies for Sustainability,
DOI 10.1007/978-94-007-1591-2_2, © Springer Science+Business Media B.V. 2012

the mobilization of nutrients and degradation of organic pollutants. Co-inoculation of plants with mycorrhizal fungi and rhizosphere bacteria is a very promising biotechnological approach for the promotion of plant growth and soil remediation. The application of microbial inoculum for the remediation of disturbed soils was tested with several plant species, e.g., fast growing tree species, but mostly on a small scale. Main reasons for the lack of field applications in a larger scale are the lack of suitable time- and cost-effective strategies for a site-specific selection, preparation and application of microbial inoculum and the strong restriction of information on on-site efficiency of inoculated microbial strains.

This chapter focuses on fundamental and applied aspects of soil microorganisms associated with the rhizosphere of plants at various disturbed sites. Major objectives are to present strategies for the promotion of phytoremediation of disturbed soils with the use of microbial inoculum and to indicate potentials and limitations of such microbial inoculation in the field.

Keywords Rhizosphere • Mycorrhiza • Bacteria • Polluted soil

2.1 Introduction

A fundamental knowledge on plants' physiological properties and their associated microorganisms in the undisturbed natural environments is necessary to understand the impact of microorganisms on the plant development in general. The existence of positive plant-microbial interactions also in disturbed soils is unquestionable, but the mechanisms are often scarcely known. Microorganisms contribute essentially to the protection of plants against unfavourable soil conditions. In this chapter a selection of possible unfavorable soil properties in disturbed soils will be focused to analyse the possible impact of associated microorganisms on plants growth and vitality. Applicability of microbial inoculum for an improved remediation of such disturbed soils will be presented.

2.2 The Rhizosphere—A Hot Spot of Microbial Activities

A narrow zone of soil affected by the presence of plant roots is defined as rhizosphere. The rhizosphere is known to be a hot spot of microbial activities. This is caused by an increased nutrient supply for microorganisms, since roots release a multitude of organic compounds (e.g., exudates and mucilage) derived from photosynthesis and other plant processes (Brimecombe et al. 2007). Therefore, rhizosphere is an environment with a high microbial diversity. An important consequence of the high diversity is an intense microbial activity with feedback effects on root development and plant growth in general. In general, the microbes serve as

2 The Potential of Rhizosphere Microorganisms to Promote the Plant Growth

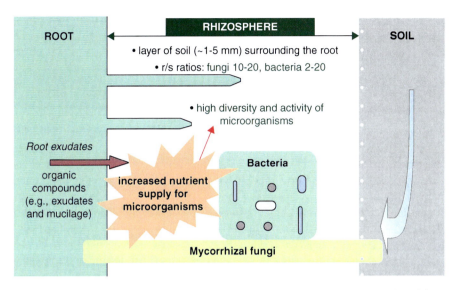

Fig. 2.1 Rhizosphere microorganisms as a critical link between plants and soil. (Adapted from Richardson et al. 2009)

intermediary between the plant (Fig. 2.1), which requires soluble mineral nutrients, and the soil, which contains the necessary nutrients but often in low concentrations and/or complex and inaccessible forms. Thus rhizosphere microorganisms provide a critical link between plants and soil (Lynch 1990).

The highest portion of microorganisms which inhabit the rhizosphere are fungi and bacteria. When considering the rhizosphere effect on their abundance, the fungal abundance is 10–20 times higher and the bacterial abundance 2–20 times higher in the rhizosphere than in the bulk soil (Morgan et al. 2005). Competition for nutrient sources in the rhizosphere is very high. Therefore, different microorganisms have developed distinct strategies, giving rise to a range of antagonistic to synergistic interactions, both among themselves and with the plant (Perotto and Bonfante 1997). A very high diversity of interactions can be assumed on the basis of the tremendous diversity of soil microorganisms and plants. The understanding of fundamentals of these interactions is critical for their use in plant growth promotion and remediation of disturbed soils.

This chapter focuses on mycorrhizal fungi and rhizosphere bacteria that are believed to play a crucial role in the proper development of plants in unfavourable soil conditions. Several examples of the role of rhizosphere microorganisms in the improvement of plant fitness in disturbed soils with unfavourable soil properties for plant growth (nutrient- and water-deficiency, soil-borne pathogens, extreme soil pH, heavy metal contamination, organic pollutants) were described. The need of site-specific selection of plant growth and soil remediation promoting rhizosphere organisms for field use was emphasized.

2.3 Role of Rhizosphere Microorganisms in the Improvement of Plant Fitness

2.3.1 *Mycorrhizal Symbiosis*

The most common mutualistic association between fungi and plant roots is the mycorrhizal symbiosis. In this association the fungal partner can provide the plant with enhanced access to water and nutrients due to the extended area for their acquisition through the extraradical hyphal network. Additionally, many fungal partners can efficiently contribute to the nutrient mobilization in the soil. They are able to produce enzymes involved in the hydrolysis of nitrogen and phosphorus compounds from the organic matter in the soil and contribute to the weathering of minerals, e.g., by the release of organic acids. Mycorrhizal fungi can alleviate abiotic (e.g., increased heavy metal concentrations) and biotic (e.g., soil-borne pathogens) stress by the increase of plant fitness through enhanced nutrient supply and in case of ectomycorrhizal fungi by covering the fine roots with a hyphal mantle. The plants, in return, provide carbohydrates for fungal growth and maintenance (Smith and Read 1997). It has been estimated that between 4 and 20% of net photosynthates could be transferred from the plant to its fungal partner (Morgan et al. 2005). The mycorrhizal symbioses usually increase the growth of the host plants in the long term, however, their effects on the plant growth might also be neutral or even parasitic if the costs for the plants exceed the advantages (Johnson et al. 1997).

The most common types of mycorrhizal associations are ectomycorrhizae (EM) and arbuscular mycorrhizae (AM).

Ectomycorrhizal fungi form associations with many woody plants ranging from shrubs to forest trees in the *Salicaceae, Pinaceae, Fagaceae, Betulaceae* and *Dipterocarpaceae* (Smith and Read 1997). Communities of ectomycorrhizal trees are dominating in boreal and temperate plant biomes and are also important in certain tropical and rain forests (Read 1993). There are about 5000–6000 ectomycorrhizal fungal species in basidiomycetes and ascomycetes described, but it was suggested that there might well be about 10,000 species (Brussaard et al. 1997). Mycelium of ectomycorrhizal fungi forms a mantle of varying thickness around the fine roots. The mantle increases the surface area of absorbing roots and often affects fine-root morphology, resulting partly in root bifurcation and clustering. Contiguous to the mantle, hyphal strands can extend into the soil and often aggregate to form rhizomorphs, specialized for long-distance transport of nutrients and water. Hyphae of ectomycorrhizal fungi also penetrate inwards between the cells of roots producing a netlike structure called the Hartig net, which constitutes the interface for the exchange of photoassimilates, soil water and nutrients between the host plant and its fungal partners. The internal organization of mantle structures and rhizomorphs with respect to hyphal differentiation and with respect to their physical relation can reveal some hints for their function (Agerer 1991).

A single plant root can be colonized by many different mycorrhizal fungi. These fungi partly are able to colonize a variety of different plant species. The natural mechanisms promoting high fungal diversity have remained unclear, however it seems likely that differential preferences for soil conditions and host plants are essential (Bruns 1995).

AM occur in the majority of herbaceous and graminaceous species of temperate and semi-arid grasslands as well as in many tree species especially of tropical and subtropical forests. Arbuscular mycorrhizal fungi are obligate biotrophs and involve a very small group of fungi in the *Glomales* (Glomeromycota). In AM, an internal mycelial phase with characteristic structure is present. In this association neither the fungal cell wall nor the host cell membrane are breached. As the fungus grows, the host cell membrane envelopes the fungus, creating a new compartment where material of high molecular complexity can be deposited. This apoplastic space prevents direct contact between the plant and fungus cytoplasm and allows efficient transfer of nutrients between the symbionts. Among commonly observed structures are found frequently: vesicles, arbuscules, hyphal coils, and internal hyphae. There are problems in extrapolating from structure to function in what now appears to be structurally and possibly also functionally diverse symbiosis (van der Heijden 2001). The term AM covers a diversity of mycorrhizal structures. AM development differs, not only over time, but also between plant species, especially with respect to the extent of development of vesicles, coils and arbuscules within the cortical cells (van der Heijden 2001). The external phase is important as well, made up of branched single hyphae that ramify through the soil, forming anastomosing networks.

Although some plants can form mycorrhizae with both arbuscular mycorrhizal and ectomycorrhizal fungi, they usually prefer one of these mycorrhizal types (Lodge 1989). However, the reasons for their preferences are still unknown. Dual mycorrhiza formation with arbuscular and ectomycorrhizal fungi were observed, e.g., in the genera *Salix*, *Alnus*, *Populus*, and *Eucalyptus*. So far it is not clear whether dual mycorrhiza is a functionally mutualistic relationship. The apparent dual mycorrhizal stage of these plants might be explained by: (i) lack of resistance (decreased plant control). This means that—arbuscular mycorrhizal fungi, with supposed higher inoculum potential, colonize seedlings, however, they are rapidly replaced by ectomycorrhizal fungi which prevent colonization of newly-formed roots by arbuscular mycorrhizal fungi; (ii) niche differentiation (increased plant benefits). This means that plants with the ability to form both types of mycorrhiza might have a selective advantage in specific environmental conditions (e.g., in flood plains) (van der Heijden 2001). It was revealed that, e.g., *Salix* spp. are able to establish AM rarely and the roots of these plants are dominated mostly by ectomycorrhizal fungi (van der Heijden 2001).

The soil surrounding mycorrhizal roots supports distinct bacterial communities compared to the bulk soil. The rhizosphere combined with the hyphosphere of mycorrhizal fungi comprises the mycorrhizosphere. Mycorrhizosphere inhabitants can include intrahyphal bacteria in ectomycorrhizal fungi (Bertaux et al. 2003), and intraspore bacteria in some arbuscular mycorrhizal fungi (Bianciotto et al. 1996). It has been shown that some mycorrhizosphere bacteria can promote mycorrhiza formation. The details of mechanisms that are involved will be described in the next paragraph.

2.3.2 Plant Growth Promoting Rhizobacteria (PGPR)

The rhizosphere is colonized with bacteria that can individually or in cooperation with mycorrhizal fungi improve the plant fitness. In the rhizosphere bacteria continuously metabolize various organic compounds from root exudates. Therefore, their activities result in quantitative and qualitative alterations of the released root exudates. Bacteria in the rhizosphere can significantly influence the nutrient supply of plants by competing for mineral nutrients and by mediating the turnover and mineralization of organic compounds. Therefore, bacteria in the rhizosphere can be a leading control of the turnover of nutrients in the soil (Robinson et al. 1989). Rhizosphere bacteria can influence plant growth also directly by releasing a variety of compounds, e.g., phytohormones or antimicrobial compounds (Perotto and Bonfante 1997).

The diversity and structure of bacterial communities is plant-specific and varies over time (Smalla et al. 2001; Barriuso et al. 2005). Diversity of bacteria is affected by the plant age, the season and the soil conditions (Hrynkiewicz et al. 2010a). Rhizosphere bacteria can have a negative, neutral or positive effects on plant fitness. Detrimental microbes include both major plant pathogens and minor parasitic and non-parasitic deleterious rhizosphere bacteria (Barea et al. 2005). Plant growth promoting rhizobacteria (PGPR), can have biofertilizing and/or biocontrol functions (Barea et al. 2005). However, the effect of rhizosphere bacteria depends mostly on the genotype of the microorganisms and plants involved as well as on the environmental conditions (Brimecombe et al. 2007). *Pseudomonas* spp. and *Bacillus* spp. belong to the largest groups of rhizosphere bacteria (Brimecombe et al. 2007).

PGPR are usually in contact with the root surface, and improve growth of plants by several mechanisms, e.g., enhanced mineral nutrition, phytohormone production, disease suppression (Tarkka et al. 2008). Two groups of PGPR were described: one group is involved in the nutrient cycling and plant growth stimulation (biofertilizers) (Vessey 2003) and the second group is involved in the biological control of plant pathogens (biopesticides) (Whipps 2001). Biofertilizers are based on living microorganisms which (when applied to seed, plant surface or soil) colonize the rhizosphere or the interior of the plant and promote growth by increasing the supply or availability of primary nutrients to the host plant (Vessey 2003). Biopesticides promote plant growth by the control of deleterious organisms, e.g., through the production of antibiotics.

In summary, bacteria may support the plant growth by several mechanisms, e.g.,: increasing the ability of nutrients in the rhizosphere (i), inducing root growth and thereby increase of the root surface area (ii), enhancing other beneficial symbioses of the host (iii) and by combination of modes of action (v) (Vessey 2003).

PGPR can increase the availability of nutrients, e.g., by enzymatic nutrient mobilization from organic matter and production of siderophores (Anderson et al. 1993; Whiting et al. 2001; Jing et al. 2007). Bacteria producing extracellular degrading enzymes are major decomposers of organic matter. They contribute essentially to the soil aggregation and nutrient availability (Johansen and Binnerup 2002). In soils with low phosphate, bacteria can release phosphate ions from low-soluble mineral P crystals and from organic phosphate sources. These bacteria exude organic acids that dissolve the P crystals and exude enzymes that split organophosphate (Vessey 2003; Tarkka et al. 2008). Some rhizosphere bacteria also produce siderophores which can be absorbed as the bacterial Fe^{3+}-siderophore complex by a number of plant species in the deficiency of iron (Vessey 2003). Microbial siderophores in the rhizosphere can significantly contribute to the biocontrol of soil-borne pathogens due to their competitive effects (Hiifte et al. 1994).

PGPR can also promote the root growth. This can be caused by the ability of most rhizobacteria to produce phytohormones, e.g. indole-3-acetic acid (IAA), cytokinins, gibberellins, ethylene which promote cell division and cell enlargement, extension of plant tissue and/or other morphological changes of roots (Salisbury 1994).

2.3.3 Mycorrhization Helper Bacteria

Mycorrhizal fungi and bacteria in the rhizosphere can interact with each other at different levels of cellular integration, ranging from apparently simple associations, through surface attachment, to intimate and obligatory symbiosis. This synergism may not only be important in promoting plant growth and health, but may also be significant to rhizosphere ecology (Perotto and Bonfante 1997).

PGPR can enhance plant growth also indirectly by stimulating the relationship between the host plant and mycorrhizal fungi. Mycorrhizae are often described as tripartite interactions, because in their natural environment, bacteria are associated with arbuscular and ectomycorrhizal fungi by colonizing the extraradical hyphae or as endobacteria living in the cytoplasm of at least some fungal taxa. For rhizobacteria that increase the ability of roots to establish symbiotic interactions with ectomycorrhizal fungi, Garbaye (1994) proposed the term "mycorrhization helper bacteria" (MHB). He suggested a number of possible mechanisms for the bacterial helper effects; for example, the production of vitamins, amino acids, phytohormones and/ or cell wall hydrolytic enzymes. Some of these effects could directly influence the germination and growth rate of fungal structures, whereas others could act on root development and on root susceptibility to fungal colonization (Garbaye 1994). Several reports have demonstrated enhanced arbuscular mycorrhizal fungal colonization in roots in the presence of MHB (e.g., Artursson et al. 2006; Hildebrandt et al. 2002). Although promotion of mycorrhiza formation and plant growth by MHB have been described (e.g., Poole et al. 2001; Artursson et al. 2006), explanations of the mechanisms are often missing.

As a possible strategy for selection of fungus-associated bacterial strains by ectomycorrhizal fungi de Boer et al. (2005) suggested exudation of soluble fungal storage sugars (usually trehalose), polyols (e.g., mannitol) or organic acids (in particular oxalic acid) which can increase the number of bacteria or exudation of inhibitory chemicals which select antibiotic-resistant bacteria. It is hardly known and a future challenge to analyze if there is a specific fungal selection for particular bacterial strains and if cooperation of these bacterial strains is restricted to given ectomycorrhizal fungi, since this would be essential to know for a successful development of joint inoculum.

The majority of MHB, which were described so far, belong to the fluorescent pseudomonads and sporulating bacilli (Garbaye and Bowen 1989; Founone et al. 2002). It seems that MHB include a variety of Gram-negative and Gram-positive species, suggesting that their activities could perhaps be found in all bacterial groups that exist in the rhizosphere (Tarkka et al. 2008). However, the beneficial effects of these bacteria are controlled by soil properties (Oliveira et al. 2005).

2.4 Effect of Rhizosphere Microorganisms on the Plant Growth in Disturbed Soils

Soils can be disturbed by a wide range of factors concerning unfavorable agricultural management or industrial activities. Fertility of various soils observed in the last decades decreased at an alarming rate due to loss of organic matter as a result of erosion, oxidation, compaction, biological impoverishment as well as wide range of pollutants. Plant growth affects the physical (e.g., formation of new soil pores) and chemical (e.g., formation of soil organic matter) quality of soils this is why plant growth at disturbed sites can increase soil fertility substantially. However, in many cases unfavourable conditions in disturbed soils may cause a lack of any vegetation or a diminished vegetation development. Rhizosphere microorganisms are especially critical for plant colonization of unfavourable soils, since they can alleviate biotic and abiotic stress of plants. In this chapter we will review, the present knowledge on plant growth promotion by rhizosphere inhabiting microorganisms in disturbed soils.

2.4.1 Nutrient- and Water-Deficiency

The world population is expanding rapidly and will likely be 10 billion by the year 2050 (Cakmak 2002). The expected increases in world population will result in a serious pressure on the existing agricultural land *via* intensification of crop production. The projected increase in food production must be accomplished on the existing cultivated areas because the expansion of new land is limited due to environmental concerns, urbanization and increasing water scarcity. In the same time, soil productivity is decreasing globally due to enhanced soil degradation in the form of erosion, nutrient depletion water scarcity, acidity, salinisation, depletion of organic matter and poor drainage (Cakmak 2002). Nearly 40% of the agricultural land on the world has been affected by soil degradation (e.g., 25% of European, 38% of Asian, 65% of African, 74% of Central American agricultrural land) (Scherr 1999). Agricultural production must be increased on the existing land, and therefore crop production must be intensified per unit of agricultural land. Mineral nutrients are the major contributor to enhancing crop production, and in maintaining soil productivity and preventing soil degradation. Generally, improving the nutritional status of plants by maintaining soil fertility is the critical step in the doubling of food production of the world. Impaired soil fertility by continuous cropping with low supply of mineral nutrients is considered a major risk for food production and ecosystem viability (Pinstrup-Andersen et al. 1999; Tillman 1999). Reduced soil fertility and crop production results in an increased pressure to bring more land into crop production at the expense of forests and marginal lands. Such areas are generally poor in fertility and sensitive to rapid degradation when cultivated (Cakmak 2002). Since especially these habitats are often very important for rare plant and animal species, it would be much better to find solutions for the remediation and later use of disturbed soils, like former sewage fields or ash dumps, if not for food at least for biomass production. A promising strategy is the integration of plant nutrition research with development of new biotechnological strategies to promote the plant growth through applications of suitable microorganisms (biofertilization). An increased nutrient use efficiency through improved soil management become an important challenge, particularly for the elements nitrogen and phosphorus.

A group of especially promising rhizosphere organisms for applications on nutrient-deficient disturbed soils are mycorrhizal fungi. They can biotrophically colonize the root cortex and develop an extramatrical mycelium which helps the plant to acquire mineral nutrients and water from the soil. They play a key role in nutrient cycling in ecosystems and their external mycelium, in association with other soil organisms, form water-stable aggregates necessary for a good soil quality (Azcón-Aguilar and Barea 1997). Moreover, it was demonstrated that arbuscular mycorrhizal fungi produce glomalin—a glycoprotein, which has been suggested to contribute to hydrophobicity of soil particles and participate in the initiation of soil aggregates (Barea et al. 2002).

Mycorrhizal fungi are known to enhance the nutrient supply of their host plants especially in nutrient-deficient conditions, e.g., by a very efficient soil exploitation and by excretion of significant quantities of phosphatases (Tibbett et al. 1998; Tibbett and Sanders 2002). Mechanisms of ectomycorrhizal fungi for adaptation on N- and P-limitation in unfavourable soil conditions, e.g., by the utilization of seed protein N were demonstrated by Tibbett et al. (1998). Ectomycorrhizal fungal species vary in their ability to acquire specific nutrients from soil (Leake and Read 1997; Erland and Taylor 2002) and this differential efficiency can be considered as niche partitioning. They may use the same substrate but extract different components. This idea could maintain the theory of high ectomycorrhizal fungal diversity in the nutrient-deficient soils. Unfortunately with the exception of N, there is little information available on the effects of specific nutrients upon ectomycorrhizal fungal diversity (Erland and Taylor 2002).

2.4.2 Extreme Soil pH

The soil pH is correlated with various biological and other chemical soil properties. About 40% of cultivated soils globally have acidity problems leading to significant decreases in crop production despite adequate supply of mineral nutrients such as N, P and K (Herrera-Estrella 1999; von Uexküll and Mutuert 1995). In acid soils major constraints to plant growth are toxicities of hydrogen (H), aluminium (Al) and manganese (Mn) and deficiencies of P, calcium (Ca) and magnesium (Mg). Among these constraints Al toxicity is the most important yield-limiting factor (Marschner 1991). Availability of P to plant roots is limited both in acidic and alkaline soils, mainly, due to formation of sparingly soluble phosphate compounds with Al and Fe in acidic and Ca in alkaline soils (Marschner 1995). Plant species have evolved adaptive mechanisms to improve their ability to cope with soils having low levels of available P by the formation of mycorrhizal association (Marschner 1995; Dodd 2000). Mycorrhizal colonization of plants enhances their ability to explore the soil for P through the action of the fungal mycelium. This results in increased exploration of the soil for available nutrients and delivers more mineral nutrients, particularly P, to plant roots (Dodd 2000; George and Marschner 1996; Marschner 1998). It is estimated that the extent of fungal mycelium may be in the range of 10–100 m per cm root or per gram of soil under field conditions in P-poor soils (McGonigle and Miller 1999). In general, the contribution of mycorrhizal associations to the plant nutrient supply is larger in soils with poor availability of mineral nutrients than in soils rich in nutrients. In pot experiments, mycorrhizal colonization contributed to the total P uptake with between 70 and 80% and to the total Zn and Cu uptake with 50 and 60% in white clover (*Trifolium repens*) (Li et al. 1991).

Mycorrhizal fungi vary with regard to their pH optima for growth and root colonization potential (Erland and Taylor 2002). Changes in soil pH can alter the enzymatic activities of some fungi, since at least some of the enzymes produced by EM have rather narrow pH optima (Leake and Read 1997). The colonization density of the ectomycorrhizal fungal species *Cenococcum geophilum* increased on beech (*Fagus sylvatica* L.) with decreasing soil pH (Kumpfer and Heyser 1986). So far, little is known on the preferences of ectomycorrhizal fungi to alkaline soils, e.g., on fly ash deposits.

Beside mycorrhizal fungi, bacteria can essentially improve the adaptation of plants to an extreme soil pH, although their distribution itself is controlled primarily by the soil pH (Fierer and Jackson 2006). Most prokaryotes grow at relatively narrow pH ranges close to neutrality. A general adaptation to extreme pH levels is to regulate the intracellular pH and keep it close to neutral. Some enzymes found in the bacterial outer membrane tend to have low pH optima, whereas all known cytoplasmic enzymes have pH optima from pH 5–8 (Torsvik and Øvreås 2002). Also in disturbed arable and landfill soils the pH was a leading control of the density of cultivable bacteria in the rhizosphere of willows (*Salix* spp.) (Hrynkiewicz et al. 2010a).

Re-vegetation of fly ash dumps, installed for final storage of this principal byproduct of coal-fired power stations, is hampered by their unfavourable chemical and physical properties for plant growth in general and in detail often by an unfavourable soil pH (often higher than 8.5) (Hrynkiewicz et al. 2009). This pH leads to a strong deficiency of essential nutrients (usually N and P), high soluble concentrations of trace elements and is often combined with the presence of compacted and cemented layers (Selvam and Mahadevan 2000). The fly ash is composed of small particles (<200 μm) and its physical composition is very uniform (Pillman and Jusaitis 1997). Beside the high soil pH, high hydrophobicity of the particle surfaces of fly ash cause water deficiency and low additionally plant growth and also microbial colonization. However, it was revealed that site-adapted rhizosphere microorganisms could significantly promote the plant establishment and growth even on fly ash (Hrynkiewicz et al. 2009). The promotion of autochtonous ectomycorrhizal fungi with site-adapted inoculated rhizosphere bacteria promoted the growth of willows (*Salix* spp.) in the mentioned study significantly.

2.4.3 Soil-Borne Pathogens

Soil is a reservoir for many potential plant pathogens and especially plants with a decreased vitality, like in unfavourable soil conditions, can be infected by them. Primary conditions that promote infection of roots can be poor sanitation, inadequate drainage and improper irrigation. Moreover, intensified production in agriculture, connected with increased use of agrochemicals may cause several negative effects e.g., development of pathogen resistance to the applied agents and their unforeseen impacts on the environment (Compant et al. 2005). Decreasing the vegetation diversity leads to pauperization of soil inhabitants, decreasing of interconnectedness and functional interchangeability (van Bruggen et al. 2006). An extremely simplified vegetation, such as a monoculture, selects a specific microbial community, including plant pathogenic microorganisms and sometimes also their parasites or antagonists (Bruggen et al. 2006). Population levels of soil-borne pathogens, include bacteria, fungi and some viruses.

Some root associated micororganisms, like mycorrhizal fungi can significantly increase the resistance of plants to soil-borne pathogens. Most of soil-borne plant diseases of roots (e.g., root rot or wilting) are caused by fungi such as *Rhizoctonia*, *Fusarium*, or *Verticillium* spp. or by oomycetes (e.g., root rot) including *Phytophtora*, *Pythium*, and *Aphanomyces* spp. (Whipps 2004). Alleviation of deleterious effects by mycorrhizal fungi were also observed in case of parasitic nematodes (de la Peña et al. 2006) and phytophagous insects (Gange 2006). Mycorrhizal effects on aboveground diseases largely rely on the lifestyle and challenge strategy of the attacker (Pozo and Azcón-Aguilar 2007). However, the ability to enhance resistance/tolerance differs among mycorrhizal isolates (i), the protection is not effective for all pathogens (ii), and the protection is modulated by environmental conditions (iii) (Pozo and Azcón-Aguilar 2007). Different mechanisms can participate in plant protection from root pathogens by mycorrhizal fungi: improved nutrient status of the host plant (i), competitive interactions with pathogenic fungi (ii), anatomical or architectural changes in the root system (iii), microbial community changes in the rhizosphere (iv) and activation of plant defence mechanisms (v) (Wehner et al. 2010).

Root pathogens can cause a considerable loss of tree seedlings in nurseries and are generally difficult to control using conventional methods. It was already well documented that inoculation of tree seedlings in nurseries with EM fungi may provide some suppression of pathogens (Duchesne 1994). Ectomycorrhizal fungi have additionally a mechanical barrier effect against infection of roots with pathogens provided by their hyphal mantle (Duchesne et al. 1987), can produce antibiotics (Duchesne 1994; Schelkle and Peterson 1996) or acidify alkaline soils (Rasanayagam and Jeffries 1992).

Also plant associated bacteria can be used sucessfully for the biocontrol of soil-borne pathogens (e.g., Sturz et al. 2000; Compant et al. 2005). Biocontrol mechanisms mediated by bacteria are: competition for an ecological niche or substrate (i), production of inhibitory allelochemicals (ii), and induction of systemic resistance in host plants to a broad spectrum of pathogens and/or abiotic stress (iii) (Compant et al. 2005).

Biocontrol effects of bacteria and mycorrhizal fungi can be combined successfully. For example, *Bacillus subtilis* inhibited the growth of root pathogens like *Fusarium* and *Cylindrocarpon* spp. in co-inoculation with ectomycorrhizal fungi (*Laccaria proxima*, *Suillus granulatus*) (Schelkle and Peterson 1996). Suppression of pathogenic fungi by helper bacteria may be direct by production of antibiotics (Malajczuk 1988) or indirect by stimulation of mycorrhizae formation on the plant roots. Both mycorrhizal fungi and bacteria may synthesise siderophores (Hrynkiewicz et al. 2010a, b) which are involved in the inhibition of pathogenic fungi (Neidhardt et al. 1990).

2.4.4 Heavy Metal Contamination

Especially ectomycorrhizal fungi can promote the establishment of plant species at heavy metal contaminated sites by immobilizing heavy metals in the soil, thereby reducing the availability of metals to plants (Chanmugathas and Bollag 1987; Fomina et al. 2005). As a result, plants colonized with these fungi show a higher tolerance to toxic metal concentrations (Brown and Wilkins 1985). Mechanisms of ectomycorrhizal fungi for amelioration of heavy metal stress are (i) sorption of metals in the hyphal sheath, (ii) reduction of their apoplastic mobility as a result of hydrophobicity of the fungal sheath, (iii) exudation of chelating substances or (iv) sorption of metals on the external mycelium (Jentschke and Godbold 2000; Turnau et al. 1996). Chelating substances of fungal origin which are able to bind metals are for example organic acids or slimes at the surface of mycelia (Jentschke and Godbold 2000). The efficiency of plant protection by ectomycorrhizal fungi differs between distinct isolates and different toxic metals (Meharg and Cairney 2000). A high intra-specific heterogeneity in metal tolerance was found in an *in vitro* screening with 49 strains of 5 species of ectomycorrhizal fungi from polluted and non-polluted sites with increased Cd, Cu, Ni and Zn contents (Blaudez et al. 2000). There can be also strong differences in the metal tolerance even between ectomycorrhizal plants of the same family. The reasons for particularly successful protection of host plants by distinct communities of ectomycorrhizal fungi are still unknown (Abler 2004). Toxic metals are believed to affect fungal populations by reducing the abundance, species diversity and selection for a tolerant population (Duxbury 1985). It has been hypothesized that tolerant EM forming plants are protected by well-adapted ectomycorrhizal fungi and that such fungi might be advantageous to remediate polluted sites more effectively by a general plant growth promotion.

It was shown, that mycorrhizal fungi, in spite of their possible metal restraining, can increase the total heavy metal uptake of their host plants through an increased plant growth (Jentschke and Godbold 2000; Schützendübel and Polle 2002). Mycorrhizal fungi are especially effective in plant protection against abiotic stress in increased heavy metal concentrations (Meharg and Cairney 2000; Schützendübel and Polle 2002). Phytoremediation, the remediation of contaminated soils and water with plants, is a promising and relatively cheap clean-up strategy and suitable in case of moderate and low contaminations (Pilon-Smits 2005). Enhanced phytoremediation can be achieved by inoculation of the plant rhizosphere with selected microorganisms (Siciliano and Germida 1999). Beside mycorrhizal fungi, rhizobacteria can contribute to increase the plant growth and metal uptake of plants on polluted soils substantially, however less is known on their mechanisms of metal protection of plants.

2.4.5 Organic Pollutants

Organic pollutants can be degraded by plants through biochemical reactions taking place within the plants and in the rhizosphere. The remediation of soils containing diverse organic pollutants, including organic solvents, pesticides, explosives and petroleum is possible with the use of plants and their rhizosphere processes (phytodegradation) (Mirsal 2004). Phytodegradation of organic pollutants may be enhanced by bacterial activities. In this process, plants interact with soil microorganisms by providing nutrients in the rhizosphere which leads to an increased microbial activity and degradation of toxic pollutans (Mirsal 2004).

Mycorrhizal fungi and rhizobacteria were demonstrated to promote plant growth and degradation of pollutants in soils with increased pollutant concentrations. For example, EM associations can display considerable resistance against toxic organic compounds such as m-toluate (Sarand et al. 1999), petroleum (Sarand et al. 1998), or polycyclic aromatic hydrocarbons (Leyval and Binet 1998; Wenzel 2009). Densely packed mycorrhizal sheaths and phenolic inter-hyphal material can protect plant roots from direct contact with the pollutant (Ashford et al. 1988).

Degradation of organic pollutants was also revealed by rhizosphere (Rentz et al. 2005) and endophytic bacteria (Wang and Dai 2011).

In summary, very specific adaptations of microorganisms on different unfavourable soil conditions are required and the adaptation is even more difficult through the fact that several unfavourable soil conditions are often combined, e.g. low soil pH with high metal mobility and low P availability. Thus, in soils with long-term stress a selection according to the site-specific needs can be supposed. The investigation of site-adapted cultivable microorganisms in unfavourable soils will contribute to identify leading controls of site-specific populations and to provide fundamental knowledge and strain collections for subsequent selections and applications of plant growth and site remediation promoting microbial strains (Hrynkiewicz et al. 2010b).

2.5 Selection and Use of Microorganisms for the Promotion of Plant Growth and Soil Remediation

Positive effects of mycorrhizae and/or rhizobacteria on plant growth and health as biostimulators, biofertilizers and/or bioprotectors in sustainable agriculture and horticulture were described by many authors (e.g., Azcón-Aguilar and Barea 1997; Barea et al. 2002, 2005; Compant et al. 2005; Ryan et al. 2009). A rising portion of disturbed soils worldwide leads to an urgent need of successful remediation strategies. It is correlated with an intense search for microorganisms, which are site-adapted and able to promote the plant growth on disturbed soils and in this way the phytoremediation of such sites.

Mycorrhizal fungi and bacteria can be well adapted to harsh soil conditions and promote the remediation of disturbed soils directly and by plant growth promotion (Schützendübel and Polle 2002; Fomina et al. 2005; Baum et al. 2006; Zimmer et al. 2009; Wenzel 2009). However, their field application is still very limited caused e.g., by the lack of the knowledge how to calculate the biological and economical efficiency and by the lack of fast and effective site-specific selection procedures. Maximum on-site benefits will only be obtained from inoculation with efficient fungi and/or bacteria in compatible host/microorganism/site combinations. Compatible highly-effective microorganisms can contribute significantly to the biological degradation (organic pollutants) or removal (heavy metals) of contaminants. This is why the selection of microbial inoculum is a main support of successful biotechnological application in the field.

2.5.1 Selection Criteria of Microorganisms for Inoculation of Unfavourable Soils

Applications of inoculation with microorganisms provide a great challenge in the future to increase crop production, cure problems with nutrient uptake, control plant

2 The Potential of Rhizosphere Microorganisms to Promote the Plant Growth   49

pathogens (Bashan 1998) and remediate disturbed soils. However, several obstacles must be overcome to achieve the successful commercialization of such treatments.

On the one hand, microbial inoculum must be economically mass-produced, formulated into a cost-effective and readily applicable product to be commercially successful (Bashan 1998). Microbial inoculum should be relatively universal for various plants and soils and its effectiveness should be relatively easy to evaluate on a standard scale. One serious problem is, that many experiments point to plant- and soil-specificity instead of being universal (Vessey 2003). Most relevant and valuable for the investigation of microorganisms for subsequent applications in remediation of polluted soils are experiments and observations made in natural conditions. However, only detailed investigations (e.g., physiological and molecular characterization) on individual microorganisms with high precision and in controlled environmental conditions can explain basic causal mechanisms of their effects on plants and soil. Therefore, all levels of methodological hierarchy are important for detection, explanation and subsequent controlled use of interactions between plants and associated soil microorganisms (Read 2002).

2.5.1.1 Selection of Microorganisms Naturally Adapted to Unfavourable Soil Conditions

Information on the diversity of microorganisms at polluted sites is supposed to be especially valuable for a future selection of microbial inoculum for such sites (e.g., Derry et al. 1998; Liu et al. 2000; Hrynkiewicz et al. 2008). Usually, at such sites a decline in the number, diversity and activity of microorganisms was observed (Schloter et al. 2003; Maila et al. 2006; Labud et al. 2007). The relationships between microbial diversity and ecosystem sustainability are still poorly understood. Information on microbial diversity and activity may provide evidence of ecosystem degradation, but might be also a valuable source of information on -site-adapted microorganisms for future application as microbial inoculum. Molecular techniques, e.g. denaturating gradient gel electrophoresis (DGGE), terminal restriction fragment length polymorphism (T-RFLP) or PCR-single-strand conformation polymorphism (PCR-SSCP), have provided detailed information on the taxonomic and phylogenetic relationships within the major groups of mycorrhizal fungi and bacteria. They can contribute to describe the host-specificity and co-evolution between plants and mycorrhizal fungi and describe the mycorrhizal community structure in the field (Read 2002). Progress towards a fundamental understanding how mycorrhizae and associated bacteria influence the vegetation development and effects at polluted sites is still a big challenge.

2.5.1.2 Properties of Successful Microbial Inoculum

The majority of soil microorganisms (95–99%) is known to be at least so far non-culturable (Torsvik and Øvreås 2002). However, the basic criterium for subsequent selection and later application of microbial inoculum useful for plant-growth promotion is cultivability and fast multiplication of microorganisms. Information of critical factors influencing plant-microbe-pollutant interactions in soils could lead to an improved selection of microbial inoculum for a microbial-assisted bioremediation. A fundamental condition for subsequent on-site applications of selected microorganisms is their safety for the environment and for humans. Therefore, before field applications, all selected microorganisms have to be precisely identified and toxicologically assessed.

Very few different microbial taxa have been tested so far for their capability to promote plant growth at disturbed and polluted soils and little is known on the microbial spectrum which might be especially relevant to promote plant species in disturbed soils. In general, numerous species of mycorrhizal fungi and soil bacteria which inhabit the rhizosphere can promote plant growth (Vessey 2003; Compant et al. 2005), e.g., by enzymatic nutrient mobilisation from organic matter (mostly P and N) and production of siderophores (Whiting et al. 2001; Jing et al. 2007) and might be promising also for disturbed or polluted soils. They can contribute essentially to soil aggregation and nutrient availability (Johansen and Binnerup 2002), which is often especially important for disturbed soils.

Therefore, enzyme activities can be suitable selection criteria for microbial inoculum for plant growth promotion in disturbed soils. Microbial enzyme activities in the soil were predominantly measured as total potential activities rather than at the level of isolates within a community (e.g., Khan et al. 2007). However, investigations of single strains are necessary for the selection of potential inoculum (Hrynkiewicz et al. 2010b).

Acid phosphatases contribute to the P mobilisation from organic matter. *In vitro* synthesis of extracellular phosphatases by ectomycorrhizal fungi was investigated by Colpaert and Van Laere (1996). These enzymes cause the release of phosphate from a range of substrates as inositol phosphate, polyphosphates, phosphorylated sugars into the soil solution (Tibbett et al. 1998). The production of these enzymes is species- and strain-dependent and often stimulated by deficiency of mineral phosphate (Dighton 1991). It was revealed that a strain of the ectomycorrhizal fungus *Paxillus involutus*, which synthesized significantly higher amounts of acid phosphatases than an other tested strain of this species, promoted the mycorrhiza formation and biomass production of willows (*Salix* spp.) also more successfully (Hrynkiewicz et al. 2010b).

In the present case, the strain-specific phosphatase activity *in vitro* was a suitable criterion for the selection of plant growth promoting candidates. However the relevance of single enzymes for selection might differ site-specific, since the growth limiting elements might be different.

Beside the phosphatase activity, cellulolytic and pectolytic activities have been used for selection of microorganisms for promotion of plant growth and mycorrhiza formation. High cellulolytic and pectolytic activities of mycorrhizal fungi and rhizosphere bacteria allow the disintegration of living and dead plant tissue (Wood 1960; Bateman and Miller 1966) and, consequently, can enable microorganisms to enter roots. High cellulolytic and pectolytic activities were detected among mycorrhizal fungi (Garbaye 1994) and their helper bacteria (Duponnois and Plenchette 2003). Therefore, also cellulolytic and pectolytic activities might be suitable selection criteria.

Furthermore, lipolytic activities might be relevant for the selection of microorganisms especially for biodegradation, since they can improve not only the N supply of plants but also promote the biodegradation of organic pollutants (e.g., petroleum-derived wastes) in soils (Chaturvedi et al. 2006).

A further suitable selection criterion for plant growth promoting microorganisms for disturbed soils is the ability to produce siderophores (Burd et al. 1998; Kuffner et al. 2008). Siderophores are complexing compounds released to improve iron acquisition, which is an essential nutrient of plants, but relatively insoluble in soil solution (Hu and Boyer 1996; Hrynkiewicz et al. 2010a, b). In rhizosphere microbial communities siderophore synthesis might be especially important for successful competition of rhizosphere microorganisms in disturbed soils with extremely low nutrient concentrations. Beside their direct effects on the iron supply of plants, siderophores can contribute additionally to the suppression of pathogens in the rhizosphere through their withhold from iron supply (Buyer and Leong 1986).

Furthermore, auxins are recognized as highly active plant growth stimulators, and indole-3-acetic acid (IAA) is a key substance (Woodward and Bartel 2005). Indole-3-acetic acid (IAA) production is widespread among soil microorganism, mostly ectomycorrhizal fungi (e.g., Rudawska and Kieliszewska-Rokicka 1997; Niemi et al. 2004; Niemi and Scagel 2007; Hrynkiewicz et al. 2010a). Several authors revealed that fungal strains with high IAA-synthesizing activity induce stronger growth of fine roots and significantly higher numbers of mycorrhizae compared to strains with low activity of IAA (Normand et al. 1996; Rudawska and Kieliszewska-Rokicka 1997; Karabaghli et al. 1998).

In summary, enzyme activities, production of siderophores and auxins can be used successfully for the selection of highly active microbial strains for the promotion of plant growth in disturbed soils. However, beside these criteria, the selection of suitable combinations of host plants and microbial inoculum is necessary.

Specificity of combinations of mycorrhizal fungal and bacterial strains as well as host plants for the remediation of disturbed soils is rarely known. It is still in discussion if a specific fungal selection of particular bacterial strains exists and whether cooperation of these bacterial strains is restricted to given ectomycorrhizal fungi. In several previous published works (e.g., Baum et al. 2006; Hrynkiewicz et al. 2009; Zimmer et al. 2009) it was demonstrated that interactions of mycorrhizal fungi and bacteria can be significantly growth promoting even in situations when the microorganisms used as inoculum does not originate from the same host plant and site. Also several previous studies (e.g., Bianciotto et al. 1996; Jana et al. 2000; Xavier and Germida 2003) revealed a low specialization of bacterial strains to mycorrhizal fungi and their host plants. This feature of inoculum might assure a broader spectrum for practical applications of microbial inoculum. As a possible mechanism for selection of fungus-associated bacterial strains by ectomycorrhizal fungi de Boer et al. (2005) suggested exudation of soluble fungal storage sugars (usually treha-

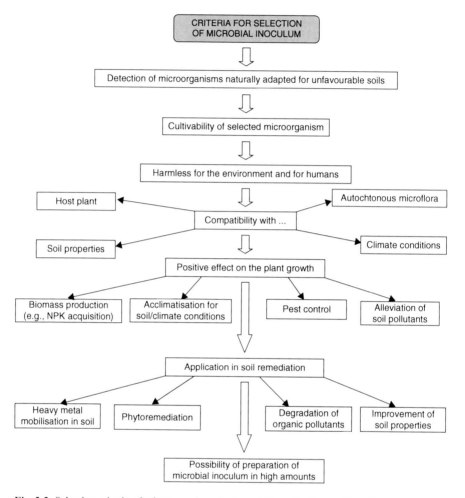

Fig. 2.2 Selection criteria of microorganisms for inoculation of unfavourable soils

lose), polyols (e.g., mannitol) or organic acids (in particular oxalic acid) which can increase the number of bacteria or exudation of inhibitory chemicals which select antibiotic-resistant bacteria (Fig. 2.2).

2.5.2 Chances and Risks of In Vitro Selection for Field Applications

Special attention should be payed to the complexity of interactions in plant-microbe-soil-pollutant systems in natural conditions. Emphasis should be layed on the evaluation of results obtained form *in vitro* and pot experiments in comparison

2 The Potential of Rhizosphere Microorganisms to Promote the Plant Growth

to the results from heterogenous, varied polluted field sites and the functioning of phyto-/rhizoremediation systems under various ecological conditions (Wenzel 2009). Moreover, some isolated microorganisms can effectively degrade single pollutants *in vitro*, but when introduced into actual field conditions with varied combinations of pollutants, they cease to function as anticipated (Quan et al. 2003; Singer et al. 2005).

In addition, introduced strains may not compete well with the indigenous microorganisms in the soil to remain dominant or viable (Bouchez et al. 2000; Das and Mukherjee 2007; Mohanty and Mukherji 2008; Supaphol et al. 2006). Bioremediation efficiency is thus a function of the ability of the inoculated microbial degraders to remain active in the natural environment (Alexander 1999).

The experimental scale and the bioavailability of the pollutants on the sites have to be considered, since some microbial treatments which were successful *in vitro* and in pot experiments failed in long-term contaminated soils in the field (Wenzel 2009). However, the scaling up of processes in the plant-soil-microbe system (as for example the bioremediation of industrially contaminated sites) represents one of the greatest challenges facing environmental scientists and yet is essential for sustainable land management worldwide (Standing et al. 2007). In this connection, investigations also on the microscale will be essential furthermore, since they provide basic information for subsequent tests in the macroscale. Enhanced degradation capabilities of inoculated microorganisms may be obtained by the selection, breeding and engineering of plants that exude specific carbon substrates (e.g., opines, flavonoides) that can be preferentially used by the microbial degrader strains/populations present in the polluted soil (Wenzel 2009). Additionally, e.g., root exudate compounds (phenylpropanoids) were identified, which created a nutritional bias in favor of enhanced polychlorinated biphenyl (PCB) degradation (Narasimhan et al. 2003). Furthermore, several experiments were carried out with genetically modified rhizosphere bacteria which were able to enhance the rhizodegradation of organic pollutants, e.g., *Pseudomonas fluorescens* expressing enzymes for toluene degradation (Yee et al. 1998), *Rhizobium tropici* expressing enzymes for degradation of dioxine-like substances (Saiki et al. 2003). Bacterial biosensors, which were designed for the detection of chemical, physical or biological signals via the production of a suitable reporter protein (e.g., green fluorescent protein, GFP), can be used successfully for fundamental research on bioremediation of polluted soils. However, such experiments so far are restricted to the lab scale, since ecological and public consequences have to be tested before their application in the field might be permitted (Wenzel et al. 2009).

2.5.3 Status and Perspectives of Commercialisation of Microbial Inoculum

Use of microbial inoculum instead of indigenous microbial populations might be preferable or even necessary to diminish plant stress or increase the biomass production at disturbed sites, with partly low or functionally-disturbed diversity of microorganisms in the soil. Applications of microbial inocula might provide a great challenge to improve the nutrient supply of plants and to control plant pathogens (Bashan 1998) or to alleviate drought stress of plants (Boomsma and Vyn 2008). However, before microbial inoculation becomes a viable technology several obstacles must be overcome. Microbial inoculum to be commercially successful must be economically mass-produced, formulated into a cost-effective and readily applicable product (Bashan 1998). Mycorrhizal and/or bacterial inoculum should be relatively universal for various plants and soils and its effectiveness should be relatively easy to evaluate on a standard scale (Vessey 2003). Solution of these and similar problems is challenging, since research of these problems is rather limited to relatively few microbial and plant taxa (Table 2.1).

2.6 Concluding Remarks and Outlook

Microbial activity in the rhizosphere contributes significantly to the sustainability of agriculture and forestry as well as to the remediation of disturbed soils. Selective promotion of desirable rhizosphere processes requires a fundamental understanding of the complex microbial interactions in the rhizosphere. Especially, mycorrhizal fungi and rhizobacteria belong to the microorganisms in the rhizosphere, which contribute essentially to increase the soil fertility and remediate physically and chemically disturbed soils.

Inoculation of soils with selected plant growth and soil remediation promoting microorganisms has the capacity to improve the plant fitness in unfavourable conditions of polluted soils and increase the plant uptake of heavy metals and the degradation of organic pollutants. The great challenge is the successful use of such inoculum in the field with natural environmental conditions and competition by autochthonous microorganisms. Complex and integrated approaches for the rhizosphere management are required, since disturbed soils are usually characterised by a high complexity and heterogeneity. An improved fundamental knowledge of physiological traits of rhizosphere microorganisms and their impact on rhizosphere processes, which are especially relevant for the remediation of disturbed soils will be essential to allow an increased and successful use of microbial inoculum in the field. The development of microbial inoculum for an improved remediation of disturbed soils should focus on such sites, where less complicated alternatives for growth promotion, like e.g. crop-selection or fertilization, failed and the lack of valuable autochthonous microorganisms was assumed. At present it seems necessary to use always site-specific selections of inoculum, since a general suitability of inoculum for diverse site conditions seems rather unlikely.

Table 2.1 Microbial inoculum with the criteria of selection and suggested mechanisms of action

Group of microorganisms	Criteria of selection	Suggested mechanisms of action (activity) and effect	References
EM fungi and/or PGPR bacteria	Improvement of plant fitness in unfavourable soil conditions (e.g., with high heavy metal concentrations): – increased biomass production (shoot and root); – increased shoot lengths	– Mutalistic benefit from mycorrhiza formation; – Physical protection of plant roots (fungal mantle) from direct contact with pollutants	Baum et al. (2006), Wu et al. (2006), Zaidi et al. (2006), Zimmer et al. (2009) and Hrynkiewicz et al. (2009, 2010a)
EM fungi	Morphological features of EM: – emanating hyphe and rhizomorphes, – thick and densely packed mycorrhizal sheaths with – phenolic inter-hyphal compounds	– Physical protection of plant roots (fungal mantle) from direct contact with the pollutant and impeded pollutant transport through increased soil hydrophobicity – Large surface increase efficiency in nutrient and water availability by plants; – Large cation exchange capacity of extramatrical hyphae reduce bioavailable concentrations of pollutants, through their substantial adsorption capacities	Ashford et al. (1988), Colpaert and Asche (1993), Hartley et al. (1997), Marschner (1998), Meharg and Cairney (2000) and Agerer (2001)
AM fungi	Increase, no effect or decrease of the uptake of metals by plants	Mechanisms not specified caused by the high diversity of effects	Wenzel et al. (2009)
EM fungi and bacteria	Broad spectrum of enzymatic activity	– Increased in plant nutrient supply (NPK); – Utilization of rhizodeposition-specific C sources – Increased metal uptake by the plants	Baum et al. (2006) and Hrynkiewicz et al. (2010a, b)
Bacteria	– N_2-fixing (e.g. *Azotobacter chroococcum*); – P-solubilising (e.g., *Bacillus megaterium*); – K-solubilising (e.g., *Bacillus mucilaginosus*)	– Increased plant nutrient supply – Accelerated phytoremediation of contaminated soils	Wu et al. (2006)

Table 2.1 (continued)

Group of microorganisms	Criteria of selection	Suggested mechanisms of action (activity) and effect	References
EM fungi and bacteria	Synthesis of chelators: – metallothioneins, – siderophores, – organic acids	Detoxification of pollutants by the following strategies: – increased solubility and changes in speciation of metals/metalloids through the production of organic ligands via microbial decomposition of soil organic matter; – exudation of metabolites (e.g., organic acids) and siderophores that can complex cationic metals or desorb anionic species (e.g., arsenate) by ligand exchange; – immobilization of cationic metals (e.g., Cd, Cu, Zn) by microbial siderophores	Mehra and Winge (1991), Neubauer et al. (2002), Gadd (2004), Hrynkiewicz et al. (2010a, b) and Wenzel (2009)
Bacteria	Synthesis of biosurfactants (e.g., rhamnolipids used in remediation of crude oil from the soil)	Mobilization of hydrophobic pollutants from soil particle surfaces, enabling their transport to sites of high degradation activity	Urum et al. (2004)
Ericoid and EM fungi, bacteria	Changes in the pH (reduction and oxidation processes) of the rhizosphere	Modification of the solubility of metals and metalloids in the soil	Martino et al. (2003), Fomina et al. (2005) and Gadd (2004)
EM fungi and bacteria	Synthesis of secondary compounds: – auxin-like substances – vitamins	– Plant growth promotion in polluted soils, – accelerated phytoremediation of polluted soils	Hrynkiewicz et al. (2010a, b) and Zaidi et al. (2006)
Bacteria	Synthesis of ACC (1-aminocyclopropane-1-carboxylate) deaminase	Increased tolerance of plants to high heavy metal concentrations in the soil: – high ethylene concentrations produced by plant roots in response to abiotic stress; – Regulation of ethylene concentrations in plants by ACC deaminase via metabolisation of the ethylene precursor ACC into a-ketobutyric acid and amonia	Burd et al. (1998), Belimov et al. (2001), Glick (2005), Idris et al. (2004) and Arshad and Frankenberger (2002)

Table 2.1 (continued)

Group of microorganisms	Criteria of selection	Suggested mechanisms of action (activity) and effect	References
EM fungi and bacteria	Degradation of soil pollutants (e.g., hydrocarbons)	– Decrease of the concentrations of pollutants and alleviation of abiotic stress for plants	Meharg and Cairney (2000) and Chaineau et al. (2005)
EM fungi	Accumulation of heavy metals in the fungal biomass	Alleviation of abiotic stress for plants	Berthelsen et al. (2000), Blaudez et al. (2000), Leyval and Joner (2000) and He et al. (2000)
EM fungi	Presence of antioxidative systems (e.g., increased activity of Mn-SOD in Cd treated fungi)	Alleviation of abiotic stress for plants	Jacob et al. (2001)
EM fungi and bacteria	Tolerance to high concentrations of diverse pollutants (e.g., heavy metals, organic compounds)	High microbial ability to tolerate abiotic stress	Wenzel (2009)

References

Abler RAB (2004) Trace metal effects on ectomycorrhizal growth, diversity, and colonisation of host seedlings. Doctoral Thesis, Blackburg, Virginia

Agerer R (1991) Ectomycorrhizae of *Sarcodon imbricatus* on Norway spruce and their chlamydospores. Mycorrhiza 1:21–30

Agerer R (2001) Exploration types of ectomycorrhizae. A proposal to classify ectomycorrhizal mycelial systems according to their patterns of differentiation and putative ecological importance. Mycorrhiza 11:107–114

Alexander M (1999) Biodegradation and Bioremediation, 2nd edn. Academic Press, New York

Anderson TA, Guthrie EA, Walton BT (1993) Bioremediation in the rhizosphere: plant roots and associated microbes clean contaminated soil. Environ Sci Technol 27:2630–2636

Arshad M, Frankenberger WT (2002) Ethylene: agricultural sources and applications. Kluwer Academic, New York

Artursson V, Finlay RD, Jansson JK (2006) Interactions between arbuscular mycorrhizal fungi and bacteria and their potential for stimulating plant growth. Environ Microbiol 8:1–10

Ashford AE, Peterson CA, Carpenter JL, Cairney JWG, Allaway WG (1988) Structure and permeability of the fungal sheath in the Pisonia mycorrhiza. Protoplasma 147:149–161

Azcón-Aguilar C, Barea JM (1997) Applying mycorrhiza biotechnology to horticulture: significance and potentials. Scientia Horticulturae 68:1–24

Barea JM, Azcón R, Azcón-Aguilar C (2002) Mycorrhizosphere interactions to improve plant fitness and soil quality. Antonie van Leeuwenhoek 81:343–351

Barea JM, Pozo MJ, Azcon R, Azcon-Aquilar C (2005) Microbial co-operation in the rhizosphere. J Exp Bot 56:1761–1778

Barriuso J, Pereyra MT, Lucas Garcia JA, Megias M, Gutierrez Manero FJ, Ramos B (2005) Screening for putative PGPR to improve establishment of the symbiosis *Lactarius deliciosus-Pinus* sp. Microb Ecol 50:82–89

Bashan Y (1998) Inoculants of plant growth-promoting bacteria for use in agriculture. Biotechnol Adv 16:729–770

Bateman DF, Miller RL (1966) Pectic enzymes in tissue degradation. Ann Rev Phytopathol 4:119–146

Baum C, Hrynkiewicz K, Leinweber P, Meißner R (2006) Heavy-metal mobilization and uptake by mycorrhizal and nonmycorrhizal willows (*Salix×dasyclados*). J Plant Nutr Soil Sci 169:516–522

Belimov AA, Safronova VI, Sergeyeva TA, Egorova TN, Matveyeva VA, Tsyganov VE, Borisov AY, Tikhonovich IA (2001) Characterisation of plant growth-promoting rhizobacteria isolated from polluted soils and containing 1-aminocyclopropane-1-carboxylate deaminase. Can J Microbiol 47:642–652

Bertaux J, Schmid M, Prevost-Boure NC, Churin JL, Hartmann A, Garbaye J, FreyKlett P (2003) In situ identification of intracellular bacteria related to *Paenibacillus* spp. in the mycelium of the ectomycorrhizal fungus *Laccaria bicolor* S238N. Appl Environ Microb 69:4243–4248

Berthelsen BO, Lamble GM, MacDowell AA, Nicholson DG (2000) Analysis of metal speciation and distribution in symbiotic fungi (ectomycorrhiza) studied by micro X-ray absorption spectroscopy and X-ray fluorescence. In: Gobran GR, Wenzel WW, Lombi E (eds) Trace elements in the rhizosphere. CRC, Boca Raton, FL, pp 149–164

Bianciotto V, Bandi C, Minerdi D, Sironi M, Tichy HV, Bonfante P (1996) An obligately endosymbiotic mycorrhizal fungus itself harbors obligately intracellular bacteria. Appl Environ Microb 62:3005–3010

Blaudez D, Jacob C, Turnau K, Colpaert JV, Ahonen-Jonnarth U, Finlay R, Botton B, Chalot M (2000) Differential response of ectomycorrhizal fungi to heavy metals *in vitro*. Mycol Res 104:1366–1371

de Boer W, Folman LB, Summerbell RC, Boddy L (2005) Living in a fungal world: impact of fungi on soil bacterial niche development. FEMS Microbiol Rev 29:795–811

2 The Potential of Rhizosphere Microorganisms to Promote the Plant Growth

Boomsma CR, Vyn TJ (2008) Maize drought tolerance: potential improvements through arbuscular mycorrhizal symbiosis? Field Crops Res 108:14–31

Bouchez T, Patureau D, Dabert P, Juretschko S, Dorés J, Delgenes P, Moletta R, Wagner M (2000) Ecological study of a bioaugmentation failure. Environ Microbiol 2:179–190

Brimecombe MJ, De Leij FAAM, Lynch JM (2007) Rhizodeposition and microbial populations. In: R Pinton, Z Varanini, P Nannipieri (eds) The rhizosphere: biochemistry and organic substances at the soil-plant interface. CRC Press, Taylor & Francis Group, Boca Raton, London, New York, pp 73–109

Brown MT, Wilkins DA (1985) Zinc tolerance of mycorrhizal *Betula*. New Phytol 99:101–106

van Bruggen AHC, Semenom AM, van Diepeningen AD, de Vos OJ, Blok WJ (2006) Relation between soil health, wave-like fluctuations in microbial populations, and soil-borne plant disease management. Eur J Plant Pathol 115:105–122

Bruns TD (1995) Thoughts on the processes that maintain local species diversity of ectomycorrhizal fungi. Plant Soil 170:63–73

Brussaard L, Behan-Pelletier VM, Bignell DE, Brown VK, Didden WAM, Folgarait PJ, Fragoso C, Freckman DW, Gupta VVSR, Hattori T, Hawksworth DL, Klopatek C, Lavelle P, Malloch D, Rusek J, Söderström B, Tiedje JM, Virginia RA (1997) Biodiversity and ecosystem functioning in soil. Ambio 26:563–570

Burd GI, Dixon DG, Glick BR (1998) A plant growth promoting bacterium that decreases nickel toxicity in plant seedlings. Appl Environ Microbiol 64:3663–3668

Buyer JS, Leong J (1986) Iron transport mediated antagonism between plant growth-promoting and plant-deleterious *Pseudomonas* strains. J Biol Chem 261:791–794

Cakmak I (2002) Plant nutrition research: Priorities to meet human needs for food in sustainable ways. Plant Soil 247:3–24

Chanmugathas P, Bollag J (1987) Microbial role in immobilisation and subsequent mobilisation of cadmium in soil suspensions. Soil Sci Soc Am J 51:1184–1191

Chaineau CH, Rougeux G, Yepremian C, Oudot J (2005) Effects of nutrient concentration on the biodegradation of crude oil and associated microbial populations in the soil. Soil Biol Biochem 37:1490–1497

Chaturvedi S, Chandra R, Rai V (2006) Isolation and characterization of *Phragmites australis* (L.): rhizosphere bacteria from contaminated site for bioremediation of colored distillery effluent. Eco Eng 27:202–207

Colpaert JV, Van Laere A (1996) A comparison of the extracellular enzyme activities of two ectomycorrhizal and leaf-saprotrophic basidiomycete colonizing beech leaf litter. New Phytol 134:133–141

Colpaert JV, Asche JA (1993) The effects of cadmium on ectomycorrhizal *Pinus sylvestris* L. New Phytol 123:325–333

Compant S, Duffy B, Nowak J, Clèment C, Barka EA (2005) Use of plant growth-promoting bacteria for biocontrol of plant diseases: principles, mechanisms of action, and future prospects. Appl Environ Microbiol 71:4951–4959

Das K, Mukherjee AK (2007) Comparison of lipopeptide biosurfactants production by *Bacillus subtilis* strains in submerged and solid state fermentation systems using a cheap carbon source: Some industrial applications of biosurfactants. Process Biochem 42:1191–1199

de la Peňa E, Rodriguez-Echeverria S, Putten WH, Freitas H, Moens M (2006) Mechanism of control of root-feeding nematodes by mycorrhizal fungi in the dune grass *Ammophila arenaria*. New Phytol 169:829–840

Derry AM, Staddon WJ, Trevors JT (1998) Functional diversity and community structure of microorganisms in uncontaminated and creosote contaminated soils as determined by sole-carbon-source-utilization. World J Microbiol Biotechnol 14:571–578

Dighton J (1991) Acquisition of nutrients from organic resources by mycorrhizal autotrophic plants. Experientia 47:362–369

Dodd JC (2000) The role of arbuscular mycorrhizal fungi in agro- and natural ecosystems. Outlook Agric 29:55–62

Duchesne LC (1994) Role of ectomycorrhizal fungi in biocontrol. In: Pfleger FL, Linderman RG (eds) Mycorrhizae and plant health. APS Press, St. Paul, MN, pp 27–45

Duchesne LC, Peterson RL, Ellis BE (1987) The accumulation of plant-produced antimicrobial compounds in response to ectomycorrhizal fungi: a review. Phytoprotection 68:17–27

Duponnois R, Plenchette C (2003) A mycorrhiza helper bacterium enhances ectomycorrhizal and endomycorrhial symbiosis of Australian Acacia species. Mycorrhiza 13:85–91

Duxbury T (1985) Ecological aspects of heavy metal responses in microorganisms In: Marshall KC (ed) Advances in microbial ecology. Plenum Press, New York, pp 185–235

Erland S, Taylor FS (2002) Diversity of ecto-mycorrhizal fungal communities in relation to the abiotic environment. In: van der Heijden MGA, Sanders IR (eds) Mycorrhizal ecology. Springer, Berlin, Heidelberg, New York, pp 163–200

Fierer N, Jackson RB (2006) The diversity and biogeography of soil bacterial communities. Proc Natl Acad Sci USA 103:626–631

Fomina MA, Alexander IJ, Colpaert JV, Gadd GM (2005) Solubilization of toxic metal minerals and metal tolerance of mycorrhizal fungi. Soil Biol Biochem 37:851–866

Founone H, Duponnois R, Ba AM, Sall S, Branget I, Lorquin J, Neyra M, Chotte JL (2002) Mycorrhiza helper bacteria stimulated ectomycorrhizal symbiosis of *Acacia holosericea* with *Pisolithus alba*. New Phytol 153:81–89

Gadd GM (2004) Microbial influence on metal mobility and application for bioremediation. Geoderma 122:109–119

Gange AC (2006) Insect–mycorrhizal interactions: patterns, processes, and consequences. In: Ohgushi T, Craig TP, Price PW (eds) Indirect interaction webs: nontrophic linkages through induced plant traits. Cambridge University Press, Cambridge , pp 124–144

Garbaye J (1994) Helper bacteria: a new dimension to the mycorrhizal symbiosis. New Phytol 128:197–210

Garbaye J, Bowen GD (1989) Stimulation of mycorrhizal infection of Pinus radiate by some micro-organisms associated with the mantle of ectomycorrhizas. New Phytol 112:383–388

George E, Marschner H (1996) Nutrient and water uptake by roots of forest trees. Z. Pflanzenernähr. Bodenkd 159:11–21

Glick BR (2005) Modulation of plant ethylene levels by the bacterial enzyme ACC deaminase. FEMS Microbiol Lett 251:1–7

Hartley J, Cairney JWG, Meharg AA (1997) Do ectomycorrhizal fungi exhibit adaptive tolerance to potentially toxic metals in the environment? Plant Soil 189:303–319

He LM, Neu MP, Vanderberg LA (2000) *Bacillus lichenformis* γ-glutamyl exopolymer: physicochemical characterization and U(VI) interaction. Environ Sci Technol 34:1694–1701

Herrera-Estrella L (1999) Transgenic plants for tropical regions: Some considerations about their development and their transfer to the small farmer. Proc Natl Acad Sci USA 96:5978–5981

Hiifte M, Vande Woestyne M, Verstraete W (1994) Role of siderophores in plant growth promotion and plant protection by fluorescent pseudomonads. In: Manthey JA , Crowley DE , Luster DG (eds) Biochemistry of metal micronutrients in the rhizosphere. Lewis Publishers, Boca Raton, FL, pp 81–92

Hildebrandt U, Janetta K, Bothe H (2002) Towards growth of arbuscular mycorrhizal fungi independent of a plant host. Appl Environ Microbiol 68:1919–1924

Hrynkiewicz K, Haug I, Baum C (2008) Ectomycorrhizal community structure under willows at former ore mining sites. Eur J Soil Biol 44:37–44

Hrynkiewicz K, Baum C, Niedojadło J, Dahm H (2009) Promotion of mycorrhiza formation and growth of willows by the bacterial strain *Sphingomonas* sp. 23L on fly ash. Biol Fertil Soil 45:385–394

Hrynkiewicz K, Baum C, Leinweber P (2010a) Density, metabolic activity and identity of cultivable rhizosphere bacteria on *Salix viminalis* in disturbed arable and landfill soils. J Plant Nutr Soil Sci 173:747–756

Hrynkiewicz K, Ciesielska A, Haug I, Baum C,(2010b) Conditionality of ectomycorrhiza formation and willow growth promotion by associated bacteria: role of microbial metabolites and use of C sources. Biol Fertil Soils 46:139–150

2 The Potential of Rhizosphere Microorganisms to Promote the Plant Growth

Hu X, Boyer GL (1996) Siderophore-mediated aluminium uptake by *Bacillus megaterium* ATCC 19213. Appl Environ Microbiol 62:4044–4048

Idris R, Trifonova R, Puschenreiter M, Wenzel WW, Sessitsch A (2004) Bacterial communities associated with flowering plants of the Ni hyperaccumulator *Thlaspi goesingense*. Appl Environ Microbiol 70:2667–2677

Jacob C, Courbot M, Brun A, Steinman HM, Jacquot JP, Botton B, Chalot M (2001) Molecular cloning, characterization and regulation by cadmium of a superoxide dismutase from the ectomycorrhizal fungus *Paxillus involutus*. Eur J Biochem 268:3223–3232

Jana TK, Srivastava AK, Csery K, Aroran DK (2000) Influence of growth and environmental conditions on cell surface hydrophobicity of *Pseudomonas fluorescens* in non-specific adhesion. Can J Microbiol 46:28–37

Jentschke G, Godbold DL (2000) Metal toxicity and ectomycorrhizas. Physiol Plant 109:107–116

Jing Y, He Z, Yang X (2007) Role of soil rhizobacteria in phytoremediation of heavy metal contaminated soils. J Zhejiang Univ Sci B 8:192–207

Johnson NC, Graham J-H, Smith FA (1997) Functioning of mycorrhizal associations along the mutualism-parasitism continuum. New Phytol 135:575–585

Johansen JE, Binnerup SJ (2002) Contribution of *Cytophaga*-like bacteria to the potential of turnover of carbon, nitrogen, and phosphorus by bacteria in the rhizosphere of barley (*Hordeum vulgare* L.). Microb Ecol 43:298–306

Karabaghli C, Frey-Klett P, Sotta M, Bonnet M, Le Tacon F (1998) In vitro effects of *Laccaria bicolor* S238N and *Pseudomonas fluorescens* strain BBc6 on rooting of de-rooted shoot hypocotyls of Norway spruce. Tree Physiol 18:103–111

Khan MS, Zaidi A, Wani PA (2007) Role of phosphate solubilizing microorganisms in sustainable agriculture: a review. Agron Sustain Dev 27:29–43

Kuffner M, Puschenreiter M, Wieshammer G, Gorfer M, Sessitsch A (2008) Rhizosphere bacteria affect growth and metal uptake of heavy metal accumulating willows. Plant Soil 304:35–44

Kumpfer W, Heyser W (1986) Effects of stem flow Beech (*Fagus sylvatica* L.). In: Gianinazzi-Pearson V, Gianinazzi S (eds) Physiological aspects and genetical aspects of mycorrhizae. Proceedings of the 1st European Symposium on Mycorrhizae. Dijon, 1–5 July 1985, INRA, pp 745–750

Labud V, Garcia C, Hernandez T (2007) Effect of hydrocarbon pollution on the microbial properties of a sandy and a clay soil. Chemosphere 66:1863–1871

Leake JR, Read DJ (1997) Mycorrhizal fungi in terrestrial ecosystems. In: Wicklow D, Soderström B (eds) The Mycota IV. Experimental and microbial relationships. Springer, Berlin, pp 281–301

Leyval C, Binet P (1998) Effect of polyaromatic hydrocarbons (PAHs) in soil on arbuscular mycorrhizal plants. J Environ Qual 27:402–407

Leyval C, Joner EJ (2000) Bioavailability of metals in the mycorhizosphere. In: Gobran GR, Wenzel WW, Lombi E (eds) Trace elements in the rhizosphere. CRC, Boca Raton, USA, pp 165–185

Li XL, Marschner H, George E (1991) Acquisition of phosphorus and copper by VA-mycorrhizal hyphae and root-to-shoot transport in white clover. Plant Soil 136:49–57

Liu A, Hamel C, Hamilton RI, Ma BL, Smith DL (2000) Acquisition of Cu, Zn, Mn and Fe by mycorrhizal maize (*Zea mays* L.) grown in soil at different P and micronutrient levels. Mycorrhiza 9:331–336

Lodge DJ (1989) The influence of soil moisture and flooding on formation of VA-endo- and ectomycorrjizae in *Populus* and *Salix*. Plant Soil 117:255–262

Lynch JM (1990) Introduction: some cosequences of microbial rhizosphere competence for plant and soil. In JM Lynch (ed) The rhizosphere. John Wiley and Sons, Chichester, UK, p 1

Maila MP, Randima P, Dronen K, Cloete TE (2006) Soil microbial communities: influence of geographic location and hydrocarbon pollutants. Soil Biol Biochem 38:303–310

Marschner H (1991) Mechanisms of adaptation of plants to acid soils. Plant Soil 134:1–20

Marschner H (1995) Mineral nutrition of higher plants. Academic Press, London

Marschner H (1998) Soil-root interface: biological and biochemical process. In: Soil chemistry and ecosystem health. SSSA Special Publication No 52. Madison, WI, pp 191–231

McGonigle TP, Miller MH (1999) Winter survival of extraradical hyphae and spores of arbuscular mycorrhizal fungi in the field. Appl Soil Ecol 12:41–50

Maila MP, Randimaa P, Drønenb K, Cloete TE (2006) Soil microbial communities: Influence of geographic location and hydrocarbon pollutants. Soil Biol Biochem 38:303–310

Malajczuk N (1988) Interaction between *Phytophthora cinnamomi* zoospores and micro-organisms on non-mycorrhizal and ectomycorrhizal roots of *Eucalyptus marginata*. Trans Br Mycol Soc 90:375–382

Martino E, Perotto S, Parsons R, Gadd GM (2003) Solubilization of insoluble inorganic zinc compounds by ericoid mycorrhizal fungi derived from heavy metal polluted sites. Soil Biol Biochem 34:133–141

Meharg AA, Cairney JWG (2000) Ectomycorrhizas—extending the capabilities of rhizosphere remediation? Soil Biol Bioch 32:1475–1484

Mehra RK, Winge DR (1991) Metal ion resistance in fungi: molecular mechanisms and their regulated expression. J Cell Biochem 45:30–40

Mirsal I (2004) Soil pollution: origin, monitoring and remediation. Springer, New York

Mohanty G, Mukherji S (2008) Biodegradation rate of diesel range n-alkanes by bacterial cultures *Exiguobacterium aurantiacum* and *Burkholderia cepacia*. Int Biodeterior Biodegrad 61:240–250

Morgan JAW, Bending GD, White PJ (2005) Biological costs and benefits to plant–microbe interactions in the rhizosphere. J Exp Bot 56:1729–1739

Narasimhan K, Basheer C, Bajic VB, Swarup S (2003) Enhancement of plant–microbe interactions using a rhizosphere metabolomics-driven approach and its application in the removal of polychlorinated biphenyls. Plant Physiol 132:146–153

Neidhardt FC, Ingraham JL, Schaechter M (eds) (1990) Physiology of the bacterial cell. Sinauer Associates, Sunderland

Neubauer SC, Emerson D, Megonigal JP (2002) Life at the energetic edge: kinetics circumneutral iron oxidation by lithotrophic iron-oxidizing bacteria isolated from the wetland-plant rhizosphere. Appl Environ Microbiol 68:3988–3995

Niemi K, Scagel C, Haggman H (2004) Application of ectomycorrhizal fungi in vegetative propagation of conifers. Plant Cell Tiss Organ Cult 78:83–91

Niemi K, Scagel CF (2007) Root induction of Pinus sylvestris L. hypocotyls cuttings using specific ectomycorrhizal fungi in vitro. In: Jain SM, Haggman H (eds) Protocols for micropropagation of woody trees and fruits. Springer, Berlin, pp 147–152

Normand L, Bartschi H, Debaud JC, Gay G (1996) Rooting and aclimatization of micropropagated cuttings of *Pinus pinaster* and *Pinus sylvestris* are enhanced by the ectomycorrhizal fungus *Hebeloma cylindrosporum*. Physiol Plant 98:759–766

Oliveira RS, Castro PML, Dodd JC, Vosátka M (2005) Synergistic effect of *Glomus intraradices* and *Frankia* spp. on the growth and stress recovery of *Alnus glutinosa* in an alkaline anthropogenic sediment. Chemosphere 60:1462–1470

Perotto S, Bonfante P (1997) Bacterial associations with mycorrhizal fungi: close and distant friends in the rhizosphere. Trends Microbiol 5:496–501

Pillman A, Jusaitis M (1997) Nutrition revegetation of waste fly ash lagoons II. Seedling transplants and plant nutrition. Waste Manag Res 15:307–321

Pilon-Smits EAH (2005) Phytoremediation. Annu Rev Plant Biol 56:15–39

Pinstrup-Andersen P, Pandya-Lorch R, Rosegrant MW (1999) World food propects: Critical issues for the early twenty-first century. 2020 Vision Food Policy Report, International Food Policy Research Institute, Washington, DC

Poole EJ, Bending GD, Whipps JM, Read DJ (2001) Bacteria associated with *Pinus sylvestris-Lactarius rufus* ectomycorrhizas and their effects on mycorrhiza formation in vitro. New Phytol 151:741–753

Pozo MJ, Azcón-Aguilar C (2007) Unraveling mycorrhiza-induced resistance. Curr Opin Plant Biol 10:393–398

Quan X, Shi H, Wang J, Qian Y (2003). Biodegradation of 2,4 dichlorophenol in sequencing batch reactors augmented with immobilized mixed culture. Chemosphere 50:1069–1074

2 The Potential of Rhizosphere Microorganisms to Promote the Plant Growth

Rasanayagam S, Jeffries P (1992) Production of acid is responsible for antibiosis by some ectomycorrhizal fungi. Mycol Res 96:971–976

Read D (1993) Appendix C: Mycorrhizas. In: Anderson JM, Ingram JSI (eds) Tropical soil biology and fertility, a handbook of methods. CAB International, Wallingford, UK, pp 121–131

Read DJ (2002) Towards ecological relevance—progress and pitfalls in the path towards an understanding of mycorrhizal functions in nature. In: van der Heijden MGA, Sanders IR (eds) Mycorrhizal ecology. Springer, Berlin, pp 3–29

Rentz JA, Alvarez PJJ, Schnoor JL (2005) Benzo[a]pyrene cometabolism in the presence of plant root extracts and exudates. Implications for phytoremediation. Environ Pollut 136:477–484

Richardson AE, Barea JM, McNeill AM, Prigent-Combaret C (2009) Acquisition of phosphorus and nitrogen in the rhizosphere and plant growth promotion by microorganisms. Plant Soil 321:305–339

Robinson D, Griffiths B, Ritz K, Wheatley R (1989) Root-induced nitrogen mineralisation: a theoretical analysis. Plant Soil 117:185–193

Rudawska M, Kieliszewska-Rokicka B (1997) Mycorrhizal formation by *Paxillus involutus* strains in relation to their IAA-synthesizing activity. New Phytol 137:509–517

Ryan PR, Dessaux Y, Thomashow LS, Weller DM (2009) Rhizosphere engineering and management for sustainable agriculture. Plant Soil 321:363–383

Saiki Y, Habe H, Yuuki T, Ikeda M, Yoshida T, Nojiri H et al (2003) Rhizoremediation of dioxine-like compounds by a recombinant Rhizobium tropici strain expressing carbazole 1,9a-dioxigenase constitutively. Biosci Biotechnol Biochem 67:1144–1148

Salisbury FB (1994) The role of plant hormones. In: Wilkinson RE (ed) Plant–environment interactions. Marcel Dekker, New York, USA, pp 39–81

Sarand I, Timonen S, Nurmiaho-Lassila E-L, Koivila T, Haahtela K, Romantschuk M et al (1998) Microbial biofilms and catabolic plasmid harbouring degradative fluorescent pseudomonads in Scots pine ectomycorrhizospheres developed on petroleum contaminated soil. FEMS Microbiol Ecol 27:115–126

Sarand I, Timonen S, Koivula T, Peltola R, Haahtela K, Sen R et al (1999) Tolerance and biodegradation of m-toluate by Scots pine, a mycorrhizal fungus and fluorescent pseudomonads individually and under associative conditions. J Appl Microbiol 86:817–826

Schelkle M, Peterson RL (1996) Suppression of common root pathogens by helper bacteria and ectomycorrhizal fungi in vitro. Mycorrhiza 6:481–485

Scherr SJ (1999) Soil degradation, a threat to developing-country food security by 2020? Food, Agriculture, and the Environmental Discussion Paper 27. International Food Policy Research Institute. Washington, DC

Schloter M, Dilly O, Munch JC (2003) Indicators for evaluating soil quality. Agric Ecosyst Environ 98:255–262

Schützendübel A, Polle A (2002) Plant responses to abiotic stresses: heavy metal-induced oxidative stress and protection by mycorrhization. J Exp Bot 53:1351–1365

Selvam A, Mahadevan A (2000) Reclamation of ash pond of Neyveli Lignite Corporation, Neyveli, India. Minetech 21:81–89

Siciliano SD, Germida JJ (1999) Enhanced phytoremediation of chlorobenzoates in rhizosphere soil. Soil Biol Biochem 31:299–305

Singer AC, van der Gast CJ, Thompson IP (2005) Perspectives and vision for strain selection in bioaugmentation. Trends Biotechnol 23:74–77

Smalla K, Wieland G, Buchner A, Zock A, Pary J, Kaiser S, Roskot N, Heuer H, Berg G (2001) Bulk and rhizosphere soil bacterial communities studied by denaturing gradient gel electrophoresis: plant-dependent enrichment and seasonal shifts reveald. Appl Environ Microbiol 67:4742–4251

Smith SE, Read DJ (1997) Mycorrhizal symbiosis, 2nd edn. Academic Press, London

Standing D, Baggs EM, Wattenbach M, Smith P, Killham K (2007) Meeting the challenge of scaling up processes in the plant-soil-microbe system. Biol Fert Soil 44:245–257

Sturz AV, Christie BR, Nowak J (2000). Bacterial endophytes: potential role in developing sustainable systems of crop production, Crit Rev Plant Sci 19:1–30

Supaphol S, Panichsakpatana S, Trakulnaleamsai S, Tungkananuruk N, Roughjanajirapa P, Gerard O'Donnell A (2006) The selection of mixed microbial inocula in environmental biotechnology: example using petroleum contaminated tropical soils. J Microbiol Method 65:432–441

Tarkka M, Schrey S, Hampp R (2008) Plant associated micro-organisms. In: Nautiyal CS, Dion P (eds) Molecular mechanisms of plant and microbe coexistence. Springer, New York, pp 3–51

Tibbett M, Sanders FE, Cairney JWG (1998) The effect of temperature and inorganic phosphorus supply on growth and acid phosphatase production in arctic and temperate strains of ectomycorrhizal *Hebeloma* spp. in axenic culture. Mycol Res 102:129–135

Tibbett M, Sanders FE (2002) Ectomycorrhizal symbiosis can enhance plant nutrition through improved access to discrete organic nutrient patches of high resource quality. Ann Bot 89:783–789

Tillman D (1999) Global environmental impacts of agricultural expansion: the need for sustainable and efficient practices. Proc Natl Acad Sci USA 96:5995–6000

Torsvik V, Øvreås L (2002) Microbial diversity and function in soil: from genes to ecosystems. Curr Opin Microbiol 5:240–245

Turnau K, Kottke I, Dexheimer J (1996) Toxic elements filtering in *Rhizopogon roseolus*—*Pinus sylvestris* mycorrhizas collected from calamine dumps. Mycol Res 100:16–22

Urum K, Pekdemir T, Copur M (2004) Surfactants treatment of crude oil contaminated soils. J Colloid Interf Sci 276:456–464

van der Heijden EW (2001) Differential benefits of arbuscular mycorrhizal and ectomycorrhizal infection of *Salix repens*. Mycorrhiza 10:185–193

von Uexküll HR, Mutuert W (1995) Global extent, development and economic impact of acid soils. In: Date RA, Grundon NJ, Rayment GE, Probert ME (eds) Plant–soil interactions at low pH: principles and management. Kluwer Academic Publishers, Dordrecht, pp 5–19

Vessey JK (2003) Plant growth promoting rhizobacteria as biofertilizers. Plant Soil 255:571–586

Wang Y, Dai CC (2011) Endophytes: a potential resource for biosynthesis, biotransformation, and biodegradation. Ann Microbiol 61:207–215

Wehner J, Antunes PM, Powell JR, Mazukatow J, Rillig MC (2010) Plant pathogen protection by arbuscular mycorrhizas: A role for fungal diversity? Pedobiologia 53:197–201

Wenzel WW (2009) Rhizosphere processes and management in plant-assisted bioremediation (phytoremediation) of soils. Plant Soil 321:385–408

Whipps JM (2001) Microbial interactions and biocontrol in the rhizosphere. J Exp Bot 52:487–511

Whipps JM (2004) Prospects and limitations for mycorrhizas in biocontrol of root pathogens. Can J Bot 82:1198–1227

Whiting SN, de Souza MP, Terry N (2001) Rhizosphere bacteria mobilize Zn for hyperaccumulation by *Thlaspi caerulescens*. Environ Sci Technol 35:3144–3150

Wood RKS (1960) Pectic and cellulolytic enzymes in plant disease. Ann Rev Plant Physiol 11:299–322

Woodward AW, Bartel B (2005) Auxin: regulation, action and interaction. Ann Bot 95:707–735

Wu SC, Cheung KC, Luo YM, Wong MH (2006) Effects of inoculation of plant growth-promoting rhizobacteria on metal uptake by *Brassica juncea*. Environ Pollut 140:124–135

Xavier LJC, Germida JJ (2003) Bacteria associated with *Glomus clarum* spores influence mycorrhizal activity. Soil Biol Biochem 35:471–478

Yee DC, Maynard JA, Wood TK (1998) Rhizoremediation of trichloroethylene by e recombinant, root-colonizing Pseudomonas fluorescens strain expressing toluene ortho-monooxygenese constitutively. Appl Environ Microbiol 64:112–118

Zaidi S, Usmani S, Singh BR, Musarrat J (2006) Significance of *Bacillus subtilis* strain SJ-101 as a bioinoculant for concurrent plant growth promotion and nickel accumulation in *Brassica juncea*. Chemosphere 64:991–997

Zimmer D, Baum C, Leinweber P, Hrynkiewicz K, Meissner R (2009) Associated bacteria increase the phytoextraction of cadmium and zinc from a metal-contaminated soil by mycorrhizal willows. Int J Phytorem 11:200–213

Chapter 3
Sustainable Solutions for Agro Processing Waste Management: An Overview

C. M. Ajila, Satinder K. Brar, M. Verma and U. J. S. Prasada Rao

Contents

3.1	Introduction	66
3.2	Agro Processing Waste: Problems and Management Opportunities	67
3.3	Sources and Characterization of Agro By-products	70
3.4	Utilization of Agro-Wastes	72
	3.4.1 Food/Feed Ingredients	72
	3.4.2 Carbon Source for Growing Microorganisms for Production of Valuable Chemicals and Enzymes	75
	3.4.3 Use for the Production of Fertilizer	86
	3.4.4 Use for Energy Production	87
	3.4.5 Recovery of Value-Added Products	90
	3.4.6 Use as Metal Adsorbent for the Environmental Pollutants	98
3.5	Conclusion and Future Outlook	98
References		99

Abstract Technological revolution in the field of agriculture has tremendously increased the agriculture production. The net impact by the revolution in agriculture has resulted in fast development on food processing industries all over the world. As a result of this rapid development, significant quantities of agricultural products are subjected to processing to make them suitable for consumption, increased storage stability, improved nutrition and sensory quality. Food industrialization has generated large quantity of food products, provided employment to large number of people and uplifted the economic status, at the same time; it generated waste in huge quantities causing environmental pollution. Pollution has not only scientific aspects but also sociological and economical, causing adverse impacts on human beings and its environment. The food wastes can be classified into different categories, such as crop waste and residues; fruits and vegetables by-products; sugar, starch and confectionary industry by-products; oil industry by-products; grain and legumes by-products; distilleries and breweries by-products. Food industry wastes and by-

---

S. K. Brar (✉)
INRS-ETE, Université du Québec, 490, Rue de la Couronne, G1K 9A9 Québec, Canada
e-mail: satinder.brar@ete.inrs.ca

A. Malik, E. Grohmann (eds.), *Environmental Protection Strategies for Sustainable Development*, Strategies for Sustainability,
DOI 10.1007/978-94-007-1591-2_3, © Springer Science+Business Media B.V. 2012

products are geographically scattered comprising large volume and low nutritional value. Consequently, collection, transportation and processing cost of the by-products can exceed the selling price. If we could produce valuable products from food industry by-products through new scientific and technological methods, these by-products could be converted into products with a higher economic value than the main products. Different ways of utilization of by-products from food processing industry can be mainly classified into five categories, such as source for food/feed ingredients, as a carbon source for growing useful microorganisms, as fertilizer by composting, as a source for direct energy generation/biogas production and as a source for high value-added products. This chapter provides a brief discussion on the utilization of agro-processing wastes as a source of nutrients, phytochemicals, and fermentable substrate.

Keywords Agro-processing wastes • Agro by-products • Nutrients • Phytochemicals • Carbon source

3.1 Introduction

Agriculture is one of the most widely practiced entities in the world. With the advent of modern civilization, agriculture has been commercialized to a greater extent in which the process from production of agriculture commodities to consumption are highly mechanized and modernized with most sophisticated technologies. Globalization in the agricultural industry is increasingly competitive. The modern agriculture should be able to tackle a diversity of challenges, such as affectively linking more clientele to domestic and international markets, promoting environmental conservation and resource management and natural resources management and also securing food security and food sufficiency. The modernization and advent of industrialization has resulted into generation of huge quantities of diverse wastes. The accumulation of wastes and ill effects on the environment are enormous. This leads to a huge threat for humans to consider and handle such wastes in an integrated manner for creating a better environment. It was estimated that approximately 87 million tonnes of waste was produced per year in UK. European countries produce approximately 700 million tonnes of agricultural waste per year (Eurostat data 2009). However there is also a wide variation in the amount of by-products generated depending on the country. France produces 400 million tonnes per year and Austria produces less than 1 million tonnes per year (www.ace.mmu.ac.uk). Globally, 998 million tonnes of agricultural wastes are produced per year (www.uncrd.or.jp). It has been reported that every year approximately 35 million tonnes of dried distilled grain in North America is wasted. Globally 500 million tonnes of sugar cane bagasse and 12.2 million tonnes of grape skin and seeds are generated every year (Briens 2009).

The management of agro-industrial residues is one of the complicated problems in agriculture and in agro-industry and has an impact on its economy and its day to day operations. Agro by-products are one of the most abundant and renewable sources on the earth. Every year, enormous biomass as agro by-products are accumulated in large quantity and causes environmental pollution. This also causes loss of potentially important materials which can be used for the production of number of valuable products, such as food, feed, fuel and variety of chemicals and bioactive compounds. The protection of environment towards the aim of sustainability is one of the mostly discussed topics in the area of waste management. The sustainable solutions for waste management range from waste prevention, waste minimization, cleaner production and also zero emissions systems. The sustainable solutions for waste management include environmentally sound management of waste, applying zero emission industrial ecosystems, agro based industrial systems and most importantly reuse, recycling, composting, bio-digestion, bio-extraction, bio-refineries, among others. The chapter summarizes recent research on sustainable management of agro-by-products with special emphasis on plant derived by-products and its different ways of utilization with innovative-end user products.

3.2 Agro Processing Waste: Problems and Management Opportunities

Agro-industrial wastes can be generally organized into different categories, such as food processing wastes, energy crops and biofuel production wastes and crop residues. Agricultural wastes comprise almost 15% of total waste generated by each country (Hsing et al. 2001). If the agriculture waste is not managed properly, it can pollute the environment resulting in impacts to water quality and a general loss of aesthetics. Waste from agriculture and food processing can become one of the most serious sources of pollution (Di Blasi et al. 1997). Due to cheap energy and raw materials following World War II, the role of utilizing by-products was not well studied till recently. An increase in the number of industrial plants in turn increased the volume of by-products, which led to considerations on the treatment and environmental friendly disposal of the processing by-products. Earlier agricultural technology started using by-products as fodder and for oil production from food processing industry. The disposal of agro-industrial waste can often be difficult because of the following reasons:

1. **Biological stability and potential growth of pathogens**: Many types of food processing waste contain a large number of microorganisms and will be altered quickly through microbiological activity. If the waste by-products are not processed properly, it will lead to hygienically unacceptable conditions through maggots, microorganisms and moulds. The breakdown of protein is characterized by evolution of strong odours.

2. **Water content:** The water content of the fruit and vegetable processing by-products lies between 70 and 95% by mass. High water content increases the transport costs of the waste. The mechanical removal of water can lead to further problems with water disposal, due to high level of organic material in the water.

3. **Rapid auto-oxidation**: The waste rich in high fat content is susceptible to oxidation, which leads to the release of foul smelling fatty acids.
4. **Changes due to enzymatic activity**: In many types of by-products from fruit and vegetable processing industry, enzymes are still active which accelerate the reaction involved in spoilage (Westendorf and Wohlt 2002).

Agricultural and food industry residues and wastes constitute a major proportion of world agricultural productivity. It has been estimated that approximately 30% of global agricultural products are becoming residues and refuses. Large volumes of solid and liquid wastes are generated from the agro-processing industries.

The agro wastes are now being handled in different ways as represented in Fig. 3.1. The waste management processes employed in most of the countries so far are given below:

1. Landfill and open-dumping sites: Landfill and dumping in open sites are common practices for disposal of wastes. Generally, wastes are dumped in swamp lands and in low lying areas. Approximately 60–80% of the wastes are disposed in this manner in many countries (Ngoc and Schnitzer 2009). This method has become one of the major sources of environmental pollution as the capacity of landfill is surpassed due to lack of environmental planning as well as due to lack of space following increased pressure on land.

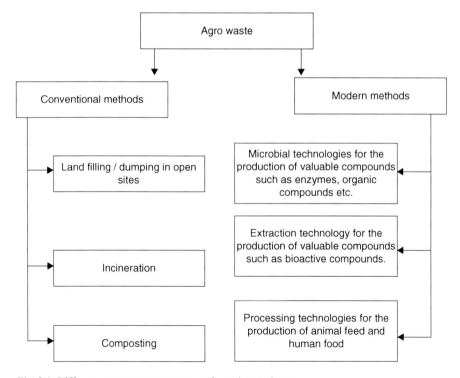

Fig. 3.1 Different management processes of agro by-products

3 Sustainable Solutions for Agro Processing Waste Management: An Overview

Table 3.1 Quantity of waste generated from agro by-products in different countries. (Sources: Eurostat and Organisation for Economic Co-operation and Development (OECD))

Year	Country	Quantity ($t \times 1000$)
2007	Japan	90 430
2007	New Zealand	150
2007	Belgium	1150
2007	Czech Republic	460
2007	Finland	860
2007	Iceland	50
2007	Ireland	60 107
2007	Italy	440
2007	Netherland	2390
2007	Norway	160
2007	Slovakia	4490
2003	Bulgaria	50
1999	United kingdom	84 000
1990	Austria	880
2007	France	400

2. Incineration: Incineration is another method of waste treatment in most of the countries. Even though this option seems to be attractive, the operating efficiency depends on the characteristics and composition of the waste. These methods need high financial start up and operating capital requirements.
3. Composting: Composting is a biological treatment in which microorganisms decompose and stabilize organic material. Composting is a low-technology approach for waste reduction. Composting of organic waste leads into highly nutritive organic manure. However, this method has limitation due to its high operational and maintenance costs and low cost of organic manure compared to the commercial fertilizers.
4. Recycling or recovery: Recycling of the solid wastes have been carried out in many developed countries. Approximately 44% of solid wastes are recycled in the developed countries and the percentage of recycling in the under developed countries are 12% and 8–11% in other low income countries (Ngoc and Schnitzer 2009). However, wastes used for recycling are mainly composed of plastic, paper, glass, rubber, ferrous etc. used for the further production of new products. The quantities of waste generated from different countries are given in Table 3.1.

Growing global demand for environmentally sustainable methods of production, pollution prevention and economical motives have changed a lot in the waste management system. Wastes are considered as a new source of resources for the production of value-added compounds. Better management of agricultural wastes can become an asset to the agro-industry. Generally employing management system for the agriculture is given in the Fig. 3.2. Energy rich agro-by-products can improve the agricultural productivity and increase resource utilization efficiency by developing and deploying appropriate technologies for the processing and reuse of the same. The efficient utilization of agro-waste can also lead to improve the environmental health by decreasing the pollution caused by the agro wastes.

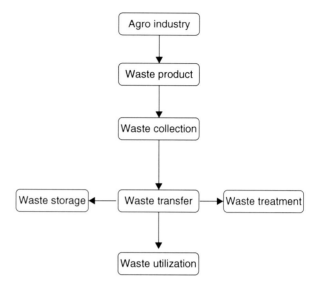

Fig. 3.2 Agricultural waste management system process

3.3 Sources and Characterization of Agro By-products

Agro wastes are of wide variety. The handling and technologies used for its processing are mainly based on the type of agro-wastes. The agro wastes can be classified into different categories, such as crop waste and residues; by-products from fruits and vegetables, sugar, starch and confectionary industry by-products; by-products from grain and legumes; by-products from distilleries and breweries, by-products from milk, poultry, fish products and egg industry.

The crop waste residues mainly include materials remaining after the harvesting and processing including the straw, husk or stubble from barley, wheat, rice, beans, oats, rye stalks or stovers from corn, sorghum, cotton etc. Fruit and crop residues include apricots, almonds, apples, avocados, grapes, lemons, limes, olives, oranges, plums etc. The vegetable crop residues consist mainly of vines, peels and leaves that remain after harvesting and processing. The major source of the waste in the fruit and vegetable processing industry are illustrated in the Table 3.2. During apple processing, pomace is the major by-product, which consists of crushed flesh, stalks, peels, seeds etc (Rahmat et al. 1995). The type of waste from mango processing industry is mainly peel (15–20%), coarse fibrous pulpy waste (5–10%) and kernel (15–20%) (Larrauri et al. 1996). The waste from starch industry, such as tapioca produce waste in the form of tapioca rind or peeling, spent pulp. Rice husk is a by-product obtained during rice milling. The major wastes from sugar cane industry are bagasse, molasses and sugar cane press mud. Wine making industry produces grape pomace as a by-product consists of skin, seed and stem in an estimated amount of 13% by weight of grapes (Torres et al. 2002). The distillery wastes mainly consist of yeast sludges. The spent grain is the most abundant brewery by-product corresponding to approximately 85% of total by-products generated (Reinold 1997). The

3 Sustainable Solutions for Agro Processing Waste Management: An Overview   71

Table 3.2 Percentage of Fruit and vegetable processing by-products

Fruit/vegetables	Nature of waste	Approx. waste (%)
Mango	Peel, stones,	45
Banana	Peel	35
Citrus	Peel, rag, seed	50
Pineapple	Skin, core	33
Grape	Stem, skin, seed	20
Guava	Peel, core, seed	10
Pea	Shell	40
Tomato	Skin, core, seed	20
Potato	Peel	15
Onion	Outer leaves	10
Apple	Peel, pomace, seed	25

spent grain accounts for an average for 31% of original malt weight representing 20kg/100L of beer produced (Reinold 1997). Brewers spent grain is available at very low cost throughout the year. Brazil, world's largest beer producer, generated around 1.7 million tonnes of spent grain in 2002 (Berto 2003). Brewers spent grain is a lignocellulosic material containing about 17% cellulose, 28% non-cellulosic polysaccharides, chiefly arabinoxylans and 28% lignin as represented in Table 3.3.

Whey is the major by-product from dairy industry which is a source of several nutrients and bioactive peptides and the chemical constituents in whey are presented in the Table 3.4. In meat industry, approximately 55% of the whole animal becomes the by-product after processing of the meat. These wastes are rich in proteins and can be used for biological secondary treatment. Marine and sea food industry is another source for by-products and related pollution. It was reported by FAO that the annual discard from world fisheries were estimated to be 20 million tonnes (25% of total production) per year (FAO 2005). Marine by-products are rich source of proteins, chitin, chitosan, gelatin, among others.

Agro-industrial wastes contains organic substances, such as polysaccharides, proteins, sugars, lipids, acids, starch and other nutrients, such as vitamins, minerals, fibers etc. The major organic compound present in most of the agro-wastes is lignocelluloses which is a compact, partly crystalline structure consisting of linear and crystalline polysaccharides cellulose, branched non-cellulosic and non crystalline heteropolysaccharide, hemicelluloses and branched non-crystalline lignin (Glasser

Table 3.3 Chemical composition of brewers spent grain (BSG) and germinated barley foodstuff (GBF)

Component (% dry wt)	BSG[a]	BSG[b]	GBF[a]	GBF[c]
Cellulose	25.4	16.8	8.9	9.1
Arabinoxylan	21.8	28.4	17.0	19.2
Lignin	11.9	27.8	8.2	6.7
Protein	24.0	15.2	46.9	48.0
Lipid	10.6	Nd	10.2	9.2

[a] Kanauchi et al. (2001)
[b] Mussatto and Roberto (2005)
[c] Fukuda et al. (2002)
nd, not determined

Table 3.4 Chemical constituents of whey. (Evans et al. 2010)

Constituents	Concentration
Water (%)	93–94
Dry matter (%)	5.8–6.8
Lactose (%)	4.2–4.7
Albumin (%)	0.8–1.0
Fat (%)	0.2–0.4
Biotin (µg/kg Dry matter)	325–599
Iron (mg/kg Dry matter)	9.16
Zinc (mg/kg Dry matter)	32.5
Copper (mg/kg Dry matter)	0.32
Cobalt (mg/kg Dry matter)	0.052
Molybdenum (mg/kg Dry matter)	0.573
Total protein (%)	0.8–1.0
Whey protein (%)	0.6–0.75
Citric acid (%)	0.1

et al. 2000). Cellulose is made up of a linear polymer chain that consists of a series of hydroglucose units in glucan chains with 1–4 glycosidic linkages. Hemicellulose is made up of various polysaccharides, such as xylose, galactose, mannose and arabinose. Hemicellulose is composed of linear and branched heteropolymers of L-arabinose, D-galactose, D-glucose, D-mannose, and D-xylose. Methyl or acetyl groups are attached to the carbon chain to various degrees. Hemicellulose and cellulose constitute 13–39% and 36–61% of the total dry matter, respectively. Lignin is made by an oxidative coupling of three major C6–C3 phenypropanoid units, namely sinapyl alcohol, coniferyl alcohol and p-coumaryl alcohol. These are arranged in a random, irregular three dimensional network that provide strength and structure and are consequently very resistant to enzymatic degradation.

3.4 Utilization of Agro-Wastes

Agro-industry waste and by-products are substances that originated during processing and can be further utilized in other ways. The agro-wastes have been used for many value-added products as shown in the Table 3.5.

3.4.1 Food/Feed Ingredients

Many by-products of agro industry can be fed to animals directly as such without any modification or can be used after fermentation of the agro-residues. Generally, by-products to be used as feedstuff should be economical, rich in nutrients and free of toxins or other substances that may be unhealthy to animals. Recovering by-products for use as animal feed can help agro industry save money by reducing

Table 3.5 Different ways of utilization of agro wastes

Agro waste	Utilization
Rice husk ash and charcoal	Additive in cement mixes
Rice husk	Water glass manufacture
Banana peel and sugarcane fibers	Active carbon
Oil palm, empty fruit bunch (EFB)	Electricity production
Oil palm stems, rubber wood	Paper making pulp
Onion skin, groundnut husk	Mulching, organic fertilizer
Husk, bagasse	Particleboard, softwood furniture
Bagasse, banana fruit reject	Heavy metal removal
Husk, straw, cow dung	Mushroom cultivation
Sunflower stalk, corn stalk	Ethanol production, animal feed
Bagasse fibers	Biogas production, electricity generation
Animal waste (dung)	Compost, fertilizer

waste discharges and can cut waste management costs and also can prevent environmental pollution. Most of the agro residues offer a less expensive source of nutrients than traditional feeds and also support acceptable animal performance.

Feeding by-products of the crop and food processing industries to livestock is a practice as old as the domestication of animals. It has two important advantages, can reduce the dependence of live stock on grains that can be used as animal feed and also can eliminate the need for costly waste management programs (Dhiman et al. 2003; Firkins et al. 2002; Gallo et al. 2001). The use of conventional feed stuffs for small ruminant is generally inefficient and expensive (Al-Jassim et al. 1997). Ruminant feeding by agro residues are often a practical alternative because the rumen microbial ecosystem can utilize this agro residues which often contain high levels of structural fibre required for their maintenance, growth, reproduction and production. Many by-products can be used as animal feed. Table 3.6 gives an idea about some commonly available by-products that are used as animal feed and they have been categorized according to the principal nutrients present in them.

During the last decades, the scientific research has been focused on many aspects of agro by-products as animal feed, in particular optimizing their nutritive value, characterization and quantification of phenolic compounds and fatty acids and their potential effects on the diet, effects on animal performance and product quality from feeding theses materials. There are many factors such as types and proportions of by-products generated, variability in moisture and nutrient content, storage of the material, potential for the presence of physical and microbial contaminants and toxins, handling characteristics etc should be considered for evaluating the suitability of by-products as animal feed (Crickenberge and Carawan 1996). It also depends on the target animal, processing and handling of the by-products, volume of material, cost versus benefits, effects on feed consumption, safety concerns etc.

Table 3.6 Principal nutrients in commonly used agro-byproducts as animal feed

Nutrient	By-products
Protein	Brewer's grains, Distiller's grains, Cull beans, Feather meal
Protein and energy	Brewer's grains, Distiller's grains, Corn gluten feed, Peanut screenings, Wheat bran
Energy	Bakery meal, Hominy feed, Snack food waste, Soyhulls, Vegetable, fruit processing waste
Roughage sources	Apple pomace, Corn cobs, Cottonseed hulls, Peanut hulls, Rice by-products

Brewery waste has been used as animal feed because of its high protein content and fibre. Brewery waste is a good feed ingredient for ruminants with other inexpensive sources such as urea which can provide all the essential amino acids (Huige 1994). Brewery waste was also reported to promote increased milk production without affecting the animal fertility (Reinold 1997). The use of agricultural by-products as animal feed has been investigated for a range of animals including pig, poultry, fish, and cattle among other agro by-products as given in Table 3.7.

Citrus by-products, such as fresh citrus pulp, citrus silage, dried citrus pulp, citrus meal and fines, citrus molasses, citrus peel and citrus activated sludge have been used as a alternative feed for ruminates in different growth stages (Bampidis and Robinson 2006). Olive by-products were also evaluated as ruminate animal feed with respect to their composition, digestion, degradation, ruminal fermentation and their impact on animal performance and product quality (Molina-Alcaide and Yáñez-Ruiz 2008). The soyhulls were also used as animal feed and the effects of feeding on ruminal fermentation, nutrient digestion and utilization and performance of dairy cows (Ipharraguerre and Clarka 2003) were investigated.

The brewery spent grain has been evaluated for the manufacture of flakes, whole wheat bread, biscuits and snacks because of its low cost and high nutritive value (Öztürk et al. 2002). Protein rich flour has been prepared from brewery spent grain and used in different bakery products including breads, muffins, cookies, mixed grain cereals, fruit and vegetable loaves, cakes, waffles, pancakes, tortillas, snacks,

Table 3.7 Use of agro by-products as an animal feed

Agro waste	Animal	References
Brewery waste, wheat bran, rye bran	Hamsters	Zhang et al. (1990) and Zhang et al. (1992)
Brewery waste	Chicken	Gondwe et al. (1999)
Brewery waste	Cows	Dhiman et al. (2003), Firkins et al. (2002) and Gallo et al. (2001)
Brewery waste	Fish	Kaur and Saxena (2004) and Muzinic et al. (2004)
Brewery waste	Pigs	Dung et al. (2002)
Citrus by-products	Ruminates	Bampidis and Robinson (2006)
Olive by-products	Sheep & goat	Molina-Alcaide and Yáñez-Ruiz (2008)
Soyhulls	Cows	Ipharraguerre and Clarka (2003)
Mango by-products	Fish	Mahadevaswamy and Venkataraman (1990)

3 Sustainable Solutions for Agro Processing Waste Management: An Overview

doughnuts and brownies etc (Huige 1994). The carrot pomace has been incorporated into bread, cakes and dressing (Ohsawa et al. 1995). A protein rich fibrous food called germinated barley has been made from brewer's yeast grain. The effluent from biogas production from mango processing waste has been utilized for the production of fresh water fishes, such as carp, rohu etc (Mahadevaswamy and Venkataraman 1990). Milling by-products, such as cereal bran and oil cakes obtained after oil extraction are commonly used as animal feed. The utilization of food processing by-products is comparatively very low when compared to the quantity of by-products generated.

An improvement in animal feeding is one of the important and basic conditions for the better management of farming of animals. It was recognized that poor quality of the feed was mainly responsible for the poor animal performance. Most of the animal feed are quiet expensive for most of the farming practices. Therefore, adequate and good quality feed supplies is the most important factor in farm management. The approach has been part of traditional agriculture for centuries. The production of animal feed is one of the most logical ways for utilizing a substantial portion of the enormous potential agro–residues. The production of animal feed could be the best profitable way of utilization at both industrial and also small scale level. The production of animal feed from agro- residues represents one of the highest cash return due to the fact that the demand for animal feed is always stable and huge. The marketing is also relatively easy and the technologies involved are not too complicated. The technologies involved for the production of animal feed can be handled in small scale industries also. Thus the production of animal feed from agro residues will be one of the most sustainable technologies for better way for attaining income to the agro community and also for the better management of environmental pollution.

3.4.2 Carbon Source for Growing Microorganisms for Production of Valuable Chemicals and Enzymes

By-products from the food processing industry as a whole can be used in a number of ways especially for biomass production. Microorganisms are grown on food processing by-products. An animal feed from apple pomace has been produced and evaluated (Joshi and Sandhu 1996). It was reported that *Kloceckera apiculata* and *Candida utilis* could transform apple pomace into an improved stock feed by solid-state fermentation (Rahmat et al. 1995). Various by-products can be used to grow microorganisms to produce enzymes, single cell protein, amino acids, lipids, carbohydrates and organic acids, which have applications in the field of animal feed and food processing. This aspect has been done by fermentative utilization of agro by-products as shown in Table 3.8.

Various fruit and vegetable wastes from the tomato, grape, apple, cabbage, carrot, beetroot and watermelon are used as the substrate for the lysine production by *Brevibacterium* spp. (Trifonova et al. 1993). Protein enrichment has been obtained

Table 3.8 Microbial utilization of agro by-products for production of various products

Products	Wastes
Ethanol	Citrus industry waste, apple pomace, peach waste, cashew apple pomace, pineapple waste, pear cuttings
Animal feed	Apple pomace, olive waste, brewery spent grain
Biogas	Waste from fruit and vegetable industry, fermentation industry
Single cell protein	Apple pomace, peach waste, cashew apple
Pectin, fibers	Citrus waste, apple pomace
Cider, beer and vinegar	Apple pomace, citrus waste
Bakers yeast	Apple pomace, brewery waste
Citric acid	Apple pomace, pine apple waste
Color	Vinery waste and distillery waste
Flavors/xanthan gum	Apple pomace and grape pomace

by growing *Aspergillus niger* on mango peel, orange peel, green immature banana and carrot wastes in solid state fermentation (Davy et al. 1981; Garg et al. 2000).

3.4.2.1 Single Cell Protein

Agriculture and food processing waste can also be used for the production of single cell protein having a potential use in the food and feed industries. Single cell protein can be used as a protein supplement in human foods, functional foods, food ingredients and protein supplement for livestock feeding. Wastes from orange, grape, pineapple and apple processing industries have been utilized for single cell protein production (Nigam 2000). The production of single cell protein will be another way for the better utilization of agro- by-products, since the composition of by-products will be similar to the ideal medium for the growth of microorganisms. The vinases from ethanol distillation of beet and cane molasses fermentation is rich in organic matter and inorganic salts and used for the culture of *spirulina*, since the by-product is similar to the medium of the organisms (González et al. 2009). Many agroby-products are being employed for the production of single cell protein as presented in Table 3.9.

Baggase has been used for the production of single cell protein and protein enriched animal feed. A unique process has been developed for mixed cultures with simultaneous saccharification and fermentation (Pandey et al.1998). Mixed cultures of two microorganisms on bagasse were used for the production of a biomass with 35.5% protein and 69.8% digestibility (Azzam 1992). Mixed cultures were used by many researchers for single cell protein production on bagasse (Rodriguez-Vazquez and Diazcervantes 1994). Moo Young et al. (1993) developed food and fodder grade mycoprotein by cultivating food grade fungus on bagasse.

Brewery waste has been used as substrate for cultivation of species of *Pleurotus, Agrocybe* and *Lentinus* for the production of single cell protein. Brewery waste was also reported for the production of single cell protein using the microorganism *Pleurotus ostreatus* (Wang et al. 2001). Protein fraction from brewery waste has

3 Sustainable Solutions for Agro Processing Waste Management: An Overview 77

Table 3.9 Single cell production by agro by-products by using different microorganisms

Agro by-products	Microorganism	Use of single cell protein	References
Brewery waste	*Pleurotus* spp., *Agrocybe* spp., *Lentinus* spp.	As single cell protein, protein rich biomass	Schildbach et al. (1992)
Brewery waste	*Pleurotus ostreatus*	As single cell protein	Wang et al. (2001)
Bagasse	*Cellulomonas flavigena and Xanthomonas* spp.	Single cell protein	Rodriguez-Vazquez and Diazcervantes (1994)
Cassava wastes	*Saccharomyces cerevisiae, Lactobacillus* spp., *Rhizopus* spp., *Aspergillus niger Streptomyces, Pleurotus* spp.	Single cell protein animal feed and food	Ubalua (2007) and Oboh and Elusiyan (2007)
Coffee pulp, coffee husk other coffee wastes	*Candida utilis*	Protein rich biomass	Orozco et al. (2008)
Defatted rice polishing	*Cryptococcus curvatus*	Single cell protein, protein enriched biomass	Rajoka et al. (2006)
Beet molasses and corn gluten meal	*Microsphaeropsis* spp; *Streptomyces cyaneus*	Single cell oil	El-Fadaly et al. (2009)
Wheat bran, straw, buck wheat, sugar beet pulp	*Phanerocheate chrysosporium*	Single cell protein, feed and food ingredients	Orozco et al. (2008) and Salmones et al. (2005)

been used as a medium for enhanced growth and sporulation of soil actinobacteria, especially *Streptomyces* (Szponar et al. 2003).

The protein supply and demand pattern is not a balanced condition because of the population growth rate. This increases the requirement of protein and better quality of food. To improve the supply of animal protein for human consumption, the production of animal stuff should be increased. Presently, animal feed production is mainly based on fish waste and plant protein source. Animal feed is relatively high cost item in the production of milk, meat, egg and broilers. This led to search for a new source of animal feed. The new source must have a high nutritional value, should not be a competitive factor for food for human consumption, should be economically feasible and most importantly it should be locally available. An urgent and immediate way to solve this problem is to develop and implement the use of single cell protein for animal feeding. Production of single cell protein can replace some of the usual protein sources on feed and food stuff. Single cell protein has more protein content than conventional feed. The protein content is ranging from 40 to 80% of their dry weight on a crude protein basis. The production of single cell protein from agro residues by microbial technologies is relevant to rural development and particularly to increase the per capita income of the community. The use of agro residues for the production of single cell protein has many advantages such as abundant and year round supply, contribution to the development of small scale and medium industries, potential for to help the low income farmers and unemployment in rural communities. The most important factors are the sustainable management of environment by the reuse of agro wastes with production of better quality products which provide new income to the community.

3.4.2.2 Organic Acids

Organic acids have numerous applications in food, beverage, chemical, pharmaceutical and cosmetics industry. The major characteristics of the organic acids are its solubility at room temperature, hygroscopic, buffering and chelating nature which provides the special characteristics for its industrial application. The major organic acids in industrial uses are citric, acetic, tartaric, malic, gluconic and lactic acids. Organic acids are obtained as the end product or as an intermediate product from biochemical cycle. Most of the organic acids can be produced by microbial technology. The agricultural residues and by-products are the ideal substrates for the production of organic acids as listed in the Table 3.10.

Citric acid is the most widely produced organic acid because of its wide range of applications. Agricultural residues, such as wheat bran, rice bran, potato fibrous residues have been employed in Japan for the production of citric acid by solid state fermentation. Agricultural residues, such as molasses, beet pulp residues, sugar cane bagasse, orange waste, apple pomace, grape pomace etc were also used for the production of citric acid by solid state fermentation using *Aspergillus niger* (Soccol et al. 2004). The pine apple peel, apple pomace, kiwi fruit peel, wheat bran and rice bran are utilized as substrates for the production of citric acid by

Table 3.10 Production of organic acids by solid state fermentation of agro-byproducts by using different microorganisms

Agro by-products	Micro-organisms used	Organic acid	References
Molasses, beet pulp residues, sugar cane bagasse, orange waste apple and grape pomace	*Aspergillus niger*	Citric acid	Soccol et al. (2004)
Pine apple waste, apple pomace	*Aspergillus niger*	Citric acid	Hang and Woodams (1986)
Pine apple waste	*Aspergillus niger*	Citric acid	De Lima et al. (1995)
Pine apple waste Grape pomace	*Yarrowia lipolytica*	Citric acid	Imandi et al. (2008)
Cassava, sugar cane bagasse	*Aspergillus niger and Gluconobacter oxidans*	Gluconic acid	Buzzini et al. (1993)
Citrus peel	*Rhizopus oryzae*	Lactic acid	Tay and Yang (2002)
	Debaryomyces courdertii	Pyruvic acid	Moriguchi (1982)
Apple pomace	Yeast	Oxalic acid	Kennedy (1994)

solid state fermentation using *Aspergillus niger* (Hang and Woodams 1986). About 99% of world's production of citric acid is by the process of microbial fermentation. Ikramul et al. (2003) have used sugar industry by-product molasses for citric acid production. The optimization of citric acid production was recently studied by Moeller et al. (2007) and Maria (2007). The comparative study on the efficiency of acid, alkaline and urea pre-treatment for the production of citric acid using *Aspergillus niger* by solid state fermentation of bagasse has been done recently. The production of citric acid by the solid state fermentation of pine apple waste has been reported with *Aspergillus niger* (De Lima et al. 1995) and *Yarrowia lipolytica* (Imandi et al. 2008).

Gluconic acid (GA) is a multifunctional carbonic acid regarded as an important ingredient in the food, feed, beverage, textile, pharmaceutical, and construction industries. The favored production process is submerged fermentation by *Aspergillus niger* utilizing glucose as a major carbohydrate source, which accompanied product yield of 98%. Gluconic acid has been produced by fermentation of agricultural by-products, such as grape pomace using *Aspergillus* spp and *Gluconobacter oxidans* (Buzzini et al. 1993). Grape by-products and banana must are used as sources for carbon for gluconic acid production by using *Aspergillus niger* ORS-4.410 under submerged fermentation. Crude grape must (GM) and banana-must (BM) resulted in significant levels of gluconic acid production i.e. 62.6 and 54.6 g/L, respectively (Singh et al. 2005).

Lactic acid is another organic acid with numerous industrial applications. Food industry wastes have been used for lactic acid synthesis using integrated glucoamylase production (Wang et al. 2008). Production of lactic acid by *Rhizopus oryzae* in oat bran has been recently developed by Koutinas et al. (2007). Agricultural residues, such as cassava, sugar cane bagasse, carrot processing waste and starch can be used as a substrate for the production of lactic acid using the microorganisms *Rhizopus oryzae* (Tay and Yang 2002). Production of lactic acid and butyric acid has been reported from sugar cane molasses (Kanwar 1995). Continuous production of lactic acid from molasses by *Sporolactobacillus cellulosolvens* has been reported (Kanwar et al.1995). Citrus peel as a sole source of carbon is used for the production of pyruvic acid by yeast (Moriguchi 1982).

Organic acids are widely used in industrial applications, such as foods, beverages, pharmaceuticals, cosmetics, detergents, plastics, resins, and other chemical and biochemical products. The production of organic acids is mainly based on two methods, such as fermentation and chemical methods. From the view point of sustainable development and human health, the fermentation methods are generally employed and most preferred method for the production of organic acids. Fermentation uses agro residues, such as feed stock, silages, molasses and cheese whey for the production of organic acids. The use of agro residues for the production of organic acids by fermentation technology provides a cheaper source of resource materials which in turn reduces the cost of production and also produces final products with higher safety degree compared to products from chemical methods.

3.4.2.3 Enzymes

Most of the chemical changes that occur in living tissues are regulated by enzymes. In recent years, there has been a renewed interest in fermentation processes for the production of enzymes by using many microorganisms. Many industrially important enzymes are produced from agro- by-products by using microorganisms. Agricultural by-products and wastes can often be used as substrates in fermentation process because of their complex composition containing carbon, nitrogen and mineral supplies suitable for the growth of microorganisms. Different fungal and bacterial strains are used in the fermentation process especially in solid state fermentation. Fermentation of agro-by-products is mainly used for the production of biomass or its metabolic products, such as enzymes (Zheng and Shetty 2000), organic acids (Pandey et al. 2000), flavour and aroma compounds (Soares et al. 2000) or pigments (Silveira et al. 2008). As most of the agricultural by-products are rich in natural polymers, such as cellulose, lignin, pectin, starch etc, microorganisms produce large amount of different types of enzymes to degrade these substrates during fermentation. Malted barley waste from the brewery industry was used as medium for the production of laccase and manganese peroxidase by using *Lentinus edodes* (Hatvani and Mècs 2001). Commercial production of enzymes is generally carried out by submerged (SmF) and solid state fermentation. The composition and concentration of media and fermentation conditions largely affect the growth and production of extracellular enzymes from microorganisms. Agro-residues have been used as an alternative to synthetic basal media for the production of amylase (Gangadharan et al. 2008). The production of different enzymes by fermentation of agro- residues by using various microorganisms is given in Table 3.11.

Amylolytic enzyme production has been mainly carried out by using wheat bran, rice bran, rice husk, oil cakes, tea waste, cassava, cassava bagasse, sugarcane

Table 3.11 Production of enzymes by solid state fermentation of agro-by-products by using different microorganisms

Agro by-products	Micro- organisms used	Organic acid	References
Fishery waste, brewery waste, apple pomace and paper industry sludge	*Phanerochate chrysoporium*	Lignolytic enzymes	Fatma et al. (2010)
Wheat bran	*Aspergillus foetidus*	Xylanase	Chapla et al. (2010)
Sugar beer waste	*Aspergillus heteromorrphus* *Phanerchate chrysosporium*	Laccase, manganese peroxidase	Vassileva et al. (2009)
Lemon pulp waste	*Aspergillus niger and Trichoderma viridae*	Pectinase	De Gregorio et al. (2002)
Grape pomace	*Aspergillus awamari*	Pectinase	Botella et al. (2007)
Corn cob and oat	*Pencillium janthinellum*	Xylanase and pectinase	Oliveira et al. (2006)

3 Sustainable Solutions for Agro Processing Waste Management: An Overview

bagasse etc (Mulimani et al. 2000). Spent brewery grain was found to be a good substrate for the production of amylase by *A.oryzae* under solid-state fermentation (Francis et al. 2003). Paddy husk was reported to enhance the nutrient utilization when mixed with the substrates, such as rice bran, corn flakes, soya flour and soy meal powder by *A. niger* CFTRI 1105 during solid state fermentation thereby increasing glucoamylase production (Arasaratnam et al. 2001). Gonzalez et al. (2008) determined the optimal nutritional and operative conditions for amylolytic enzymes production by *S. fiuligera*. Banana waste has been exploited as solid state fermentation substrate for amylase production by *B. subtilis* (Krishna and Chandrasekaran 1996). Amylase has been produced from bean waste, banana waste, wheat and maize bran using microorganisms (Hang et al. 1986).

Cellulolytic enzymes are the third most important industrial enzymes due to their versatile applications in various industries, such as paper and pulp, textile and detergent industry. Agro-industrial residues have been widely used for cellulase production by employing variety of microorganisms that are the rich source of cellulose. Solid substrate fermentation can be proposed as a better technology for commercial production of cellulases considering the low cost input and ability to utilize naturally available sources of cellulose as substrate. *T reesei* has been exclusively studied as a microbial source of extracellular cellulase capable of hydrolyzing native cellulose. Agro-residues, such as straw, spent hulls of cereals and pulses, rice or wheat bran, rice or wheat straw, sugarcane bagasse, water hyacinth, paper industry waste and other cellulosic biomass have been used for the production of cellulose enzymes (Belghith et al. 2001; Tengerdy and Szakacs 2003). Cellulase has been produced from cabbage waste, apple pomace, wheat bran, sugar beet, rice straw, bagassae etc by using microorganisms such as *Trichoderma harzianum, Aspergillus ustus, Bortrytes* spp. etc. The production of cellulolytic enzymes by *Aspergillus niger* in submerged culture with millet, guinea corn straw, rice husks and maize straw as substrates was also reported (Milala et al. 2005).

Pectinases are a group of enzymes that contribute to the degradation of pectin by various mechanisms. In nature, pectinases are important for plants as they help in cell wall extension and fruit ripening. Sugarcane bagasse has been successfully used in solid state fermentation for the production of pectinases by *Aspergillus niger, Thermoascus aurantiacus, Moniliella* spp., *Penicillium* spp. (Jacob and Prema 2008). Sugar beet pulp has been used as raw material for pectinase production by *Aspergillus niger* (Bai et al. 2004). Sugar beet pulp has also been used as the carbon source as well as the pectinase inducer to produce extracellular alkaline pectinase, by *Bacillus gibsoni*, under solid state fermentation (Li et al. 2005). Microorganisms, such as *Thermoascus aurantiacus, Penicillium* spp., *Moniliella* spp., *Penicillium viridicatum* proved to be suitable for the production of pectinases by utilizing orange bagasse as raw material for solid state fermentation (Silva et al. 2005). Wheat bran has been studied extensively for pectinase production by solid-state fermentation (Ghildyal et al. 1981). Apple pomace has been effectively used for the production of pectinases by *Aspergillus niger* (Joshi et al. 2006). Grape pomace was also found to be a good source for pectinase production by *Aspergillus awamori* (Botella et al. 2007). Pectinolytic enzyme production by using wheat bran was achieved by various microbial cultures, such as *Aspergillus niger, Aspergillus carbonarius, Streptomyces* spp., *Streptomyces lydicus, Thermoascus auriantacus, Penicillium viridicatum, Fusarium moniliforme, Bacillus* sp. (Dinu et al. 2007)

Ligninolytic enzymes are involved in the degradation of complex and recalcitrant polymer lignin. These groups of enzymes are highly versatile in nature and they find application in a wide variety of industries. The term lignin degrading enzymes encompasses mainly three oxidative enzymes; lignin peroxidase (LiP), manganese peroxidase (MnP) and laccase. The production of laccase and manganese peroxidase in solid state fermentation by *Trametes versicolor* and laccase alone by *Flammulina velutipes* by using bagasse has been reported (Pradeep and Datta 2002). Wheat straw is one of the best substrates for the production of ligninolytic enzymes. Laccase and MnP activity has been reported during the growth of *Trametes versicolor* on wheat straw (Schlosser et al. 1997). Wheat straw has served as a best substrate for the production of LiP, MnP and laccase from several other fungi, too (Arora et al. 2002). Wheat straw has also been used for the production of ligninolytic enzymes under solid state fermentation by *Streptomyces strains* (Berrocal et al. 1997). Production of all the three ligninolytic enzymes by *Streptomyces psammoticus* using rice straw in submerged fermentation has been reported (Niladevi and Prema 2005) and it was the best substrate for laccase production in solid state fermentation as compared to other agro-industrial residues (Niladevi et al. 2007). Wheat bran has been reported to be the best substrate for laccase production in SmF by *Ganoderma lucidum* (Songulashvili et al. 2007). Wheat bran has been used for the production of laccase by *Ganoderma* strain under solid state fermentation. Production of MnP and laccase from wheat bran has also been reported (Papinutti et al. 2003). Laccase production has been carried out from rice bran in submerged fermentation and solid state fermentation using the basidiomycete fungus, *Coriolus versicolor* and it was observed that rice bran was a better substrate for laccase production in both submerged and solid state fermentation as compared to other substrates like wheat bran and rice straw meal (Chawachart et al. 2004). Recently, production of ligninolytic enzymes by solid-state cultures of *Phanerochaete chrysosporium* BKM-F-1767 was investigated using different agro-industrial wastes, such as fishery residues, brewery waste, apple waste (pomace) and pulp and paper industry sludge (Fatma et al. 2010). Banana skin has been used as a support-substrate for the production of extracellular laccase by the white-rot fungus *Trametes pubescens* (Osmaa et al. 2006). Production of cellulolytic and ligninolytic enzymes from banana waste under solid state fermentation conditions has been investigated to produce high level of ligninolytic enzymes (Shah et al. 2005).

Lipase was produced from copra waste (coconut oil cake) and rice bran by using *Candida rugosa* (Benzamin and Pandey 1996). Production of xylanase has been reported on dried apple pomace (Bhalla and Joshi 1993). Recently, kiwi fruit waste has been used as substrate for production of laccase by using white rot fungi (Rosales et al. 2002). Fruit processing by-products such as apple pomace, cranberry pomace, and strawberry pomace were used as substrates for production of polygalacturonase by *Lentinus edodes* through solid-state fermentation (Zheng and Shetty 2000). *Bacillus firmus* produced very high level of polygalacturonase by solid state fermentation of potato wastes (Bayoumi et al. 2008). Different enzymes such as α and β amylase, cellulase, pectinase, lipase, esterase and peroxidase were produced by the solid state fermentation with different micro-organisms such as fungi, bacteria and yeast by using agro-industrial orange peel and pulp waste (Attyia and Ashour 2002). Recently, Sangeetha et al. (2004) have studied the production of fructosyl transferase by *A. oryzae* while employing a great variety of agricultural by-products as substrates: cereal brans (wheat bran, rice bran and oat bran), corn products (corn cob, corn bran, corn germ, corn meal, corn grits and whole corn powder), coffee and tea-processing by-products (coffee husk, coffee pulp, spent coffee and spent tea), sugarcane bagasse and cassava bagasse.

Enzymes are one of the most important biological catalysts for many industrial and medical applications. Global enzyme demand will rise 6.3% annually through 2013 depending on the strong demand in the speciality enzyme segment and good growth in different industrial applications (www.freedoniagroup.com). North America and Western Europe have the healthy enzyme market and developing countries also show a faster growth in the same sector. The cost and availability of substrates also play an important role in the enzyme production and development of an efficient process in addition to the manufacturing and purification technologies. The feasibility of agro residues for the commercial production of enzymes has been well explored. Reduction in the cost of enzyme production and improvement in the efficiency of production are major goals for future research. Agro industrial by-products can be successfully utilized for solid state fermentation to produce the microbial production enzymes. These residues are abundant low cost materials and are easily available.

3.4.2.4 Flavor Compounds

Flavour compounds comprise major part of the world market for food additives. Most of the flavouring compounds are produced via chemical synthesis or by extraction from natural materials. Recently there has been a growing interest in the production of flavor and other food additives of plant origin. There are various groups of aroma compounds and the classification of these compounds are mainly based on their chemical structure, physicochemical properties, sensorial properties and also the chemical nature of the compounds. Lipid derived flavor/aroma compounds are one of the major important families which include volatile acids or esters, lactones, aldehydes, alcohols, ketones and carotenoid derived aroma compounds. Many microorganisms, including bacteria and fungi, are currently known for their ability to synthesize different aroma compounds.

Several researchers have studied the production of aroma compounds by solid state fermentation by using several microorganisms, such as *Neurospora* spp. (Pastore et al. 1994), *Zygosaccharomyces rouxii*, *Aspergillus* spp. (Ito et al. 1990), using pre-gelatinized rice, miso and cellulose fibres, respectively.

Abundant research has been reported on the production of flavor compounds by solid state fermentation of agro-byproducts as a substrate as represented in the Table 3.12.

Table 3.12 Production of flavor compounds by using different agro-industrial wastes

Agro-by-product	Microorganisms	Flavor compounds	References
Cassava bagasse, apple pomace and soybean	*Ceratocystis fimbriate*	Fruity flavour	Bramorski et al. (1998)
Cofffe husks	*Ceratocystis fimbriate*	Pine apple flavour	Soares et al. (2000)
Tropical agro-by-products	*Rhizopus oryzae*	acetaldehyde and 3-methylbutanol	Bramorski et al. (1998) and Christen et al. (2000)
Cassava bagasse and giant palm bran	*Kluveromyces marxianus*	Fruity aroma	Medeiros et al. (2001)
Soybean by-products	*Bacillus subtilis*	2,5-Dimethyl-pyrazine and etramethylpyrazine	Besson et al. (1997) and Larroche et al. (1999)
Semi solid maize	*Pediococcus pentosaceus Lactobacillus acidophilus*	Dairy flavour compounds	Escamilla-Hurtado et al. (2005)
Copra fat by-products	*Aspergillus niger*	Methyl ketones	Allegrone et al. (1991)
Wheat bran	*E.coli* JM109 (pBBI)	Vanillin	Gioia et al. (2007)

Cassava bagasse, apple pomace, giant palm bran and coffee husk have been used as substrates for the aroma production in solid state fermentation (Bramorski et al. 1998; Soares et al. 2000; Medeiros et al. 2001). Production of aroma compounds by agro-industrial by-products, such as wheat bran, cassava bagasee and sugar cane bagasse were reported using the fungus *Ceratocystis* (Christen et al. 1997). Fruity aroma was produced by solid state fermentation of coffee husk by using the organism *Ceratocystis fimbriata*. It has been reported that the flavor development depends on the amount of glucose added in the medium. High levels of addition of glucose reduced the aroma intensity (Soares et al. 2000).

Fruity aroma was produced by *Ceratocystis fimbriata* in solid-state cultures by using several agro-industrial wastes, such as cassava bagasse, apple pomace, amaranth and soybean (Bramorski et al. 1998). Strong pine apple flavor was produced by solid state fermentation by using coffee husk as substrates by using the strain *Ceratocystis fimbriata* (Soares et al. 2000). Volatile compounds, such as acetaldehyde and 3-methylbutanol were also produced by the edible fungus *Rhizopus oryzae* during solid state fermentation on tropical agro by-products (Bramorski et al. 1998; Christen et al. 2000). *Kluyveromyces marxianus* produced aroma compounds, such as monoterpene alcohols and isoamyl acetate (responsible for fruity aromas), by solid state fermentation of cassava bagasse or giant palm bran as a substrate (Medeiros et al. 2001). Volatile compounds with fruity characteristics, such as ethanol, acetaldehyde, ethyl acetate, ethyl propionate and isoamyl acetate were produced by *Ceratocystis fimbriata* in two different bioreactors: columns (laboratory scale) and horizontal drum (semi-pilot scale) by using coffee husk as a substrate (Medeiros et al. 2006). Recently attempts have been made for the production of vanillin by the bioconversion of ferulic acid derived from wheat bran by genetically engineered *E.coli* strain JM109 (pBBI) with a bioconversion yield of 50% (Gioia et al. 2007).

Presently, most of the aroma compounds are obtained by traditional methods, such as chemical methods and extraction from natural products. Increasing awareness and interest for natural products has forced the aroma industry to seek new methods to obtain aroma compounds naturally. An alternative way for the natural biosynthesis is mainly based on microbial biosynthesis or bioconversion. The increasing demand for natural flavours in the food industry has made remarkable efforts towards the development of biotechnological process for the production of aroma compounds. The production of aroma compounds for the food processing industry by the microbial technology offers several advantages over traditional methods. The solid state fermentation of agro-residues leads to tremendous improvement in the economical feasibility of the production process and application of aroma compounds and also can give higher yields. Moreover, the use of agro residues as the starting material can reduce the cost of production in terms of raw material which can lead to the development of a low cost technology.

3.4.2.5 Pigments

An alternative route for the production of the natural food colorants is through the application of biotechnological tools by employing the microorganisms. There is a growing demand for natural colours because of their presumed safety. The selection of substrate for the solid state fermentation process depends upon several factors, mainly related with cost and availability. In the solid state fermentation process, the solid substrate supplies the nutrients to the microbial culture growing in it. There is an increased interest in natural pigments to replace some currently used synthetic dyes, since the latter have been associated with toxic effects in foods (Mapari et al. 2005). The red pigment of the fungus *Monascus* is widely used in Asia for centuries as a food colorant (Kim et al. 2002).

Traditionally, *Monascus* has been cultivated on rice (forming ang-kak or red rice), although several other media, have been tested for the pigment production. Rice is the natural substrate which gives the best production, compared to other typical cereals, tubers and leguminous plants (Carvalho et al. 2003). However, some of the other substrates used also presented good biopigment production, especially corn, wheat and cassava. Cassava bagasse gave a low pigment yield, but being an agro-industrial residue, its low price might compensate for its low yield. Very recently, jackfruit seed powder has been identified as a potent substrate for pigment production (Babitha et al. 2007). Pigment production by *Monascus purpureus* in submerged fermentation by using grape waste as growth substrate was optimized by employing factorial design and response surface techniques (Silveira et al. 2008).

There has been a great interest in the use of microbial technology for the production of color from food processing by-products, such as *Rhodotorula, Cryptococcus, Phatlia, Rhodozyma, Monosaus purpureus, Bacillus* spp., *Xanthomonas campestris* (Bilanovic et al. 1984).

There is an increased interest and demand on edible natural pigments to replace some currently used synthetic colouring agents, which have potential for carcinogenicity and teratogenicity. There is a growing need for the development of low cost technologies for the production of natural pigments due to the high cost of technology for the production of pigments on industrial scale. The production of natural pigments by solid state fermentation by using agro-industrial residues as substrates will be very cost effective and also will add value to the industry.

3.4.3 Use for the Production of Fertilizer

Agro-by-products can be beneficially used as a soil conditioner or fertilizer. The by-product characteristics of interest include the moisture content, BOD, calcium carbonate, C:N ratio, fat and oils, odors, pathogens, pH, soluble salt and toxicity. Agro-industrial wastes from agriculture; food processing or any cellulose based industries remain largely unutilized and often cause environmental pollution. These wastes could be converted into potential renewable source of energy, if managed sustainably and scientifically. In the last few decades, composting technology has been arising as a sustainable tool for the efficient utilization of the agro-industrial processing wastes and to convert them into value-added products. The usability of the process depends upon several factors, such as raw material, various process conditions like pH, temperature, moisture, aeration etc. Composting is a natural aerobic biochemical process in which thermophilic microorganisms transform organic material into a stable soil like product. It is a managed biological conversion of waste material, under controlled conditions into hygienic humus rich, relatively bio stable product that conditions soils and nourishes plants. Organic wastes from fruit and vegetable processing industry used for composting include peelings, pulping, outer skins, pomace, cores, leaves, fruit twigs and sludge from the processing and packaging of various products.

The by-product from sugarcane industry can be used for the production of fertilizer by composting. The potential of these winery wastes in vermicomposting has been investigated by using *Eisenia andrei* (Nogales et al. 2005). The post-harvest residues of some crops, e.g. wheat, millets, and pulse were utilized to recycle through vermicomposting by *Eudrilus eugeniae* (Suthar 2008). Gobi et al. (2001) studied the vermicomposting of coir pith by *Eudrilus eugeniae* and found that the NPK values increased significantly from its original value after vermicomposting. The ability of *E. fetida* pre-adapted to coffee pulp was tested to transform coffee pulp into vermicompost under different experimental conditions in outdoor containers (Orozco et al. 1996).

Composting technology is a suitable tool for efficient conversion of agro industrial processing wastes, which serves as a rich source of plant nutrients. These waste materials are packed with a tremendous source of energy, protein and nutrients, which would otherwise be lost if they are disposed as such in the open dumps and landfills. Moreover, with the use of compost as organic amendments in the agriculture, recycling of the nutrients back to the soil takes place, in turn, maintaining the sustainability of the ecosystem. Therefore, the production of fertilizer has enormous potential in agro-industrial waste management in a sustainable and decentralized manner, as it yields rich organic fertilizer, safely disposes the organic waste and helps tackle environmental problems, such as landfill and reduces the expense of collecting and transporting this waste.

3.4.4 Use for Energy Production

Over the last century, energy consumption has increased beyond control as a result of growing world population and industrialization (Sun and Cheng 2002). Renewable energy sources, such as ethanol, methane and bio-hydrogen can be produced by fermentation of sugars unlike the fossil fuels. Owing to diminishing natural oil and gas resources, interest in the bioconversion of renewable cellulosic biomass into fuel ethanol as an alternate to petroleum is rising around the world (Stevenson and Weimer 2002). Biomass is the earth's most attractive alternative among fuel sources and sustainable energy resource.

Bio-fuels can be broadly classified into two major types, gaseous and liquid biofuels. Purification of the conventional biogas into methane-enriched biofuel led to the development of biomethane. Biohydrogen is a relatively new type of gaseous biofuel, which is produced by anaerobic fermentation of agro-industrial wastes by the synergistic action of a consortium of methanogenic, acidogenic and hydrogenic bacteria (Fountoulakis and Manios 2009.) Liquid biofuels have recently been classified into bioethanol and biodiesel. While bioethanol has recently gained rejuvenated importance in the wake of present energy crisis worldwide, biodiesel occupied the centre stage as a potential substitute for petroleum diesel in the last two decades (Sumathi et al. 2008; Demirba 2009).

In developing countries, production of biogas from waste is an appropriate solution. Biogas generation through anaerobic fermentation is considered to be a simple and economical system of waste treatment and a number of processes have been developed and commercially exploited. Cuzi and Labat (1992) studied methanogenic fermentation of cassava peel with mean methane content of 50%. Fruit and vegetables wastes subjected to anaerobic degeneration produced 0.12 m^3 biogas/kg (Viswanath et al. 1992). Anaerobic digestion of mango peel resulted in biogas production of 0.33 m^3/kg with 53% methane content (Somayaji 1992). The biomethanation of banana peel and pine apple processing waste suggested their potential and suitability for economically viable waste treatment by technologically anaerobic digester (Bardiya et al. 1996).

3.4.4.1 Ethanol

Ethanol can be produced from agro-byproducts which are rich in sugars/starch by the microbial technology; it may evolve as an alternative to our limited and renewable source of energy. Ethanol production by solid state fermentation from agro industrial wastes, such as apple pomace, grape pomace has been reported (Hang et al. 1982, 1986). Ethanol production from grape and sugar beet pomaces has been developed using solid state fermentation (Rodrígueza et al. 2010). Ethanol has been produced from food wastes such as citrus, molasses, apple pomace, orange waste, rice husk and banana waste (Shankaranand and Lonsane 1993). It has been estimated that 2 million metric tonnes of dry agro wastes can produce 809.2 million L of ethanol (Borup and Fenhaus 1990).

Bioethanol is a biofuel used as a petrol substitute, produced by simple fermentation processes involving cheaper and renewable agricultural carbohydrate feedstock and yeasts as biocatalysts. A variety of common sugar feed stocks including sugarcane stalks; sugar beet tubers and sweet sorghum are used. The fermentation process was mediated by two enzymes, invertase and zymase, produced by the yeast cells. The overall process for the conversion of bioethanol production from cellulose and hemicelluloses are shown in Fig. 3.3.

Ethanol produced from renewable and cheap agricultural products reduces the green house gas emissions, such as CO_2, NO_2 and SO_2 and eliminate smog from the environment. Agricultural residues and wastes have several advantages as they do not require any large agricultural and forestry facilities. Some of the agro-industrial residues and waste materials abundantly available are mentioned in Table 3.13

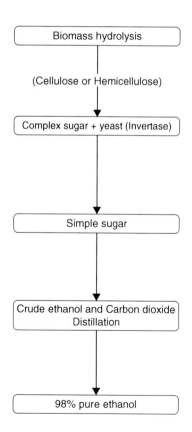

Fig. 3.3 Basic ethanol production process flow chart

3 Sustainable Solutions for Agro Processing Waste Management: An Overview 89

Table 3.13 Agro-industrial residues and plant waste materials used for biofuel production

Agro industrial waste	Biofuel	References
Brewer's yeast; autolyzate and	Bioethanol	Ruanglek et al. (2006)
Fish soluble waste	Bioethanol	Liimatainen et al. (2004)
Waste Potato	Bioethanol	Ruanglek et al. (2006)
Corn steep liquor	Biogas	Fountoulakis et al. (2008)
Olive oil waste, Vinery waste	Biogas	Kivaisi and Rubindamayugi (1996)
Cereals waste, sugar bagasse, coffee husk		

which is used as potential substance for ethanol production. Fermentation of banana waste to ethanol by co-culture fermentation by using *C. thermocellum* in combination with *C. thermosaccharolyticum* and *T. ethanolicus* bacterial species was recently reported (Reddy et al. 2010).

3.4.4.2 Biodiesel

Limited and increasing demand for diesel all over the world led researchers to find some alternative sources. Transesterification is the most widely used process in which triglycerides react with an alcohol (mainly methanol) in the presence of chemical (acid or alkali) or biological (enzyme) catalysts to produce mono alkyl esters, popularly known as biodiesel. Some alkali catalyzed batch processes have been commercialized.

Biodiesel is generally produced from vegetable oils or animal fats. Various oils, such as palm oil, soybean oil, sunflower oil, rice bran oil, rapeseed oil etc. are used for biodiesel. There is also some research going on the production of biodiesel from agro wastes. However, bioethanol produced from agro-industrial residues can in turn be used for the trans-esterification of vegetable oils to produce mono-ethyl esters of fatty acids as biodiesel (Peng and Chen 2007). Whey concentrate and tomato waste hydrolysate, which contain more than 1 g/L total organic nitrogen, in turn produce 14.3 and 39.6%, lipid respectively can be used for gamma-linolenic acid production. The amount of gamma-linolenic acid produced from these wastes can be used as two good raw materials for biodiesel production (Fakasa et al. 2008). Recently, suitability and potential of oil mill waste has been studied for the production of biodiesel. Evaluation of bio-waste materials by Nuclear Magnetic Spectrum revealed the potential of this tool to identify waste-oil sources in a cost-effective and quick manner (Willson et al. 2010).

3.4.4.3 Biogas

The conventional biogas, which is produced in biogas plants by employing anaerobic digestion of organic wastes including manure by mixed microbial cultures, is primarily composed of methane (typically 55–70% by volume) and carbon dioxide (typically 30–45%) and may also include smaller amounts of hydrogen sulfide (typically, 50–2000 ppm), water vapor (saturated), oxygen, and various trace hydrocarbons (Amigun et al. 2008). Due to its lower methane content (and therefore lower heating value) compared to natural gas, biogas use is generally limited to engine-generator sets and boilers (Krich et al. 2005).

Biomethane is upgraded or sweetened biogas after the removal of the bulk of the carbon dioxide, water, hydrogen sulfide and other impurities from raw biogas. Biomethane is extremely similar to natural gas (which contains 90% methane) except that it comes from renewable sources (Krich et al. 2005). Biogas can also be purified and upgraded and used as a vehicle fuel. Over a million vehicles are now using biogas and fleet operators have reported savings of 40–50% in vehicle maintenance costs (Parawira 2004). Table 3.13 gives some recent reports on the production of biogas from agro-residues.

The production and use of biofuels and bioenergy are in a new era of global growth. The depletion in the non renewable energy sources force us to search for new source of energy. The development in biofuel production is mainly due to many factors such as development of more conversion technologies, introduction of global policies on fuels and environmental issues, and the rising in the price of conventional energy and fuel sources. The reduction of petroleum products and emissions of carbon dioxide and other gases that are contributing to global warming can be controlled by the increase in use of biofuels. Different types of biofuels are already in the global market. The two conventional biofuels that are primarily in use are ethanol and biodiesel. Currently, ethanol accounts for more than 90% of total biofuel production and the biodiesel makes up the remaining percentage. The development of low cost technologies with the utilization of low cost raw materials as new energy source will have tremendous impact on the world fuel economy and for fuel crisis. The economic feasibility of biofuel production is mainly based on its cost and availability of raw materials. Agro residues offer a cheaper option as raw material for biofuel production. The use of agro industrial residues and wastes with the help of modern biotechnological tools has impact on many factors, such as effective reduction in emission of toxic pollutants and green house gases, environmental management and a control in fuel rise.

3.4.5 Recovery of Value-Added Products

In recent years, attempts have been made to convert food by-products into a variety of valuable products. The recycling of by-products and minimizing wastes are crucial aspects of this strategy. The recovery of high value- added compounds are of special interest.

There is an increasing consumer appreciation of natural products as alternative to synthetic compounds in a variety of goods ranging from food to personal care formulations. Consumers have increasing awareness of the diet related health problem, therefore, demanding natural ingredients, which are expected to be safe and health promoting. By-products of plant food processing were found to be promising sourc-

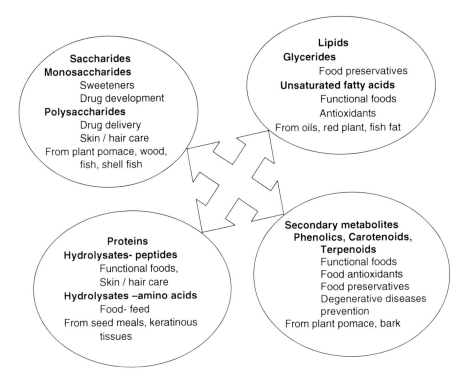

Fig. 3.4 Major bioactive compounds, their sources and applications

es of biological active compounds, which may be used because of their favourable nutraceutical properties. There are large variety of value-added compounds in the by-products and wastes of biological origin. These products may be used as such or may serve as a starting material for the preparation of novel compounds. The characteristics of commercial interest include antioxidants, carbohydrates, dietary fibers, fat and oils, pigments, proteins and starch (Torres et al. 2002). Figure 3.4 summarizes the bioactive compounds, their sources and their applications. Agricultural products are good source of anti-mutagens that repress the genotoxic potency or activity of specific chemical mutagens or carcinogens (De Flora et al. 1993). Recently, isoflavones, such as genistin, genistein, daidzin, daidzein, glycitin and glycitien have been isolated from soybean and corn processing by-products and showed anti-genotoxic activity and anti-cancer property (Plewa et al. 2001).

Polysaccharides, such as pectins from fruit by-products have different applications mostly related to their film and pharmaceutical properties. Oligosaccharides from fruit processing wastes have been used as prebiotic components of functional foods and monosaccharide as starting material for drug development. Seeds obtained from fruit beverage and fruit canning industries are rich in oils of high content in unsaturated fatty acids and glycerides of possible direct application in cosmetics and can also be used as starting material for the preparation of novel chemicals, such as biocompatible food preservatives (Biermann et al. 2000). By-products from tomato processing industry contain polyunsaturated antioxidant carotenoids, such as lycopene, which can use nutraceuticals. Protein rich meals are by-products from the oil seed processing industry. These proteins and their hydrolysates may be used in animals and poultry feed and in human functional nutrition (Baysal et al. 2002). A great variety of secondary metabolites, such as phenolic compounds, carotenoids and other bioactive compounds, such as dietary fiber, enzymes and vitamins with biological activities can be isolated from agro- processing by-products as shown in Table 3.14. They have been given increasing significance as functionally active components of food and drink and possess many nutraceutical properties (Diplock et al. 1998).

3.4.5.1 Biofibers

Agricultural and forestry residues are one of the major sources of cellulose in the world. Approximately 2×10^{11} t of lignocellulosics are produced every year (Mohanty et al. 2000). Primary ligonocellulosic by-products that are easily available in huge quantity at low cost are corn stover, wheat, rice, barley straw, sorghum stalks,

Table 3.14 Bioactive compounds from food processing industry by-products

By-products	Bioactive compounds	References
Fruits		
Apple peel	Flavonoids and anthocyanins	Wolfe et al. (2003)
	Pectin and phenolics	Carle et al. (2001)
Citrus peel	Phenolics, dietary fiber	Gorinstein et al. (2001)
Banana peel	Phenolic compounds	Someya et al. (2002)
Banana bract	Anthocyanins	Pazmiño-Durán et al. (2001)
Apple pomace	Flavonoids and anthocyanins	Sheiber et al. (2003)
Grape skin and seed	Phenolics, anthocyanin, dietary fiber	Torres et al. (2002)
Peach pomace	Pectin	
Mango seed	Phenolics and phospholipids	Arogba (2000)
Mango peel	Phneolics and dietary fiber	Ajila et al. (2007a, b, 2008)
Pine apple waste	Antioxidants	Wrolstad and Ling (2001)
Vegetable		
Carrot pomace	Carotenoids	Stoll et al. (2001)
Onion waste	Quercetin	Waldron (2001)
Red beet waste	Betalins	Kujala et al. (2000)
Potato peel	Phenolic acids	Rodriguez et al. (1994)
Tomato peel	Lycopene	Baysal et al. (2002)
Others		
Soybean hull	Peroxidase	Sessa (2003)
Black gram husk	Phenolic acids	Ajila et al. (2009)
Almond hulls	Phenolic compounds	Duh and Yen (1995)
Pea nut hulls	Phenolic compounds	Ge et al. (2002)
Wheat germ	Vitamin E	

coconut husks, sugarcane bagasse and pine apple and banana leaves. These crop residues used for industrial applications could be a better source of revenue for farmers without adversely affecting the soil fertility. Lignocellulosics from agro by-products possess many industrial applications depending on their composition and physical properties. Wheat and rice straw have been used in pulp and paper industry and production of cellulose fibers as alternative to wood for cellulose based materials (Focher et al. 1998; Majumdar and Chanda 2001; Lim et al. 2001). Recently, many agricultural residues, such as coir, pine apple waste, banana leaves, corn husk and stalks have been used as natural cellulose fibers for textiles, composites, paper and other industrial applications (Focher et al. 1998; Ganjyal et al. 2004; Reddy and Yang 2004). In Brazil, agro by-products from pine apple industry, pine apple leaves can produce about 19,600 t of fibers (Sivam et al. 2006). Currently, cellulose microfibers have been isolated, purified and characterized from bagasse which has many industrial applications (Bhattacharya et al. 2008). Natural cellulose fibers are extracted from the lignocellulosic agro by-products by different methods, such as microbial, chemical and mechanical methods. The retting, extraction of cellulose, by microbial method has been done by using different microorganisms, such as bacteria and fungi. *Bacillus* and *Clostridium* species are mainly used bacteria for the extraction of cellulose. Fungi, such as *Rhizomucor pusillus*, *Fusarium lateritium* are also used for the extraction of cellulose (Henriksson et al. 1997). The chemical extraction of cellulose has been carried out with mild acids, alkali and enzymes. Sodium hydroxide is the commonly used alkali for the extraction of cellulose. Acids, such as sulphuric acid, oxalic acid in combination with detergent have also been used for fiber extraction (Henriksson et al.1997). Combination of enzymes, such as pectinases, hemicellulases and cellulases are generally used for enzymatic fiber extraction. Multi-enzyme complex has been recently developed for the extraction of fiber (Reddy and Yang 2005). Mechanical separation of fiber is carried out using different methods, such as decorticating machines, steam explosion, tilby process, ammonia fiber extraction etc (Focher et al. 1998; Gollapalli 2002).

The major constituents of lignocellulosic sources are cellulose, hemicellulose and lignin as shown in Table 3.15 and the proportion of these components in a fiber depends on age, source of the fiber and the extraction conditions used for the fiber recovery (Batra 1985). Cellulose is the major component that provides strength and stability to the plant cell wall and the fiber. The content of cellulose in fiber influences the properties, economics of fiber production and industrial application of fiber. Hemicellulose is composed of multiple polysaccharide polymers with a degree of polymerization and orientation less than that of cellulose and mainly consists of glucose, xylose, galactose, arabinose and mannose. It acts as filler between cellulose and lignin. Lignin is a highly cross-linked molecular complex with amorphous structure and act as glue between individual cells and the fibrils forming the cell wall (Majumdar and Chanda 2001). Lignin provides strength and stiffens the cell wall of the fibres to protect the carbohydrate from physical and chemical damage (Saheb and Jog 1999). The structure, morphology, flexibility and rate of hydrolysis of fibres mainly depend on the lignin content. Fibers from lignocellulosic sources find various applications, such as particle boards, building materials,

Table 3.15 Composition of biofibers in agro-by-products

By-products	Cellulose (%)	Liginin (%)	Hemicellulose (%)	References
Bagasse	50	25	25	Pandey et al. (2000)
Banana	60–65	6–8	5–10	Majumdar and Chanda (2001)
Barely straw	31–45	27–38	14–19	Rowell et al. (1997)
Corn stover	38–40	28	7–21	Reddy and Yang (2004)
Coir	36–43	0.15–0.25	41–45	Majumdar and Chanda (2001)
Pineapple by-products	70–82	18	5–12	Majumdar and Chanda (2001)
Rice straw	28–36	23–28	12–14	Rowell et al. (1997)
Sorgham stalks	27	25	11	Rowell et al. (1997)
Wheat straw	33–38	26–32	17–19	Gressel and Zilberstein (2003)

textiles, human food, animal feed, cosmetics, and medicine and for other biopolymers production and fine chemicals (Majumdar and Chanda 2001). Cellulose and hemicelluloses can be used for the production of single cell protein, cellulolytic enzymes, mushrooms etc. It can also be used for the production of resins, plastics, pharmaceuticals, herbicides etc. Fibers can also used for the production of paper and paper boards (Sundstrom and Klei 1982; Ebringerova and Heinze 2000). Recently, a novel process for the production of industrially important cellulose acetate from agro by-products, such as rice straw, wheat hull and corn fibre by removing hemicellulosic sugars was developed (Biswas et al. 2006).

An increasing population and depletion of natural resources leads to increased utilization of agro based fibers compared to natural fibers. Sustainable management of natural resources becomes more complex due to increase in population and increased demands on the natural resources. The utilization of agro residues as raw material for the production of natural fibers with the development of modern technologies will lead to a sustainable management of depleting natural resources and environmental pollution. This will also lead to production of natural fibers in a cost effective manner.

3.4.5.2 Antioxidant Compounds

Antioxidants are widely used as ingredients in dietary supplements in the hope of maintaining health and preventing diseases such as cancer and coronary heart disease. The major antioxidant compounds present in agro- residues are phenolic compounds and carotenoids.

Phenolic compounds have a major role in growth and reproduction, providing protection against pathogens and predators besides contributing towards the color and sensory characteristics of fruits and vegetables (Alasalvar et al. 2001). Phenolic compounds exhibit a wide range of physiological properties, such as anti-allergenic, anti-inflammatory, antimicrobial, antioxidant, antithrombotic, cardio protective and vasodilatory effects. Several beneficial effects derived from phenolic compounds have been attributed to their antioxidant activity (Heim et al. 2002). The antioxidant properties of phenolics are mainly due to their redox properties, which allow them to act as reducing agents, hydrogen donors and singlet oxygen quenchers. They also act as chelators of metal ions, preventing metal catalyzed formation of free radical species (Salah et al. 1995). Phenolic antioxidants interfere with the oxidation of lipid and other molecules by rapid donation of hydrogen atom to radicals. The phenoxy radical intermediates are relatively stable; therefore, a new chain reaction is not easily possible. The phenoxy radical intermediates also act as terminator of the propagation route by reacting with other free radical (Shahidi and Naczk 1995). Structurally, phenolic compounds comprise an aromatic ring, bearing one or more hydroxyl substituents, and range from simple phenolic molecules to highly polymerized compounds (Bravo 1998). Most naturally occurring phenolic compounds are present as conjugates with mono and polysaccharides, linked to one or more of the phenolic groups, and also may occur as functional derivatives, such as esters and methyl esters (Shahidi and Naczk 1995; Harborne et al. 1999). Agro- byproducts are rich sources of phenolic compounds as shown in Table 3.14. Recently antioxidant properties and cyto protective activity of phenolic compounds from mango peel has been reported (Ajila et al. 2007a, b, 2008).

Carotenoids are a class of natural fat-soluble pigments. Among the pigments present in living organisms, carotenoids are the most widely distributed in nature. They are found throughout the plant kingdom in both photosynthetic and non-photosynthetic tissues, in bacteria, fungi and animals. They consist of eight isoprenoid units joined in such a manner that the arrangement of isoprenoid units is reversed at the center of the molecule so that two central methyl groups are in 1, 6-position relationship and the remaining non-terminal methyl groups are in 1, 5-position relationship. All carotenoids may be formally derived from the acyclic $C_{40}H_{56}$ structure having a long central chain of conjugated double bonds by hydrogenation, dehydrogenation, cyclization or oxidation. Some 600 different carotenoids are known to occur naturally (Ong and Tee 1992). In general, carotenoids can be classified into two different groups: carotene, which are strictly hydrocarbons and xanthophylls, which are derived from the former and contain oxygenated functional groups. Structurally, the carotenoids are acyclic or contain a five or six number cyclic ring at one or both ends of the molecules. Carotenoids have also been classified as primary and secondary compounds. Primary carotenoids are the compounds required by plants for photosynthesis (carotene, violoxanthin, neoxanthin), where as secondary carotenoids are localized in fruits and flowers (carotene, cryptoxanthin, zeaxanthin, anthrxanthin, capsanthin, capsorubin) (Litchenthaler 1987).

Carotenoids are lipophyllic substances except in certain cases where highly polar functional groups are present, as in norbixin, a carotenoid with dicarboxyl acid structures. The polyene chain makes the carotenoid molecule extremely susceptible to isomerizing and oxidizing conditions, such as light, heat or acids. In green plants, beta-carotene was found to be the major one followed by xanthophylls, lutein, iolaxanthin and neoxanthin. Zeaxanthin, β-carotene, cryptoxanthin and anthrxanhin are found in small amounts. In the case of fruits, xanthophylls are the major carotenoids. In maize, the predominant pigments are lutein and zeaxanthin, while in mango and perissmon, the major pigments are cryptoxanthin and zeaxanthin. In tomato, the major carotenoid is lycopene. Capsanthin and capsorubin are found almost exclusively in ripe capsicum fruits and are also responsible for their attractive red color (Davis 1976; Goodwin 1976). The presence of most common carotenoid pigments found in agro residues are shown in Table 3.14. Lycopene is a fat soluble carotenoid with 11 conjugated double bonds in the molecule and is the precursor of β carotene with a very good antioxidant activity twice that of β carotene (Di Mascio et al. 2002). Lycopenes were also reported to have many pharmacological and anti-carcinogenic properties (Livny et al. 2002). Natural products, such as tomato, water lemon, red peppers etc are found to be a good source of lycopene. The by-products derived from tomato processing industries are growing annually. In Europe, it was evaluated that almost 10 million tons of tomatoes are processed by the food industry and the by-products derived from the industry is almost 0.1 million tons of tomatoes (Naviglio et al. 2008). The tomato processing by-product is a very good source of pure lycopene. An innovative eco-friendly, and cost effective extraction process for pure lycopene from tomato waste with an average recovery of 14% w/w and 98% purity was developed using an Extractor Naviglio with water as solvent for extraction (Naviglio et al. 2008).

Consumers are increasingly interested in the health benefits of foods such as diseases prevention, health enhancing and antioxidant properties beyond their normal nutritional benefits. The functional foods and nutraceuticals have an important role in improving the health, reduce health care costs and support economic development in rural communities. Antioxidants play an important role in nutraceutical industry. Global demand for nutraceutical industries will grow 5.8% annually through 2010 (www.freedoniagroup.com). Recently, there has been plenty of research on the production of nutraceuticals from many agro residues. This leads to the development of nutraceuticals in a cost effective manner for the industry and also availability of cheap and worthy nutraceutical products to the end user. The use of agro residues for nutraceuticals leads to better management of these residues which is one of the main causes of environmental pollution.

3.4.5.3 Dietary Fiber

Dietary fiber includes the compounds found in plant foods, which are non-digestible in the mammalian small intestine. These compounds include cellulose, non-cellulose polysaccharides (Southgate 1982). It plays an important role in the prevention and treatment of diseases. Cereals, legumes, vegetables, fruits, lentils, nuts and seeds are found to be a rich source of dietary fiber. Dietary fibers do not constitute a defined chemical group, but are a combination of chemically heterogeneous substances, such as cellulose, hemicellulose, pectins, lignins, gums and polysaccharides (Asp et al. 1992). The most widely accepted definition of dietary fiber is a physiological one, in which 'dietary fibers' correspond to the vegetable cell wall residues that are resistant to enzymatic hydrolysis in the small intestine. A chemical definition describes dietary fibers as non-starch polysaccharides. The most commonly used definition is dietary fiber are oligosaccharides, polysaccharides and the derivatives which cannot be digested by the human digestive enzymes to absorbable components in the upper alimentary tract (Trowell et al. 1976). Dietary fibers can be generally classified into two main categories, insoluble and soluble dietary fiber depending on their solubility in water. Insoluble dietary fiber (IDF) is a coarse material that does not dissolve in water. The chemical components of insoluble fiber include cellulose, lignin and hemicellulose found in primary and secondary cell wall of plants. The cellulose in the cell wall is chemically bound to hemicellulose and lignin (Salisbury and Ross 1992). The hemicellulose and lignin act to give significant rigidity and resistance to the action of enzymes and acids. Soluble dietary fiber (SDF) is made up of sticky substances, such as gums and dissolves in water. Each fraction has different physiological effects (Schneeman 1987). The insoluble part is related to both water absorption and intestinal regulation, whereas the soluble fraction is associated with the reduction of cholesterol in blood and the decrease in intestinal absorption of glucose. In terms of health benefits, both fibers complement each other and a 50–70% insoluble and 30–50% soluble dietary fiber is considered a well-balanced proportion (Schneeman 1987). In cereals, the SDF content is quiet low whereas in fruits, the ratio between soluble and insoluble DF fractions is more balanced (Saura-Calixto 1993). SDF mainly consists of hemicelluloses, pectic substances, gums and mucilages. Hemicellulose includes xylan, mannans and xyloglucans (Aspinall 1980). The pectic substances comprise the second major category of SDF and include the arabinans and galactans (Schneeman 1987). The third major category in SDF are gums and are considered to be a hydrophilic polymeric material, generally polysaccharides in nature, that can be dissolved or dispersed in water to give a gelling effect (Whistler 1973). The monomeric units found in gums include the neutral sugars, uronic acids and other acid groups (Schneeman 1987). The physical and chemical properties of the components of dietary fiber appear to be important in determining the physiological response to sources of fiber in the diet. Four major properties have been associated with physiological responses to various sources including fermentability by bacteria, binding to organic compounds, ion-exchange capacity and water holding capacity which has also been associated with viscosity and solubility of various fiber sources (Schneeman 1986). Agro-residues are a rich source of dietary fibers (Table 3.14).

Dietary fiber is one of the major components in functional foods. Dietary fibers are known to reduce cholesterol levels, digestive complaints, make feeling of satiety and stimulate bacterial fermentation in the colon, prevent chronic diseases such as cancer, diabetes, obesity etc. Increased attention to dietary fiber resulted in the introduction of many food products with the incorporation of dietary fiber products and dietary fiber enriched products. The U.S. food fiber industry earned revenue of $ 193.1 million in 2004 and is expecting a growth of $ 495.2 million in 2011 (Frost and Sullivan 2005). Technologies for the extraction of dietary fibers from the agro residues lead to a new economical technology for the utilization of the agro residues with simultaneous value addition.

3.4.6 Use as Metal Adsorbent for the Environmental Pollutants

Removal of heavy metals ions from the aqueous streams by agricultural waste materials is an innovative and promising technology. Agricultural materials rich in cellulose have metal absorption capacity. The cellulose, hemicelluloses, lignin, sugars, proteins, and starch contain a variety of functional groups that facilitate metal complexion which in turn helps in the sequestration of heavy metals (Hashem et al. 2007). Most of the agricultural wastes seem to be a viable option for heavy metal adsorption due to their unique chemical composition, low cost and availability in abundance etc. Various agricultural wastes materials, such as rice bran, rice husk, wheat bran, wheat husk, husks, coconut shells, apple, banana, peels etc have been reported to possess metal adsorption capacity (Annadurai et al. 2002; Hashem et al. 2007). The metal adsorption using aqueous stream by using agricultural materials is based on metal desorption. The process of metal biosorption involves a solid phase (sorbent) and a liquid phase (solvent) containing dissolved species to be absorbed. The high affinity of the sorbent for the metal ion species, the metal ions are attracted and bound by a complex process involving different mechanisms, such as chemisorption, complexation, adsorption on surfaces and pores, ion exchange, chelation, adsorption by physical forces, entrapment in inter and intrafibrillar capillaries and spaces of the structural polysaccharides network (Basso et al. 2002). The functional groups, such as acetamido groups, carbonyl, phenolic, structural polysaccharides, amido, amino, sulphydryl, carboxyl groups, alcohols and esters present in agro by-products possess the affinity for metal complexation (Gupta and Ali 2000). Some biosorbents are specific for certain types of metals depending upon their chemical composition and some are non- specific. Agricultural wastes, such as hazelnut shells, orange peels, maize combs, peanut shells have been reported to have nickel removal efficiency (Kurniawan et al. 2006).

Current treatment technologies employed for the removal of heavy metals are not economical and generate huge quantity of toxic chemical sludge and other toxic by-product compounds. Recently, biosorption has emerged as a alternative method to the existing conventional methods for the removal and remediation of metal ions from the environment. The major advantages of metal adsorption by agro residues over conventional treatment methods are low cost, minimization of chemical and biological sludge, regeneration of biosorbents, high efficiency, and easy availability of the raw material. Hence, the use of agro residues for metal remediation will not only reduce the heavy metal pollution but also mitigate the biological pollution resulting from agro-residues.

3.5 Conclusion and Future Outlook

Global perception of wastes in general and agricultural wastes in particular is changing rapidly in response to need for environmental conservation, sustainable agricultural productivity and global food security. This trend has increased the need for

3 Sustainable Solutions for Agro Processing Waste Management: An Overview

appropriate technologies for the re-processing of such wastes. Protein enrichment of wastes for use in animal nutrition offers opportunities for the reuse of abundant agricultural wastes and refuse. Biotechnological processes, such as solid substrate fermentation and ensiling offer immense opportunities for the reuse of abundant agricultural wastes for the production of valuable compounds. Modern research and technologies will help to improve confidence in the final products derived from these processes, as well as drive development of the application of biotechnology for the valorization of agricultural waste. Fermentation technologies are widely used in many environmental friendly and economical industrial sectors. These technologies offer several advantages of low cost substrates, high product specificity and low energy consumption.

Acknowledgements The authors are sincerely thankful to the Natural Sciences and Engineering Research Council of Canada (Discovery Grant 355254, Canada Research Chair), FQRNT (ENC 125216) and MAPAQ (No. 809051) for financial support. The views or opinions expressed in this article are those of the authors.

References

Ajila CM, Bhat SG, Prasada Rao UJS (2007a) Valuable components of raw and ripe peels from two Indian mango varieties. Food Chem 102:1006–1011

Ajila, CM, Naidu KA, Bhat SG et al (2007b) Bioactive compounds and antioxidant potential of mango peel extract. Food Chem 105:982–988

Ajila CM, Prasada Rao UJS (2008) Protection against hydrogen peroxide induced oxidative damage in rat erythrocytes by *Mangifera indica* L. peel extract. Food Chem Toxicol 46:303–309

Ajila CM, Prasada Rao UJS (2009) Purification and characterization of black gram (*Vigna mungo*) husk peroxidase. J Mol Catal B: Enzym 60:36–44

Alasalvar CM, Grigor JM, Zhang D et al (2001) Comparison of volatiles, phenolics, sugars, antioxidant vitamins, and sensory quality of different colored carrot varieties. J Agri Food Chem 49:1410–1416

Al-Jassim RAM, Awadeh FT, Abodabos A (1997) Supplementary feeding value of urea-treated olive cake when fed to growing Awasi lambs. Anim Feed Sci Technol 64:287–292

Allegrone G, Barbeni M, Cardillo R et al (1991) On the steric course of the microbial generation of (Z6)-gamma-dodecenolactone from (10R,S) 10-hydroxyoctadeca-(E8,Z12)-dienoic acid. Biotechnol Lett 13:765–768

Amigun B, Sigamoney R, von Blottmitz H (2008) Commercialization of biofuels industry in Africa. A review article. Renewable Sustainable Energy Rev 12:690–711

Annadurai G, Juang RS, Lee DJ (2002) Adsorption of heavy metals from water using banana and orange peels. Water Sci Technol 47:185–190

Arasaratnam V, Mylvaganam K, Balasubramaniam K (2001) Glucoamylase production by *Aspergillus niger* in solid state fermentation with paddy husk as support. J Food Sci Technol 38:334–338

Arogba SS (2000) Mango (*Mangifera indica*) kernel: chromatographic analysis of the tannin, and stability study of the associated polyphenol oxidase activity. J Food Com Anal 13:149–156

Arora DS, Chander M, Gill PK (2002) Involvement of lignin peroxidase, manganese peroxidase and laccase in degradation and selective ligninolysis of wheat straw. Int Biodeterior Biodegrad 50:115–120

Asp NG, Mattsson B, Onning G (1992) Variation in dietary fibre, β-glucan, starch, protein, fat, and hull content of oats grown in Sweden 1987–1989. Euro J Clini Nutri 46:31–47

Aspinall GO (1980) Carbohydrate: structure and function, Vol. 3. In: J Preiss (ed) "The Biochemistry of Plants". Academic Press, New York

Attyia SH, Ashour SM (2002) Biodegradation of agro-industrial orange waste under solid state fermentation and natural environmental conditions. Egy J Biol 4:23–30

Azzam AM (1992) Pre-treatment of cane bagasse with alkaline hydrogen peroxide for enzymatic hydrolysis of cellulose and ethanol fermentation. J Environ Sci Health Part B 24:421–433

Babitha S, Soccol CR, Pandey A (2007) Solid-state fermentation for the production of *Monascus* pigments from jackfruit seed. Biores Technol 98:1554–1560

Bai ZH, Zhang HX, Qi HY et al (2004) Pectinase production by *Aspergillus niger* using waste water in solid state fermentation for eliciting plant disease resistance. Bioresour Technol 95:49–52

Bampidis VA, Robinson PH (2006) Citrus by-products as ruminant feeds: A review. Ani Feed Sci Tech 128:175–217

Bardiya N, Somayaji D, Khanna S (1996) Biomethanation of banana peel and pineapple waste. Biores Tech 58:73–76

Basso MC, Cerrella EG, Cukierman AL (2002) Lignocellulosic materials as potential biosorbents of trace toxic metals from wastewater. Ind Eng Chem Res 41:3580–3585

Batra SK (1985) Other long vegetable fibers. In: Dekker M (ed) Handbook of fiber science and technology (Vol IV: Fiber Chemistry)

Bayoumi RA, Yassin HM, Swelim MA et al (2008) Production of Bacterial Pectinase(s) from Agro-Industrial Wastes Under Solid State Fermentation Conditions. J App Sci Res 4:1708–1721

Baysal T, Ersus S, Starmans DA (2002) Supercritical CO_2 extraction of β-carotene and lycopene from tomato paste waste. J Agri Food Chem 8:5507–5511

Belghith H, Ellouz-Chaabouni S, Gargouri A (2001) Biostoning of denims by *Penicillium occitanis* (Pol6) cellulases. J Biotechnol 89:257–262

Benzamin S, Pandey A (1996) Lipase production by *Candida rugosa* on Copra Waste Extract. Indian J Microbiol 36:201–204

Berrocal MM, Rodriguez J, Ball AS, Perez-Leblic MI, Arias ME (1997) Solubilization and mineralization of [14C] lignocellulose from wheat straw by *Streptomyces cyaneus* CECT 3335 during growth in solid state fermentation. Appl Microbiol Biotechnol 48:379–384

Berto D (2003) Panorama do mercado de bebidas. cerveja, a bebida alcoólica mais consumida no país. Food Ingredients 23:36–39

Besson I, Creuly C, Gros JB et al (1997) Pyrazine production by Bacillus subtilis in solid state fermentation on soybeans. Appl Microbiol Biotechnol 47:489–495

Bhalla TC, Joshi M (1993) Production of cellulase and xylanase by *Trichoderma viridae* and *Aspergillus* sp. on apple pomace. Indian J Microbiol 3:253–255

Bhattacharya D, Germinario LT, Winter WT (2008) Isolation, preparation and characterization of cellulose microfibers obtained from bagasse. Carbohyd Polym 73:371–377

Biermann U, Friedt W, Lang S, Lühs W, Machmüller G, Metzger JO, Klaas MR, Schäfer HJ, Schneider MP (2000) New syntheses with oils and fats as renewable raw materials for the chemical industry. Angew Chem Int Ed 39:2206–2224

Bilanovic D, Shelef G, Green M (1984) Xanthan fermentation of citrus waste. Bioresource Technol 48:169–172

Biswas A, Saha BC, Lawton JW, Shogren RL, Willett JL (2006) Process for obtaining cellulose acetate from agricultural by-products. Carbohydrate Polymers 64:134–137

Borup MB, Fenhaus SL (1990) Food processing wastes. Res J Water Pollut Control Fed 62:461–465

Botella C, Diaz A, de Ory I, Webb C, Blandino A et al (2007) Xylanase and pectinase production by *Aspergillus awamori* on grape pomace in solid state fermentation. Process Biochem 42:98–101

Bramorski A, Christen P, Ramirez M et al (1998) Production of volatile compounds by the fungus *Rhizopus oryzae* during solid state cultivation on tropical agroindustrial substrates. Biotechnol Lett 20:359–362

Bravo L (1998) Polyphenols: chemistry, dietary sources, metabolism, and nutritional significance. Nutri Rev 7:317–333

3 Sustainable Solutions for Agro Processing Waste Management: An Overview 101

Briens (2009) Biomass pyrolysis. http://www.agri-therm.com/TCBiomass2009_Pyrolysis_ CBriens.pdf)

Buzzini P, Gobbertti M, Rossi J et al (1993) Utilization of grape must and concentrated rectified grape must to produce gluconic acid by *Aspergillus niger* in batch fermentations. Biotechnol Lett 15:151–156

Carle R, Keller P, Schieber A, Rentschler et al (2001) Method for obtaining useful materials from the by-products of fruit and vegetable processing. Patent application WO 01/78859 A1

Carvalho JC, Pandey A, Babitha S, Soccol CR (2003) Production of *Monascus* biopigments: an overview. Agro Food Ind Hi-tech 14:37–42

Chapla D, Divecha J, Madamwar D et al (2010) Utilization of agro-industrial waste for xylanase production by *Aspergillus foetidus* MTCC 4898 under solid state fermentation and its application in saccharification. Biochem Eng J 49:361–369

Chawachart N, Khanongnuch C, Watanabe T, Lumyong S (2004) Rice bran as an efficient substrate for laccase production from thermotolerant basidiomycete *Coriolus versicolor* strain RC3. Fungal Divers 15:23–32

Christen P, Meza JC, Revah S (1997) Fruity aroma production in solid state fermentation by *Ceratocystis fimbriata*: Influence of the substrate type and the presence of precursors. Mycol Res 101:911–919

Christen P, Bramorski A, Revah S et al (2000) Characterization of volatile compounds produced by Rhizopus strains grown on agro-industrial solid wastes. Bioresour Technol 71:211–215

Crickenberge RG, Carawan RE (1996) Using food processing by-products for animal feed. Water quality and waste management. North Carolina co-operative extension services. http://www. bae.ncsu.edu/programs/extension/publicat/wqwm/cd37.html

Cuzi N, Labat M (1992) Reduction of cyanide levels during anaerobic digestion of cassava. Int J Food Sci Technol 27:329–336

Davis BH (1976) Carotenoids. In: Goodwin TW (ed) Chemistry and biochemistry of plant pigments. Academic Press, New York

Davy C, Eng AEC, Chem E (1981) In: Herzka A, Booth RG (eds) Food industry wastes: Disposal and recovery. Applied Science Publishers, London

De Flora S, Izzotti A, Bennicelli C (1993) Mechanisms of antimutagenesis and anticarcinogenesis: role in primary prevention. In: Bronzetti G, Hayatsu H, De Flora S, Waters MD, Shankel DM (eds) Antimutagenesis and Anticarcinogenesis Mechanisms III. Plenum Press, New York

De Gregorio A, Mandalari G, Arena N et al (2002) SCP and crude pectinase production by slurry-state fermentation of lemon pulps. Bioresource Techn 83:89–94

De Lima VAG, Stamford TM, Salgueiro AA (1995) Citric acid production from pine apple waste by solid state fermentation using *Aspergillus niger*. Arq Biol Technol 38:777–783

Demirba A (2009) Biofuels securing the planet's future energy needs. Energy Conversion Manag 50:2239–2249

Dhiman TR, Bingham HR, Radloff HD (2003) Production response of lactating cows fed dried versus wet brewers' grain in diets with similar dry matter content. J Dairy Sci 86:2914–2921

Di Blasi C, Tanzi V, Lanzetta MA (1997) Study on the production of agricultural residues in Italy. Biomass Bioenergy 12:321–331

Di Mascio P, Kaiser SP, Sies H (2002) Lycopene as the most efficient biological carotenoid singlet oxigen quencher. Arch Biochem Biophys 274:179–185

Dinu D, Nechifor MT, Stoian G, Costache M, Dinischiotu A et al (2007) Enzymes with new biochemical properties in the pectinolytic complex produced by *Aspergillus niger* MIUG 16. J Biotechnol 131:128–137

Diplock AT, Charleux JL, Crozier-Willi G et al (1998) Functional food science and defence against reactive oxidative species. Br J Nutri 80:S77–S112

Duh PD, Yen GC (1995) Antioxidant activity and methanolic extracts of peanut hulls from various cultivars. J Am Oil Chem Soc 72:1065–1067

Dung NNX, Manh LH, Uden P (2002) Tropical fibre sources for pigs—digestibility, digesta retention and estimation of fibre digestibility in vitro. Anim Feed Sci Technol 102:109–124

Ebringerova A, Heinze T (2000) Xylan and xylan derivatives – biopolymers with valuable properties. Macromol Rapid Comm 21:542–556

El-Fadaly HA, El-Ahmady El-Nagga N, El-Sayed MM (2009) Single cell oil production by an oleaginous yeast strain in a low cost cultivation medium. Res J Microbiol 4:301–313

Escamilla-Hurtado ML, Valdes-Martinez SE et al (2005) Effect of culture conditions on production of butter flavor compounds by *Pediococcuspentosaceus* and *Lactobacillus acidophilus* in semisolid maize-based cultures. Int J Food Microbiol 105:305–316

Eurostat (2009) Environment and energy: Wim Kloek, Karin Blumenthal. Statistics in Focus 30/2009. http://epp.eurostat.ec.europa.eu/cache/ITY_OFFPUB/KS-SF-09-030/EN/KS-SF-09-030-EN.PDF. Accessed in March 2010

Evans J, Zulewska J, Newbold M et al (2010) Comparison of composition and sensory properties of 80% whey protein and milk serum protein concentrates. J Dairy Sci 93:1824–1843

Fakasa S, Čertikb M, Papanikolaoua S et al (2008) γ-Linolenic acid production by *Cunninghamella echinulata* growing on complex organic nitrogen sources. Bioresource Technol 99:5968–5990

FAO (2005) The State of the World's Fisheries

Fatma G, Satinder KB, Tyagi RD, Verma M, Surampalli RY (2010) Screening of agro-industrial wastes to produce ligninolytic enzymes by *Phanerochaete chrysosporium*. Biochem Eng J 49:388–394

Firkins JL, Harvatine DI, Sylvester IT et al (2002) Lactation performance by dairy cows fed wet brewers grains or whole cottonseed to replace forage. J Dairy Sci 85:2662–2668

Focher B et al (1998) Regenerated and graft copolymer fibers from stem-exploded wheat straw: characterization and properties. J Appl Poly Sci 67:961–974

Fountoulakis MS, Drakopoulou S, Terzakis S et al (2008) Potential for methane production from typical Mediterranean agro-industrial by-products. Biomass Bioenergy 32:155–161

Fountoulakis MS, Manios T (2009) Enhanced methane and hydrogen production from municipal solid waste and agro-industrial by-products co-digested with crude glycerol. Bioresource Technol 100:3043–3047

Francis F, Sabu A, Nampoothiri KM et al (2003) Use of response surface methodology for optimizing process parameters for the production of α-amylase by *Aspergillus oryzae*. Biochem Eng J 15:107–115.

Frost & Sullivan (2005) New dietary guidelines create increased potential for the United States Food Fiber Industry. News Release A940-88. Retrieved August 16, 2007, from www.prnewswire.co.uk/cgi/news/release?id=159753

Fukuda M, Kanauchi O, Araki Y et al (2002) Prebiotic treatment of experimental colitis with germinated barley foodstuff: a comparison with probiotic or antibiotic treatment. Int J Mole Med 9:65–70

Gallo M, Sommer A, Mlynar R et al (2001) Effect of dietary supplementation with brewery draff on rumen fermentation and milk production in grazing dairy cows. J Farm Anim Sci 34:107–113

Gangadharan D, Sivaramakrishnan S, Nampoothiri KM et al (2008) Response surface methodology for the optimization of alpha amylase production by *Bacillus amyloliquefaciens*. Bioresour Technol 99:4597–4602

Ganjyal GM, Reddy N, Yang YQ et al (2004) Biodegradable packaging foams of starch acetate blended with corn stalk fibers. J Appl Poly Sci 93:2627–2633

Garg N, Tandon DK, Kalra SK (2000) Protein enrichment of mango peel through solid state fermentation. Indian Food Packer 54:62–65

Ge Y, Yan H, Hui B et al (2002) Extraction of natural vitamin E from wheat germ by supercritical carbon dioxide. J Agri Food Chem 50:685–689

Ghildyal NP, Ramakrishna SV, Nirmala Devi P et al (1981) Large scale production of pectolytic enzyme by solid state fermentation. J Food Sci Technol 19:248–251

Gioia DD, Sciubba L, Setti L et al (2007) Production of biovanillin from wheat bran. Enzyme Micro Technol 41:498–505

Glasser WG, Kaar WE, Jain RK, Sealey JE (2000) Isolation options for noncellulosic heteropolysaccharides (Hetps). Cellulose 7:299–317

3 Sustainable Solutions for Agro Processing Waste Management: An Overview

Gobi M, Balamurugan V, Vijayalakshmi GS (2001) Studies on the composting a fungi of coir waste by using *Phanerochaete chrysopsporium* and the earthworms, *Eudrilus eugineae*. J Environ Pollun 8:45–48.

Gollapalli LE (2002) Predicting digestibility of ammonia fiber explosion (AFEX)-treated rice straw. Appl Biochem Biotechnol 98–100:23–35

Gondwe TNP, Mtimuni JP, Safalaoh ACL (1999) Evaluation of brewery by-products replacing vitamin premix in broiler finisher diets. Indian J Anim Sci 69:347–349

Gonzalez CF, Farina JI, de Figueroa LIC (2008) Optimized amylolytic enzymes production in *Saccharomycopsis fiuligera* DSM-70554 – an approach to efficient cassava starch utilization. Enzyme Microb Technol 42:272–277

González G, Barrocala V, Boladoa S et al (2009) Valorisation of by-products from food industry, for the production of single cell protein (SCP) using microalgae. New Biotechnol 25:S262–S265

Goodwin TW (1976) In: Goodwin TW (ed) Chemistry and biochemistry of plant pigments. Academic Press, London

Gorinstein S, Zachwieja Z, Folta M, Barton H, Piotrowicz J, Zember M, Weisz M, Trakhtenberg, S, Martın-Belloso O (2001) Comparative content of dietary fiber, total phenolics, and minerals in persimmons and apples. J Agri Food Chem 49:952–957

Gressel J, Zilberstein A (2003) Let them eat (GM) straw. Trends Biotechnol 21:525–530

Gupta VK, Ali I (2000) Utilization of bagasse fly ash (a sugar industry waste) for the removal of copper and zinc from wastewater. Separation Purification Technol 18:131–140

Hang YD, Woodams EE (1986) Utilisation of grape pomace for citric acid production by solid state fermentation. Am J Enol Vitic 34:426–428

Hang YD, Lee CY, Woodams EE (1982) A solid state fermentation system for production of ethanol from apple pomace. J Food Sci 47:1851–1852

Hang YD, Lee CY, Woodams EE (1986) Solid-state fermentation of grape pomace for ethanol production. Biotechnol Lett 8:53–56

Harborne JB, Baxter H, Moss GPA (1999) Handbook of bioactive compounds from plants. Taylor and Francis, London

Hashem MA, Abdelmonem RM, Farrag TE (2007) Human hair as a biosorbent to uptake some dyestuffs from aqueous solutions. Alexandria Eng J 1:1–9

Hatvani N, Mècs I (2001) Production of laccase and manganese peroxidase by *Lentinus edodes* on malt-containing by-product of the brewing process. Process Biochem 37:491–496

Heim K, Tagliaferro A, Bobilya D (2002) Flavonoid antioxidants: chemistry, metabolism and structure–activity relationships. J Nutri Biochem 13:572–584

Henriksson G, Akin DE, Hanlin RT et al (1997) Identification and retting efficiencies of fungi isolated from dew-retted flax in the United States and Europe. Appl Environ Microb 63:3950–3956

Hsing H, Wang W, Chiang P et al (2001) Hazardous wastes transboundary movement management – case study in Taiwan. J Resource Conserv Recycl 40:329–342

Huige NJ (1994) Brewery by-products and effluents. In: Hardwick WA (ed) Handbook of brewing. Marcel Dekker, New York

Ikramul H, Ali S, Qadeer M et al (2003) Stimulatory effect of alcohols on citric acid productivity by 2-deoxy d-glucose resistant culture of *Aspergillus niger* GCB-47. Bioresour Technol 86:227–233

Imandi SB, Bandarua VVR, Somalanka SR, Bandaru SR, Garapati HR (2008) Application of statistical experimental designs for the optimization of medium constituents for the production of citric acid from pineapple waste. Bioresour Technol 99:4445–4450

Ipharraguerre IR, Clarka JH (2003) Soyhulls as an alternative feed for lactating dairy cows: a review. J Dairy Sci 86:1052–1073

Ito K, Yoshida K, Ishikawa T, Kobayashi S et al (1990) Volatile compounds produced by fungus *Aspergillus oryzae* in rice koji and their changes during cultivation. J Ferment Bioengin 70:169–172

Jacob N, Prema P (2008) Novel process for the simultaneous extraction and degumming of banana fibers under solid-state cultivation. Braz J Microbiol 39:115–120

Joshi VK, Sandhu DK (1996) Preparation and evaluation of an animal feed byproduct produced by solid-state fermentation of apple pomace. Biores Technol 56:251–255

Joshi VK, Parmar M, Rana NS et al (2006) Pectin esterase production from apple pomace in solid-state and submerged fermentations. Food Technol Biotechnol 44:253–256

Kanauchi O, Mitsuyama K, Araki Y (2001) Development of a functional germinated barley foodstuff from brewers' spent grain for the treatment of ulcerative colitis. J Am Soci Brewing Chem 59:59–62

Kanwar SS, Tewari HK, Chadha BS, Punj V, Sharma VK (1995) Lactic acid production from molasses by *Sporolactobacillus cellulosolvens*. Acta Microbiol Immunol Hung 42:331–338

Kaur VI, Saxena PK (2004) Incorporation of brewery waste in supplementary feed and its impact on growth in some carps. Biores Technol 91:101–104

Kim HJ, Kim HJ, Oh HJ et al (2002) Morphology control of Monascus cell and scale-up of pigment fermentation. Process Biochem 38:649–655

Kivaisi AK, Rubindamayugi MST (1996) The potential of agro-industrial residues for production of biogas and electricity in Tanzania. Renew Energ 9:917–921

Koutinas AA, Malbranque F, Wang RH et al (2007) Development of an oat-based biorefinery for the production of lactic acid by *Rhizopus oryzae* and various valueadded co-products. J Agric Food Chem 55:1755–1761

Krich K, Augenstein D, Batmale JP et al (2005) Biomethane from dairy waste: a sourcebook for the production and use of renewable natural gas in California

Krishna C, Chandrasekaran M (1996) Banana waste as substrate for amylase production by *Bacillus sublitis* (CBTK 106) under solid state fermentation. Appl Microbiol Biotechnol 46:106–111

Kujala TS, Loponen JM, Kika KD, Pihlaja K (2000) Phenolics and betacyanins in red beetroot (*Beta vulgaris*) root: distribution and effect of cold storage on the content of total phenolics and three individual compounds. J Agri Food Chem 48:5338–5342

Kurniawan TA, Chan GY, Lo GH et al (2006) Comparisons of low-cost adsorbents for treating wastewaters laden with heavy metals: A review. Sci Total Environ 366:409–426.

Larrauri JA, Goñi IJ, Martin-Carrón N et al (1996) Measurement of health promoting properties in fruit dietary fibres: Antioxidant capacity, fermentability and glucose retardation index. J Sci Food Agri 71:515–519

Larroche C, Besson I et al (1999) High pyrazine production by *Bacillus subtilis* in solid substrate fermentation on ground soy-beans. Process Biochem 34:67–74

Li Z, Bai Z, Zhang B et al (2005) Newly isolated *Bacillus gibsonii* S-2 capable of using sugar beet pulp for alkaline pectinase production.World J Microbiol Biotechnol 21:1483–1486

Lichtenthaler HK (1987) Chlorophylls and carotenoids: pigments of photosynthetic biomembranes. Methods Enzymol 148:350–382

Liimatainen H, Kuokkanen T, Käärïainen J et al (2004) Development of bio-ethanol production from waste potatoes. Proceedings of the Waste Minimization and Resources Use Optimization Conference, June 10, 2004, University of Oulu, Finland. Oulu University Press, Oulu

Lim SK et al (2001) Novel regenerated cellulose fibers from rice straw. J Appl Poly Sci 82:1705–1708

Livny O, Kaplan I, Reifen R et al (2002) Lycopene inhibits proliferation and enhances gap-junction communication of KB-1 human oral tumor cells. J Nutr 132:3754–3759

Mahadevaswamy M, Venkataraman LV(1990) Integrated utilization of fruit-processing wastes for biogas and fish production. Biological Wastes 32:243–251

Majumdar P, Chanda S (2001) Chemical profile of some lignocellulosic crop residues. Indian J Agric Biochem 14:29–33

Mapari SAS, Nielsen KF, Larsen TO et al (2005) Exploring fungal biodiversity for the production of water-soluble pigments as potential natural food colorants. Curr Opin Biotechnol 16:231–238

Maria P (2007) Advances in citric acid fermentation by *Aspergillus niger*: Biochemical aspects, membrane transport and modeling. Biotechnol Adv 25(3):244–263

Medeiros ABP, Pandey A, Christen RJS (2001) Aroma compounds produced by *Kluyveromyces marxianus* in solid-state fermentation on packed bed column bioreactor. World J Microbiol Biotechnol 17:767–771

Medeiros ABP, Pandey A, Vandenberghe LPS, Pastore GMP (2006) Production and recovery of aroma compounds produced by solid-state fermentation using different adsorbents. Food Technol Biotechnol 44:47–51

Milala MA, Shugaba A, Gidado A et al (2005) Studies on the use of agricultural wastes for cellulase enzyme production by *Aspegillus niger*. Res J Agri Biol Sci 1:325–328

Moeller L, Strehlitz B, Aurich A et al (2007) Optimization of citric acid production from glucose by *Yarrowia lipolytica*. Eng Life Sci 7(5):504–511

Mohanty AK, Misra M, Hinrichsen G (2000) Biofibres, biodegradable polymers and biocomposites: an overview. Macromol Mater Eng 276–277:1–24

Molina-Alcaide E, Yáñez-Ruiz DR (2008) Potential use of olive by-products in ruminant feeding: a review. Anim Feed Sci Technol 147:247–264

Moo Young M, Chisti Y, Vlach D (1993) Fermentation of cellulosic materials to mycoprotein foods. Biotechnol Adv 11:469–479

Moriguchi M (1982) Fermentative production of pyruvic acid from citrus peel extract by *Debaryomyces courdertii*. Agric Biol Chem 46:955–961

Mulimani VH, Patil GN, Ramalingam (2000) Amylase production by solid state fermentation: a new practical approach to biotechnology courses. Biochem Educ 28:161–163

Mussatto SI, Roberto IC (2005) Acid hydrolysis and fermentation of brewers spent grain to produce xylitol. J Sci Food Agri 85:2453–2460

Muzinic LA, Thompson KR, Morris A et al (2004) Partial and total replacement of fish meal with soybean meal and brewers grains with yeast in practical diets for Australian red claw crayfish *Cherax quadricarinatus*. Aquaculture 230:359–376.

Naviglio D, Pizzolongo B, Ferrara L et al (2008) Extraction of pure lycopene from industrial tomato waste in water using the extractor Naviglio®. Afri J Food Sci 2:37–44

Ngoc UN, Schnitzer H (2009) Sustainable solutions for solid waste management in Southeast Asian countries. Waste Manag 29:1982–1995

Nigam JN (2000) Cultivation of *Candida langeronii* in sugar cane bagasse hemicellulosic hydrolyzate for the production of single cell protein. World J Microbiol Biotechnol 16:367–372

Niladevi KN, Prema P (2005) Mangrove actinomycetes as the source of ligninolytic enzymes. Actinomycetol 19:40–47

Niladevi KN, Sukumaran RK, Prema P (2007) Utilization of rice straw for laccase production by *Streptomyces psammoticus* in solid-state fermentation. J Ind Microbiol Biotechnol 34:665–674

Nogales R, Celia C, Benitez E (2005) Vermicomposting of winery wastes: a laboratory study. J Environ Sci Health 40:659–673

Oboh G, Elusiyan CA (2007) Changes in the nutrient and anti nutrient content of Micro-fungi fermented cassava flour prpoduced from low and medium-cyanide variety cassava tubers. Afr J Biotechnol 6:2150–2157

Ohsawa K, Chinen C, Tajanami S, Kuribayashi T, Kurukouchi K (1995) Studies on effective utilisation of carrot pomace. I. Effective utilisation to cake, dressing and pickles. Res Report of the Nagano State. Laboratory of Food Technol 23:15–18

Oliveira LA, Porto ALF, Tambourgi EB (2006) Production of xylanase and protease by *Penicillium janthinellum* CRC 87M-115 from different agricultural wastes. Biores Technol 97:862–867

Ong ASH, Tee ES (1992) Natural sources of carotenoids in plants and oils. In: L Packer (ed) Carotenoids. Part A. Chemistry, Separation, Quantitation, and Antioxidation, vol. 213, pp 142–167. Academic Press, London

Orozco FH, Cegarra J, Trujillo LM, Roig A (1996) Vermicomposting of coffee pulp using the earthworm *Eisenia foetida*: effects on C and N contents and the availability of nutrients. Biol Fertil Soil 22:162–166

Orozco AL, Perez MI, Guevara O et al (2008) Biotechnological enhancement of coffee pulp residues by solid-state fermentation with *Streptomyces*. J Anal Appl Pyrolysis 81:247–252

Osmaa JF, Toca Herreraa JL et al (2006) Banana skin: a novel waste for laccase production by *Trametes pubescens* under solid-state conditions. Application to synthetic dye decolouration. Dyes Pigments 75:32–37

Öztürk S, Özboy O, Cavidoglu I et al (2002) Effects of brewers spent grain on the quality and dietary fibre content of cookies. J Inst Brew 108:23–27

Pandey A, Nigam P, Soccol CR et al (1998) Bioconversion of sugar cane bagasse in solid state fermentation by micro-organisms. Paper presented in the International Symposium on Microbial Biotechnology for Sustainable Development and Applications, 14–16 November, Jabalpur, India

Pandey A, Soccola CR, Nigam P et al (2000) Biotechnological potential of agro-industrial residues. I: sugarcane bagasse. Biores Technol 74:69–80

Papinutti VL, Diorio LA, Forchiassin F (2003) Production of laccase and manganese peroxidase by *Fomes sclerodermeus* grown on wheat bran. J Ind Microb Biotechnol 30:157–160

Parawira W (2004) Doctoral Dissertation, Biotechnology Department, Lund University, Sweden

Pastore GM, Park YK et al (1994) Production of a fruity aroma by *Neurospora* from beiju. Mycol Res 98:25–35

Pazmiño-Durán EA, Giusti MM, Wrolstad RE et al (2001) Anthocyanins from banana bracts (*Musa paradisiaca*) as potential food colorants. Food Chem 73:327–332

Peng XW, Chen HZ (2007) Single cell oil production in solid-state fermentation by *Microsphaeropsis* sp. from steam-exploded wheat straw mixed with wheat bran. Biores Residues Biomass Bioenergy 26:361–375

Plewa MJ, Berhow MA, Vaughn SF et al (2001) Isolating antigenotoxic components and cancer cell growth suppressors from agricultural by-products. Mutation Res 480–481:109–120

Pradeep V, Datta M (2002) Production of ligninolytic enzymes for decolorization by cocultivation of white-rot fungi *Pleurotus ostreatus* and *Phanerochaete chrysosporium* under solid-state fermentation. Appl Biochem Biotechnol 102:109–118

Rahmat H, Hodge RA, Manderson GJ et al (1995) Solid substrate fermentation of *Kloeckera apiculata* and *Candida utilis* on apple pomace to produce an improved stock-feed. World J Microbial Biotechnol 11:168–170

Rajoka MI, Khan SH, Jabbora MA et al (2006) Kinetics of batch single cell protein production from rice polishings with *Candida utilis* in continuously aerated tank reactors. Biores Technol 97:1934–1944

Reddy N, Yang Y (2004) Structure of novel cellulosic fibers from cornhusks. Polymer Preprints (American Chemical Society, Division of Polymer Chemistry) 45:411

Reddy N, Yang Y (2005) Biofibers from agricultural byproducts for industrial applications. Trends Biotechnol 23:22–27

Reddy HKY, Srijana M, Reddy MD, Reddy G (2010) Co-culture fermentation of banana agro-waste to ethanol by cellulolytic thermophilic *Clostridium thermocellum* CT2. Afri J Biotec 9:1926–1934

Reinold MR (1997) Manual prático de cervejaria, first ed. Aden Editora eComunicac¸ões Ltda, São Paulo

Rodriguez-Vazquez R, Diazcervantes D (1994) Effect of chemical solutions sprayed on sugarcane bagasse pith to produce single cell protein physical and chemical analysis of pith. Biores Technol 47:159–164

Rodriguez D, Hadley M, Holm ET (1994) Phenolics in aqueous potato peel extract: extraction, identification and degradation. J Food Sci 59:649–651

Rodrígueza LA, Toroa ME, Vazqueza F (2010) Bioethanol production from grape and sugar beet pomaces by solid-state fermentation. Int J Hydrogen Energy 35:5914–5917

Rosales E, Couto RS, Sanromán MÁ (2002) New uses of food waste: application to laccase production by *Trametes hirsuta*. Biotechnol Lett 24:701–704. doi:10.1023/A:1015234100459

Rowell RM et al (1997) In paper and composites from agro-based resources. CRC Press, London

Ruanglek V, Maneewatthana D, Tripetchkul S (2006) Evaluation of Thai agro-industrial wastes for bio-ethanol production by *zymomonas mobilis*. Process Biochem 41:1432–1437

Saheb ND, Jog JP (1999) Natural fiber polymer composites: a review. Adv Polym Tech 18:351–363

Salah N, Miller NJ, Paganga G et al (1995) Polyphenolic flavonols as scavengers of aqueous phase radicals and as chain-breaking antioxidants. Arch Biochem Biophy 322:339–346

3 Sustainable Solutions for Agro Processing Waste Management: An Overview 107

Salisbury FB, Ross CW (1992) Plant physiology, 4th edn. Wadsworth Publishing Company, Belmont, CA

Salmones D, Mata G, Waliszewski KN (2005) Comparative culturing of *Pleurotus spp.* on coffee pulp and wheat straw: biomass production and substrate biodegradationm. Bioresource Technol 96:537–544

Sangeetha PT, Ramesh MN, Prapulla SG (2004) Production of fructosyl transferase by *Aspergillus oryzae* CFR 202 in solidstate fermentation using agricultural by-products. Appl Microbiol Biotechnol 65:530–537

Saura-Calixto F (1993) Ethanolic precipitation: a source of error in dietary fibre determination. Food Chem 47:351–355

Schieber A, Hilt P, Streker P, Endre HU, Rentschler C, Carle R (2003) A new process for the combined recovery of pectin and phenolic compounds from apple pomace. Inno Food Sci Emerging Technol 4:99–107

Schildbach R, Ritter W, Schmithals K et al (1992) New developments in the environmentally safe disposal of spent grains and waste kieselguhr from breweries. Proc Convention—Institute of Brewing (Asia Pacific Section) 22:139–143

Schlosser D, Grey R, Fritsche W (1997) Patterns of ligninolytic enzymes in *Trametes versicolor.* Distribution of extra- and intracellular enzyme activities during cultivation on glucose, wheat straw and beech wood. Appl Microbial Biotechnol 47:412–418

Schneeman B (1986) Dietary fiber: physical and chemical properties, methods of analysis and physiological effects. Food Technol 40:104

Schneeman BO (1987) Soluble vs insoluble fiber–different physiological responses. Food Tech 47:81

Sessa DJ (2003) Processing of soybean hulls to enhance the distribution and extraction of value-added proteins. J Sci Food Agri 84:75–82

Shah MP, Reddy GV, Banerjee R, Ravindra Babu P, Kothari IL (2005) Microbial degradation of banana waste under solid state bioprocessing using two lignocellulolytic fungi (*Phylosticta spp.* MPS-001 and *Aspergillus spp.* MPS-002). Process Biochem 40:445–451

Shahidi F, Naczk M (1995) Food phenolics: sources, chemistry, effects and applications. Technomic Publishing Company, Lancaster

Shankaranand VS, Lonsane BK (1993) Sugarcane press mud as a novel substrate for the production of citric acid by soild state fermentation. World J Microbiol Biotechnol 9:377–380

Silva D, Tokuioshi K, Martins ED, Da Silva R, Gomes E et al (2005) Production of pectinase by solid state fermentation with *Penicillium viridicatum* RFC3. Process Biochem 40:2885–2889

Silveira ST, Daroit DJ, Brandelli A (2008) Pigment production by *Monascus purpureus* in grape waste using factorial design. LWT – Food Sci Technol 41:170–174

Singh OV, Kapur N, Singh RP (2005) Evaluation of agro-food byproducts for gluconic acid production by *Aspergillus niger* ORS-4.410. World J Microbiol Biotechnol 21:519–524

Sivam RL, Alexandre ME, Carvalho LH (2006) Mechanical properties of composites made of pineapple leaf fiber (PALF) and polyester. In: Bandopadhyay S, Zheng Q, Brendt CC et al (eds) Proceedings of ACUN-5-developments in composites: advanced, infrastructural, natural and nano-composites, UNSW, Sydney (2006) ISBN 0 7334 2363 9:452–461

Soares M, Christen P, Pandey A et al (2000) Fruity flavour production by *Ceratocystis fimbriata* grown on coffee husk in solid-state fermentation. Process Biochem 35:857–861

Soccol CR, Prado FC, Vandenberghe LPS et al (2004) Organic acids: production and application. Citric acid. In: Pandey A (ed) Concise encyclopedia of bioresource technology. The Howarth Reference Press

Somayaji D (1992) PhD thesis, Gulbarga University, Gulbarga

Someya Y, Yoshiki, Okubo K (2002) Antioxidant compounds from bananas (*Musa cavendish*). Food Chem 79:351–354

Songulashvili G, Elisashvili V, Wasser SP, Nevo E, Hadar Y (2007) Basidiomycetes laccase and manganese peroxidase activity in submerged fermentation of food industry wastes. Enzyme Microb Technol 41:57–61

Southgate DAT (1982) In: Vahouny GV, Kritchevesky D (eds) Dietary fiber in health and diseases. Plenum Press, New York

Stevenson DM, Weimer PJ (2002) Isolation and characterization of a *Trichoderma* strain capable of fermenting cellulose to ethanol. Appl Microbiol Biotechnol 59:721–726

Stoll T, Schieber A, Carle R (2001) In: Pfannhauser W, Fenwick GR, Khokhar S (eds) Biologically active phytochemicals in food: analysis, metabolism, bioavailability and function. Royal Society of Chemistry, Cambridge

Sumathi S, Chani SP, Mohamed AR (2008) Utilization of oil palm as a source of renewable energy in Malaysia. Renew Sustain Energ Rev 12:2404–2421

Sun Y, Cheng J (2002) Hydrolysis of lignocellulosic materials for ethanol production: a review. Bioresour Technol 83:1–11

Sundstrom DW, Klei HE (1982) Uses of by-product lignins from alcohol fuel processes. Biotechnol Bioeng Symp 12:45–56

Suthar S (2008) Bioconversion of post harvest crop residues and cattle shed manure into value added products using earthworm *Eudrilus eugeniae* Kinberg. Ecol Eng 32:206–214

Szponar B, Pawlik KJ, Gamian A et al (2003) Protein fraction of barley spent grain as a new simple medium for growth and sporulation of soil Actinobacteria. Biotechnol Lett 25:1717–1721

Tay A, Yang S (2002) Production of L (+)-lactic acid from glucose and starch by immobilized cells of *Rhizopus oryzae* in a rotating fibrous bed bioreactor. Biotechnol Bioeng 80(1):1–12

Tengerdy RP, Szakacs G (2003) Bioconversion of lignocelluloses in solid-state fermentation. Biochem Eng J 13:169–179

Torres JL, Varela B, García MT et al (2002) Valorization of grape (*Vitis vinifera*) byproducts. Antioxidant and biological properties of polyphenolic fractions differing in procyanidin composition and flavonol content. J Agri Food Chem 50:7548

Trifonova VV, Ignotova NI, Milyukova TB et al (1993) Possibility of utilizing various types of fruit and vegetable raw material for microbial synthesis of lysine. Appl Biochem Microbiol 29:429–432

Trowell HC, Southgate DAT, Wolever TMS et al (1976) Dietary fiber redefine. Lancet 1:967

Ubalua AO (2007) Cassava wastes: treatment options and value addition alternatives. Afr J Biotechnol 6:2065–2073

Vassilev N, Requena AR, Nieto LM, Nikolaeva L, Vassileva N (2009) Production of manganese peroxidase by *Phanerochaete chrysosporium* grown on medium containing agro-wastes/rock phosphate and biocontrol properties of the final product. Ind Crop Prod 30(1):28–32

Viswanath P, Devi SS, Nand K (1992) Anaerobic digestion of fruit and vegetable processing wastes for biogas production. Biores Technol 40:43–48

Waldron K (2001) Useful ingredients from onion waste. Food Sci Technol 15:38–41

Wang XQ, Wang QH, Ma HZ, Yin W (2008) Lactic acid fermentation of food waste using integrated glucoamylase production. J Chem Technol Biotechnol. online publication. (www.interscience.com) DOI 10.1002/jctb.2007

Wang D, Sakoda A, Suzuki M (2001) Biological efficiency and nutritional value of *Pleurotus ostreatus* cultivated on spent beer grain. Bioresour Technol 78:293–300

Westendorf ML, Wohlt JE (2002) Brewing by-products: Their use as animal feeds. Vet Clin Food Anim Pract 18:233–252

Whistler RL (1973) Solubility of polysaccharides and their behavior in solution. ACS Adv Chem Ser 117:245–255

Willson RM, Wiesman Z, Brenner A (2010) Analyzing alternative bio-waste feedstocks for potential biodiesel production using time domain (TD)-NMR. Waste Manag doi:10.1016/j.wasman.2010.03.008

Wolfe K, Xianzhong WU, Liu RH (2003) Apple peels as a value-added food ingredient. J Agri Food Chem 51:1676–1683

Wrolstad RE, Ling W (2001) Natural antibrowning and antioxidant compositions and methods for making the same. Patent application US6224926, 2001

Zheng ZX, Shetty K (2000) Solid-state production of polygalacturonase by *Lentinus edodes* using fruit processing wastes. Process Biochem 35:825–830

Zhang JX, Bergman F, Hallmans G et al (1990) The influence of barley fibre on bile composition, gallstone formation, serum cholesterol and intestinal morphology in hamsters, APMIS: Acta Pathologica, Microbiologica et Immunologica Scandinavica 98:568–574

Zhang JX, Lundin E, Hallmans G et al (1992) Dietary effects of barley fibre, wheat bran and rye bran on bile composition and gallstone formation in hamsters, APMIS: Acta Pathologica, Microbiologica et Immunologica Scandinavica 100:553–557

Chapter 4
Dyes—Environmental Impact and Remediation

Luciana Pereira and Madalena Alves

Contents

4.1	Introduction	112
4.2	Dye Structures and Properties	113
4.3	Dye Applications	117
4.4	Environmental Impact	117
4.5	Wastewater Remediation	121
	4.5.1 Sorption	122
	4.5.2 Coagulation-Flocculation-Precipitation	130
	4.5.3 Membrane Filtration	130
	4.5.4 Electrochemical Wastewater Treatment	131
	4.5.5 Advanced Oxidation Processes	132
	4.5.6 Bioremediation	135
4.6	Products Identification and Mechanisms of Dye Degradation	152
4.7	Conclusions	153
4.8	Future Perspectives	154
References		154

Abstract Dyes are an important class of synthetic organic compounds used in many industries, especially textiles. Consequently, they have become common industrial environmental pollutants during their synthesis and later during fibre dyeing. Textile industries are facing a challenge in the field of quality and productivity due to the globalization of the world market. As the highly competitive atmosphere and the ecological parameters become more stringent, the prime concern of the textile processors is to be aware of the quality of their products and also the environmental friendliness of the manufacturing processes. This in turn makes it essential for innovations and changes in these processes, and investigations of appropriate and environmentally friendly treatment technologies or their residues. The large-scale production and extensive application of synthetic dyes can cause considerable environmental pollution, making it a serious public concern. Legislation on the limits of

---

M. Alves (✉)
IBB-Instituto Biotecnologia e Bioengenharia, Centro Engenharia Biológica, Universidade do Minho, Campus de Gualtar, 4710-057, Braga, Portugal
e-mail: madalena.alves@deb.uminho.pt

A. Malik, E. Grohmann (eds.), *Environmental Protection Strategies for Sustainable Development,* Strategies for Sustainability,
DOI 10.1007/978-94-007-1591-2_4, © Springer Science+Business Media B.V. 2012

colour discharge has become increasingly rigid. There is a considerable urgent need to develop treatment methods that are effective in eliminating dyes from their waste. Physicochemical and biological methods have been studied and applied, although each has its advantages and disadvantages, with the choice being based on the wastewater characteristics, available technology and economic factors. Some industrial-scale wastewater treatment systems are now available; however, these are neither fully effective for complete colour removal nor do they address water recycling.

This chapter outlines the background of dye chemistry, the application areas and the impact of dyeing effluents in the environment. The processes/techniques being implemented and developed for wastewaters remediation are revisited.

Keywords Dye • Textile industry • Decolourisation • Physico-chemical treatment • Bioremediation

4.1 Introduction

Environmental pollution is one of the major and most urgent problems of the modern world. Industries are the greatest polluters, with the textile industry generatings high liquid effluent pollutants due to the large quantities of water used in fabric processing. In this industry, wastewaters differing in composition are produced, from which coloured water released during the dyeing of fabrics may be the most problematic since even a trace of dye can remain highly visible. Other industries such as paper and pulp mills, dyestuff, distilleries, and tanneries are also producing highly coloured wastewaters. It is in the textile industry that the largest quantities of aqueous wastes and dye effluents are discharged from the dyeing process, with both strong persistent colour and a high biological oxygen demand (BOD), both of which are aesthetically and environmentally unacceptable (Wang et al. 2007). In general, the final textile waste effluent can be broadly categorized into 3 types, high, medium and low strength on the basis of their COD content (Table 4.1).

The textile industry plays a major role in the economy of Asian and other countries. In India, it accounts for the largest consumption of dyestuffs at ~80% (Mathur et al. 2003), taking in every type of dye and pigment produced, this amounts to close to 80 000 tonnes. India is the second largest exporter of dyestuffs, after China. Worldwide, $~10^6$ tons of synthetic dyes are produced annually, of which $1-1.5 \times 10^5$ tons are released into the environment in wastewaters (Zollinger 1987). This release is because not all dye binds to the fabric during the dyeing processes; depending on the class of the dye, the losses in wastewaters can vary from 2% for basic dyes to as high as 50% for reactive dyes, leading to severe contamination of

Table 4.1 Some characteristics of typical textile effluents

Wastewater type	COD ($mg \cdot L^{-1}$)	Conductivity (μScm^{-1})
High strength	1500	2900
Medium strength	970	2500
Low strength	460	2100

4 Dyes—Environmental Impact and Remediation

surface and ground waters in the vicinity of dyeing industries (O'Neill et al. 1999). It is estimated that globally 280 000 tons of textile dyes are discharged in textile industrial effluent every year (Jin et al. 2007). Apart from the aesthetic point of view, dyes are undesirable because they can affect living creatures in the water discharged as effluent into the environment. Industrial effluents containing synthetic dyes reduce light penetration in rivers and thus affect the photosynthetic activities of aquatic flora, thereby severely affecting the food source of aquatic organisms. The thin layer of discharged dyes that can form over the surfaces of the receiving waters also decreases the amount of dissolved oxygen, thereby affecting the aquatic fauna. Furthermore, dye-containing effluents increase biochemical oxygen demand. Dyes are in general stable organic pollutants that persist in the environment, and concern has been raised that such artificial compounds are xenobiotic. Therefore, methods for their degradation have been increasingly explored and development. Despite the number of successful systems employing various physicochemical and biological processes, economical removal of colour from effluents remains a major problem. Since these concerns about the environment are gaining momentum, it is necessary to develop better economically and environmentally friendly treatment technologies. Among the current pollution control technologies, biodegradation of synthetic dyes by various microbes is emerging as an effective and promising approach. The bioremediation potential of microbes and their enzymes acting on synthetic dyes has been demonstrated, with others needing to be explored in the future as alternatives to conventional physicochemical approaches (Husain 2006; Ali 2010). It is obvious that each process has its own constraints in terms of cost, feasibility, practicability, reliability, stability, environmental impact, sludge production, operational difficulty, pre-treatment requirements, the extent of the organic removal and potential toxic by-products. Also, the use of a single process may not completely decolourise the wastewater and degrade the dye molecules. Even when some processes are reported to be successful in decolourising a particular wastewater, the same may not be applicable to other types of coloured wastewaters. Certainly, the effective removal of dye from industrial coloured wastewater is a challenge to the manufactures and researchers, as some of the processes are neither economical nor effective. The amount of water consumed in textile industries must also be considered because the traditional textile finishing industry consumes ~100 L of water in the processing of a kg of textile material. Consequently the potential of water re-use should be an objective when applying a particular wastewater treatment.

In this chapter, all these issues will come under focus and discussion, based on the theoretical and practical aspects of each of them.

4.2 Dye Structures and Properties

The textile dyeing industry has been in existence for over 4000 years. In ancient times, dyes were obtained from natural sources and not everyone could possess coloured fabrics. For example, during the early Roman Empire period, only kings

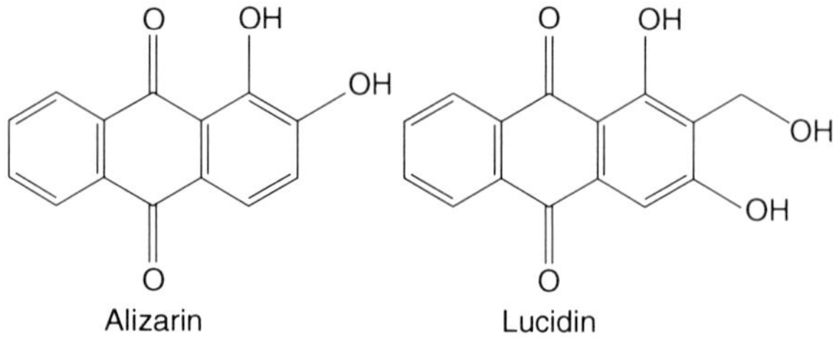

Fig. 4.1 Example of two natural extracts obtained from madder-rot: Alizarin and Lucidin

Alizarin   Lucidin

and priests could wear purple dyed fabrics while in the middle-ages, scarlet dyed fabrics were reserved exclusively for important members of the clergy. Natural colouring agents are mainly of inorganic origin (clays, earths, minerals, metal salts, and even semi-precious stones, such as malachite) or organic dyestuffs traditionally divided into 2 groups, one of animal and the other of plant origin (Ackacha et al. 2003). Undoubtedly, botanical sources were the most important, but a wide variety of other organisms was used, including lichens, insects and shellfish. Organic dyes present a broad spectrum of compounds with different physical and chemical properties. Among them, anthraquinone red colorants (e.g. cochineal, lac dye or madder root) are of special interest. Madder root has a long tradition as a dyestuff because of its bright red colour. The red pants of Napolean's army and the red coats of the English soldiers in the 18/19th century were dyed with madder. However, extracts of Madder root contain mainly alizarin (1,2-dihydroxy-anthraquinone) and several by-products, in which lucidin (1,3-dihydroxy-2-hydroxymethyl-anthraquione) is of the special concern because it has proved to have mutagenic character, severely constraining the use of Madder root (Fig. 4.1). Moreover, not every shade is directly available from a natural source. Synthetic dyes quickly replaced the traditional natural dyes. They cost less, offered a vast range of new colors, and imparted improved properties to the dyed materials.

In 1856, William Henry Perkin accidentally discovered the world's first commercially successful synthetic dye. By the end of the 19th century, 10 000 new synthetic dyes had been developed and manufactured. Nowadays, India, the former USSR, Eastern Europe, China, South Korea and Taiwan consume ~600 thousand tons (kt) of dyes per annum. Since 1995, China has been the leading producer of dyestuffs, exceeding 200 kt per annum (Wesenberg et al. 2003).

A large variety of dyestuffs is available, which can be natural or synthetic substances, but synthetic dyes are commonly used for textile fibres, whereas natural dyes tend to be reserved for the food industry.

A dye can generally be described as a coloured substance that has an affinity for the substrate to which it is being applied. It is a coloured because it absorbs in the visible range of the spectrum at a certain wavelength (Table 4.2). In general, a small amount of dye in aqueous solution can produce a vivid colour, which is related with the high molar extinction coefficients. Colour can be quantified by spectrophotometry (visible spectra), chromatography (usually high performance liquid, HPLC) and high performance capillary electrophoresis (Fig. 4.2).

4 Dyes—Environmental Impact and Remediation

Table 4.2 Colours of the visible spectrum: wavelengths and frequencies intervals

Colour	Wavelength interval (nm)	Frequency interval (THz)
Red	~700–635	~430–480
Orange	~635–590	~480–510
Yellow	~590–560	~510–540
Green	~560–490	~540–610
Blue	~490–450	~610–670
Violet	~450–400	~670–750

The major structure element responsible for light absorption in dye molecules is the **chromophore** group, i.e., a delocalized electron system with conjugated double or simple bonds (Gomes 2001). Chromophores frequently contain heteroatoms as N, O, and S, with non-bonding electrons. Common chromophores include –N=N– (azo), =C=O (carbonyl), =C=C=, C=NH, –CH=N–, NO or N–OH (nitroso), –NO$_2$ or NO–OH (nitro) and C=S (sulphur). As a complement to the electron acceptors action are the groups called **auxochromes**, which are electron donors generally on the opposite side of the molecule and their basic function is to increase colour. Indeed, the basic meaning of the word auxochrome is *colour enhancer*. Some auxochromes include –NH$_3$, –COOH, HSO$_3$ and –OH. These groups also have the important property of giving a higher affinity to the fibre. The chromogen, which is an aromatic structure (normally benzene, naphtalene or anthracene rings), is part of a chromogen-chromophore structure along with an auxochrome. Synthetic dyes exhibit considerable structural diversity and thus possess very different chemical and physical properties. The chemical classes of dyes more frequently employed on an industrial scale are azo, anthraquinone, indigoid, xanthene, arylmethane and phthalocyanine derivatives (Fig. 4.3). Nevertheless, it needs to be emphasized that the overwhelming majority of synthetic dyes in current use are **azo** derivatives. The colour range obtained with this class of dyes is very wide (Gomes 2001). More than one azo group can be present in the dye structure, dyes then being classified as azo, disazo, trisazo or poliazo as they have one, two, three or more groups. The

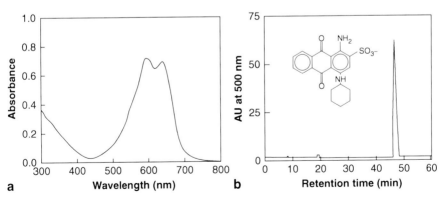

Fig. 4.2 Visible spectra (a) and HPLC chromatogram (b) of 1 mM Acid Blue 62 (Pereira et al. 2009b)

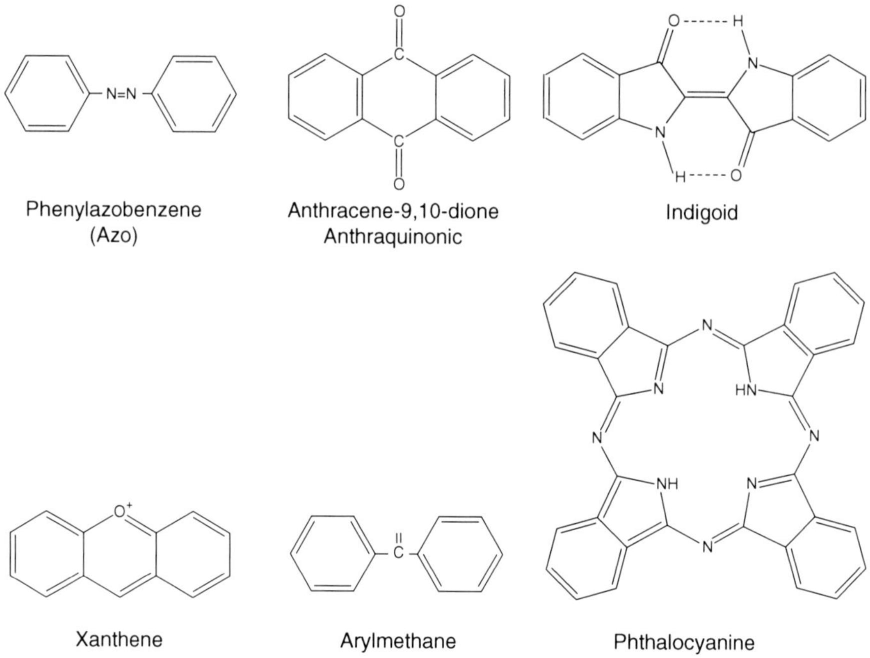

Fig. 4.3 Different chromophores: Azo, anthraquinonic, indigoid, xanthene, arylmethane and phthalocyanine dye structures

anthraquinonic dyes are also widely used and comprise of a carbonyl group associated with a conjugated system of 2 benzene rings. As with the azo dyes, substitution groups in the aromatic rings are required to intensify the colour. The major difference is their need only of electron donors once the carbonyl groups are in the uniquely possible position to act as electron acceptors.

Dyes are usually classified by their Colour Index (CI), developed by the Society of Dyes and Colourist (1984), which is edited every three months. It lists dyes firstly by a generic name based on its application and colour, then by assigning a 5-digit CI number based on its chemical structure, if known (O´Neill et al. 1999). Examples include Acid Blue 120 (26400), Reactive Red 4 (18105), and Mordant Yellow 10 (14010). They can be grouped in different classes: acid, basic, direct, disperse, metallic, mordant, pigment, reactive, solvent, sulphur and vat dyes, which reflects their macroscopic behaviour and also their prevailing functionalities. They are used in accordance to their compatibility with the type of textile substrate being processed (Gomes 2001). Acid, direct and reactive dyes are water-soluble anionic dyes; basic dyes are cationic, whereas disperse, pigment and solvent dyes are non-ionic (Hao et al. 2000; Gomes 2001). Disperse dyes are sparing soluble in water for application in hydrophobic fibres from a aqueous dispersion. They are often of anthraquinone and sulfide structure, with many $-C=O$, $-NH-$ and aromatic groups (Fu and Viraraghavan 2001). Most of the mordant dyes are anionic, but some cationic ones also exist. In aqueous solution, anionic dyes carry a net charge due to the presence of sulphonate

4 Dyes—Environmental Impact and Remediation

(SO^{3-}) groups, while cationic dyes carry a net positive charge due to protonated amine or sulfur containing groups. Sulphonic groups in the molecule provide solubility. Disperse Vat dyes (of which indigo and woad are the most important examples) are water-insoluble; however, under reducing conditions, they can be converted into a 'leuco' form (soluble in alkaline aqueous solutions), which penetrates the fibres during dyeing. Metal-complex dyes exhibit higher light and wash fastness due to the presence of transition metals, such as chromium, copper, nickel or cobalt that modify the surface chemistry between the dye molecule and the fabric (Hao et al. 2000; Gomes 2001).

4.3 Dye Applications

Approximately 40 000 different synthetic dyes and pigments are used industrially, and about 450 000 tons of dyestuffs are produced worldwide. Azo dyes are the largest and more versatile class of dyes, accounting for up to 50% of the annual production (Zollinger 1987). They are extensively used in many fields of up-to-date technology, in e.g., various branches of the textile industry, the leather tanning industry, paper production, food, colour photography, pharmaceuticals and medicine, cosmetic, hair colourings, wood staining, agricultural, biological and chemical research, light-harvesting arrays, and photoelectrochemical cells (Kuhad et al. 2004; Couto 2009). Moreover, synthetic dyes have been employed for the efficacious control of sewage and wastewater treatment, for the determination of specific surface area of activated sludge for ground water tracing, etc. (Forgacs et al. 2004).

The largest consumer of these dyes is the textile industry, accounting for 2/3rds of its market. Different classes of dyes are used according to the fibres to which they can be applied. Reactive dyes are most commonly used as they can be applied to both in natural (wool, cotton, silk) and synthetic (modified acrylics) fibres (O'Neill et al. 1999). Reactive dyes differ from other class of dyes in that their molecules contain one or more reactive groups capable of forming a covalent bond with a compatible fibre group. They have become very popular due to their high wet-fastness, brilliance and range of hues (Hao et al. 2000). Their use has increased as synthetic fibres became more abundant. Acid and basic dyes are used for dyeing all natural fibres (wool, cotton, silk) and some synthetics (polyesters, acrylic and rayon). Direct dyes are classified this way because they are applied directly to cellulose fibres. Furthermore, they are used for colouring rayon, paper, leather and to a small extent nylon. The application of mordant dyes is limited to the colouring of wool, leather, furs and anodised aluminium. Solvent dyes are used for colouring inks, plastics, and wax, fat and mineral oil products.

4.4 Environmental Impact

Colour is usually the first contaminant to be recognized in a wastewater because a very small amount of synthetic dyes in water (< 1 ppm) are highly visible, affecting the aesthetic merit, transparency and gas solubility of water bodies. They adsorb

Table 4.3 Colour concentrations limits and quantum of water generated from industries (adapted from Anjaneyulu et al. 2005)

Industry	Quantum of water generated standards (m³/Ton)	Colour concentration (hazen units)	Colour limits (hazen units)	
			USPHS	BIS
Textile	120 m³/Ton$_{fibre}$	1100–1300	0–25	20
Pulp & Paper				
• Large	175 m³/Ton$_{paper}$	100–600	0–10	5–101
• Small	150 m³/Ton$_{paper}$			
Tannery	28 m³/Ton$_{raw\ hide}$	400–500	10–50	25
Kraft mill	40 m³/Ton	2100–2300	10–40	20
Sugar	0.4 m³/Ton$_{cane}$	150–200	5–10	20

and reflect the sunlight entering water, thereby interfering with the aquatic species growth and hindering photosynthesis. Additionally, they can have acute and/or chronic effects on organisms depending on their concentration and length of exposure. Removal of colour from dye-containing wastewater is the first and major concern, but the point of degrading dyes is not only to remove colour, but to eliminate, or substantially decrease, the toxicity (i.e. detoxification).

Government legislation regarding the removal of dyes from industrial effluents is becoming increasingly stringent, especially in the more developed and developing countries (Robinson et al. 2001). Enforcement of the law will continue to ensure that textile and other dye-utilizing industries treat their dye-containing effluent to the required standards. In India, colour limits in industrial waters have also been set and have been made more stringent in the last few years. Table 4.3 presents the colour concentrations, their limits and the quantity of water generated from textile and other industries in United States and India (Anjaneyulu et al. 2005). European Community (EC) regulations are also becoming more stringent (O'Neill et al. 1999). A large variety of dyes can be found in real effluents. It has been estimated that ~9% (or 40 000 tons) of the total amount (450 000 tons) of dyestuffs produced in the world are discharged in textile wastewaters (O'Neill et al. 1999). Desirable criteria when producing those dyes are their fixation degree to fibre and fastness (i.e. high stability in light and washing) and resistant to microbial attack. Indeed, dyes are design to resist to very harsh conditions, difficulting colour removal from textile wastewaters by the conventional wastewater treatments. The degree of fixation of an individual dye varies with the type of fibre, shade and dyeing parameters. Dye fixation rate values are useful in giving an idea of the amount released, but can only be approximated. These losses are < 2–10% for basic, disperse and direct dyes, but can reach 50% for reactive dyes (Al-Degs et al. 2000; Hao et al. 2000). This high degree for reactive dyes is due to the hydrolyzed form of reactive dyes which has no affinity for the fibres (Fig. 4.4). As nowadays, reactive dyes are the most commonly used in the textile industries, and there is a need for finding an efficient method dye removal with special attention to this class. Moreover, once in the effluents and due to their high stability, they may remain in the environment for a long time (~50 years). Because of their commercial importance, the impact and toxicity of dyes released in the environment have been extensively studied (Pinheiro et al.

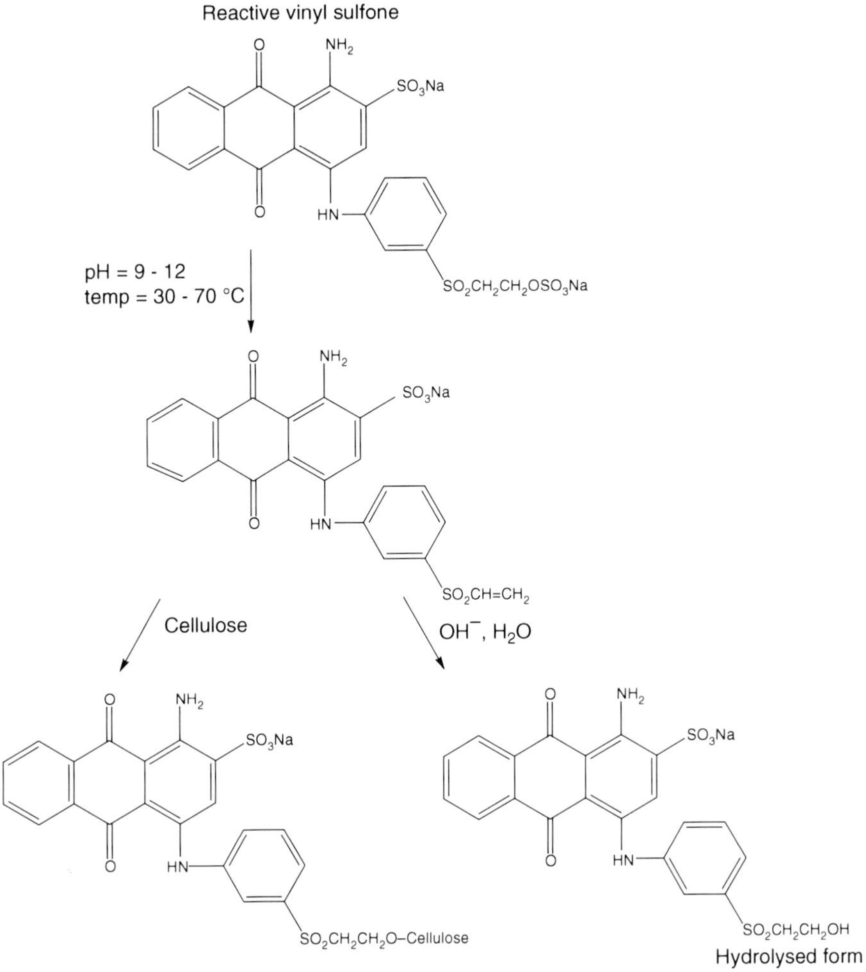

4 Dyes—Environmental Impact and Remediation

Reactive vinyl sulfone

Fig. 4.4 Reactive dye undergoing hydrolysis (Hao et al. 2000)

2004; Mathur et al. 2003; Puvaneswari et al. 2006; Pereira et al. 2009a, b). As several thousand different synthetic dyes are employed, exhibit various biological activities, it is understandable that our knowledge concerning their behaviour in the environment and health hazards involved in their use, remain incomplete (Forgacs et al. 2004). In general, dyes have low toxicity in mammals and aquatic organisms (O′Neill et al. 1999), but products formed by their biodegradation, mainly aromatic amines from the anaerobic reduction of azo dyes (see "Dye biodegradation" section), can be harmful (Razo-Flores et al. 1997; Pinheiro et al. 2004). Definitely, azo dyes that constitute the largest group of synthetic colorants used are consequently the most common synthetic dyes released into the environment (Zhao and Hardin 2007; Ali 2010). Some have been linked to bladder cancer, splenic sarcomas, and hepatocarcinomas, producing nuclear anomalies in experimental animals and chromosomal aberrations in cultured mammalian cells. An increased incidence of bladder cancer in dye workers exposed to large quantities of azo dyes has been reported (Puvaneswari et al. 2006). Assessment of the toxicity of dyes is therefore of the utmost importance. Various short-term screening methods have been developed to detect mutagenic/carcinogenic substances; these have played important roles not only in screening suspected chemicals, but also in studying the mechanisms of mutagenesis and carcinogenesis, thereby providing useful information for assessing the genetic effects of chemicals in man. Microorganisms have several attributes that make them attractive for use in quick screening of effluents and chemicals for toxicity. In the review of Hao et al. (2000) results on the toxicity valuation for the single cell alga, *Selenastrum capricornutum,* and for the fathead minnow, *Pimephales promelas,* are tabled. Other examples include the effect of the azo dye Sudan Orange G and the anthraquinonic dye Acid Blue 62 before and after an enzymatic treatment on the yeast *Saccharomyces cerevisiae* (Pereira et al. 2009a, b). The Ames test is a common, well implemented method for the evaluation of mutagenic potential of many compounds (Ames et al. 1975). The mutagenic potential of the locally (Indian) available and used textile dyes has been evaluated by Mathur et al. (2003) by the Ames test using the TA 100 strain of *Salmonella typhimurium.* Among the seven dyes tested, only one showed absence of mutagenic activity. The remaining six dyes were all positively mutagenic.

Also of particular concern are more specific compounds, used throughout the wet-processing steps, that can be toxic to aquatic life. Those include heavy metals, surfactants (wetting agents), fabric rinsing and/or washing detergents and other additives such as salts, sodium sulphate, sulphuric acid and dispersive agents). Additionally, dyeing baths often use extreme pH values (either acidic or alkaline, depending on the dye) and high temperatures; they have high BOD and chemical oxygen demand (COD), solid, oil and possibly toxic organics that include phenols (Shrestha and Kazama 2007; Tüfekci et al. 2007). These compounds will change the effluent water, causing a variety of physiological and biochemical disturbance. It is noteworthy that each fibre being processed produces effluents of its own distinctive characteristics, and for all textiles mill processing the same fibre, effluent characteristics are broadly similar, although quantities will vary. Differences can also arise between different plants processing the same fibre due to variations in the production technology. The impact of dyeing factories on plants and fishes can be found in the review of Puvaneswari et al. (2006).

Pollution prevention programs also need to focus on reduction in water and energy consumption by introducing new technologies and the reuse of dyeing water after the treatment processes. The textile industry is a high consumer of water (Allegre et al. 2004), with an average of 200 L water per kg of fibre. Another case of colour pollution is so-called "red-water", which results from trinitrotoluene (TNT) production generated in the purification stage (Hao et al. 1994). Other coloured wastewaters result from other industrial processes, including the bleaching of pulp, paper and textile fibres.

4.5 Wastewater Remediation

The textile finishing industry has been put under immense pressure to reduce use of harmful substances, especially mutagenic, carcinogenic, and allergenic effects of textile chemicals and textile dyes. There are regulations regarding the colour limits in effluents, which vary in different countries.

Textile dye wastewater remediation is based not only in colour removal (decolourisation), but also in the degradation and mineralization of the dye molecules. Indeed, decolourisation occurs when the molecules are removed from the solution or when the chromophore bond is broken, but the molecule in the first case, and the major fragments in the second, remain intact. The absorption of light by the associated molecules shifts from the visible to the ultraviolet or infrared region of the electromagnetic spectrum.

A wide range of technologies has been developed for the removal of synthetic dyes from waters and wastewaters to decrease their environmental impact. These include: **physical methods** such as membrane-filtration processes (nanofiltration, reverse osmosis, electrodialysis) and sorption techniques; **chemical methods** such as coagulation or flocculation combined with flotation and filtration, precipitation-flocculation with $Fe(II)/Ca(OH)_2$, electroflotation, electrokinetic coagulation, conventional oxidation methods (e.g. with ozone), irradiation or electrochemical processes; and **biological methods**, aerobic and anaerobic microbial degradation, and the use of pure enzymes. All of these procedures have advantages and disadvantages.

Traditional wastewater treatment technologies are markedly ineffective for handling wastewater of synthetic textile dyes because of the chemical stability of the pollutants (Forgacs et al. 2004). Additionally, they do not address the water recycling issue (Soares et al. 2004). The major disadvantage of physicochemical methods is primarily the high cost, low efficiency, limited versatility, need for specialized equipment, interference by other wastewater constituents, and the handling of the generated waste (van der Zee and Villaverde 2005; Kaushik and Malik 2009). Physical methods can effectively remove colour, but the dye molecules are not degraded, becoming concentrated and requiring proper disposal. With the chemical techniques, although the dyes are removed, accumulation of concentrated sludge can create a disposal problem. There is also the possibility that a secondary pollution problem arises because of the excessive amounts of chemicals involved. Recently, other emerging techniques—advanced oxidation processes, which are based on the generation of very powerful oxidizing agents such as hydroxyl radicals—have been applied with success in pollutant degradation (Arslan and Balcioğlu 1999; Zhou and He 2007). Although these methods are efficient for the treatment of waters contaminated with pollutants, they are very costly and commercially unattractive. The high electrical energy demand and the consumption of chemical reagents are common problems. The development of efficient, economic and environmentally friendly technologies to decrease dye content in wastewater to acceptable levels at affordable cost is of utmost importance (Couto 2009). Biological methods are generally considered environmentally friendly because they can lead to complete mineralization of organic pollutants at low cost (Pandey et al. 2007). They also remove BOD, COD and suspended solids. The main limitation can be related in some cases to the toxicity of some dyes and/or their degradation products to the organisms used in the process. Indeed, the removal of dyes depends on their physical and chemical characteristics, as well as the selected treatment method, with no technology in use today having universal application. However, some of the processes do not satisfactorily remove the colour and others are costly (Mondal 2008). The technologies used and those in development for the dye removal have been discussed in several reports (Vandevivere et al. 1998; Robinson et al. 2001; Forgacs et al. 2004; Anjaneyulu et al. 2005; Hai et al. 2007; Hao et al. 2007; Mondal 2008).

4.5.1 Sorption

Sorption of synthetic dyes on inexpensive and efficient solid supports has been considered a simple and economical method for their removal from water and wastewater, producing high quality of water (Forgacs et al. 2004; Allen and Koumanova 2005), making it an attractive alternative for the treatment of contaminated waters, especially where the sorbent is inexpensive and does not require a pre-treatment step before its application. It is superior to other techniques for water re-use in terms of initial cost, flexibility and simplicity of design, ease of operation, lower interference with diurnal variation, and insensitivity to toxic pollutants. Some of the advantages of applying sorption (see Weber et al. 1978; Mckay et al. 1999; Amin 2008) include: (1) less land area (half to quarter of what is required for a biological system); (2) lower sensitivity to diurnal variation; (3) not being affected by toxic chemicals; (4) greater flexibility in design and operation, and (5) superior removal of organic contaminants. Furthermore sorption does not result in the formation of harmful substances.

Decolourisation by sorption is a result of two mechanisms, sorption and ion exchange, and is influenced by many physicochemical factors, such as, dye/sorbent interaction, sorbent surface area, particle size, temperature, pH, and contact time (Robinson et al. 2001; Dhodapkar et al. 2006; Jovančić and Radetić 2008). The nature of the bound between the dye and the adsorbent during sorption is important for its effectiveness. Two types of sorption occur: **physical sorption**, when the interparticle bonds between the adsorbate and adsorbent are weak (van der Waals, hydrogen, and dipole-dipole); and **chemical sorption**, characterized by strong interparticle bonds due to an exchange of electrons (covalent and ionic bonds). The first is usually a reversible process and the second irreversible (Allen and Koumanova 2005).

An effective sorption model requires an accurate equilibrium isotherm, kinetic/mass transfer relationships and coupling equations (Mckay 1998). The mass transfer stage usually assumes a three-step model: (i) external film diffusion across the boundary layer, (ii) sorption at a surface site, and (iii) internal mass transfer within the particle, based on a pore or solid surface diffusion mechanism (Mckay and Swee-

4 Dyes—Environmental Impact and Remediation

ney 1980). To explain the sorption mechanism and predict sorption uptake rates by adsorbent pellets, various assumptions have been applied to equilibrium data: linear isotherm (Dryden and Kay 1954), irreversible isotherm (Liapis 1987) and nonlinear isotherm (Tien 1961; Weber and Rummer 1965). Specific cases of nonlinear isotherms include the Langmuir equation (Langmuir 1918), the Freundlich equation (Freundlich 1906) and the Redlich-Peterson equation (Redlich and Peterson 1959). The **Langmuir and Freundlich sorption isotherms** (Eqs. 4.1 and 4.2, respectively) are the most commonly used to quantify the amount of adsorbate adsorbed by an adsorbent (Annadurai and Krishnan 1997; Al-Degs et al. 2000, 2008).

$$q_e = Q_{max} K_L C_e/(1 + K_L C_e) \qquad \text{(Langmuir)} \qquad (1)$$
$$q_e = K_F C_e^n \qquad \text{(Freundlich)} \qquad (2)$$

Where, q_e (mmol·g_{AC}^{-1}) is the surface concentration of dye at equlibrium; C_e (mmol·dm^3) is the equilibrium concentration of dye in solution; Q_{max} (mmol·g_{AC}^{-1}) is the amount of dye adsorbed at a complete monolayer coverage; K_L (dm^3.mmol^{-1}) is a constant that relates to the heat of sorption; K_F [mmol·g^{-1}(mmol·dm^3)$^{-n}$] represents the sorption capacity when the dye equilibrium concentration (C_e) equals one unit, and n represents the degree of dependence of sorption on the equilibrium concentration.

Sorption is also related to the molecular size of the dye and the number of sulfonic groups. Smaller dyes have higher sorption, whereas bigger molecules are more difficult to adsorb due to diffusion limitations (Allen and Koumanova 2005; Tsang et al. 2007).

Sorption methods, independently of the inorganic or organic character of the supports, have certain drawbacks. Since sorption processes are generally non-selective, other components of the wastewater can also be adsorbed by the support. Competition between the sorbents can influence the dye-binding capacity of supports in an unpredictable manner. Moreover, a sorption process removes the synthetic dyes from wastewater by concentrating them on the surface, without structurally changing them. When the support is regenerated, the fate of the resulting concentrated solution of dyes presents a problem that has not yet been satisfactorily solved.

Large-scale applications based on the sorption process have to take all these factors into consideration. This technique can be successfully applied as a polishing step at the end of the treatment stage, to meet discharge colour standards. The reviews of Allen and Koumanova (2005) and Mondal (2008) outline the fundamental principles of dye sorption and also evaluate a number of different adsorbents used in the removal of colorants from wastewaters.

4.5.1.1 Activated Carbon

Activated Carbon (AC) is the most widely used sorbent and therefore will be described in more detail, but some of the fundamental principles outlined are also applicable to other adsorbents.

Fig. 4.5 Image of Activated Carbon both in powder and block form (AC)

Commercial ACs are available (Fig. 4.5), with different physical forms depending on their application. The type of carbon sorbent and its mode of preparation exert a marked influence on the sorption capacity (Pereira et al. 2003; Forgacs et al. 2004), and both by their texture and surface chemistry determine the performance (Figueiredo et al. 1999). The diversity of surface groups on AC (of acid and base character) makes it much more versatile than other adsorbents (Fig. 4.6) and the nature and concentration of surface functional groups can be modified by suitable chemical or thermal treatments. It is also possible to prepare carbons with different proportions of micro, meso and macropores (Rodriguez-Reinoso 1998; Figueiredo et al. 1999). This becomes advantageous once AC is modified physical and chemically, for specific applications, in order to optimise its performance. Liquid (HNO_3) and gas oxidation (O_2) produce samples with a higher amount of surface oxygen-containing groups. Nitric treatment increases in particular the carboxylic acid groups and the gas oxidation treatment is the most effective way to introduce less acidic ones, such as phenols and carbonyl/quinone groups (Figueiredo et al. 1999; Pereira et al. 2003; Faria et al. 2005). No significant changes in the texture of the carbon are likely to occur with liquid phase treatments, while an increase of the micropore volume, mesopore surface area and average of the micropores with gas phase oxidation is usual; these changes increase as the degree of oxidation itself increases (Figueiredo et al. 1999). Thermal treatment at high temperature produce materials with a low amount of oxygen-containing groups and high basicity, resulting mainly from the ketonic groups remaining on the surface, fewer acidic groups, and the delocalisation of π-electrons of the carbon basal planes (Moreno-Castilha et al. 2000; Faria et al. 2005, 2008).

Activated carbon samples are amphoteric in nature and therefore the pH of a dye solution plays an important role in the sorption process. As already mentioned, acidity and basicity is related to the chemical groups at the AC surface. The charge thereon is correlated with the pH of the solution. A convenient index of the propensity of a surface to became either positively or negatively charged depends on the pH required to give a zero net surface charge (pH_{pzc}). In a solution at $pH < pH_{pzc}$,

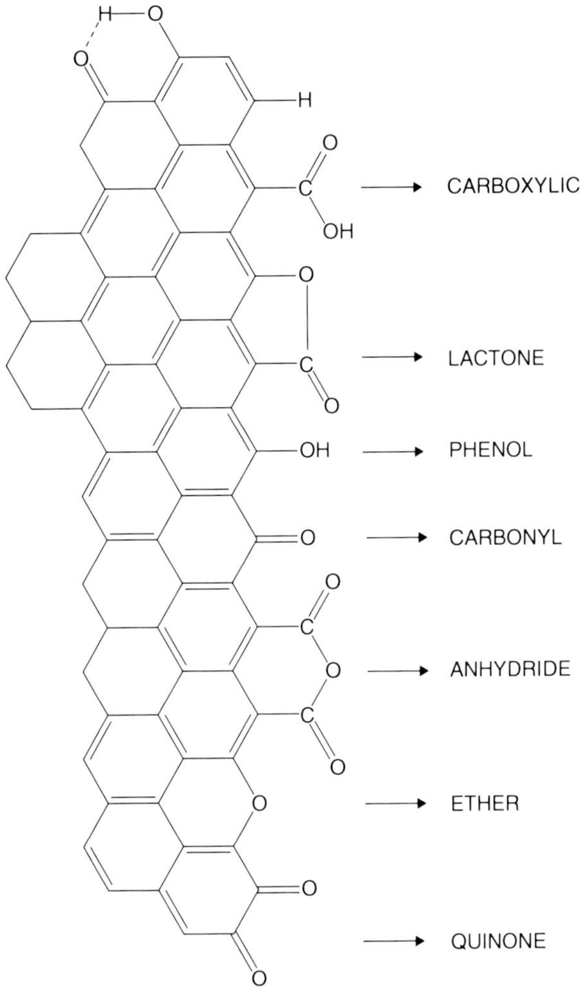

4 Dyes—Environmental Impact and Remediation

Fig. 4.6 Possible surface groups at carbon surface (adapted from Figueiredo et al. 1999)

AC has a net positive surface charge and at $pH > pH_{pzc}$ a net negative charge. High sorption on AC is expected at high pH for cationic dyes and at low pH for anionic dyes, due to electrostatic interaction and attractive forces (**Coulomb's law**). In contrast, when the pH of the solution is not in the ideal range, the electrostatic repulsive forces repel the dyes from the AC surface and it will be covered by the solvent. Some experimental results also indicate that sorption capacity increases with a decrease in AC particle size (Mckay and Sweeney 1980; Al-Degs et al. 2000). This can be attributed to the large molecular diameter and chemical nature of the adsorbate species. Decolourisation is also dependent of other parameters, such as the molecular structure, pK_a and redox potential of the dye. Dye ionization (protonation/deprotonation) is also dependent on the pH of the solution. A variety of experimental techniques have been used to characterize the AC surfaces, including chemical titration, temperature-programmed desorption (TPD), X-ray photoelectron spectroscopy (XPS), mass spectrometry, NMR, infra-red spectroscopy (FTIR, DRIFTS) (Rodriguez-Reinoso 1987; Figueiredo et al. 1999; Pereira et al. 2003; Faria et al. 2004; Shen et al. 2008; Klein et al. 2008).

Carbon-based sorbents have excellent sorption properties for a considerable number of synthetic dyes (Pereira et al. 2003; Malik 2004; Faria et al. 2005, 2008; Al-Degs et al. 2008). The possibility of using of AC as a redox mediator in dye biological degradation has more recently, also been proposed (van der Zee et al. 2001; Pereira et al. 2010; see also Sect. 4.5.6.4). However, preparation of carbon sorbents is generally energy-consuming, making commercially available products relatively expensive. Since a large amount of carbon sorbent is needed for the removal of dyes from a large volume of effluent, the high cost can hamper its application (Fu and Viraraghavan 2001; Forgacs et al. 2004). In addition, the technology for manufacturing AC of good quality is not fully in place in developing countries. This has prompted a growing research interest in the production of low-cost alternatives to AC from a range of carbonaceous and mineral precursors.

4.5.1.2 Other Adsorbents

Research to find cheaper alternatives to AC has been discussed by Gupta and Suhas (2009). Sorption capacity and selectivity are also factors important when choosing the material for certain applications. Many materials have proved to be good candidates, including inorganic and organic materials, some examples with their advantages and disadvantages being given in Table 4.4. Many of the starting materials for these replacement adsorbents are agricultural or industrial by-products; hence their reuse as secondary adsorbents minimizes waste. But applicability of the sorption process is largely dependent on the availability of cheap adsorbents and thus recent initiatives in sorption process have sought economically sound adsorbents.

4.5.1.3 Biosorption/Biomaterials

Biosorption is defined as the accumulation and concentration of pollutants from aqueous solutions using biological materials, thus allowing the recovery and/or environmentally acceptable disposal of pollutants. Biosorption, a property of both living and dead organisms (and their components), has been heralded for a number of years as a promising biotechnology for pollutant removal from solution and/or pollutant recovery, because of its efficiency and simplicity, as an analogous operation to conventional ion-exchange technology, and the availability of biomass (Aksu 2005; Gadd 2009). Biosorption for dyes can also be adopted for the treatment of textile effluents since a wide variety of microorganisms including algae, yeasts, bacteria and fungi are capable of efficiently decolourising a huge range of dyes (Fu and Viraraghavan 2001; Padmesh et al. 2005; Prigione et al. 2008a, b; Bergsten-

Table 4.4 Example of dye adsorbents, major advantages and disadvantages of their use

Adsorbent	Advantages	Disadvantages	References
Alumina	• can be modified in order to be improved as sorbent • can be regenerated • has high affinity for cationic dyes	• has lower affinity for anionic dyes	Adak et al. (2005), Adak and Pal (2006)
Coal (high silica content)	• achieve equilibrium in short time as compared with activated carbon • is effective for cationic dyes (basic dyes) due to the low pH_{pzc} • not being a pure material is suggested to have a variety of surface properties and sorption properties	• not effective for anionic dyes (acid, direct and reactive), which are the most commonly use and highly released to wastewaters;	Mohan et al. (2002), Karaca et al. (2004)
Fly ash	• high availability • after its use in construction, the safe disposal of this material is problematic; its use on sorption will solve also this environmental problem • low cost material • has hydrophilic surface and porous structure • remove both anionic and cationic dyes • inexpensive by-product management technology is needed for its re-use	• some fly ash have unstable composition and properties (depending on their origin) • can contain high level of toxic metals	Mohan et al. (2002), Wang and Wu (2006)
Silica-based and clay (high silica content) $pH_{pzc} \sim 2$	• low cost • available in abundance • good sorption properties	• not effective for the anionic dyes (acid, direct and reactive), which are the most commonly use and highly released to wastewaters • side reactions	Kannan et al. (2008)
Diatomite (siliceous rock from the skeletons of aquatic plants, diatoms); $pH_{pzc} = 6.2$	• presence of high reactive groups (silanol); • it has a unique combination of physical and chemical properties, which make it applicable as a sorbent; • high selectivity for basic and reactive dyes; • high permeability; • high porosity, low density and high surface area	• low selectivity for anionic dyes • naturally occurring diatomice has a lesser ability to adsorb dyes compared with the chemically modified diatomice	Al-Ghouti et al. (2003), Allen and Koumanova (2005), Badii et al. (2010)

Table 4.4 (continued)

Adsorbent	Advantages	Disadvantages	References
Functional granular polymers	• high selectivity for the removal of cationic dyes • exhibits relatively high surface area and porosity • its ease of regeneration	• solubility in water can be low • high cost	Chowdhury et al. (2004)
Agricultural, industrial and domestic waste by-products	• widespread availability • economically attractive for dye removal • because they are so cheap, regeneration is not necessary	• usually can be used just once	Nigam et al. (2000), Forgacs et al. (2004)
Peat (low-grade carbonaceous fuel containing lignin, cellulose and humic acids)	• widely available biosorbent • costs much less than activated carbon • has adsorption capabilities for a variety of pollutants • good adsorption for cationic dyes (basic) at high pH • does nor requires activation • the exhausted peat adsorbent may be disposed of by burning and the heat used or steam generation • can also be modified with some chemical pretreatment to improve its sorption properties and selectivity	• low mechanical strength • high affinity for water • poor chemical stability • tendency to shrink and/or swell, and to leach fulvic acid • influenced by the pH of solution • low capacity for acid dyes (anionic); • specific surface area for adsorption is lower than that of activated carbon	Brown et al. (2000)
Zeolite (mezoporous material)	• natural zeolites have excellent ion exchange properties and high surface area • well defined pore structure in the microporous range • surface can be modified by quaternary amine surfactants enhancing the adsorption capacity	• smaller dyes (such many azo) are excluded from zeolite structure due to the large pore size; • natural zeolites are not suitable sorbents for reactive azo dyes	Alpat et al. (2008), Armagan et al. (2004), Ozdemir et al. (2004)

Torralba et al. 2009). Although the mechanisms of biosorption are not yet fully explained, it seems to take place essentially on the cell wall.

Textile dyes vary greatly in their chemistries and, therefore, their interactions with microorganisms/biomaterials depends on a particular dye and also on the specific chemistry of microbial biomass/biomaterial (Robinson et al. 2001; Erdem et al. 2005). The use of biomass for wastewater treatment is increasing because it is available in large quantities at a low price. The major advantages of biosorption technology are its effectiveness in reducing the concentration of dyes down to very low levels, and the use of cheap biosorbent material. Fungal biomass can be produced economically using relatively simple fermentation techniques and cheap growth media (Fu and Viraraghavan 2002). Generally, it can be eluted and regenerated by some solvents, such as methanol and ethanol, certain surfactants (e.g. non-ionic Tween), or NaOH solution. Biosorption as an emerging technology also attempts to overcome the selectivity disadvantage of conventional sorption processes. The use of dead rather than live biomass eliminates the problems of toxic waste and nutrient requirements. In spite of good sorption properties and high selectivity, the sorption process is slow. Clogging in some bioreactors has also been a limitation. Biosorption processes are particularly suitable for the treatment of solutions containing dilute (toxic) dye concentrations. Biomass has a high potential as a sorbent because of its physico-chemical characteristics. Biosorption is strongly influenced by the functional groups in the fungal biomass, the specific surface properties, and the initial pH of the dye solution. Its performance also depends on external factors such as salts and ions in solution that may be competing with the dye.

Not only microorganisms, but other biomaterials are being increasingly used for economical and ecofriendly remediation of textile dye from effluents (Crini 2006; Allen and Koumanova 2005; Mondal 2008). **Chitin/Chitosan**-based absorbents present a new group of bioabsorbents that can remove dyes from wastewater (Chiou and Li 2002; Chatterjee et al. 2005). **Chitin** is a naturally occurring derivative of cellulose and **chitosan** is a dervative of chitin. Chitosan is the deacylated form of chitin, which is a linear polymer of acylamino-D-glucose. Chitosan has the attraction of having a high content of amino and hydroxyl functional groups, giving it a high potential for dye sorption (Yoshida et al. 1991, 1993; Juang et al. 1997; Chiou and Li 2002). These materials can contain amine or amide nitrogen in varying proportions, and have high sorption capacity for anionic dyes without significant sorption for cationic dyes. Van der Walls attraction, hydrogen bonding and Coulombic attraction are the main ones involved. Sorption capacity of chitosan increases with a decrease in pH: the $-NH_2$ group is easily protonated in acid solution to become NH_3^+, thereby creating electrostatic attraction to anionic dyes. Chitin contains the amide group $-CO-NH-$ which cannot easily be protonated, and hence less electrostatic attraction between the chitin and the dye molecules occurs. Some of the useful features of chitosan include its abundance (it is found in the skins or shells of anthropods), hydrophilicity, biodegradability and nontoxicity.

4.5.2 Coagulation-Flocculation-Precipitation

Coagulation is the destabilization of electrostatic interactions that exist between the molecules of reactive hydrolyzed dyes (or auxiliaries) and water through the addition of a chemical reagent, a coagulant (Allegre et al. 2004). Coagulation is used together with flocculation or sedimentation, with an efficiency that depends on the type of flocculant and the pH of the medium (Allegre et al. 2004; Mondal 2008). In practice, flocculation reagents are large synthetic polymers with a linear structure used along with $FeCl_3$, $FeSO_4$, $AlCl_3$ or $Al_2(SO_4)_3$. There have been reports of the use of several co-polymers, such as pentaethylene, hexamine and ethylediene dichloride, as flocculants for the decolourizing of dye effluents (Anjaneyulu et al. 2005). Colour removal is accomplished by aggregation/precipitation and sorption of colouring substances onto the polynuclear coagulant species and hydrated flocks. The co-polymers have large pore diameters (up to ~ 400 μm), thereby enhancing the process of sorption on flocs.

Although these methods are advantageous in eliminating insoluble dyes (such as disperse), a large amount of sludge is produced leading to extra costs for its treatment (Hao et al. 2000). Acid, direct, vat, mordant and reactive dyes usually coagulate, but the resulting floc is of poor quality and does not settle well, yielding mediocre results. Cationic dyes simply do not coagulate.

Some features of the coagulation process include (Soares et al. 2004):

- Short detention times;
- Good removal efficiencies;
- Quick response to operational factors, making automation simple;
- Equipment requiring less space than a biological lagoon (but nevertheless expensive);
- Removal of 90% of suspended solids and ~ 10–50% COD;
- Performance dependent on the final floc formation and its setting quality
- Overdoses of polyelectrolyte leading to residual concentrations of it in the effluent, and thus a detrimental effect on the nitrification process;
- Dye colour in solid phase remaining a problem, although removal from the aqueous phase can be up to 90%;
- Sludge production dependent on the nature of the flocculant used

The high cost of chemicals for precipitation and pH adjustments, problems associated with dewatering and disposing of generated sludge, and the high concentration of residual cation levels left in the supernatant are some of the limitations of this method.

4.5.3 Membrane Filtration

Membrane filtration is another important process for separating dyes from aqueous solution. A membrane is a permeable or semi-permeable phase, often known as a

4 Dyes—Environmental Impact and Remediation

thin polymeric solid, which restricts the motion of certain species. A membrane is a barrier that allows one component of a mixture to permeate freely while hindering permeation of another component. Membrane techniques offer the appeal of recovering and reusing chemicals (dyes) for producing reusable water. This separation process has the ability to clarify, concentrate and, most importantly, to separate dye continuously from effluent. The technique effectively removes of all type of dyes, but produces highly concentrated sludge. Dissolved solids are not removed by this technique. Selection of a specific membrane depends on the e type of dye or wastewater composition, and on the process operation. Membranes with varying pore size can retain solutes according to their different molecular weight cutoffs (MWCO) and are classified as: **reverse osmosis** (< 1000 MWCO), **nanofiltration** (500–$15\ 000$ MWCO) and **ultrafiltration** membranes (1000–$100\ 000$ MWCO). [Nanofiltration (NF) membranes are a new class of membranes, which have properties between those of ultrafiltration (UF), and reverse osmosis (RO) membranes.]

The major drawbacks are the high costs of labour and of membrane replacement, since membranes are prone to clogging and fouling. The need to be regenerated or changed at regular intervals, which entails high capital and energy costs, and this is not always technical and economically viable. Additionally, adsorbed dye molecules are not degraded but concentrated, and subsequently need to be disposed of. Membrane filtration is therefore more effective for low concentrated waters. Membrane blockage due to high dye concentration was observed by Ahmad et al. (2002). When the feed concentration of dye increased from 0.5 to 10 gL^{-1}, the flux decreased dramatically from 1904 to 208 $L/m^2 h$, and the average percentage rejection increased from 21 to 91%.

4.5.4 *Electrochemical Wastewater Treatment*

The electrochemical treatment of wastewater is a potential powerful method of pollution control, offering high removal efficiency. The process includes **electrooxidation** and **electrocoagulation,** which are relatively non-specific, and thus applicable to a variety of contaminants.

The principle of electrochemical techniques is to send an electric current through electrodes resulting in different chemical reactions (Fig. 4.7). The reducing agent, instead of being added in the typical conventional method is replaced by an innovative cathodic *electron transfer* (electrons are employed, instead of chemicals). An electrochemical cell is used to perform the reduction/oxidation process. The dyeing apparatus is coupled to the electrochemical cell and the dye bath gets effectively reduced through an electrochemical process. The factors affecting electrochemical performance include current intensity, design geometry, stratification, type, number and spacing of electrodes used, pH, temperature, nature of electrolyte, surface tension, flow rate of wastewater and the properties of the dye (Hao et al. 2000). Higher degrees of decolourisation can be achieved by this method, although this does not imply a high COD reduction. Depending on electrolysis time and the applied poten-

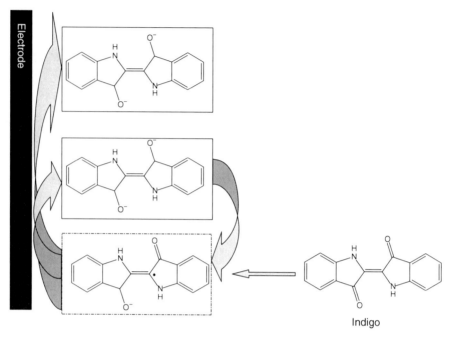

Fig. 4.7 Mechanism of the direct electrochemical reduction of indigo (Adapted from Roessler et al. 2001)

tial, total decolourisation can be attained. The electrochemical method of oxidation for colour removal is more efficient in the treatment of textile wastewater for the dyeing stage than for the total dyeing and finishing stages. Some of the advantages are non-hazardous resulting products, the little need for additional chemicals, and the low temperature required compared with other treatments (Kim et al. 2002b; Esteves and Silva 2004); furthermore no sludge is formed. In terms of apparatus, it does not require much space. Electrochemical processes generally have lower temperature requirements than those of other equivalent non-electrochemical treatments, and no need for additional chemicals. The equipment and operations needed are generally simple, with the controls being easy and the electrochemical reactors compact. The high cost of the electricity requirement is the main limitation (Kim et al. 2002a).

4.5.5 Advanced Oxidation Processes

Oxidation is the most commonly used chemical decolouration processes because it is simple. Alternative to the traditional methods are the Advanced Oxidation Processes (AOPs) based on the generation of highly reactive species, such as hydroxyl radicals (•OH), that can oxidize quickly and non-selectively a broad range of or-

4 Dyes—Environmental Impact and Remediation

ganic pollutants. Furthermore, several investigations have demonstrated that AOPs effectively remove colour and, in part, some organic content of dyestuffs. AOPs include (see Rauf and Ashraf 2009):

1. photolysis (UV);
2. hydrogen peroxide, such as H_2O_2+UV, Fenton ($H_2O_2+Fe^{2+}/Fe^{3+}$), Fenton-like reagents ($H_2O_2+Fe^{2+}$-solid/Fe^{3+}-solid) and photo-Fenton ($H_2O_2+Fe^{2+}/Fe^{3+}+UV$);
3. ozone (ozonation, photo-ozonation, ozonation + catalysis, and $O_3+H_2O_2$ and O_3+Fe^{2+}/Fe^{3+}); and
4. photocatalysis (semiconductor-mediated photocatalysis and TiO_2+CdS+combinations).

AOPs have common principles in terms of the participation of hydroxyl radicals generally assumed to participate in the reaction. Although other species are also thought to be involved, the active species responsible for the destruction of contaminants in most cases is the hydroxyl radical ($\bullet OH$), which is unstable and therefore very reactive. Hydroxyl radicals may attack organic molecules by extracting a hydrogen atom from the molecule under attack (Hao et al. 2000). The common pathway for the degradation of organics by the hydroxyl radical is as follows:

$$\bullet OH + RH \rightarrow H_2O + R\bullet$$

$$R\bullet + H_2O_2 \rightarrow ROH + \bullet OH$$

$$R\bullet + O_2 \rightarrow ROO\bullet$$

$$ROO\bullet + RH \rightarrow ROOH + R\bullet$$

The oxidation potential of $\bullet OH$ is higher than that of other oxidizing agents (Table 4.5). Hydrogen peroxide (H_2O_2) is a widely used agent that needs to be activated by, for example, UV light. Many methods of chemical decolourisation vary depending on the method by which H_2O_2 becomes activated. H_2O_2 alone is often not effective and fails to react with a particular azo dye over a pH range of 3–9.5 (Hao et al. 2000). For light-mediated oxidation of dye molecules, UV of different wavelengths and visible light can be used. As expected, UV alone may not be applicable to wastewaters with high colour intensity. But the combination of UV and H_2O_2 can effectively decolourise dyes, and also reduce TOC and COD, and hence to a potential mineralization.

Table 4.5 Oxidation potential for several oxidizing agents (Hao et al. 2000)

Process	E_0 (V)
$O_2+4\,H^++2\,e^-\rightarrow 2\,H_2O$	1.23
$H_2O_2+2\,H^++2\,e^-\rightarrow 2\,H_2O$	1.78
$O_3+2\,H^++2\,e^-\rightarrow H_2O_2$	2.07
$\bullet OH+H^++e^-\rightarrow H_2O$	2.28

In Fenton's reactions, the combination of ferrous and H_2O_2 serves two functions: oxidation of dye molecules by •OH and coagulation with iron ions. After addition of the iron and hydrogen peroxide, they react together to generate hydroxyl radicals in the following manner:

$$Fe^{2+} + H_2O_2 \rightarrow Fe^{3+} + \bullet OH + OH^-$$

$$Fe^{3+} + H_2O_2 \rightarrow Fe^{2+} + \bullet OOH + OH^+$$

The generated hydroxyl radicals will in turn react with the pollutants and oxidize them. With the formation of Fe^{3+} and its complexes, the Fenton process can precipitate the dissolved solutes. Fenton's reagent is effective in decolourising both soluble and insoluble dyes, but leads to the formation of a sludge that can present disposal problems. The pH of wastewater needs to be acidified for effective utilization of Fenton's reagent; if it is is too high, the iron precipitates as $Fe(OH)_3$ and decomposes the H_2O_2 to oxygen. The use of ozone, pioneered in the early 1970s, is due its high instability, making it a good oxidizing agent. Ozonation, as an effective oxidation process, has found application in the decolourisation of synthetic dyes; although it is ineffective in dispersing dyes and needs high pH values. With no residual or sludge formation and no toxic metabolites, ozonation leads to uncoloured effluent and low COD, suitable for discharge into aqueous systems. One major advantage is that ozone can be used in its gaseous state and therefore does not increase the volume of wastewater and sludge. The disadvantage of ozonation is the short half-life (typically being 20 min) of ozone, demanding continuous application and thereby making it a costly process. Operating costs for ozone have proved higher than for electrochemical treatment giving the same level of colour removal.

Photocatalytic degradation is a part of AOP which has proven to be a promising technology for degrading organic compounds (Rauf and Ashraf 2009). Commercial dyes are designed to resist photodegradation, and hence the selection of optimal photocatalytic conditions for the decolourisation of dyes requires considerable expertise. Direct and indirect photocatalytic pathways are the 2 suggested mechanisms for a given photocatalytic reaction. The effective and economic performance of the process is strongly dependent on the electrode materials, and many researchers have investigated electrochemical oxidation for azo dye degradation through operating parameter optimization using a variety of anodes. With the advancement of experimental techniques, semiconductors have been tested for their efficiencies in dye degradation, including TiO_2, V_2O_5, ZnO, WO_3, CdS, ZrO_2 and their impregnated forms (Rauf and Ashraf 2009). Other factors affecting photochemical reactions include pH, light intensity, temperature and initial dye concentration.

Ultraviolet photolysis combined with hydrogen peroxide (UV/H_2O_2) is one of the most appropriate AOP technologies for removing toxic organics from water because it probably occurs in nature itself (Schrank et al. 2007). While the UV/H_2O_2 process appears to be very slow, costly and weakly effective for possible full-scale application, the combination of UV/TiO_2 is more promising (Dominguez et al. 2005). TiO_2 during photocatalysis generates electron hole pairs when irradiated by the light of <380 nm wavelength. Organic pollutants are thus oxidized via direct hole transfer or, in most cases, attacked by the •OH radical formed in the irradiated TiO$_2$ (Xu 2001).

One of the disadvantages of AOP's is that they may also produce undesirable toxic products and release aromatic amines. In aqueous solution, photochemical degradation is likely to progress slowly, but there is no sludge production.

4.5.6 Bioremediation

Bioremediation is defined as the biologically mediated breakdown of chemical compounds: microorganisms (filamentous fungi, yeasts, bacteria, actinomycetes and algae) or enzymes are applied to assist in the removal of xenobiotics (synthetic organic compounds, which are not found in nature, and are thus foreign and new to the biota) from polluted environments. Biological processes are design to take advantage of the biochemical reactions that are carried out in living cells and/or via the enzymes synthesised by them. They are energy-dependent processes and usually involve the breakdown of the pollutant into a number of by-products.

Biodegradation of synthetic dyes has been identified as an effective and environmentally friendly solution (Fig. 4.8). Synthetic dyes are not commonly present in the environment, so they may not be readily biodegradable. Observations indicate that dyes themselves are not biologically degradable, since microorganisms do

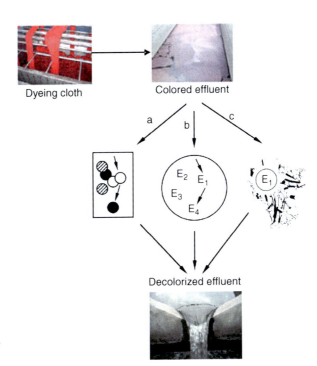

Fig. 4.8 Dye bioremediation by: (a) mixed cultures, (b) isolated organisms, and (c) isolated enzymes (e) (Adapted from Kandelbauer and Guebitz 2005)

not use colour constituents as a source of food. For bioremediation, an additional carbon and energy source has to be present (Nigam et al. 1996). In addition, synthetic dyes are designed in such a way that they can be made resistant to microbial degradation under aerobic conditions. Also the high solubility in water of, especially, sulfonate dyes and their high molecular weight inhibit permeation through biological membranes. Bio-decolourisation of dyes may occur either by sorption on growing/living and dead microbial cells, biosorption, or biodegradation of the dyes by the cells (Fu and Viraraghavan 2001). In biosorption, the original structure of the dyes remains intact and the environmental pollution problem is not eradicated because the pollutant is not destroyed but instead becomes entrapped in the microbial biomass (Ali 2010). In contrast, in biodegradation the original dye structure is destroyed, and the pollutant is split into fragments by the microbial cells, sometimes achieving complete mineralization, i.e., conversion of the xenobiotic into CO_2, H_2O and some salts of inorganic origin. Although many organic molecules are degraded, many others are recalcitrant due to their complex chemical structure, and to their synthetic origin and xenobiotic nature. The rates of dye decolourisation/biodegradation are usually assessed by Monod-type kinetics. Zero, first and second-order reaction kinetics have been used to estimate colour removal rate constants using Eqs. 4.3, 4.4 and 4.5, respectively:

$$\mathbf{C_t} = \mathbf{C_0} - \mathbf{k_0 t} \qquad \text{(zero order)} \tag{3}$$

$$\mathbf{C_t} = \mathbf{C_0} \mathbf{e}^{-klt} \qquad \text{(first order)} \tag{4}$$

$$\mathbf{C_t} = 1 / \mathbf{C_0} + \mathbf{k_2 t} \quad \text{(second order)} \tag{5}$$

where C_0 and C_t are the initial and time of reaction concentrations, respectively; t is the reaction time, and k is the reaction rate constant (time^{-1}).

Biological treatment is often the most economical alternative compared with other physical and chemical processes. An advantage of biological treatment over certain physicochemical treatment methods is that >70% of the organic material measured by the COD test can be converted to biosolids (Forgacs et al. 2004).

Development of an efficient dye degradation biotechnology requires application of a suitable selected strain and its use under favourable conditions to optimize its degradation potential (Novotný et al. 2004; Lucas et al. 2008). A number of microorganisms are capable of decolourising textile dyes, including bacteria, filamentous fungi and yeasts (Stolz 2001). The isolation of new strains or the adaptation of existing ones to the decomposition of dyes will probably increase the efficacy of bioremediation of dyes in the near future. The addition of activators (e.g. Tween 80, veratryl alcohol and manganese (IV) oxide) to the culture medium of *P. chrysosporium* for the production of lignolytic enzymes increased the decomposition rate of the dye Poly R-478 (Couto et al. 2000). In many cases, it is preferable that suitable organisms excrete the active enzymes into the medium, otherwise transport into the cells becomes limiting for bio-elimination. Processes using immobilized growing cells or immobilized enzymes seem to be more promising than those involving free cells, since immobilization allows repeated use of the microbial cells.

4 Dyes—Environmental Impact and Remediation

Some disadvantages of biological treatments are the requirement for a large land area, the sensitivity toward diurnal variation, the toxicity of some chemicals, and less flexibility in design and operation.

4.5.6.1 Decolourisation by Mixed Cultures

Utilization of microorganism consortia offers considerable advantages over the use of pure cultures in the degradation of synthetic dyes. Furthermore, mixed culture studies may be more comparable to practical situations. With the increasing complexity of a xenobiotic, one cannot expect to find complete catabolic pathways in a single organism; a higher degree of biodegradation and even mineralization might be accomplished when co-metabolic activities within a microbial community complement one another (Nigam et al. 1996; Khadijah et al. 2009). Using mixed cultures instead of pure cultures, higher degrees of biodegradation and mineralization can be achieved due to synergistic metabolic activities of the microbial community (Ramalho et al. 2004; Khehra et al. 2005; Ali 2010). The individual strains can attack dye molecules at different positions, yielding metabolic end products that may be toxic; these can be further metabolised as nutrient sources to carbon dioxide, ammonia and water by another strain. Other species present may not be involved in bioremediation at all, but can stabilise the overall ecosystem (Kandelbauer and Gübitz 2005). This type of mineralization is the safest way to assure that no potentially harmful and unrecognized intermediate degradation products are released into the environment. Mixed consortia usually do not require sterile conditions and have greater stability towards changes in the prevailing conditions (pH, temperature and feed composition) compared with pure cultures (Ramalho et al. 2004). Therefore, the use of mixed cultures is a good strategy for bioreactors.

Biological activated sludge is the most common type of treatment system using mixed cultures. However, activated sludge technology has several inherent disadvantages, such as low biomass concentration and easy washout. It should be stressed that the composition of mixed cultures can change during the decomposition process due to the metabolism, which can interfere with the control of technologies using mixed cultures.

4.5.6.2 Decolourisation by Isolated Organisms and Their Enzymes

The use of an isolated culture system ensures that the data are reproducible and more readily interpreted. The detailed mechanisms of biodegradation can be understood in terms of biochemistry and molecular biology, and these disciplines can also be used to upregulate enzyme system to generate modified strains with enhanced biodegradation activities. The quantitative analysis of the kinetics of azo-dye decolourisation by a particular bacterial culture can be meaningfully valuable asset. Also, the response of the system to changes in operational parameters can be examined. Many strains and enzymes that can degrade dyes or other pollutants now been commercialized. Enzymes presently under investigation are still expensive

Trametes hirsuta   Pycnoporus sanguineus

Fig. 4.9 Examples of white-rot-fungi: *Trametes hirsuta* and *Pycnoporus sanguineous*

due to the high cost of isolation, purification and production. Some commercial fungal enzymes and their costs are listed in table 2 of the review by Durán and Esposito (2000).

Filamentous Fungi

By far the most efficient single class of microorganisms in breaking down synthetic dyes is the white-rot fungi (WRF; Fig. 4.9). White-rot fungi are a class of microorganisms that produce non-specific extracellular ligninolytic enzymes capable of extensive aerobic depolymerization and mineralization of lignin (Couto 2009). Lignin is a complex chemical compound abundant in wood, being an integral part of the cell walls of plants. Its degradation is a rate-limiting step in carbon recycling (Ohkuma et al. 2001). Microbial ability to metabolize lignin and its components is one of the plausible evolutionary origins of the degrading pathway of aromatic xenobiotics and/or environmental pollutants. The main extracellular enzymes participating in lignin degradation are lignin peroxidase (ligninase, LiP, EC 1.11.1.14), manganese peroxidase (MnP, EC 1.11.1.13), and the Cu-containing laccase (benzenediol:oxygen oxidoreductase, EC 1.10.3.2). A new group of ligninolytic heme-containing peroxidases, combining structural and functional properties of the LiPs and MnPs, are the versatile peroxidases (VPs). In addition, enzymes involved in hydrogen peroxide production such as glyoxal oxidase (GLOX) and aryl alcohol oxidase (AAO) (EC 1.1.3.7) can be considered to belong to the ligninolytic system (Wesenberg et al. 2003). It is the non-specificity of these enzymes synthesized by WRF that makes them efficient degraders of a wide range of xenobiotics under aerobic conditions, including dyes (Ohkuma et al. 2001; Wesenberg et al. 2003; Couto 2009). *Phanerochaete chrysosporium* was the first species identified that could degrade polymeric synthetic dyes (Glenn and Gold 1983). In addition to enzymatic biodegradation, fungi may decolourise solutions by biosorption (Fu and

4 Dyes—Environmental Impact and Remediation

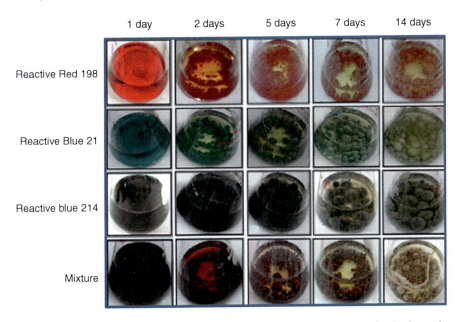

Fig. 4.10 Visual observation of different reactive dyes and a mixture decolourisation by *P. simplicissimum* at increasing incubation times (adapted from Bergsten-Torralba et al. 2009)

Viraraghavan 2002; Zeroual et al. 2006; Prigione 2008a, b). The cell-adsorbed dye can be further enzymatically decolourised.

Long growth cycles and moderate decolourization rates limit the performance of fungal systems (Banat et al. 1996). Fungi growth takes normally from three to seven days. They can be cultivate in different culture media, such as yeast extract, malt extract, agar, starch and potato dextrose using low pHs normally ranging from 3 to 5 (Fu and Viraraghavan 2001). The optimal growth temperature depends on the fungi, but is in the range of 20–35°C. In general, better decolourisation and biodegradation activities are also obtained at acidic or neural pH, and at temperatures between 20 and 35°C. The process can take hours, but more often several days (Ali 2010). For many recent publications on fungi decolourisation of synthetic dyes, see table 2 in Ali (2010). Figure 4.10, is an example of decolourisation by the fungus, *P. Simplicissimum,* of several different single reactive dyes and their mixtures (Bergsten-Torralba et al. 2009). Various type of reactor systems using fungi have been described (Fu and Viraraghavan 2001; Kapdan and Kargi 2002), although, processes using immobilized growing cells look more promising than those using free cells, since the former can be repeatedly and continuously used in the remediation process (Zhang et al. 1999). Furthermore, immobilized cultures are more resilient to environmental perturbations, e.g. pH or exposure to toxic chemicals concentrations, than suspension cultures (Kuhad et al. 2004; Couto 2009). Two interesting reviews on the use of immobilized fungi for dye decolourisation come from Couto (2009) and Mazmanci (2010).

Bacteria

Efforts to isolate bacterial cultures capable of degrading azo dyes started in the 1970s with reports of *Bacillus subtilis* (Horistsu et al. 1977), followed by *Aeromonas hydrophilia* (Idaka and Ogewa 1978) and *Bacillus cereus* (Wuhrmann et al. 1980). Numerous other bacteria capable of dye decolourisation have been found in the interim (Banat et al. 1996; Dave and Dave 2009). Khadijah et al. (2009) isolated 1540 bacteria and screened them for the ability to degrade azo dyes; from the initial screening in microtitre plates, 220 isolates showed decolourisation potential, of which 37 showed higher decolourised zones on dye-incorporated agar plates. In the final screening in liquid medium, 9 proved capable of degrading a wide spectrum of dyes. Bacteria degrade azo dyes reductively under anaerobic conditions to give colourless aromatic amines. These in turn need to be further degraded due to their possible toxic, mutagenic and/or carcinogenic character in humans and animals (Chen 2006). Human intestinal bacteria can also degrade azo dyes to carcinogenic aromatic amines (Chen 2006), which presents a public health problem where low amounts of dyes might be ingested. Anthraquinonic dyes are less susceptible to anaerobic reduction. Whole-cell biodegradation is often carried out by a number of enzymes working sequentially; however, as with other microorganism, only a few of the expressed bacterial enzymes are directly involved in dye biotransformation. The bacterial enzymes involved in the reductive azo bound cleavage are usually azoreductases, whose actions may depend on the presence of other substances such as cofactors, co-substrates or mediators. To avoid the formation of carcinogenic amines, aerobic conditions are preferable in aromatic amine degradation, but it should also be noted that some of them may be auto-oxidized to polymeric structures in the presence of oxygen (Kudlich et al. 1999). Undeniably, the isolation of bacteria capable of aerobic decolourisation and mineralization of dyes has attracted interest, although, especially for sulfonated azo dyes, things have proven difficult (McMullan et al. 2001). Contrary to the unspecific mechanism of azo dye bacterial reduction under anaerobic conditions, aerobic bacteria usually need to be specifically adapted to achieve a significant reductive process. This adaptation involves long-term aerobic growth in continuous culture in the presence of a very simple azo compound. Induction leads to the bacteria synthesis of azoreductases, specific for the reduction of the inducer azo compound, or even others related compounds, in the presence of oxygen (Stolz 2001). Some strains of aerobic bacteria can degrade azo groups by special oxygen-tolerant azo reductases, but they have limited substrate range (Zimmermann et al. 1982; Nachiyar and Rajakumar 2005; Chen 2006). The concepts of anaerobic and aerobic biodegradation will be described in Sects. 4.5.5.4 and 4.5.5.5.

The use of bacteria is influenced by factors at the level of the cell, which in turn will influence the permeability and diffusion of dye molecules. Parameters, such as cell density, enzymes per cell, enzymatic catalytic efficiency, substrate charge and even cell permeability, can be modeled in order to achieve the highest removal rate (Martinez et al. 1999). Generally, unlike fungi, bacteria show better decolourisation

4 Dyes—Environmental Impact and Remediation

and biodegradation activities at basic pH. In comparison to fungi, bacterial decolourisation tends to be faster (Kalyani et al. 2009).

For dye degradation in a bioreactor, immobilization of bacterial cells is also preferable. The advantages of using intact or immobilized cells for biocatalysis is that there is no need to recover and purify the enzymes involved in the process; in addition enzymes encapsulated in the cells may be more resistant to the operating conditions in the long term, in particular for decolourising model or real wastewaters that are rich in salts, additives, surfactants, detergents and others compounds. In addition, costs associated with enzyme purification are negated. Cell immobilization is an effective way to maintain continuous substrate degradation with concomitant cell growth for the treatment of toxic materials. Compared with suspension cells, the main advantage of immobilized cells includes the retention of microorganisms in the reactor and hence protection of cells against toxic substances. The biocatalysts (cells) can be used in repeated cycles, which is of great importance when applied on an industrial scale. Indeed, Advanced Immobilized Cell Reactor technology has been developed specifically to attend for a cost effective treatment system that would accommodate shock load applications and be extremely flexible in its operation. One disadvantage of immobilization is the increased resistance of substrates and products to diffusion through matrices used for immobilization. Owing to the low solubility of oxygen in water and the high local cell density, oxygen transfer often becomes the rate-limiting factor in the performance of aerobic immobilized cell systems. Thus, when aerobic cells are used, aeration technique becomes a very important consideration in bioreactor design technology. Immobilization commonly is accomplished using natural high molecular hydrophilic polymeric gel, such as calcium or sodium alginate, carrageenan and agarose (Palmieri et al. 2005) or many synthetic polymers, such as poly-acrylamide (PAM), polyvinyl alcohol (PVA) (Zhou et al. 2008). Activated carbon has also been used, especially in high performance bioreactors (Walker and Weatherley 1999; van der Zee and Villaverde 2005). The recently developed BIOCOL process (Conlon and Khraisheh 2002) is a commercially available option for the treatment of azo dyes and uses a bacterium isolated from soil contaminated with textile wastewater as the biocatalyst. In this process, the bacterial cells are grown and immobilised on an activated carbon support material that adsorbs the target dye molecules and the potentially toxic amine breakdown products for biodegradation. As well as acting as an absorbent, the activated carbon can potentially promote the degradative activity of the biocatalyst (van der Zee et al. 2001)

Yeast

Very little work has been devoted to the study of the decolourising ability of yeast, most often mentioning sorption as the main cause (Meehan et al. 2000; Donmez 2002). Nevertheless, there are some reports on biodegradation by yeast strains, such as *Candida zeylanoides* (Martins et al. 1999), *Candida zeylanoides* and *Issatchenkia occidentalis* (Ramalho et al. 2002 and 2004, respectively). Yeast cells, like bacteria, are capable of azo dye reduction to the corresponding amines. Testing adapted and unadapted cultures, Ramalho et al. (2004) found that the azo dye reduction activity was due to a constitutive enzyme and that activities were dependent on intact, active cells. Moreover, they noted that *I. occidentalis* (a dye reducer strain) has an absolute requirement of oxygen. Compared to bacteria and filamentous fungi, yeasts have some of the advantages of both: they not only grow rapidly like bacteria, but like filamentous fungi, they also have the ability to resist unfavourable environments (Yu and Wen 2005). In yeast, the ferric reductase system participates in the extracellular reduction of dyes (Ramalho et al. 2005; Chen 2006). In general, yeasts show better decolourisation and biodegradation activities at acidic or neutral pH.

4.5.6.3 Decolourisation by Isolated Enzymes

Independent of the organism to which they belong, fungi, bacteria or yeast, the degradation of some compounds is catalysed by specific enzymes. Indeed, there is a growing recognition that enzymes can be used in many remediation treatments to target specific pollutants. In this direction, recent biotechnological advances have allowed the production of cheaper and more readily available enzymes through better isolation and purification procedures. At the same time, fundamental studies on enzyme structures and enzymatic mechanisms have been conducted.

The main enzymes involved in dye degradation are the lignin-modifying extracellular enzymes (laccase, lignin peroxidase, phenol oxidase, Mn-dependent peroxidase and Mn-independent peroxidase) secreted by WRF, and the bacterial azoreductases. Because of their high biodegradation capacity, they are of considerable biotechnological interest and their application in the decolourisation process of wastewaters has been extensively investigated (Young and Yu 1997). Several enzymes have been isolated and characterised (Grigorious et al. 2000); however, azo reductases have little activity in vivo (Rus et al. 2000). From an environmental point of view, the use of enzymes instead of chemicals or microorganisms presents several advantages, including the potential for scale-up, with enhanced stability and/or activity, and at a lower cost through using recombinant-DNA technology. Another advantage of using pure enzymes instead of the microorganism is that the expression of enzymes involved in dye degradation is not constant with time, but dependent on the growth phase of the organisms and is influenced by inhibitors that may be present in the effluent. Synthetic or natural redox mediators have, to be added many times to the enzymatic bath in order to achieve the total capacity of the enzyme(s) or even to make their work possible (see Sect. 4.5.6.4).

The major drawback of using enzyme preparations is that once the enzymes become inactivated, it is of no use. Because enzymes can be inactivated by the presence of the other chemicals, it is likely that enzymatic treatment will be most effective in streams that have the highest concentrations of target contaminants but the lowest concentration of other compounds that could interfere with their action. In order to increase the potential use of enzymes in a wastewater bioremediation

4 Dyes—Environmental Impact and Remediation

process, their immobilisation is recommended for biochemical stability and reuse, thereby reducing the cost (Durán and Esposito 2000; Kandelbauer et al. 2004).

4.5.6.4 Anaerobic Dye Decolourisation

Anaerobic processes convert the organic contaminants principally into methane and carbon dioxide. They usually occupy less space, can treat wastes containing up to 30 000 mg L^{-1} of COD, have lower running costs, and produce less sludge.

Azo dye degradation occurs preferentially under anaerobic or oxygen limited concentrations, acting as final electron acceptors during microbial respiration. Oxygen, when it is present, may compete with the dyes. In many cases the decolourisation of reactive azo dyes under anaerobic conditions is a co-metabolic reaction (Stolz 2001). Several mechanisms have been proposed for the decolourisation of azo dyes under anaerobic conditions (Rus et al. 2000). One of these is the reductive cleavage of the azo bond by unspecific cytoplasmic azo reductases using flavoproteins ($FMNH_2$ and $FADH_2$) as cofactors. A second proposed mechanism is an intracellular, non-enzymatic reaction consisting of a simple chemical reduction of the azo bond by reduced flavin nucleotides. These reductive cleavages, with the transfer of 4 electrons and the respective aromatic amine formation (Fig. 4.11), usually occur with low specific activities but are extremely nonspecific with regard to the organism involved and the dyes converted. Transport of the reduction equivalents from the cellular system to the azo compounds is also important, because the most relevant azo compounds are either too polar and/or too large to pass through the cell membrane (Rau and Stolz 2003). Mediators generally enable or accelerate the electron transfer of reducing equivalents from a cell membrane of a bacterium to the terminal electron acceptor, the azo dye (Kudlich et al. 1997; Rus et al. 2000; Keck et al. 2002; Van der Zee et al. 2001). Such compounds can either result from the aerobic metabolism of certain substances by bacteria themselves or be added to the medium. They are enzymatically reduced by the cells and these reduced mediator compounds in turn reduce the azo group in a purely chemical reaction (Stolz 2001). It has been suggested that quinoide redox mediators with standard redox potentials (E0') between −320 and −50 mV could function as effective redox mediators in the microbial reduction of azo dyes. The first example of an anaerobic cleavage of azo dyes by redox mediators which are naturally formed during the aerobic metabolism of xenobiotic compound was reported by Keck et al. (1997). Dos Santos et al. (2005) studied the impact of different redox mediators on colour removal of

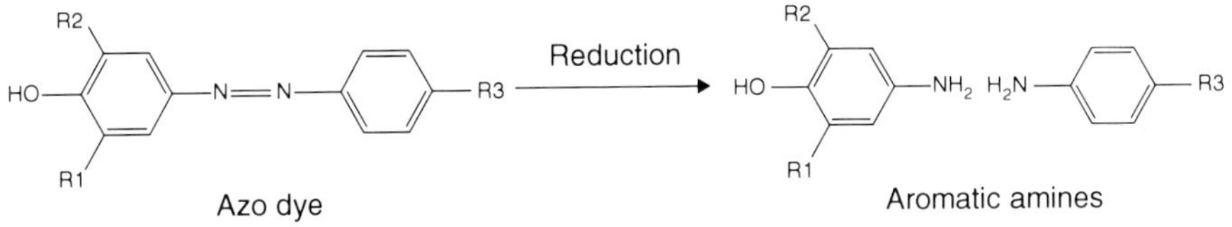

Fig. 4.11 Mechanism of an azo dye reduction

azo dye model compound; up to an eightfold increase in decolourisation rates were achieved compared with mediator-free incubations. Many authors have reported increase of azo dye decolourisation rates in the presence of such molecules, either by membrane-bound respiratory chain or cytosolic enzymes (Kudlich et al. 1997; Rau et al. 2002; 2003; Ramalho et al. 2004; Pearce et al. 2006). The need in some cases for external addition of chemical mediators may increase the cost of the process. In crude extracts and crude enzyme preparations, low molecular weight compounds may be naturally present that act as natural enhancing compounds. The physiology of the possible reactions that result in a reductive cleavage of azo compounds under anaerobic conditions differs significantly from the situation in the presence of oxygen because the redox active compounds rapidly react either with oxygen or azo dyes (Stolz 2001). Therefore, under aerobic conditions, oxygen and dyes compete for the reduced electron carriers. In bioremediation, an anaerobic treatment or pretreatment step can be a cheap alternative compared with an aerobic system because expensive aeration is omitted and the problem with bulking sludge is avoided.

In addition to azo dyes, other classes of dyes have been degraded under anaerobic conditions (McMullan et al. 2001), but the mechanisms are less well described.

4.5.6.5 Aerobic Dye Decolourisation

In aerobic pathways, azo dyes are oxidized without the cleavage of the azo bond through a highly non-specific free radical mechanism, forming phenolic type compounds (Fig. 4.12). This mechanism avoids the formation of the toxic aromatic amines arising under reductive conditions (Chivukula and Renganahathan 1995; Chen 2006; Pereira et al. 2009a).

The main organisms involved in the oxidative degradation of dyes are the fungi WRF by the so-called lignolitic enzymes (Sect. 5.5.2.1). The application of these organisms or their enzymes in dye wastewater bioremediation have attracted increasing scientific attention because they are able to degrade a wide range of organic pollutants, including various azo, heterocyclic and polymerise dyes (Abadulla et al. 2000; Wesenberg et al. 2003). It is also noteworthy that they require mild conditions, with better activities and stability in acidic media and at temperatures from 20 to 35°C. Oxidases can also be found in some bacteria, plants and animals.

Peroxidases, including LiP and MnP, use hydrogen peroxide to promote the one-electron oxidation mechanism of chemicals to free radicals (Figs. 4.13 and 4.14, respectively). Those enzymes have an important role on the cellular detoxification

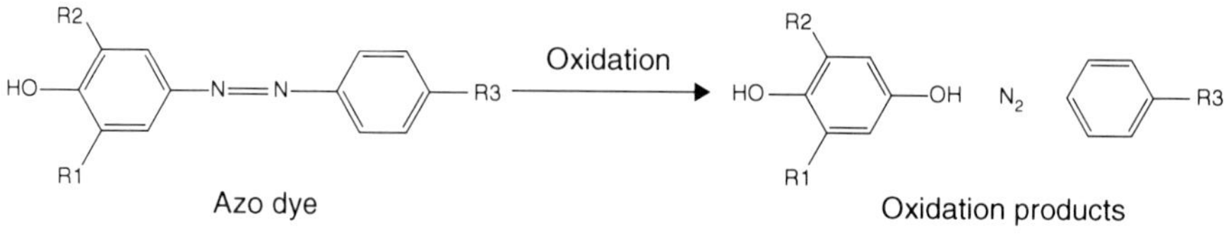

Fig. 4.12 Mechanism of an azo dye oxidation

4 Dyes—Environmental Impact and Remediation

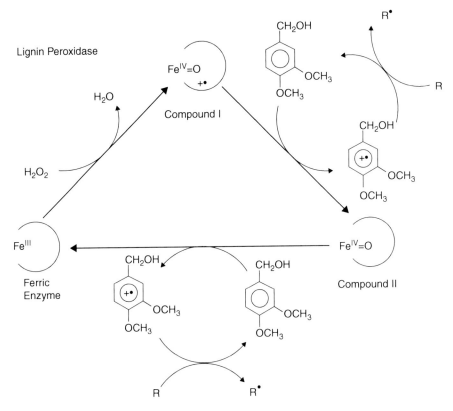

Fig. 4.13 Catalytic cycle of a manganese peroxidase (Cameron et al. 2000)

of organisms by eliminating hydrogen peroxide. They are hemoproteins belonging to the oxidoreductase enzymes that catalyze the oxidation of phenols, biphenols, anilines, benzidines and related hetereoaromatic compounds, for which H_2O_2 is the final electron acceptor (Durán and Esposito 2000). The dye does not need to bind to the enzyme; instead oxidation occurs through simple electron transfer, either directly or through the action of low molecular weight redox mediators (Eggert et al. 1996). Lignin peroxidase was discovered earlier than MnP and exhibits the common peroxidase catalytic cycle. It interacts with its substrates via a ping-pong mechanism, i.e. firstly it is oxidized by H_2O_2 through the removal of two electrons that give compound I, and further oxidized through the removal of one electron to give compound II, which oxidizes its substrate, returning to the resting enzyme. The active intermediates of LiP (i.e. componds I and II) have considerably higher reduction potentials than the intermediates of other peroxidises, extending the number of chemicals that can be oxidised. MnP mechanism differs from that of LiP in using Mn^{+2} as a mediator. Once Mn^{+2} has been oxidized by the enzyme, Mn^{+3} can oxidize organic substrate molecules. MnP compounds I and II can oxidize Mn^{+2} to Mn^{+3}, but compound I can also oxidize some phenolic substrates.

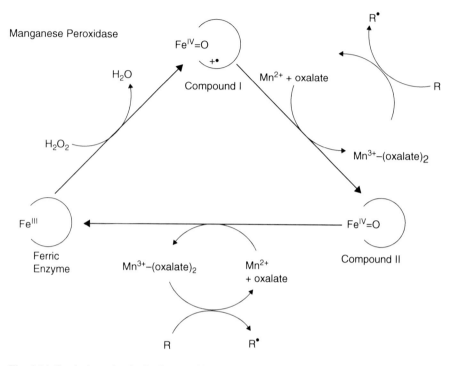

Fig. 4.14 Catalytic cycle of a lignin peroxidase (Cameron et al. 2000)

The use of pure peroxidases for dye degradation, in place of the organism, requires the addition of hydrogen peroxide. The pathways for the degradation of two sulfonated azo dyes by the peroxidise and ligninase of *Phanerochaete chrysosporium* and by the peroxidise of *Streptomyces chromofuscust* have been proposed by Goszczynski et al. (1994). More recently, López et al. (2004) have proposed a degradation pathway of Orange II by MnP enzyme.

Laccases are another group of oxidoreductases, but these use oxygen as the final electron acceptor. This may be advantageous when using the pure enzyme, relatively to peroxidases, since H_2O_2 and cofactors are not needed. Laccases belong to the small group of named multicopper blue oxidase enzymes, and the mechanism of oxidative reactions catalysed by them involves the transfer of 4 single electrons from the substrate to the final acceptor. Their catalytic centres consist of 3 structurally and functionally distinct copper centres (Solomon et al. 1996; Stoj and Kosman et al. 2005). T1 copper ("blue copper") is a mononuclear centre involved in the substrate oxidation, whereas T2 and T3 form a trinuclear centre involved in the reduction of oxygen to water (Fig. 4.15). Owing to their high relative nonspecific oxidation capacity, laccases have proven useful for diverse biotechnological applications including degradation of dyes (Abadulla et al. 2000; Husain 2006) and organic synthesis (Eggert et al. 1996; Riva 2006; Schroeder et al. 2007). The first described laccase had plant origin, the Japanese lacquer tree, *Rhus vernicefera*

4 Dyes—Environmental Impact and Remediation

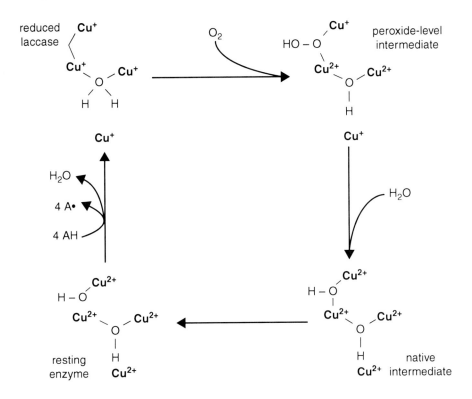

Fig. 4.15 Catalytic cycle of a laccase (Wesenberg et al. 2003)

(Yoshida and Takemori 1997). Nowadays, most of the known laccases have fungal (e.g. white-rot fungi) or plant origins, although a few have recently been identified and isolated from bacteria (Claus 2003; Gianfreda et al. 1999). The ability of a bacterial laccase (CotA) to degrade a wide range of azo and anthraquinonic dyes was reported for the first time by Pereira et al. (2009a, b). The mechanism of azo dye degradation has been described by some authors (Chivukula and Renganahathan 1995: Zille et al. 2005; Pereira et al. 2009a). Antraquinonic dyes are also an important class of dyes, although there is limited information on their physicochemical or biological degradation, and even less on the molecular mechanisms of transformation (Pereira et al. 2009b). Based on the characterisation of intermediates and final products of the reaction, Pereira et al. (2009b) described the mechanistic pathway for the biotransformation of the anthraquinonic dye AB62 by the bacterial CotA-laccase.

The mechanism of Indigo dye (the most important dye in the manufacturing of blue jeans) degradation by a laccase has also been described (Campos et al. 2001). The dye is cleaved under laccase catalyzed electron transfer to give isatin, which, after further decarboxylation, is catalyzed to the final stable oxidation product, anthranilic acid (Fig. 4.16). Likewise, in peroxidise-catalyzed decolourisation of Indi-

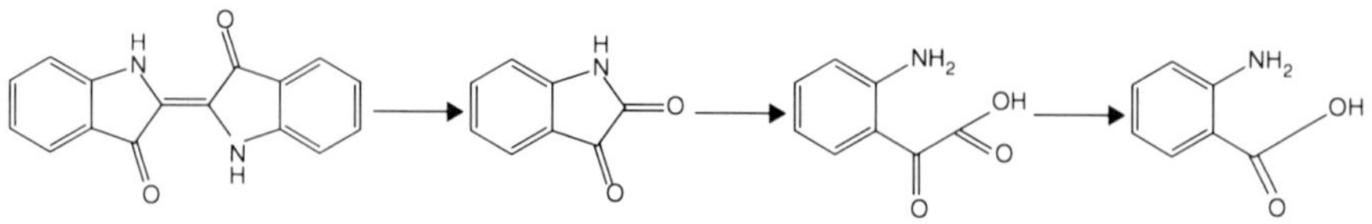

Indigo   Isatin   Anthranilic acid

Fig. 4.16 Mechanism of Indigo oxidation by purified laccases from *Trametes hirsuta* and *Sclerotium rolfsii* (Campos et al. 2001)

go Carmine, isatin sulfonic acid is formed, although a stable red oxidation product, probably a dimeric condensation product, was obtained (Podgornik et al. 2001).

4.5.6.6 Combination of Anaerobic/Aerobic Processes for Dye Treatment

Many dyes used in textile industry cannot be degraded aerobically as the enzymes involved are dye-specific. Anaerobic reduction of azo dyes is generally more satisfactory than aerobic degradation, but the intermediate products (carcinogenic aromatic amines) must be further degraded. These colourless amines are, however, very resistant to further degradation under anaerobic conditions, and therefore aerobic conditions are required for complete mineralisation (Melgoza et al. 2004; Forgacs et al. 2004; Pandey et al. 2007). For the most effective wastewater treatment, 2-stage biological wastewater treatment systems, are then necessary in which an aerobic treatment is introduced after the initial anaerobic reduction of the azo bond (Sponza and Işik 2005; Van der Zee and Villaverde 2005; see Fig. 4.17). The balance between the anaerobic and aerobic stages in this treatment system must be carefully controlled because it may become darker during re-aeration of a reduced dye solution. This is to be expected, since aromatic amines are spontaneously unstable in the presence of oxygen. Oxidation of the hydroxyl and amino groups to quinines and quinine imines can occur and these products can also undergo dimerisation or polymerisation, leading to the development of new darkly coloured chromophores (Pereira et al. 2009a, b). However, with the establishment of correct operating conditions, many strains of bacteria are capable of achieving high levels of decolourisation when used in a sequential anaerobic/aerobic treatment process (Steffan et al. 2005; van der Zee and Villaverde 2005). Aerobic biodegradation of many aromatic amines has been extensively studied (Brown and Laboureur 1983; Pinheiro et al. 2004; van der Zee and Villaverde 2005), but these findings may not apply to all aromatic amines. Specially sulfonated aromatic amines are often difficult to degrade (Razo-Flores et al. 1996; Tan and Field et al. 2000; Tan et al. 2005). Aromatic amines are commonly not degraded under anaerobic conditions. Melgoza et al. (2004) studied the fate of Disperse Blue 79 in a two-stage anaerobic/aerobic process; the azo dye was biotransformed to amines in the anaerobic stage and an increase of toxicity was obtained; the toxic amines were subsequently mineralized in the aerobic phase, resulting in the detoxification of the effluent.

4 Dyes—Environmental Impact and Remediation

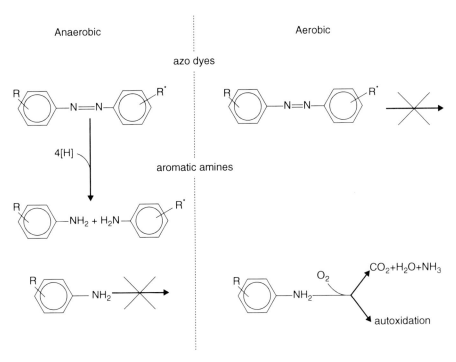

Fig. 4.17 Azo dye degradation in a two-step (aerobic/anaerobic) process (van der Zee and Villaverde 2005)

The 2-stage treatment systems can be the result of many different combinations, such as 2 biological or 2 physicochemical processes, but also of combinations of both these types. The choice for the best treatments is related to many other factors, such as the type of effluent, the availability of the methods, the locality where it is to be applied, and others.

4.5.6.7 Factors Affecting Biodegradation of Dyes

Ecosystems are dynamic environments with variable abiotic conditions, e.g. pH, temperature, dissolved oxygen, nitrate concentration, metals, salts, etc. Microorganisms are affected by changes in these parameters, and consequently their decomposing activities are also affected. Other non-dye related parameters are the type and source of reduction equivalents, bacteria consortium, and cell permeability. Enzymes synthesised by microorganisms, when isolated, are also sensitive to the medium culture conditions in which they are applied. Usually, they proceed better in environments more suitable for the original microorganism. Textile wastewaters result from different classes of dyes, and consequently vary in their composition and pH. It is therefore important, while evaluating the potential of different microorganisms for dye degradation, to consider the effects of other present compounds. Optimization of abiotic conditions will greatly help in the development of industrial-scale bioreactors for bioremediation. Optimal pH and temperatures are always related to the environment where the organisms were collected, but usually fungi work better, either for growth or compounds degradation, in acidic or neutral media, whereas bacteria prefer alkaline conditions. Dye decolourisation proceeds better under acidic conditions for fungi and fungal enzymes (Abadulla et al. 2000; Kandelbauer et al. 2004; Almansa et al. 2004; Zille et al. 2005), but under alkaline conditions for bacteria and bacterial enzymes (Pereira et al. 2009a). The optimal temperatures of microorganisms usually range from 20 to 35°C, but there are also others that tolerate higher values, although the stability may be compromised. Beyond the optimum temperature, the degradation activities of the microorganisms decrease because of slower growth, reproduction rate and the deactivation of enzymes responsible for degradation.

Biodegradation of azo dyes and textile effluents can be affected by dye related parameters, such as class and type of azo dye, reduction metabolites, dye concentration, dye side-groups and organic dye additives. Microbial activity can decrease with increasing dye concentration, which can be attributed to the toxicity of the dyes to the growing microbial cells at higher concentrations (in the biodegradation) and/or cell saturation (biosorption). In the most enzymatic decolourisation studies, the kinetics are described by Michaelis-Menten model and an increase of the rate with increase the dye concentration is observed up to a certain concentration (saturation). At dye amounts higher than the optimal, the rates usually remain constant due to saturation, but there are also some cases of inhibition at concentrations higher than the optimal. The inhibition concentration of dye is not the same for all the microorganisms, and for the same organisms, the inhibition concentration will depend on the dye. Adaptation of a microbial community to the compound is very useful in improving the rate of decolourisation process, due to the natural expression of genes encoding for enzymes responsible for its degradation, when previously exposed (Ramalho et al. 2004). The fact that some dyes are biodegraded and others not, even under the same conditions, is explained by the role of the chemical structure of the dye on the process. Even when belonging to the same class and type, dyes differ in their structure and present different p_{K_a} and potential redox. Zille et al. (2004) studied the biodegradation under aerobic conditions of azo dyes by yeasts with reducing activity and by an oxidative enzyme, laccase, with or without mediator; they compared these 2 approaches on the basis of the electrochemical properties of dyes and bioagents. A linear increase of dye decolourisation with decreasing redox potential of dye was obtained with laccase and laccase/mediator systems; in the reductive approach, they observed that the less negative the redox potential of the azo dye, the more favourable (and faster) its reduction. The redox potential should reportedly be below –450 to –500 mV for azo dye reduction to occur (Delée et al. 1998). It is worth mentioning that the redox potential is influenced by other external factors, such as the pH of the solution. The redox potential of the enzymes may also be involved, although in many cases, enzymes with redox potential lower than that of the dyes can decolourise them. These facts can be understood in the light of the Nernst equation; any redox reaction is dependent on the formal redox potential, and on the concentrations of the reduced and oxidized species (Zille et al. 2004).

4 Dyes—Environmental Impact and Remediation

Little information is available in the literature describing quantitatively the effects of dye chemical structure on the reactivity towards laccase oxidation. Almansa et al. (2004) have synthesized 22 model azo-dyes of chemically very similar structure and studied the effect of substituents on the enzymatic kinetics of their decolourisation. The model dyes only differed in nature and position of the substituent on the phenyl ring, which carried either a single methyl ($-CH_3$), trifluoromethyl ($-CF_3$), fluoro ($-F$), chloro ($-Cl$), bromo ($-Br$), nitro ($-NO_2$), or hydroxy ($-OH$) group in the ortho, meta or para-position with respect to the azo linkage. Without the assistance of an electron mediator, enzymatic degradation took place only with the hydroxy-substituted model azo-dyes. All other dyes were only degraded in the presence of mediator, except those substituted with trifluoromethyl that were not degraded at all. The electron withdrawing effect of 3 fluorine atoms proved strong enough to completely prevent the reaction. In general, they found that electron-withdrawing substitutents diminish reaction rates whereas electron-donating groups enhanced the susceptibility of the dye towards oxidative attack. Similar conclusions were previous reported by Chivukula and Renganathan (1995) and Kandelbauer et al. (2004).

Wastewaters from textile processing and dyestuff manufacture industries contain also substantial amounts of salts in addition to azo dye residues. Thus, microbial species capable of tolerating salt stress will be beneficial for treating such wastewaters. High salt concentrations can also inhibit enzyme activity.

All the referred factors have to be considered and studied for a full-scale application. The influence of the flow rate of the effluent, the nature (concentration) of the effluent, the pursued extent of treatment, the location, the climatic conditions and the configuration of the reactor are all of great importance, not only for the success of the process itself, but also in terms of making it cost-effective.

4.5.6.8 Genetic Engineering of Dye Degrading Organisms

The vast majority of the current publications in the field of the synthetic dyes removal from waters have dealt with the various aspects of the application of microbiological methods and techniques, with the search for new microorganisms providing improved decomposition rates, and with the elucidation of the principal biochemical and biophysical mechanisms underlying the decolourisation process of dyes. Bioprocesses, whether involving the microorganisms themselves or their enzymes, are sufficiently versatile to be customised. Identification, isolation, and transfer of genes encoding degradative enzymes can greatly help in designing microbes with enhanced degradation capabilities. Thus, acclimatization and genetic engineering can both help in designing "super-degraders". Of the two approaches, acclimatization is natural, since in this case the built-in genetic setup of the microorganism is not disturbed; only some components are enabled. On the other hand, in genetic engineering, the natural genetic set-up of a microorganism is changed by incorporating a new gene or genes. Therefore, many scientists—especially environmentalists—are skeptical about the usefulness of genetically modified organisms. They fear that such modified organisms may create new environmental problems

(Ali 2010). However, there is evidence that gene manipulation for the creation of recombinant strains with higher biodegradation capacity will be applied in the future. By cloning and transferring genes encoding for dye degrading enzymes, organisms could be designed that combine the abilities of mixed cultures within a single specie. A number of genes conferring the ability of dye decolourising have been identified. Successful decolourisation of an azo dye using *Escherichia coli* carrying the azoreductase gene from a wild-type *Pseudomonas luteola* has been reported (Chang et al. 2000; Chang and Kuo 2000). This approach could become a useful alternative for shortening the extended time-periods otherwise needed to adapt appropriate cultures and isolated strains, respectively. CotA-laccase, a bacterial enzyme from *Bacillus subtilis* cloned and over-expressed in *E. coli*, has proved to be a thermoactive and intrinsically thermostable enzyme with a high capacity for the decolourisation of azo and anthraquinonic dyes (Pereira et al. 2009a, b). The expression level of CotA-laccases in different *E. coli* host strains, growing under different culture conditions, was compared and a high-throughput screenings for the oxidation of dyes with high potential redox developed by Brissos et al. (2009).

4.6 Products Identification and Mechanisms of Dye Degradation

We have already pointed out that pollutant degradation, including dyes, vary according to non-related and related dye parameters. Identification of the products from synthetic dyes biodegradation is most helpful in determining the mechanistic pathways involved. Such findings are not only important in the knowledge about the fate of organic pollutants, but indirectly in the assessment of the toxicity of the intermediates and main products, and also in describing the microbial system and/or enzymatic activities. Different microorganisms/enzymes may have different pathways of degradation depending on the particular dye structure; thus the strategy of the microbial system for dye degradation and many other factors has to be studied. Dyes, as colourants, absorb in the visible region of the spectra and each one has a maximal wavelength, depending on its visible colour (see Sect. 4.2); therefore, the easiest way to monitor dye degradation is by means of spectrophotometry, following the decrease in its absorbance. By this technique, all the molecules present are quantified, and intermediates and degradation products will contribute to the spectra absorbance. Various basic and advanced instrumental techniques of chromatography such as gas chromatography (GC), high performance liquid chromatography (HPLC), nuclear magnetic resonance spectroscopy (NMR), mass spectrometry (Ion-trap, MALDI) and capillary electrophoresis (CE) are available to assist in the isolation and characterization of the intermediates and products of dye degradation, thereby giving new insight into the mechanism of biodegradation. Prior procedures of extraction of the aqueous sample with an organic solvent or filtration are adopted

4 Dyes—Environmental Impact and Remediation

when a heterogeneous catalyst or solid reactant is employed, or when a pre-separation is needed.

Quantification of CO_2 and NH_3 produced in incubated culture media can also provide additional information. Recent studies on the on the mechanisms and pathways using those techniques have been published by López et al. (2004), Vanhulle et al. (2008); Bafana et al. (2009) and Pereira et al. (2009a, b).

4.7 Conclusions

The management of textile industrial effluents is a complicated task, taking into consideration the complexity of the waste compounds that may be present, in addition to the dyes, and the numerous established options for treatment and reuse of water. Wide ranges of water pH, temperature, salt concentration and in the chemical structure of numerous dyes in use today add to the complication. Economical removal of colour from effluents remains an important problem, although a number of successful systems employing various physicochemical and biological processes have been successfully implemented. Regulatory agencies are increasingly interested in new, efficient, and improved decolourisation technologies. Solid and evolving scientific knowledge and research is of the utmost relevance for the effective response to current needs. In view of the requirement for a technically and economically satisfactory treatment, a flurry of emerging technologies are being proposed and are at different stages of being tested for commercialization. A broader validation of these new technologies and the integration of different methods in the current treatment schemes will be most likely in the near future, rendering them both efficient and economically viable.

Conventional physicochemical treatments are not always efficient. The high cost, the generation of sludge and of other pollutants, and the need for sophisticated technologies are limiting factors as well. Bioremediation of textile effluents is still seen as an attractive solution due to its reputation as a low-cost, sustainable and publicly acceptable technology. Microorganisms are easy to grow and the use of their isolated enzymes for textile dyes degradation is not expensive in relative terms because there is no need for high purity levels in treating effluents. Many microorganisms and enzymes have been isolated and explored for their ability and capacity to degrade dyes. Others have been modified by the genetic engineering tools to obtain "super and faster degraders". Biological processes may require two stages, especially for azo dyes, in which dyes are reduced in the first anerobic step to their respective aromatic amines, which are then oxidized and mineralized in a final aerobic step. In some cases, a combination of biological with physical—such as adsorption or filtration, or chemical such as coagulation/oxidation—processes may be necessary to achieve the desirable goal. Several low cost and efficient sorbents including natural wastes are very promising, not only due to the lower cost and high availability, but also a new utility is granted to those wastes.

4.8 Future Perspectives

The increasing manufacture and application of synthetic dyes, taking into account their impact in the environment, needs an effective response in terms of modern and viable treatment processes of coloured effluents, prior to their discharge as waste into waterways. Biodegradation of synthetic dyes using different microorganisms and isolated enzymes offer a promising approach by themselves or in combination with conventional treatments.

The complexity of dyes degradation and the existence of an immense variety of structurally different dyes, indicates the need for more research. Moreover, the most of the available studies on dye degradation are referent to azo dyes; the studies shall be extended to the other classes: anthraquinone, indigoid, xanthene, arylmethane and phthalocyanine derivatives. The pathways for dye degradation are also still not totally understood; walk in that direction has high relevance for the development of future modern technology. The increasing research in microbiology, molecular biology, chemistry and genetic fields associated with the degradation technology, are fundamental for that knowledge. Additionally, the effect of coloured substances and their products either in the environment or during the treatment processes will be better evaluated and understood. That better understand, in turn, will leads to a technology improvement and application of more efficient, either from existent or new, treatment processes. New microorganisms and enzymes, with broader substrate specificity and higher activity, will also be found, isolated and studied for their ability and capacity as key agents in pollution remediation. Through the benefits of genetic engineering, random or selective modification of the microorganisms and enzymes can greatly help in designing microbes with higher catalytic power for a wider range of compounds. Optimization of the remediation process in terms of time, efficiency, stabilization and, as consequence, in costs will gain from this. The knowledge and evolution in biochemical, biological and process engineering is also essential for the establishment of processes at industrial scale. Promising processes seem to be the combination of more than one treatment, either biologic or chemical, for the complete mineralization and detoxification of the coloured effluent.

The study and implementation of the new treatments shall not only be focused on pollution reduction, but also in the reuse of water and exploitation of the final by-products for other applications.

Acknowledgements This work was supported by the PTDC/AMB/69335/2006 project grants. L. Pereira holds a Pos-Doc fellowship (SFRH/BPD/20744/2004).

References

Abadulla E, Tzanov T, Costa S, Robra K-H, Cavaco-Paulo A, Gübitz G (2000) Decolorization and detoxification of textile dyes with a laccase from *Trametes hirsute*. Appl Environ Microbiol 66:3357–3362

4 Dyes—Environmental Impact and Remediation

Ackacha MA, Połeć-Pawlak K, Jarosz M (2003) Identification of anthraquinone coloring matters in natural red dyestuffs by high performance liquid chromatography with ultraviolet and electrospray mass spectrometric detection. J Sep Sci 26:1028–1034, doi: 10.1002/jssc.200301484

Adak A, Bandyopadhyay M, Pal A (2005) Removal of crystal violet dye from wastewater by surfactant-modified alumina. Sep Purif Tech 44:139–144, doi:10.1016/j.seppur.2005.01.002

Adak A, Pal A (2006) Removal kinetics and mechanism for crystal violet uptake by surfactant-modified alumina. J Environ Sci Heal A, 41:228–2297, doi: 10.1080/10934520600872953

Ahmad AL, Harris WA, Syafiie, Seng OB (2002) Removal of dye from wastewater of textile industry using membrane technology. J Technol 36:31–44

Aksu Z (2005) Application of biosorption for the removal of organic pollutants: a review. Proc Biochem 40:9–1026

Al-Degs YS, Khraishen MAM, Allen SJ, Ahmad MN (2000) Effect of carbon surface chemistry on the removal of reactive dyes from textile effluent. Water Res 34:927–935

Al-Degs YS, El-Barghouthi MI, El-Sheikh AH, Walker GM (2008) Effect of solution pH, ionic strength, and temperature on adsorption behaviour of reactive dyes on activated carbon. Dyes Pigments 77:16–23

Al-Ghouti MA, Khraisheh MAM, Allen SJ, Ahmad MN (2003) The removal of dyes from textile wastewater: a study of the physical characteristics and adsorption mechanisms of diatomaceous Earth. J Environ Manage 69:229–238

Allen SJ, Koumanova B (2005) Decolourisation of water/wastewater using adsorption. J Univ Chem Technol Metall 43(3):175–192

Ali H (2010) Biodegradation of synthetic dyes—a review. Water Air Soil Pollut doi:.1007/s11270-010-0382-4

Allegre C, Maisseub M, Charbita F, Moulin P (2004) Coagulation—flocculation—decantation of dye house effluents: concentrated effluents. J Hazard Mater B116:57–64, doi:10.1016/j.jhazmat.2004.07.005

Almansa E, Kandelbauer A, Pereira L, Cavaco-Paulo A, Gubtiz G (2004) Influence of structure on dye degradation with laccase mediator systems. Biocat Biotransf 22:315–324, doi: 10.1080/10242420400024508

Alpat SK, Ozbayrak O, Alpat S, Akcay H (2008) The adsorption kinetics and removal of cationic dye, Toluidine Blue O, from aqueous solution with Turkish zeolite. J Hazard Mater 151:213–220

Ames B, McCann J, Yamasaki E (1975) Methods for detecting carcinogens and mutagens with the Salmonella/mammalian-microsome mutagenicity test. Mut Res 31:347–364

Amin NK (2008) Removal of reactive dye from aqueous solutions by adsorption onto activated carbons prepared from sugarcane bagasse pith. Desalination 223:152–161, doi:10.1016/j.desal.2007.01.203

Anjaneyulu Y, Chary NS, Raj DSN (2005) Decolourization of industrial effluents—available methods and emerging technologies—a review. Rev Environ Sci Biotechnol 4:245–273, doi: 10.1007/s11157-005-1246-z

Annadurai G, Krishnan MRV (1997) Batch equilibrium adsorption of reactive dye onto natural biopolymer. Iranian Polymer Jovial 6:169–175

Armagan B, Turan M, Celik MS (2004) Equilibrium studies on the adsorption of reactive azo dyes into zeolite. Desalination 170:33–39

Arslan I, Balcioğlu IA (1999) Degradation of commercial reactive dyestuffs by heterogenous and homogenous advanced oxidation processes: a comparative study. Dyes Pigments 43:95–108

Badii K, Ardejani FD, Saberi MA, Limaee NY, Shafaei ZS (2010) Adsorption of acid blue 25 dye on diatomite on aqueous solutions. Indian J Technol 17 (1):7–16

Bafana A, Chakrabarti T, Muthal P, Kanade G (2009) Detoxification of benzidine-based azo dye by *E. gallinarum*: time-course study. Ecotoxicol Environ Saf 72:960–964

Banat IM, Nigam P, Singh D, Marchant R (1996) Microbial decolorization of textile-dye-containing effluent: a review. Biores Technol 58:217–227

Bergsten-Torralba LR, Nishikawa MM, Baptista DF, Magalhães DP, Da Silva M (2009) Decolourisation of different textile dyes by *Penicillium Simplicis* and toxicity evaluation after fungal treatment. Braz J Microbiol 40:808–817

Brissos V, Pereira L, Munteanu F-D, Cavaco-Paulo A, Martins LO (2009) Expression system of CotA-laccase for directed evolution and high-throughput screenings for the oxidation of high redox potential dyes. J Biotechnol 4:1–6, doi: 10.1002/biot.200800248

Brown D, Laboureur P (1983) The aerobic biodegradability of primary aromatic amines. Chemosphere 12:405–414

Brown PA, Gill SA, Alle SJ (2000) Metal removal from wastewater using peat. Water Res 34 (16):3907–3916

Cameron MD, Timofeevski S, Aust SD (2000) Enzymology of *Phanerochaete chrysosporium* with respect to the degradation of recalcitrant compounds and xenobiotics. Appl Microbiol Biotechnol 54:751–758

Campos R, Kandelbauer A, Robra KH, Cavaco-Paulo A, Gübitz GM (2001) Indigo degradation with purified laccases from *Trametes hirsuta* and *Sclerotium rolfsii*. J Biotechnol 89:131–139

Chowdhury AN, Jesmeen SR, Hossain MM (2004) Removal of dyes from water by conducting polymeric adsorbent. Polym Adv Technol 15:633–638

Chang JS, Kuo TS (2000) Kinetics of bacterial decolorization of azo dye with *Escherichia coli* NO3. Bioresour Technol 75:107–11

Chang JS, Kuo TS, Chao YP, Ho JY, Lin PJ (2000) Azo dye decolorization with a mutant *Escherichia coli strain*. Biotechnol Lett 22:807–812

Chatterjee S, Chatterjee S, Chatterjee BP, Das AR, Guha AK (2005) Adsorption of a model anionic dye, eosin Y, from aqueous solution by chitosan hydrobeads. J Colloid Interface Sci 288(1):30–35

Chen H (2006) Recent advances in azo dye degrading enzyme research. Curr Protein Peptide Sci 7:101–111

Chiou M-S, Li H-Y (2002) Equilibrium and kinetic modeling of adsorption of reactive dye on cross-linked chitosan beads. J Hazard Mater 3:233–248

Chivukula M, Renganathan V (1995) Phenolic azo dye oxidation by laccase from *Pyricularia oryzae*. Appl Environ Microbiol 61:4374–4377

Claus H (2003) Laccases and their occurrence in prokaryotes. Arch Microbiol 179:145–150

Conlon M, Khraisheh M (2002) Bioadsorption process for the removal of colour from textile effluent. GB Patent WO0242228

Couto SR (2009) Dye removal by immobilised fungi. Biotechnol Adv 27:227–235, doi:10.1016/j.biotechadv.2008.12.001

Couto SR, Rivela I, Muñoz MR, Sanromán A (2000) Ligninolytic enzymes production and the ability of decolourisation of Poly R-478 in packed-bed bioreactors by *Phanerochaete chrysosporiu*. Bioprocess Eng 23:287–93

Crini G (2006) Non-conventional low-cost adsorbents for dye removal: A review. Biores Technol 97:1061–1085, doi:10.1016/j.biortech.2005.05.001

Dave SR, Dave RH (2009) Isolation and characterization of *Bacillus thuringiensis* for Acid red 119 dye Decolourisation. Biores Technol 100:49–253, doi:10.1016/j.biortech.2008.05.019

Delée W, O'Neill C, Hawkes FR, Pinheiro HM (1998) Anaerobic treatment of textile effluents: a review. J Chem Technol Biotechnol 73:323–335

Dhodapkar R, Rao NN, Pande SP, Kaul SN (2006) Removal of basic dyes from aqueous medium using a novel polymer: Jalshakti. Biores Technol 97:877–885, doi:10.1016/j.biortech.2005.04.033

Dominguez JR, Beltran J, Rodriguez O (2005) Vis and UV photocatalytic detoxification methods (using TiO2, TiO2/ H2O2, TiO2/O3, TiO2/S2O8–2, O3, H2O2, S2O8–2, Fe3+/H2O2 and Fe3+/H2O2/C2O4–2) for dyes treatment. Catalysis Today 101:389–395

Donmez G (2002) Bioaccumulation of the reactive textile dyes by *C. tropicalis* growing in molasses medium. Enzyme Microb Technol 30:363–366

Dos Santos AB, Madrid MP, Stams AJM, Van Lier JB, Cervantes FJ (2005) Azo dye reduction by mesophilic and thermophilic anaerobic consortia. Biotechnol Prog 21:1140–1145

Dryden CE, Kay WB (1954) Kinetics of batch adsorption and desorption. Ind Eng Chem 46:2294–2297

Durán N, Esposito E (2000) Potential applications of oxidative enzymes and phenoloxidase-like compounds in wastewater and soil treatment: a review. Appl Catal B: Environ 28:83–99

4 Dyes—Environmental Impact and Remediation

Eggert C, Temp U, Eriksson KEL (1996) Laccase-producing white-rot-fungus lacking lignin peroxidase and manganese peroxidase. In: Jefries TW, Viikari L (eds) ACS Symposium Series 655. Enzymes for Pulp and Paper Processing, Washington, DC, pp 130–150

Erdem E, Çölgeçen G, Donat R (2005) The removal of textile dyes by diatomite earth. J Coll Interf Sci 282:314–319, doi:10.1016/j.jcis.2004.08.166

Esteves MF, Silva JD (2004) Electrochemical degradation of reactive blue 19 dye in textile wastewater, 4th World Textile Conference AUTEX 2004, Roubaix, June , 2004

Faria PCC, Órfão JJM, Pereira MFR (2004) Adsorption of anionic and cationic dyes on activated carbons with different surface chemistries. Water Res 38(8):2043–2052

Faria PCC, Orfão JJM, Pereira MFR (2005) Mineralisation of coloured aqueous solutions by ozonation in the presence of activated carbon. Water Res 39:1461–1470

Faria PCC, Orfão JJM, Figueiredo JL, Pereira MFR (2008) Adsorption of aromatic compounds from the biodegradation of azo dyes on activated carbon. Appl Surface Sci 254:3497–3503

Figueiredo JL, Pereira MFR, Freitas MMA, Orfão JJM (1999) Modification of the surface chemistry of activated carbons. Carbon 37:1379–1389

Forgacs E, Cserháti T, Oros G (2004) Removal of synthetic dyes from wastewaters: a review. Environt Int 30:953–971, doi:10.1016/j.envint. 2004.02.001

Freundlich HZ (1906) Over the adsorption in solution. J Phys Chem 57:384–470

Fu Y, Viraraghavan T (2001) Fungal decolorization of dye wastewaters: a review. Biores Technol 79:51–262

Fu Y, Viraraghavan T (2002) Removal of Congo red from an aqueous solution by fungus *Aspergillus niger*. Adv Environ Res 7(1):239–247

Gadd GM (2009) Biosorption: critical review of scientific rationale, environmental importance and significance for pollution treatment. J Chem Technol Biotechnol 84:3–28, doi: 10.1002/jctb.1999

Gianfreda L, Xu F, Bollag J-M (1999) Laccases: a useful group of oxidoreductive enzymes. Bioremediation J 3:1–25

Glenn JK, Gold MH (1983) Decolorization of several polymeric dyes by the lignin-degrading basidiomycete *Phanerochaete chrysosporium*. Appl Environ Microbiol 45:1741–1747

Gomes JR (2001) Estrutura e Propriedades dos Corantes. Barbosa e Xavier Lda. Braga, Portugal

Goszczynski S, Paszczynski A, Pasti-Grigsby MB, Crawford RL, Crawford DL (1994) New pathway for degradation of sulfonated azo dyes by microbial peroxidases of *Phanerochaete chrysosporium* and *Streptomyces chromofuscust*. J Bacteriol 176:1339–1347

Grigorious D, Aline E, Patrick P, René B (2000) Purification and characterisation of the first bacterial laccase in the Rhizospheric bacterium *Azospirillum lipoferum*. Soil Biol Biochem 32:919–927

Gupta VK, Suhas (2009) Application of low-cost adsorbents for dye removal—a review. J Environ Manag 90:2313–2342, doi:10.1016/j.jenvman.2008.11.017

Hai FI, Yamamoto K, Fukushi K (2007) Hybrid treatment systems for dye wastewater. Crit Rev Environ Sci Technol 37:315–377

Hao OJ, Phull KK, Chen JM (1994) Wet oxidation of TNT red water and bacterial toxicity of treated water. Water Res 28:283–290

Hao OJ, Kim H, Chiang P-C (2000) Decolorization of watewater. Crit Rev Environ Sci Technol 30:449–505, doi.org/10.1080/10643380091184237

Hao J, Song F, Huang F, Yang C, Zhang Z, Zheng Y, Tian X (2007) Production of laccase by a newly isolated deuteromycete fungus *Pestalotiopsis* sp. and its decolorization of azo dye. J Indust Microbiol Biotechnol 34(3):233–240

Horistsu H, Takada M, Idaka E, Tomoyda M, Ogewa T (1977) Degradation of P-amino azo benzene by *Bacillus subtilis*. Eur J Appl Microbiol 4:217–224

Husain Q (2006) Potential applications of the oxidoreductive enzymes in the decolourisation and detoxification of textile and other synthetic dyes from polluted water: a review. Crit Rev Biotechnol 26:201–221, doi: org/10.1080/07388550600969936

Idaka E, Ogewa Y (1978) Degradation of azo compounds by *Aeromonas hydraphilia* var. 2413. J Soc Dye Colorist 94:91–94

Jin X, Liu G, Xu Z, Yao W (2007) Decolorization of a dye industry effluent by *Aspergillus fumigatus* XC6. Appl Microbiol Biotechnol 74:239–243

Jovančić P, Radetić M (2008) Advanced sorbent materials for treatment of wastewaters. Hdb Environ Chem 5:239–264, doi: 10.1007/698_5_097

Juang R, Tseng R, Wu F, Lee S (1997) Adsorption behaviour of reactive dyes from aqueous solution on chitosan. J Chem Technol Biotechnol 70:391–399

Kandelbauer A, Guebitz GM (2005) Bioremediation for the decolorization of textile dyes, a review. In: Lichtfouse E, Schwarzbauer J, Robert D (eds) Environmental chemistry. Springer-Verlag, Heidelberg, pp 269–288

Kandelbauer A, Maute O, Kessler RW, Erlacher A, Gübitz GM (2004) Study of dye decolorization in an immobilized laccase enzyme-reactor using online spectroscopy. Biotechnol Bioeng 87:552–563

Kalyani DC, Telke AA, Dhanve RS, Jadhav JP (2009) Ecofriendly biodegradation and detoxification of Reactive Red 2 textile dye by newly isolated *Pseudomonas* sp. SUK1. J Hazard Mater 163:35–742

Kannan C, Sundaram T, Palvannan T (2008) Environmentally stable adsorbent of tetrahedral silica and non-tetrahedral alumina for removal and recovery of malachite green dye from aqueous solution. J Hazard Mater 157:137–145, doi:10.1016/j.jhazmat.2007.12.116

Kapdan IK, Kargi F (2002) Biological decolorization of textile dyestuff containing wastewater by *Coriolus versicolor* in a rotating biological contactor. Enzyme Microbiol Technol 30:195–199

Karaca S, Gürses A, Bayrak R (2004) Effect of some pretreatments on the adsorption of methylene blue by Balkaya lignite. Energy Convers Manag 45:1693–1704

Kaushik P, Malik A (2009) Fungal dye decolorization: recent advances and future potential. Environ Int 35:127–141

Keck A, Klein J, Kudlich M, Stolz A, Knackmuss HJ, Mattes R (1997) Reduction of azo dyes by redox mediators originating in the naphthalenesulfonic acid degradation pathway of *Sphingomonas* sp. strain BN6. Appl Environ Microbiol 63:3684–3690

Keck A, Rau J, Reemtsma T, Mattes R, Stolz A, Klein J (2002) Identification of quinoide redox mediators that are formed during the degradation of naphthalene-2-sulfonate by *Sphingomonas xenophaga* BN6. Appl Environ Microbiol 68:4341–4349

Khadijah O, Lee KK, Abdullah MFF (2009) Isolation, screening and development of local bacterial consortia with azo dyes decolourising capability. Malaysian J Microbiol 5(1):25–32

Khehra MS, Saini HS, Sharma DK, Chadha BS, Chimni SS (2005) Decolorization of various azo dyes by bacterial consortium. Dyes Pigments 67:55–61

Kim T-H, Park C, Leec J, Shin E-B, Kim S (2002a) Pilot scale treatment of textile wastewater by combined process (fluidized biofilm process—chemical coagulation—electrochemical oxidation). Water Res 36:3979–3988

Kim T-H, Park C, Shin E-B, Kim S (2002b) Decolorization of disperse and reactive dyes by continuous electrocoagulation process. Desalination 150:165–175

Klein KL, Melechko AV, McKnight TE, Retterer ST, Rack PD, Fowlkes JD, Joy DC, Simpson ML (2008) Surface characterization and functionalization of carbon nanofibers. J Appl Phys 103:1–23

Kudlich M, Keck A, Klein J, Stolz A (1997) Localization of the enzyme system involved in anaerobic reduction of azo dyes by *Sphingomonas* sp. Strain BN6 and effect of artificial redox mediators on the rate of azo dye reduction. Appl Environ Microbiol 63:3691–3694

Kudlich M, Hetheridge MJ, Knackmuss H-J, Stolz A (1999) Autooxidation reactions of different aromatic *ortho*-aminohydroxynaphthalenes which are formed during the anaerobic reduction of sulfonated azo dyes. Environ Sci Technol 33:896–901, doi: 10.1021/es9808346

Kuhad RC, Sood N, Tripathi KK, Singh A, Ward OP (2004) Developments in microbial methods for the treatment of dye effluents. Adv Appl Microbiol 56:185–213

Langmuir I (1918) The adsorption of gases on plane surfaces of gases, mica and platinium. J Am Chem Soc 40:1361–1403

Liapis AI (1987) Fundamentals of adsorption. Engineering Foundation, New York

4 Dyes—Environmental Impact and Remediation 159

López C, Valade AG, Combourieu B, Mielgo I, Bouchon B, Lema JM (2004) Mechanism of enzymatic degradation of the azo dye Orange II determined by ex situ 1H nuclear magnetic resonance and electrospray ionization-ion trap mass spectrometry. Anal Biochem 335:135–149

Lucas M, Mertens V, Corbisier A-M, Vanhulle S (2008) Synthetic dyes decolourisation by white-rot fungi: development of original microtitre plate method and screening. Enzyme Microbial Technol 42:97–106

Mazmanci MA (2010) Decolorization of azo dyes by immobilized fungi. In: Atacag Erkurt H (ed) Biodegradation of azo dyes. Hdb Env Chem Springer/Verlag Berlin, Heidelberg, 9 pp 169–181, doi: 10.1007/698_2009_51

McKay G (1998) Application of surface diffusion model to the adsorption of dyes on bagasse pith. Adsorption 4:361–372

McKay G, Sweeney AG (1980) Principles of dye removal from textile effluent. Water Air Soil Pollut 14:3–11

Mckay G, Porter JF, Prasad GR (1999) The removal of dye colours from aqueous solutions by adsorption on low-cost materials. Water Air Soil Pollut 114:423–438

McMullan G, Meehan C, Conneely A, Kirby N, Robinson T, Nigam P, Banat IM, Marchant R, Smyth WF (2001) Microbial decolourisation and degradation of textile dyes. Appl Microbiol Biotechnol 56:81–87, doi: 10.1007/s002530000587

Malik PK (2004) Dye removal from wastewater using activated carbon developed from sawdust: adsorption equilibrium and kinetics. J Hazard Mater B 113:81–88

Martinez MB, Flickinger MC, Nelsestuen GL (1999) Stady-state enzyme kinetics in the *Escherichia coli* perisplam: a model of a whole cell biocatalyst. J Biotechnol 71:59–66

Martins MA, Cardoso MH, Queiroz MJ, Ramalho MT, Oliveira-Campos AM (1999) Biodegradation of azo dyes by the yeast *Candida zeylanoides* in batch aerated cultures. Chemosphere 38:2455–2460

Mathur N, Bhatnagar P, Bakre P (2003) Assessing mutagenicity of textile dyes from Pali(Rajasthan) using Ames bioassay. Appl Ecol Environ Res 4(1):111–118

Melgoza RM, Cruz A, Buitron G (2004) Anaerobic/aerobic treatment of colorants present in textile effluents. Water Sci Technol 50:149–155

Meehan C, Banat IM, McMullan G, Nigam P, Smyth F, Marchant R (2000) Decolorization of remazol black-B using a thermotolerant yeast, *K. marxianus* IMB3. Environ Int 26:75–79

Mohan D, Singh KP, Singh G, Kumar K (2002) Removal of dyes from wastewater using flyash, a low-cost adsorbent. Ind Eng Chem Res 41:3688–3695

Mondal S (2008) Methods of dye removal from dye house effluent—an overview. Environ Eng Sci 25:383–396, doi: 10.1089/ees.2007.0049

Moreno-Castilla C, López-Ramón MV, Carrasco-Marín F (2000) Changes in surface chemistry of activated carbons by wet Oxidation. Carbon 38:1995–2001

Nachiyar CV, Rajakumar GS (2005) Purification and characterization of an oxygen insensitive azoreductase from *Pseudomonas aeruginosa*. Enzyme Microb Technol 36:503–509

Nigam P, McMullan G, Banat IM, Marchant R (1996) Decolourisation of effluent from the textile industry by a microbial consortium. Biotechnol Lett 18:117–120

Nigam P, Armour G, Banat IM, Singh D, Marchant R (2000) Physical removal of textile dyes and solid state fermentation of dye adsorbed agricultural residues. Bioresour Technol 72:219–226

Novotný C, Svobodová K, Kasinath A, Erbanová P (2004) Biodegradation of synthetic dyes by *Irpex lacteus* under various growth conditions. Int Biodet Biodeg 54:215–223, doi: 10.1016/j.ibiod.2004.06.003

Ohkuma M, Maeda Y, Johjima T, Kudo T (2001) Lignin degradation and roles of white rot fungi: Study on an efficient symbiotic system in fungus-growing termites and its application to bioremediation. Riken Review, Focused Ecomol Sci Res 42:39–42

O'Neill C, Hawkes FR, Hawkes DL, Lourenço ND, Pinheiro HM, Delée W (1999) Colour in textile effluents-sources, measurement, discharge consents and simulation: a review. J Chem Technol Biotechnol 74:1009–1018

Ozdemir O, Armagan B, Turan M, Celik MS (2004) Comparison of the adsorption characteristics of azo-reactive dyes on mezoporous minerals. Dyes Pigments 62:49–60

Padmesh TVN, Vijayaraghavan K, Sekaran G, Velan M (2005) Batch and column studies on biosorption of acid dyes on fresh water macro alga *Azolla filiculoides*. J Hazard Mater 125:121–129, doi: 10.1016/j.jhazmat.2005.05.014

Palmieri G, Cennamo G, Sannia G (2005) Remazol Brilliant Blue R decolourization by the fungus *Pleurotus ostreatus* and its oxidative enzymatic system. Enzyme Microbial Technol 36:17–24

Pandey A, Singh P, Iyengar L (2007) Bacterial decolorization and degradation of azo dyes. Int Biodet Biodeg 59:73–84

Pearce CI, Lloyd JR, Guthrie JT (2006) The removal of colour from textile wastewater using whole bacterial cells: a review. Dyes Pigments 58:179–196

Pereira MFR, Soares SF, Orfão JJM, Figueiredo JL (2003) Adsorption of dyes on activated carbons: influence of surface chemical groups. Carbon 41:811–821

Pereira L, Coelho AV, Viegas CA, Santos MMC, Robalo MP, Martins LO (2009a) Enzymatic biotransformation of the azo dye Sudan Orange G with bacterial CotA-laccase. J Biotechnol 139:68–77, doi:10.1016/j.jbiotec.2008.09.001

Pereira L, Coelho AV, Viegas CA, Ganachaud C, Iacazio G, Tron T, Robalo MP, Martins LO (2009b) On the mechanism of biotransformation of the anthraquinonic dye acid blue 62 by laccases. Adv Synt Catal 351:1857–1865, doi: 10.1002/adsc.200900271

Pereira L, Pereira R, Pereira MFR, Van der Zee FP, Cervantes FJ, Alves MM (2010) Thermal modification of activated carbon surface chemistry improves its capacity as redox mediator for azo dye reduction. J Hazard Mater 183:931–939

Pinheiro HM, Touraud E, Thomas O (2004) Aromatic amines from azo dye reduction: status review with emphasis on direct UV spectrophotometric detection in textile industry wastewaters. Dyes Pigments 61:121–139

Podgornik H, Poljansek I, Perdih A (2001) Transformation of indigo carmine by *Phanerochaete chrysosporium* ligninolytic enzymes. Enzyme Microb Technol 29:166–172

Prigione V, Varese GC, Casieri L, Marchisio VF (2008a) Biosorption of simulated dyed effluents by inactivated fungal biomasses. Bioresour Technol 99:3559–3567

Prigione V, Tigini V, Pezzella C, Anastasi A, Sannia G, Varese GC (2008b) Decolorisation and detoxification of textile effluens by fungal biosorption. Water Res 42:2911–2920

Puvaneswari N, Muthukrishnan J, Gunasekaran P (2006) Toxicity assessment and microbial degradation of azo dyes. Indian J Exp Biol 44:618–626

Ramalho PA, Scholze H, Cardoso MH, Ramalho MT, Oliveira-Campos AM (2002) Improved conditions for the aerobic reductive decolorisation of azo dyes by *Candida zeylanoides*. Enzyme Microb Technol 31:848–854

Ramalho PA, Cardoso MH, Cavaco-Paulo A, Ramalho MT (2004) Characterization of azo reduction activity in a novel ascomycete yeast strain. Appl Environ Microbiol 70:2279–2288

Ramalho PA, Saraiva S, Cavaco-Paulo A, Casal M, Cardoso MH, Ramalho MT (2005) Azo reductase activity of intact *Saccharomyces cerevisiae* cells is dependent on the fre1p component of plasma membrane ferric reductase. Appl Environ Microbiol 71:3882–3888

Rau J, Knackmuss HJ, Stolz A (2002) Effects of different quinoid redox mediators on the anaerobic reduction of azo dyes by bacteria. Environ Sci Technol 36:1497–1504

Rau J, Stolz A (2003) Oxygen-insensitive nitroreductases NfsA and NfsB of *Escherichia coli* function under anaerobic conditions as lawsone-dependent azo reductase. Appl Environ Microbiol 69:3448–3455, doi: 10.1128/AEM.69.6.3448-3455.2003

Rauf MA, Ashraf SS (2009) Fundamental principles and application of heterogeneous photocatalytic degradation of dyes in solution. Chem Eng J 151:10–18, doi:10.1016/j.cej.2009.02.026

Razo-Flores E, Donlon BA, Field JA, Lettinga G (1996) Biodegradability of N-substituted aromatics and alkylphenols under methanogenic conditions using granular Sludge. Water Sci Technol 33:47–57

Razo-Flores E, Luijaten M, Donlon B, Lettinga G, Field A (1997) Complete biodegradation of the azo dye azodisalicylate under anaerobic conditions. Environ Sci Technol 31:2098–2103

Redlich O, Peterson DL (1959) A useful adsorption isotherm. J Phys Chem 63:1024–1026

Riva S (2006) Laccases: blue enzymes for green chemistry. Trends Biotechnol 24 (5):219–226, doi:10.1016/j.tibtech.2006.03.006

4 Dyes—Environmental Impact and Remediation 161

Robinson T, McMullan G, Marchant R, Nigam P (2001) Remediation of dyes in textile effluent: a critical review on current treatment technologies with a proposed alternative. Biores Technol 77:247–255

Rodriguez-Reinoso F, Martin-Martinez JM, Prado-Burguete C, McEnaney B (1987) A standard adsorption isotherm for the characterization of activated carbons. J Phys Chem 91:515–516

Rodriguez-Reinoso F (1998) The role of carbon materials in the heterogeneous catalysis. Carbon 36 (3):159–175

Roessler A, Dossenbach O, Meyer U, Marte W, Rys P (2001) Direct electrochemical reduction of indigo. Chimia 55:879–882

Rus R, Rau J, Stolz A (2000) The function of cytoplasmatic reductases in the reduction of azo dyes by bacteria. Appl Environm Microbiol 66:1429–1434

Schrank SG, dos Santos JN, Souza DS, Souza EES (2007) Decolourisation effects of Vat Green 01 textile dye and textile wastewater using H2O2/UV process. J Photochem Photobiol A: Chem 186:125–129, doi:10.1016/j.jphotochem.2006.08.001

Schroeder M, Pereira L, Couto SR, Erlacher A, Schoening K-U, Cavaco-Paulo A, Guebitz GM (2007) Enzymatic synthesis of Tinuvin. Enzyme Microb Technol 40:1748–1752, doi:10.1016/j.enzmictec.2006.11.026

Shrestha S, Kazama F (2007) Assessment of surface water quality using multivariate statistical techniques: a case study of the Fuji river basin, Japan. Environ Model Software 22:464–475

Shen W, Li Z, Liu Y (2008) Surface chemical functional groups modification of porous carbon. Recent Patents Chem Eng 1:27–40

Soares GMB, Miranda TM, Oliveira-campos AMF, Hrdina R, Costa-Ferreira M, Amorim MTP (2004) Current situation and future perspectives for textile effluent decolourisation. Paper presented at the Textile Institute 83rd World Conference, Shanghai, China, 23–7 May , 2004

Society of Dyes and Colourist (1984) Colour Index, vol. 1, 3rd ed., S.D.C., A. Ass. Tex. Chem. Colour

Solomon EI, Sundaram UM, Machonkin TE (1996) Multicopper oxidases and oxygenases. Chem Rev 96:2563–2605

Sponza DT, Işik M (2005) Reactor performance and fate of aromatic amines through decolorization of Direct Black 38 dye under anaerobic/aerobic sequentilas. Process Biochem 40:35–44

Stoj CS, Kosman DJ (2005) Copper proteins: Oxidases. In: King RB (ed) Encyclopedia of inorganic chemistry, Vol II, 2nd edn. John Wiley and Sons, New York, pp 1134–1159

Steffan S, Bardi L, Marzona M (2005) Azo dye biodegradation by microbial cultures immobilized in alginate beads. Environ Int 31:201–205, doi:10.1016/j.envint.2004.09.016

Stolz A (2001) Basic and applied aspects in the microbial degradation of azo dyes. Appl Microbiol Biotechnol 56:69–80, doi: 10.1007/s002530100686

Tan NCG, Field JA (2000) Biodegradation of sulfonated aromatic compounds. In: Environmental Technologies to Treat Sulfur Pollution Principles and Engineering. IWA Publishing, London, pp 377–392

Tan NCG, van Leeuwen A, van Voorthuinzen EM, Slenders P, Prenafeta-Boldú FX, Temmink H, Lettinga G, Field JA (2005) Fate and biodegradability of sulfonated aromatic amines. Biodegradation 16:527–537

Tien C (1961) Adsorption kinetics of a non-flow system with nonlinear equilibrium relationship. Aiche J 7:410–419

Tsang DCW, Hu J, Liu MY, Zhang W, Lai KCK, Lo IMC (2007) Activated carbon produced from waste wood pallets: adsorption of three classes of dyes. Water Air Soil Pollut 184:141–155, doi:.1007/s11270-007-9404-2

Tüfekci N, Sivril N, Toroz I (2007) Pollutants of textile industry wastewater and assessment of its discharge limits by water quality standards. Turkish J Fish Aquatic Sci 7:97–103

Van der Zee FP, Bouwman RHM, Strik DPBTB, Lettinga G, Field JA (2001) Application of redox mediators to accelerate the transformation of reactive azo dyes in anaerobic bioreactors. Biotechnol Bioeng 75:691–701

Van der Zee FP, Villaverde S (2005) Combined anaerobic-aerobic treatment of azo dyes—a short review of bioreactors studies. Water Res 39:1425–1440

Van der Zee FP, Cervantes FJ (2009) Impact and application of electron shuttles on the redox (bio) transformation of contaminants: a review. Biotechnol Adv 27:256–277

Vandevivere PC, Bianchi R, Verstraete W (1998) Treatment and reuse of wastewater from the textile wet-processing industry: review of emerging technologies. J Chem Technol Biotechnol 72:289–302

Vanhulle S, Enaud E, Trovaslet M, Billottet L, Kneipe L, Jiwan J-LH, Corbisier A-M, Marchand-Brynaert J (2008) Coupling occurs before breakdown during biotransformation of acid blue 62 by white rot fungi. Chemosphere 70:1097–1107, doi:10.1016/j.chemosphere.2007.07.069

Walker GM, Weatherley LR (1999) Biological activated carbon treatment of industrial wastewater in stirred tank reactors. Chem Eng J 75:201–206

Wang S, Wu H (2006) Environmental-benign utilisation of fly ash as low-cost adsorbents. J Hazard Mater B136:482–501, doi:10.1016/j.jhazmat.2006.01.067

Wang XJ, Gu XY, Lin DX, Dong F, Wan XF (2007) Treatment of acid rose dye containing wastewater by ozonizing-biological aerated filter. Dyes Pigments 74(3):736

Weber Jr WJ, Rummer RR (1965) Intraparticle transport of sulphonated alkylbenzenes in a porous solid: diffusion and non-linear adsorption. Water Resour Res 1:361–365

Weber WJ, Pirbazari M, Melson GL (1978) Biological growth on activated carbon: an investigation by scanning electron microscopy. Environ Sci Tech 127:817–819

Wesenberg D, Kyriakides I, Agathos SN (2003) White-rot fungi and their enzymes for the treatment of industrial dye effluents. Biotechnol Adv 22:161–187, doi:10.1016/j.biotechadv.2003.08.011

Wuhrmann K, Mechsner KI, Kappeler TH (1980) Investigation on rate determining factors in the microbial reduction of azo dyes. Environ J Appl Microbiol 9:325–338

Xu Y (2001) Comparative studies of the Fe3+/2+UV, H2O2-UV, TiO2-UV vis system for the decolorization of a textile dye X-3B in water. Chemosphere 43:1103–1107

Yoshida H, Fukuda S, Okamoto A, Kataoka T (1991) Recovery of direct dye and acid dye by adsorption on chitosan fiber—equilibria. Wat Sci Technol 23(7/9):1667–1676

Yoshida H, Okamoto A, Kataoka T (1993) Adsorption of acid dye on cross-linked chitosan fibers: equilibria. Chem Eng Sci 48:2267–2272

Yoshida H, Takemori T (1997) Adsorption of direct dye on cross-linked chitosan fiber: breakthrough curve. Water Sci Technol 35:29–37

Young H, Yu J (1997) Ligninax catalyzed decolorization of synthetic dyes. Water Res 31(5):1187–1193

Yu Z, Wen X (2005) Screening and identification of yeasts for decolorizing synthetic dyes in industrial wastewater. Int Biodet Biod 56:109–114

Zeroual Y, Kim BS, Kim CS, Blaghen M, Lee KM (2006) A comparative study on biosorption characteristics of certain fungi for bromophenol blue dye. Appl Biochem Biotechnol 134:51–60

Zhang FM, Knapp JS, Tapley KN (1999) Development of bioreactor systems for decolorization of orange II using white-rot fungus. Enzyme Microb Technol 24:48–53

Zhao X, Hardin I (2007) HPLC and spectrophotometric analysis of biodegradation of azo dyes by *Pleurotus ostreatus*. Dyes Pigments 73:322–325

Zhou XF, Zhang YL, Xu DQ, Cao WH, Dai CM, Qiang ZM, Yang Z, Zhao JF (2008) Treatment of succinonitrile wastewater by immobilized high efficiency microorganism strains. Water Sci Technol 58:911–918

Zhou M, He J (2008) Degradation of azo dye by three clean advanced oxidation processes: wet oxidation, electrochemical oxidation and wet electrochemical oxidation-A comparative study. Electrochimica Acta 53:1902–1910

Zille A, Ramalho P, Tzanov T, Millward R, Aires V, Cardoso MH, Ramalho MT, Gübitz GM, Cavaco-Paulo A (2004) Predicting dye biodegradation from redox potentials. Biotechnol Prog 20:1588–1592

Zille A, Górnacka B, Rehorek A, Cavaco-Paulo A (2005) Degradation of Azo Dyes by *Trametes villosa* Laccase under long time oxidative conditions. Appl Environ Microbiol 71:6711–6718

Zimmermann T, Kulla HG, Leisinger T (1982) Properties of purified Orange II azoreductase, the enzyme initiating azo dye degradation by Pseudomonas KF46. Eur J Biochem 129:197–203

Zollinger H (1987) Colour chemistry—synthesis, properties and applications of organic dyes and pigments. VCH, New York, p 92

Chapter 5
Molecular Detection of Resistance and Transfer Genes in Environmental Samples

Elisabeth Grohmann and Karsten Arends

Contents

5.1	Introduction	164
5.2	Molecular Detection Methods	165
	5.2.1 DNA Hybridization	166
	5.2.2 PCR (Simple and Multiplex PCR)	166
	5.2.3 Quantitative PCR	166
	5.2.4 DNA Microarray	167
5.3	Antibiotic Resistance	167
	5.3.1 Mechanisms of Antibiotic Resistance	168
	5.3.2 Antibiotic Resistance Genes in the Environment	169
5.4	Heavy Metal Resistance	176
	5.4.1 Mechanisms of Heavy Metal Resistance	176
	5.4.2 Heavy Metal Resistance Genes in the Environment	177
5.5	Key Transfer Genes	180
	5.5.1 Detection of Transfer Genes in the Environment	181
5.6	Conclusions	184
5.7	Perspectives	184
References		185

Abstract Horizontal plasmid transfer is the most important means of spreading resistance to antibiotics and heavy metals, as well as virulence genes, to closely and remotely related microorganisms thereby increasing the horizontal gene pool in so diverse habitats such as soils, wastewater, aquifer recharge systems and glacier ice. An overview about the currently used molecular tools to detect and quantify the abundance of antibiotic and heavy metal resistance and transfer genes in aquatic and terrestrial environments is provided. Habitats studied range from nutrient rich environments such as manured agricultural soils to oligotrophic habitats such as drinking water or glaciers in the Antarctic. The state of the art in antibiotic and heavy metal resistance mechanisms and monitoring of conjugative transfer factors

E. Grohmann (✉)
Department of Infectious Diseases, University Hospital Freiburg, Hugstetter Strasse 55,
79106 Freiburg, Germany
e-mail: elisabeth.grohmann@uniklinik-freiburg.de

A. Malik, E. Grohmann (eds.), *Environmental Protection Strategies for Sustainable Development*, Strategies for Sustainability,
DOI 10.1007/978-94-007-1591-2_5, © Springer Science+Business Media B.V. 2012

to assess the transmissibility of the resistance factors is summarized. The chapter ends with perspectives on emerging new molecular monitoring tools suited to rapid, reliable and high throughput analysis of environmental samples differing in origin and level of pollution.

Keywords Antibiotic resistance • Heavy metal resistance • Plasmid transfer • Contamination • PCR • Microarray • Terrestrial habitats • Aquatic habitats

5.1 Introduction

Antibiotics are widely used to protect the health of humans and animals or to increase the growth rate of animals in livestock breeding as food additives. The majority of antibiotics are released unchanged into the environment after application in humans and animals. Therefore, concerns about the potential impact of antibiotic residues in the environment keep growing (Sarmah et al. 2006; Wright 2007; Kemper 2008; Zhang et al. 2009). In surface water it is difficult to find an area where antibiotics are not detected, except for pristine sites in the mountains before the rivers enter urban or agricultural areas (Yang and Carlson 2003). Some antibiotics can even be found in groundwater deeper than 10 m (Batt et al. 2006). Apart from chemical pollution of the environment by antibiotics, the application of antibiotics may also accelerate the development of antibiotic resistant bacteria presenting a health threat to humans and animals. These bacteria might be transmitted from the environment to humans via direct or indirect contact (Iversen et al 2004; Kim et al. 2005; Zhang et al. 2009). Considering the increasing evidence that clinical resistance is closely linked to environmental Antibiotic Resistance Genes (ARG) and bacteria (e.g. Prabhu et al. 2007; Abriouel et al. 2008) it is clear that research activities have to be extended to put more emphasis on the study of non pathogenic and indigenous bacteria (Zhang et al. 2009) to understand the mechanisms of resistance development and spread in the environment.

Mercury (Hg) is the most toxic of the heavy metals and it has no biological function described up to now (Nies 1999). Hg is not abundant in the soil crust, but can be found in ores, the dominant being cinnabar (HgS) (Oregaard and Sørensen 2007). Hg is emitted to the atmosphere by natural sources, e.g. soil erosion and volcanic eruptions and by anthropogenic sources, e.g. coal-fuelled power plants (Nriagu and Pacyna 1988; Oregaard and Sørensen 2007). Hg resistance is found in both *Bacteria* (Barkay et al. 2003) and *Archaea* (Schelert et al. 2004). Hg resistance in *Bacteria* is encoded by the *mer* operon (Oregaard and Sørensen 2007). Hg resistance genes have provided a model system for a wide range of molecular ecological studies, such as investigations of gene diversity and evolution within bacterial communities in the environment (e.g. Barkay et al. 1989; Rochelle et al. 1991; Bruce 1997).

Copper (Cu) is an essential micronutrient for eukaryotes and prokaryotes but it is also a toxic element in high concentrations, leading to detrimental effects on the metabolism of the cell (Brown et al. 1992; Lejon et al. 2007). Cu appears to be widespread in the environment, in particular in agricultural soils, due to applica-

5 Molecular Detection of Resistance and Transfer Genes in Environmental Samples

tion of copper-based pesticides to control plant diseases, or because of its presence as a contaminant in organic and mineral amendments (Tom-Petersen et al. 2003; Lejon et al. 2007). Cu resistance is displayed by several species belonging to the Gram-negative bacteria. In most cases, Cu resistance is conferred by *copA* or *copA* homologous resistance genes (e.g. Rensing and Grass 2003; Lejon et al. 2007).

Arsenic (As) is considered to be a semimetal with metallic and non-metallic properties. It is not only toxic to bacteria, but also to other domains of life. As is present in diverse environments, it is released either by natural weathering of rocks or by anthropogenic processes, e.g., by mining industries and agricultural practices (Pechrada et al. 2010). Operons coding for As resistance genes (*ars*) have been detected on the chromosomes and on conjugative plasmids in a variety of Gram-positive and Gram-negative bacteria (Pechrada et al. 2010).

Heavy metal resistance genes and ARGs are readily exchanged within bacterial communities by conjugative plasmids, or by mobilizable elements horizontally transferred by the help of self-transmissible Mobile Genetic Elements (MGE). Key protein factors of horizontal plasmid transfer have been explored; they include, depending on the origin of the MGE, two or three conjugative ATPases and a relaxase, the enzyme that initiates conjugative transfer by a local enzymatic attack at the origin of transfer (*oriT*) of the conjugative plasmid. These key transfer genes are required for horizontal transmission of virulence genes, ARGs and heavy metal resistance genes.

In the recent decades many valuable molecular tools have been developed which enable to detect, and if desired to quantify, the presence of resistance genes in virtually all environments without prior cultivation of the bacteria. These techniques present powerful tools for the assessment of the environmental pollution by emerging contaminants such as ARGs. The presence of key transfer factors in the habitats contaminated by ARGs and/or heavy metal resistance genes indicates the horizontal transmissibility of these resistance factors in the environment.

The current knowledge in the presence of resistance factors in diverse aquatic and terrestrial environments and their transmissibility within the microbial community is explored. Emphasis will be placed on molecular detection tools currently used and emerging tools with perspectives in the near future. Perspectives on molecular fingerprinting techniques that might enable quick and efficient monitoring of environmental pollution with tendency to complement and eventually substitute the classical culture-dependent hygienic monitoring techniques will be described.

5.2 Molecular Detection Methods

So far, the methods used for detection, typing, and characterization of ARGs, heavy metal resistance genes and transfer genes have covered, but have not been limited to, specific and multiplex PCR, quantitative real-time PCR (qRT-PCR), DNA sequencing, and DNA-DNA or DNA-RNA hybridization-based techniques including microarray (Zhang et al. 2009).

5.2.1 DNA Hybridization

Molecular hybridization has been applied to detect ARGs for 30 years (Mendez et al. 1980). But during this time many improvements have been made on molecular hybridization, in particular in probe design and synthesis, so that dot blot and Southern blot are still often applied to distinguish different ARGs in one group from each other (Roberts and Kenny, 1986; Levy et al. 1999), to identify the presence of specific resistance genes in the environment (e.g., Agersø and Petersen 2007) or to verify the results obtained by gene-specific PCR (e.g. Ansari et al. 2008; Malik et al. 2008). With a number of non-radiolabeled systems commercially available, of which the best known and most often used is digoxigenin, there is no more need at present for radioactive labelling of probes.

5.2.2 PCR (Simple and Multiplex PCR)

PCR assays have been widely used in environmental samples for the detection of specific ARGs and heavy metal resistance genes, less frequently for the detection of transfer genes. Environmental target DNA or RNA at low concentrations can be amplified and detected by PCR-based methods. However, false-positive results are often obtained in the PCR assay. Southern hybridization of PCR products labelled and used as DNA probes to plasmid or chromosomal DNA samples from strains harbouring target genes or to total DNA isolated from the environment of interest can avoid false-positive PCR results (Ahmed et al. 2006; Akinbowale et al. 2007; Zhang et al. 2009). Furthermore, DNA sequencing is another common method used to verify the PCR products of certain ARGs (Thompson et al. 2007).

To save time and effort, multiplex PCR protocols have been developed and often applied for simultaneous detection of more than one environmental ARG (e.g. Bell et al. 1998; Jensen et al. 2002; Agersø and Petersen 2007; Ramachandran et al. 2007). Multiplex PCR saves considerable time and cost when different target genes are investigated simultaneously, but as a result of all the reactions taking place at the same conditions, some amplifications can be inhibited and false-negative results might be obtained. Another draw-back of multiple PCR is that primer dimer formation between the different primer pairs can interfere with experimental results and result in poor sensitivity (Markoulatos et al. 2002; Zhang et al. 2009). Despite of these disadvantages, multiplex PCR is still considered as a rapid and convenient method for the detection of multiple ARGs in isolated bacteria or environmental DNA (e.g. Gilbride et al. 2006; Agersø and Petersen 2007).

5.2.3 Quantitative PCR

qRT-PCR is usually used to quantify target DNA on the basis that the initial target gene concentration can be estimated according to the increase of the PCR product concentration with amplification cycles (Zhang and Fang 2006). SYBR Green is the most common used fluorescent dye to quantify ARGs in environmental samples (e.g. Yu et al. 2005; Mackie et al. 2006; Pei et al. 2006; Auerbach et al. 2007; Zhu 2007). TaqMan probes have also often been applied to quantify ARGs in environmental samples (Smith et al. 2004; Volkmann et al. 2004; Mackie et al. 2006; Pei et al. 2006; Auerbach et al. 2007; Chen et al. 2007; Böckelmann et al. 2009).

5.2.4 DNA Microarray

The DNA microarray technique is a genomic analysis technique with high throughput and high-speed (Zhang et al. 2009). For detection of antibiotic resistances, DNA microarray can provide detailed information on the presence or absence of a large number of ARGs simultaneously in a single assay (Gilbride et al. 2006). Microarray allows detection of ARGs within several hours and can be used as a time-saving and convenient tool supporting conventional resistance detection assays (Antwerpen et al. 2007). Although microarrays have been successfully used to determine the antibiotic resistances of clinical samples, few reports have been published using this technique to detect ARGs in environmental samples. The most important factor hampering its application in environmental samples is the low detection limit of the method. Nevertheless, microarray coupled with PCR can enhance the detection limit for environmental ARGs (Gilbride et al. 2006; Zhang et al. 2009). Patterson et al. (2007) successfully developed a microarray system based on PCR amplification of 23 tetracycline resistance genes and ten erythromycin resistance genes to screen environmental samples for the presence of these ARGs. Another reason for poor application of microarrays in environmental samples is the complexity of the samples and their pre-treatment. The presence of undesirable contaminants in environmental samples inhibits DNA extraction and/or target gene amplification, thus, a complicated pre-treatment of environmental samples is often necessary and crucial to get satisfactory and reproducible detection results (Call 2005; Zhang et al. 2009). Recently, new microarrays aimed to detect ARGs, pathogens and to monitor bacterial horizontal gene flux have been developed (e.g., Stabler et al. 2008; Frye et al. 2010). It seems evident that these arrays have the potential for application on environmental samples.

5.3 Antibiotic Resistance

Application of antibiotics in human medicine, veterinary medicine, and agriculture for nearly 60 years has exerted a tremendous impact on microbial communities, resulting in a multitude of (combined) resistances to the antibiotics. Most of them are genetically controlled by ARGs. The development of antibiotic resistance has been studied in great detail in recent decades, whereas resistance mechanisms towards heavy metals have not been explored to the same extent.

5.3.1 Mechanisms of Antibiotic Resistance

Antibiotic resistance is mainly caused by the following mechanisms: (1) target bypass: inaccessibility of the antibiotics to their target enzyme due to mutation or loss of the gene encoding the target enzyme (e.g. Happi et al. 2005); (2) efflux pumps: reduction of intracellular concentrations of antibiotics by conformational alteration of the cell membrane (e.g. Kumar and Schweizer 2005); (3) antibiotic inactivation, in most cases performed by specific enzymes (e.g. Wright 2005) or (4) target modification: modification of the action/binding site of the antibiotic (e.g. Lambert 2005).

5.3.1.1 Tetracycline Resistance Genes

There have been at least 38 different tetracycline resistance (*tet*) genes and three oxytetracycline resistance genes (*otr*) characterized to date (e.g. Thompson et al. 2007; Zhang et al. 2009). These genes include 23 genes, which code for efflux proteins, 11 genes for ribosomal protection proteins, three genes for an inactivating enzyme and one gene with unknown resistance mechanism (e.g. Roberts 2005; Zhang et al. 2009). Most environmental *tet* genes encode transport proteins, which pump the antibiotic out of the bacterial cell and keep the intracellular concentrations low to make ribosomes function normally (Roberts 2002).

5.3.1.2 Aminoglycoside Resistance Genes

The most important mechanism of aminoglycoside resistance is direct inactivation of the antibiotic by enzymatic modification (Shakil et al. 2008). More than 50 modification enzymes have been found so far (Ramón-García et al. 2006; Zhang et al. 2009). These enzymes are divided into three groups based upon their biochemical actions on the aminoglycoside, including acetyltransferases, phosphotransferases and nucleotidyltransferases. Different aminoglycoside-modifying enzymes have been detected in a broad range of polluted and natural water environments (for a summary, see Zhang et al. 2009).

5.3.1.3 ARGs Conferring Resistance to Macrolide-Lincosamide-Streptogramin (MLS), Chloramphenicol and Vancomycin

Although structurally unrelated to each other, macrolides, lincosamide, and streptogramin, are often investigated simultaneously for microbial resistance, since some macrolide resistance genes (*erm*) encode resistance to two or all three of these compounds (Roberts et al. 1999). In total, more than 60 different genes conferring resistance to one or more of the MLS antibiotics have been identified (Zhang et al.

5 Molecular Detection of Resistance and Transfer Genes in Environmental Samples 169

2009), including genes associated with rRNA methylation, efflux, and inactivation. The mechanisms responsible for resistance to chloramphenicol and florfenicol include chloramphenicol acetyltransferases, specific exporters and multidrug exporters (Schwarz et al. 2004). So far, six types of vancomycin resistance genes have been identified (Messi et al. 2006).

5.3.1.4 ARGs Conferring Resistance to Sulphonamides and Trimethoprim

Sulphonamides target dihydropteroate synthase. Trimethoprim competitively inhibits dihydrofolate reductase. Both enzymes are partly responsible for folate biosynthesis (Brochet et al. 2008). Resistances to sulphonamides and trimethoprim are often encoded by mutations located on highly conserved areas of dihydropteroate synthase (DHPS) genes and dihydrofolate reductase genes (Sköld 2000, 2001). DHPS catalyses an essential step in the biosynthesis of folic acid and is the target for the sulphonamide group of antimicrobial drugs. The enzyme binds its substrates in a fixed order: 6-hydroxylmethyl-7,8-dihydropterin pyrophosphate (DHPP) binds first, followed by the second substrate, pABA (*p*-aminobenzoic acid). Binding of pyrophosphate also allows the enzyme to recognize pABA or sulphonamide drugs, which act as pABA analogues. Mutations on the DHPS gene which confer resistance to sulphonamide drugs have a substantial effect on sulphonamide recognition and binding (Levy et al., 2008). Sulphonamide and trimethoprim resistance genes have been detected in both aquatic and terrestrial environments, in particular in those containing faecal pollution (Srinivasan et al. 2005; Moura et al. 2007; da Silva et al. 2007).

5.3.1.5 ARGs Conferring Resistance to β-Lactams

β-lactams are the most widely used antibiotics, and resistance to these antibiotics is a severe threat because they have low toxicity and are used to treat a broad range of infections (Zhang et al. 2009). The mechanisms of β-lactam resistance include inaccessibility of the antibiotics to their target enzymes, modifications of target enzymes, and/or direct inactivation of the antibiotics by β-lactamases (Zhang et al. 2009). In Gram-negative bacteria, the major resistance mechanism is enzymatic inactivation through the cleavage of the β-lactam ring by β-lactamases. More than 400 different β-lactamases encoded by hundreds of ARGs (*bla*) have been identified (Li et al. 2007). A variety of *bla* genes have been identified in diverse aquatic and soil environments (Zhang et al. 2009; and in this chapter, see below).

5.3.2 Antibiotic Resistance Genes in the Environment

Antibiotic resistance has been classified by the World Health Organization as one of the three major public health threats of our century (Lachmayr et al. 2009). Most efforts to control and study the epidemic have focused on pathogens in the clinical setting. However, ARGs also occur in non-pathogenic indigenous bacteria, which can then be passed on by Horizontal Gene Transfer (HGT) (Levy et al. 1976; Levy 2002). Thus, it is important to understand the environmental distribution of ARGs and the influence of anthropogenic inputs on the development and dissemination of ARGs in the environment.

5.3.2.1 Antibiotic Resistance in Aquatic Habitats

Clearly, most studies on the presence of ARGs in aquatic environments have been performed on the occurrence of *tet* genes. *Tet* genes have been detected in virtually all investigated aquatic habitats reaching from pristine rivers, glaciers, lagoons downstream of animal feeding operations to wastewater treatment plants. Some recent examples on molecular detection of *tet* genes in various aquatic environments are summarized below: Storteboom et al. (2010) detected *tet*(H), *tet*(Q), *tet*(S), and *tet*(T) genes in association with animal feeding operations; *tet*(C), *tet*(E), and *tet*(O) in association with wastewater treatment plants. In contrast, the source of the pristine Cache La Poudre River in Colorado, was dominated by *tet*(M) and *tet*(W) genes. Pei et al. (2006) applied PCR for the detection of five *tet* families in the sediments of the mixed-landscape Cache La Poudre River; the *tet*(O) gene families were further quantified by qRT-PCR. The La Poudre River downstream is known to contain high concentrations of antibiotics due to urban and agricultural activities. Pei et al. (2006) demonstrated that the quantities of *tet* genes normalized to the bacterial 16S rRNA gene copy number were significantly different between the sites investigated, with higher *tet* concentrations in the sludge at the impacted sites than at the pristine sites. Böckelmann et al. (2009) frequently detected *tet*(O) in reclaimed water from two artificial groundwater recharge systems. Börjesson et al. (2009a) applied qRT-PCR for the detection of *tet*(A) and *tet*(B) in wastewater-associated environments, namely soil from an overland flow area treating landfill leachates, biofilm from a municipal wastewater treatment plant, and sludge from a hospital wastewater pipeline. As expected, the highest concentration of both genes was observed in the hospital pipeline and the lowest in the overland flow system. *Tet*(A) was detected in all the environments.

Bacterial isolates from aquaculture sources in Australia were found to harbour a variety of *tet* genes, which can be horizontally transferred to other bacteria (Akinbowale et al. 2007). By *tet* specific PCR analysis, Akinbowale et al. (2007) demonstrated that *tet*(M) was the most common determinant (50%), followed by *tet*(E) (45%) and *tet*(A) (35%) in the tetracycline resistant bacterial isolates from aquaculture sources. To assess the diversity of antibiotic-resistant bacteria and their resistance genes in typical maricultural environments, Dang et al. (2007) performed both a culture-based and PCR-based study on a mariculture farm of China. Oxytetracycline-resistant bacteria were abundant in all the water samples. 16S rDNA sequence analyses demonstrated that the typical resistant isolates belonged to marine *Vibrio*, *Pseudoalteromonas* or *Alteromonas* species. For oxytetracycline resistance, *tet*(A),

5 Molecular Detection of Resistance and Transfer Genes in Environmental Samples 171

tet(B) and *tet*(M) genes were detected in some multidrug-resistant isolates, with *tet*(D) being the most common *tet* determinant. Dang et al. (2007) concluded that there is risk of multidrug-resistant bacteria contamination in mariculture environments, and marine *Vibrio* and *Pseudoalteromonas* species can serve as reservoirs of resistance determinants.

Nikolakopoulou et al. (2008) studied the presence of selected *tet* genes in different Greek seawater habitats, originated from wastewater treatment facilities, fish farm, and coastal environments. The molecular methods employed included assessment of the presence of twelve *tet* gene clusters by PCR, followed by hybridization with specific probes, in total DNA extracted from the habitats. The molecular analysis showed that *tet*(A) and *tet*(K) were detected in all the habitats, whilst *tet*(C) and *tet*(E) were present in fish farm and wastewater effluent samples, and *tet*(M) was found in fish farm and coastal samples. *tet*(A), *tet*(C), *tet*(K), and *tet*(M) were detected in 60 of the 89 isolates screened. The isolates were identified as pertaining to the *Stenotrophomonas*, *Acinetobacter*, *Pseudomonas*, *Bacillus* or *Staphylococcus* genera. The occurrence of *tet* genes in 15% of the bacterial isolates coincided with the presence of putative conjugative IncP plasmids. A habitat-specific dissemination of IncPα plasmids in wastewater effluent isolates and of IncPβ plasmids in fish farm isolates was observed. Isolated plasmids were shown to carry *tet*(A), *tet*(C), *tet*(E), and *tet*(K) genes. Nikolakopoulou et al. (2008) concluded that *tet* genes are widespread in the seawater habitats studied and are often encoded on conjugative broad-host-range plasmids.

Chlortetracycline is a commonly applied antimicrobial used for growth promotion and prophylaxis in swine production in the United States (Jindal et al. 2006). Chlortetracycline resistance was measured throughout the waste treatment processes at five swine farms by culture-based and molecular methods. Conventional farm samples had the highest levels of resistance with both culture-based and molecular methods and had similar levels of resistance despite differences in antimicrobial usage. All of the feed samples tested, including those from an organic farm, where no antimicrobials were used, contained *tet* genes (Jindal et al. 2006). Generally, the same *tet* genes and frequency of detection were found in the manure and lagoon samples for each commercial farm. The level of tetracycline resistance remained high throughout the waste treatment processes, suggesting that the potential impact of land application of treated wastes on the environmental resistance level should be extensively investigated (Jindal et al. 2006).

Koike et al. (2007) determined the occurrence of *tet* genes in groundwater underlying two swine confinement operations. Groundwater and lagoon samples were collected from both sites, total DNA was extracted, and PCR was used to detect seven *tet* genes, *tet*(M), *tet*(O), *tet*(Q), *tet*(W), *tet*(C), *tet*(H), and *tet*(Z). *Tet* gene concentrations were quantified by qRT-PCR. All seven *tet* genes were continually detected in groundwater during the 3-year monitoring period at both sites. Elevated detection frequency and concentration of *tet* genes were observed in the wells located down-gradient of the lagoon. The authors concluded that resistance genes in groundwater are affected by swine manure, but they are also part of the indigenous gene pool (Koike et al. 2007).

Ushida et al. (2010) investigated the prevalence of ARGs in glacier ice samples from various locations in the northern hemisphere (Alaska, Central Asia) and Antarctica by qRT-PCR with a Taqman probe system. They tested 15 different *tet* genes. Only *tet*(D), *tet*(E) and *tet*(G) were detected in low concentrations in the Alaskan glacier, the glacier most highly affected by anthropogenic pollution.

The spread of methicillin-resistant *Staphylococcus aureus* (MRSA), in which the *mecA* gene mediates resistance, threatens the treatment of staphylococcal diseases. Börjesson et al. (2009b) determined the effect of wastewater treatment processes on *mecA* gene concentrations, and the prevalence of *S. aureus* and MRSA over time in a municipal wastewater treatment plant. Water samples were collected for one year, at eight sites in the plant, and investigated by qRT-PCR for the presence of *mecA*, *S. aureus* and MRSA. *MecA* and *S. aureus* were detected throughout the year at all sampling sites. MRSA was also detected, but mainly in the early treatment steps. The authors concluded that the wastewater treatment process reduced the *mecA* concentrations, which can partly be explained by removal of biomass, but did not completely eliminate the *mecA* gene from the effluent (Börjesson et al. 2009b). Barker-Reid et al. (2010) studied the occurrence of ARGs in water used for irrigation in the Werribee River Basin, Australia, including river water and reclaimed effluent water. Samples were collected over a one-year period and screened for the presence of ARGs by PCR. Of the river water samples collected, 4% were positive for methicillin, and 4% for sulphonamide resistance genes, while 9% of the reclaimed water samples were positive for methicillin and 45% for sulphonamide. The low detection of ARGs in river water indicated that, regardless of its poor water quality, the river has not yet been severely contaminated with ARGs. The authors also concluded that the greater prevalence of ARGs in reclaimed water indicates that this important agricultural water source will need to be extensively monitored in the future (Barker-Reid et al. 2010).

Picão et al. (2008) searched for plasmid-mediated quinolone resistance genes (*qnr*) among waterborne environmental *Aeromonas* spp. isolated from the Lugano lake in Switzerland. Isolates resistant to nalidixic acid or ciprofloxacin were screened for *qnr* genes by multiplex PCR followed by sequencing. Isolated plasmids were transformed, and further analysis of the genetic structures surrounding the *qnrS2* gene was performed by PCR and sequencing. A *qnrS2* gene was identified from an *Aeromonas allosaccharophila* isolate, as part of a mobile insertion cassette located on a broad host range plasmid that co-harboured a class1 integron containing further ARG cassettes. Picão et al. (2008) pointed out that these findings further strengthen the role of *Aeromonas* spp. as a reservoir of ARGs in the environment.

Lachmayr et al. (2009) investigated the presence of TEM β-lactamases, conferring resistance to β-lactam antibiotics in a wastewater stream. They designed consensus DNA probes for all known TEM β-lactamase genes, and applied them in Taqman qRT-PCR assays to quantify these genes in environmental samples. The Taqman assays were used to study whether sewage, both treated and untreated, contributes to the spread of these genes in receiving waters. Lachmayr et al. (2009) found that while modern sewage treatment technologies reduce the concentration of the TEM β-lactamase genes, the ratio of TEM β-lactamase genes to 16S rRNA genes increases with treatment, suggesting that bacteria harbouring these resistance genes are more likely to survive the treatment process. The Authors concluded that TEM β-lactamase genes are introduced into the environment in significantly higher concentrations than those occurring naturally, creating reservoirs of increased resistance potential.

The prevalence of kanamycin and ampicillin resistant bacteria has also been investigated in public drinking water from the Lahore Metropolitan, in Pakistan (Samra et al. 2009). Bacteria resistant to kanamycin and ampicillin were tested by PCR for the detection of the corresponding resistance genes, and further characterized by colony hybridization and transformation studies. Among the 625 drinking water samples, 400 contained kanamycin and ampicillin resistant bacteria. Transformation of *Escherichia coli* DH5α with plasmids isolated from kanamaycin and ampicillin resistant bacteria confirmed that the ARGs were encoded on plasmids. These results showed that, due to poor sanitary conditions, the Lahore Metropolitan drinking water is contaminated with kanamycin and ampicillin resistant bacteria (Samra et al. 2009). Chen et al. (2007) developed six qRT-PCR assays for the quantification of six classes of *erm* genes (classes A through C, F, T, and X) that encode the major mechanism of resistance to macrolides-lincosamides-streptogramin B (MLS$_B$). These qRT-PCR assays were validated and used in quantifying the six *erm* classes in five types of environmental samples, including samples from bovine manure, swine manure, compost of swine manure, swine waste lagoons, and an upflow biofilter system treating hog house effluents. The bovine manure samples contained much smaller reservoirs of each of the six *erm* classes than the swine manure samples. In comparison to the swine manure samples, the composted swine manure samples had strongly reduced *erm* gene concentrations (up to 7.3 logs), whereas the lagoon or the biofilter had similar *erm* gene abundances (Chen et al. 2007). These data suggest that the methods of manure storage and treatment have a significant impact on the persistence and decrease of MLS$_B$ resistance originating from animals. It is likely that they affect the dissemination of ARGs into the environment (Chen et al. 2007). The abundances of the *erm* genes appeared to be positively correlated with those of the *tet* genes previously determined in these samples (Yu et al. 2005).

Böckelmann et al. (2009) frequently detected *ermB* in reclaimed water from two European artificial groundwater recharge systems by qRT-PCR.

Castiglioni et al. (2008) selected the resistance gene *marA*, mediating resistance to multiple antibiotics, and investigated its distribution in sediment and water samples from surface and sewage treatment waters. *MarA* was found in almost all environmental samples, and was confirmed by PCR in antibiotic-resistant bacterial isolates. The majority of the resistant isolates belonged to the genus *Bacillus*, not previously known to possess the regulator *marA*. Phylogenetic analysis indicated the possible horizontal acquisition of *marA* by *Bacillus* from Gram-negative *Enterobacteriaceae* revealing a novel *mar* homolog in *Bacillus*. qRT-PCR assays demonstrated that the frequency of *marA* in anthropogenic environments seems to be related to bacterial exposure to water-borne antibiotics.

5.3.2.2 Antibiotic Resistance in Terrestrial Habitats

Patterson et al. (2007) developed a macroarray system to screen environmental samples for the presence of specific *tet* and *erm* genes. The macroarray was loaded with PCR amplicons of 23 different *tet* and ten *erm* genes. Total bacterial genomic DNA was extracted from soil and animal faecal samples collected from different European countries. Intensively farmed pig herds had a significantly higher number of resistant bacteria than pigs from organic herds reared without antibiotic use. The relative proportions of the different ARGs were constant across the different countries (Patterson et al. 2007). Ribosome protection type *tet* genes were the most common ARGs in animal faecal samples, with the *tet*(W) gene the most abundant. Different ARGs were present in soil samples, where *erm*(V) and *erm*(E) were the most prevalent followed by the efflux type *tet* genes (Patterson et al. 2007).

Guardabassi and Agersø (2006) studied the occurrence of D-Ala:D-Lac ligase genes homologous to the glycopeptide resistance gene *vanA* in samples of agricultural and garden soil by PCR. Sequencing of 25 clones from an *E. coli* TOP10 library with insertions of the amplified *vanA* homologous genes revealed 23 novel sequences with 86–99% identity with the *vanA* gene in enterococci. *VanA* homologous sequences were recovered from all agricultural samples, as well as from two garden samples with no history of organic fertilization. Guardabassi and Agersø (2006) concluded that soil might be a rich and assorted reservoir of ARGs closely related to those conferring vancomycin resistance in clinical bacteria.

Malik et al. (2008) investigated the presence of the *ampC*, *tet*(O), *ermB*, *bla-SHV5*, and *vanA* ARGs by PCR in five different soils, three Indian soils with wastewater irrigation history, one soil from an abandoned sewage field in Germany and one from a German urban park. All the Indian soil samples irrigated with wastewater contained the *ampC* gene, whereas the other ARGs were not found in any of the samples.

To investigate the ARGs among uncultured bacteria in an undisturbed soil environment, Allen et al. (2009) undertook a functional metagenomic analysis of a remote Alaskan soil. They found that this soil is a reservoir for β-lactamases that function in *E. coli*, including divergent β-lactamases and the first bifunctional β-lactamase. These important findings suggested that even in the absence of selective pressure, the soil microbial community in an unpolluted site harbours unique and ancient β-lactam resistance determinants.

Binh et al. (2008) studied the prevalence and diversity of transmissible antibiotic resistance plasmids in piggery manure. Samples from manure storage tanks of 15 farms in Germany were investigated, representing diverse sizes of herds, meat or piglet production. Antibiotic resistance plasmids from manure bacteria were captured by exogenous isolation in *gfp*-tagged *E. coli* and further characterized. 228 transconjugants were captured from 15 manures using selective media supplemented with amoxicillin, sulfadiazine or tetracycline. Replicon probing demonstrated that 28 of the plasmids belonged to IncN, one to IncW, 13 to IncP-1 and 19 to the recently discovered pHHV216-like plasmids. The amoxicillin resistance gene *bla-*TEM was detected on 44 plasmids and the sulphonamide resistance genes *sul1, sul2*

5 Molecular Detection of Resistance and Transfer Genes in Environmental Samples 175

and/or *sul3* on 68 plasmids. Hybridization of replicon-specific sequences amplified from community DNA revealed that IncP-1 and pHHV216-like plasmids were found in all manures, while IncN and IncW plasmids were less frequent. Binh et al. (2008) showed that 'field-scale' piggery manure might be a reservoir of broad host range plasmids conferring multiple ARGs.

Heuer et al. (2009) treated two soils with manure, either with or without the sulphonamide antibiotic sulfadiazine. In both soils a significant increase in copy number of the sulphonamide resistance gene *sul2* was detected by qRT-PCR. All *sul2*-carrying elements, captured in *E. coli* from soil, belonged to a novel class of conjugative plasmids (Heuer et al. 2009).

The prevalence of three sulphonamide resistance genes, *sul1*, *sul2* and *sul3* and sulphachloropyridazine resistance were studied in bacterial isolates from manured agricultural clay soils and slurry samples in the United Kingdom over a 2-year period by Byrne-Bailey et al. (2009). Slurry from tylosin-fed pigs amended with sulphachloropyridazine and oxytetracycline was used for manuring. Phenotypic resistance to sulphachloropyridazine was significantly higher in isolates from pig slurry and post-application soil than in those from pre-application soil. Of 531 isolates, 23% carried *sul1*, 18% *sul2* and 9% *sul3* only. Two percent of the isolates contained all three *sul* genes. Byrne-Bailey et al. (2009) reported for the first time sulphonamide resistance in *Psychrobacter*, *Enterococcus* and *Bacillus* spp. Thus, this study provided the first description of the genotypes *sul1*, *sul2*, and *sul3* outside the *Enterobacteriaceae* and in the soil environment.

Zhou et al. (2010) evaluated the effects of land application of swine manure on the levels of tetracycline, macrolide, and lincosamide antibiotics and on MLS_B resistance in field soil samples and laboratory soil batch tests. MLS_B and tetracycline antimicrobials were quantified by liquid chromatography-tandem mass spectrometry, the prevalence of the ribosomal modification responsible for MLS_B resistance in the samples was quantified by fluorescence in situ hybridization (FISH). Macrolide antibiotics were not detected in soil samples, while tetracyclines were found, suggesting that the tetracyclines persist in soil. No significant differences in MLS_B resistance were detected when amended and unamended field soils were compared. Only a transient (< 20-day) MLS_B resistance increase was documented in most batch tests. In contrast to other studies, Zhou et al. (2010) did not detect a persistent increase in the occurrence of MLS_B resistance due to land application of treated swine manure.

Mindlin et al. (2008) generated a collection of antibiotic resistant bacterial strains isolated from arctic permafrost subsoil sediments of various age and genesis. The collection included approximately 100 strains of Gram-positive and Gram-negative bacteria resistant to aminoglycoside antibiotics (gentamycin, kanamycin and streptomycin), chloramphenicol and tetracycline. Multidrug resistant strains were found for the first time in "ancient" bacteria. In studies of the molecular nature of the streptomycin resistance, two different resistance determinants were detected: *strA-strB* genes coding for aminoglycoside phosphotransferases and *aadA* genes encoding aminoglycoside adenylyltransferases. These genes proved to be highly homologous to those of contemporary bacteria (Mindlin et al. 2008).

Knapp et al. (2010) speculated that antibiotic resistance levels might be apparent in historic soil archives as evidenced by ARG abundances over time. To this goal, DNA was extracted from five long-term soil-series (soil chronosequence) from different locations in The Netherlands differing in soil type, irrigation type and fertilizer use that spanned 1940–2008, and concentrations of 16S rRNA gene and of 18 ARGs from different major antibiotic classes were quantified by qRT-PCR. Knapp et al. (2010) showed that ARGs from all classes of antibiotics tested have significantly increased since 1940, but especially for the tetracyclines, with some individual ARG being >15 times more abundant now than in the 1970s. This is alarming, because waste management procedures have broadly improved and stricter rules on non-therapeutic antibiotic use in agriculture are being promulgated. These data suggest that basal environmental levels of ARGs might still increase, requiring counteractive measures worldwide (Knapp et al. 2010).

5.4 Heavy Metal Resistance

A variety of heavy metals are found in different concentrations in the environment, in pristine as well as in anthropogenic sites. Many bacteria have developed resistance mechanisms against heavy metals or they can even use them as electron acceptors or electron donors. The most important resistance mechanisms are briefly summarized.

5.4.1 Mechanisms of Heavy Metal Resistance

Of all the heavy metals, Hg is the most toxic and with no biological function (Nies, 1999). Hg resistance in *Bacteria* is conferred by the *mer* operon. Several *mer* operon-encoded proteins are involved in transport of inorganic oxidized Hg into the cytosol, where the *merA*-encoded mercuric reductase protein reduces Hg^{2+} to volatile, less reactive elemental Hg^0 (Oreggard and Sørensen 2007; for a review on bacterial Hg resistance see Barkay et al. 2003). Hg resistance has been recognized in many bacterial phyla, including *Firmicutes*, *Actinobacteria* and *Proteobacteria*, of both clinical and environmental origin. It is considered to be an ancient resistance mechanism (Osborn et al. 1997).

As is present in the pentavalent As (V) (arsenate) and trivalent As (III) (arsenite) forms in the environment (Cullen and Reimer 1989). Arsenite is more toxic than arsenate and has been shown to inhibit several dehydrogenases (Ehrlich 1996; Jareonmit et al. 2010). Microbial As (III) oxidation and As (V) reduction activities are cellular strategies for either detoxification or for generating energy. Microorganisms can use As compounds as electron donors, or electron acceptors, or they encode one or more As detoxification mechanisms (e.g. Ahmann et al. 1994; Jareonmit et al. 2010). These mechanisms include (i) minimizing the uptake of arsenate through the phosphate uptake system, (ii) peroxidation reactions with membrane lipids, and (iii) microbial As detoxification pathways involving the *ars* operon (summarized in Jareonmit et al. 2010). Operons encoding analogous arsenic resistance genes (*ars*) have been found on the chromosome and conjugative plasmids of a wide variety of Gram-positive and Gram-negative bacteria (Jareonmit et al. 2010). The *ars* operons generally consist of either the *arsRBC* or *arsRDABC* genes (Silver and Phung 1996).

Pseudomonas syringae, E. coli and *Xanthomonas campestris* were the first bacterial species reported to survive in the presence of high Cu concentrations (Cooksey 1994). The high-level resistance to Cu resulted from the presence of the *cop* or *pco* operons (Cooksey 1994). These operons are genetically homologous and encode a Cu resistance mechanism that may involve Cu sequestration and/or enzymatic transformation (Cooksey 1994; Rensing and Grass 2003). In the recent decade, several bacterial strains such as *Cupriavidus metallidurans* CH34 (Borremans et al. 2001), *Ralstonia solanacearum* (Salanoubat et al. 2002), *Pseudomonas aeruginosa* PAO1 (Stover et al. 2000) and *Caulobacter crescentus* (Nierman et al. 2001) have been shown to be resistant to Cu and to encode all or part of these operons. The Cu resistance of these strains has always implied the presence of *copA* homologous genes encoding a periplasmic inducible multi-copper oxidase (Nies 1999; Rensing and Grass 2003). Thus, *copA* may represent a relevant marker to resolve the diversity of the Cu-resistant functional microbial communities (Lejon et al. 2007).

5.4.2 Heavy Metal Resistance Genes in the Environment

Studies on the microbial impact on the fate of minerals and geologically important compounds of mining areas can lead to a better understanding of biogeochemical cycles (Haferburg and Kothe 2007). Whereas metabolic processes of microorganisms are the cause for the dissolution of minerals, microbial metabolism can contribute to the formation of certain ore deposits over geological time. The adaptation to heavy metal rich environments is resulting in microorganisms that show activities in biosorption, bioprecipitation, extracellular sequestration, transport mechanisms, and/or chelation. These resistance mechanisms are the basis for the application of microorganisms in bioremediation approaches (Haferburg and Kothe 2007). Most molecular studies on the detection of heavy metal resistance genes in the environment have been performed on Hg and As resistance genes.

5.4.2.1 Heavy Metal Resistance in Aquatic Habitats

Only a few studies have been carried out on the detection of heavy metal resistance genes in water and river sediments. Smalla et al. (2006) analyzed river sediment samples from two Hg-polluted and two non-polluted or less-polluted areas of the river Nura in Kazakhstan by PCR for the presence and abundance of Hg resistance genes and of broad-host-range plasmids. The study revealed that Hg pollution corresponded to an increased abundance of Hg resistance genes and of IncP-1β replicon-specific sequences detected in total community DNA. Without selection for Hg resistance, three different IncP-1β plasmids (pTP6, pTP7, and pTP8) were captured directly from contaminated sediment slurry in *Cupriavidus necator* based on their ability to mobilize the IncQ plasmid pIE723. These plasmids hybridized with the *merRTΔP* probe and conferred Hg resistance to their host (Smalla et al. 2006).

To understand how contaminants affect microbial community diversity, heterogeneity, and functional structure, Waldron et al. (2009) investigated six ground water monitoring wells from a Field Research Centre in the U.S., with a wide range of pH, nitrate, and heavy metal contamination. DNA from the bacterial groundwater community was analyzed with a functional gene array containing 2006 probes to detect genes involved in metal resistance, sulphate reduction, organic contaminant degradation, and carbon and nitrogen cycling. Microbial diversity declined in relation to the contamination levels of the wells. Highly contaminated wells had lower gene diversity but greater signal intensity than the pristine wells (Waldron et al. 2009). Metal-resistant and metal-reducing microorganisms were detected in both contaminated and pristine wells, suggesting the potential for successful bioremediation of metal-contaminated ground waters (Waldron et al. 2009).

5.4.2.2 Heavy Metal Resistance in Terrestrial Habitats

Bruce (1997) performed a profiling of *mer* gene subclasses in Hg-polluted and pristine natural environments by Fluorescent-PCR-restriction fragment length polymorphism (FluRFLP). For FluRFLP, PCR products were amplified from individual *mer* operons in Hg-resistant bacterial isolates and from DNA directly isolated from bacteria in soil and sediment samples. The primers used to amplify DNA were designed from consensus sequences of the major subclasses of archetypal Gram-negative *mer* operons. One of the oligonucleotide primers was labelled at the 5′end with a green (TET) fluorescent dye. *Mer* PCR products amplified from DNA extracted directly from soil and sediment bacteria were investigated to determine the profiles of the major *mer* subclasses present in each natural environment. Profiles generated were highly similar for samples taken within the same soil type. The profiles, however, changed significantly on crossing from one soil type to another, with gradients of the different groupings of *mer* genes identified (Bruce 1997).

Oregaard and Sørensen (2007) studied the diversity of bacterial mercuric reductase genes from surface and sub-surface floodplain soil. DNA was extracted from different soil depths (0–5, 45–55 and 90–100 cm below surface) sampled at a floodplain in Oak Ridge, USA. The presence of *merA* genes was examined by PCR targeting *Actinobacteria*, *Firmicutes* or β/γ-*Proteobacteria*. β/γ-*Proteobacteria* were amplified from all soils, whereas *Actinobacteria* were amplified only from surface soil (Oregaard and Sørensen 2007). Moreover, β/γ-*Proteobacteria merA* sequences showed high diversity in all soils, but limited vertical similarity. Less than 20% of the operational taxonomic units (OTU) (DNA sequence ≥95% identical) were shared between the different soils. Due to the high functional diversity of *mer* genes and the limited vertical distribution of shared OTU, Oregaard and Sørensen suggested HGT as one of the mechanisms of bacterial adaptation to Hg resistance.

de Lipthay et al. (2008) studied the acclimation to Hg of bacterial communities of different depths from contaminated and non-contaminated floodplain soils by molecular methods. The level of Hg tolerance of the bacterial communities from the contaminated site was higher than that of the reference site. The level of Hg tolerance and functional versatility of bacterial communities in contaminated soils initially were higher in surface soil, compared with the deeper soils. However, following new Hg exposure, no differences between bacterial communities were observed, indicating a high adaptive potential of the subsurface communities. Four different Hg-resistance plasmids, all belonging to the IncP-1β group, were isolated from the contaminated soils in different soil depths (de Lipthay et al. 2008). The abundance of *merA* and IncP-1 plasmid carrying populations increased, after new Hg exposure, which could be the result of selection, as well as HGT. The data suggest a role for IncP-1 plasmids in the adaptation to Hg of surface as well as subsurface soil microbial communities (de Lipthay et al. 2008).

Cai et al. (2009) investigated the distribution and diversity of arsenite-resistant bacteria in three different As-contaminated soils and further studied the As(III) resistance levels and corresponding resistance genes of these microbes. 58 arsenite-resistant bacteria were identified from soils with three different arsenic-contamination levels. Highly arsenite-resistant bacteria (MIC >20 mM) were only isolated from the highly contaminated site and belonged to various *Acinetobacter*, *Agrobacterium*, *Arthrobacter* and *Pseudomonas* species (Cai et al. 2009). Five arsenite-oxidizing bacteria that belonged to *Achromobacter*, *Agrobacterium* and *Pseudomonas* were identified and showed a higher average arsenite resistance level than the non-arsenite oxidizers. Five *aoxB* genes encoding arsenite oxidase and 51 arsenite transporter genes were amplified from these strains by PCR. The *aoxB* genes were specific for the arsenite-oxidizing bacteria. Strains containing both an *aoxB* gene and an arsenite transporter gene (*ACR3* or *arsB*) displayed higher arsenite resistance than those carrying an arsenite transporter gene only. Cai et al. (2009) discussed the potential impact of HGT of *ACR3* and *arsB* genes on the bacterial adaptation to highly arsenite-contaminated habitats.

Jareonmit et al. (2010) investigated the microbial community structure in Thailand soils from an old tin mine, contaminated with low and high levels of As, using denaturing gradient gel electrophoresis (DGGE). Band pattern analysis indicated that the bacterial community was not significantly different in the two soils. Two hundred and sixty-two bacterial isolates were obtained from As-contaminated soils. The majority of the As-resistant isolates belonged to the Gram-negative bacteria. The As-resistant isolates were further investigated by PCR for the presence of *ars* genes: 30% were found positive for an *arsC* arsenate reductase gene (Jareonmit et al. 2010).

Lejon et al. (2007) developed a molecular fingerprinting assay to assess the diversity of *copA* genes, conferring bacterial resistance to Cu. PCR detection resulted in the identification of a novel *copA* determinant in *Pseudomonas fluorescens* (Lejon et al. 2007). The *copA* DNA fingerprinting assay was optimized for DNA directly extracted from soils differing in physicochemical characteristics and in organic status. Particular *copA* genes were obtained for each studied soil. The molecular phylogeny of the *copA* gene confirmed that particular *copA* gene clusters are specific for each soil organic status. The molecular fingerprinting assay demonstrated to be sensitive to short-term responses of *copA* gene diversity to Cu additions to soil samples. Thus, microbial community adaptation appeared to be preferentially controlled by the diversity of the indigenous *copA* genes rather than by the bioavailability of the metal (Lejon et al. 2007).

Abou-Shanab et al. (2007) tested 47 bacterial cultures, from the rhizosphere of yellowtuft (*Alyssum murale*), and from Ni-rich serpentine soil, for their ability to tolerate As (V), Cd, Cr, Zn, Hg, Pb, Co, Cu and Ni in their growth medium. All cultures contained multiple metal-resistant bacteria, with heptametal resistance as the major pattern (29%). Five of the cultures were tolerant to nine different metals (Abou-Shanab et al. 2007). PCR in combination with DNA sequencing was applied to investigate the metal resistance mechanism in some of these Gram-positive and Gram-negative bacteria that were highly resistant to Hg, Zn, Cr and Ni. The *czc*, *chr*, *ncc* and *mer* genes responsible for resistance to Zn, Cr, Ni and Hg, respectively were amplified in these bacteria by PCR. Thus, Hg, Zn, Cr, and Ni resistance genes are widely distributed in both Gram-positive and Gram-negative bacteria isolated from *A. murale* rhizosphere and Ni-rich soils (Abou-Shanab et al. 2007).

5.5 Key Transfer Genes

Transfer regions of conjugative plasmids are considered as a subtype of type IV secretion systems (T4SS) which have evolved to transport DNA/protein complexes from one bacterial cell to another instead of protein effectors to eukaryotic target cells (Grohmann et al. 2003; Alvarez-Martinez and Christie 2009). These T4SS encode for 15–30 proteins, which are all involved in transfer regulation, processing of the transferred substrate (DNA or protein(s)), contact formation with the target cell and formation of the multi-protein transport channel in the cell envelope. The T4SS transfer factors can be divided into four families, the Dtr family (DNA transfer and replication), the energy-supply family, the transport channel family, and the surface protein or adhesin family. Of these families, the most conserved T4SS proteins belong to (i) the Dtr family, the so called relaxases which initiate DNA transfer by a site- and strand-specific cleavage in the origin of transfer (*oriT*) and (ii) to the energy supply family which comprises three types of ATPases, a VirB4-like, a VirB11-type and the VirD4-type. The VirD4-type ATPases are better known as T4SS coupling proteins that link the relaxasome (the DNA-protein complexes formed at the *oriT*) with the Mpf (mating-pair formation) complex consisting of the assembled multi-protein transport complex.

Plasmid relaxases belong to six different protein families (for a review see Francia et al. 2004; and Garcillán-Barcia et al. 2009). Thus, they are good marker genes

5 Molecular Detection of Resistance and Transfer Genes in Environmental Samples   181

for the classification of plasmid families. Two proteins present in practically all T4SS involved in DNA transport are the VirB4 and VirD4 family proteins. In contrast, VirB11 family proteins are absent from most T4SS of *Firmicutes* and *Archaea* and also from model plasmids such as F (Alvarez-Martinez and Christie 2009). Therefore, best suited marker genes for the detection of mobilizable and/or self-transmissible plasmids and assessment of resistance and virulence factor transfer in the environment are the relaxase genes and the genes encoding the T4SS ATPases, *virB4* and *virD4*. The only DNA sequence indispensable for DNA transfer in *cis* (on mobilizable and conjugative plasmids or transposons) is the *oriT*.

5.5.1 Detection of Transfer Genes in the Environment

As indicated above, the most appropriate transfer genes for molecular detection of MGE in the environment include those of the relaxase families, the VirB4 and VirD4 ATPases as well as, in some cases, the genes of the VirB1-like lytic transglycosylases which locally open the peptidoglycan to facilitate assembly of the T4SS protein complex and/or T4SS substrate transport through the bacterial cell envelope. Some authors also searched for the presence of *oriT*s, which can be divided in *oriT* families (Francia et al. 2004) enabling classification of the detected MGE. The method of choice for the detection of transfer genes in the environment is gene-specific PCR, followed by Southern or dot blot hybridization, if higher sensitivity is required.

An example for the design of a specific primer pair for the detection of a sub-group of relaxases belonging to the MOB$_V$ relaxase family abundant in Gram-positive bacteria is given in Fig. 5.1.

5.5.1.1 Detection of Plasmids and Transfer Genes in Aquatic Habitats

Most molecular studies on the presence of plasmids or transfer genes in aquatic environments have been performed in wastewater treatment plants. The presence and diversity of IncP-1 plasmids (most of them are conjugative) in the influent of a Danish wastewater treatment plant was investigated by Bahl et al. (2009). They performed PCR amplification of the IncP-1 replication gene, *trfA*, in community DNA from the wastewater influent, followed by DNA sequencing. All five subgroups of the IncP1 plasmids, α, β, γ, δ, and ε were present in the wastewater treatment plant influent. These results confirmed that wastewater constitutes a reservoir for the conjugative IncP-1 plasmids, which often harbour multiple ARGs.

To investigate the mobile gene pool present in wastewater environments, Moura et al. (2010) isolated total community DNA from two distinct raw effluents: urban and slaughterhouse wastewaters. The bacterial community structure was evaluated by DGGE analysis of 16S rRNA gene fragments. Detection of broad-host-range plasmid sequences was carried out by PCR and Southern hybridization. *TrfA* of

```
                            ───────▶
pAMalpha1_AF503772    3263  ATGAGTTATGCAGTTTGTAGAATGCAAAAAGTGAAATCAGCTGGACTAAAAGGCATGCAA  3204
pBC16_U32369          1074  ATGAGTTATGCAGTTTGTAGAATGCAAAAAGTGAAATCAGCTGGACTAAAAGGCATGCAA  1015
pUB110_M19465         1074  ATGAGTTATGCAGTTTGTAGAATGCAAAAAGTGAAATCAGCTGGACTAAAAGGCATGCAA  1015
pGO1_FM207042        20733  ATGAGTTATGCAGTTTGTAGAATGCAAAAAGTGAAATCAGCTGGACTAAAAGGCATGCAA  20674
pSK41_AF051917       20733  ATGAGTTATGCAGTTTGTAGAATGCAAAAAGTGAAATCAGCTGGACTAAAAGGCATGCAA  20674
pKKS2187_FM207105     4680  ATGAGTTATGCAGTTTGTAGAATGCAAAAAGTGAAATCAGCTGGACTAAAAGGCATGCAA  4739
SAP016A_GQ900381     37987  ATGAGTTATGCAGTTTGTAGAATGCAAAAAGTGAAATCAGCTGGACTAAAAGGCATGCAA  37928
pIP1714_AF015628      2463  ATGAGTTATGCAGTTTGTAGAATGCAAAAAGTGAAATCAGCTGGACTAAAAGGCATGCAA  2522
pGR71_X15503           112  ATGAGTTATGCAGTTTGTAGAATGCAAAAAGTGAAATCAGCTGGACTAAAAGGCATGCAA  171
pTB19_M63891          3622  ATGAGTTATGCAGTTTGTAGAATGCAAAAAGTGAAATCAGCTGGACTAAAAGGCATGCAA  3681
pTB53_D14852           506  ATGAGTTATGCAGTTTGTAGAATGCAAAAAGTGAAATCAGCTGGACTAAAAGGCATGCAA  565
                           *************************************************************


pAMalpha1_AF503772    2903  TATGCAACAGTTCATAATGATGAGCAAACCCCTCACATGCATTTAGGTGTTGTGCCTATG  2844
pBC16_U32369           714  TATGCAACAGTTCATAATGATGAGCAAACCCCTCACATGCATTTAGGTGTTGTGCCTATG  655
pUB110_M19465          714  TATGCAACAGTTCATAATGATGAGCAAACCCCTCACATGCATTTAGGTGTTGTGCCTATG  655
pGO1_FM207042        20373  TATGCAACAGTTCATAATGATGAGCAAACCCCTCACATGCATTTAGGTGTTGTGCCTATG  20314
pSK41_AF051917       20373  TATGCAACAGTTCATAATGATGAGCAAACCCCTCACATGCATTTAGGTGTTGTGCCTATG  20314
pKKS2187_FM207105     5040  TATGCAACAGTTCATAATGATGAGCAAACCCCTCACATGCATTTAGGTGTTGTGCCTATG  5099
SAP016A_GQ900381     37627  TATGCAACAGTTCATAATGATGAGCAAACCCCTCACATGCATTTAGGTGTTGTGCCTATG  37568
pIP1714_AF015628      2823  TATGCAACAGTTCATAATGATGAGCAAACCCCTCACATGCATTTAGGTGTTGTGCCTATG  2882
pGR71_X15503           472  TATGCAACAGTTCATAATGATGAGCAAACCCCTCACATGCATTTAGGTGTTGTGCCTATG  531
pTB19_M63891          3982  TATGCAACAGTTCATAATGATGAGCAAACCCCTCACATGCATTTAGGTGTTGTGCCTATG  4041
pTB53_D14852           866  TATGCAACAGTTCATAATGATGAGCAAACCCCTCACATGCATTTAGGTGTTGTGCCTATG  925
                           *************************************************************


pAMalpha1_AF503772    2843  CGTGATGGAAAACTGCAAGGAAAAAATGTGTTTAATCGTCAAGAACTGTTATGGCTACAA  2784
pBC16_U32369           654  CGTGATGGAAAACTGCAAGGAAAAAATGTGTTTAATCGTCAAGAACTGTTATGGCTACAA  595
pUB110_M19465          654  CGTGATGGAAAACTGCAAGGAAAAAATGTGTTTAATCGTCAAGAACTGTTATGGCTACAA  595
pGO1_FM207042        20313  CGTGATGGAAAACTGCAAGGAAAAAATGTGTTTAATCGTCAAGAACTGTTATGGCTACAA  20254
pSK41_AF051917       20313  CGTGATGGAAAACTGCAAGGAAAAAATGTGTTTAATCGTCAAGAACTGTTATGGCTACAA  20254
pKKS2187_FM207105     5100  CGTGATGGAAAACTGCAAGGAAAAAATGTGTTTAATCGTCAAGAACTGTTATGGCTACAA  5159
SAP016A_GQ900381     37567  CGTGATGGAAAACTGCAAGGAAAAAATGTGTTTAATCGTCAAGAACTGTTATGGCTACAA  37508
pIP1714_AF015628      2883  CGTGATGGAAAACTGCAAGGAAAAAATGTGTTTAATCGTCAAGAACTGTTATGGCTACAA  2942
pGR71_X15503           532  CGTGATGGAAAACTGCAAGGAAAAAATGTGTTTAATCGTCAAGAACTGTTATGGCTACAA  591
pTB19_M63891          4042  CGTGATGGAAAACTGCAAGGAAAAAATGTGTTTAATCGTCAAGAACTGTTATGGCTACAA  4101
pTB53_D14852           926  CGTGATGGAAAACTGCAAGGAAAAAATGTGTTTAATCGTCAAGAACTGTTATGGCTACAA  985
                           *************************************************************
                                   ◀───────
```

Fig. 5.1 DNA sequence alignment of MOB_V relaxase family members using ClustalW2 (http://www.ebi.ac.uk/Tools/clustalw2/). Primers were designed using primer3 (http://frodo.wi.mit.edu/primer3/) and primer specificity was determined by BLAST search against Nucleotide Collection (nr/nt). GenBank accession numbers, AF503772 – *Enterococcus faecalis* pAMalpha1, *mobB*; U32369 – *Bacillus cereus* pBC16, *mob*; M19465 – *Staphylococcus aureus* pUB110, beta protein gene; FM207042 – *S. aureus* pGO1, *pre*; AF051917 – *S. aureus* pSK41, *pre*; FM207105 – *S. aureus* pKKS2187, *pre*; GQ900381 – *S. epidermidis* plasmid SAP016A, SAP016A_047; AF015628 – *S. cohnii* pIP1714, *pre*; X15503 – *B. subtilis* pGR71, K1 delta1 gene; M63891 – *B. stearothermophilus* pTB19, *mob*; D14852 – *Bacillus.* sp. pTB53, *preT*. 5′ and 3′ PCR primer sequences of the MOB_V relaxase family are indicated in bold letters. The 395-bp PCR product is flanked by arrows

IncP-1 and replication genes of the IncN, IncQ, and IncW plasmid families were present in both effluents. In agreement with Bahl et al. (2009), Moura et al. (2010) concluded that wastewater environments bring together various types of MGE that may play a major role in bacterial adaptation and evolution.

Schlüter et al. (2008) analysed the plasmid metagenome sequence data obtained by the ultrafast 454-pyrosequencing technology from wastewater treatment plant bacteria. The sequence dataset was analysed for genetic diversity and composition by a newly developed bioinformatic pipeline based on assignment of environmental gene tags (EGTs) to protein families in the Pfam database. Many sequences represented genes with predicted functions in plasmid replication, stability and mobility, which indicates that wastewater treatment plant bacteria harbour stable and mobile plasmids. The data of Schlüter et al. (2008) confirmed a high diversity of plasmids residing in

5   Molecular Detection of Resistance and Transfer Genes in Environmental Samples   183

wastewater treatment plant bacteria, with the mobile organic peroxide resistance plasmid pMAC from *Acinetobacter baumannii* as the most abundant replicon type.

Sobecky and Hazen (2009) summarized recent insights gained from different methodological approaches used to characterize the biodiversity and ecology of MGEs in marine environments and their contributions to HGT. Recent advances in genome sequencing technology have significantly increased the number of plasmids characterized as a part of genome-sequencing projects. For example, the 48.5-kb plasmid of *Vibrio vulnificus* (Chen et al. 2003), the 40.4-kb plasmid pA of *Nitrosococcus oceani* (Klotz et al. 2006) and four plasmids in the size range of 5.8–106.5 kb from *Roseobacter denitrificans* (Swingley et al. 2007) were all characterized as part of genome-sequencing studies while studies characterizing the contribution of MGEs to HGT and the diversity of plasmids from deep sea systems are still missing (Sobecky and Hazen 2009).

5.5.1.2   Detection of Plasmids and Transfer Genes in Terrestrial Habitats

Disqué-Kochem et al. (2001) were one of the first to develop PCR primers for the detection of *virD4*- and *virB11*-like sequences in bacterial soil isolates. They designed degenerate PCR primers to detect transfer regions from plasmids belonging to different incompatibility groups that are often found in the environment. In several of the isolates conjugative plasmids were identified which were able to mobilize a derivative of the non-conjugative IncQ plasmid RSF1010 into recipient strains (Disqué-Kochem et al. 2001).

Coombs (2009) published an interesting book chapter on the potential of HGT in microbial communities of the terrestrial subsurface where several plasmids have been documented. The incidence of plasmids among "culturable" isolates from two subsurface sites with different sediment composition varied from 1.8% (Ogunseitan et al. 1987) to 33% (Fredrickson et al. 1988). Based on these data and further studies (Coombs 2009) the occurrence of plasmids at subsurface sites appeared to be within the range of plasmid incidences in other environments. Subsurface strains may carry single (Feng et al. 2007) or multiple (Brockman et al. 1989) plasmids, with large plasmids appearing to dominate (Fredrickson et al. 1988). The occurrence of very large (>150 kb) likely conjugative plasmids in deep subsurface bacteria has been found to be greater than the frequency from shallower subsurface soils (Ogunseitan et al. 1987; Coombs, 2009) or from marine environments (e.g. Sobecky et al. 1997).

Ansari et al. (2008) searched for conjugative plasmids in multiple antibiotic resistant bacterial isolates from contaminated Indian soils (steel industry, Ghaziabad) by PCR assays, detecting *oriT*s and replication functions of conjugative plasmids from different incompatibility groups. All the isolates gave PCR products with *trfA2* and *oriT* primers of the IncP group. Ansari et al. (2008) concluded that the conjugative IncP plasmids are prevalent in these soils, and may be mainly responsible for the antibiotic resistance transfer in these anthropogenic environments.

Total community DNA of the heavy-metal-contaminated soil from Lucknow, India (pesticide industry) was isolated and investigated for the presence of different plasmid specific sequences such as *tfrA2*, IncP- *oriT*, IncW- *oriT*, IncN- *rep*, *oriV*,

IncQ-*oriT* and pMV158-*oriT* by gene-specific PCR (M.I. Ansari, K. Schiwon and E. Grohmann, unpubl. data). In agreement with the data obtained by Ansari et al. (2008), only IncP specific sequences were detected in the Lucknow soil. Some of the isolates were also tested for the presence of transfer genes, *virB4*-, *virD4*- and *virB1*-like sequences derived from conjugative plasmids from Gram-positive bacteria, in particular from enterococci and staphylococci. Three isolates harboured the *orf5* gene, the *virB4* homolog of the broad-host-range conjugative plasmid pIP501, one of the isolates additionally harboured the *orf7* gene, encoding the VirB1-like lytic transglycosylase from plasmid pIP501. The pSK41/pGO1 relaxase (Mob_V relaxase family, Fig. 5.1) was not detected in any of the isolates.

Hu et al. (2009) developed a PCR-based strategy to screen for the presence of the *Bacillus anthracis* pXO1- and/or pXO2-like replicons, as well as for the presence of transfer modules related to those of the pXO2-like conjugative plasmid pAW63. pXO1- and pXO2-like replicons were present in ca. 6.6% and 7.7% of random environmental samples originating from soil, plants, insects, mammals, and water, respectively. Only ca. 1.54% of the strains were positive for pXO2-like transfer module genes.

5.6 Conclusions

The development of molecular detection methods applicable to complex environmental samples of aquatic and terrestrial origin has made tremendous progress in the recent twenty years. These techniques are not considered any more as costly sophisticated techniques only useful for the analysis of pure cultures and clinical samples, but they have achieved broad acceptance in microbial ecology. As the majority of environmental microorganisms are not culturable under laboratory conditions, on standard media, molecular techniques are often the sole option to characterize the physiology, core and accessory genome(s) and mobile gene pool of microbial consortia. This chapter has summarized the current molecular methodology successfully applied on environmental samples. It has also provided an overview on the occurrence and frequency of antibiotic resistance and heavy metal resistance genes in different aquatic and terrestrial habitats, with emphasis on environments suffering anthropogenic impact. The analysis of the occurrence of conjugative plasmids and transfer genes in the environment should help assessing the transmissibility of resistance and virulence factors within and among the microbial populations in the respective habitat.

5.7 Perspectives

The development of super fast next generation high throughput sequencing technologies at affordable costs with application on environmental samples has already revolutionized genetic analysis of environmental samples of diverse origins.

5 Molecular Detection of Resistance and Transfer Genes in Environmental Samples 185

Metagenome analyses of anthropogenic environments as well as of remote habitats have provided first insights on the global picture of the indigenous microbial communities, their physiology and genetic potential with respect to adaptability to stress factors and HGT. Many large-scale metagenome projects on diverse terrestrial and aquatic habitats (e.g. Alaskan soil metagenome, marine metagenome from coastal waters, wastewater metagenome) have been recently finished or are still in progress (e.g., Barz et al., 2010; Biddle et al., 2011; Schlüter et al., 2008).

qRT-PCR is getting more and more popular, and less costly, in the application on complex environmental samples. In the near future, we will not rely any more on qualitative or semi quantitative data with respect to environmental pollution by pathogenic microbes or resistance factors, but we will be continuously provided with reliable quantitative data on the environmental pollution factor(s) of interest.

The design of molecular signatures, such as those developed by Storteboom et al. (2010) for ARGs in pristine rivers, animal feeding operations lagoons and wastewater treatment plants will help trace anthropogenic sources of pollution and unambiguously distinguish between the various putative sources of the pollutants. Thus, they will speed up the identification of the pollutant sources and consequently the opportunity to take appropriate measures to reduce, to avoid or to prevent the pollution.

Acknowledgements We sincerely thank Miquel Salgot for critical reading of the manuscript. We regret that not all valuable contributions of colleagues in the field could have been included in this review due to space limitation.

References

Abou-Shanab RA, van Berkum P, Angle JS (2007) Heavy metal resistance and genotypic analysis of metal resistance genes in gram-positive and gram-negative bacteria present in Ni-rich serpentine soil and in the rhizosphere of *Alyssum murale*. Chemosphere 68:360–367

Abriouel H, Omar NB, Molinos AC, López RL, Grande MJ, Martínez-Viedma P, Ortega E, Cañamero MM, Galvez A (2008) Comparative analysis of genetic diversity and incidence of virulence factors and antibiotic resistance among enterococcal populations from raw fruit and vegetable foods, water and soil, and clinical samples. Int J Food Microbiol 123:38–49

Agersø Y, Petersen A (2007) The tetracycline resistance determinant Tet 39 and the sulphonamide resistance gene *sulII* are common among resistant *Acinetobacter* spp. isolated from integrated fish farms in Thailand. J Antimicrob Chemother 59:23–27

Ahmann D, Roberts AL, Krumholz LR, Morel FM (1994) Microbe grows by reducing arsenic. Nature 371(6500):750

Ahmed AM, Furuta K, Shimomura K, Kasama Y, Shimamoto T (2006) Genetic characterization of multidrug resistance in Shigella spp. from Japan. J Med Microbiol 55:1685–1691

Akinbowale OL, Peng H, Barton MD (2007) Diversity of tetracycline resistance genes in bacteria from aquaculture sources in Australia. J Appl Microbiol 103:2016–2025

Allen HK, Moe LA, Rodbumrer J, Gaarder A, Handelsman J (2009) Functional metagenomics reveals diverse beta-lactamases in a remote Alaskan soil. ISME J 3:243–251

Alvarez-Martinez CE, Christie PJ (2009) Biological diversity of prokaryotic type IV secretion systems. Microbiol Mol Biol Rev 73:775–808

Ansari MI, Grohmann E, Malik A (2008) Conjugative plasmids in multi-resistant bacterial isolates from Indian soil. J Appl Microbiol 104:1774–1781

Antwerpen MH, Schellhase M, Ehrentreich-Foerster E, Witte W, Nuebel U (2007) DNA microarray for detection of antibiotic resistance determinants in *Bacillus anthracis* and closely related *Bacillus cereus*. Mol Cell Probes 21:152–160

Auerbach EA, Seyfried EE, McMahon KD (2007) Tetracycline resistance genes in activated sludge wastewater treatment plants. Water Res 41:1143–1151

Bahl MI, Burmølle M, Meisner A, Hansen LH, Sørensen SJ (2009) All IncP-1 plasmid subgroups, including the novel epsilon subgroup, are prevalent in the influent of a Danish wastewater treatment plant. Plasmid 62:134–139

Barkay T, Miller SM, Summers AO (2003) Bacterial mercury resistance from atoms to ecosystems. FEMS Microbiol Rev 27:355–384

Barkay T, Liebert C, Gillman M (1989) Hybridization of DNA probes with whole-community genome for detection of genes that encode microbial responses to pollutants: *mer* genes and Hg^{2+} resistance. Appl Environ Microbiol 55:1574–1577

Barker-Reid F, Fox EM, Faggian R (2010) Occurrence of antibiotic resistance genes in reclaimed water and river water in the Werribee Basin, Australia. J Water Health 8:521–531

Barz M, Beimgraben C, Staller T, Germer F, Opitz F, Marquardt C, Schwarz C, Gutekunst K., Vanselow KH, Schmitz R, LaRoche J, Schulz R, Appel J 2010. Distribution analysis of hydrogenases in surface waters of marine and freshwater environments. PLoS One 5:e13846

Batt AL, Snow DD, Aga DS (2006) Occurrence of sulfonamide antimicrobials in private water wells in Washington County, Idaho, USA. Chemosphere 64:1963–1971

Bell JM, Paton JC, Turnidge J (1998) Emergence of vancomycin-resistant enterococci in Australia: phenotypic and genotypic characteristics of isolates. J Clin Microbiol 36:2187–2190

Biddle JF, White JR, Teske AP, House CH (2011) Metagenomics of the subsurface Brazos-Trinity Basin (IODP site 1320): comparison with other sediment and pyrosequenced metagenomes. ISME J 5(6):1038–1047. doi:10.1038/ismej.2010.199

Binh CT, Heuer H, Kaupenjohann M, Smalla K (2008) Piggery manure used for soil fertilization is a reservoir for transferable antibiotic resistance plasmids. FEMS Microbiol Ecol 66:25–37

Böckelmann U, Dörries H-H, Ayuso-Gabella MN, Salgot de Marçay M, Tandoi V, Levantesi C, Masciopinto C, Van Houtte E, Szewzyk U, Wintgens T, Grohmann E (2009) Quantitative real-time PCR monitoring of bacterial pathogens and antibiotic resistance genes in three European artificial groundwater recharge systems. Appl Environ Microbiol 75:154–163

Börjesson S, Dienues O, Jarnheimer PA, Olsen B, Matussek A, Lindgren PE (2009a) Quantification of genes encoding resistance to aminoglycosides, beta-lactams and tetracyclines in wastewater environments by real-time PCR. Int J Environ Health Res 19:219–230

Börjesson S, Melin S, Matussek A, Lindgren PE (2009b) A seasonal study of the *mecA* gene and *Staphylococcus aureus* including methicillin-resistant *S. aureus* in a municipal wastewater treatment plant. Water Res 43:925–932

Borremans B, Hobman JL, Provoost A, Brown NL, van der Lelie D (2001) Cloning and functional analysis of the *pbr* lead resistance determinant of *Ralstonia metallidurans* CH34. J Bacteriol 183:5651–5658

Brochet M, Couvé E, Zouine M, Poyart C, Glaser P (2008) A naturally occurring gene amplification leading to sulfonamide and trimethoprim resistance in *Streptococcus agalactiae.* J Bacteriol 190:672–680

Brockman FJ, Denovan BA, Hicks RJ, Fredrickson JK (1989) Isolation and characterization of quinoline-degrading bacteria from subsurface sediments. Appl Environ Microbiol 55:1029–1032

Brown NL, Rouch DA, Lee BTO (1992) Copper resistance determinants in bacteria. Plasmid 27:41–51

Bruce KD (1997) Analysis of *mer* gene subclasses within bacterial communities in soils and sediments resolved by fluorescent-PCR-restriction fragment length polymorphism profiling. Appl Environ Microbiol 63:4914–4919

Byrne-Bailey KG, Gaze WH, Kay P, Boxall AB, Hawkey PM, Wellington EM (2009) Prevalence of sulfonamide resistance genes in bacterial isolates from manured agricultural soils and pig slurry in the United Kingdom. Antimicrob Agents Chemother 53:696–702

5 Molecular Detection of Resistance and Transfer Genes in Environmental Samples 187

Cai L, Liu G, Rensing C, Wang G (2009) Genes involved in arsenic transformation and resistance associated with different levels of arsenic-contaminated soils. BMC Microbiol 9:4

Call DR (2005) Challenges and opportunities for pathogen detection using DNA microarrays. Crit Rev Microbiol 31:91–99

Castiglioni S, Pomati F, Miller K, Burns BP, Zuccato E, Calamari D, Neilan BA (2008) Novel homologs of the multiple resistance regulator *marA* in antibiotic-contaminated environments. Water Res 42:4271–4280

Chen CY, Wu KM, Chang YC, Chang CH, Tsai HC, Liao TL, Liu YM, Chen HJ, Shen AB, Li JC, Su TL, Shao CP, Lee CT, Hor LI, Tsai SF (2003) Comparative genome analysis of *Vibrio vulnificus*, a marine pathogen. Genome Res 13:2577–2587

Chen J, Yu Z, Michel FC Jr, Wittum T, Morrison M (2007) Development and application of real-time PCR assays for quantification of *erm* genes conferring resistance to macrolides-lincos-amides-streptogramin B in livestock manure and manure management systems. Appl Environ Microbiol 73:4407–4416

Cooksey DA (1994) Molecular mechanisms of copper resistance and accumulation in bacteria. FEMS Microbiol Rev 14:381–386

Coombs JM (2009) Potential for horizontal gene transfer in microbial communities of the terrestrial subsurface. In: Gogarten MB (ed) Horizontal gene transfer: genomes in flux, vol. 532, Humana Press, New York

Cullen WR, Reimer KJ (1989) Arsenic speciation in the environment. Chem Rev 89:713–764

Dang H, Zhang X, Song L, Chang Y, Yang G (2007) Molecular determination of oxytetracycline-resistant bacteria and their resistance genes from mariculture environments of China. J Appl Microbiol 103:2580–2592

da Silva MF, Vaz-Moreira I, Gonzalez-Pajuelo M, Nunes OC, Manaia CM (2007) Antimicrobial resistance patterns in *Enterobacteriaceae* isolated from an urban wastewater treatment plant. FEMS Microbiol Ecol 60:166–176

de Lipthay JR, Rasmussen LD, Oregaard G, Simonsen K, Bahl MI, Kroer N, Sørensen SJ (2008) Acclimation of subsurface microbial communities to mercury. FEMS Microbiol Ecol 65:145–155

Disqué-Kochem C, Battermann A, Strätz M, Dreiseikelmann B (2001) Screening for *trbB*- and *traG*-like sequences by PCR for the detection of conjugative plasmids in bacterial soil isolates. Microbiol Res 156:159–168

Ehrlich HL (1996) Geomicrobial interactions with arsenic and antimony, In: Ehrlich HL (ed) Geomicrobiology, 3rd edn. Dekker Inc., New York, NY, pp 276–293

Feng L, Wang W, Cheng J, Ren Y, Zhao G, Gao C, Tang Y, Liu X, Han W, Peng X, Liu R, Wang L (2007) Genome and proteome of long-chain alkane degrading *Geobacillus thermodenitrificans* NG80-2 isolated from a deep-subsurface oil reservoir. Proc Natl Acad Sci USA 104:5602–5607

Francia MV, Varsaki A, Garcillán-Barcia MP, Latorre A, Drainas C, de la Cruz F (2004) A classification scheme for mobilization regions of bacterial plasmids. FEMS Microbiol Rev 28:79–100

Fredrickson JK, Hicks RJ, Li SW, Brockman FJ (1988) Plasmid incidence in bacteria from deep subsurface sediments. Appl Environ Microbiol 54:2916–2923

Frye JG, Lindsey RL, Rondeau G, Porwollik S, Long F, McClelland M, Jackson CR, Englen MD, Meinersmann RJ, Berrang ME, Davis JA, Barrett JB, Turpin JB, Thitaram SN, Fedorka-Cray PJ (2010) Development of a DNA microarray to detect antimicrobial resistance genes identified in the National Center for Biotechnology Information database. Microb Drug Resist 16:9–19

Garcillán-Barcia MP, Francia MV, de la Cruz F (2009) The diversity of conjugative relaxases and its application in plasmid classification. FEMS Microbiol Rev 33:657–687. Review

Gilbride KA, Lee DY, Beaudette LA (2006) Molecular techniques in wastewater: understanding microbial communities, detecting pathogens, and real-time process control. J Microbiol Methods 66:1–20

Grohmann E, Muth G, Espinosa M (2003) Conjugative plasmid transfer in gram-positive bacteria. Microbiol Mol Biol Rev 67:277–301

Guardabassi L, Agersø Y (2006) Genes homologous to glycopeptide resistance *vanA* are widespread in soil microbial communities. FEMS Microbiol Lett 259:221–225

Haferburg G, Kothe E (2007) Microbes and metals: interactions in the environment. J Basic Microbiol 47:453–467. Review

Happi CT, Gbotosho GO, Folarin OA, Akinboye DO, Yusuf BO, Ebong OO, Sowunmi A, Kyle DE, Milhous W, Wirth DF, Oduola AM (2005) Polymorphisms in *Plasmodium falciparum dhfr* and *dhps* genes and age related *in vivo* sulfadoxine-pyrimethamine resistance in malaria-infected patients from Nigeria. Acta Trop 95:183–193

Heuer H, Kopmann C, Binh CT, Top EM, Smalla K (2009) Spreading antibiotic resistance through spread manure: characteristics of a novel plasmid type with low %G+C content. Environ Microbiol 11:937–949

Hu X, Van der Auwera G, Timmery S, Zhu L, Mahillon J (2009) Distribution, diversity and potential mobility of extra-chromosomal elements related to the *Bacillus anthracis* pXO1- and pXO2 virulence plasmids. Appl Environ Microbiol 75: 3016–3028

Iversen A, Kühn I, Rahman M, Franklin A, Burman LG, Olsson-Liljequist B, Torell E, Möllby R (2004) Evidence for transmission between humans and the environment of a nosocomial strain of *Enterococcus faecium*. Environ Microbiol 6:55–59

Jareonmit P, Sajjaphan K, Sadowsky MJ (2010) Structure and diversity of arsenic resistant bacteria in an old tin mine area of Thailand. J Microbiol Biotechnol 20:169–178

Jensen LB, Agerso Y, Sengelov G (2002) Presence of *erm* genes among macrolide-resistant Gram-positive bacteria isolated from Danish farm soil. Environ Int 28:487–491

Jindal A, Kocherginskaya S, Mehboob A, Robert M, Mackie RI, Raski L, Zilles JL (2006) Antimicrobial use and resistance in swine waste treatment systems. Appl Environ Microbiol 72:7813–7820

Kemper N (2008) Veterinary antibiotics in the aquatic and terrestrial environment. Ecol Indic 8:1–13

Kim SH, Wei CI, Tzou YM, An HJ (2005) Multidrug-resistant *Klebsiella pneumoniae* isolated from farm environments and retail products in Oklahoma. J Food Prot 68:2022–2029

Klotz MG, Arp DJ, Chain PS, El-Sheikh AF, Hauser LJ, Hommes NG, Larimer FW, Malfatti SA, Norton JM, Poret-Peterson AT, Vergez LM, Ward BB (2006) Complete genome sequence of the marine, chemolithoautotrophic, ammonia-oxidizing bacterium *Nitrosococcus oceani* ATCC 19707. Appl Environ Microbiol 72:6299–6315

Knapp CW, Dolfing J, Ehlert PA, Graham DW (2010) Evidence of increasing antibiotic resistance gene abundances in archived soils since (1940). Environ Sci Technol 44:580–587

Koike S, Krapac IG, Oliver HD, Yannarell AC, Chee-Sanford JC, Aminov RI, Mackie RI (2007) Monitoring and source tracking of tetracycline resistance genes in lagoons and groundwater adjacent to swine production facilities over a 3-year period. Appl Environ Microbiol 73:4813–4823

Kumar A, Schweizer HP (2005) Bacterial resistance to antibiotics: active efflux and reduced uptake. Adv Drug Deliv Rev 57:1486–1513

Lachmayr KL, Kerkhof LJ, Dirienzo AG, Cavanaugh CM, Ford TE (2009) Quantifying nonspecific TEM beta-lactamase (*bla*TEM) genes in a wastewater stream. Appl Environ Microbiol 75:203–211

Lambert PA (2005) Bacterial resistance to antibiotics: Modified target sites. Adv Drug Deliv Rev 57:1471–1485

Lejon DP, Nowak V, Bouko S, Pascault N, Mougel C, Martins JM, Ranjard L (2007) Fingerprinting and diversity of bacterial *copA* genes in response to soil types, soil organic status and copper contamination. FEMS Microbiol Ecol 61:424–437

Levy SB (2002) Factors impacting on the problem of antibiotic resistance. J Antimicrob Chemother 49:25–30

Levy SB, FitzGerald GB, Macone AB (1976) Changes in intestinal flora of farm personnel after introduction of a tetracycline-supplemented feed on a farm. N Engl J Med. 295: 583–588

Levy SB, McMurry LM, Barbosa TM, Burdett V, Courvalin P, Hillen W, Roberts MC, Rood JI, Taylor DE (1999) Nomenclature for new tetracycline resistance determinants. Antimicrob Agents Chemother 43:1523–1524

Levy C, Minnis D, Derrick JP (2008) Dihydropteroate synthase from *Streptococcus pneumoniae*: structure, ligand recognition and mechanism of sulfonamide resistance. Biochem J 412: 379–388

5 Molecular Detection of Resistance and Transfer Genes in Environmental Samples 189

Li XZ, Mehrotra M, Ghimire S, Adewoye L (2007) β-lactam resistance and β-lactamases in bacteria of animal origin. Vet Microbiol 121: 197–214

Mackie RI, Koike S, Krapac I, Chee-Sanford J, Maxwell S, Aminov RI (2006) Tetracycline residues and tetracycline resistance genes in groundwater impacted by swine production facilities. Anim Biotechnol 17:157–176

Malik A, Celik E-K, Bohn C, Boeckelmann U, Knobel K, Grohmann E (2008) Molecular detection of conjugative plasmids and antibiotic resistance genes in anthropogenic soils from Germany and India. FEMS Microbiol Lett 279:207–216

Markoulatos P, Siafakas N, Moncany M (2002) Multiplex polymerase chain reaction: a practical approach. J Clin Lab Anal 16:47–51

Mendez B, Tachibana C, Levy SB (1980) Heterogeneity of tetracycline resistance determinants. Plasmid 3:99–108

Messi P, Guerrieri E, de Niederhäusern S, Sabia C, Bondi M (2006) Vancomycin-resistant enterococci (VRE) in meat and environmental samples. Int J Food Microbiol 107:218–222

Mindlin SZ, Soina VS, Ptrova MA, Gorlenko ZhM (2008) Isolation of antibiotic resistance bacterial strains from East Siberia permafrost sediments. Genetika 44:36–44

Moura A, Henriques I, Ribeiro R, Correia A (2007) Prevalence and characterization of integrons from bacteria isolated from a slaughterhouse wastewater treatment plant. J Antimicrob Chemother 60:1243–1250

Moura A, Henriques I, Smalla K, Correia A (2010) Wastewater bacterial communities bring together broad-host range plasmids, integrons and a wide diversity of uncharacterized gene cassettes. Res Microbiol 161:58–66

Nierman WC, Feldblyum TV, Laub MT et al (2001) Complete genome sequence of *Caulobacter crescentus*. Proc Natl Acad Sci USA 98:4136–4141

Nies DH (1999) Microbial heavy-metal resistance. Appl Microbiol Biotechnol 51:730–750

Nikolakopoulou TL, Giannoutsou EP, Karabatsou AA, Karagouni AD (2008) Prevalence of tetracycline resistance genes in Greek seawater habitats. J Microbiol 46:633–640

Nriagu JO, Pacyna JM (1988) Quantitative assessment of worldwide contamination of air, water and soils by trace metals. Nature 333:134–139

Ogunseitan OA, Tedford ET, Pacia D, Sirotkin KM, Sayler GS (1987) Distribution of plasmids in groundwater bacteria. J Ind Microbiol 1:311–317

Oregaard G, Sørensen SJ (2007) High diversity of bacterial mercuric reductase genes from surface and sub-surface floodplain soil (Oak Ridge, USA). ISME J 1:453–467

Osborn AM, Bruce KD, Strike P, Ritchie DA (1997) Distribution, diversity and evolution of the bacterial mercury resistance (*mer*) operon. FEMS MicrobiolRev 19:239–262

Patterson AJ, Colangeli R, Spigaglia P, Scott KP (2007) Distribution of specific tetracycline and erythromycin resistance genes in environmental samples assessed by macroarray detection. Environ Microbiol 9:703–715

Pechrada J, Sajjaphan K, Sadowsky MJ (2010) Structure and diversity of arsenic-resistant bacteria in an old tin mine area of Thailand. J Microbiol Biotechnol 20:169–178

Pei R, Kim SC, Carlson KH, Pruden A (2006) Effect of river landscape on the sediment concentrations of antibiotics and corresponding antibiotic resistance genes (ARG). Water Res 40:2427–2435

Picão RC, Poirel L, Demarta A, Silva CS, Corvaglia AR, Petrini O, Nordmann P (2008) Plasmid-mediated quinolone resistance in *Aeromonas allosaccharophila* recovered from a Swiss lake. J Antimicrob Chemother 62:948–950

Prabhu DIG, Pandian RS, Vasan PT (2007) Pathogenicity, antibiotic susceptibility and genetic similarity of environmental and clinical isolates of *Vibrio cholerae*. Indian J Exp Biol 45:817–823

Ramachandran D, Bhanumathi R, Singh DV (2007) Multiplex PCR for detection of antibiotic resistance genes and the SXT element: application in the characterization of *Vibrio cholerae*. J Med Microbiol 56:346–351

Ramón-García S, Otal I, Martín C, Gómez-Lus R, Aínsa JA. (2006) Novel streptomycin resistance gene from *Mycobacterium fortuitum*. Antimicrob Agents Chemother 50:3920–3922

Rensing C, Grass G (2003) *Escherichia coli* mechanisms of copper homeostasis in a changing environment. FEMS Microbiol Rev 27:197–213

Roberts MC (2002) Resistance to tetracycline, macrolide-lincosamide-streptogramin, trimethoprim, and sulphonamide drug classes. Mol Biotechnol 20:261–283

Roberts MC (2005) Update on acquired tetracycline resistance genes. FEMS Microbiol Lett 245:195–203

Roberts MC, Kenny GE (1986) Dissemination of the *tetM* tetracycline resistance determinant to *Ureaplasma urealyticum*. Antimicrob Agents Chemother 29:350–352

Roberts MC, Sutcliffe J, Courvalin P, Jensen LB, Rood J, Seppala H (1999) Nomenclature for macrolide and macrolide-lincosamide-streptogramin B resistance determinants. Antimicrob Agents Chemother 43:2823–2830

Rochelle PA., Wetherbee MK, Olson BH (1991) Distribution of DNA sequences encoding narrow- and broad-spectrum mercury resistance. Appl Environ Microbiol 57:1581–1589

Salanoubat M, Genin S, Artiguenave F et al (2002) Genome sequence of the plant pathogen *Ralstonia solanacearum*. Nature 415:497–502

Samra ZQ, Naseem M, Khan SJ, Dar N, Athar MA (2009) PCR targeting of antibiotic resistant bacteria in public drinking water of Lahore metropolitan, Pakistan. Biomed Environ Sci 22:458–463

Sarmah AK, Meyer MT, Boxall ABA (2006) A global perspective on the use, sales, exposure pathways, occurrence, fate and effects of veterinary antibiotics (VAs) in the environment. Chemosphere 65:725–759

Schelert J, Dixit V, Hoang V, Simbahan J, Drozda M, Blum P (2004) Occurrence and characterization of mercury resistance in the hyperthermophilic archaeon *Sulfolobus solfataricus* by use of gene disruption. J Bacteriol 186:427–437

Schlüter A, Krause L, Szczepanowski R, Goesmann A, Pühler A (2008) Genetic diversity and composition of a plasmid metagenome from a wastewater treatment plant. J Biotechnol 136:65–76

Schwarz S, Kehrenberg C, Doublet B, Cloeckaert A (2004) Molecular basis of bacterial resistance to chloramphenicol and florfenicol. FEMS Microbiol Rev 28:519–542

Shakil S, Khan R, Zarrilli R, Khan AU (2008) Aminoglycosides versus bacteria- a description of the action, resistance mechanism, and nosocomial battleground. J Biomed Sci 15:5–14

Silver S, Phung LT (1996) Bacterial heavy metal resistance: new surprises. Annu Rev Microbiol 50:753–789

Sköld O (2000) Sulfonamide resistance: mechanisms and trends. Drug Resist Updat 3:155–160

Sköld O (2001) Resistance to trimethoprim and sulfonamides. Vet Res 32:261–273

Smalla K, Haines AS, Jones K, Krögerrecklenfort E, Heuer H, Schloter M, Thomas CM (2006) Increased abundance of IncP-1beta plasmids and mercury resistance genes in mercury-polluted river sediments: first discovery of IncP-1beta plasmids with a complex *mer* transposon as the sole accessory element. Appl Environ Microbiol 72:7253–7259

Smith MS, Yang RK, Knapp CW, Niu Y, Peak N, Hanfelt MM, Galland JC, Graham DW (2004) Quantification of tetracycline resistance genes in feedlot lagoons by real-time PCR. Appl Environ Microbiol 70:7372–7377

Sobecky PA, Hazen TH (2009) Horizontal gene transfer and mobile genetic elements in marine systems. In: Gogarten MB et al (eds) Horizontal gene transfer: genomes in flux, vol. 532. Humana Press, New York

Sobecky PA, Mincer TJ, Chang MC, Helinski DR (1997) Plasmids isolated from marine sediment microbial communities contain replication and incompatibility regions unrelated to those of known plasmid groups. Appl Environ Microbiol 63:888–895

Srinivasan V, Nam HM, Nguyen LT, Tamilselvam B, Murinda SE, Oliver SP (2005) Prevalence of antimicrobial resistance genes in *Listeria monocytogenes* isolated from dairy farms. Foodborne Pathog Dis 2:201–211

Stabler RA, Dawson LF, Oyston PC, Titball RW, Wade J, Hinds J, Witney AA, Wren BW (2008) Development and application of the active surveillance of pathogens microarray to monitor bacterial gene flux. BMC Microbiol 8:177

Storteboom H, Arabi M, Davis JG, Crimi B, Pruden A (2010) Identification of antibiotic-resistance-gene molecular signatures suitable as tracers of pristine river, urban, and agricultural sources. Environ Sci Technol 44:1947–1953

5 Molecular Detection of Resistance and Transfer Genes in Environmental Samples 191

Stover CK, Pham XQ, Erwin AL et al (2000) Complete genome sequence of *Pseudomonas aeruginosa* PAO1, an opportunistic pathogen. Nature 406:959–964

Swingley WD, Sadekar S, Mastrian SD, Matthies HJ, Hao J, Ramos H, Acharya CR, Conrad AL, Taylor HL, Dejesa LC, Shah MK, O'huallachain ME, Lince MT, Blankenship RE, Beatty JT, Touchman JW (2007) The complete genome sequence of *Roseobacter denitrificans* reveals a mixotrophic rather than photosynthetic metabolism. J Bacteriol 189:683–690

Thompson SA., Maani EV, Lindell AH, King CJ, McArthur JV (2007) Novel tetracycline resistance determinant isolated from an environmental strain of *Serratia marcescens.* Appl Environ Microbiol 73:2199–2206

Tom-Petersen A, Leser TD, Marsh TL, Nybroe O (2003) Effects of copper amendment on the bacterial community in agricultural soil analyzed by the T-RFLP technique. FEMS Microbiol Ecol 46:53–62

Ushida K, Segawa T, Kohshima S, Takeuchi N, Fukui K, Li Z, Kanda H (2010) Application of real-time PCR array to the multiple detection of antibiotic resistant genes in glacier ice samples. J Gen Appl Microbiol 56:43–52

Volkmann H, Schwartz T, Bischoff P, Kirchen S, Obst U (2004) Detection of clinically relevant antibiotic-resistance genes in municipal wastewater using real-time PCR (TaqMan). J Microbiol Methods 56:277–286

Waldron PJ, Wu L, Van Nostrand JD, Schadt CW, He Z, Watson DB, Jardine PM, Palumbo AV, Hazen TC, Zhou J (2009) Functional gene array-based analysis of microbial community structure in groundwaters with a gradient of contaminant levels. Environ Sci Technol 43:3529–3534

Wright GD (2005) Bacterial resistance to antibiotics: enzymatic degradation and modification. Adv Drug Deliv Rev 57:1451–1470

Wright GD (2007) The antibiotic resistome: the nexus of chemical and genetic diversity. Nat Rev Microbiol 5:175–186

Yang S, Carlson K (2003) Evolution of antibiotic occurrence in a river through pristine, urban and agricultural landscapes. Water Res 37:4645–4656

Yu Z, Michel FC Jr, Hansen G, Wittum T, Morrison M (2005) Development and application of real-time PCR assays for quantification of genes encoding tetracycline resistance. Appl Environ Microbiol 71:6926–6933

Zhang T, Fang HHP (2006) Applications of real-time polymerase chain reactions for quantification of microorganisms in environmental samples. Appl Microbiol Biotechnol 70:281–289

Zhang X-X, Zhang T, Fang HP (2009) Antibiotic resistance genes in water environment. Appl Microbiol Biotechnol 82: 397–414

Zhou Z, Raskin L, Zilles JL (2010) Effects of swine manure on macrolide, lincosamide, and streptogramin B antimicrobial resistance in soils. Appl Environ Microbiol 76:2218–2224

Zhu B (2007) Abundance dynamics and sequence variation of neomycin phosphotransferase gene (*npt*II) homologs in river water. Aquat Microbiol Ecol 48:131–140

Chapter 6
Key Biochemical Attributes to Assess Soil Ecosystem Sustainability

Vito Armando Laudicina, Paul G. Dennis, Eristanna Palazzolo
and Luigi Badalucco

Contents

6.1 Introduction 194
6.2 Microbial Biomass 195
6.3 Microbial Activity 198
    6.3.1 Soil Enzyme Activities 199
    6.3.2 Carbon Mineralisation (Soil Respiration) 202
    6.3.3 Nitrogen Mineralisation and Net Nitrification 204
6.4 Microbial Community Diversity and Function 205
    6.4.1 Nucleic Acid-Based Techniques 205
    6.4.2 Fatty Acid-Based Methods (PLFAS, FAMES, ELFAS) 211
6.5 Soil Quality Indexes 213
    6.5.1 Simple Indexes 213
    6.5.2 Complex Indices 215
6.6 Concluding Remarks 216
References 217

Abstract Soil is not a renewable resource, at least within the human timescale. In general, any anthropic exploitation of soils tends to disturb or divert them from a more "natural" development which, by definition, represents the best comparison term for measuring the relative shift from soil sustainability. The continuous degradation of soil health and quality due to abuse of land potentiality or intensive management occurs since decades. Soil microbiota, being 'the biological engine of the Earth', provides pivotal services in the soil ecosystem functioning. Hence, management practices protecting soil microbial diversity and resilience, should be pursued. Besides, any abnormal change in rate of innumerable soil biochemical processes, as mediated by microbial communities, may constitute early and sensitive warning of soil homeostasis alteration and, therefore, diagnoses a possible risk for soil sustainability. Among the vastness of soil biochemical processes and related attributes (bioindicators) potentially able to assess the sustainable use of soils, those

---

L. Badalucco (✉)
Dipartimento dei Sistemi Agro-Ambientali, University of Palermo, Viale delle Scienze, Ed. 4, 90128 Palermo, Italy
e-mail: badalucc@unipa.it

A. Malik, E. Grohmann (eds.), *Environmental Protection Strategies for Sustainable Development*, Strategies for Sustainability,
DOI 10.1007/978-94-007-1591-2_6, © Springer Science+Business Media B.V. 2012

related to mineralisation-immobilisation of major nutrients (C and N), including enzyme activity (functioning) and composition (community diversity) of microbial biomass, have paramount importance due to their centrality in soil metabolism. In this chapter we have compared, under various pedoclimates, the impact of different agricultural factors (fertilisation, tillage, etc.) under either intensive and sustainable managements on soil microbial community diversity and functioning by both classical and molecular soil quality indicators, in order to outline the most reliable soil biochemical attributes for assessing risky shifts from soil sustainability.

Keywords Soil quality • Soil enzymes • C and N mineralisation-immobilisation • Microbial diversity • Nucleic acid- and fatty acid-based indicators

6.1 Introduction

Soils perform a wide range of ecosystem functions that are crucial for the majority of terrestrial life. Soils provide microbial habitat spaces of diverse size and architecture, as well as reservoirs of chemicals such as nutrients, and ecosystem services such as water filtration and storage. In addition, soils support biological activities such as decomposition and recycling of dead organic matter, and play a major role in mitigating climate change through the sequestration of carbon.

Many human activities strongly influence soil functioning. Generally, exploitation of soils tends to disturb or divert them from a more "natural" and, as much as possible, undisturbed development which, by definition, represents the best comparison term for measuring any relative shift from soil operational sustainability. Agriculture management practices have been developed to manipulate soil system to facilitate planting and the production of food, forage and fibre, i.e. to improve the overall plant productivity.

Developing sustainable land management systems is complicated by the need to consider their utility to humans, their efficiency of resource use, and their ability to maintain a balance with the environment that is favourable both to humans and most other species, in order to preserve biodiversity (Harwood 1990). Although considerable research activity is aimed at the development and evaluation of sustainable management systems, these efforts are hampered by disagreements on what may constitute a reliable measure of sustainability. Gregorich et al. (2001) have defined sustainability as "the ability of an ecosystem to maintain ecological processes and functions, biological diversity and productivity over time".

Soil quality, being conceptualized as the major linkage between the strategies of conservation management practices and the achievement of the major goals of sustainable agriculture (Acton and Gregorich 1995; Parr et al. 1992), it is defined as the "continued capacity of soil to (1) function as a vital living system, within ecosystem and land use boundaries, (2) sustain biological productivity, (3) promote the quality of air and water environments, and (4) maintain plant, animal and human health" (Doran and Safley 1997). Hence, the assessment of soil quality or health, and direction of change with time, is the prerequisite for sustainable land management.

Therefore, soil quality indicators refer to measurable soil attributes that influence the capacity of soils to perform crop production and/or environmental functions. Among the vastness of chemical, physical and biological properties of soil involved in its functioning, Doran and Parkin (1996) have proposed a minimum data set to be used for soil quality assessment which includes physical (texture, rooting depth, infiltration rate, bulk density, water retention capacity), chemical (pH, total carbon content, electrical conductivity, nutrients level) and biological (C and N microbial biomass, potentially mineralisable N, soil respiration) as basic indicators for an initial characterisation of soil quality. Beyond those identified as basic, they suggested that other supplementary soil properties could be used to evaluate soil quality on the basis of specific climatic, geographic and socio-economic conditions. As the physical and physico-chemical parameters are altered only when the soil undergo a really drastic change (Filip 2002), many authors (Klein et al. 1985; Nannipieri et al. 1990) have proposed several biological and biochemical parameters as sensitive parameters also to the slight modifications that the soil can undergo under the action of any disturbing agent. Since soil microorganisms due to their quick metabolism can respond rapidly, they may reflect a hazardous environment and should be, therefore, preferentially considered when monitoring soil status.

Reliable soil microbiological and biochemical indicators to determine soil quality would be simple to measure, should work equally well in all environments and reveal which problems exist wherever. It is unlikely that a sole indicator can be defined with a single measurement because of the multitude of microbiological components and biochemical pathways. Microbial indicators of soil health cover a diverse set of microbial measurements due to the multi-functional properties of microbial communities in the soil ecosystem. Therefore, the basic indicators and the number of estimated measures are still under debate. Bloem et al. (2006) suggested that national and international programs for monitoring soil quality should include microflora and respiration measurements and also nitrogen mineralization, microbial diversity and functional groups of soil fauna.

In this chapter we intend to review and discuss both classical (microbial biomass and activity) and modern (microbial community structure and functioning) soil biochemical attributes widely accepted within the soil science scientific community for assessing the sustainability of the soil ecosystem.

6.2 Microbial Biomass

Soil microbial biomass (MB) can be defined as the portion of soil organic matter that constitutes living microorganisms smaller than 5–10 μm^3 (Jenkinson and Ladd 1981). Microbial biomass constitutes approximately 1–4% of the total organic carbon (MBC; Anderson and Domsch 1989; Sparling 1992) and 2–6% of the total organic nitrogen (MBN; Jenkinson 1988). Typically MBC ranges from 100 to >1000 mg C kg^{-1} soil (Paul et al. 1999). MB has a turnover time less than 1 year (Paul 1984) and, therefore, responds to stress/disturbance factors more rapidly than the whole soil organic matter, the content of which may need decades to appreciably change. Due to its dynamic nature, MB content at any one time cannot indicate whether soil organic matter (generally soil quality) is increasing, decreasing or at equilibrium. When using MB as soil quality indicator, it is important to consider its seasonal variability, which shows clear patterns depending partly on soil temperature and moisture as well as organic and inorganic inputs (Kandeler et al. 1999a; Emmerling et al. 2001; Ge et al. 2010; Feng and Simpson 2009). Hence, MB monitoring through the time is required to infer considerations on the changes in amount and nutrient content of the MB (Rice et al. 1996).

MB is typically measured indirectly using the following methods: fumigation-incubation (FI), substrate-induced respiration (SIR), fumigation-extraction (FE) and/or measurement of ATP content (Jenkinson and Powlson 1976; Anderson and Domsch 1978; Jenkinson and Ladd 1981; Brookes et al. 1985; Vance et al. 1987). These methods have greatly improved measurements of MB and its associated nutrient pools. However, Rice et al. (1996) recommend the fumigation techniques since they are inexpensive, easy to use, and facilitate assessment of the mineralisable C and N fractions of soil organic matter. Of the two fumigation methods, Paul et al. (1999) recommend the FE method as it is faster than FI and gives more reproducible results. FE involves two key steps: (1) destruction of microbial cell membranes by chloroform fumigation, and (2) extraction and analysis of cell constituents (Paul et al. 1999). Following FE, MBC and MBN are routinely determined but also MB phosphorus (MBP; Brookes et al. 1982), sulphur (MBS; Wu et al. 1994) and potassium (MBK; Lorenz et al. 2010) can be analysed. However, as Badalucco et al. (1997)) reported, when using the FE method the efficiency of chloroform in lysing microbial cells seems to be inversely related to soil structural stability.

Microbial biomass acts as both a nutrient reservoir and as a catalyst for organic matter decomposition. Consideration of MB is crucial, therefore, to understand nutrient fluxes within and between ecosystems (Smith and Paul 1990). Jenkinson et al. (1987) defined the MB as "the eye of the needle through which all organic matter needs to pass".

The effects on MB of tillage, crop rotation, inorganic/organic fertiliser applications and other agricultural practices have been extensively studied. Recently, Gonzalez-Chavez et al. (2010) reported the impact of 28-years of tillage on MBC, MBN and MBP in Texas. They found that MBC and MBN were nearly doubled, whereas MBP was 2.5 times higher in no-tillage compared to conventional tillage treatments. Balota et al. (2004) investigated the long-term effects of tillage and crop rotation on MB in a subtropical environment and found that no-tillage increased MBC soybean/wheat, maize/wheat and cotton/wheat rotations (from 11 to 98%) across all soil depths (0–5, 5–10 and 10–20 cm) when compared to conventional tillage. Doran (1980) suggests that, in tropical or arid/semiarid environments, the larger MB in no-tilled or minimally tilled soils when compared with soils that are conventionally tilled is due to the fact that surface litter can lower soil temperature and increase water content, soil aggregation and carbon content. Thus, no-tilled soils are not only high in available substrates but also wetter, cooler and less variable with respect to their temperature and moisture regimes. These conditions stim-

6 Key Biochemical Attributes to Assess Soil Ecosystem Sustainability 197

ulate the growth of soil microorganisms, particularly in surface layers (Balota et al. 2004; Ferreira et al. 2000).

The relationship between MBC and soil quality as a function of crop rotations has proven more difficult to understand. Crop rotation systems that include leguminous crops encourage a higher MBC/total organic carbon (TOC) ratio (microbial quotient) than monoculture systems (Anderson and Domsch 1989). This difference is attributed to the input of a high organic residue variety under crop rotation systems (Anderson and Domsch 1989). Franchini et al. (2007) observed increases in MBC in soybean fields previously cultivated with legumes (lupins) in comparison with those previously cultivated with wheat, and found that crop rotations that included a higher ratios of legume to non-legume resulted in higher microbial quotient values. Positive effects of legumes on MB were not observed in many studies, however. For this reason, it has been proposed that differences in microbiological parameters resulting from crop rotation are detectable in long-term trials only, even under no-tillage (Franchini et al. 2007). Moreover, results from crop rotations may be related to shifts in rhizodeposition of organic compounds (Matson et al. 1997) that either stimulate or suppress microbial activities. For example, with legumes flavonoids are released into the rhizosphere and facilitate rhizobial and mycorrhizal symbioses (Hungria and Stacey 1997; Ferreira et al. 2000). There are also reports on the effects of crop rotations on the suppression of soil-borne plant pathogens in wheat (Santos et al. 2000), maize (Denti and Reis 2001), and soybean (Hoffmann et al. 2004). There is, in conclusion, strong evidence that crop rotations generally lead to qualitative positive changes in microbial communities, also with positive overall effects on crop productivity (Nogueira et al. 2006; Pereira et al. 2007).

With regards to organic fertiliser amendments, there is growing and clear evidence that organic agricultural systems exhibit improved soil quality characterised by higher biological activity than conventional systems (Drinkwater et al. 1995; Droogers and Bouma 1996; Ge et al. 2010; Zhong et al. 2010). Microbial biomass and activity are greatly stimulated by the addition of manure through improvement of soil physical characteristics and addition of readily available sources of carbon and nitrogen (Reganold 1988; Fraser et al. 1988). Fließbach and Madër (2000) compared the effects on MB of three different long-term management systems: organic, biodynamic and conventional. They found that MBC and MBN, as well as their ratios to the total and light fraction C and N pools in soils of the organic systems, were higher than in conventional systems. This was interpreted as an enhanced decomposition of the easily available light fraction pool of soil organic matter with increasing amounts of microflora. The positive changes in MB and its activity are probably due to the addition of more labile organic substrates (Araujo et al. 2008).

Long-term inorganic fertilizer applications (NPK), typical in intensive agriculture, can have either deleterious or beneficial effects on soil microbial biomass and activity (Biederbeck et al. 1996; Simek et al. 1999; Böhme et al. 2005; Deng et al. 2006; Weigel et al. 1998; Kandeler et al. 1999a; Fließbach et al. 2000; Thirukkumaran and Parkinson 2000; Rampazzo and Mentler 2001; Ge et al. 2010). Recently, Badalucco et al. (2010) reported that reversing agriculture from intensive to sustainable improved soil quality under semiarid conditions.

MB is a sensitive indicator of toxicity attributable to pesticides, heavy metals, and other pollutants. Pollutants can affect microorganisms directly by causing toxic effects or indirectly by, for example, decreasing the availability of substrates such as plant root exudates. Thus the decreased energy available to the microbes could also result in a smaller population (Brookes 1995; Kizilkaya et al. 2004; Perez-de-Mora et al. 2006).

Soil MB is also affected by soil salinity and/or sodicity, which are serious land degradation issues worldwide (Laudicina et al. 2009). Rietz and Haynes (2003) found a significant negative relationship between MBC and soil electrical conductivity, and sodium adsorption ratio and exchangeable sodium percentage indicated an adverse effect of salinity and sodicity on MB. These findings are in agreement with other studies that demonstrate that MB is negatively correlated with total soluble salts in naturally saline soils (Mallouhi and Jacquin 1985; Ragab 1993; Garcia et al. 1994). Wong et al. (2008), however, observed a positive relationship between MB and salinity in a laboratory-based study. They attributed this observation to additional substrate becoming available with increasing salt concentration, with salinity effects being stronger than those attributable to sodicity. Besides, they have suggested that, when exchangeable sodium percentage is $\geq 6\%$, the associated increase in substrate availability could be related to the dominance of sodic processes such as aggregate dispersion. Under saline conditions, however, substrate availability results from an increase in the solubility of SOM (Wong et al. 2008).

6.3 Microbial Activity

Soil microbial activities are of critical importance for biogeochemical cycles. Microbial activity is regulated by many factors including: nutrient, oxygen and water availability, temperature, and soil pH. Soil microbial activity can be measured under either field or laboratory conditions. In the field, variations in meteorological conditions during the experiment are inevitable, i.e. soil aeration, moisture and temperature will change and may strongly influence the results (Madsen 1996). Furthermore, field measurements are often difficult to be interpreted. For example, soil respiration determined in the field is due to activity of microorganisms and other organisms such as macrofauna and plants, which vary significantly in different systems and throughout the season (Dilly et al. 2000). Laboratory procedures are usually carried out on sieved and stabilised soil samples at standardized temperature and water content. Such measurements generally include assays of enzyme activities, C and N mineralisation. These, and eventually other microbial activity measurements, may be helpful to evaluate effects of soil management, land use and specific environmental conditions (Burns 1977, 1978) on microbial activity. Laboratory methods allow the standardization of environmental factors and, thus, the comparison of results from soils of different geographical locations, environmental conditions and even different laboratories. However, laboratory measurements generally represent microbial potential activities, as they are determined under op-

6 Key Biochemical Attributes to Assess Soil Ecosystem Sustainability   199

timized conditions (Nannipieri et al. 1990). Field measurements of soil microbial activity are more complex than laboratory measurements and the interpretation and synthesis of results require considerable care, but they are needed in studies at ecosystem level.

6.3.1 Soil Enzyme Activities

Soil enzyme activities are reliable indicators of different soil management practices as they rapidly respond to different environmental conditions (Albiach et al. 1999; Benítez et al. 2006; Melero et al. 2007). The rationale behind using soil enzyme activities as soil quality indicators is that enzyme activities: (i) are often related to, and thus reflect, other soil parameters such as organic matter decay and turnover, soil physico-chemical properties and microbial activity or biomass (Dick 1994); (ii) can change more rapidly (1 or 2 years) than more stable soil properties (e.g. total organic C), thus providing an early indication of soil quality with changes in soil management practices or environmental conditions; (iii) can be integrated to give soil biological indices of past soil management; and (iv) involve simple assay methods that can be performed routinely by most laboratories (Dick et al. 1996).

Taylor et al. (2002) suggested two reasons for measuring soil enzyme activities: (1) for use as indicators of process diversity and, therefore, the biochemical potential of the soil, also in view of the manipulation of the soil system, and the possible resilience; (2) for use as indicators of soil quality, in the sense that changes in key functions and activities can provide information about the progress of specific remediation operations or the sustainability of particular types of land management.

The information obtained from enzyme assays should, however, be considered with caution (Pettit et al. 1977; Burns 1978; Nannipieri et al. 1990). For example, soil enzyme assays generally provide a measure of the potential activity, that is encoded in the whole "soil genotype", but this will rarely ever be expressed. In addition, information from soil enzyme assays may represent the redundancy of the soil biochemical system and as such is an aspect of resilience. Some soil enzyme assays attempt to measure real activity, i.e. a phenotypic property, but are rarely successful. By considering soil enzymes as indicators of soil quality, the question that arises is which enzymes are important. A case can be argued for at least 500 enzymes with critical roles in the cycling of C or N or both, but this is currently unfeasible for most laboratories. If there is genuine redundancy in enzymatic functions in soil, the loss of activity of a specific "keystone" enzyme should not have a major effect. If, on the other hand, changes in the activity of some "keystone" enzymes provide an early indication of changes in process diversity, soil enzyme measurements have a clear role in the assessment of soil quality.

The type of enzyme assay to use depends on the goal of the experiment. Dehydrogenase has potential to act as an indicator of viable soil microbial activity. It has a long history as a biological indicator and is often closely related to the average activity of microbial populations because it exists solely as an integral part of viable microorganisms (Skujins 1973). Dick et al. (1996) suggested that enzymes correlating with microbial activity may be less suited to predict long-term changes in soil quality because they would reflect recent management or seasonal (climatic) effects that may be transitory. Therefore, enzyme activities that remain after air drying and closely correlate with organic matter content may be better indicators of permanent changes in soil quality because such enzymes are probably associated to surviving microorganisms and/or complexed and protected against inactivation by soil humic- or clay-complexes. The rationale behind these considerations is that soil management practices that promote stabilisation of enzymes also promotes stabilisation of organic matter and associated structural properties. In general, hydrolytic enzymes are a good choice as indices of both (i) soil fertility (because they are involved in nutrient cycling), and (ii) soil quality (because organic residue-decomposing organisms are probably the major contributors to soil enzyme activity (Speir 1977; Speir and Ross 1976). The most widely assayed enzymes are those involved in the degradation of cellulose and lignin, which are the most abundant components of plant litter (Allison et al. 2007). Other commonly measured enzymes activities relate to hydrolysis of proteins, chitin and peptidoglycan, which are the principal reservoirs of organic N (Caldwell 2005). Phosphatases are of interest for their role in mineralizing P from nucleic acids, phospholipids and other ester phosphates (Turner et al. 2002; Toor et al. 2003). The structural heterogeneity of biopolymers requires the interaction of several classes of enzymes to reduce them to constituent monomers available for microbial consumption (Ljungdahl and Eriksson 1985; Kirk and Farrell 1987; Sinsabaugh et al. 2005). However, most studies of soil enzyme activities are limited to those that catalyse the production of the terminal monomers, because the kinetics are easier to study and the reactions produce nutrient mineralisation end products that are important in plant nutrition (Allison et al. 2007). Currently, the following soil enzyme assays are of widespread use:

1. β-glucosidases catalyse the hydrolysis of various β-glucosides, mainly cellobiose to glucose; they are active during the decomposition of organic matter and release products which are important energy source for microorganisms in soil (Tabatabai 1982);
2. Urease catalyses the hydrolysis of urea to carbon dioxide and ammonium (Tabatabai and Bremner 1972);
3. Proteases catalyse the hydrolysis of proteins to oligopeptides, and oligopeptides to aminoacids (Ladd and Butler 1972);
4. Phosphomonoesterases catalyse the hydrolysis of organic phosphomonoesters releasing phosphate and esters (Tabatabai and Bremner 1969);
5. Arylsulphatases catalyse the hydrolysis of estersulphates releasing sulphate and esters (Tabatabai and Bremner 1970).

Using the soil enzyme assays listed above as well as a few other, many authors have investigated soil fertility and/or soil quality changes following different management practices. However, despite the extensive bibliography, the results obtained by different researchers are often contradictory, with regard to both the effects of different agricultural practices and to changes shown by different enzyme activities

(Gil-Sotres et al. 2005). Trasar-Cepeda et al. (2008) pointed that there are different possible explanations for the lack of standard patterns of soil biochemical activity, particularly enzyme activities, in response to soil use and management. One possibility is that these differences are due to a lack of standardized protocols for determining these enzyme activities. Another reason for the contradictory results may be the small number of soil samples that are usually analysed in each study. Biochemical properties usually display a high degree of both spatial and temporal variability; therefore, a large number of samples may be required to overcome the statistical uncertainty, and this does not always occur. Another possible explanation for the random response of enzyme activities to soil use and management is the small number of enzyme activities that are usually analysed, and it is not uncommon to find studies in which only one enzyme activity is measured (Skujins 1978). Conclusions are reached, therefore, about the type of variation suffered by soil enzyme activities, even though the behaviour of a single enzyme activity hardly reflects the responses of other enzyme activities, as they may be associated with different cycles and participate in different stages of degradation and thus they will not necessarily respond in the same way (Sinsabaugh et al. 1991; Trasar-Cepeda et al. 2007). There is also a problem related to the lack of agreement about which kind of soil represents the maximum quality and can be used, therefore, as reference soil with which to compare soils affected by use (Gil-Sotres et al. 2005). Two contrasting options are usually considered: (1) quality is linked to health, and soils with maximum quality are the natural ones developing under climax vegetation without any human interference (Fedoroff 1987; Rasmussen et al. 1989; Trasar-Cepeda et al. 1998); (2) quality is linked to crop yield, and soils with maximum quality are highly productive but cause the least environmental impact (Doran et al. 1994; Jackson 2002). Finally, it is also possible that the different behaviour of hydrolytic enzymes in response to soil use and management may be because their activities are not good indicators and their validity as soil quality indicators may have been overestimated (Trasar-Cepeda et al. 2008). To overcome doubts about the information that soil enzyme assays provide, it is important to keep in mind that (i) a single enzymes activity cannot represent the overall soil quality (Nannipieri et al. 1990), and (ii) it is needed to measure several chemical, physical and other biological properties in combination with enzyme activity assays (Dick et al. 1996).

Recently, García-Ruiz et al. (2008) determined alkaline and acid phosphatase, arylsulphatase, β-glucosidase and dehydrogenase activities in combination with potential nitrification rates to investigate the short-to-medium term effect of organic management on soil quality. They found that these enzyme activities discriminated between organic and conventional management system as they were higher under organic management system than in the conventional one. This was not surprising as organic inputs increase microbial biomass which, in turn, may increase microbial activity. In addition, they calculated, and used as a soil quality index, the geometric mean of the assayed enzyme activities. The applied algorithm indicates high soil quality in organic amended soils (García-Ruiz et al. 2008), as well as low quality in plots receiving different doses of heavy-metal enriched sludge (Hinojosa et al. 2004).

Soil enzyme activities can be interpreted from the perspective of total and/or specific activities. Total enzyme activity gives an estimate of the rate at which the product of activity is made available in the soil. Indeed, being linked to soil microflora and to organic matter, soil enzyme comparisons among different pedo-climatic areas could be difficult or could provide misleading information. Hence, to normalize enzyme activities across differences in standing microflora or organic C, the total enzyme activity may be transformed to specific activity by dividing it by the microbial biomass C or organic C. In the first case, the specific activity of the enzyme may be considered as a measure of the physiological capacity, assuming the enzymes assayed represent the activities of the currently or recently viable microbial community (Landi et al. 2000). This information coupled with data on microbial community composition may help to link microbial diversity and function in soil (Kandeler et al. 1999b; Waldrop et al. 2000). In the second case, specific enzyme activities provide insight of how suitable the organic matter is to degradation by each specific enzyme, thus providing a measure of organic matter quality (Sinsabaugh et al. 2008).

6.3.2 Carbon Mineralisation (Soil Respiration)

Carbon dioxide (CO_2) release from soils, or soil respiration, is a multi-component process and, according to Kuzyakov (2006), it can depend on five sources:

1. SOM-derived CO_2 in root- and plant residue-free soil, frequently referred to as "basal respiration";
2. Additional SOM-derived CO_2 in root- and plant residue-affected soil, frequently referred to as "priming effect";
3. Microbial degradation of dead plant residues;
4. Microbial decomposition of rhizodepositions, frequently referred to as "rhizomicrobial respiration", and
5. Plant root respiration.

Being the first four CO_2 sources produced by soil microorganisms, they are also collectively named "microbial respiration" or "heterotrophic respiration".

Microbial respiration (MR; soil respiration minus root respiration) is a measure of the total metabolic activity of soil microorganisms that are decomposing organic matter (Haynes 2005). Hence MR, provides an indication of C available to soil microbes (Robertson et al. 1999) and can act as an index of soil organic matter quality (Haynes 2005).

Microbial respiration is measured by incubating, in sealed chambers and at a constant temperature (generally 20 or 25°C), pre-conditioned live root-free soil samples at field-moisture level or re-wetted up to 40–60% of their water holding capacity. The CO_2 accumulated in the chamber headspace is then determined by different methods, i.e. by alkali trapping, gas-chromatography, or infrared-gas analysis.

6 Key Biochemical Attributes to Assess Soil Ecosystem Sustainability 203

As it is evolved under controlled and optimal conditions, the amount of CO_2 efflux evolved during a short-term incubation (from 10 to 30 days) is also referred to as the potentially mineralisable C, and is generally expressed in mg kg^{-1} soil. The potentially mineralisable C generally coincides with the soil C fraction easily available to microflora. On the other hand, long-term incubations (up to several months) may supply information on C pools with a slow turnover. Thus soil organic C can be partitioned into various functionally distinct pools that are important components of most of the present soil C mineralisation models (Robertson et al. 1999). The short- and long-term release of CO_2 can, in fact, be used mathematically to indicate the functional pools of soil organic C commonly referred to as active (labile) and passive (recalcitrant) fraction soil organic C (Riffaldi et al. 1996; Robertson et al. 1999), respectively.

Microbial respiration is a sensitive indicator of the response of the biotic soil component to management practices. Indeed, studies have shown that MR responds to a multitude of factors such as soil temperature, moisture content, irrigation practices, tillage systems, presence of organic matter and nutrients, soil aeration, microbial processes and soil diffusivity (Edwards 1975; Mielnick and Dugas 1999).

Jabro et al. (2008) found that irrigation increased soil respiration from the soil surface to the atmosphere and, therefore, reduced the amount of C sequestered by the soil. Furthermore, the increase in soil temperature was accompanied by increase in soil respiration fluxes. A positive, high correlation was found between soil temperature and the CO_2 emission from the soil surface, which was well described by an exponential function.

With regard to crop management effects, Haynes (2005) reported a disproportionately greater increase in MR than total organic C in response to decreases in the amount of fallow in cereal rotations, and cropping with grasses rather than cereals. Tillage may be considered as the management practice most affecting soil respiration. Tillage loosens up the soil, increases the exposure of the previously protected soil organic matter and speeds up microbial oxidation processes, intensifying soil CO_2 emission to the atmosphere and reducing organic C (Beare et al. 1994). Tillage accelerates soil CO_2 emission by improving soil aeration, disaggregating soil, increasing the contact between soil and crop residue, and speeding organic C decomposition (Logan et al. 1991; Angers et al. 1993; Al-Kaisi and Yin 2005). Therefore, reduced or no-tillage systems decrease CO_2 emissions, thereby increasing the storage of carbon in the soil (Kern and Johnson 1993; Ball et al. 1999; Jabro et al. 2008). So et al. (2001) found that adoption of conservation tillage greatly reduced CO_2 emission from the soil relative to conventional tillage, effectively increasing the potential for C retention.

Interpretation of soil respiration results in terms of soil quality is, however, problematic and is still under debate. On the one hand, rapid decomposition of soil organic matter is not necessarily desirable because stable organic matter plays an important role in soil physical and chemical characteristics. On the other hand, the decomposition of organic residues to release nutrients at time when plants have a high demand is a desirable characteristic (Parkin et al. 1996; Sparling 1997).

6.3.3 Nitrogen Mineralisation and Net Nitrification

Almost all soil nitrogen (N) is present in the form of organic compounds that cannot be used directly by plants and is not susceptible to loss through leaching. The soil's capacity to transform organic N into inorganic N, i.e. its N mineralisation potential, is often used as an index of the N available to plant (Robertson et al. 1999; Nannipieri and Paul 2009). It is perhaps the most common and best means available to asses N fertility (e.g. Keeney 1980; Binkley and Hart 1989; Palm et al. 1993), as it is related to both the size of the labile soil organic N pool and the activity of the organisms responsible for the mineralization processes. Mineralisation potentials, i.e. the net production of inorganic N released from the mineralisable organic fraction in soil under constant moisture and temperature conditions, are better than inorganic soil N concentrations (pool size) as an indicator of site fertility, because the supply rate of a limiting nutrient affects more its availability than its instantaneous concentration. Most mineralisation assays are designed to exclude plant uptake and leaching but include microbial immobilisation and denitrification, thus providing net mineralisation potentials.

Nitrogen mineralisation assays usually refer to the net increase in both ammonium (NH_4^+) and nitrate (NO_3^-) in soil, since any nitrate formed must be derived from ammonium. The term "net" refers to the difference between the gross N mineralisation and the gross N immobilisation. While other forms of inorganic N are also produced during mineralisation assays (e.g. NO_2^-, N_2O and NO_x), in most soils their appearance is highly transient and relative pools are quickly converted to another form (NO_2^-) or their fluxes are inconsequential relative to increases in the NH_4^+ and NO_3^- pools (N_2O, NO_x). Nitrification refers specifically to the conversion (chemolithoautotrophic oxidation) of ammonium to nitrate by nitrifiers, which are microorganisms that oxidize ammonium to nitrite and then to nitrate. Nitrification assays also usually measure the net flux, excluding the nitrate that may be immobilised into microflora (Davidson et al. 1991) or denitrified to N gas (Robertson and Tiedje 1985) during the course of the assay.

While both net N mineralisation and nitrification assays have their limitations as measures of N availability, they can nevertheless provide substantial insight into soil fertility and ecosystem functioning at many sites, and they are used widely as soil quality indicators (Sparling 1997).

Large differences between sites or experimental treatments, for example, imply large differences in plant-available N, as well as large differences in the potential loss of N from an ecosystem. Nitrate, for example, is more readily lost from most ecosystems than ammonium, so large potential nitrification rates at a site can indicate a higher likelihood of nitrogen loss, all else being equal. It is hard to interpret small differences in mineralisation or nitrification between sites or treatments because concomitant processes can occur during the assays.

Drinkwater et al. (1996) suggest that the interpretation of net N mineralisation data requires to consider not only mineral N but also organic N and C pools. Short-term incubations are useful for estimating labile soil N but they exclude the large and stable pools contribution to the active soil N. Greater mineralisation potentials

6 Key Biochemical Attributes to Assess Soil Ecosystem Sustainability   205

have been found in a variety of cropping systems under organic than conventional management (Doran et al. 1987; Drinkwater et al. 1995; Gunapala and Scow 1997) reflecting the increased role of decomposers in determining N availability. The ratio of N mineralised to total organic N can be a sensitive indicator of differences in soil organic matter. The percent of total soil N mineralised in short-term anaerobic incubations was more than two-fold greater in soils with organic compared to conventional management, indicating significant qualitative differences in soil organic matter (Drinkwater et al. 1995). Microbial biomass N, N mineralisation potentials and mineral N pools can serve as indicators on the status of N dynamics in the soil. High mineralisation potentials in conjunction with high concentration of mineral N, especially during times of reduced crop uptake, could indicate susceptibility to N losses through leaching. Organically managed soils are sometimes characterised by high microbial activity and potentially mineralisable N, together with a smaller concentration of mineral N when compared with soils receiving conventional mineral fertiliser (Drinkwater et al. 1995). The combination of low mineral N concentration and enhanced microbial activity in the organic soils are indicative of a more tightly coupled N cycle (Sprent 1987; Jackson et al. 1989; Jenkinson and Parry 1989), with higher turnover rates of mineral N pools than in conventional soils.

6.4 Microbial Community Diversity and Function

Soil functioning depends on the composition and activities of microbial communities which generally respond rapidly to changes in their environment. They are, therefore, excellent indicators of soil quality. Only a small proportion of microbial populations can be cultivated under laboratory conditions; however, all populations can be characterised using molecular techniques. The methods generally target nucleic acids or fatty acid methyl esters (FAMES). DNA-base techniques can determine the presence and abundance of genes, but the expression of genes must be evaluated using mRNA or proteomic analyses. FAME-based methods facilitate assessment of active vs. inactive microbial types but only at coarse phylogenetic resolution. The following section provides an overview of the most widely used and emerging molecular methods for phylogenetic and functional analyses of microbial communities.

6.4.1 Nucleic Acid-Based Techniques

6.4.1.1 Phylogenetic Marker Genes

Phylogenetic analyses generally focus on ribosomal RNA (rRNA) and/or ribosomal DNA (rDNA or rRNA genes; regions of the genome that encode for rRNA). The rRNA operon includes genes that encode for structural rRNAs, which are important constituents of ribosomes, and internal transcribed spacer (ITS) regions that lie between the structural rRNA genes. Ribosomes are critical for cell function as they are the sites of protein synthesis. The structural rRNAs act as a scaffold around which ribosomal proteins can be positioned in a precise manner. If alterations occur in rDNA sequences at locations that ultimately determine the secondary structure of rRNAs, the organism may be unable to function. Some sections of the rDNA encoding for structural rRNAs are, therefore, highly conserved across broad phylogenetic distances; however, between these conserved regions are sections that have little or no effect on the conformation of the rRNA. These regions are highly variable between different microbial types because they have evolved at rate that is disproportionate to that of those that are constrained by the requirement to maintain ribosome function (Smit et al. 2007). Likewise, the ITS regions are also highly variable between different microbial types as they have no effect on the structure of rRNAs. In fact, they are more variable than structural rRNAs, and as such, ITS regions are often used to differentiate between microbial types at a sub-species level (Boyer et al. 2001). Post transcription, ITS regions are degraded prior to translatation. Analysis of ITS regions is very common for differentiating between fungal types; however, analysis of the structural 18S rRNA (gene) small subunit is also common in fungal diversity studies. In bacterial and archaeal diversity studies, analysis of the structural 16S rRNA (gene) small subunit is by far the most common target and extensive databases of identified sequences are available online. As rRNA is less stable than rDNA, information derived from rRNA analyses are often considered to reflect the active components of a community, whereas information derived from rDNA analyses reflects the entire community whether dead or alive.

6.4.1.2 Functional Genes

Both phylogenetic marker and functional genes contain regions that are highly conserved and variable between microbial types. Differentiation of these variable regions provides insight into the diversity of organisms involved in specific ecosystem functions (e.g. denitrification, nitrification, and nitrogen fixation). Current evidence indicates that these 'functional guilds' are immensely diverse (Zehr et al. 1998; Hugenholtz et al. 1998; DeLong and Pace 2001; Braker et al. 2001) and exhibit functional redundancy. The extent to which the diversity of these guild is related to the productivity and stability of functions is poorly understood even for large organisms. However, macroecologists have reported evidence indicating that the productivity and stability of ecosystem functions is linked to species richness and equitability (evenness) as well as community composition (Hooper et al. 2005; Cardinale et al. 2006; Wittebolle et al. 2009). The diversity of functional guilds is, therefore, a good indicator of soil quality and sustainability. Another attractive feature of functional guilds is that they are easier to study as their members constitute only a fraction of the total 'species' in a community. There are two ways to investigate the diversity of functional genes: (1) analysis of genomic DNA (gDNA); and (2) analysis of messenger RNA (mRNA). Analysis of gDNA provides insight into

6 Key Biochemical Attributes to Assess Soil Ecosystem Sustainability 207

the whole community whether dead or alive and analysis of mRNA provides insight into the genes that are being transcribed. Analysis of mRNA is focussed, therefore, entirely on active organisms.

6.4.1.3 Gene Amplification and Discrimination

Using polymerase chain reaction (PCR) and oligonucleotide primers that are complimentary to conserved regions that flank a region of sequence variability, it is possible to make millions of copies of the variable regions of phylogenetic and functional genes, which can then be discriminated by a variety of methods. This PCR step is generally needed to obtain sufficient DNA for detection but introduces a range of biases that tend to lead to overestimation of the most dominant sequences (Suzuki and Giovannoni 1996; Crosby and Criddle 2003). Another bias, in phylogenetic analyses, is related to the fact that different microbial types harbour multiple rRNA operons (Rainey et al. 1996; Klappenbach et al. 2000). For bacteria the copy number of rRNA operons varies between 1 and 15 (Rainey et al. 1996). This results in a quantitative bias towards organisms with higher copy numbers (Crosby and Criddle 2003). Despite these caveats, rRNA (genes) are the focal point of most microbial diversity studies. Techniques that discriminate between PCR amplicons of differing sequence can be broadly grouped into fingerprinting, sequencing or probe hybridisation methods (e.g. microarrays).

Fingerprinting Methods

In microbial diversity studies, Terminal Restriction Fragment Length Polymorphism (T-RFLP) and Denaturing Gradient Gel Electrophoresis (DGGE) are the most common fingerprinting methods. For T-RFLP, PCR products are generated using fluorescent-primers, digested with restriction enzymes and then terminal restriction fragments are detected by capillary electrophoresis (Avaniss-Aghajani et al. 1994). Denaturing Gradient Gel Electrophoresis involves the electrophoretic separation of PCR products according to their base sequence and composition using a denaturing gradient (Muyzer et al. 1993). Both methods enable large numbers of samples to be analysed and provide similar results (Smalla et al. 2007). However, these techniques facilitate detection of the most dominant amplicon types only, meaning that they poorly represent the diversity of rare 'species' or genes. This is a problem for assessment of microbial diversity as the majority of microbial types are present at low abundance (Bent and Forney 2008; Forney et al. 2004). For many studies, however, being able to detect relative changes in structure is sufficient and T-RFLP and DGGE are reliable for this purpose (Hartmann and Widmer 2008). In soils, these techniques have been used to provide insight into the diversity of total microbial communities (Ogilvie et al. 2008; Yergeau et al. 2007) as well as that of functional guilds such as nitrogen fixing bacteria, nitrifiers and denitrifiers (Gamble et al. 2009; Zhang et al. 2008). Arguably, the main advantage of T-RFLP over DGGE is that it is more suitable for high-throughput analyses as samples can be processed and analysed in multi-well format. The key advantage of DGGE is the ease with which bands can be cut from gels and identified by sequencing.

Sequencing

The sequence of a PCR product can be compared against known sequences compiled in online databases to facilitate identification of its origin. Traditionally, sequencing involves the construction of clone libraries, in which individual PCR products are cloned into a vector and then sequenced. The number of microbial populations is inferred from the number of unique sequences observed. The construction and analysis of clone libraries is laborious and costly. As such, most libraries contain approximately 1,000 clones, which is insufficient to observe the vast majority of microbial 'species'. It is estimated that up to 100,000 clones would need to be sequenced in order to accurately represent microbial diversity in soils (Dunbar et al. 2002); however, this is beyond the scope of most laboratories. Of course, the amount of sampling effort required to accurately reflect the diversity of a system is proportional to its richness. If the aim of a study was to characterise the number of bacterial 'species' in a typical soil, the amount of sampling effort would be very high, reflecting the immense richness of soil bacterial communities. If, however, the aim of a study was to characterise the diversity of soil bacteria involved in specific ecosystem functions the amount of sampling effort required would be much lower as the richness of the corresponding bacteria is relatively small.

Recently, new platforms have become available that significantly decrease the costs and time requirements of sequencing. Of these platforms, 454 Life Science's GS FLX Titanium platform is particularly appropriate for high-throughput analysis of PCR products as a single run generates approximately 10^6, 500 base pair reads (1 kb reads forthcoming). It is possible to run multiple samples in parallel using sample specific barcoded primers that enable the sequences to be affiliated with the samples from which they derive post sequencing (Binladen et al. 2007; Hamady et al. 2008). This technology has revolutionised our ability to characterise diverse communities and brings accurate estimation of phylogenetic and functional gene diversity a step closer to reality (Roesch et al. 2007; Elshahed et al. 2008; Teixeira et al. 2010; Lee et al. 2010). It is important, however, to properly account for the occurance of sequencing errors as these may lead to significant over-estimation of richness (Kunin et al. 2009; Huse et al. 2007; Quince et al. 2009).

Microarrays

Microarrays are slides that have thousands of single stranded DNA (ssDNA) probes, complimentary to phylogenetic and/or functional gene sequences, arrayed on their surface. In a microarray experiment, a DNA sample is fragmented, labelled and

6 Key Biochemical Attributes to Assess Soil Ecosystem Sustainability

then incubated on the surface of the microarray. If a sample sequence is complementary to a DNA probe on the microarray surface it will hybridise to the probe and fluoresce when excited at an appropriate wavelength. Fluorescence is measured for the whole array simultaneously; therefore, the relative abundance of thousands of sequences can be assessed in parallel. The most common microarray chips used for soil microbial ecology studies are the Phylochip, which has 5×10^5 16S rRNA gene ssDNA probes covering 8741 bacterial and archaeal taxa (Brodie et al. 2006; DeSantis et al. 2007), and the Geochip, which arrays 10,000 genes covering more than 150 functional groups (He et al. 2007). The main drawback of microarrays is that it is necessary to have *a priori* knowledge of the sequences that are expected in an environment. Any sequences that are present in an environmental sample that do not have a complimentary ssDNA probes on the microarray will be undetected irrespective of their abundance. Therefore, in environments that contain many novel sequences a microarray designed for another environment may poorly assess the diversity present.

6.4.1.4 Metagenomics

Metagenomics is the study and characterisation of entire core and accessory genomes from mixed microbial communities. This facilitates simultaneous insight into both the phylogenetic and function diversity of communities. Whole genomes from mixed communities can be fragmented and cloned into vectors for sequencing (shotgun sequencing), or fragmented and then sequenced using next-generation sequencing platforms. Due to the relatively small number of reads that are feasible using shotgun sequencing, the cloning-sequencing approach is considered to be acceptable only for low diversity systems with a few dominant community members, e.g. biofilms associated with acid mine drainage (Tyson et al. 2004). The cloning step is also known to introduce biases due to the toxicity of cloned genes to *E.coli* (Sorek et al. 2007). The modern high-thoughput sequencing platforms give much greater sequencing depth, which is important when characterising more diverse communities. A particular strength of the metagenomics approach is the potential for discovering novel genes. For example, proteorhodopsin proteins, which are light-driver proton pumps, were first identified in a metagenomic study of marine bacterioplankton, and have since been found in a diverse range of microorganisms (Béjà et al. 2000a, b). It is now thought that these proteins may represent an important source of energy flow in the world's oceans.

6.4.1.5 Gene Expression Analyses

In assessing soil quality and sustainability it may be of interest to characterise which genes are being expressed at a population (transcriptomics) or community level (metatranscriptomics). For these analyses it is necessary to extract mRNA, reverse transcribe it to cDNA and then analyse it using a variety of methods. Functional gene microarrays, such as the Geochip, are particularly useful for rapid, high-throughput analysis of cDNA; however, only genes that have complimentary ssDNA probes will be detected using this approach. Alternatives include real-time quantitative PCR (RT-qPCR) and sequencing. RT-qPCR is the most sensitive method available but primers must be designed and reactions conducted for each gene. Sequencing allows multiple gene transcripts to be characterise in parallel. The traditional cloning and sequencing approach can be adopted; however, new sequencing platforms circumvent the cloning step and provide far greater numbers of reads than would be feasible with Sanger sequencing. A range of free, online, open-source databases exist for identification of sequences post analysis such as the NCBI non-redundant database and the *curated* pathway databases such as KEGG, COG and SEED.

6.4.1.6 Linking Specific Organisms/Genes with Specific Substrates

Currently, methods for determining which populations are actively catabolising specific compounds within their environment involve the use of isotope tracers. One such example is stable isotope probing (SIP; Manefield et al. 2002a, b; Dumont and Murrell 2005). In SIP experiments, specific or broad-range substrates are highly enriched with a stable isotope such as ^{13}C or ^{15}N such that microbial cells utilising the substrate become isotopically labelled. After sampling, isotope labelled and unlabelled molecules such as DNA/RNA can then be separated by buoyant density gradient centrifugation. During this procedure labelled and unlabelled molecules are stained to facilitate their visualisation. Separated bands are then isolated and can be analysed using a wide range of molecular techniques.

Alternative techniques include fluorescence in-situ hybridisation-microautoradiography (FISH-MAR; Lee et al. 1999; Ouverney and Fuhrman 1999) and isotope arrays (Adamczyk et al. 2003). Using high intensity radiotracers, MAR has a resolution of 0.5–2 µm (Wagner et al. 2006) and can, therefore, be used to detect radio-labelled bacteria, or nucleic acids hybridised to a microarray chip. When combined with FISH, which is a technique that enables cells with specific DNA sequences to be indentified by hybridisation of fluorescent probes, MAR can be used to link key physiological features to targeted phylogenetic groups. When combined with microarrays, MAR facilitates detection of radio-labelled spots representing functional genes or known 'species'. The main drawback of these approaches is the use of radio-isotopes; however, by using secondary ion mass spectrometry (SIMS) instead of MAR, stable isotopes may be used. Secondary ion mass spectrometry can be used to image and quantify enrichment of multiple isotopes in single cells or nucleic acids hybridised to microarray chips. This greatly increases the scope of these techniques as microbial interactions with nitrogen-containing compounds, as well as other elements for which there are no useable radioisotopes, can also be investigated. Furthermore the fact that SIMS facilitates simultaneous detection of multiple isotopes means that more complicated experimental systems can be designed. A key advantage of using isotope tracers is that it is possible to focus on a defined assemblage of organisms. As this assemblage would represent only a frac-

6 Key Biochemical Attributes to Assess Soil Ecosystem Sustainability

tion of the entire community, it would be easier to recognise relationships between diversity, ecosystem function and stability. The use of isotopes could be particularly appropriate where soil quality is defined as the ability of a community to utilise specific resources (e.g. reclamation of soils contaminated with organic pollutants).

6.4.2 Fatty Acid-Based Methods (PLFAS, FAMES, ELFAS)

The use of microbial lipids to identify microorganisms and characterise microbial communities in natural systems constitutes a well established method. Indeed, fatty acids are the key component (about 40%) of cellular membranes of all living cells. Phospholipids consist of a single molecule of glycerol, in which two OH groups of the glycerol are ester-bonded with two fatty acid chains and the third OH group is ester-bonded with a phosphate group, that in turn, has one OH group ester-bonded with an aminoalcohol.

The most widely used extraction and separation method to obtain fatty acids derived from phospholipids is that proposed by Bligh and Dyer (1959) and modified by White et al. (1979). Briefly, the soil sample is added to a single-phase mixture of chloroform:methanol:buffer solution (1:2:0.8, v/v/v) for total lipid extraction, lipids are then separated into neutral-, glyco- and phospho-lipids (PLFAs) in a silicic acid column. PLFAs are converted, by alkaline methanolysis, to fatty acid methyl esters that are analyzed by gas chromatography to determine the types and quantities of each one. A sample profile consists of the abundance of each of the extracted PLFAs, which are chains of carbon atoms with varying numbers and positions of double bonds and side chains.

The concentration of total PLFA (expressed as nanomoles per gram of dry soil) provides quantitative insights into the viable soil microflora because they are rapidly degraded after cell death and are not found in the storage products (White et al. 1979; Federle 1986). Bailey et al. (2002) found a significant linear relationship between microbial biomass estimated by the chloroform-fumigation extraction method and PLFAs. They proposed that 1 nmol of microbial PLFA corresponds approximately to an extra-flush of 2.4 μg K_2SO_4-extractable C following chloroform-fumigation. As the suite of PLFAs identified in different laboratories varies, for simplicity, they considered only the saturated PLFAs, the monounsaturated PLFAs, and the polyunsaturated PLFAs less than 20, 18, and 20 C atoms in length, respectively. Leckie et al. (2004) compared the PLFA and chloroform-fumigation extraction methods in forest humus and found that 1 nmol of microbial PLFA corresponds approximately with an extra- flush of 3.2 μg K_2SO_4-extractable C following chloroform-fumigation. Besides, Olsson (1999) proposed a single PLFA to calculate arbuscular mycorrhizal fungal biomass C. They calculated that 1 mg of AM fungal biomass C correspond to 38 nmol of PLFA 16:1ω5, whilst Frostegård and Bååth (1996) proposed the use of 18:2ω6,9 only as a relative indicator of the fungal biomass.

PLFAs are above all useful biomarkers or signatures for fingerprinting the soil microbial community because of relative abundance of certain PLFAs, which differ

Table 6.1 Fatty acids used as biomarkers

Fatty acids	Biomarkers	References
i15:0, a15:0, 15:0, i16:0, 16:1ω7c, i17:0, a17:0, 17:0, cy17:0, 18:1ω7c, cy19:0	Total bacterial biomass	Frostegård and Bååth (1996), Bailey et al. (2002)
18:2ω6,9c	Fungal biomass	
i15:0, a15:0, i16:0, i17:0, a17:0	Gram-positive bacteria	O'Leary and Wilkinson (1988), Zogg et al. (1997)
16:1ω7; 18:1ω7; cy17:0; cy19:0;	Gram-negative bacteria	Zelles (1999), Waldrop et al. (2000)
18:1ω9c; 16:1ω5	Arbuscular mycorrhizal fungal biomass	Madan et al. (2002), Olsson (1999)
10Me16:0; 10Me17:0; 10Me18:0	Actinomycetes (recently renamed actinobacteria)	Lechevalier (1977)

considerably among the specific group of microorganisms. This approach, based on the variability of fatty acids present in cell membranes of different organisms, allows a phenotypic fingerprinting of soil microbial communities (Table 6.1).

Furthermore, the following relative abundance ratios of fatty acids are used as indexes of environmental stress: cyclopropyl / precursor (cy17:0 / 16:1ω7 and cy19:0/18:1ω7); iso/anteiso [(i15:0 + i17:0)/(a15:0 + a17:0)]; total saturated/total monounsaturated and trans/cis monounsaturated fatty acids (e.g. 16:1ω7, 18:1ω7) (Hedlund 2002; McKinley et al. 2005; Pettersson and Bååth 2003; Wu et al. 2009). The increase in the 16:1ω7 and 18:1ω7 trans-to-cis isomers ratios have been associated with stress and starvation conditions for bacteria (Guckert et al. 1986; Heipieper et al. 1996). The trans-to-cis ratio of monounsaturated fatty acids can be affected by high temperature, toxicity by organic compounds, starvation, osmotic stress, low pH and heavy metal toxicity (Frostegard et al. 1993; Heipieper et al. 1996; Pietikainen et al. 2000). Cyclopropyl PLFAs have been shown to increase relative to their monoenoic precursors during prolonged stationary growth phase of some bacteria, during growth under low carbon and oxygen availability, low pH, and high temperature (Guckert et al. 1986; Ratledge and Wilkinson 1988; Asuming-Brempong et al. 2008). Increases in cy17:0 and cy19:0 concentrations, relative to their respective metabolic precursors, 16:1ω7c and 18:1ω7 may, therefore, indicate physiological stress rather than a change in the composition of the community.

Recently, microbial fatty acids can be directly extracted from soil by a simple method that reduces the costs and analysis time. It consists of a mild alkaline reagent to lyse cells (KOH in methanol) and release fatty acids from lipids (ester-linked fatty acids; ELFAs) once the ester bonds are broken (Schutter and Dick 2000). Ritchie et al. (2000) used the ELFA method to characterise microbial communities of several cropped soils and found that the ELFA method grouped communities in a similar way as DNA-based methods. The ELFA method has been successfully used to investigate microbial communities in compost (Steger et al. 2003), in long-term tilled soil with different crop rotation (Gonzalez-Chavez et al. 2010) and in Antarctic dry valley soils (Hopkins et al. 2008).

6.5 Soil Quality Indexes

A soil quality index could be defined as the minimum set of parameters that, when interrelated, provides numerical data on the capacity of a soil to carry out one or more functions (Acton and Padbury 1993). Therefore, a soil quality index is a composite measure that reflects the individual measures relative to each other. This results in a single-digit index and threshold values can be established for the index rather than for the individual indicators. The drawbacks of the index approach are that all information on the relationships between indicators are lost and that weighting of the individual indicators may be subjective (Stenberg 1999). Furthermore, there is no direct relationship between an index value and a specific function or indicator, which may cause problems when interpreting the reasons for a high or low index values (Stenberg et al. 1998; Sojka and Upchurch 1999).

All studies on soil quality indexes point to the complexity since a diversity of physical, chemical, microbiological and biochemical properties need to be integrated to establish such quality (Papendick and Parr 1992; García et al. 1994; Halvorson et al. 1996). In general, the physical and physico-chemical parameters are of little use as they alter only when the soil undergoes a really dramatic change (Filip 2002). On the contrary, biological and biochemical parameters are sensitive to the slight modifications that the soil can undergo in the presence of any stressing or disturbing agent (Klein et al. 1985; Nannipieri et al. 1990; Yakovchenko et al. 1996). Therefore, whenever the sustainability of soil natural functions and the impact of soil different uses have to be evaluated, key indicators must include biological and biochemical parameters.

Regarding the use of both general and specific biochemical parameters to estimate soil quality, three approaches are widely used: (i) the use of individual properties (see above); (ii) the use of simple indexes; or (iii) the use of complex indexes.

6.5.1 Simple Indexes

The most straightforward simple indexes used to evaluate soil quality are the metabolic quotient (respiration to microbial biomass ratio, qCO_2), and the microbial quotient (the percentage of total organic C present as microbial biomass C, MBC/TOC; Bastida et al. 2008; Laudicina et al. 2009; Badalucco et al. 2010). Physiologically, the qCO_2 describes the substrate mineralised per unit of microbial biomass carbon per unit of time. The qCO_2, conceptually based on Odum's theory of ecosystem succession, has been widely used as an indicator of either ecosystem development (during which it supposedly declines; Insam and Haselwandter 1989), and ecosystem disturbance (supposedly increases; Anderson and Domsch 1985). Higher qCO_2 of microbial communities from young sites have been observed and compared to matured sites (Insam and Domsch 1988). Anderson and Domsch (1990) observed a decrease of qCO_2 in soils under monoculture when compared to soils under continuous crop rotations, suggesting that the richness of organic C substrates from different crops benefits respiration. In addition, this ratio has been widely used as an indicator of the alterations that take place in soil due to heavy metal contamination (Brookes 1995; Liao and Xiao 2007), deforestation (Bastida et al. 2006), changes in temperature (Joergensen et al. 1990) or soil management practices (Dilly et al. 2003). The qCO_2 is also affected by plant dry matter production and different tillage and cropping practices (Alvarez et al. 1995; Anderson 2003). However, this index has also received criticism for its insensitivity to certain disturbances and to the ecosystem's development whenever stress increases along successional gradients (Wardle and Ghani 1995; Nannipieri et al. 2003). Therefore, the significance of its increase in disturbed ecosystems is unclear since it may result from many reasons such as a drop in efficiency of substrate utilisation by the microflora, a response of the microflora to adverse conditions (monoculture, acidification, high heavy metal concentrations, etc.), the predominance of zymogene flora (r-strategists) over the autochthonous flora (K-strategists), or the alteration of the bacteria/fungi ratio since they have different carbon use strategies (Dilly and Munch 1998).

The microbial quotient has been proposed as a more sensitive index of soil changes than total organic C, since the microbial biomass of a soil responds more rapidly to changes than organic matter (Anderson and Domsch 1989, 1990; Powlson and Jenkinson 1981). This means that if a soil is in a degradation process, this degradation could be primarily detected by microbial changes whereas changes in organic matter would not be detected at an early degradation state.

For example, it is well known that intensive tillage, on one hand, favours C losses by speeding up microbial oxidation of soil organic matter but, on the other, may expose to drying the microflora. Thus, likely there may be a greater reduction of microbial biomass C than total organic C. However, conflicting results are reported: Roldán et al. (2003) found a higher MBC/TOC ratio under no tillage plot amended with residue cover or planted with legumes compared to conventional tilled plots, while Balota et al. (2004) did not find any significant change due to tillage and crop rotations.

The percentage of total organic C present as microbial biomass C in agricultural and forest soils at neutral pH is very similar and in the range between 2.0 and 4.4%, depending on nutrient status and soil management (Anderson 2003). Jenkinson and Ladd (1981) also proposed that for cultivated soils, a value of 2.2% reflects a good equilibrium between microbial biomass C and total organic carbon. However, Balota et al. (2004), who investigated the long-term tillage and crop rotation effects on microbial biomass C and N mineralisation under tropical/subtropical conditions, found in no tillage treatments an increase in organic C with a microbial quotient equal or less than 1.7%. They suggested that the microbial quotient under tropical/subtropical conditions may have a different threshold as an indicator of C accumulation than in temperate regions.

Despite some conflicting results, the microbial quotient is of great help in comparing the effects of different soil management practices also among soils with different organic matter contents as, being a percentage, avoids the problem of working with absolute values (Sparling 1997).

6.5.2 Complex Indices

Complex indices are derived from combinations of different biochemical properties of soils or deduced on the base of statistical procedures. Multivariate statistical tools, such as principal component, discriminant, factor and covariance analyses, simplify the interpretation of the large amounts of data and can be used to develop these indices. These analyses generally, reduce the data into a smaller number of indices (principal components, factors) which are linear combinations of the original values, representing most of the variation in the data set. These indices can be combined into a soil quality index by the multiple variable indicator transformation procedure (Smith et al. 1993). By this procedure, data on several soil properties at one location are combined together into a single binary indicator value, the multiple variable indicator transformation. Combined with geostatistics and kriging, soil maps can be calculated based on specified threshold values of each individual indicator. If the threshold values adequately reflect soil quality, then the kriging can produce maps of the probability of a soil being of good or bad quality.

The first attempt to calculate complex indices was carried out by Stefanic et al. (1984) who proposed the biological fertility index and the enzymatic activity number, respectively. Both these indices were polynomial formulae based on enzyme activities. Since then, many other soil indices, based on enzymatic activities have been developed. Among them, that of Sinsabaugh et al. (1994), the lignocellulosic factor, which is based on a cascade of enzyme activities which are able to degrade plant residues. The latter, thought to be one of the best biochemical index of litter degradation kinetics (Nannipieri et al. 2002), is based on the concept that if one of the selected enzymes does not work well the decomposition process will be short-circuited, which leads to a loss in the soil degrading function and, therefore, soil quality. A criticism against the lignocellulosic factor is that only considers enzymes of the C cycle, ignoring the enzymes from the cycles of the rest of the biophilic elements (N, P and S). Consequently, it can behave as a good indicator of the soil's capacity to degrade lignocellulosic materials, but not as an indicator of the global capacity of the soil to degrade organic compounds.

A list of other soil quality indices based on chemical and biochemical soil properties is reported by Bastida et al. (2008). These authors stressed the fact that many of the indices are based on pool size (e.g. microbial biomass C) and on microbial activity, whereas only few are also based on microbial diversity. The only index that considers microbial diversity is that proposed by Puglisi et al. (2005) who established a soil alteration index based on phospholipid fatty acid (PLFA) analysis. This index is rapid and easy to determine, and provides a profile of numerous fatty acids by gas chromatography, together with information about size of microflora and abundance of specific microbial groups (fungi, bacteria Gram+ and Gram–, actinomycetes, etc.), but it provides little information on the alteration (and, with it, quality or degradation) of a given soil because of the scant number of parameters (only PLFAs) used. As mentioned before for the lignocellulosic factor, one sole indicator is not sufficient to evaluate the state of a soil. In addition, PLFA, as a sole technique, is not enough to explain the processes and high-resolution microbial diversity of a soil.

However, according to Gil-Sotres et al. (2005) the use of complex equations in which various biochemical properties are involved appears to be a promising approach. The inclusion of different biochemical properties makes it possible to better reflect the complexity of the soil system, and complex equations also seem to be suitable for evaluating soil quality, at least for the conditions in which they have been designed. The main drawback of these complex indices is that they have not been tested in locations or under conditions other then those for which they were designed. Consequently, there is no way of verifying whether they are universally valid. This tell us that in order to progress in the development of a general expression for the evaluation of soil quality it will be necessary to carry out further intense and coordinated work at international level.

6.6 Concluding Remarks

The choice of bioindicators depends on economic resources available. The soil quality and functioning coincide with the biochemical potential of the microbial communities inhabiting soil. There is no single method able to describe activity, size and composition of soil microbial communities. Soil respiration, organic matter degradation, and microbial biomass are determined by classic methods. As such, they are often included in soil sustainability monitoring programs and can almost be considered as baseline parameters. No consensus on which indicators to include exists and probably never will, partly due to differences in policy targets. Hence, a selection must be made among the many microbial indicators and it should not neglect the following considerations: (1) standardized (ISO or others) and transparent techniques should be used; (2) as many techniques as possible should be used owing to no single technique so far available yields a full picture of the microbial community and soil quality; (3) when reduction of biochemical/microbial indicators is necessary, measures of microbial biomass, respiration, N mineralization, and a method of profiling the microbial community structure (either by nucleic acid- or fatty acid-based techniques) should be included.

Ongoing discussion of the ecological relevance of different techniques is inevitable and healthy for the scientific development of techniques. Only through continued research and application of the mentioned and developing techniques (e.g., techniques based on gene expression) will our understanding of the diversity and functions of biochemical attributes/microbial communities and their significance for soil quality be increased.

References

Acton DF, Gregorich EG (1995) The health of our soils. In: Acton DF, Gregorich EG (eds) Towards sustainable agriculture in Canada. Agriculture and Agri-food Canada, Ottawa

Acton DF, Padbury GA (1993) A conceptual framework for soil quality assessment and monitoring. A program to assess and monitor soil quality in Canada. Soil quality evaluation summary. Res Branch Agric. Ottawa, Canada

Adamczyk J, Hesselsoe M, Iversen N, Horn M, Lehner A, Nielsen PH, Schloter M, Roslev P, Wagner M (2003) The isotope array, a new tool that employs substrate-mediated labelling of rRNA for determination of microbial community structure and function. Appl Environ Microbiol 69:6875–6887

Albiach R, Canet R, Pomares F, Ingelmo F (1999). Structure, organic components and biological activity in citrus soils under organic and conventional management. Agrochimica 43:235–241

Al-Kaisi MM, Yin X (2005) Tillage and crop residue effects on soil carbon and carbon dioxide emission in corn–soybean rotations. J Environ Qual 34:437–445

Allison SD, Gartner T, Holland K, Weintraub M, Sinsabaugh RL (2007) Soil enzymes: linking proteomics and ecological process. In: Hurst CJ, Crawford RL, Garland JL, Lipson DA, Mills AL, Stetzenbach LD (eds) Manual of environmental microbiology, 3rd edn. ASM Press, Washington

Alvarez R, Díaz RA, Barbero N, Santanatoglia OJ, Blotta L (1995) Soil organic carbon, microbial biomass and CO_2-C production from three tillage systems. Soil Till Res 33:17–28

Anderson JPE, Domsch KH (1978) A physiological method for measurement of microbial biomass in soils. Soil Biol Biochem 10:215–221

Anderson JPE, Domsch KH (1989) Ratios of microbial biomass carbon to total carbon in arable soils. Soil Biol Biochem 21:471–479

Anderson TH (2003) Microbial eco-physiological indicators to asses soil quality. Agr Ecosyst Environ 98:285–293

Anderson TH, Domsch KH (1985) Determination of ecophysiological maintenance requirements of soil micro-organisms in a dormant state. Biol Fertil Soils 1:81–89

Anderson JPE, Domsch KH (1989) Ratios of microbial biomass carbon to total carbon in arable soils. Soil Biol Biochem 21:471–479

Anderson TH, Domsch KH (1990) Application of ecophysiological quotients (qCO2 and qD) on microbial biomass from soils of differing cropping histories. Soil Biol Biochem 25:393–395

Angers DA, N'dayegamiya AN, Cote D (1993) Tillage induced difference in organic matter of particle-size fractions and microbial biomass. Soil Sci Soc Am J 57:512–516

Araújo ASF, Santos VB, Monteiro RTR (2008) Response of soil microbial biomass and activity for practices of organic and conventional farming systems in Piauí state, Brazil. Eur J Soil Biol 44:225–230

Asuming-Brempong S, Gantner S, Adiku SGK, Archer G, Edusei V, Tiedje JM (2008) Changes in the biodiversity of microbial populations in tropical soils under different fallow treatments. Soil Biol Biochem 40:2811–2818

Avaniss-Aghajani E, Jones K, Chapman D, Brunk C (1994) A molecular technique for identification of bacteria using small sub-unit ribosomal RNA sequences. Biotechniques 17:144–149

Badalucco L, De Cesare F, Grego S, Landi L, Nannipieri P (1997) Do physical properties of soil affect chloroform efficiency in lysing microbial biomass? Soil Biol Biochem 29:1135–1142

Badalucco L, Rao M, Colombo C, Palumbo G, Laudicina VA, Gianfreda L (2010) Reversing agriculture from intensive to sustainable improves soil quality in a semiarid South Italian soil. Biol Fertil Soils 46:481–489

Bailey VL, Peacock AD, Smith JL, Bolten Jr H (2002) Relationships between soil microbial biomass determined by chloroform fumigation-extraction, substrate-induced respiration, and phospholipid fatty acid analysis. Soil Biol Biochem 34:1385–1389

Ball BC, Scott A, Parker JP (1999) Field N_2O, CO_2 and CH_4 fluxes in relation to tillage, compaction and soil quality in Scotland. Soil Till Res 53:29–39

Balota EL, Colozzi-Filho A, Andrade DS, Dick RP (2004) Long-term tillage and crop rotation effects on microbial biomass and C and N mineralization in a Brazilian Oxisol. Soil Till Res 77:137–145

Bastida F, Moreno JL, Hernández T, García C (2006) Microbiological activity in a soil 15 years alter its devegetation. Soil Biol Biochem 38:2503–2507

Bastida F, Zsolnay A, Hernández T, García C (2008) Past, present and future of soil quality indices: a biological perspective. Geoderma 147:159–171

Beare MH, Hendrix PF, Coleman DC (1994) Water-stable aggregates and organic matter fractions in conventional and no-tillage soils. Soil Sci Soc Am J 58:777–786

Béjà O, Suzuki MT, Koonin EV, Aravind L, Hadd A, Nguyen LP, Villacorta R, Amjadi M, Garrigues C, Jovanovich SB, Feldman RA, DeLong EF (2000a) Construction and analysis of bacterial artificial chromosome libraries from a marine microbial assemblage. Environ Microbiol 2:516–529

Béjà O, Aravind L, Koonin EV, Suzuki MT, Hadd A, Nguyen LP, Jovanovich SB, Gates CM, Feldman RA, Spudich JL, Spudich EN, DeLong EF (2000b) Bacterial rhodopsin: evidence for a new type of phototrophy in the sea. Science 289:1902–1906

Benitez E, Nogales R, Campos M, Ruano F (2006) Biochemical variability of olive-orchard soils under different management systems. Appl Soil Ecol 32:221–231

Bent SJ, Forney LJ (2008) The tragedy of the uncommon: understanding limitations in the analysis of microbial diversity. ISME J 2:689–695

Biederbeck VO, Campbell CA, Ukrainetz H, Curtin D, Bouman OT (1996) Soil microbial and biochemical properties after ten years of fertilization with urea and anhydrous ammonia. Can J Soil Sci 76:7–14

Binkley D, Hart SC (1989) The components of nitrogen availability assessments in forest soils. Adv Soil Sci 10:57–112

Binladen J, Gilbert MTP, Bollback JP, Panitz F, Bendixen C (2007) The use of coded PCR primers enables high-throughput sequencing of multiple homolog amplification products by 454 parallel sequencing. PLoS ONE 2:e197

Bligh EG, Dyer WJ (1959) A rapid method of total lipid extraction and purification. J Biochem Physiol 37:911–917

Bloem J, Schouten AJ, Sørensen SJ, Rutgers M, van der Werf A, Breure AM (2006) Monitoring and evaluating soil quality. In: Bloem J, Benedetti A, Hopkins DW (eds) Microbiological methods for assessing soil quality. Wallingford, United Kingdom

Böhme L, Langer U, Böhme F (2005) Microbial biomass, enzyme activities and microbial community structure in two European long-term field experiments. Agr Ecosyst Environ 109:141–152

Boyer SL, Flechtner VR, Johansen JR (2001) Is the 16S–23S rRNA internal transcribed spacer region a good tool for use in molecular systematics and population genetics? A case study in Cyanobacteria. Mol Biol Evol 18:1057–1069

Braker G, Ayala-del-Rio HL, Devol AH, Fesefeldt A, Tiedje JM (2001) Community structure of denitrifiers, Bacteria, and Archaea along redox gradients in Pacific Northwest marine sediments by terminal restriction fragment length polymorphism analysis of amplified nitrite reductase (nirS) and 16S rRNA genes. Appl Environ Microbiol 67:1893–1901

Brodie EL, DeSantis TZ, Joyner DC, Baek SM, Larsen JT, Andersen GL, Hazen TC, Richardson PM, Herman DJ, Tokunaga TK, Wan JM, Firestone MK (2006) Application of a high density oligonucleotide microarray approach to study bacterial population dynamics during uranium reduction and reoxidation. Appl Environ Microbiol 72:6288–6298

Brookes PC (1995) The use of microbial parameters in monitoring soil pollution by heavy-metals. Biol Fertil Soils 19:269–279

Brookes PC, Landman A, Pruden G, Jenkinson DS (1985) Chloroform fumigation and the release of soil nitrogen: a rapid direct extraction method to measure microbial biomass nitrogen in soil. Soil Biol Biochem 17:837–842

Brookes PC, Powlson DS, Jenkinson DS (1982) Measurement of microbial biomass phosphorus in soil. Soil Biol Biochem 14:319–329

Burns RG (1977) Soil enzymology. Sci Prog 64:275–285

6 Key Biochemical Attributes to Assess Soil Ecosystem Sustainability 219

Burns RG (1978) Enzyme activity in soil: some theoretical and practical considerations. In: Burns RG (ed) Soil enzymes. Academic Press, London

Caldwell B (2005) Enzyme activities as a component of soil biodiversity: a review. Pedobiologia 49:637–644

Cardinale BJ, Srivastava DS, Duffy EJ, Wright JP, Downing AL, Sankaran M, Jouseau C (2006) Effects of biodiversity on the functioning of trophic groups and ecosystems. Nature 443:989–992

Crosby LD, Criddle CS (2003) Understanding bias in microbial community analysis techniques due to rrn operon copy number heterogeneity. Biotechniques 34:790–794

Davidson EA, Hart SC, Shanks CA, Firestone MK (1991) Measuring gross nitrogen mineralisation, immobilisation and nitrification by 15N isotopic pool dilution in intact soil cores. J Soil Sci 42:335–349

DeLong EE, Pace NR (2001) Environmental diversity of Bacteria and Archaea. Syst Biol 50:470–478

Deng SP, Parham JA, Hattey JA, Babu D (2006) Animal manure and anhydrous ammonia amendment alter microbial carbon use efficiency, microbial biomass, and activities of dehydrogenase and amidohydrolases in semiarid agroecosystems. Appl Soil Ecol 33:258–268

Denti EA, Reis EM (2001) Efeito da rotação de culturas, da monocultura e da densidade de plantas na incidência das podridões da base do colmo e no rendimento de grãos do milho. Fitopatologia Brasileira 26:635–639

DeSantis TZ, Brodie EL, Moberg JP, Zubieta IX, Piceno YM, Andersen GL (2007) High-density universal 16S rRNA microarray analysis reveals broader diversity than typical clone library when sampling the environment. Microb Ecol 53:371–383

Dick RP (1994) Soil enzyme activities as indicators of soil quality. In: Doran JW (ed) Defining soil quality for sustainable environment. Soil science society of America, special publication 35. SSSA-ASA, Madison

Dick RP, Breakwell DP, Turco RF (1996) Soil enzyme activities and biodiversity measurements as integrative microbiological indicators. In: Doran JW, Jones AJ (eds) Methods for assessing soil quality. Soil science society of America, Madison

Dilly O, Bach HJ, Buscot F, Eschenbach C, Kutsch WL, Middelhoff U, Pritsch K, Munch JC (2000) Characteristics and energetic strategies of the rhizosphere in ecosystems of the Bornhöved Lake district. Appl Soil Ecol 15:201–210

Dilly O, Blume HP, Sehy U, Jiménez M, Munich JC (2003) Variation of stabilised, microbial and biologically active carbon and nitrogen soil under contrasting land use and agricultural management practices. Chemosphere 52:557–569

Dilly O, Munch JC (1998) Ratios between estimates of microbial biomass content and microbial activity in soils. Biol Fertil Soils 27:374–379

Doran JW (1980) Soil microbial and biochemical changes associated with reduced tillage. Soil Sci Soc Am J 44:764–771

Doran JW, Fraser DG, Culik MN, Liebhardt WC (1987) Influence of alternative and conventional agriculture management on soil microbial processes and nitrogen availability. Am J Altern Agric 2:99–106

Doran JW, Parkin TB (1996) Quantitative indicators of soil quality: a minimum data set. In: Doran JW, Jones AJ (Eds) Methods for assessing soil quality. Soil Sci Soc Am Spec Public, Madison

Doran JW, Safley M (1997) Defining and assessing soils health and sustainable productivity. In: Pankhurst CE, Doube BM, Gupta VVSR (eds) Biological indicators of soil health. CAB International, Wallingford

Doran JW, Sarrantonio M, Jaure R (1994) Strategies to promote soil quality and health. In: Pankhurst CE, Doube BM, Gupta VVSR, Grace PR (eds) Soil biota: management in sustainable farming systems. CSIRO, Melbourne

Drinkwater LE, Cambardella CA, Reeder JD, Rice CW (1996) Potentially mineralizable nitrogen as an indicator of biologically active soil nitrogen. In: Doran JW, Jones AJ (eds) Methods for assessing soil quality. Soil Science Society of America, Madison

Drinkwater LE, Letoumeau DK, Workneh F, van Bruggen AHC, Shennan C (1995) Fundamental differences between conventional and organic tomato agroecosystems in California. Ecol Appl 5:1098–1112

Droogers P, Bouma J (1996) Biodynamic versus conventional farming effects on soil structure expressed by simulated potential productivity. Soil Sci Soc Am J 60:1552–1558

Dumont MG, Murrell JC (2005) Stable isotope probing—Linking microbial identity to function. Nat Rev Microbiol 3:499–504

Dunbar J, Barns SM, Ticknor LO, Kuske CR (2002) Empirical and theoretical bacterial diversity in four Arizona soils. Appl Environ Microbiol 68:3035–3045

Edwards NT (1975) Effects of temperature and moisture on carbon dioxide evolution in a mixed deciduous forest floor. Soil Sci Soc Am J 39:361–365

Elshahed MS, Youssef NH, Spain AM, Sheik C, Najar FZ, Sukharnikov LO, Roe BA, Davis JP, Schloss PD, Bailey VL, Krumholz LR (2008) Novelty and uniqueness patterns of rare members of the soil biosphere. Appl Environ Microbiol 74:5422–5428

Emmerling C, Udelhoven T, Schröder D (2001) Response of soil microbial biomass and activity to agriculture de-intensification over a 10 year period. Soil Biol Biochem 33:2105–2114

Federle TW (1986) Microbial distribution in soil—new techniques. In: Megusar F, Gantar M (eds) Perspectives in Microbial Ecology. Ljubljana Slovene Society for Microbiology, Ljubljana

Fedoroff N (1987) The production potential of soils. Part 1. Sensitivity of principal soil types to the intensive agriculture of north-western Europe. In: Barth E, L'Hermite P (eds) Scientific basis for soil protection in the European community. Elsevier, London

Feng XJ, Simpson MJ (2009) Temperature and substrate controls on microbial phospholipid fatty acid composition during incubation of grassland soils contrasting in organic matter quality. Soil Biol Biochem 41:804–812

Ferreira MC, Andrade DS, Chueire LMO, Takemura SM, Hungria M (2000) Effects of tillage method and crop rotation on the population sizes and diversity of bradryhizobia nodulating soybean. Soil Biol Biochem 32:627–637

Filip Z (2002) International approach to assessing soil quality by ecologically-related biological parameters. Agr Ecosyst Environ 88:169–174

Fließbach A, Madër P (2000) Microbial biomass and size-density fractions differ between soils of organic and conventional agricultural systems. Soil Biol Biochem 32:757–768

Fließbach A, Mäder P, Niggli U (2000) Mineralization and microbial assimilation of 14C-labeled straw in soils of organic and conventional agricultural systems. Soil Biol Biochem 32:1131–1139

Forney LJ, Zhou X, Brown CJ (2004) Molecular microbial ecology: land of the one-eyed king. Curr Opin Microbiol 7:210–220

Franchini JC, Crispino CC, Souza RA, Torres E, Hungria M (2007) Microbiological parameters as indicators of soil quality under various soil management and crop rotation systems in southern Brazil. Soil Till Res 92:18–29

Fraser DG, Doran JW, Sahs WW, Lesoing GW (1988) Soil microbial populations and activities under conventional and organic management. J Environ Qual 17:585–590

Frostegård A, Bååth E (1996) The use of phospholipid fatty acid analysis to estimate bacterial and fungal biomass in soil. Biol Fertil Soils 22:59–65

Frostegård A, Bååth E, Tunlid A (1993) Shifts in the structure of soil microbial communities in limed forests as revealed by phospholipids fatty acid analysis. Soil Biol Biochem 25:723–730

Gamble MD, Bagwell CE, LaRocque J, Bergholz PW, Lovell CR (2009) Seasonal variability of diazotroph assemblages associated with the rhizosphere of the salt marsh cordgrass, spartina alterniflora. Microb Ecol 59:253–265

Garcia C, Hernandez T, Costa F (1994) Microbial activity in soils under Mediterranean environmental conditions. Soil Biol Biochem 26:1185–1191

García-Ruiz R, Ochoa V, Hinojosa MB, Carreira JA (2008) Suitability of enzyme activities for the monitoring of soil quality improvement in organic agricultural systems. Soil Biol Biochem 40:2137–2145

6 Key Biochemical Attributes to Assess Soil Ecosystem Sustainability 221

Ge G, Li Z, Fan F, Chu G, Hou Z, Liang Y (2010) Soil biological activity and their seasonal variations in response to long-term application of organic and inorganic fertilizers. Plant Soil 326:31–44

Gil-Sotres F, Trasar-Cepeda C, Leiròs MC, Seoan S (2005) Different approaches to evaluating soil quality using biochemical properties. Soil Biol Biochem 37:877–887

Gonzalez-Chavez MA, Aitkenhead-Peterson JA, Gentry TJ, Zuberer D, Hons F, Loeppert R (2010) Soil microbial community, C, N, and P responses to long-term tillage and crop rotation. Soil Tillage Res 106:285–293

Gregorich EG, Turchenek LW, Carter MR, Angers DA (2001) Soil and environmental science, dictionary. In: Gregorich EG, Turchenek LW, Carter MR, Angers DA (eds) Canadian society of soil science. CRC Press, Washington

Guckert JB, Hood MA, White DC (1986) Phospholipid esterlinked fatty acid profile changes during nutrient deprivation of Vibrio cholerae: increases in the trans/cis ratio and proportions of cyclopropyl fatty acids. Appl Environ Microbiol 52:794–801

Gunapala N, Scow K (1997) Dynamics of soil microbial biomass and activity in conventional and organic farming systems. Soil Biol Biochem 30:805–816

Halvorson JJ, Smith JL, Papendick RI (1996) Integration of multiple soil parameters to evaluate soil quality: a field experiment example. Biol Fertil Soils 21:207–214

Hamady M, Walker JJ, Harris JK, Gold NJ, Knight R (2008) Error-correcting barcoded primers for pyrosequencing hundreds of samples in multiplex. Nat Methods 5:235–237

Hartmann M, Widmer F (2008) Reliability for detecting composition and changes of microbial communities by T-RFLP genetic profiling. FEMS Microbiol Ecol 63:249–260

Harwood RR (1990) A history of sustainable agriculture. In: Edwards CA, Lal R, Madden JP, Miller RH, House G (eds) Sustainable agricultural systems, Ankeny IA, soil and water conservation society. Ankeny, USA

Haynes RJ (2005) Labile organic matter fractions as central components of the quality of agricultural soils: an overview. Adv Agron 85:221–268

He Z, Gentry TJ, Schadt TW, Wu L, Liebich J, Chong SC, Huang Z, Wu W, Gu B, Jardine P, Criddle C, Zhou J (2007) GeoChip: a comprehensive microarray for investigating biogeochemical, ecological and environmental processes. ISME J 1:67–77

Hedlund K (2002) Soil microbial community structure in relation to vegetation management on former agricultural land. Soil Biol Biochem 34:1299–1307

Heipieper HJ, Meulenbeld G, Oirschot QV, de Bont JAM (1996) Effect of environment factors on trans/cis ratio of unsaturated fatty acids in Pseudomonas putida S12. Appl Environ Microbiol 62:2773–2777

Hinojosa MB, García-Ruiz R, Vinegla B, Carreira JA (2004) Microbiological rates and enzyme activities as indicators of functionality in soils affected by the Aznalcòllar toxic spill. Soil Biol Biochem 36:1637–1644

Hoffmann LL, Reis EM, Forcelini CA, Panisson E, Mendes CS, Casa RT (2004) Efeito da rotação de cultura, de cultivares e da aplicação de fungicida sobre o rendimento de grãos e doenças foliares em soja. Fitopatol Bras 29:245–251

Hooper DU, Chapin FS, Ewel JJ, Hector A, Inchausti P, Lavorel S, Lawton JH, Lodge DM, Loreau M, Naeem S, Schmid B, Setala H, Symstad AJ, Vandermeer J, Wardle DA (2005) Effects of biodiversity on ecosystem functioning: a consensus of current knowledge. Ecol Monogr 75:3–35

Hopkins DW, Sparrow AD, Shillam LL, English LC, Dennis PG, Novis P, Elberling B, Gregorich EG, Greenfield LG (2008) Enzymatic activities and microbial communities in an Antarctic dry valley soil: responses to C and N supplementation. Soil Biol Biochem 40:2130–2136

Hugenholtz P, Goebel BM, Pace NR (1998) Impact of culture-independent studies on the emerging phylogenetic view of bacterial diversity. J Bacteriol 180:4765–4774

Hungria M, Stacey G (1997) Molecular signals exchanged between host plants and rhizobia: basic aspects and potential application in agriculture. Soil Biol Biochem 29:819–830

Huse SM, Huber JA, Morrison HG, Sogin ML, Welch MD (2007) Accuracy and quality of massively parallel DNA pyrosequencing. Genome Biol 8: R143

Insam H, Domsch KH (1988) Relationship between soil organic-carbon and microbial biomass on chronosequences of reclamation sites. Microb Ecol 15:177–188

Insam H, Haselwandter K (1989) Metabolic quotient of the soil microflora in relation to plant succession. Oecologia 79:174–178

Jabro JD, Sainju U, Stevens WB, Evans RG (2008) Carbon dioxide flux as affected by tillage and irrigation in soil converted from perennial forages to annual crops. J Environ Manag 88:1478–1484

Jackson LE, Schimel JP, Firestone MK (1989) Short-term partitioning of ammonium and nitrate between plants and microbes in an annual grassland. Soil Biol Biochem 21:409–415

Jackson W (2002) Natural systems agriculture: a truly radical alternative. Agr Ecosyst Environ 88:111–117

Jenkinson DS (1988) Determination of microbial biomass carbon and nitrogen in soil. In: Wilson JR (ed) Advances in nitrogen cycling in agricultural ecosystem. CAB int., Wallingford

Jenkinson DS, Hart PBS, Rayner JN, Parry LC (1987) Modelling the turnover of organic matter in long-term experiments at Rothamsted. Intecol Bull 15:1–8

Jenkinson DS, Ladd JN (1981) Microbial biomass in soil: measurement and turnover. In: Paul EA, Ladd JN (eds) Soil Biochemistry. Dekker, New York

Jenkinson DS, Parry LC (1989) The nitrogen cycle in the Broadbalk Wheat Experiment: A model for the turnover of nitrogen through the soil microbial mass. Soil Biol Biochem 21:535–541

Jenkinson DS, Powlson DS (1976) The effect of biocidal treatment on metabolism in soil. V. A method for measuring soil biomass. Soil Biol Biochem 8:209–213

Joergensen RG, Brookes PC, Jenkinson DS (1990) Survival of the microbial biomass at elevated-temperatures. Soil Biol Biochem 22:1129–1136

Kandeler E, Stemmer M, Klimanek EM (1999a) Response of soil microbial biomass, urease and xylanase within particle size fractions to long-term soil management. Soil Biol Biochem 31:261–273

Kandeler E, Palli S, Stemmer M, Gerzabek MH (1999b) Tillage changes microbial biomass and enzyme activities in particle-size fractions of a Haplic Chernozem. Soil Biol Biochem 31:1253–1264

Keeney DR (1980) Prediction of soil nitrogen availability in forest ecosystems: a literature review. For Sci 26:159–171

Kern JS, Johnson MG (1993) Conservation tillage impacts on national soil and atmospheric carbon levels. Soil Sci Soc Am J 57:200–210

Kirk TK, Farrell RL (1987) Enzymatic "combustion": the microbial degradation of lignin. Annu Rev Microbiol 41:465–505

Kizilkaya R, Aşkin T, Bayrakli B, Sağlam M (2004) Microbiological characteristics of soils contaminated with heavy metals. Eur J Soil Biol 40:95–102

Klappenbach JA, Dunbar JM, Schmidt TM (2000) rRNA operon copy number reflects ecological strategies of bacteria. Appl Environ Microbiol 66:1328–1333

Klein DA, Sorensen DL, Redente EF (1985) Soil enzymes: a predictor of reclamation potential and progress. In: Tate RL, Klein DA (eds) Soil reclamation processes. Microbiological analyses and applications. Marcel Dekker, New York

Kunin V, Engelbrektson A, Ochman H, Hugenholtz P (2009) Wrinkles in the rare biosphere: pyrosequencing errors can lead to artificial inflation of diversity estimates. Environ Microbiol 12:118–123

Kuzyakov Y (2006) Sources of CO_2 efflux from soil and review of partitioning methods. Soil Biol Biochem 38:425–448

Ladd JN, Butler HA (1972) Short-term assays of soil proteolytic enzyme activities using proteins and dipeptide derivatives as substrates. Soil Biol Biochem 4:19–30

Landi L, Renella G, Moreno JL, Falchini L, Nannipieri P (2000). Influence of cadmiuum on the metabolic quotient, L-:D-glutamic acid respiration ratio and enzyme activity:microbial biomass ratio under laboratory conditions. Biol Fertil Soils 32:8–16

Laudicina VA, Hurtado Bejarano MD, Badalucco L, Delgado A, Palazzolo E, Panno M (2009) Soil chemical and biochemical properties of a salt-marsh alluvial Spanish area after long-term reclamation. Biol Fertil Soils 45:691–700

6 Key Biochemical Attributes to Assess Soil Ecosystem Sustainability 223

Lechevalier MP (1977) Lipids in bacterial taxonomy—a taxonomist's view. Crit Rev Microbiol 5:109–210

Leckie SA, Prescott CE, Grayston SJ, Neufeld JD, Mohn WW (2004) Comparison of chloroform fumigation-extraction, phospholipid fatty acid, and DNA methods to determine microbial biomass in forest humus. Soil Biol Biochem 36:529–532

Lee N, Nielsen PH, Andreasen KH, Juretschko S, Nielsen JL, Schleifer KH, Wagner M (1999) Combination of fluorescent in situ hybridization and microautoradiography—a new tool for structure-function analyses in microbial ecology. Appl Environ Microbiol 65:1289–1297

Lee TK, Van Doan T, Yoo K, Choi S, Kim C, Park J (2010) Discovery of commonly existing anode biofilm microbes in two different wastewater treatment MFCs using FLX Titanium pyrosequencing. Appl Microbiol Biotech 87:2335–2343

Liao M, Xiao XM (2007) Effect of heavy metals on substrate utilization pattern, biomass, and activity of microbial communities in a reclaimed mining wasteland of red soil area. Ecotoxicol Environ Saf 66:217–223

Ljungdahl LG, Eriksson KE (1985) Ecology of microbial cellulose degradation. Adv Microb Ecol 8:237–299

Logan TJ, Lal R, Dick WA (1991) Tillage systems and soil properties in North America. Soil Till Res 20:241–270

Lorenz N, Verdell K, Ramsier C, Dick RP (2010) A rapid assay to estimate soil microbial biomass potassium in agricultural soils. Soil Sci Soc Am J 74:512–516

Madan R, Pankhurst C, Hawke B, Smith S (2002) Use of fatty acids for identification of AM fungi and estimation of the biomass of AM spores in soil. Soil Biol Biochem 34:125–128

Madsen EL (1996) A critical analysis of methods for determining the composition and biogeochemical activities of soil microbial communities in situ. In: Stotzky G, Bollag JM (eds) Soil biochemistry. Marcel Dekker, New York

Mallouhi N, Jacquin F (1985) Essai de correlation entre proprietes biochimiques d'un sol salsodique et sa biomasse. Soil Biol Biochem 17:23–26

Manefield M, Whiteley AS, Griffiths RI, Bailey MJ (2002a) RNA stable isotope probing, a novel means of linking microbial community function to phylogeny. Appl Environ Microbiol 68:5367–5373

Manefield M, Whiteley AS, Ostle N, Ineson P, Bailey MJ (2002b) Technical considerations for RNA-based stable isotope probing: an approach to associating microbial diversity with microbial community function. Rapid Commun Mass Spec 16:2179–2183

Matson PA, Parton WJ, Power AG, Swift MJ (1997) Agricultural intensification and ecosystem properties. Science 277:504–509

McKinley VL, Peacock AD, White DC (2005) Microbial community PLFA and PHB responses to ecosystem restoration in tallgrass prairie soils. Soil Biol Biochem 37:1946–1958

Melero S, Madejòn E, Ruiz JC, Herencia JF (2007) Chemical and biochemical properties of a clay soil under dryland agriculture system as affected by organic fertilization. Eur J Agron 36 26:327–334

Mielnick PC, Dugas WA (1999) Soil CO_2 flux in a tallgrass prairie. Soil Biol Biochem 32:221–228

Muyzer G, Dewaal EC, Uitterlinden AG (1993) Profiling of complex microbial-populations by denaturing gradient gel-electrophoresis analysis of polymerase chain reaction-amplified genescoding for 16S ribosomal-RNA. Appl Environ Microbiol 59:695–700

Nannipieri P, Ascher J, Ceccherini MT, Landi L, Pietramellara G, Renella G (2003) Microbial diversity and soil functions. Eur J Soil Sci 54:655–670

Nannipieri P, Ceccanti B, Grego S (1990) Ecological significance of biological activity in soil. In: Bollag JM, Stotzky G (eds) Soil biochemistry. Marcel Dekker, New York

Nannipieri P, Kandeler E, Ruggiero P (2002) Enzyme activities and microbiological and biochemical processes in soil. In: Burns RG, Dick RP (eds) Enzymes in the environment. Marcel Dekker, New York

Nannipieri P, Paul E (2009) The chemical and functional characterization of soil N and its biotic components. Soil Biol Biochem 41:2357–2369

Nogueira MA, Albino UB, Brandão-Junior O, Braun G, Cruz MF, Dias BA, Duarte RTD, Gioppo NMR, Menna P, Orlandi JM, Raimam MP, Rampazzo LGL, Santos MA, Silva MEZ, Vieira FP, Torezan JMD, Hungria M, Andrade G (2006) Promising indicators for assessment of agroecosystems alteration among natural, reforested and agricultural land use in southern Brazil. Agr Ecosyst Environ 115:237–247

O'Leary WM, Wilkinson SG (1988) Gram-positive bacteria. In: Ratledge C, Wilkindon SC (eds) Microbial lipids. Academic Press, London

Ogilvie LA, Hirsch PR, Johnston AWB (2008) Bacterial diversity of the Broadbalk 'classical' winter wheat experiment in relation to long-term fertilizer inputs. Microb Ecol 56:525–537

Olsson PA (1999) Signature fatty acids provide tools for determination of the distribution and interactions of mycorrhizal fungi in soil. FEMS Microbiol Ecol 29:303–310

Ouverney CC, Fuhrman JA (1999) Combined microautoradiography-16S rRNA probe technique for determination of radioisotope uptake by specific microbial cell types *in situ*. Appl Environ Microbiol 65:1746–1752

Palm C, Robertson GP, Vitousek PM (1993) Nitrogen availability. In: Anderson JM, Ingram JSI (eds) Tropical soil biology and fertility: a handbook of methods, 2nd edn. CAB International, Oxford

Papendick RI, Parr JF (1992) Soil quality—the key to a sustainable agriculture. Am J Altern Agric 7:2–3

Parkin TB, Doran JW, Franco-VizCaino E (1996) Field and laboratory tests of soil respiration. In: Doran JW, Jones AJ (eds) Methods for assessing soil quality. Soil Science Society of America, Madison

Parr JF, Papendick RI, Hornick SB, Meyer RE (1992). Soil quality: attributes and relationship to alternative and sustainable agriculture. Am J Alter Agr 7:5–11

Paul EA (1984) Dynamics of organic matter in soils. Plant Soil 76:275–285

Paul EA, Harris D, Klug MJ, Ruess WR (1999) The determination of microbial biomass. In: Robertson GP, Coleman DC, Bledsoe CS, Sollins P (eds) Standard soil methods for long-term ecological research. Oxford University Press, New York

Pereira AA, Hungria M, Franchini JC, Kaschuk G, Chueire LMO, Campo RJ, Torres E (2007) Variações qualitativas e quantitativas na microbiota do solo e na fixação biológica do nitrogênio sob diferentes manejos com soja. Rev Bras Ciência Solo 31:1397–1412

Perez-de-Mora A, Burgos P, Madejon E, Cabrera F, Jaeckel P, Scholter M (2006) Microbial community structure and function in a soil contaminated by heavy metals: effects of plant growth and different amendments. Soil Biol Biochem 38:327–341

Pettersson M, Bååth E (2003) The rate of change of a soil bacterial community after liming as a function of temperature. Micr Ecol 46:177–186

Pettit NM, Gregory LJ, Freedman RB, Burns RG (1977) Differential stabilities of soil enzymes. Assay and properties of phosphatase and arylsulphatase. Acta Biochim Biophys 485:357–366

Pietikainen J, Hiukka R, Fritze H (2000) Does short term heating of forest humus change its properties as a substrate for microbes? Soil Biol Biochem 32:277–288

Powlson DS, Jenkinson DS (1981) A comparison of the organic matter, biomass, adenosine-triphosphate, and mineralizable nitrogen contents of ploughed and direct drilled soils. J Agric Sci 97:713–721

Puglisi E, Nicelli M, Capri E, Trevisan M, Del Re AAM (2005) A soil alteration index based on phospholipid fatty acids. Chemosphere 61:1548–1557

Quince C, Lanzen A, Curtis TP, Davenport RJ, Hall N, Head IM, Read LF, Sloan WT (2009) Accurate determination of microbial diversity from 454 pyrosequencing data. Nat Methods 6:639–641

Ragab M (1993) Distribution pattern of soil microbial population in salt affected soils. In: Lieth H, Al-Masoom AA (eds) Towards rational use of high salinity tolerant plants, Deliberations about high salinity tolerant plants and ecosystems. Kluwer, Dordrecht

Rainey FA, Ward-Rainey NL, Janssen PH, Hippe H (1996) Clostridium paradoxum DSM7308(T) contains multiple 16S rRNA genes with heterogenous intervening sequences. Microbiol 142:2087–2095

6 Key Biochemical Attributes to Assess Soil Ecosystem Sustainability 225

Rampazzo N, Mentler A (2001) Influence of different agricultural land-use on soil properties along the Austrian–Hungarian border. Bodenkultur 52:89–115

Rassmussen PE, Collins HP, Smiley RE (1989) Long-term management effects on soil productivity and crop yields in semi-arid regions of eastern Oregon. Station Bulletin 675. USDA-ARS and Oregon State University, Pendleton

Ratledge C, Wilkinson SG (1988) Microbial lipids. Academic Press, London

Reganold JP (1988) Comparison of soil properties as influenced by organic and conventional farming systems. Am J Alt Agr 3:144–155

Rice CW, Moorman TB, Beare M (1996) Role of microbial biomass carbon and nitrogen in soil quality. In: Doran JW, Jones AJ, (eds) Methods for assessing soil quality. Soil Science Society of America Special Publication 49, Madison

Rietz DN, Haynes RJ (2003) Effects of irrigation induced salinity and sodicity on soil microbial activity. Soil Biol Biochem 35:845–854

Riffaldi R, Saviozzi A, Levi-Minzi R (1996) Carbon mineralization kinetics as influenced by soil properties. Biol Fertil Soils 22:293–298

Ritchie NJ, Schutter ME, Dick RP, Myrold DD (2000). Use of length heterogeneity-PCR and FAME to characterize microbial communities in soil. Appl Environ Microbiol 66:1668–1675

Robertson GP, Tiedje JM (1985) Denitrification and nitrous oxide production in successional and old growth Michigan forest. Soil Sci Soc Am J 48:383–389

Robertson GP, Wedin D, Groffman PM, Blair JM, Holland EA, Nadelhoffer KJ, Harris D (1999) Soil carbon and nitrogen availability. In: Robertson GP, Coleman DC, Bledsoe CS, Sollins P (eds) Standard soil methods for long-term ecological research. Oxford University Press, New York

Roesch LFW, Fulthorpe RR, Riva A, Casella G, Hadwin AKM, Kent AD, Daroub SH, Camargo FA, Farmerie WG, Triplett EW (2007) Pyrosequencing enumerates and contrasts soil microbial diversity. ISME J 1:283–290

Roldán A, Caravaca F, Hernández MT, Garcia C, Sánchez-Brito C, Velásquez M, Tiscareño M (2003) No-tillage, crop residue additions, and legume cover cropping effects on soil quality characteristics under maize in Patzcuaro watershed (Mexico). Soil Till Res 72:65–73

Santos HP, Lhamby JCB, Prestes AM, Lima MR (2000). Efeito de manejo de solo e de rotação de culturas de inverno no rendimento e doenças de trigo. Pesquisa Agropecuária Brasileira 35:2355–2361

Schutter ME, Dick RP (2000) Comparison of fatty acid methyl ester (FAME) methods for characterizing microbial communities. Soil Sci Soc Am J 64:1659–1668

Simek M, Hopkins DW, Kalcik J, Picek T, Santruckova H, Stana J, Travnik K (1999) Biological and chemical properties of arable soils affected by long-term organic and inorganic fertilizer applications. Biol Fertil Soils 29:300–308

Sinsabaugh RL, Antibus RK, Linkins AE (1991) An enzymic approach to the analysis of microbial activity during plant litter decomposition. Agr Ecosyst Environ 34:43–54

Sinsabaugh RL, Moorhead DL, Linkins AE (1994) The enzymatic basis of plant litter decomposition: emergence of an ecological process. Appl Soil Ecol 1:97–111

Sinsabaugh RL, Gallo ME, Lauber C, Waldrop M, Zak DR (2005) Extracellular enzyme activities and soil carbon dynamics for northern hardwood forests receiving simulated nitrogen deposition. Biogeochemistry 75:201–215

Sinsabaugh RL, Lauber CL, Weintraub MN, Ahmed BA, Steven D, Crenshaw C, Contosta AR, Cusack D, Frey S, Gallo ME, Gartner TB, Hobbie SE, Holland K, Keeler BL, Powers JS, Stursova M, Takacs-Vesbach C, Waldrop MP, Wallenstein MD, Zak DR, Zeglin LH (2008) Stoichiometry of soil enzyme activity at global scale. Ecol Lett 11:1252–1264

Skujins J (1973) Dehydrogenase: an indicator of biological activities in arid soils. Bull Ecol Res Commun 17:235–241

Skujins J (1978) Hystory of abiontic soil enzyme research. In: Burns RG (ed) Soil enzymes. Academic Press, London

Smalla K, Oros-Sichler M, Milling A, Heuer H, Baumgarte S, Becker R, Neuber G, Kropf S, Ulrich A, Tebbe CC (2007) Bacterial diversity of soils assessed by DGGE, T-RFLP and SSCP fingerprints of PCR-amplified 16S rRNA gene fragments: do the different methods provide similar results? J Microbiol Meth 69:470–479

Smit S, Widmann J, Knight R (2007) Evolutionary rates vary among rRNA structural elements. Nucleic Acids Res 35:3339–3354

Smith JL, Paul EA (1990) The significance of soil microbial biomass estimations. In: Bollag JM, Stotzky G (eds) Soil Biochemistry, vol 6. Dekker, New York, pp 359–396

Smith JL, Halvorson JJ, Papendick RI (1993) Using multiple-variable indicator kriging for evaluating soil quality. Soil Sci Soc Am J 57:743–749

So HB, Kirchhof G, Bakker R, Smith GD (2001) Low input tillage/cropping system for limited resource areas. Soil Till Res 61:109–123

Sojka RE, Upchurch DR (1999) Reservations regarding the soil quality concept. Soil Sci Soc Am J 63:1039–1054

Sorek R, Zhu YW, Creevey CJ, Francino MP, Bork P, Rubin EM (2007) Genome-wide experimental determination of barriers to horizontal gene transfer. Science 318:1449–1452

Sparling GP (1992) Ratio of microbial biomass to soil organic carbon as a sensitive indicator of changes in soil organic matter. Aust J Soil Res 30:195–207

Sparling GP (1997) Soil microbial biomass, activity and nutrient cycling as indicators of soil health. In: Pankhurst C, Doube BM, Gupta VVSR (eds) Biological Indicators of Soil Health. CAB International, Wallingford

Speir TW (1977) Studies on a climosequence of soils in tussock grassland. 11. Urease, phosphatase and sulphatase activities of topsoils and their relationships with other properties including plant available sulfur. N Z J Sci 20:159–166

Speir TW, Ross DJ (1976) Studies on a climosequence of soils in tussock grassland. 9. Influence of age of Chionochloa rigida on enzyme activities. N Z J Sci 19:389–396

Sprent N (1987) The ecology of the nitrogen cycle. Cambridge University Press, Cambridge

Stefanic G, Eliade G, Chinorgeanu I (1984) Researches concerning a biological index of soil fertility. In: Nemes MP, Kiss S, Papacostea P, Stefanic G, Rusan M (eds) Fifth symposium on soil biology. Roman National Society of Soil Science, Bucharest

Steger K, Jarvis Å, Smårs S, Sundh I (2003) Comparison of signature lipid methods to determine microbial community structure in compost. J Microbiol Meth 55:371–382

Stenberg B (1999) Monitoring soil quality of arable land: microbiological indicators. Acta Agric Scand 49:1–24

Stenberg B, Pell M, Torstensson L (1998) Integrated evaluation of variation in biological, chemical and physical soil properties. Ambio 27:9–15

Suzuki MT, Giovannoni SJ (1996) Bias caused by template annealing in the amplification of mixtures of 16S rRNA genes by PCR. Appl Environ Microbiol 62:625–630

Tabatabai MA (1982) Soil enzymes. In: Page AL (ed) Methods of soil analysis, part 2. Chemical and microbiological properties, 2nd edn. American Society of Agronomy, Madison

Tabatabai MA, Bremner JM (1969) Use of p-nitrophenyl phosphate for assay of soil phosphatase activity. Soil Biol Biochem 1:301–307

Tabatabai MA, Bremner JM (1970) Arylsulphatase activity of soils. Soil Sci Soc Am Proc 34:427–429

Tabatabai MA, Bremner JM (1972) Assay of urease activity in soils. Soil Biol Biochem 4:479–487

Taylor JP, Wilson M, Mills S, Burns RG (2002) Comparison of microbial numbers and enzymatic activities in surface soils and subsoils using various techniques. Soil Biol Biochem 34:387–401

Teixeira LCRS, Peixoto RS, Cury JC, Sul WJ, Pellizari VH, Tiedje J, Rosado AS (2010) Bacterial diversity in rhizosphere soil from Antarctic vascular plants of Admiralty Bay, maritime Antarctica. ISME J. doi:10.1038/ismej.2010.35

Thirukkumaran CM, Parkinson D (2000) Microbial respiration, biomass, metabolic quotient and litter decomposition in a lodgepole pine forest floor amended with nitrogen and phosphorous fertilizers. Soil Biol Biochem 32:59–66

Toor GS, Condron LM, Di HJ, Cameron KC, Cade-Menum BJ (2003) Characterisation of organic phosphorus in leachate from a grassland soil. Soil Biol Biochem 35:1317–1323

Trasar-Cepeda C, Gil-Sotres F, Leirós MC (2007) Thermodynamic parameters of enzymes in grassland soils of Galicia, NW Spain. Soil Biol Biochem 39:311–319

Trasar-Cepeda C, Leiros MC, Gil-Sotres F (2008) Hydrolytic enzyme activities in agricultural and forest soils. Some implications for their use as indicators of soil quality. Soil Biol Biochem 40:2146–2155

Trasar-Cepeda C, Leirós MC, Gil-Sotres F, Seoane S (1998) Towards a biochemical quality index for soils: an expression relating several biological and biochemical properties. Biol Fertil Soils 26:100–106

Turner BL, McKelvie ID, Haygarth PM (2002) Characterisation of water extractable soil organic phosphorus by phosphatase hydrolysis. Soil Biol Biochem 34:27–35

Tyson GW, Chapman J, Hugenholtz P, Allen EE, Ram RJ, Richardson PM, Solovyev VV, Rubin EM, Rokhsar DS, Banfield JF (2004) Community structure and metabolism through reconstruction of microbial genomes from the environment. Nature 428:37–43

Vance ED, Brookes PC, Jenkinson DS (1987) An extraction method for measuring soil microbial biomass C. Soil Biol Biochem 19:703–707

Wagner M, Nielsen PH, Loy A, Nielsen JL, Daims H (2006) Linking microbial community structure with function: fluorescence in situ hybridization-microautoradiography and isotope arrays. Curr Opin Biotech 17:83–91

Waldrop MP, Balser TC, Firestone MK (2000) Linking microbial community composition to function in tropical soil. Soil Biol Biochem 32:1837–1846

Wardle DA, Ghani A (1995) A critique of the microbial metabolic quotient qCO_2 as a bioindicator of disturbance and ecosystem development. Soil Biol Biochem 27:1601–1610

Weigel A, Klimanek EM, Körschens M, Mercik S (1998) Investigations of carbon and nitrogen dynamics in different long-term experiments by means of biological soil properties. In: Lal R, Kimble JM, Follet RF, Stewart BA (eds) Soil processes and the carbon cycle, papers from a symposium entitled "Carbon Sequestration in Soil". CRC Press, Boca Raton

White DC, Davies WM, Nickels JS, King JD, Bobbie RJ (1979) Determination of the sedimentary microbial biomass by extractable lipid phosphate. Oecologia 40:51–62

Wittebolle L, Marzorati M, Clement L, Balloi A, Daffonchio D, Heylen K, De Vos P, Verstraete W, Boon N (2009) Initial community evenness favours functionality under selective stress. Nature 458:623–626

Wong VNL, Dalal RC, Greene RSB (2008) Salinity and sodicity effects on respiration and microbial biomass of soil. Biol Fertil Soils 44:943–953

Wu J, O'Donnell AG, He ZL, Syers JK (1994) Fumigation-extraction method for the measurement of soil microbial biomass-S. Soil Biol Biochem 26:117–125

Wu Y, Ding N, Wang G, Xu J, Wu J, Brookes PC (2009) Effects of different soil weights, storage times and extraction methods on soil phospholipid fatty acid analyses. Geoderma 150:171–178

Yakovchenko VI, Sikora LJ, Rauffman DD (1996) A biologically based indicator of soil quality. Biol Fertil Soils 21:245–251

Yergeau E, Bokhorst S, Huiskes AHL, Boschker HTS, Aerts R, Kowalchuk GA (2007) Size and structure of bacterial, fungal and nematode communities along an Antarctic environmental gradient. FEMS Microbiol Ecol 59:436–451

Zehr JP, Mellon MT, Zani S (1998) New nitrogen-fixing microorganisms detected in oligotrophic oceans by amplification of nitrogenase (nifH) genes. Appl Environ Microbiol 64:3444–3450

Zelles L (1999) Fatty acid patterns of phospholipids and lipopolysaccharides in the characterisation of microbial communities in soil: a review. Biol Fertil Soils 29:111–129

Zhang YY, Dong JD, Yang ZH, Zhang S, Wang YS (2008) Phylogenetic diversity of nitrogen-fixing bacteria in mangrove sediments assessed by PCR-denaturing gradient gel electrophoresis. Arch Microbiol 190:19–28

Zhong W, Gu T, Wang W, Zhang B, Lin X, Huang Q, Shen W (2010) The effects of mineral fertilizer and organic manure on soil microbial community and diversity. Plant Soil 326:511–522

Zogg GP, Zak DR, Ringelberg DB, MacDonald NW, Pregitzer KS, White DC (1997) Compositional and functional shifts in microbial communities due to soil warming. Soil Sci Soc Am J 61:475–481

Chapter 7
Methods for Genotoxicity Testing of Environmental Pollutants

Farhana Masood, Reshma Anjum, Masood Ahmad and Abdul Malik

Contents

7.1 Introduction 230
7.2 Objectives of the Genetic Toxicology 232
7.3 Methods for Genotoxicity Testing 234
    7.3.1 Bacterial Test Methods 234
    7.3.2 Tests Employing Eukaryotic Cells and Organisms 239
    7.3.3 Other Assays 248
7.4 Recent Developments 250
    7.4.1 Green Screen HC assay 250
    7.4.2 Engineered Cells and Cell Lines 251
7.5 Future Prospects 252
7.6 Conclusions 253
References 253

Abstract Genetic hazard assessment deals with changes in genetic material of organisms, either human or other natural origin. Although considered an important element of the basic mechanisms of evolution, mutations often have a more detrimental effect on individuals and their offspring, and may adversely affect populations. There is consensus about a close association of DNA damage, mutations and the induction of various types of cancer. In eco-genotoxicity, possible effects of mutagenic/genotoxic substances on populations and ecosystems are investigated. Mutagenicity testing has been performed with all types of organisms. For monitoring purposes higher organisms (eukaryotes) are exposed to the environmental compartment "*in situ*" or in laboratory tests "*in vivo*". Mutagenicity represents permanent changes to single genes or chromosomes, while genotoxicity focuses on primary damage of DNA. The bacterial Ames, umuC and SOS chromo assays have been predominantly used. Tests with eukaryotic cells or organisms might be more relevant for human and ecological risk assessment, but generally they are

---

A. Malik (✉)
Department of Agricultural Microbiology, Faculty of Agricultural Sciences, Aligarh Muslim University, Aligarh-202002, India
e-mail: ab_malik30@yahoo.com

A. Malik, E. Grohmann (eds.), *Environmental Protection Strategies for Sustainable Development,* Strategies for Sustainability,
DOI 10.1007/978-94-007-1591-2_7, © Springer Science+Business Media B.V. 2012

much more time-consuming. Several tests have been developed using the integrity of DNA as a non-specific endpoint of genotoxicity e.g. comet assay, alkaline DNA-elution assay, DNA alkaline unwinding assay, UDS-assay; the comet assay probably the most cost-efficient test among them. Most eukaryotic genotoxicity tests detect macro damage of chromosomes in the visible light microscope following appropriate staining (chromosomal aberration, micronucleus assay, SCE assay). Plants, amphibians, fish and water mussels as well as permanent mammalian cell lines such as V79, CHO or CHL have been used as the test organisms. Newer technologies such as transcriptomics, proteomics and metabolomics provide the opportunity to gain insight into genotoxic mechanisms and also to provide new markers *in vitro* and *in vivo*. There is also an increasing number of animal models with relevance to genotoxicity testing. These types of models will undoubtedly have an impact on genotoxicity testing in the future.

Keywords Genotoxicity • Mutagenicity • DNA • Environmental pollutants

7.1 Introduction

Thousands of chemicals are released by industrial activity, agricultural practices, domestic activity *etc.* and find their ways into various compartments of the environment, like air, soil, groundwater and surface water. Numerous genotoxic chemicals have been detected in both the particulate and gas phases of outdoor air, particularly in densely populated urban regions. Combustion of fossil fuels for power generation or transportation in industrial facilities, power plants and motor vehicles is thought to be a major source of these genotoxic compounds. Among the plethora of genotoxic chemicals that are released into the environment by combustion process, several compounds are thought to be formed from primary combustion products via chemical and photochemical reaction in the outdoor environment. Most of these genotoxic substances eventually descend to the ground, and thereby contaminating it. It was reported that some industries, e.g. pulp and paper mills, steel foundries and organic chemical manufacturing facilities, discharge wastes of significant genotoxic potency. When improperly handled and disposed of, these industrial wastes and effluents also contaminate the soil with their genotoxic compounds. For agricultural land, naturally occurring genotoxic compounds in cultivated plants may be ploughed into the soil by tillage. Various chemicals are also applied to agricultural land as fertilizers, pesticides and herbicides. Moreover soil microflora may also convert nongenotoxic compounds into genotoxic derivatives.

Exposure of a living organism to the genotoxicant usually disrupts the normal cellular processes that eventually results in structural modifications to the DNA which may cause subsequent problems of diverse nature. For instance, exposure of human individuals to various environmental genotoxic chemicals has been associated with the occurrence of cardiovascular diseases, premature aging and the emergence of neoplasia (de Vizcaya-Ruiz et al. 2008; Farmer and Singh 2008; Garinis et al. 2008).

7 Methods for Genotoxicity Testing of Environmental Pollutants

An organism's inability, whether transient or permanent to cope with this type of stress and to maintain its structural integrity provides the investigator the opportunity to test for the genotoxicity of agents (chemical and physical) in the environment.

Physical–chemical analyses are often conducted in order to detect the presence of chemical agents potentially hazardous to the environment and to human health. However, the chemical identity of many substances released into the environment, as well as their resulting metabolites, is still not well characterized, making it necessary to assess the potential impact of these chemical agents (Van der Oost et al. 2003). The characterization of biological consequences of toxic exposure based only on chemical procedures is almost impossible due to their limitations in predicting synergistic and antagonistic effects of contaminants, in complex environmental mixtures (Helma et al. 1998). Biological assays are capable of characterizing the effects of environmental contaminants caused by chronic and/or acute exposure, without prior knowledge of the chemical components present in the environment. Genetic toxicity testing is routinely performed to identify potential genotoxic carcinogens and germ cell mutagens. With regard to the identification of genotoxic carcinogens, all the minimal batteries of genetic toxicology tests recommended by regulatory agencies include at least two or three test procedures, generally an Ames test, a mammalian cell chromosome damage test, and in some cases a mammalian cell mutation assay (Fig. 7.1). Depending on the responses in the tests, the types of substances tested, and on their intended uses (e.g., pharmaceuticals, pesticides, chemicals, cosmetics, etc.), one or more *in vivo* rodent tests (e.g., bone marrow micronucleus; liver UDS) also have to be conducted (ICH 1997; Dearfield et al. 2002; Kirkland et al. 2005).

The standard batteries of tests are selected to address two types of genetic damage of concern, i.e., gene mutations and chromosome damage. Some more recent

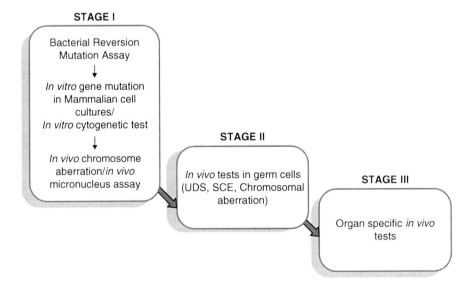

Fig. 7.1 Different bioassays for genotoxicity testing

guidelines have suggested inclusion of the *in vitro* micronucleus test to detect chromosome loss (Thybaud et al. 2007). Additional tests may be needed to clarify the substances activity, or to determine if the activity seen in the initial testing is relevant. Such additional testing may include investigation of aneuploidy, chromosome non-disjunction, DNA interaction, and/or primary DNA damage. The various kinds of tests in the battery are selected because they measure different genotoxic effects and different genotoxic mechanisms of action, thereby providing important mechanistic information. For example, generation of chromosomal aberrations involves strand breakage and rejoining, reversion of the Ames *Salmonella* tester strains containing specific base substitutions requires base mutations that may arise from DNA alkylation or mis-repair of bulky adducts, whereas reversion of the Ames' strains containing frameshift mutations requires induction of a second frameshift that is characteristic of intercalating agents and other classes, but not of small molecular weight alkylating agents. These types of information should be used to guide the interpretation of the results and the selection of follow-up tests. *In vivo* endpoints should be chosen to reflect the types and the mechanisms of damage found in the initial screening battery preferably of different trophic levels.

Agents that are positive in multiple tests with different endpoints, especially if they are positive *in vivo*, impart the highest level of concern and would require extensive investigations to develop a sufficient weight of evidence to establish conditions under which human exposure might be permissible. Selection for choosing an *in vivo* assay as a follow-up test to an initial finding includes choice of the appropriate tissue(s) in which evaluations should be performed. This depends on anticipated route(s) of exposure, tissue distribution (often known from other toxicology or pharmacokinetic studies), and metabolic degradation and/or activation in various tissues as well as target organ in long-term toxicity studies. It was noted that such considerations might dictate the use of nonstandard studies (i.e., not in the regulatory initial battery or standard follow-up studies but sufficiently validated) and that in those cases the use of non-standard studies would be preferable to a standard *in vivo* assay in which the endpoint or target tissue is not relevant. It was further noted that when metabolic modification of an effect is observed to occur, consideration should be given to human metabolism in relation to the laboratory models being considered. Likewise, the relationship between exposure-response information in laboratory models to blood and tissue levels from human exposures is important, and it is desirable to determine this relationship whenever possible.

7.2 Objectives of the Genetic Toxicology

In the field of genetic toxicology, all kinds of changes to the genetic material of an organism are evaluated in order to identify genetic risks after exposure to chemical substances or certain environmental conditions (e.g. solar radiation). Mutagenic agents might be man-made or of natural origin. Especially plants have developed chemical interactions with their environment and many of these substances have

7 Methods for Genotoxicity Testing of Environmental Pollutants

mutagenic properties. Besides exogenous sources, the endogenous agents as reactive intermediates of cell metabolism also contribute to the induced DNA damage though arbitrarily classified under the spontaneous damage. Therefore, complex systems of DNA repair have evolved for detecting and eliminating the damage at the affected DNA strand during the processes of replication and transcription. However, all these DNA repair processes are to a certain degree error-prone, so there still remains a chance that induced DNA damage might lead to permanent changes (mutations) of the genetic make-up of an organism.

Although considered as an important part of the basic mechanisms of evolution, mutations more often have a detrimental effect for individuals and their offspring. Furthermore, increased mutation rates, e.g. due to environmental pollution, might even negatively affect populations. There is consensus about a correlation among DNA damage, mutations and the induction of various kinds of cancer. It is the dominant paradigm in genetic toxicology that the ability of a chemical to cause mutation presages its ability to cause cancer. Even though carcinogenesis is a complex, multi-step process, that is still not fully unravelled, growing evidence shows that it involves multiple mutations eventually leading to uncontrolled cell proliferation.

In genetic toxicology two testing strategies can be distinguished by the endpoint used:

1. mutagenicity testing wherein mutation is the endpoint (single gene, chromosome, or genome mutations);
2. genotoxicity testing with different endpoints representing primary DNA damage such as DNA strand breaks, DNA adducts, induction of the SOS repair system, and chemically altered DNA bases.

Mutagenicity testing has been done with all kinds of organisms including bacteria, invertebrates, mammals, fishes and plants. For monitoring purposes higher organisms (eukaryotes) were exposed to the environmental compartment "*in situ*" or in the laboratory "*in vivo*". *In vitro* test systems usually use bacteria or unicellular eukaryotes (e.g. yeast), primary cultures of tissues, blood cells as well as permanent cell lines originating from eukaryotic organisms. Although mutation represents the indisputable endpoint, the proof is often difficult and time-consuming in higher eukaryotic test systems. As primary DNA damage is one of the important prerequisites for exogenously induced mutations, results from genotoxicity tests can be used as an indicator for an interaction of test substance and DNA with the potential to induce mutations.

The main objectives of genetic toxicology testing are as under:

1. identifying mutagenic/genotoxic substances in order to minimize the risk of exposure to these compounds with suspected carcinogenic properties;
2. genotoxic/mutagenic substances may also induce hereditary defects through mutations in germ cells and they often exhibit teratogenic properties;
3. in an ecological context, mutagenic/genotoxic compounds might induce substantial reproductive loss in exposed populations and could further influence individual fitness by a toxicity-related phenomenon described as genotoxic disease syndrome (Kurelec 1993).

7.3 Methods for Genotoxicity Testing

7.3.1 Bacterial Test Methods

There are many asays for detecting the mutagenicty/ genotoxicity of environmental pollutants, but the utilization of bioassays with bacteria has proven to be very effective for monitoring because these assays are sensitive, inexpensive, reliable, and can be performed in a short period of time with relatively low cost. Genotoxicity data using the bacterial assays are summarized in Table 7.1. All bacterial test methods have some common characteristics. Most tester strains contain mutations which increase sensitivity to genotoxins. The *rfa* mutation for example causes a partial loss of cell wall and therefore increases permeability to larger molecules such as benzo[a]pyrene. The *uvrB* mutation of most Ames-tester strains deletes a gene coding for the DNA excision repair system and therefore hinders the repair of DNA damage. Often a test battery of several tester strains is applied in order to characterize specific genotoxic spectra or get hints on the origin of genotoxins. As bacteria do not possess the metabolic capacity of eucaryotes the tests are usually performed in the absence and the presence of S9 liver homogenate (supernatant of rat liver extract centrifuged at 9000 *g*).

7.3.1.1 Ames Test

Among the microbial bioassays, the *Salmonella* mutagenicity test is undoubtedly the most popular bioassay in environmental mutagenesis research, particularly for the analysis of complex mixtures such as organic extracts of soil, air, and water (Siddiqui and Ahmad 2003; Claxton et al. 2004; Fatima and Ahmad 2006; Ansari and Malik 2009a; Chakraborty and Mukherjee 2009). The standard version of the assay, known as the plate incorporation assay, is a reverse mutation test that quantifies the frequency of reversion from histidine auxotroph to wild-type following a 48 to 72 h incubation with the test substances (Ames et al. 1973; Maron and Ames 1983). Several tester strains of *Salmonella typhimurium* are available, carrying a variety of *his* mutations. The most popular tester strains, TA98 and TA100, carry the *hisD3052* and *hisG46* alleles, respectively. The former is a −1 frameshift mutation reverted to wild type by frameshift mutagens (e.g., ICR-191, nitrosamines etc). The latter carries a base-substitution mutation that is reverted by base-pair substitutions at a GC pair in a proline codon. These strains have been extensively employed for the detection of environmental mutagens including PAHs, nitroarenes, aromatic amines (e.g., N containing heterocyclics), S-containing heterocyclics, and phenyl-benzotriazoles (White and Claxton 2004). The dose response can be quantified by varying the sample concentrations and counting revertant colonies per plate at each concentration.

S. typhimurium YG1021 and YG1026, strains that possess high nitroreductase levels, were developed by introducing plasmids containing the nitroreductase gene

Table 7.1 Various genotoxicity bioassays in bacteria

Sample source	Assay method/strain	Suspected mutagen	Reference
Composite soil from agricultural fields irrigated with industrial and domestic wastewaters (Aligarh City, India)	Ames assay/TA97a, TA102 and TA104; Differential survival in recA, lexA, polA mutants of E. coli K-12	Organochlorines and Organophosphorus pesticides	Aleem and Malik (2003)
Industrial, surface and ground waters from Aligarh (India)	Ames test and fluctuation test/TA97a, TA98, TA100, TA102 and TA104	Heavy metals and pesticides	Siddiqui and Ahmad (2003)
Industrial waste waters of Ghaziabad and Aligarh (India)	Ames test and fluctuation test/TA97a, TA98, TA100, TA102 and TA104	Heavy metals and pesticides	Fatima and Ahmad (2006)
Soils contaminated with industrial effluents (France)	SOS Umu	Hydrocarbons and Metals	Xiao et al. (2006)
Surface soil (Kyoto, Japan)	Ames assay/TA98, TA100	Nitroarenes: 1, 6-DNP1, 8-DNP, 1, 3, 6-TNP, 3, 9-DNF, 3, 6-DNBeP	Watanabe et al. (2008)
Contaminated soils and waste waters (Germany)	automated Umu-test	PAH, PCB, Metals	Brinkmann and Eisentraeger (2008)
Soil samples not contaminated by industrial wastes and discharges	Ames assay/TA98	–	Courty et al. (2008)
Agricultural soils near industrial area (Ghaziabad, India)	TA97a, TA98, TA100,TA102 and TA104; Differential survival in recA, lexA, polA mutants of E. coli K-12; Microscreen induction assay	Organochlorines and Organophosphorus pesticides	Ansari and Malik (2009a)
Soil from uranium mine area (Mangualde, Portugal)	Ames assay/TA98, TA100	Heavy metals	Pereira et al. (2009)
Draining water from dredged sediments (India)	Ames assay/TA98, TA100; Mutatox test (Vibrio fischeri)	Heavy metals, PAHs, PCB	Mouchet et al. (2005)
Nitroarene contaminated industrial sludge (Karlskoga, Sweden)	umuC assay	Nitroarenes	Gustavsson and Engwall (2006)

Table 7.1 (continued)

Sample source	Assay method/strain	Suspected mutagen	Reference
Leachates from dry wastes of metal, tannery, dye industries (Uttar Pradesh, India)	Spot test/Ames assay/TA98,TA100, TA97a, TA102	Metals and dyes	Singh et al. (2007)
Industrial Wastewater from Ghaziabad (India)	TA97a, TA98, TA100,TA102 and TA104 Differential survival in *recA, lexA, polA* mutants of *E. coli* K-12 Microscreen induction assay	Organochlorines and Organo-phosphorus pesticides	Ansari and Malik (2009b)
Coal fly ash water leachate (India)	Ames assay/TA97a, TA102	Metals	Chakraborty and Mukherjee (2009)
Ground water samples (Sao Paulo, Brazil)	Ames assay/TA98, TA100, YG1041 and YG1042	Metals and Nitrocompounds	Valente-Campos et al. (2009)
Surface sediments from Taihu Lake (China)	Ames assay/TA 98, TA100	Organochlorine pesticides	Zhao et al. (2010)
Dimethylamino-2-ethylazide (DMAZ) N,N,N′,N′-tetramethylethanediamine (TMEDA)	Ames assay/TA98, TA100, TA1535, TA1537 and *E. coli,* WP2 *uvrA*)	–	Reddy et al. (2010)
Polychlorinated butadienes (Cl_4–Cl_6)	Ames assay	–	Brüschweiler et al. (2010)
PM 2.5 (Italy)	Ames assay/TA98, TA98NR,YG1021	Nitrocompounds	Traversi et al. (2009)

7 Methods for Genotoxicity Testing of Environmental Pollutants 237

from *S. typhimurium* TA1535 into TA98 and TA100, respectively. These strains have been shown to detect various kinds of mutagenic nitro compounds much more efficiently than TA98 and TA100 (Watanabe et al. 1989). Watanabe et al. (1990) successfully developed some new tester strains, for instance *S. typhimurium* YG1024 and YG1029 strains derived from TA98 and TA100 respectively that show remarkably high sensitivity to both nitroarenes and aromatic amines. *S. typhimurium* YG1041 and YG1042 (Hagiwara et al. 1993), derived from TA98 and TA100, respectively, have enhanced levels of both nitroreductase and O-acetyltransferase and are, consequently, highly sensitive to nitroarenes and aromatic amines.

The simple reason for the greater popularity of the *Salmonella* assay amidst other assays is due to the ease and cost-effectiveness of the *Salmonella* assay compared with other assays. In addition, the *Salmonella* assay has been used more than any other test for evaluating complex mixtures. The coupling of this assay with chemical analysis i.e., bioassay-directed chemical fractionation enhanced the utility of this bioassay and permitted the isolation and identification of defined chemical fractions that contain genotoxic activity. These defined fractions can then be subjected to further analyses in an effort to characterize the precise chemical composition. The *Salmonella*/microsome assay has been used to detect mutagens in samples of cigarette smoke (Roemer et al. 2004), diesel exhaust (Seagrave et al. 2005), surface waters (Siddiqui and Ahmad 2003; Tabrez and Ahmad 2009; Valente-Campos et al. 2009), sewage sludge (Perez et al. 2003), and soils (Xiao et al. 2006; Brinkmann and Eisentraeger 2008). This bioassay has also been used to identify and characterize various mutagens such as heterocyclic amines in cooked foods (Shishu and Kaur 2003; Sugimura et al. 2004).

The liquid version of the Ames test, Ames fluctuation test is also used. Though it could not gain such popularity despite its high sensitivity, ability of automation, possibility of using hepatocytes for metabolic activation and its better sensitivity for aqueous samples containing low levels of mutagens. Recently in the authors laboratory a cooperative analysis between the solid (plate incorporation) and liquid (fluctuation) version of Ames testing has been conducted, wherein they concluded that both the tests were equally efficacious in their own right but fluctuation test is better adapted for aqueous samples with low levels of mutagenic potentials (Siddiqui and Ahmad 2003; Fatima and Ahmad 2006).

7.3.1.2 umuC-assay

The umuC-assay was originally developed by Oda et al. (1985). A microplate version of the test is available. The assay is based on the use of a genetically modified *Salmonella typhimurium* strain TA1535 that contains the plasmid pSK1002. Here the *umuC* gene, as a part of the SOS system, is fused to a reporter gene, *lacZ*, that encodes for ß-galactosidase. If genotoxins induce the SOS function, the reporter gene is also activated and the formation of ß-galactosidase is quantified photometrically at 420 nm by its ability to form a yellow-coloured metabolite (Oda et al. 1985). The test is carried out with and without S9. Bacterial growth is measured as turbidity at 600 nm and biomass factors are considered in the test results. A reduction of cell growth by more than 50% is considered as a toxic effect hence ß-galactosidase activity is not measured for those wells.

The *umu* test is specific for determining the mutagenic effect of nitroarenes. These chemicals are widespread in the environment because of their rapid formation from products of incomplete combustion of polycyclic aromatic hydrocarbons and oxides of nitrogen (Rosenkranz et al. 1980). Some of them are reported to be potent mutagens and carcinogens (Rosenkranz and Mermelstein 1985; Weisburger 1988). They are activated to mutagens by reduction to arylhydroxylamines by nitroreductase (NR) and then these arylhydroxylamines are activated by O-acetyltransferase (O-AT) to form the ultimate reactive electrophiles, nitronium ions that bind to DNA and cause mutagenesis (Wild and Dirr 1989). Since nitroarenes are an important risk factor for humans, the studies concerning their mutagenicity and genotoxicity are important. Several other tester strains which overexpress specific activation enzymes (acetyltransferase, nitroreductase) have been developed in order to increase the sensitivity against specific genotoxins like nitroarenes and/or aromatic amines (Oda et al. 1992, 1993, 1995). The application of a fluorometric *umu*-test system has been developed in order to increase the sensitivity of the test for the detection of genotoxic compounds in surface waters (Reifferscheid and Zipperle 2000), contaminated soils and waste waters (Brinkmann and Eisentraeger 2008).

7.3.1.3 SOS Chromoassay

The SOS chromotest originally was developed by Quillardet et al. (1982, 1985). The test employs a variant of *Escherichia coli* (strain PQ37) in which the production of β- galactosidase is under the express control of the SOS response to DNA damage, and SOS induction is monitored calorimetrically. Test results are expressed as SOS induction factor (IF), the ratio of toxicity-corrected SOS induction in the samples relative to the solvent control, with sample potency usually expressed as the SOS inducing potency (SOSIP), the initial slope of the concentration response relationship. There is some evidence that the *umu* test detects lower genotoxic responses than the SOS chromotest for two reasons: firstly, the outer wall of the *Salmonella* tester strain used is made more permeable to genotoxins, and secondly, the *umuC* reporter gene is placed on a multicopy plasmid while in the SOS chromotest it is placed on a single bacterial chromosome (de Maagd and Tonkes 2000).

Rojiéková et al. (1998) used the SOS chromotest solid phase test (i.e., SOS chromotest pad) to analyze soils amended with a variety of petroleum products including kerosene, used motor oil, and crude petroleum. The chromotest offers the convenience of miniaturization and the test can be performed in only 2–3 h. Moreover, sample sterility and the survival of the tester strain are also not required. The microplate version of the SOS chromotest/*umu*-test was developed as a rapid and sensitive screening tool, for the detection of genotoxins in surface waters (Jolibois and Guerbet 2005) and contaminated soils (Xiao et al. 2006).

7 Methods for Genotoxicity Testing of Environmental Pollutants

7.3.1.4 Microscreen Phage-Induction Assay

The *E. coli* prophage induction assay, also known as the microscreen prophage induction assay was developed by Rossmann et al. (1984). The activation of the SOS system results in the release of lytic phages from *E. coli* [WP2s], which are detected following their infection of a second (indicator) *E. coli* strain [TH-008]. The genotoxic potency is evaluated by counting the plaques in the bacterial layer. The DNA-repair assay with *E. coli* K12 strains enables the detection of (repairable) DNA-damage by comparison of the differential survival of strains differing in their DNA-repair capacity.

Several studies have shown the microscreen phage-induction assay to be an appropriate analytic methodology to detect halogenated compounds, organochlorine compounds and metal components, with a low level of detection by the *Salmonella* assay (Vargas et al. 2001; Ansari and Malik 2009a, b). In addition, this test was considered a good screening assay for genotoxic compounds present in small concentrations in environmental samples.

7.3.1.5 Mutatox Assay

The Mutatox assay uses a non luminescent variant of the luminescent saltwater bacteria *Vibrio fischeri* (*Photobacterium phosphoreum*), which is also used for the determination of acute bacterial toxicity. Genotoxic damage induces the re-establishment of luminescence, which indicates the degree of genotoxicity. In contrast to the SOS chromotest and the umuC test where the activation of the SOS pathway occurs, the formation of a protease is measured, that degrades a repressor protein of the lux pathway thus leading to luminescence (de Maagd and Tonkes 2000). The intensity of the luminescent signal at a given concentration is directly related to the reversion frequency and the mutagenic activity of the test sample. The test responds to a variety of base-pair and frame-shift mutagens, crosslinking agents, intercalating agents, and DNA synthesis inhibitors (Kwan et al. 1990). It has also been employed in examining the mutagenicity of soil and water samples (Frische 2002; Mouchet et al. 2005).

7.3.2 Tests Employing Eukaryotic Cells and Organisms

In vitro and *in vivo* testings of genotoxicity at a higher level of biological organization with eukaryotic cells or organisms might be more relevant for human and ecological risk assessment. Several genotoxicity tests have been developed which use the integrity of DNA as a non specific endpoint of genotoxicity. Among them different techniques are used to measure DNA fragmentation as a result of DNA strand breaks. Alkaline DNA denaturating conditions are added to the test protocols in order to detect, besides double strand breaks, single strand breaks, alkali-labile sites and repair-enzyme-mediated incisions. Genotoxicity data using the eukaryotic cells are summarized in Table 7.2. A major drawback of these tests is their relatively more time-consuming nature compared with the bacterial tests.

7.3.2.1 ^{32}P-postlabeling Assay

The majority of chemicals that are genotoxic exert their effects only after metabolic conversion to chemically reactive forms. Under certain conditions, and depending upon the chemical properties of the genotoxicant, the reactive entity may bind covalently to cellular macromolecules, including nucleic acids and proteins. The product of this reaction with DNA is termed a DNA adduct. Some genotoxicants are so chemically reactive that metabolic activation is not necessary and direct adduction to cellular molecules occurs without metabolic assistance. The presence of a specific adduct in the DNA is an indicator of exposure to a genotoxicant, and may be used to identify, unequivocally, the genotoxicant of concern. Currently, methods of varying sensitivity are available to detect and quantify the DNA adducts. The most prominent among them is the ^{32}P-postlabeling technique while other methods employ the high performance liquid chromatography (HPLC), fluorescence spectrophotometry and immunoassays using adduct-specific antibodies (Qu et al. 1997).

In the ^{32}P-postlabeling assay, DNA is hydrolyzed enzymatically to 3′-monophosphates and DNA adducts are enriched by the selective removal of normal nucleotides. The DNA adducts are then labelled with [^{32}P] phosphate and resulting ^{32}P-labeled DNA adducts are usually separated by thin-layer chromatography or high performance liquid chromatography. Radioactivity of DNA adducts is detected by autoradiography and measured by liquid scintillation counter. The ^{32}P-postlabeling technique is the most sensitive method for the detection of a wide range of large hydrophobic compounds bound to DNA, and can potentially detect even a single DNA adduct, such as those derived from polycyclic aromatic compounds (PACs), in 10^9–10^{10} bases. The ^{32}P-postlabeling methodology for the detection of adducted chemicals to DNA is finding increasing applications in environmental monitoring studies where the genotoxicity is of special concern. The salient features of this methodology are sensitivity and selectivity.

7.3.2.2 Alkaline Unwinding Assay

Alkaline unwinding technique is used to detect and quantitate DNA strand breaks caused by physical factors such as radicalizations, and DNA damages induced by organic compounds. This method can be employed for the measurement of DNA damage not only in the cultured cells *in vitro*, but also in the organs, tissues and the whole organism under *in vivo* conditions. DNA strand separation occurs in alkali due to destruction of hydrogen bonds between two strands of the double helix. The rate of unwinding in alkali depends on the covalent length of the DNA strands

Table 7.2 Various genotoxicity bioassays in eukaryotes

Tested agents	Organism/cell	Organ/tissue	Bioassay employed	Response	Reference
Industrial waste waters from Aligarh and Ghaziabad (India)	*Allium cepa*	Root	CA	+	Fatima and Ahmad (2006)
Caí river (Rio Grande do Sul State)	*Pimephales promelas*	Erythrocytes	MN (*in vivo*)	+	de Lemos et al.(2007)
River of Tucuman (Argentina) receiving industrial and surface water	*Allium cepa*	Root	CA MN	+ +	Gana et al. (2008)
Water soil leachates (Slovenia)	Caco-2 and HepG2 cells *Tradescantia*		Comet assay (*in vitro*) MN	+ +	Lah et al. (2008)
Coastal water (Goa)	*Cronia contracta*	–	Alkaline unwinding assay (*in vivo*)	+	Sarkar et al. (2008)
Coal fly ash water leachate (India)	Human blood cells *Nicotiana*		Comet assay (*in vitro*) Comet assay (*in vivo*)	+ +	Chakraborty and Mukherjee (2009)
Treated urban sewage sludge	*Wistar* rats	Peripheral blood Bone marrow	Comet assay (*in vivo*) MN (*in vivo*)	Not significant	Marzo Solano et al. (2009)
Danube River (Europe)	Rainbow-trout *Barbus barbus*	Liver Erythrocytes	Comet assay (*in vitro*) MN (*in vitro*) MN (*in vivo*)	+ + +	Boettcher et al. (2010)
Water accommodated fraction of petroleum hydrocarbons (WAF-P)	*Chaetoceros tenuissimus* *Skeletonema costatum*	–	Alkaline unwinding assay (*in vivo*)	+	Deasi et al. (2010)
Coking wastewater (China)	*Vicia faba* *Hordeum vulgare*	Root	MN (*in vivo*) SCE (*in vivo*)	+	Dong and Zhang (2010)
Soils spiked with benzo[*a*]pyrene	–	–	^{32}P-postlabelling (*in vivo*)	+	Hua et al. (2009)
Urban air (Italy)	*Tradescantia* clone # 443 *Nicotiana tabacum*	Buds Leaves	Trad-MN Comet assay	+ +	Villarini et al. (2009)
Phenanthrene (Phe)	*Liza aurata*	Gill Liver	Alkaline unwinding assay (*in vivo*)	+	Oliveira et al. (2007)

Table 7.2 (continued)

Tested agents	Organism/cell	Organ/tissue	Bioassay employed	Response	Reference
Polyphenols	*Unio timidus*	Erythrocytes	Comet assay (*in vivo*)	+	Labieniec et al. (2007)
bisphenol A (BPA)	Mice	Liver Mammary Cells	^{32}P-postlabelling (*in vivo*)	+	Izzotti et al. (2009)
3-Nitrobenzanthrone (3-NBA),	NMRI mouse	Skin	^{32}P-postlabelling (*in vivo*)	+	Schmeiser et al. (2009)
Chloroacetic acid (CAA) and chloro-benzene (CB)	*Rattus norvegicus*	Bone marrow cells	CA (*in vivo*) MN (*in vivo*)	+	Siddiqui et al. (2006)
Benzoic acid	Human lymphocytes	–	CA (*in vitro*) MN (*in vitro*) SCEs (*in vitro*)	+	Yilmaz et al. (2009)
dimethylamino-2-ethylazide (DMAZ), N,N,N′,N′-tetramethylethanediamine (TMEDA)	–	CHO cells	CA (*in vivo*)	–DMAZ +TMEDA	Reddy et al. (2010)
Fluconazole	Mouse bone marrow cells Human lymphocytes	–	CA (*in vivo*) CA (*in vitro*) MN (*in vitro*) SCE (*in vitro*)	Not significant + + +	Yüzbaşıoğlu et al. (2008)
AMPA	Hep-2cells Human lymphocytes Mice		Comet assay (*in vitro*) CA (*in vitro*) MN (*in vivo*)	+ + +	Mañas et al. (2009)
Glyphosate	Hep-2 cells Human lymphocytes Mice		Comet assay (*in vitro*) CA assay (*in vitro*) MN (*in vivo*)	+ – +	Mañas et al. (2009)
Pirimicarb, Aficida® (50% pirimicarb)	–	CHO-K1 cells	CA (*in vitro*) SCE (*in vitro*)	+ +	Soloneski and Larramendy (2010)

CA chromosome aberration; *MN* micronucleus; *SCE* sister chromatid exchange; (+) positive response; (−) negative response

7 Methods for Genotoxicity Testing of Environmental Pollutants 243

where the breaks act as starting points for unwinding. Therefore, the amount of double stranded DNA leftover after a period of unwinding is inversely related to the number of strand breaks. Alkaline unwinding assay involves following three steps, isolation and purification of DNA, alkaline unwinding, and fluorometric analysis of DNA unwinding.

Alkaline unwinding assay has been widely used to analyze cultured cells and other biological samples, and to detect possible strand breakage caused by potential chemical carcinogens and mutagens. Siu et al. (2003) exposed green-lipped mussels (*Perna viridis*) to PCB, and monitored the levels of DNA strand breaks in the mussel hepatopancreas using an alkaline unwinding assay. *in situ* investigations for the detection of genotoxic potential in selected surface water with the DNA alkaline unwinding assay have been reported using fish cells, early life stages of fish, crustaceae and mussels (Wittekindt et al. 2000). At present, the application of alkaline unwinding assay has expanded from small ecological environments to real contaminated area. This method has been found to be quite effective in evaluation on genetic damage on organisms caused by environmental pollutants.

7.3.2.3 Alkaline DNA Elution Assay

Alkaline DNA elution assay was first reported by Kohn et al. (1976). This assay actually detects both the single and double strand breaks in the DNA. The test measures the rate of elution of DNA through a membrane filter after the digestion of treated tissue and DNA denaturation in a buffer with protease and detergents respectively. The content of DNA in the filtrate is measured by fluorimetry. The elution rate increases with the reduction of the molecular weight of the DNA fragments. The test has been applied with wild living clams (*Corbicula fluminea*) for the assessment of genotoxic potentials in native surface waters (Waldman et al. 2000) and with the Chinese hamster cell line V79 for assessing the genotoxicity of the waste water from a paraquat manufacturing plant (Kuo and Lin 1993).

7.3.2.4 DNA-Repair Synthesis (UDS-Assay)

The unscheduled DNA synthesis assay measures the incorporation of radioactively labelled nucleosides usually tritium-labelled thymidine in cells that are not undergoing scheduled (S-phase) DNA synthesis. The test therefore detects DNA repair synthesis after excision and removal of a DNA stretch damaged by mutagenic agents. The uptake of thymidine usually is determined by autoradiography. Garry et al. (2003) employed DNA repair synthesis UDS test in rat's primary hepatocytes to assess genotoxicity of benzo[a]pyrene. Despite various merits, equipment cost and the experimental time for test performance are high. Currently, the unscheduled DNA synthesis (UDS) assay and the *in vivo* bone marrow MN assay are the most internationally accepted *in vivo* genotoxicity assays.

7.3.2.5 Random Amplified Polymorphic DNA (RAPD) Assay

RAPD assay evaluates the genetic variability directly at the genome level. RAPD provides a less biased genomic sample and is able to generate a nearly unlimited number of markers, given that DNA fragments are flanked by sequences that are complementary to a specific primer and located on different template strands that are within 3 kb of each other (Fritsch and Rieseberg 1996). Generally, RAPD reactions are performed with a single 10 bp primer and amplified fragments are visualized by agarose or polyacrylamide gel electrophoresis and subsequent staining with ethidium bromide. The resulting DNA profiles may differ among individuals, depending on the (1) presence/absence (p/a) of priming sites, (2) priming complementary completeness/incompleteness, or (3) the distance between priming sites. Hence, RAPD bands are lost or gained when point mutations, inversions, deletions, additions or gross chromosomal rearrangements affect the p/a of primer sites, their complementarity to primers and/or the distance between priming sites. RAPD bands of different molecular weight are interpreted as separate loci which are scored on a present (amplification)—absent (non-amplification) basis (Fritsch and Rieseberg 1996). Additionally, band intensity differences may be interpreted as well through visual inspection or by using image analysis software.

The first study measuring genotoxic effects using the RAPD assay was performed by Savva et al. (1994). In this study, the RAPD profiles generated from rats exposed to benzo(a)pyrene revealed the appearance and disappearance of bands in comparison to control patterns. These changes observed in the fingerprints of exposed animals may be due to the presence of DNA adducts, mutations or DNA strand breaks. Since, the RAPD method was also successfully used to detect 'DNA effects' induced by benzo(a)pyrene (Castano and Becerril 2004), metals such as lead, manganese, cadmium and copper (Atienzar et al. 2001; Liu et al. 2005; Enan 2006), UV, X-ray, gamma radiations (Kuroda et al. 1999; Atienzar et al. 2000; Atak et al. 2004). DNA effects include DNA damage (e.g. DNA adduct, DNA breakage) as well as mutations (point mutations and large rearrangements) and possibly other effects (e.g. structural effects) at the DNA level that can be induced by chemical or physical agents that directly and/or indirectly interact with genomic DNA. RAPD assay can detect mutations if they occur in at least 2% of the tested DNA (Jones and Kortenkamp 2000). Yang et al. (2000) reported that agrochemicals (mainly triadimefon and ammonium bicarbonate and their metabolites) affect the diversity of soil microbial community at DNA level by the RAPD assay.

Compared with other mutagenicity tests such as the Ames test or the hprt (hypoxanthine phosphoribosyl transferase) test in mammalian cell lines, RAPD assay is less sensitive (Jones and Kortenkamp 2000). However, in these well-validated mutagenicity assays, a relatively small target is investigated whereas RAPD examines mutations in the whole genome. Until now, there has been very little data evaluating the potential of RAPD assay to detect genotoxic effects with some well-known genotoxicity tests. Nevertheless, studies comparing the RAPD method with routinely used genotoxicity/mutagenicity assays are necessary to further evaluate the potential of the RAPD assay for the detection of DNA damage and mutations.

7.3.2.6 Comet Assay

In recent years the comet assay has gained broad attention, because this test is relatively easy to handle and can be applied with cells from different organisms and tissues. The alkaline version of the comet assay has been developed by Singh et al. (1988). In general cells are mixed with low-melting agarose, placed on microscope slides and lysed by an alkaline buffer with ionic detergents. The liberated DNA is resolved in an electrophoresis chamber, stained and evaluated by fluorescence microscopy. Cells with increased DNA damage display increased migration from the nuclear region towards the anode. The resulting comet-like structure is quantified by measuring the length of the tail and/or the tail moment (the intensity of the migrated DNA multiplied by the respective tail length (integral) with respect to the nuclear DNA). The distance and or amount of DNA migration is indicative of the number of strand breaks.

The test has been applied to a broader range of aquatic organisms such as algae (Erbes et al. 1997; Aoyama et al. 2003), mussels (Nagarajappa et al. 2006; Rank et al. 2007), amphibians (Mouchet et al. 2006; Huang et al. 2007) and fish (Masuda et al. 2004; Kosmehl et al. 2008). The advantages of the comet assay include the following: (a) genotoxic damage is detected at the single cell level; (b) most eukaryotic cell types are suitable for the assay; (c) only a small number of cells are required; (d) it is generally faster to conduct and more sensitive than other available methods for the assessment of strand breaks; (e) DNA strand breaks form quickly following genotoxic exposure, allowing for an early response evaluation on biota. On the other hand there are still no standard test protocols and a certain degree of handling skills is a necessary prerequisite to routinely performing the test. Moreover, owing to very high sensitivity of this test, the problem of false positivity should also be taken care of.

Single cell gel electrophoresis (SCGE) is currently the most widely employed method to detect DNA lesions in eco-genotoxicology. The production of DNA strand breaks correlates well with the mutagenic and carcinogenic properties of environmental pollutants with diverse structures. Chromosomal damage expressed after cell replication represents an accumulated effect associated with long-term exposure. Cavas and Könen (2007) using the alkaline comet assay, evaluated the genotoxic potential of the herbicide roundup (R) containing isopropylamine salt of glyphosate in goldfish (*Carassus auratus*). The comet assay is a tool widely used to detect genotoxicity in aquatic environments (Kim and Hyun 2006; Frenzilli et al. 2009). Andrade et al. (2004) reported that this test can be recommended for screening of complex mixture contaminants in biomonitoring studies.

7.3.2.7 Chromosome Aberration Assays

Chromosomal aberration represents a macrodamage of chromosomes that include structural aberrations such as fragments or intercalations and numerical aberrations formed as a result of (unequal segregation of homologous chromosomes during cell divisions, which leads to a loss or surplus of chromosomes (aneuploidy and polyploidy). Cytogenetic effects can be studied either in whole animals (*in vivo*) or in cells grown in culture (*in vitro*). Generally the cell culture is exposed to the test substance and then treated with a metaphase-arresting substance (Colcimide). Following suitable staining the metaphase cells are analysed microscopically for the presence of chromosomal aberrations. Although not currently used in cytogenetic testing for regulatory submissions, fluorescence *in situ* hybridization (FISH) staining techniques have been recently developed for human and mouse chromosomes, in which each chromosome can be differentially stained, revealing chromosomal rearrangements not apparent with conventional staining techniques. When FISH staining is translated from a research approach to a testing protocol, it may be possible to reduce the number of chromosomes to be analyzed and, hence, the time for chromosomal aberration tests. Plants have been especially used for the evaluation of chromosome aberration (Gana et al. 2008), fish have been applied only occasionally. Most of the experience gained is available with permanent cell lines such as Chinese hamster lung cells (V79) (Jung et al. 2001) and Chinese hamster ovary cells (CHO) (Soloneski and Larramendy 2010). The detection of aberrations in fish cells is difficult due to the high number and small size of chromosomes in most species.

7.3.2.8 *Allium cepa* Anaphase Aberration Assay

Onion bulbs are briefly germinated and 1–2 cm long roots are then exposed to soils, soil extracts, soil slurries, or soil leachates for 2–24 h (usually one mitotic cycle). Roots are then fixed, stained and 100 anaphase or telophase cells scored for aberrations including chromosome fragments and bridges. Some researchers also score the frequency of vagrant chromosomes and multipolar cells that are presumed to be the result of c-mitotic events (Kovalchuk et al. 1998). *Allium cepa* is an efficient test system routinely used to evaluate the genotoxic potential of chemicals in the environment, due to its sensitivity and good correlation with mammalian test systems (Chauhan et al. 1999). *Allium cepa* chromosome aberration test, a standardized test for cyto- and genotoxicity, is routinely used for studying the effects of toxic substances (Fiskesjo 1985). The anaphase–telophase chromosome aberration assay has been shown to be highly reliable in genotoxicity testing (Konuk et al. 2007; Yildiz and Arikan 2008; Yildiz et al. 2009).

7.3.2.9 Sister Chromatid Exchange (SCE) Assay

The sister chromatid exchange (SCE) assay detects reciprocal exchanges of DNA segments between two sister chromatids of a duplicating chromosome. SCEs represent the interchange of DNA at apparently homologous loci. This process involves DNA breakage and repair but as this process does not necessarily lead to permanent mutations some researchers classify the SCE assay as a genotoxicity test. Although little is known about its molecular basis, the SCE frequency is elevated under the

7 Methods for Genotoxicity Testing of Environmental Pollutants

influence of mutagenic agents and therefore serves as a model for mutagenicity. The detection of sister chromatids is achieved by incorporation of unusual precursors like bromodesoxyuridine into chromosomal DNA for two cell cycles followed by fluorescence microscopy. For pesticides genotoxicity assessment SCE assays have been performed by many researchers (Zeljezic and Garaj-Vrhovac 2002; Ergene et al. 2007; Soloneski and Larramendy 2010). Usually SCEs are performed with mussels, fish cells and mammalian cells (Chinese hamster lung, CHL, Chinese hamster ovary cells, CHO). SCEs are useful markers of cytogenetic effects *in vitro* and *in vivo* (Yilmaz et al. 2009; Dong and Zhang 2010), however, their utility as biomarkers of adverse health effect is disputed (Hagmar et al. 2004; Bonassi et al. 2004).

7.3.2.10 Micronucleus Induction

Micronuclei are chromosome fragments or whole chromosomes that were not incorporated in the daughter cell nuclei and appear in the cytoplasm. For the measurement of micronuclei cell division must be allowed to continue up to the interphase. A mammalian erythrocyte micronucleus assay with bone marrow has been standardized by OECD and EC. Micronucleus formation along with the sister chromatid exchanges and chromosome aberration assays is considered as a clastogenic endpoint. In principle flow cytometric measurement of micronuclei is possible (Sánchez et al. 2000) but equipment costs are high. The micronucleus assay is a widely used cytogenetic assay for the assessment of *in vivo* or *in vitro* chromosomal damage. In general, the chromosomes of fish and other aquatic organisms are relatively small in size and/or high in number. Therefore, the metaphase analysis of chromosomal aberrations using these organisms is difficult. However, small size and large chromosome number does not affect the performance of the micronucleus assay, so it can be easily applied to fish or other aquatic organisms. The micronucleus test in the peripheral blood of erythrocytes in fish is widely used and recommended for studies on chronic exposure to different types of environmental pollutants with clastogenic and aneugenic properties (Udroiu 2006). The genotoxicity of a number of metals such as chromium has been demonstrated by several authors by means of the micronucleus test. Zhu et al. (2004) found positive results in erythrocytes of *Cyprinus carpio* exposed to concentrations of hexavalent chromium ranging between 1 and100 µg/mL.

Environmental biomonitoring with micronuclueus assays usually has been performed *in vivo* by exposure of relevant aquatic organisms for several days followed by microscopic analysis of erythrocytes, gill cells (animals) or roots (plants). But fish and human derived permanent cell lines (RTG-2) have also been used "*in vitro*" (Kohlpoth et al. 1999, Sánchez et al. 2000). *In vitro* micronucleus tests are currently under development in a number of laboratories as a less subjective and more economical alternative to *in vitro* chromosomal aberration tests. For these approaches, cytochalasin B is used to arrest cell division (cytokinesis) but not nuclear division, and up to 1000 binucleate cells are examined for the presence or absence of micronuclei. However, because the *in vitro* micronucleus tests have yet to be validated and shown to be at least as effective as tests for chromosomal aberrations *in vitro,* none is currently recommended for regulatory submissions. *In vivo* micronucleus tests are justified for regulatory submissions for assessing chromosomal breakage and aneuploidy in an environment including *in vitro* metabolic reactions. Micronuclei are readily observed microscopically in stained preparations of (otherwise anucleate) polychromatic erythrocytes (PCEs) from the bone marrow of rats or mice or from the peripheral blood of mice; the latter because, in mice, the spleen does not remove micronucleated cells from the blood. Bone marrow cells, which give a more informative index of toxicity, are routinely used for the micronucleus test as well as for *in vivo* chromosomal aberration assays. Because micronuclei can be evaluated more rapidly and economically than chromosomal aberrations, the micronucleus test in rodents is now used more extensively than the rodent bone marrow chromosomal aberration test.

7.3.2.11 *Tradescantia* Micronucleus Test

The *Tradescantia* MN assay or Trad-MN, an assay originally developed for the assessments of gaseous and airborne mutagens (Van't Hof and Schairer 1982; White and Claxton 2004), is often used for investigations of contaminated aqueous media such as surface waters or wastewaters (Lah et al. 2008) and complex environmental samples (Meireles et al. 2009). The assay involves exposure of 15–30 cuttings per treatment to the test material, and buds are subsequently fixed and early stage tetrads (i.e., meiotic products of spore mother cells) are stained and scored for MN. Generally, 300 tetrads are scored from each experimental group and the results are expressed as MN per 100 tetrads (Ma et al. 1994). Exposure times vary widely depending on the nature of the study and the samples being examined.

7.3.3 Other Assays

7.3.3.1 *Tradescantia* Stamen Hair Mutation

The stamen hair mutation system is carried out using sterile hybrid clones such as *Tradescantia* clone 4430, an interspecific hybrid of *T. hirsutiflora* and *T. subcaulis.* The sterility of the clones is a convenient feature that ensures genetic homogeneity in the absence of mutagenic effects. The stamen hair mutation assay or Trad-SHM is based on the fact that stamen hair cells in clone 4430, and other clones such as BNL02, are heterozygous for phenotypically visible flower colour markers (i.e., blue-dominant and pink-recessive) (Ma et al. 1994). Cuttings of *Tradescantia* clones heterozygous for the alleles controlling stamen hair colour are exposed to aqueous extracts, organic extracts diluted in an aqueous medium, whole soils, or soil slurries for up to several days. Following a lag period of up to 14 days (depend-

7 Methods for Genotoxicity Testing of Environmental Pollutants

ing on treatment time) stamen filaments are microscopically examined and scored for pink mutations. The results are usually expressed as mutation events per 1000 stamen hairs. Ferreira et al. (2007) reported mutagenic potential of contaminated atmosphere of Ibirapuera park (Brazil) by applying the Trad-SHM assay. Similarly Česnienė et al. (2010) evaluated the genotoxicity of contaminated soil from military and urban territories.

7.3.3.2 *Saccharomyces cerevisiae* Assay

Recently, *Saccharomyces cerevisiae* tester strains have been developed for the detection of genotoxic (Walmsley et al. 1997; Afanassiev et al. 2000) and combined geno- and cytotoxic potential (Lichtenberg-Frate et al. 2003). The *Saccharomyces cerevisiae* reporter assay is based on transcriptional activation of the yeast optimized version of the green fluorescent protein (Gfp) of *Aequorea victoria* (Cormack et al. 1996) by the DNA damage inducible RAD54 promoter in a sensitized yeast strain devoid of endogenous efflux transporters. Like the *rfa* mutation in the *Salmonella typhimurium*-based *umu* test where the *rfa* gene-cluster encodes the lipopolysaccharides (LPS) core oligosaccharide biosynthesis, deletions of ABC-type transporters in the *Saccharomyces cerevisiae* strain, aim to enhance test sensitivity. In yeast, the activities of these ABC-type extrusion systems are involved in drug extrusion and the continuous removal of the relevant compounds from the cell can compromise the sensitivity of any assay (Schmitt et al. 2004). Since the cell's own DNA damage assessment apparatus is being monitored, the entire genome is used as the target for DNA damage. This is in contrast to mutation assays which monitor damage at a particular genetic locus, such as the *Salmonella* HIS operon in the Ames test. Many DNA-damaging agents affect other cellular targets and, depending on the exposure level, this can result in a reduction in proliferative potential as well as increasing mortality. In this assay relative total growth is assessed by comparing the extent of proliferation of treated cells with that of untreated cells. Yeast cells have the same advantages as bacteria in terms of high throughput, ease of culture and manipulation, fast growth and low cost. Unlike bacterial systems, DNA damage-inducible genes in yeast cells respond to a much broader spectrum of DNA lesions. While retaining the simplicity of unicellular organisms, they are truly eukaryotes and therefore closer to mammalian cells. This system allows the detection of expression levels similar to the minimal values detectable by quantitative RT-PCR, below one mRNA molecule per cell.

7.3.3.3 Mutagenicity Testing with Mammalian Cell Systems

Mutagenicity testing in mammalian cells is conducted to confirm whether a chemical is mutagenic for higher animals. Testing in mammals requires to acknowledge the fact that higher degrees of DNA repair and xenobiotic metabolism are markedly different from those of bacterial systems. Consequently, several assays incorporate various continuous cell lines and primary cultures of animal and human origin. While primary cultures maintain some biotransformation capability, continuous cell lines have limited potential. Cell culture systems thus permit a closer examination of the mechanisms of action responsible for gene mutation and chromosome aberrations.

Mammalian forward mutation assays, such as the thymidine kinase (Tk) assay or the hypoxanthine-guanine phosphoribosyl transferase (Hprt) assay, detect mutations at the heterozygous Tk or hemizygous Hprt gene (MacGregor et al. 2000). Cells such as L5178Y mouse lymphoma cells (Tk locus), several Chinese hamster cell lines (Hprt locus), and human lymphoblastoid cells (Tk locus) are most commonly used. Mutations are selected by incubation of the cell cultures with the selective agents trifluorothymidine (Tk assay) or 6-thioguanine (Hprt assay). Cells having forward mutations at the TK or Hprt genes survive in the presence of the selective agent, while wild-type cells accumulate a toxic metabolite and do not proliferate. Comparison of the mutant frequency of the treatment groups with the concurrent negative control group allows the identification of a mutagenic chemical.

The development of the hprt assay employing human peripheral blood lymphocytes has contributed significantly to our understanding of *in vivo* somatic cell mutagenesis in the human (Strauss and Albertini 1979). In the mouse, rat, and human hprt systems, promutagens, direct acting mutagens and X-irradiation readily induce positive responses (Cochrane and Skopek 1994; Klarmann et al. 1995; Aidoo et al. 1997) indicating that rodent lymphocytes effectively detect the induction of genotoxic damage *in vivo*. The results of these studies indicate the ability of rat lymphocytes to detect the carcinogenic potential of many tissue- or organ-specific mutagens.

Other *in vivo* single gene mutation assays include human cell assays based on the glycophorin A (GPA) surface glycoprotein on the human red blood cell, the lymphocyte T-cell receptor in mice and humans, and the detection of hemoglobin variants (MacGregor 1994). New mouse models utilize the endogenous autosomal adenine phosphoribosyltransferase (*aprt*) and *tk* genes as targets for monitoring *in vivo* mutation. These models will provide *in vivo* counterparts to the established *in vitro* cell culture assays heterozygous for the *aprt* and the *tk* genes (Clive et al. 1972, Bradley and Latovanec 1982). This will be important because results obtained *in vitro* indicate that autosomal genes may be better suited than the *hprt* gene or transgenes for detecting genetic events known to be associated with carcinogenesis.

7.4   Recent Developments

7.4.1   *Green Screen HC assay*

This assay uses human lymphoblastoid TK6 cells transfected with the GADD45α genotoxic stress-specific response gene linked to a green fluorescent protein reporter (Hastwell et al. 2006). In the initial validation study of the assay, which was

7 Methods for Genotoxicity Testing of Environmental Pollutants

carried out in microtitre trays, 31 out of 34 of the expected genotoxins gave a positive response, whilst none of the 36 non-genotoxins was positive. This group included 11 cytotoxic non-carcinogens that had given positive responses in chromosome aberration *in vitro* studies, when tested at high levels of cytotoxicity. Only direct-acting chemicals (not requiring metabolic activation) were included in the initial trial as S9 is fluorescent and absorbing at the relevant wavelengths, and can interfere with the standard protocol. However, a flow cytometric version of the assay has been developed which has allowed the detection of several compounds that require metabolic activation. New versions of the assay using metabolically competent cells are being considered. Multi-centre trials are under way to determine the robustness of this assay. The assay is not intended for more analytical studies (such as mechanism of action), however, where a genotoxin is known to give a positive response in the assay, its easy read-out lends itself to use in monitoring applications such as environmental contamination, as well as compound screening.

7.4.2 Engineered Cells and Cell Lines

Chinese hamster lung cell lines (V79) cells have been genetically engineered to express various human P450's. Several chemicals that are known to be *in vivo* genotoxic, DNA-reactive mutagenic carcinogens could be detected at very much lower concentrations than when S9 is used as the activation system (Glatt et al. 2005). This is highly relevant to overcoming the current need for very high concentrations in the conventional assays, which may lead to spurious results in some cases. It would be helpful to extend the testing database of these engineered lines to see if specificity is affected with non-genotoxic carcinogens. Several transfected lines of HepG2 have been constructed which express increased levels of phase I enzymes (such as CYP1A1, CYP1A2, CYP2E1 etc.); furthermore, cell lines are available which express human glutathione-*S*-transferases. HepG2 cells have also been used successfully as a source of S9. As for most of these possible options, there is little published information on the accuracy of tests using this cell line to find nongenotoxic carcinogens and non-carcinogens negative. There is some information that high concentrations of ascorbic acid and beta-carotene are positive in tests using these cells, which demonstrates that there may be problems with specificity (Tweats et al. 2007).

Newer technologies such as transcriptomics, proteomics and metabolomics probably have the greatest potential to provide accurate assessments of genotoxic potential of chemicals in the longer term. They provide the opportunity to gain insight into genotoxic mechanisms of carcinogenicity beyond the initial DNA-damaging events, provide new markers *in vitro* and *in vivo* and possibly increase the accuracy of discriminating carcinogens from non-carcinogens. However, at present the sheer complexity of responses to genotoxic stresses provides major difficulties in formulating a clear way forward. One area where there may be a potential advantage in the use of transcriptomic technology, however, is to distinguish mechanisms of genotoxicity, e.g. those that have a threshold from those that act without a threshold (Newton et al. 2004). The use of transcriptomics will also become more powerful as databases of expression profiles would be available populated with larger numbers of studies of genotoxins acting by a variety of mechanisms, as well as of non-genotoxins.

There appears to be a few publications on the use of proteomics to study the cellular or biomarker response against the exposure to genotoxins and carcinogens. However, proteomics has been used extensively as an aid to cancer detection and risk assessment (Tweats et al. 2007). Laser Capture Microdissection (LCM), coupled with downstream proteomics applications, such as 2D polyacrylamide gel electrophoresis and SELDI (surface enhanced laser desorption ionization) separation followed by mass spectrometry (MS) analysis, can facilitate the characterization and identification of protein expression changes that track normal and cancer phenotypes. Identification of such proteins could enable the development of protein chips allowing changes relevant to genotoxicity and carcinogenicity to be measured.

7.5 Future Prospects

The current limitations of *in vivo* genotoxicity tests, including concerns about their sensitivity, and the reasoned desire to reduce unnecessary animal use if an acceptable *in vitro* system is at hand, make plausible an important role of standard *in vitro* assays in the foreseeable future despite the problems encountered and an appreciable rate of positives of doubtful relevance (Kirkland and Müller 1999). In addition, combinatorial chemistry and other methods of design and generation of numerous compounds, call for the development of more rapid screening methods that should require smaller quantities of test material in toxicology/safety assessment. Since genotoxicity assessment is an essential factor in the development of chemicals for human use, *in vitro* genotoxicity methods are likely to be among those that will be adapted to faster and more efficient screening modes and ultimately to be incorporated into routine regulatory testing. As with all toxicological endpoints, the inability to reliably use *in vitro* data to quantitatively predict tissue- and cell-specific effects *in vivo* ultimately relegates *in vitro* data to qualitative or semi-quantitative applications. Thus, one of the most important needs in the field of genetic toxicology is to develop the efficient and fool-proof methods for better evaluation of heritable mutation rates in various tissues of mammals including humans, so that the quantitative health risks associated with genetic damage can be estimated more reliably.

It seems certain that the major advances in molecular methodologies that have been achieved in the recent past will provide improved *in vivo* models for mutagenecity testing in the near future. Several possibilities exist that include, techniques for the visualization of mutant cells *in situ*, genotypic selection techniques for direct analysis of DNA isolated from tissues *in vivo*, and improved transgenic models that include "humanized" characteristics (e.g., human target sequences, humanized metabolism and repair systems, etc.). Direct analysis of mutations via

7 Methods for Genotoxicity Testing of Environmental Pollutants

genotypic selection methods has already been proven feasible and may become practical in the near future. As experience with established *in vivo* assay systems increases and improved methodologies are introduced, it can be expected that reliance on *in vivo* mutagenesis data will become an increasingly important part of regulatory testing schemes.

7.6 Conclusions

The currently established *in vitro* tests for determining genotoxicity have good sensitivity but poor specificity for predicting rodent carcinogens. Thus all batteries composed only of *in vitro* tests would result in a large proportion of tested chemicals without carcinogenic potential being discarded unnecessarily. New models and approaches are at various stages of development including a mammalian cell model looking at induction of Gadd45α; novel cell lines and engineered cell lines with metabolic competence or boosted cellular defences against reactive oxygen species; use of the 'omic technologies; organ models particularly of the skin that could all make a valuable contribution to better prediction of genotoxicity and potential carcinogenicity. Models that measure compound induced epigenetic changes are available and may yield useful data for the prediction of carcinogenic potential. The ability to investigate the relevance of positive *in vitro* genotoxicity results for prediction of carcinogenicity in humans without the use of animals represents a significant scientific and technical challenge. A range of accepted, robust tools to investigate *in vitro* positives should be available to allow appropriate experiments to be conducted on a case-by-case basis depending on the nature of the initial observation in the standard *in vitro* tests.

References

Afanassiev V, Sefton M, Anantachaiyong T et al (2000) Application of yeast cells transformed with GFP expression constructs containing the RAD54 or RNR2 promoter as a test for the genotoxic potential of chemical substances. Mutat Res 464:297–308

Aidoo A, Morris SM, Casciano DA (1997) Development and utilization of the rat lymphocyte hprt mutation assay. Mutat Res 387:69–88

Aleem A, Malik A (2003) Genotoxic hazards of long-term application of wastewater on agricultural soil. Mutat Res 538:145–154

Ames B, Lee F, Durston WD (1973) An improved bacterial system for the detection and classification of mutagens and carcinogens. Proc Natl Acad Sci U S A 70:782–786

Andrade VM, Silva J, Silva FR et al (2004) Fish as bioindicators to assess the effects of pollution in two southern Brazilian rivers using the Comet assay and micronucleus test. Environ Mol Mutagen 44:459–468

Ansari MI, Malik A (2009a) Genotoxicity of agricultural soils in the vicinity of industrial area. Mutat Res 673:124–132

Ansari MI, Malik A (2009b) Genotoxicity of wastewaters used for irrigation of food crops. Environ Toxicol 24:103–115

Aoyama K, Iwahori K, Miyata N (2003) Application of *Euglena gracilis* cells to comet assay: evaluation of DNA damage and repair. Mutat Res 538:155–162

Atak C, Alikamanoglu S, Acik L et al (2004) Induced of plastid mutations in soybean plant (*Glycine max* L. Merrill) with gamma radiation and determination with RAPD. Mutat Res 556:35–44

Atienzar FA, Cordi B, Donkin ME (2000) Depledge, Comparison of ultraviolet-induced genotoxicity detected by random amplified polymorphic DNA with chlorophyll fluorescence and growth in a marine macroalgae *Palmaria palmate*. Aquat Toxicol 50:1–12

Atienzar FA, Cheung VV, Jha AN et al (2001) Fitness parameters and DNA effects are sensitive indicators of copper induced toxicity in *Daphnia magna*. Toxicol Sci 59:241–250

Bonassi S, Lando C, Ceppi M et al (2004) No association between increased levels of high frequency sister chromatid exchange cells (HFCs) and the risk of cancer in healthy individuals. Environ Mol Mutagen 43:134–136

Boettcher M, Grund S, Keiter S et al (2010) Comparison of *in vitro* and *in situ* genotoxicity in the Danube River by means of the comet assay and the micronucleus test. Mutat Res 700:11–17

Bradley WE, Letovanec D (1982) High-frequency non random mutational event at the adenine phosphoribosyltransferase (aprt) locus of sib-selected CHO variants heterozygous for aprt. Somatic Cell Genet 9:51–66

Brinkmann C, Eisentraeger A (2008) Completely automated short-term genotoxicity testing for the assessment of chemicals and characterisation of contaminated soils and waste waters. Env Sci Pollut Res 15: 211–217

Brüschweiler BJ, Märki W, Wülser R (2010) *In vitro* genotoxicity of polychlorinated butadienes (Cl4–Cl6). Mutat Res 699:47–54

Castano A, Becerril C (2004) *In vitro* assessment of DNA damage after short- and long-term exposure to benzo(a)pyrene using RAPD and the RTG-2 fish cell line. Mutat Res 552:141–151

Cavas T and Konen S (2007) Detection of cytogenetic and DNA damage in peripheral erythrocytes of goldfish (*Carassius auratus*) exposed to a glyphosate formulation using the micronucleus test and the comet assay. Mutagenesis 22:263–268

Česnienė T, Kleizaitė V, Ursache R et al (2010) Soil-surface genotoxicity of military and urban territories in Lithuania, as revealed by *Tradescantia* bioassays. Mutat Res 697:10–18

Chakraborty R, Mukherjee A (2009) Mutagenicity and genotoxicity of coal fly ash water leachate. Ecotoxicol Environ Saf 72:838–842

Chauhan LKS, Saxena PN, Gupta SK (1999) Cytogenetic effects of cypermethrin and fenvalerate on the root meristem cells of *Allium cepa*. Environ Exp Bot 42:181–189

Claxton LD, Matthews PP, Warren SH (2004) The genotoxicity of ambient outdoor air, a review: *Salmonella* mutagenicity. Mutat Res 567:347–399

Clive D, Flamm WG, Machesko MR (1972) A mutational assay system using the thymidine kinase locus in mouse lymphoma cells. Mutat Res 16:77–87

Cochrane JE, Skopek TR (1994) Mutagenicity of butadiene and its epoxide metabolites: II. Mutational spectra of butadiene, 1, epoxybutane and diepoxybutane at the hprt locus in splenic T-cells from exposed B6C3F1 mice. Carcinogenesis 15:719–723

Cormack BP, Valdivia RH, Falkow S (1996) FACS optimized mutants of the green fluorescent protein (GFP). Gene 173:33–38

Courty B, Le Curieux F, Belkessam L et al (2008) Mutagenic potency in *Salmonella typhimurium* of organic extracts of soil samples originating from urban, suburban, agricultural, forest and natural areas. Mutat Res 653:1–5

de Lemos CT, Rödel PM, Terra NR et al (2007) River water genotoxicity evaluation using micronucleus assay in fish erythrocytes. Ecotoxicol Environ Saf 66:391–401

de Maagd PG, Tonkes M (2000) Selection of genotoxicity tests for risk assessment of effluents. Environ Toxicol 15:81–90

de Vizcaya-Ruiz A, Barbier O, Ruiz-Ramos R et al (2008) Biomarkers of oxidative stress and damage in human populations exposed to arsenic. Mutat Res 674:85–92

Dearfield KL, Cimino MC, McCarroll NE et al (2002) Genotoxicity risk assessment: a proposed classification strategy. Mutat Res 521:121–135

7 Methods for Genotoxicity Testing of Environmental Pollutants 255

Deasi SR, Verlecar XN, Ansari ZA et al (2010) Evaluation of genotoxic responses of *Chaetoceros tenuissimus* and *Skeletonema costatum* to water accommodated fraction of petroleum hydrocarbons as biomarker of exposure. Water Res 44:2235–2244

Dong Y and Zhang J (2010) Testing the genotoxicity of coking wastewater using *Vicia faba* and *Hordeum vulgare* bioassays. Ecotoxicol Environ Saf 73:944–948

Enan MR (2006) Application of random amplified polymorphic DNA to detect genotoxic effect of heavy metals. Biotechnol Appl Biochem 43:147–154

Erbes M, Wessler A, Obst U et al (1997) Detection of primary DNA damage in *Chlamydomonas reinhardtii* by means of modified microgel electrophoresis. Environ Mol Mutagen 30:448–458

Ergene S, Çelik A, Çavaş T et al (2007) Genotoxic biomonitoring study of population residing in pesticide contaminated regions in Göksu Delta: Micronucleus, chromosomal aberrations and sister chromatid exchanges. Environ Int 33:877–885

Farmer PB, Singh R (2008) Use of DNA adducts to identify human health risk from exposure to hazardous environmental pollutants: the increasing role of mass spectrometry in assessing biologically effective doses of genotoxic carcinogens. Mutat Res 659:68–76

Fatima RA, Ahmad M (2006) Genotoxicity of industrial wastewaters obtained from two different pollution sources in northern India: A comparision of three bioassays. Mutat Res 609:81–91

Ferreira MI, Domingos M, Gomes HA et al (2007) Evaluation of mutagenic potential of contaminated atmosphere at Ibirapuera Park, São Paulo-SP, Brazil, using the *Tradescantia* stamen-hair assay. Environ Pollut 145:219–224

Fiskesjo G (1985) The *Allium* test as a standard in environmental monitoring. Hereditas 102:99–112

Frenzilli G, Nigro M, Lyons BP (2009) The Comet assay for the evaluation of genotoxic impact in aquatic environments. Mutat Res 681:80–92

Frische T (2002) Screening for soil toxicity and mutagenicity using luminescent bacteria-a case study of the explosive 2, 4, 6-trinitrotoluene (TNT). Ecotoxicol Environ Saf 51:133–144

Fritsch P, Rieseberg LH (1996) The use of random amplified polymorphic DNA (RAPD) in conservation genetics. In: Smither TB, Wayne RK (eds) Molecular Genetic Approaches in Conservation, Oxford University, pp 55–73

Gana JM, Ordonez R, Zampini C et al (2008) Industrial effluents and surface waters genotoxicity and mutagenicity evaluation of a river of Tucuman, Argentina. J Hazard Mater 155:403–406

Garinis GA, Van Der Horst GT, Vijg J et al (2008) DNA damage and ageing: new-age ideas for an age-old problem. Nat Cell Biol 10:1241–1247

Garry S, Nesslany F, Aliouat EM et al (2003) Potent genotoxic activity of benzo[a]pyrene coated onto hematite measured by unscheduled DNA synthesis *in vivo* in the rat. Mutagenesis 18:449–455

Glatt H, Schneider H, Liu Y (2005) V79-hCYP2E1-hSULT1A1, a cell line for the sensitive detection of genotoxic effects induced by carbohydrate pyrolysis products and other food-borne chemicals. Mutat Res 580:41–52

Gustavsson L, Engwall M (2006) Genotoxic activity of nitroarene-contaminated industrial sludge following large-scale treatment in aerated and non-aerated sacs. Sci Total Environ 367:694–703

Hagiwara Y, Watanabe M, Oda Y et al (1993) Specificity and sensitivity of Salmonella typhimurium YG1041 and YG1042 strains possessing elevated levels of both nitroreductase and acetyltransferase activity. Mutat Res 291:171–180

Hagmar L, Stromberg U, Tinnerberg H et al (2004) Epidemiological evaluation of cytogenetic biomarkers as potential surrogate end-points for cancer. IARC Sci Publ 157:207–215

Hastwell PW, Chai LL, Roberts KJ et al (2006) High-specificity and high sensitivity genotoxicity assessment in a human cell line: validation of the Green Screen HC GADD45a-GFP genotoxicity assay. Mutat Res 607:160–175

Helma C, Eckl P, Gottmann E et al (1998) Genotoxic and ecotoxic effects of groundwaters and their relation to routinely measured chemical parameters. Environ Sci Technol 32:1799–1805

Hua G, Lyons B, Killham K et al (2009) Potential use of DNA adducts to detect mutagenic compounds in soil. Environ Pollut 157;916–921

Huang D, Zhang Y, Wang Y et al (2007) Assessment of the genotoxicity in toad *Bufo raddei* exposed to petrochemical contaminants in Lanzhou Region, China. Mutat Res 629:81–88

ICH (International Cooperation on Harmonization) (1997) ICH Harmonized Tripartite Guideline. S2B. Genotoxicity: A Standard Battery for Genotoxicity Testing of Pharmaceuticals. http://www.ich.org. Accessed 16 July 1997

Izzotti A, Kanitz S, D'Agostini F et al (2009) Formation of adducts by bisphenol A, an endocrine disruptor, in DNA *in vitro* and in liver and mammary tissue of mice. Mutat Res 679:28–32

Jolibois B, Guerbet M (2005) Efficacy of two wastewater treatment plants in removing genotoxins. Arch Environ Contam Toxicol 48:289–295

Jones C, Kortenkamp A (2000) RAPD library fingerprinting of bacterial and human DNA: applications in mutation detection. Teratogen Carcin Mut 20:49–63

Jung Y, Youn Y, Ryu J, Surh Y (2001) Salsolinol, a naturally occurring tetrahydroisoquinoline alkaloid, induces DNA damage and chromosomal aberrations in cultured Chinese hamster lung fibroblast cells 474:25–33

Kim IY, Hyun CK (2006) Comparative evaluation of the alkaline comet assay with the micronucleus test for genotoxicity monitoring using aquatic organisms. Ecotoxicol Environ Saf 64:288–297

Kirkland D, Aardema M, Henderson L et al (2005) Evaluation of the ability of a battery of three genotoxicity tests to discriminate rodent carcinogens and non-carcinogens. I. Sensitivity, specificity and relative predictivity. Mutat Res 584:1–256

Kirkland D, Müller L (1999) Interpretation of the biological relevance of genotoxicity test results: the importance of thresholds. Mutat Res 464:137–147

Klarmann B, Wixler V, Lorenz R et al (1995) Mutant frequency at the H-2K class 1 and hprt genes in T-lymphocytes from the X-ray exposed mouse. Int J Radiat Biol 67:421–430

Kohlpoth M, Rusche B, Nusse M (1999) Flow cytometric measurement of micronuclei induced in a permanent fish cell line as a possible screening test for the genotoxicity of industrial waste waters. Mutagenesis 14:397–402

Kohn K, Ericson L, Ewig R et al (1976) Fractionation of DNA from mammalian cells by alkaline elution. Biochemistry 15:4629–4637

Konuk M, Liman R, Cigerci IH (2007) Determination of genotoxic effect of boron on *Allium cepa* root meristematic cells. Pakistan J Bot 39:73–79

Kosmehl T, Hallare AV, Braunbeck T et al (2008) DNA damage induced by genotoxicants in zebrafish (*Danio rerio*) embryos after contact exposure to freeze-dried sediment and sediment extracts from Laguna Lake (The Philippines) as measured by the comet assay. Mutat Res 650:1–14

Kovalchuk O, Kovalchuk I, Arkhipov A et al (1998) The *Allium cepa* chromosome aberration test reliably measures genotoxicity of soils of inhabited areas in the Ukraine contaminated by the Chernobyl accident. Mutat Res 415:47–57

Kuo ML, Lin JK (1993) The genotoxicity of the waste water discharged from paraquat manufacturing and its pyridyl components. Mutat Res 300:223–229

Kurelec B (1993) The genotoxic disease syndrome. Mar Environ Res 35:341–348

Kuroda S, Yano H, Koga-Ban Y et al (1999) Identification of DNA polymorphism induced by X-ray and UV irradiation in plant cells. Jpn Agric Res Quart 33: 223–226

Kwan KK, Dutka BJ, Rao SS et al (1990) Mutatox test: a new test for monitoring environmental genotoxic agents. Environ Pollut 65:323–332

Labieniec M, Biernat M, Gabryelak T (2007) Response of digestive gland cells of freshwater mussel *Unio tumidus* to phenolic compound exposure *in vivo*. Cell Biol Int 31:683–690

Lah B, Vidic T, Glasencnik E et al (2008) Genotoxicity evaluation of water soil leachates by Ames test, comet assay, and preliminary *Tradescantia* micronucleus assay. Environ Monit Assess 139:107–118

Lichtenberg-Frate H, Schmitt M, Gellert G et al (2003) A yeast based method for the detection of cyto- and genotoxicity. Toxicol *in vitro* 17:709–716

Liu W, Li PJ, Qi XM et al (2005) DNA changes in barley (*Hordeum vulgare*) seedlings induced by cadmium pollution using RAPD analysis. Chemosphere 61:158–167

7 Methods for Genotoxicity Testing of Environmental Pollutants

Ma TH, Cabrera GL, Chen R et al (1994) *Tradescantia* micronucleus bioassay. Mutat Res 310:221–230

MacGregor JT, Casciano D, Müller L (2000) Strategies and testing methods for identifying mutagenic risks. Mutat Res 455:3–20

MacGregor JT (1994) Environmental mutagenesis: past and future directions. Mutat Res 23:73–77

Mañas F, Peralta L, Raviolo J et al (2009) Genotoxicity of AMPA, the environmental metabolite of glyphosate, assessed by the Comet assay and cytogenetic tests. Ecotoxicol Environ Saf 72:834–837

Maron D, Ames BN (1983) Revised methods for the *Salmonella* mutagenicity test. Mutat Res 113:173–215

Marzo Solano M , Alves de Lima PL, João Francisco Lozano Luvizutto JFL et al (2009) *In vivo* genotoxicity evaluation of a treated urban sewage sludge sample. Mutat Res 676:69–73

Masuda S, Deguchi Y, Masuda Y et al (2004) Genotoxicity of 2-[2-(acetylamino)-4-[bis(2-hydroxyethyl)amino]-5-methoxyphenyl]-5-amino-7-bromo-4-chloro-2H-benzotriazole (PBTA-6) and 4-amino-3,30-dichloro-5,40-dinitro-biphenyl (ADDB) in goldfish (*Carassius auratus*) using the micronucleus test and the comet assay. Mutat Res 560:33–40

Meireles J, Rocha R, Neto AC (2009) Genotoxic effects of vehicle traffic pollution as evaluated by micronuclei test in *Tradescantia* (Trad-MCN). Mutat Res 675:46–50

Mouchet F, Gauthier L, Mailhes C et al (2005) Biomonitoring of the genotoxic potential of draining water from dredged sediments, using the comet and micronucleus tests on amphibian (Xenopus Laevis) larvae and bacterial assays (Mutatox® and Ames tests). J Toxicol Environ Health 68:811–832

Mouchet F, Gauthier L, Mailhes C (2006) Comparative evaluation of genotoxicity of captan in amphibian larvae (*Xenopus laevis* and *Pleurodeles waltl*) using the comet assay and the micronucleus test. Environ Toxicol 21:264–277

Nagarajappa A, Ganguly U, Goswami (2006) DNA damage in male gonad cells of Green mussel (*Perna viridis*) upon exposure to tobacco products. Ecotoxicology 15:365–369

Newton RK, Aardema M, Aubrecht J (2004) The utility of DNA microarrays for characterizing genotoxicity. Environ Health Perspect 112:420–422

Oda Y, Nakamura S, Oki I et al (1985) Evaluation of the new system (umu-test) for the detection of environmental mutagens and carcinogens. Mutat Res 147:219–229

Oda Y, Shimada T, Watanabe M et al (1992) A sensitive *umu* test system for the detection of mutagenic nitroarenes in *Salmonella typhimurium* NM1011 having high nitroreductase activity. Mutat Res 272:91–99

Oda Y, Yamazaki H, Watanabe M et al (1993) Highly sensitive umu test system for the detection of mutagenic nitroarenes in *Salmonella typhimurium* NM3009 having high acetyltransferase and nitroreductase activities. Environ Mol Mutagen 21:357–364

Oda Y, Yamazaki H, Watanabe M et al (1995) Development of high sensitive umu test system: rapid detection of genotoxicity of promutagenic aromatic amines by *Salmonella typhimurium* strain NM2009 possessing high O-acetyltransferase activity. Mutat Res 334:145–156

Oliveira M, Pacheco M, Santos MA (2007) Cytochrome P4501A, genotoxic and stress responses in golden grey mullet (*Liza aurata*) following short-term exposure to phenanthrene. Chemosphere 66:1284–1291

Pereira R, Marques CR, Ferreira MJS et al (2009) Phytotoxicity and genotoxicity of soils from an abandoned uranium mine area. Appl Soil Ecol 42:209–220

Perez S, Reifferscheid G, Eichhorn P et al (2003) Assessment of the mutagenic potency of sewage sludges contaminated with polycyclic aromatic hydrocarbons by an ames sludges for fluctuation assay. Environ Toxicol Chem 22:2576–2584

Qu SX, Bai CL, Stacey NH (1997) Determination of bulky DNA adducts in biomonitoring of carcinogenic chemical exposures: Features and comparison of current techniques. Biomarkers 2:3–6

Quillardet P, Huisman O, D′Ari R et al (1982) SOS chromotest, a direct assay of induction of an SOS function in Escherichia coli K-12 to measure genotoxicity. Proc Natl Acad Sci U S A 79:5971–5975

Quillardet P, De Bellecombe C, Hofnung M (1985) The SOS chromotest; a colorimetric bacterial assay for genotoxins: validation study with 83 compounds. Mutat Res 147:79–95

Rank J, Lehtonen KK, Strand J et al (2007) Aquatic toxicology, DNA damage, acetylcholinesterase activity and lysosomal stability in native and transplanted mussels (*Mytilus edulis*) in areas close to coastal chemical dumping sites in Denmark. Aquat Toxicol 84:50–61

Reddy G, Song J, Mecchi MS et al (2010) Genotoxicity assessment of two hypergolic energetic propellant compounds. Mutat Res 700:26–31

Reifferscheid G, Zipperle J (2000) Development and application of luminometric and fluorometric umu-test systems for the detection of genotoxic compounds in surface water. In Grummt T (ed) Erprobung, Vergleich, Weiterentwicklung und Beurteilung von Gentoxizitätstests für Oberflächenwasser. Forschungszentrum Karlsruhe GmbH, Projektträger Wassertechnologie und Entsorgung, Außenstelle Dresden, Dresden. p 217

Roemer E, Stabbert R, Rustemeier K et al (2004) Chemical composition, cytotoxicity and mutagenicity of smoke from US commercial and reference cigarettes smoked under two sets of machine smoking conditions. Toxicology 195:31–52

Rojiéková R, Marsálek B, Dutka B et al (1998) Bioassays used for detection of ecotoxicity at contaminated areas. NATO Sci Ser 2:227–232

Rosenkranz HS, Mermelstein R (1985) The genotoxicity, metabolism and carcinogenicity of nitrated polycyclic aromatic hydrocarbons. J Environ Sci Heal C 3:221–272

Rosenkranz HS, McCoy EC, Sanders DR et al (1980) Nitropyrenes: isolation, identification and reduction of mutagenic impurities in carbon black and toners. Science 209:1039–1043

Rossmann TG, Molina M, Meyer LW (1984) The genetic toxicology of metal compounds. I. Induction of lambda prophage in *E. coli* WP2s (lambda). Environ Mutagen 6:59–69

Sánchez P, Llorente MT, Castano A (2000) Flow cytometric detection of micronuclei and cell cycle alterations in fish-derived cells after exposure to three model genotoxic agents: mitomycin C, vincristine sulfate; and benzo[a]pyrene. Mutat Res 465:113–122

Sarkar A, Dipak CS, Gaitonde et al (2008) Evaluation of impairment of DNA integrity in marine gastropods (*Cronia contracta*) as a biomarker of genotoxic contaminants in coastal water around Goa, West coast of India. Ecotoxicol Environ Saf 71:473–482

Savva D, Castellani S, Mattei N et al (1994) The use of PCR and DNA fingerprints to detect the genotoxic effects of environmental chemicals. In: Varnavas SP (ed) Environmental Contamination, Proceedings of the 6th International Conference, Delphi, Greece, CEP Consultants, Edinburgh

Schmitt M, Gellert G, Ludwig J et al (2004) Phenotypic yeast growth analysis for chronic toxicity testing. Ecotoxicol Environ Saf 59:142–150

Schmeiser HH, Fürstenberger G, Takamura-Enya T et al (2009) The genotoxic air pollutant 3-nitrobenzanthrone and its reactive metabolite N-hydroxy-3-aminobenzanthrone lack initiating and complete carcinogenic activity in NMRI mouse skin. Cancer Lett 284:21–29

Seagrave J, Gigliotti A, McDonald JD et al (2005) Composition, toxicity, and mutagenicity of particulate and semivolatile emissions from heavy-duty compressed natural gas-powered vehicles. Toxicol Sci 87:232–241

Shishu, Kaur IP (2003) Inhibition of mutagenicity of food-derived heterocyclic amines by sulforaphane—a constituent of broccoli. Indian J Exp Biol 41:216–219

Siddiqui AH, Ahmad M (2003) The *Salmonella* mutagenicity of industrial, surface and groundwater water samples of Aligarh region of India. Mutat Res 541:21–29

Siddiqui MF, Ahmad R, Ahmad W (2006) Micronuclei induction and chromosomal aberrations in *Rattus norvegicus* by chloroacetic acid and chlorobenzene. Ecotoxicol Environ Saf 65:159–164

Singh NP, McCoy MT, Tice RR et al (1988) A simple technique for quantitation of low levels of DNA damage in individual cells. Exp Cell Res 175:184–191

Singh A, Chandra S, Gupta SK et al (2007) Mutagenicity of leachates from industrial solid wastes using *Salmonella* reverse mutation assay. Ecotoxicol Environ Saf 66:210–216

Siu WHL, Hung CL, Wong BJ et al (2003) Exposure and time dependent DNA strand breakage in hepatopancreas of green-lipped mussels (*Perna viridis*) exposed to Aroclor 1254, and mixtures of BαP and Aroclor 1254. Mar Pollut Bull 46:1285–1293

7 Methods for Genotoxicity Testing of Environmental Pollutants 259

Soloneski S, Larramendy ML (2010) Sister chromatid exchanges and chromosomal aberrations in Chinese hamster ovary (CHO-K1) cells treated with the insecticide pirimicarb. J Hazard Mater 174:410–415

Strauss GH, Albertini RJ (1979) Enumeration of 6-thioguanineresistant peripheral blood lymphocytes in man as a potential test for somatic cell mutations arising *in vivo*. Mutat Res 61:353–379

Sugimura T, Wakabayashi K, Nakagama H et al (2004) Heterocyclic amines: mutagens/carcinogens produced during cooking of meat and fish. Cancer Sci 95:290–299

Tabrez S, Ahmad M (2009) Toxicity, biomarker, genotoxicity and carcinogenicity of trichloroethylene (TCE) and its metabolites: A review. J Environ Sci Heal C 27:178–196

Thybaud V, Aardema M, Clements J et al (2007) Strategy for genotoxicity testing: Hazard identification and risk assessment in relation to *in vitro* testing. Mutat Res 627:41–58

Traversi D, Degan R, Marco RD et al (2009) Mutagenic properties of PM2.5 urban pollution in the Northern Italy: The nitro-compounds contribution. Environ Int 35:905–910

Tweats DJ, Scott AD, Westmoreland C, Carmichael PL (2007) Determination of genetic toxicity and potential carcinogenicity *in vitro*-challenges post the Seventh Amendment to the European Cosmetics Directive. Mutagenesis 22:5–13

Udroiu I (2006) The micronucleus test in piscine erythrocytes. Aquat Toxicol 79:201–204

Valente-Campos S, Dias CL, Barbour EDA et al (2009) The introduction of the *Salmonella*/microsome mutagenicity assay in a groundwater monitoring program. Mutat Res 675:17–22

Van der Oost R, Beyer J, Vermeulen NPE (2003) Fish bioaccumulation and biomarkers in environmental risk assessment: a review. Environ Toxicol Pharmacol 13:57–149

Van't Hof J, Schairer LA (1982) *Tradescantia* assay system for gaseous mutagens. A report of the US Environmental Protection Agency Gene-Tox Program. Mutat Res 99:303–315

Vargas VM, Migliavacca SB, de Melo AC et al (2001) Genotoxicity assessment in aquatic environments under the influence of heavy metals and organic contaminants. Mutat Res 490:141–158

Villarini M, Fatigoni C, Dominici L et al (2009) Assessing the genotoxicity of urban air pollutants using two *in situ* plant bioassays. Environ Pollut 157:3354–3356

Waldman P, Kramer M, Metz I et al (2000) The alkaline elution with the clam *Corbicula fluminea*: a rapid and sensitive method for the detection of genotoxic potentials in native surface waters-data of two laboratories. Pages 217 in T. Grummt, editor. Erprobung, Vergleich, Weiterentwicklung und Beurteilung von Gentoxizitätstests für Oberflächenwasser. Forschungszentrum Karlsruhe GmbH, Projektträger Wassertechnologie und Entsorgung, Außenstelle Dresden, Dresden

Walmsley RM, Billinton N, Heyer WD (1997) Green fluorescent protein as a reporter for the DNA damage induced gene RAD54 in *Saccharomyces cerevisiae*. Yeast 13:1535–1545

Watanabe M, Ishidate M Jr, Nohmi T (1989) A sensitive method for detection of mutagenic nitroarenes: construction of nitroreductase-overproducing derivatives of *Salmonella typhimurium* strains TA98 and TA100. Mutat Res 216:211–220

Watanabe M, Ishidate M Jr, Nohmi T (1990) Sensitive method for detection of mutagenic nitroarenes and aromatic amines: new derivatives of *Salmonella typhimurium* tester strains possessing levated O-acetyltransfearse levels. Mutat Res 234:337–348

Watanabe T, Takahashi K, Konishi E et al (2008) Mutagenicity of surface soil from residential areas in Kyoto city, Japan, and identification of major mutagens. Mutat Res 649:201–212

Weisburger JH (1988) Past, present and future role of carcinogenic and mutagenic-N-substituted aryl compounds in human cancer causion. In: King CM et al (eds) Carcinogenic and Mutagenic Responses to Aromatic Amines and Nitroarenes. Elsevier, New York

White PA, Claxton LD (2004) Mutagens in contaminated soil: a review. Mutat Res 567:227–345

Wild D, Dirr A (1989) Mutagenic nitrenes/nitrenium ions from azidoimidoazoarenes and their structure-activity relationship. Mutagenesis 4:446–452

Wittekindt E, Saftic S, Matthess C, Fischer B, Hansen P D, Schubert J (2000) *In situ* investigations for the detection of genotoxic potential in selected surface water with the DNA alkaline unwinding assay using fish cells, Early life stage of fish, crustaceae and mussels. Pages 217 in T. Grummt, editor. Erprobung, Vergleich, Weiterentwicklung und Beurteilung von Gentoxizitätstests für Oberflächenwasser. Forschungszentrum Karlsruhe GmbH, Projektträger Wassertechnologie und Entsorgung, Außenstelle Dresden, Dresden

Xiao R, Wang Z, Wang C et al (2006) Contamination at a major industrialized city in north east China by a combination of *in vitro* and *in vivo* bioassays. Environ Sci Technol 40:6170–6175

Yang Y, Yao J, Hu S (2000) Effects of agricultural chemicals on DNA sequence diversity of soil microbial community: a study with RAPD Marker. Microb Ecol 39:72–79

Yildiz M, Arikan ES (2008) Genotoxicity testing of quizalofop-P-ethyl herbicide using the *Allium cepa* anaphase-telophase chromosome aberration assay. Caryologia 61:45–52

Yildiz M, Cigerci IH, Konuk M et al (2009) Determination of genotoxic effects of copper sulphate and cobalt chloride in *Allium cepa* root cells by chromosome aberration and comet assays. Chemosphere 75:934–938

Yilmaz S, Ünal F, Yüzbaşıoğlu D (2009) The *in vitro* genotoxicity of benzoic acid in human peripheral blood lymphocytes. Cytotechnology 60:55–61

Yüzbaşıoğlu D, Ünal F, Yilmaz S et al (2008) Genotoxicity testing of fluconazole *in vivo* and *in vitro*. Mutat Res 649:155–160

Zeljezic D, Garaj-Vrhovac V (2002) Sister chromatid exchange and proliferative rate index in the longitudinal risk assessment of occupational exposure to pesticides. Chemosphere 46:295–303

Zhao Z, Zhang L, Wu J et al (2010) Assessment of the potential mutagenicity of organochlorine pesticides (OCPs) in contaminated sediments from Taihu Lake. China 696:62–68

Zhu Y, Wang J, Bai Y et al (2004) Cadmium, chromium, and copper induce polychromatocyte micro nuclei in carp (*Cyprinus carpio*). Bull Environ Contam Toxicol 72:78–86

Chapter 8
Trends in Biological Degradation of Cyanobacteria and Toxins

Fatma Gassara, Satinder K. Brar, R. D. Tyagi and R. Y. Surampalli

Contents

8.1	Introduction	262
8.2	Cyanobacteria	263
	8.2.1 Factors Supporting the Formation of Cyanobacteria	263
8.3	Cyanotoxins	269
	8.3.1 Classification	269
8.4	Physicochemical Treatment	272
8.5	Biological Treatment	273
	8.5.1 Biological Treatment of Cyanobacteria	273
	8.5.2 Biological Treatment of Cyanotoxins	284
8.6	Future Perspectives	286
8.7	Conclusions	287
References		288

Abstract Cyanobacteria are known as blue-green algae, blue-green bacteria, and Cyanophyta. They are present in both toxic and non-toxic forms and it is actually the toxic form which proliferates in the aquatic environment. There are principally two types of toxins (neurotoxin and hepatotoxin) which lead to adverse environmental and human health impacts. Thus, the cyanobacteria and their cyanotoxins must be eliminated from fresh waters (lakes, river) to avoid contamination of drinking water and prevent other environmental adversities. Several treatment methods, such as physical and chemical treatment comprising chlorination, ozonation, photooxidation, activated carbon, and biological treatment including utilization of pure microorganisms such as bacteria, virus, fungi, protozoa, among others have been studied to ensure higher elimination of cyanobacteria. The physico-chemical treatment is the most prevalent and faster than biological treatment. However, this treatment causes the lysis of cyanobacterial cells and releases cyanotoxins and other carcinogenic and mutagenic substances in to the medium. In this context, the biological treatment is an eco-friendly option for removal of cyanobacteria and their toxins present in fresh

---

S. K. Brar (✉)
INRS-ETE, Université du Québec, 490, Rue de la Couronne, G1K 9A9 Québec, Canada
e-mail: satinder.brar@ete.inrs.ca

A. Malik, E. Grohmann (eds.), *Environmental Protection Strategies for Sustainable Development*, Strategies for Sustainability,
DOI 10.1007/978-94-007-1591-2_8, © Springer Science+Business Media B.V. 2012

waters. This mini-review is an attempt to explore different aspects of the research in the field of removal of cyanobacteria. The review presents the ecological aspects of cyanobacteria, physical-chemical treatment methodologies in short, biological treatment of cyanobacteria and cyanotoxins in details as latter are potentially more toxic.

Keywords Cyanobacteria • Cyanotoxins • Hepatototoxins • Neurotoxins • Biological treatment • Fresh waters

8.1 Introduction

Cyanobacteria are microorganisms having a diameter ranging between 3 and 10 μm and are known as blue-green algae, blue-green bacteria, and Cyanophyta. For a long time, they were called "green-blue algae" as they possess pigments and carry out photosynthesis similar to algae (Carmichael 1994; Chorus and Bartram 1999; Pitois et al. 2000). However, the absence of nucleus and intracellular organelles actually places them in the proximity of bacteria. The cyanobacteria are divided into 150 genera comprising more than 2,000 species (Duy et al. 2000). In Quebec, there are approximately 300 species (Lavoie et al. 2007). These bacteria are unicellular, filamentous or occur as colonies (Chorus and Bartram 1999). They form an important component of the marine nitrogen cycle, but are also found in habitats other than the marine environment, such as freshwater (Betsey Betsey Dexter Dyer 2003, Leão et al. 2010) and hypersaline inland lakes (Michael Hogan 2008). When there are favourable conditions, these photosynthetic prokaryotes can lead to demographic explosion in water called efflorescence, which results in the production of significant biomass. This factor enhances the cyanobacteria number in fresh water which induces production of toxins by some species (Chorus and Bartram 1999; Duy et al. 2000; Cox et al. 2005).

There are two principal types of cyanobacteria: toxic and non-toxic. The non-toxic cyanobacteria were in fact utilized for human consumption in the past. Currently, *Arthrospira* (*Spirulina*) species is being employed for production of food in specialized commercial farms. There are nearly 40 species of toxic cyanobacteria (Skulberg et al. 1993, Paldavicien et al. 2009). They can have a negative impact on the environment (anoxia, organic pollution, among others) and present a potential risk to animal and human health. This type of cyanobacteria is able to synthesize various types of toxins (hepatotoxins, neurotoxins). These cyanobacteria and their toxins are produced in large quantities in fresh water and are responsible for hepatic and neurotic problems in aquatic and terrestrial animals (Billings 1981; Falconer 1989; Carmichael 1992; Codd et al. 1997; Sivonen and Jones 1999; Carmichael 2001; Klitzke et al. 2010). The water contaminated with cyanotoxins can cause adverse health impacts on human population (liver cancer) leading to mortality (Yu 1989, 1995; Ueno et al. 1996; Jochimsen et al. 1998; Azevedo et al. 2002). Thus, cyanobacteria and their cyanotoxins must be eliminated from fresh waters (lakes, river) to avoid contamination of drinking water. The species of cyanobacteria present in Canada and which are often associated to intoxication are *Anabaena, Aphani-*

8 Trends in Biological Degradation of Cyanobacteria and Toxins

zomenon, Microcystis, Oscillatoria and *Nodularia* (Carmichael 1992). The detailed classification of cyanobacteria with their specific location is presented in Table 8.1. This review principally focuses on the toxic forms which proliferate in the environment leading to adverse environmental and human health impacts.

8.2 Cyanobacteria

8.2.1 Factors Supporting the Formation of Cyanobacteria

8.2.1.1 Factors Related to the Intrinsic Properties of Cyanobacteria

Cyanobacteria are well-equipped with certain components which enable them to colonize the fresh water ecosystems as discussed below.

Diversity of Photosynthetic Pigments

Cyanobacteria produce different types of pigments, such as chlorophyll a, chlorophyll b (Hess et al. 1996; Bumba et al. 2005) and additional pigments called photosynthetic phycobiliproteins (phycocyanin: blue pigment), phycoerythrin (red pigment, for certain species only) and allophycocyanin (blue pigment). The diversity of pigments allows cyanobacteria to possess higher photosynthetic efficiency and maintain complete photosynthetic function with a low light intensity. The cyanobacteria are able to modify the composition of pigment-proteins in their photosynthetic complexes. Hence, when the wavelength of growth medium is modified, cyanobacteria change their colour (Grossman et al. 2001).

The cyanobacteria have also developed various protection strategies against ultraviolet radiations (UV) and excessive radiation. For example, they synthesize compounds which act as a solar shield (mycosporin-like amino acids) and other UV protector pigments (carotenoid), since they eliminate the oxidizing molecules resulting from the excessive radiation. They also have effective repair mechanisms for the cellular components damaged by strong solar radiations (Vincent and Quesada 1993).

Vertical Migration

The majority of cyanobacterial species possess small gas vacuoles which enable them to migrate vertically at a speed of 50 m per day in the water column (Walsby and McCallister 1987; Oliver and Ganf 2000), when water stagnates. They can thus benefit from light radiations on the surface during the day, and migrate in-depth at the end of the day, in order to exploit the nutrients which are present in larger concentration (Oliver and Ganf 2000). The vertical migration

Table 8.1 Species of known toxic cyanobacteria and types of toxins

Order	Taxonomy	Toxin	Type of toxicity	Location
Chroococcales	*Microcystis aeruginosa*	Microcystins Anatoxin-a	Hepatotoxic Neurotoxic	Worldwide
	Microcystis botrys	Microcystins	Hepatotoxic	Denmark, Norway
	Microcystis viridis	Microcystins	Hepatotoxic	Japan
	Microcystis wesenbergii	Microcystins	Hepatotoxic	Japan
	Microcystis sp./spp.	Microcystins	Hepatotoxic	Scandinavia, Australia, USA
	Snowella lacustris	Microcystins	Hepatotoxic	Norway
	Woronichinia naegeliana	Microcystins	Hepatotoxic	Norway
Nostocales	*Anabaena circinalis*	Anatoxin-a	Neurotoxic	USA, Finland
		Toxins PSP	Neurotoxic	Australia
	Anabaena flos-aquae	Microcystins	Hepatotoxic	Scotland, Norway, Canada
		Anatoxin-a	Neurotoxic	Japan, Finland, USA,
		Anatoxin-a(s)	Neurotoxic	USA, Canada
	Anabaena lemmermannii	Anatoxin-a(s)	Neurotoxic	Denmark
		Toxins PSP	Neurotoxic	Denmark
		Microcystins	Hepatotoxic	Denmark, Norway
	Anabaena planctonica	Anatoxin-a	Neurotoxic	Italy
	Anabaena sp.	Anatoxin-a	Neurotoxic	Ireland, Germany, Finland, Japan
		Microcystins	Hepatotoxic	France, Egypt, Denmark, Finland
	Anabaena spiroides	Anatoxin-a	Neurotoxic	Japan
	Aphanizomenon flos-aquae	Toxins PSP	Neurotoxic	USA, Portugal
		Anatoxin-a	Neurotoxic	Finland
	Aphanizomenon ovalisporum	Cylindrospermopsin	Hepatotoxic	Israel
	Aphanizomenon sp.	Anatoxin-a	Neurotoxic	Finland
	Cylindrospermopsis raciborskii	Cylindrospermopsin	Hepatotoxic	Australia, Hungary
		Toxins PSP	Neurotoxic	Brazil

Table 8.1 (continued)

Order	Taxonomy	Toxin	Type of toxicity	Location
	Cylindrospermum sp.	Anatoxin-a	Neurotoxic	Finland
	Nostoc sp.	Microcystins	Hepatotoxic	England, Finland
	Lyngbya wollei	Toxins PSP	Neurotoxic	USA
	Oscillatoria sp.	Anatoxin-a	Neurotoxic	Ireland,
		Homoanatoxin-a	Neurotoxic	Scotland,
		Microcystins	Hepatotoxic	Finland, Norway, Great Britain
	Phormidium formosum	Homoanatoxin-a	Neurotoxic	Norway
		Microcystins	Hepatotoxic	Norway
Oscillatoriales	*Planktothrix agardhii*	Microcystins	Hepatotoxic	Denmark, Norway, Finland, China
		Anatoxin-a	Neurotoxic	Norway, Finland
	Planktothrix mougeotii	Microcystins	Hepatotoxic	Denmark, Norway
	Planktothrix rubescens	Microcystins	Hepatotoxic	Norway
	Planktothrix prolifica	Microcystins	Hepatotoxic	Norway

also permits cyanobacteria to adjust their position in the water column when illumination is too high and avoid damage caused by an excess of light (UV radiations).

Capacity of Fixation and Storage of Nitrogen

The majority of cyanobacterial species are able to produce an enzyme, namely nitrogenase, harbored in the heterocysts, and responsible for nitrogen fixation. Contrary to other algae, the cyanobacteria are able to store nitrogen which is in excess in fresh water, in the form of cyanophicin (a copolymer of aspartate and arginin) and phycocyanin (pigment responsible for photosynthesis) so as to use it in the event of absence of nitrogen.

Storage of Phosphorus

When present in excess concentrations, cyanobacteria store phosphorus in the granular form as polyphosphates. These intracellular reserves are used for growth and multiplication when nutrient concentrations are low (Ishikawa et al. 2002).

Temperature

Optimum temperature for cyanobacteria growth is situated in general, between 25 and 35°C (Reynolds and Walsby 1975; Robarts and Zohary 1987). However, some cyanobacteria species can survive and proliferate even at low temperatures. The capacity to support low temperatures depends on cyanobacteria species. For example, *Oscillatoria* can proliferate at 10°C (Robarts and Zohar 1987). However, Microcystis species is inhibited below 15°C (Robarts and Zohary 1987). Thus, the proliferation of cyanobacteria at different temperatures helps them to escape from biological control by other organisms whish can not grow at low temperatures.

Dormance

When the conditions are not favorable, cyanobacteria enter sporulation phase, cells are transformed into spores that can survive for a long time in the sediments, while using reserves of nutrients. When the natural conditions become favorable, the cells go back to the surface and proliferate.

Defence Against Zooplankton

For protection, cyanobacteria produce toxins and allelopathic substances (secondary metabolites) which tend to affect the physiology of many zooplankton species. The change in physiology can sometimes result in the mortality of predators (Smayda 1997). Cyanobacteria can also regroup into colony forming large aggregates which are difficult to be ingested by zooplankton (Smayda 1997).

8 Trends in Biological Degradation of Cyanobacteria and Toxins

Capacity to Support Presence of Hydrogen Sulfide and Low Oxygen Concentrations

In comparison to other algae, cyanobacteria are capable of surviving at low oxygen concentration and high concentrations of hydrogen sulphide, usually poisonous for other algal species (Cohen et al. 1986; Whitton and Potts 2000). H_2S acts as electron donor during photosynthesis of cyanobacteria (Cohen et al. 1986). The presence of H_2S in water inhibits the growth of other bacteria degrading cyanobacteria. Thus, biological control of cyanobacteria becomes limited.

8.2.1.2   Environmental Factors Related to Survival of Cyanobacteria

There are three types of environnemental factors supporting the formation of cyanobacteria, namely, physical, chemical and biological. These parameters responsible for the formation of cyanobacteria are presented in details in Table 8.2.

Physical Factors

There are several physical factors supporting the formation of cyanobacteria. High temperature supports the proliferation of cyanobacteria in water. The optimal temperature of growth for the majority of these cells is located between 25 and 35°C (Reynolds and Walsby 1975; Robarts and Zohary 1987; Health Canada 1998). Longer duration of light during the day ensures optimal growth of these algae, which benefit from the light to carry out their photosynthesis. Nevertheless, the duration of light during the day necessary to optimize the growth depends on the species. For

Table 8.2 Parameters responsible for the formation and/or proliferation of cyanobacteria

Parameters	Details
Physical	Increase in temperature
	Availability of light
	Increase in turbidity
	Stagnation of water
Chemical	Availability of phosphorus
	Availability of nitrogen
	Presence of iron allowing nitrogen fixation and photosynthesis
	Presence of molybdenum ensuring carbon fixation and nitrogen contribution
	Increase in pH of medium
	High carbonate and bicarbonate concentrations
Biological	Non digestibility of cyanobacteria that is not consumed by zooplanktons
	Low concentrations of macrophytes: absence of competition of nutrients
	Low concentration of bacteria destroying the cyanobacteria

example, *Microcystis* is more adapted to shorter days than *Anabaena*. It is probably one of the reasons that the species of *Microcystis* are dominant species in North America at the end of summer, when the days are shorter (Health Canada 1998). Increased turbidity favours cyanobacteria compared to other algae, which cannot support the turbidity of the medium, limiting the light necessary for photosynthesis. Cyanobacteria can grow in turbid medium as they can migrate towards the surface, in order to obtain maximum light (Walsby and McCallister 1987; Health Canada 1998; Oliver and Ganf 2000). However, extremely high turbidity limits the availability of phosphorus in the medium, and hence affects the growth of cyanobacteria. The stagnation of water supports the proliferation of cyanobacteria while enabling them to maintain a strategic position in the water, where the medium conditions are optimal (Health Canada 1998). In the given context, removal of turbidity by physical-chemical means as discussed later can result in reduced growth of these algae.

Chemical Factors

The availability of macroelements, such as phosphorus and nitrogen in water supports the growth of cyanobacteria. The macroelements play a key role during photosynthesis. Cyanobacteria are able to store phosphorus in an effective way as discussed above.

The availability of iron and molybdenum in water also supports the growth of cyanobacteria. Iron is directly involved in nitrogen fixation and photosynthesis and molybdenum is involved in carbon fixation and nitrogen contribution. pH is a very important factor in growth, establishment and diversity of cyanobacteria which have generally been reported to prefer neutral to slightly alkaline pH for optimum growth (Brock 1973). High pH of the medium promotes the growth of cyanobacteria due to availability of carbonate and bicarbonate. Acidic soils are stressed environments for the proliferation of these organisms and hence they are normally absent at pH values below 4 or 5, however, eukaryotic algae flourish under these conditions (Fogg et al. 1973). Thus, reduction in pH can aid in removal of cyanobacteria and this could be the use of Fenton's oxidation reaction where the reaction conditions are acidic for the treatment of these contaminated waters.

Biological Factors

The nondigestibility of cyanobacteria by zooplanktons is a factor supporting the growth of these algal species in comparison with other algae. However, the growth of cyanobacteria is controlled by the presence of macrophytes and other phytoplanktons, which are in competition with cyanobacteria for nutritive elements and light. Other aquatic bacteria can also compete with cyanobacteria for nutritive elements and metabolize their toxins which is normally explored for various biological treatment strategies.

8 Trends in Biological Degradation of Cyanobacteria and Toxins

Table 8.3 Values of acute mammalian toxicity of cyanotoxins (as measured for mice)

Cyanotoxin	LD_{50} (μg/kg)	References
Microcystin-LR	43–50	Botes et al. (1985), Rinehart et al. (1988), Krishnamyrthy et al. (1989), Watanabe et al. (1988), Svrcek and Smith (2004), Gupta et al. (2003)
Microcystin-LA	50	Kaya and Watanabe (1990), Svrcek and Smith (2004), Gupta et al. (2003)
Microcystin-YR	70–110	Botes et al. (1985), Watanabe et al. (1988), Svrcek and Smith (2004), Gupta et al. (2003)
Microcystin-RR	235–600	Kusumi et al. (1987), Painuly et al. (1988), Watanabe et al. (1988), Sivonen et al. (1992a), Svrcek and Smith (2004), Gupta et al. (2003)
(D-Asp³) microcystin-LR	50–300	Krishnamyrthy et al. (1989), Cremer and Henning (1991), Harada et al. (1990b, 1991a), Luukkainen et al. (1993), Svrcek and Smith (2004)
(D-Asp³) microcystin-RR	250	Meriluoto et al. (1989), Sivonen et al. (1992a), Svrcek and Smith (2004)
(Dha⁷) microcystin-LR	250	Harada et al. (1991b), Sivonen et al. (1992a), Luukkainen et al. (1993), Svrcek and Smith (2004)
(6Z-Adda) microcystin-LR	>1,200	Harada et al. (1990a, b), Svrcek and Smith (2004)
Nodularin	50	Svrcek and Smith (2004)
Cylindrospermopsin (pure)	2000	Svrcek and Smith (2004)
Anatoxine-a et homoanatoxin-a	200–250	Svrcek and Smith (2004)
anatoxin-a(S)	20	Svrcek and Smith (2004)
Saxitoxin	10	Svrcek and Smith (2004)

8.3 Cyanotoxins

The cyanotoxins can have adverse health impacts following cutaneous contact or ingestion (Lavoie et al. 2007). Their presence constitutes a world wide problem. In order to surmount the ecological and health problems, several countries have set standard norms. These norms comprise threshold concentrations of toxins present in drinking water. About 46 species of cyanobacteria have the potential to produce toxins (Duy et al. 2000; Ernst et al. 2005; Lavoie et al. 2007). Cyanotoxins have been known to induce acute mammalian toxicity as evident from the values enumerated in Table 8.3.

8.3.1 Classification

Cyanotoxins are classified into two molecular classes according to their physiological effects: hepatotoxins (acting on the liver) and neurotoxins (acting on the nervous system). The detailed classification of cyanotoxins is presented in Fig. 8.1.

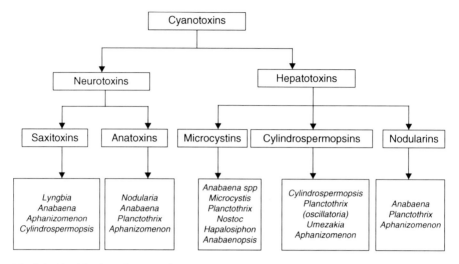

Fig. 8.1 Classification of cyanotoxins

8.3.1.1 Hepatotoxins

Hepatotoxins are cyclic peptides of low molecular weight, including the species of microcystins, nodularins and cylindrospermopsins. These compounds are mainly synthesized by certain genus, such as *Microcystis*, *Oscillatoria* or *Nostoc*, present worldwide. Hepatotoxins act on the liver, via bile duct, leading to hepatic disorders. Animals poisoned by hepatoxins, suffer signs of weakness, anoxia, paleness of mucous membranes, vomiting and diarrhoea. Mortality results after few hours to few days exposure, in general, due to liver haemorrhage (MacKintosh et al. 1990; Falconer 1991; Carmichael 1994).

Microcystins

Microcystins are heptapeptides having a molecular mass ranging from 800 to 1,100 g. mol^{-1}. The word microcystin is derived from, *"Microcystis"* genus, which produces these toxins. The microcystins are toxic cyanotoxins (Haider et al. 2003). Until date, nearly 80 variants of these toxins have been identified (Dietrich and Hoeger 2005). The structure of microcystins is characterized by the presence of a certain number of amino acids which can substitute itself in pairs into two of the basic structures. Leucine (L), arginine (R) and tyrosine (Y) are the most prevalent amino acids in the structure of microcystins (Krishnamurthy et al. 1989; Health Canada 2002). Microcystin LR is the most known and well studied mycrocystin (Health Canada 2002; Institut National de Santé Publique du Québec 2004).

8 Trends in Biological Degradation of Cyanobacteria and Toxins 271

Nodularins

Nodularins are pentapeptides synthesized primarily by *Nodularia* genus. This genus is often found in brackish waters. Nodularins, similar to microcystins, are inhibitors of proteins (phosphatases and other enzymes) which play a significant role in many cellular functions. It is considered that they act as promoters of tumours (MacKintosh et al. 1990; Falconer 1991; Carmichael 1994).

Cylindrospermopsins

Cylindrospermopsins are alkaloid cyanotoxins which have cytotoxic and hepatotoxic effects. These cyanotoxins have a tricyclic unit guanidine and a molecular mass of 415 Da (Svrcek and Smith 2004). Cylindrospermopsin and its derivatives are highly soluble in water as they occur in the form of zwitterions (AFSSET 2006). These molecules are stable when exposed to light and can resist high temperatures (50°C). However, when they are present as cellular metabolites and are exposed to the UV radiations, they quickly degrade (WHO 1999). Cylindrospermopsins inhibit protein synthesis by binding in an irreversible way with transfer RNA. Thus, they affect all tissues, but more particularly, the neurons (AFSSET 2006).

8.3.1.2 Neurotoxins

Neurotoxins are alkaloids that include anatoxins and saxitoxins. These toxins were detected in USA, Canada, Australia, Japan, Scandinavian countries, Great Britain, Italy, Portugal and Denmark. Exposure to these cyanotoxins generally causes headache, vomiting and diarrhoea. Even if the intoxicants produced by these toxins are less frequent, it should be known that a simple exposure by bathing or inhalation of water contaminated by efflorescence of *Anabaena* is enough to cause neurotoxic symptoms. The neurotoxins are often responsible for paralyzing intoxicants due to the absorption of water or ingestion of contaminated shells. These toxins cause death by affecting respiratory functioning, and by disturbing the acetylcholine/acetylcholinesterase system (Carmichael et al. 1975, 1979; Aronstam and Witkop 1981).

Anatoxins

Anatoxin is the first discovered neurotoxin. It is produced by *Anabaena, Aphanizomenon, Oscillatoria, Cylindrospermum* and *Microcystis* (Park et al. 1993; Bumke-Vogt et al. 1999; Namikoshi et al. 2003; Viaggiu et al. 2004; Ballot et al. 2005; Araóz et al. 2005; Ghassempour et al. 2005; Gugger et al. 2005). This toxin was detected in Canada, Finland, Scotland, Italy and Japan. Anatoxins act as agents that block neuromuscular post-synapses as they are fixed at the same level as receptors of acetylcholine and inhibit acetylcholinesterase. Thereafter, they cause an excessive stimulation of the muscular cells including a progressive paralysis of the respiratory muscles.

Saxitoxins

Saxitoxins are mainly synthesized by dinoflagellates. They are responsible for food poisoning resulting normally after consumption of contaminated seafood (Chevalier et al. 2001). These saxitoxins are synthesized by *Lyngbia, Anabaena, Aphanizomenon* and *Cylindrospermopsis*. They are seldom present in North America and traced only in Europe and Asia.

After understanding the classification of cyanobacteria and corresponding cyanotoxins, there is a need to explore the various treatment methods employed for the removal of the cyanobacteria and more so, the cyanotoxins which are more toxic. The presence of cyanobacteria in fresh water causes ecological and health problems (Billings 1981; Falconer 1989; Yu 1989, 1995; Carmichael 1992; Ueno et al. 1996; Codd et al. 1997; Jochimsen et al. 1998; Sivonen and Jones 1999; Carmichael 2001; Azevedo et al. 2002). Hence, these cyanobacteria should be removed. Several types of treatments, such as, physical, chemical and biological treatment are proposed, in order to ensure higher removal of cyanobacteria.

8.4   Physicochemical Treatment

The physico-chemical treatment is the most often used treatment in the elimination of toxic substances, such as phenols (Hu et al. 2002; Li et al. 2008). In the past, algae have been treated by an algicide, such as copper sulfate. However, algicidal treatment provoked the bursting of cells and the release of their cyanotoxins, which were higly toxic in fresh water (Watanabe et al. 1992; Ishii et al. 2004). Recently, the treatment by coagulation, ultrafiltration and clarification were found to be very effective in the removal of cyanobacteria (Gijsbertsen-Abrahamse et al. 2006). However, these treatments do not result in the complete neutralization of toxins, which are released following the bursting of blue green algae. On the other hand, the use of permanganate to decontaminate water infected with cyanotoxins allows complete neutralization of the microcystins (Rodriguez et al. 2007). This method was unable to eliminate other types of cyanotoxins, such as cylindrospermopsins and anatoxins. Chlorination treatment also has been used to remove cyanotoxins in solutions possessing pH lower than 8 and residual chlorine concentrations lower than 0.5 mg/L (Acero et al. 2005; Rodriguez et al. 2007). However, chlorination was accompanied by the production of toxic intermediate products which further accumulated in the sediments and would lead to secondary pollution.

Cyanobacterial toxins have also been degraded by ozonation which is able to ensure the removal of mycrocystins in contaminated water. This treatment is often

8 Trends in Biological Degradation of Cyanobacteria and Toxins

faster and more effective than the biological treatment. For example, ozonation allows the degradation of 1 mg/L of toxins in 90 s (Smith et al. 2008). In comparison, biological treatment needs hours and even days to cause the degradation of the same mass concentration of toxins. However, the ozonation method is unable to eliminate other types of cyanotoxins, such as cylindrospermopsins and anatoxins (Rodriguez et al. 2007). Other chemical treatments (such as UV treatment and H_2O_2) are endowed with low efficiency and are not completely valid for removal of cyanotoxins. Another method used for treatment of contaminated water has been activated carbon. The presence of large volume of mesopores allowed an effective elimination of cyanotoxins. Nevertheless, this treatment is not selective, as it allows the adsorption of other organic substances in addition to toxins leading to composite efficacy. The advantages and disadvantages of different physico-chemical treatment techniques for cyanobacteria and cyanotoxins are presented in details in Table 8.4.

8.5 Biological Treatment

8.5.1 Biological Treatment of Cyanobacteria

The biological breakdown is carried out by several organisms which are able to destroy the cyanobacteria (blue green algae) by using them as nutrient source. The advantages and disadvantages of different biological treatment techniques for removal of cyanobacteria and cyanotoxins are presented in Table 8.5.

8.5.1.1 Virus

Virus play a key role in the aquatic ecosystems due to their significant abundance, ubiquitous presence and their impact on the mortality of bacterial and phytoplanktonic populations (Dorigo et al. 2004; Pereira et al. 2009). There are large number of viruses present in the environment which can degrade cyanobacteria, referred to as cyanophages (Safferman and Morris 1963; Padan et al. 1967; Ohki and Fujita 1996; Breitbart et al. 2002; Venter et al. 2004; Baker et al. 2006; Li and Hayes 2008). Viruses can thus play a significant role in the biocontrol of blue green algae. In fact, the viruses posses extremely low generation times (10 h for the cyanophage LPP-DUN1) (Daft et al. 1970; Breitbart et al. 2002; Venter et al. 2004) and high burst size which is about 100 phage particles for each infected cell making up *Plectonema boryanum* filament (Daft et al. 1970). The fast appearance of highly resistant mutants of cyanophages can destroy resistant cyanobacteria, make the viral biocontrol process of cyanobacteria more advantageous and effective (Padan and Shilo 1973; Barnet et al. 1981; Gumbo et al. 2008). On the other hand, the treatment of cyanobacteria by virus has some limitations. High degree of specificity to the host, the fast appearance of resistant cyanobacteria and effect of environmental factors, such as darkness, addition of inhibitors of photosystem II and decouplers of photosynthetic electron transport, and CO_2 starvation on cyanophages, lead to complexity and unpre-

Table 8.4 Advantages and disadvantages of physico-chemical treatment techniques for cyanobacteria and cyanotoxins

Type of treatment	Parameters	Advantages	Disadvantages	References
Ozonation	$[Ozone] = 0.4$ mg/L $[microcystin-LR] = 1$ mg/L Temperature $= 20°C$ pH $= 7$ Time $= 90$ s	Method valid for degradation of microcystins	– Method valid only for residual ozone concentrations ranging from $0.05–0.1$ mg/L	Smith et al. (2008)
$H_2O_2/Fe(II)$	$[H_2O_2] = 0.005$ mg/L $[microcystin-LR] = 1$ mg/ $[Fe(II)] = 0.2$ mg/L Temperature $= 20°C$ pH $= 7$ Time $= 60$ s	Method valid for degradation of the microcystins	– Problem of secondary pollution due to the addition of Fe(II)	Smith et al. (2008)
$O_3/Fe(II)$	$[microcystin-LR] = 1$ mg/ $[Fe(II)] = 0.2$ mg/L $[Ozone] = 0.2$ mg/L Temperature $= 20°C$ Ph $= 7$ Time $= 80$ s	Degradation more than 99% of initial toxin	– Method valid only for residual ozone concentrations of $0.05–0.1$ mg/L as well as secondary pollution	Smith et al. (2008)
O_3/H_2O_2	$[microcystin-LR] = 1$ mg/ $[H_2O_2] = 0.001$ mg/L $[Ozone] = 0.5$ mg/L Temperature $= 20°C$ pH $= 7$ Time $= 80$ s	Degradation moreover than 98% of initial toxin	– Method valid only for residual ozone concentrations of $0.05–0.1$ mg/L	Smith et al. (2008)
H_2O_2	$[microcystin-LR] = 1$ mg/ $[H_2O_2] = 0.001$ mg/L Temperature $= 20°C$ pH $= 7$ Time $= 80$ s		– Degradation less than 5% of initial toxin: Not a valid method	Smith et al. (2008)

Table 8.4 (continued)

Type of treatment	Parameters	Advantages	Disadvantages	References
UV	[toxin]=2.36 µg/mL Time=48 h UVB, UVA, VIS		– Very low degradation: Not a valid method	Mazur-Marzec et al. (2006)
UV/ TiO$_2$	[microcystin-LR]=55 ng/mL [microcystin- RR]=60 ng/mL Time=20 min Type of UV radiation: UVC Water of lake	Complete decomposition of toxins with a half-life of 10 min		Shephard et al. (2002)
	P. agardhii and P. rubescens	Effective elimination of cyanobacteria	– No elimination of cyanotoxins – Lysis of algal cells and release of cyanotoxins in the medium	Gijsbertsen-Abrahamse et al. (2006)
Ultrafiltration	Molecular weight=100 KDa			
Coagulation	Average diameter of pores=30 µm			
Clarification	Surface membrane=0.8 m^2			
Addition of an algicide: copper sulphate		Effective elimination of cyanobacteria	– Release of toxin in water. – Neutralization of all algae, not only cyanobacteria	NRA Toxic Algae Task Group 1990
Activated carbon	[Toxin]=30 µg/L saxitoxins equivalent time=6 min	Presence of large volume of pores allowing an effective elimination of cyanotoxins	– Non-selective method allowing adsorption of other organic substances	Orr et al. (2004)

Table 8.4 (continued)

Type of treatment	Parameters	Advantages	Disadvantages	References
Chlorination	[cylindrospermopsin (CYN)]=1 μM [Anatoxine A]=1 μM [Chlorine]=1.5–3 mg/L pH=7 T=20°C Time=24 h	Effective elimination of the cyanotoxins	– Method valid only for pH lower than 8 and waste chlorine concentrations lower than 0.5 mg/L – Production of toxic intermediate products which accumulate in sediments	Rodriguez et al. (2007); Acero et al. (2005)
Permanganate	[cylindrospermopsin (CYN)]=1 μM [Anatoxin A]=1 μM [Chlorine]=1 mg/L [microcystin-LR]=3.2 μg/L [microcystin-RR]=7.1 μg/L [microcystin- YR]=1.1 μg/L pH=7 T=20°C Time=2–3 h	Effective method for elimination of microcystins	– Nonapplicable method for water treatment containing cylindrospermopsin and Anatoxin A as it requires high concentration of substrate	Rodriguez et al. (2007)

Table 8.5 Advantages and disadvantages of biological treatment of cyanobacteria and cyanotoxins

	Treatment	Advantages	Disadvantages	References
Phages	Cyanophages	– Effective lysis of cyanobacteria – Very short generation time and high burst size. – Fast appearance of mutants of cyanophages which attack resistant cyanobacteria	– Complexity and unpredictability of interactions cyanobacteria/phages in the medium caused by very high degree of specificity, effect of environmental factors and fast appearance of resistant cyanobacteria – Lysis of algal cells and release of cyanotoxins in the medium	Safferman and Morris. 1963; Padan et al. (1967). Ohki and Fujita (1996)
	Bacillus sp.	– Effective method for elimination of filamentous cyanobacteria.	– Lysis of algal cells and release of cyanotoxins in the medium	Reim et al. (1974), Wright and Thompson (1985), Wright et al. (1991)
	Flexibacter flexilis *Flexibacter sancti*	– Bacteria secrete a lysozyme resulting in inhibition of photosynthesis and activity of glycolate and dehydrogenase of cyanobacteria – Neutralization of cells without lysis – No release of toxin in the medium		Sallal (1994)

Table 8.5 (continued)

	Treatment	Advantages	Disadvantages	References
Bacteria	*Myxobacter*	– Adaptavity of Myxobacter in the physical conditions – Large capacity to imprison the host – Large capacity to multiply – Large choice of hosts: effectiveness on several hosts	– Degradation requires a high population of Myxobacter and a rigorous control of conditions of treatment: nutrient and environmental conditions	Daft et al. (1975), Fraleigh and Burnham (1988)
	Micrococcus xanthys strain Pco2	– Destruction of cyanobacteria by enzymes produced by *Myxobacter*		Burnham et al. (1984)
	Burkholderia sp.	– Elimination of more than 90% of microcystin LR after 43 days	– Very slow method of degradation – Degradation rate = 0.05 µg microcystin - LR/ml/day	Lemes et al. (2008)
	Sphingomonas pauemobilis strain Y2	Fast elimination of microcystins LR and RR	– The process of degradation is inhibited by presence of organic matter in the medium	Park et al. (2001)
	Sphingpoyxis sp. LH21	– Effective elimination of the microcystins and their toxicity – Absence of intermediate products of degradation	– Effective degradation only for microcystins LR and RR but not for the others	Ho et al. (2007)
	Gram negative Sphingomonas strain 7CY	– Elimination of microcystins LR and RR and nodularin after 4 days – High rate of degradation	– This bacterium is able to degrade nodularin only in the presence of microcystin RR as it does not have an enzyme for degradation of nodularin	Ishii et al. (2004)
	Biofilms	– Degradation of 1000 µg/L microcystin after 6 days		Saitou et al. (2002)

8 Trends in Biological Degradation of Cyanobacteria and Toxins

Table 8.5 (continued)

	Treatment	Advantages	Disadvantages	References
Actinomy-cetes	Actinomycetes: *Streptomyces plectonema*	– Production of extracellular products destroying cyanobacteria	– Non-selective action: the extracellular products of this strain are able to destroy fungi, bacteria and algae. – Unstable lytic agents: they lose their lytic activity quickly	
	Actinomycetes: *Streptomyces exfoliatis*	– Lysis of cyanobacteria. – Stable lytic agent.	– This agent must be used in combination with other techniques, in order to ensure the biocontrol of cyanobacteria.	Sigee et al. 1999
Protozoa	Protozoa	– Effective destruction of cyanobacteria.	– The protozoa are often consumed by higher organisms – Low density of protozoa in the medium – Low rate of predation.	Canter et al. (1990), Cole and Wynne (1974); Laybourne-Parry et al. (1987)

dictability of cyanobacteria/phage interactions in the field (Sigee et al. 1999). There would be additional difficulties of production of large quantities of active inocula for the effective use of cyanophages as biocontrol agents for cyanobacteria in the aqueous environment. This entails detailed studies on growth of these cyanophages in environmentally controlled fermenters and subsequent harvesting for application. The final field application of these harvested cyanophages would also involve several steps of separation and purification increasing the overall production costs.

The treatment of cyanobacteria by phages is a complicated process, nevertheless it can be used for the removal of blue green algae present in fresh waters. At this point from the existing literature, it is still not clear if the cyanophages can simultaneously destroy cyanobacteria and corresponding cyanotoxins.

8.5.1.2 Bacteria

Cyanobacteria can also be treated by bacteria. The first isolated strain of a bacterium was a Gram positive bacterium, *Bacillus* sp. which was able to lyse seven genera of cyanobacteria (*Anabaena* and *Microcystis*). The lysis was carried out by a complex volatile product of low molecular weight and high thermal stability (Reim et al. 1974; Wright and Thompson 1985). The product responsible for lysis was isoamylic alcohol (3-methyl-1-butanol) (Wright et al. 1991) which was a landmark in the field of cyanobacteria as it allowed better understanding of cyanotoxicity. Likewise, two strains of *Flexibacter* (*Flexibacter flexilis* and *Flexibacter sancti*) isolated from wastewater, were able to destroy the genus *Oscillatoria (Oscillatoria williamsii)* (Sallal 1994). The destruction was carried out by a metabolite (lysozyme) produced by these two bacteria that inhibited the photosynthetic reaction of transport of electrons. It also inhibited the activity of glycolate, dehydrogenase, and nitrogenase (Sallal 1994), the enzymes important in metabolism of cyanobacteria.

An isolated *Myxococcus* sp was also able to degrade cyanobacterial species (*Phormidium luridum* and *Nostoc muscorum*). Bacteria encapsulate their hosts and release lysozymes which are actually responsible for the destruction (Burnham et al. 1981; Fraleigh and Burnham 1988). Degradation is influenced by the density of predation. *Myxococcus* uses the secretions of its host to live and does not destroy it, if the number of predators is lower than the number of the hosts (Burnham et al. 1981; Fraleigh and Burnham 1988). In another study, four strains of *Myxobacter* (Cp1-4) were able to destroy 40 strains of cyanobacteria after 20 min of incubation through simple contact (Daft and Stewart 1971). The degradation of cyanobacteria by *Myxobacter* seemed to be effective (Shi et al. 2006). However, this degradation was influenced by the quantity of algae and *Myxobacter* (10^6 cells/mL of *Myxobacter* are required to remove the cyanobacterials cells at 1 to >110 plaque-forming units ml and the nature of nutrients present in the medium (Shilo 1970; Daft and Stewart 1971; Daft et al. 1975; Fraleigh and Burnham 1988). However, these studies were restricted to pure culture medium and at environmentally insignificant concentrations, thus implying no conclusion on their role and degree of removal of cyanobacteria. Another point to be raised is that these studies do not validate the simultaneous destruction of cyanotoxins with the cyanobacteria.

8 Trends in Biological Degradation of Cyanobacteria and Toxins 281

Out of the different bacteria reported for cyanobacteria control, *Myxobacter* was found to be the best potential bacterium due to three principal reasons: (a) rapid adaptability in variable environmental conditions; (b) capacity to seek and imprison the host and; (c) ability to multiply and imprison several types of hosts. This bacterium can be exploited further at large scale to combat cyanotoxin contamination.

8.5.1.3 Actinomycetes

Actinomycetes are bacteria which grow as thin filaments similar to a mold rather than as single cells. In fact, they were long thought to be fungi and were called actinomycetes. But fungi are eukaryotes and the actinobacteria are not eukaryotes (Holt et al. 1994). Many of them have turned out to be the source of valuable antibiotics (Stackebrandt et al. 1997), including streptomycin, erythromycin, and the tetracyclines. Actinomycetes have been also known to destroy cyanobacteria in fresh waters and soil environments. Different strains of *Streptomyces* have been isolated by Silvey and Wright et al. which can lyse several cyanobacteria, namely *Anabaena cylindrica, Tolypothrix tenuis*. However, the lytic agents lose their activity after four successive transfers from the laboratory to bench scale and eventual field application. Another study by Sigee et al. (1999) isolated a strain of *Streptomyces* (*Streptomyces exfoliatus*) which produced stable lytic agents that persisted in lakes and river and hence showed potential of field application. Likewise, another strain of actinomycetes AN6 was isolated from the Iraqi soil which showed a broad spectrum of biological control by destroying several strains of cyanobacteria, bacteria, and green algae. The strain possessed high potential as no resistant cyanobacteria were detected (Al-Tai 1982). The mechanism of destruction of cyanobacteria involved lysis carried out by extracellular products and secretion of antibacterial substances produced by actinomycetes. The amino acid, L-lysine, secreted by actinomycetes was found to be the cause of lysis. L-lysine caused severe damage to the cell wall of cyanobacteria. All these studies have been restricted to laboratory scale experiments and no study so far has been reported on the possibility of scale-up for field application.

The efficiency of these biocontrol agents is not the same in the laboratory and fresh waters. This may be due to the instability of lytic agents. These biological agents can be stabilized by addition of stabilizing chemicals, such as polyethylene glycol, polyvinyl alcohol or by immobilization of agents. Therefore, actinomycetes may be utilized in conjunction with other microorganisms, for example, the *Myxobacter* as discussed earlier to obtain efficient biocontrol of cyanobacteria.

8.5.1.4 Protozoa

In the ecosystem, the protozoa play significant role in the reduction of green algae and the cyanobacteria by grazing (Canter et al. 1990). Cyanobacteria also constitute appropriate food source for several genera of protozoa including the ciliate, *Nassula,* flagellate, *Ochromonas* (Cole and Wynne 1974) and amoebae, *Acanthamoeba,*

Mayorella (Laybourne-Parry et al. 1987) and *Nuclearia* (Yamamoto 1981). The predation of cyanobacteria by protozoa is prominent in the natural environment (Cook et al. 1974; Laybourn-Parry et al. 1987; Canter et al. 1990), at laboratory scale (Yamamoto 1981; Dryden and Wright 1987) and in the field (Brabrand et al. 1983). Protozoa can grow in lab scale reactors and they can efficiently control cyanobacteria (Yamamoto 1981; Dryden and Wright 1987). However, the efficiency of biocontrol by protozoa is lower under field conditions. The efficiency depends on several factors, such as growth of protozoa, flowering rate, specificity of predation, growth of cyanobacteria and rate of predation by higher organisms, such as Copepoda. Protozoa in the natural systems are superseded by the predatory carnivores and fish. Thus, the density of these protozoa is very low which does not enable them to multiply and eliminate cyanobacteria.

However, ex-situ growth of these protozoa in fermenters, proper recovery and stabilization can result in large population of protozoa which may be released into the aquatic environment. For example, the harvested protozoa can be encapsulated and released into the aquatic streams so that there is sustained release of these protozoans over a period of time in order to control the proliferation of cyanobacteria.

8.5.1.5 Fungi

Fungi are also responsible for reduction of numbers of cyanobacteria. Canter and Lund isolated fungal strain (*Rhizophidium planktonicum*) which destroyed the cyanobacteria, *Oscillatoria agardhii* var. isothrix. Another fungal strain, *Rhizophidium planktonicum* was able to destroy the cyanobacteria (*Oscillatoria agardhii* var. isothrix). However, the capacity of this strain was limited by the impossibility of production on a large scale (Safferman and Morris 1962). Fungal species, *Acremonium*, *Emericellopsis* and *Verticillium* are also able to lyse *Anabaena flos-aquae* and other species of cyanobacteria (Redhead and Wright 1978). The lysis of the cynobacteria by *Acremonium*, *Emericellopsis* is facilitated by thermostable extracellular agents which are antibiotics (Redhead and Wright 1978, 1980). However, these antibiotics are produced in very small quantities. These quantities are effective only if the fungal strains are in close contact with cyanobacteria. Hence, it is recommended to extract these antibiotics, concentrate them and use them as different formulations— crude or pure for biocontrol of cyanobacteria. High performing fungal strains which can be super-producers of antibiotics can also be developed by using the tools of genetic engineering.

Majority of the recent research in the field of biological control of cyanobacteria has been focussed on the biodegradation of cyanotoxins. In fact, the toxity of cyanobacteria is due to their toxins. If toxins are effectively removed, the toxicity of cyanobacteria will be simultaneously controlled. The detailed mechanisms of degradation of cyanobacteria are presented in Figs. 8.2, 8.3, respectively.

8 Trends in Biological Degradation of Cyanobacteria and Toxins 283

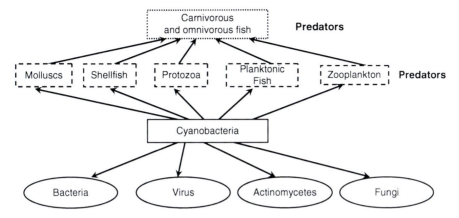

Fig. 8.2 Degradation chain of cyanobacteria

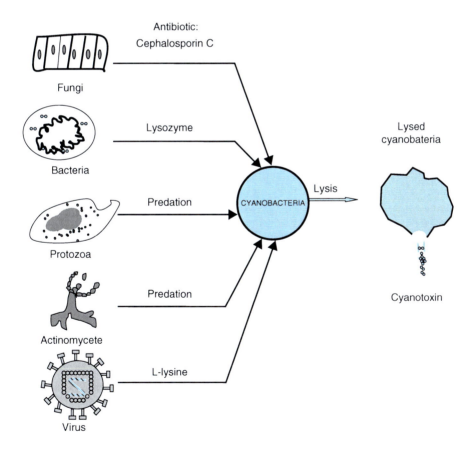

Fig. 8.3 Biological degradation mechanisms of cyanobacteria

8.5.2 Biological Treatment of Cyanotoxins

The degradation of cyanobacteria does not necessarily mean the removal of their toxins. It is an established fact that the cyanotoxins, namely nodularin, microcystin, anatoxin, saxitoxin, among others are the chief contributors, as discussed earlier. Physico-chemical treatments efficiently destroy cyanobacteria and release toxins into the medium leading to secondary toxicity. The cyanotoxins should be eliminated in order to avoid the secondary intoxication causing further contamination of water. There are a large number of bacteria present in the environment which can degrade cyanotoxins. Lemes et al. (2008) isolated a strain of *Burkholderia* from south Brazilian lagoon which could degrade microcystins. However, the rate of removal was very low. The bacterium removed less than 10% of initial microcystin concentration of 1 mg mL^{-1} after 43 days (Lemes et al. 2008). Hence, this treatment cannot be applied in the decontamination of cyanotoxin contaminated fresh water. Likewise, several microcystin degrading *Sphingomonas* strains were isolated from lagoon water (Bourne et al. 1996; Park et al. 2001; Saitou et al. 2003; Imanishi et al. 2005; Lemes et al. 2008) and lake water (Jones and Orr 1994; Cousins et al. 1996; Harada et al. 2004). The degradation resulted in intermediate metabolites which were carcinogenic, toxic and mutagenic. Thus, it is not just a question of degradation of native cyanotoxin, it is equally important to know if the by-products released are potentially more toxic as it may further create secondary toxicity.

Different strains of *Sphingomonas* sp. do not necessarily possess same degradation efficiency and mechanism of action on microcystins. *Sphingomonas sp.* Y2 isolated from lake water by Park et al. (2001), degraded microcystins more efficiently (degradation rates of microcystins-RR and -LR were 13 and 5.4 mg L^{-1} day^{-1}) as compared to another strain of *Sphingomonas* isolated from Australian drainage water (degradation rate of microcystins LR < 0. 2 mg L^{-1} day^{-1}) (Jones et al. 1994; Bournes et al. 1996). Degradation strongly depended on the species isolated and the medium of isolation. Likewise, microcystins LR were degraded by *Sphingomonas* sp. Acm-3962. This degradation was interceded by three intracellular enzymes, namely: enzyme 1 (microcystinase), which catalyzed the conversion of microcystin to linearized microcystin LR; enzyme 2, which catalyzed the conversion of linearized microcystin LR to tetra- peptide (NH$_2$- Adda- IsoGlu- Mdha- Ala- OH); and enzyme 3, which catalyzed the breakdown of tetra- peptide (NH$_2$- Adda- IsoGlu-Mdha- Ala- OH) (Bournes et al. 1996). Jones et al. (1994) demonstrated that the degradation was carried out by two bacterial populations, a population which used microcystin LR as source of carbon and energy and another, which cometabolized the residual microcystins LR present at low concentrations in the aquatic medium. A Gram negative bacterium, *Sphingomonas sp. 7CY* isolated from lake water by Ishii et al. (2004) degraded nodularin Har in the presence of microcystin RR as well as hydrophobic microcystins. However, it was unable to degrade nodularin without microcystin-RR. The cytoplasmic enzyme degrading microcystin-RR might be constitutive and not be able to decompose nodularin-Har due to their structural differences. The nodularin-Har was degraded by another bacterial enzyme(s) induced

8 Trends in Biological Degradation of Cyanobacteria and Toxins 285

by either microcystin-RR or degradation products of microcystin-RR. The synthesis of these enzymes does not take place in the presence of other sources of carbon and nitrogen (Ishii et al. 2004). Another strain of *Sphingomonas,* named *Sphingopyxis* sp. LH21 isolated by Ho et al. (2007) from a biological sand filter degraded microcystins LR after 5 h of incubation. The degradation of microcystins by this strain was faster and more efficient than by other strains of *Sphingomonas*. Another *Sphingomonas sp. B9* isolated from lake water by Imanishi et al. (2005) produced enzymes which could degrade cyclic peptides (microcystins and nodularin). B9 degraded the cyanobacterial cyclic peptides by hydrolysis of their peptide bonds.

Pseudomonas aeruginosa isolated from lake water degraded microcystins LR by secretion of an alkaline protease synthesized by this strain (Harada et al. 2004; Kato et al. 2007). These results indicated that this bacterium can degrade structurally different cyclicpeptides and may be effective for the detoxification of hazardous cyclic peptides. The reaction of microcystin-LR with cell extract of this strain produced a final degradation product Adda ((2S, 3S, 8S, 9S)-3-amino-9-methoxy-2,6,8-trimethyl-10-phenyldeca-4(E),6(E)-dienoic acid) via two intermediates, linearized microcystin-LR and a tetrapeptide (Harada et al. 2004). The final degradation product did not show any toxicity against mice or inhibition of protein phosphatase activity in contrast to the native toxin. Hence, this bacterium can be used to efficiently remove dissolved cyclic cyanotoxins in fresh waters and their toxicity. However, it is not known if this strain can degrade microcystins and nodularins located inside cyanobacterial cells. The bacterium also rapidly loses its degradation activity in lakes and rivers. To overcome these limitations, a study was conducted on immobilization of *Sphingomonas B9* on a polyester resin (Tsuji et al. 2006). The immobilization preserved the activity of B9 against microcystins for 2 months, however the release of these immobilized versions into the aquatic streams is challenging. Nevertheless, the immobilization of these microorganisms can be carried out on other microorganisms, such as actinomycetes which can act as "ghost hosts" and this can be tested for its efficacy in treatment of cyanobacteria-cyanotoxin contaminated aquatic streams.

Microcystins have also been removed by bacteria present in the ecosystem in biofilms (Saitou et al. 2002). Contrary to other biological treatments carrying out the removal of dissolved microcystins, treatment by biofilm allows complete removal of extracellular and intracellular microcystins. Hence, the biofilm systems can be effectively applied for removal of cyanobacteria and their toxins present in potable water through biological membrane reactors. This type of treatment, however, will be limited to potable water treatment and hence can treat only off-site contaminated waters.

There are several bacterial species degrading cyanotoxins in particular, hepatotoxins. When compared to the physico-chemical treatments, the biological treatment allows the removal of cyanotoxin toxicity from the contaminated media to a certain extent. In addition, the breakdown products resulting from this type of treatment are normally non-toxic. The treatment also encompasses some drawbacks. For example, the biological agents produced by bacteria quickly lose their activity against cyanotoxins. Hence, the use of these agents in the decontamination of fresh waters is limited as it needs complete stabilization. The majority of biological methods employed so far were found to remove only dissolved cyanotoxins. In lakes, rivers and lagoons, large amounts of cyanotoxins are located inside the cyanobacteria and are often suspended. Hence, it is necessary to seek an efficient treatment which can simultaneously lyse cyanobacteria and degrade their toxins. At this juncture, utilization of cyanophage cannot be overlooked and it can be used in combination with other biological control, such as actinomycetes which show broad spectrum of control, an important issue while combating cyanobacteria.

8.6 Future Perspectives

Several physico-chemical treatments have been proposed for the removal of cyanobacteria and their toxins. These treatments are simple, faster than biological treatment, and easy to apply and efficient. However, chemical treatment causes the bursting of cyanobacterial cells and release of the cyanotoxins into the medium. These cyanotoxins are chemically very stable and persist for a long time in fresh waters (Watanabe et al. 1992). Moreover, these physico-chemical treatments result in by-products that are carcinogenic (Bellar et al. 1974; Rook 1974) and mutagenic (Hemming et al. 1986) substances. The physicochemical methods also result in secondary pollution of treated water. In order to avoid these side effects, there is a need to optimize other types of treatments. At this stage, an alternative that employs sequential treatments or hybrid treatments rather than the single treatment would be a better option.

Biological breakdown is a moderate and effective treatment in the removal of cyanobacteria and their cyanotoxins. Biological control has been proven as an excellent treatment option for cyanobacteria at laboratory scale. However, the biological agents lose their efficiency when transferred to lakes and rivers. Hence, the combination of biological with chemical technology could be considered as a viable technology to give complete solution to removal of cyanobacteria present in fresh waters. The sequential treatment can be set-up in this fashion, partial chemical treatment followed by biological treatment which can take care of the residual toxins from chemical treatment. In fact, physical-chemical treatment cannot accomplish complete destruction as there are cyanotoxins which are released into the medium and they have been found to be more toxic. The field application of these hybrid technologies warrant detailed research studies which have not been carried out in the past.

At this juncture, adsorption of different organisms, for example, bacteria, virus, protozoa etc. on nanoparticles such as, SiO_2, Al_2O_3, among others can also be utilized. Nanoparticles will act as carriers and allow reduction of bacterial numbers due to increase in surface area and thus enhanced potential efficiency. In literature, various studies report the use of nanoparticles as carriers for microorganisms (Soppimath et al. 2001; Couvreur et al. 2002; Vauthier et al. 2003; Bala et al. 2004; Wissing et al. 2004). The following will be among the important technological advantages of nanoparticles as microbial carriers: high stability (i.e., long shelf life); high carrier capacity (i.e., many destroying or predator organisms can be incorporated in the particle matrix); feasibility of incorporation of both hydrophilic and hydrophobic substances. These carriers can also be designed to enable controlled (sustained) organism (to degrade cyanobacteria) release from the matrix. Thus, in comparison to other physico-chemical treatments, the use of nanoparticles in combination with biological agents may not result in secondary pollution of the treated medium. The use of large number of nanoparticle adsorbed predator organisms will lead to an effective degradation of cyanobacteria. Interestingly, these particles, even, if they are applied in large numbers, net mass remains low. And thus, residual nanoparticles concentrations will be much lower than the maximum concentrations allowed in lakes and rivers water. However, future research has to be carried out in this direction comprising following research perspectives: (1) adhesion stability of microorganisms to nanoparticles; (2) residual concentration of nanoparticles and its secondary effect on the environmental streams; and (3) treatment efficacy, more so in the removal of cyanotoxins. There is also a possibility of using genetic engineering to develop high performance microbial strains which can jointly address the contamination and/or intoxication problems of cyanobacteria as well as cyanotoxins.

8.7 Conclusions

There are several control methods available for cyanobacteria, namely physical-chemical and biological. The physical-chemical methods comprise, chlorination, ozonation, photooxidation, activated carbon, among others. The biological methods include utilization of pure organisms, such as bacteria, virus, fungi and protozoa, among others. Biological control was found to be a relatively effective option for removal of cyanobacteria. The biological treatment has the key advantage of being a more specific method vis-à-vis physical-chemical methods which are general and can also destroy the useful organisms in the ecosystem affecting the biodiversity. Nevertheless, biological control encompasses various drawbacks, namely; (1) limited duration of life due to elimination by other organisms and; (2) large scale production, storage and application are difficult to achieve. The efficiency of biological control methods can be enhanced by the use of tools of genetic engineering leading to production of microorganisms which can be super-producers of anticyanobacterial agents such as, lysozymes, antibiotics and others. The use of several biological antagonistic agents at the same time may also allow effective control of cyanobacteria due to variable mode of action. However, there is a caveat that there may be synergy or antagonism among different microorganisms which needs careful investigation. Therefore, in the current scenario, biological control must constitute only as a stage in the complete process of removal of cyanobacteria.

Acknowledgements The authors are sincerely thankful to the Natural Sciences and Engineering Research Council of Canada (Discovery Grants A4984 and 355254, STP235071, Canada Research Chair) and INRS-ETE for financial support. The views or opinions expressed in this article are those of the authors and should not be construed as opinions of the U.S. Environmental Protection Agency.

References

Acero JL, Rodriguez E, Meriluoto J et al (2005) kinetics of reactions between chlorine and the cyanobacterial toxins microcystins. Water Res 39:1628–1638

Agence française de sécurité sanitaire de l'environnement et du travail (AFSSET) et Agence française de sécurité sanitaire des aliments (AFSSA), Rapport sur l'évaluation des risques liés à la présence de cyanobactéries et de leurs toxines dans les eaux destinées à l'alimentation, à la baignade l'eau de baignade et aux autres activités récréatives, [Paris], AFSSA, juillet 2006. (http://tinyurl.com/4zzz65)

Al-Tai AMS (1982) Characteristics and microbial interactions of certain actinomycetes from Iraq. Ph.D. Thesis, University of Dundee

Araóz R, Nghiêm HO, Rippka R, Palibroda N, Tandeau de Marsac N, Herdman M et al. (2005) Neurotoxins in axenic oscillatorian cyanobacteria: coexistence of anatoxin-a and homoanatoxin-a determined by ligand-binding assay and GC/MS. Microbiology 151:1263–1273

Aronstam RS, Witkop B (1981) Anatoxin-a interactions with cholinergic synaptic molecules. P Nat Scad Sci USA 78:4639–4643

Azevedo SMFO, Carmichael WW, Jochimsen EM, Rinehart, KL, Lau S, Shaw GR, Eaglesham, GK et al (2002) Human intoxication by microcystins during renal dialysis treatment in Caruaru—Brazil. Toxicology 181–182:441–446

Baker SC, Shimizu C, Shike H, Garcia F, Van der Hoek L, Kuijper TW, Reed SL, Rowley AH, Shulman ST, Talbot HKB, Williams JV, Burns JC et al. (2006) Human coronavirus NL63 infection is not associated with acute Kawasaki disease. Adv Exp Med Biol 581:523–526

Bala I, Hariharan S, Kumar MN et al (2004) PLGA nanoparticles in drug delivery: the state of the art, Crit Rev Ther Drug Carrier Syst 21:387–422

Ballot B, Krienitz L, Kotut K, Wiegand C, Pflugmacher S et al (2005) Cyanobacteria and cyanobacterial toxins in the alkaline crater lakes Sonachi and Simbi, Kenya. Harmful Algae 11:39–50

Barnet YM, Daft MJ, Stewart WD et al. (1981) Cyanobacteria—cyanophage interactions in continuous culture. J Appl Bacteriol 5:541–552

Bellar TA, Lichtenberg JJ, Kromer RC et al (1974). The occurrence of organohalides in chlorinated drinking waters. J Am Water Works Assoc 68:703–706

Betsey Betsey Dexter Dyer B (2003) A field guide to bacteria. Cornell University Press, Cornell, ISBN 0801488540

Billings WH (1981) In: Carmichael WW (eds) The water environment: algal toxins and health. Plenum Press, New York, pp 243–250

Botes DP, Wessels PL, Kruger H, Runnegar MTC, Santikarn S, Smith RJ, Barna JCJ, Williams DH et al (1985) Structural studies on cyanoginosins-LR, -YR, -YA, and -YM, peptide toxins from *Microcystis aeruginosa*? J Chem Soc Perkin Transactions I:2747–2748

Bourne DG, Jones GJ, Blakeley RL, Jones A, Negri AP, Riddles P et al. (1996) Enzymatic pathway for the bacterial degradation of the cyanobacterial cyclic peptide toxin microcystin LR. Appl Environ Microbiol 62:4086–4094

Brabrand A, Faafeng BA, Kallquist T, Nilssen JP et al. (1983) Biological control of undesirable cyanobacteria in culturally eutrophic lakes. Oecologia 60:1–5

Breitbart M, Salamon P, Andresen B, Mahaffy JM, Segall AM et al. (2002) Genomic analysis of uncultured marine viral communities. Pro Natl Academy Sci USA 99:14250–14255

Brock TD (1973) Evolutionary and ecological aspects of the cyanophytes. In: Carr NG, Whitton BA (eds) The biology of the blue-green algae. Blackwell Scientific Publications, Oxford, pp 487–500

Bumba L, Prasil O, Vacha F et al. (2005) Antenna ring around trimeric Photosystem I in chlorophyll b containing cyanobacterium *Prochlorothrix hollandica*. Biochim Biophys Acta 1–5

Bumke-Vogt C, Mailahn W, Chorus I et al. (1999) Anatoxin-a and neurotoxic cyanobacteria in German lakes and reservoirs. Environ Toxicol 141:17–25

Burnham JC, Collart SA, Daft MJ et al. (1984) Myxococcal predation of the cyanobacterium *Phormidium luridum* in aqueous environments. Arch Microbiol 137:220–225

8 Trends in Biological Degradation of Cyanobacteria and Toxins 289

Burnham JC, Collart SA, Highison BA et al. (1981) Entrapment and lysis of the cyanobacterium *Phormidium luridum* by aqueous colonies of *Myxococcus xanthus* PCO2. Arch Microbiol 129:285–294

Canter HM, Heaney SI, Lund JW et al (1990) The ecological significance of grazing on planktonic populations of cyanobacteria by the ciliate *Nassula*. New Phytol 114:247–263

Carmichael WW (1994) The toxins of cyanobacteria. Sci Am 270:64–72

Carmichael WW (2001) A mini-review of cyanotoxins: toxins of cyanobacteria (blue-green algae). In: Koe WJ, Samson RA, van Egmond HP, Gilbert J, Sabino JM et al. (eds) Mycotoxins and phycotoxins in perspective at the turn of the millennium, de Koe WJ, Wageningen, pp 495–504

Carmichael WW, Biggs DF, Gorham PR et al (1975) Toxicology and pharmacological action of *Anabaena josaquae*. Toxin. Science 187:542–544

Carmichael WW, Biggs DF, Peterson MA et al. (1979) Pharmacology of anatoxin-a, produced by the freshwater cyanophyte Anabaena Jos-aquae NRC-44-1. Toxicon 17:229–236

Carmichael, WW (1992) Cyanobacteria secondary metabolites-cyanotoxins. J Appl Bacteriol 72:445–459

Chevalier P, Pilote R, Leclerc JM et al (2001) Risques à la santé publique découlant de la présence de cyanobactéries (algues bleues) toxiques et de microcystines dans trois bassins versants du sud-ouest québécois tributaires du fleuve Saint-Laurent. Unité de recherche en santé publique (Centre hospitalier de l'Université Laval) et Institut national de santé publique, Québec, p 151

Chorus I, Bartram J (1999) Toxic cyanobacteria in water: a guide to their public health consequences, monitoring and management. E&FN Spon, London, UK

Codd GA, Ward CJ, Bell SG (1997) Cyanobacterial toxins: occurrence, modes of action, health effects and exposure routes. Arch Toxicol 19:399–410

Cohen Y, Jorgensen BB, Revbech NP, Poplawski R et al(1986) Adaptation tohydrogen sulfide of oxygenic and anoxygenic photosynthesis among cyanobacteria. Appl Environ Microbiol 51:398–407

Cole GT, Wynne MJ (1974). Endocytosis of microcystis aeruginsoa by ochromonas danica. J Phycol 10:397–410

Cook WL, Ahearn DG, Reinhardt DJ, Reiber RJ et al (1974) Blooms of an algophorous amoeba associated with *Anabaena* in a freshwater lake. Water Air Soil Pollut 3:71–80

Cousins IT, Bealing DJ, James HA, Sutton A et al (1996) Biodegradation of microcystin-LR by indigenous mixed bacterial populations. Water Res 30:481–485

Couvreur P, Barratt G, Fattal E, Legrand P, Vauthier C et al. (2002) Nanocapsule technology: a review. Crit Rev Ther Drug Carrier Syst 19:99–134

Cox PA, Banack SA, Murch SJ, Rasmussen U,Tien G, Bidigare RR,Metcalf JS,Morrison LF,Codd GA,Bergman B (2005) Diverse taxa of cyanobacteria produce -N-methylamino-L-alanine, a neurotoxic amino acid. P Nat Scad Sci USA 102:5074–5078

Cremer J, Henning K (1991) Application of reversed-phase medium-pressure liquid chromatography to the isolation, separation and amino acid analysis of two closely related peptide toxins of the cyanobacterium *Microcystis aeruginosa* strain PCC 7806. J Chromatog 587:71–80

Daft MJ, Begg J, Stewart WD et al. (1970) A virus of bluegreen algae from fresh-water habitats in Scotland. New Phytol 69:1029–1038

Daft MJ, McCord SB, Stewart WD et al (1975) Ecological studies on algal lysing bacteria in fresh waters. Freshwat Biol 5:577–596

Daft MJ, Stewart WD (1971) Bacterial pathogens of freshwater blue-green algae. New Phytol 70:819–829

Dietrich DR, Hoeger SJ (2005) Guidance values for microcystin in water and cyanobacterial supplement products (blue-green algae supplements): a reasonable or misguided approach? Toxicol Appl Pharmacol 20:273–289

Dorigo U, Jacquet S, Humbert JF et al (2004) Cyanophage diversity inferred from g20 gene analysis in the largest natural French lakes, Lake Bourget, France. Appl Environ Microbiol 70:1017–1022

Dryden RC, Wright JC (1987) Predation of cyanobacteria by protozoa. Can J Microbiol 33:471–482

Duy TN, Lam PKS, Shaw GR, Connel DW et al (2000) Toxicologiy and risk assessment of freshwater cyanobacterial (blue-green algal) toxins in water. Rev Environ Contam Toxicol 163:113–186

Ernst B, Dietz L, Hoeger SJ, Dietrich DR et al (2005) Recovery of MC-LR in fish liver tissue. Environ Toxicol 20:449–458

Falconer IR (1989) Effects on human health of some toxic cyanobacteria (blue-green algae) in reservoirs, lakes, and rivers. Toxic Assess 4:175–184

Falconer IR (1991) Tumor promotion and liver injury caused by oral consumption of cyanobacteria. Environ Toxicol Water Qual 6:177–184

Fogg GE, Stewart WDP, Walsby AE et al (1973) The blue-green algae. Academic Press, London

Fraleigh PC, Burnham JC (1988) Myxococcal predation on cyanobacterial populations: nutrient effects. Limnol Oceanogr 33:476–483

Ghassempour A, Najafi NM, Mehdinia A, Davarani SSH, Fallahi M, Nakhshab M et al (2005) Analysis of anatoxin-a using polyaniline as a sorbent in solid-phase microextraction coupled to gas chromatography-mass spectrometry. J Chromatogr A 1078:120–127

Gijsbertsen-Abrahamse AJ, Schmidt W, Chorus I, Heijman SGJ et al (2006) Removal of cyanotoxins by ultrafiltration and nanofiltration. J Membr Sci 276:252–259

Grossman AR, Bhaya D, He Q et al (2001) Tracking the light environment by cyanobacteria and the dynamic nature of light harvesting. J Biol Chem 276:11449–11452

Gugger M, Lenoir S, Berger C, Ledreux A, Druart J, Humbert JF et al (2005) First report in a river in France of the benthic cyanobacterium *Phormidium favosum* producing anatoxin-a associated with dog neurotoxicosis. Toxicon 45:919–928

Gumbo R. J, Ross G, Cloete ET et al (2008) Biological control of Microcystis dominated harmful algal blooms. African J Biotech 25:4765–4773

Gupta N, Pant SC, Vijayaraghavan R, Lakshmana Rao PV et al (2003) Comparative toxicity evaluation of cyanobacterial cyclic peptide toxin microcystin variants (LR, RR, YR) in mice. Toxicology 188:285–296

Haider S, Naithani V, Viswanathan PN, Kakkar P et al (2003) Cyanobacterial toxins: a growing environmental concern. Chemosphere 52:1–21

Harada KI, Imanishi S, Kato H, Mizuno M, Ito E, Tsuji K et al (2004) Isolation of Adda from microcystin-LR by microbial degradation. Toxicon 44:107–109

Harada KI, Matsuura K, Suzuki M, Watanabe MF, Oishi S, Dahlem AM, Beasley VR, Carmichael WW et al (1990b) Isolation and characterization of the minor components associated with microcystins-LR and -RR in the cyanobacterium (bluegreen algae). Toxicon 28:55–64

Harada KI, Ogawa K, Matsuura K, Murata H, Suzuki M, Watanabe MF, Itezono Y, Nakayama N et al (1990a) Structural determination of geometrical isomers of microcystins-LR and -RR from cyanobacteria by two-dimensional NMR spectroscopic techniques. Chem Res Toxicol 3:473–481

Harada KI, Ogawa K, Matsuura K, Nagai H, Murata H, Suzuki M, Itezono Y, Nakayma N, Shirai M, Nakano M et al (1991b) Isolation of two toxic heptapeptide microcystins from an axenic strain of Microcystis aeruginosa, K-139. Toxicon 29:479–489

Hemming J, Holombim B, Reunannen M, Kronberg L et al (1986) Determination of the strong mutagen 3-chloro-4-(dichloromethyl)- 5-hydroxy-2(5H)-furanone in the chlorinated drinking and humic waters. Chemosphere 15:549–556

Hess WR, Partensky F, van der Staay GW, Garcia- Fernandez JM, Borner T, Vaulot D et al (1996) Proc Nat Acad Sci USA 93:11126–11130

Ho L, Gaudieux AL, Fanok S, Newcombe G, Humpage AR et al (2007) Bacterial degradation of microcystin toxins in drinking water eliminates their toxicity. Toxicon 50:438–441

Holt JG, Krieg NR, Sneath PHA, Staley JT, Williams ST (1994) Bergey's manual of determinative bacteriology, 9th ed. The Williams and Wilkins Co, Baltimore

Hu JY, Aizawa T, Ookubo S et al (2002) Products of aqueous chlorination of bisphenol A and their estrogenic activity. Environ Sci Technol 36:1980–1987

Imanishi S, Kato H, Mizuno M, Tsuji K, Harada KI et al (2005) Bacterial degradation of microcystins and nodularin. Chem Res Toxicol 18:591–598

Institut National de Santé Publique du Québec. Groupe Scientifique sur l'eau. Fiche cyanobactéries et cyanotoxines, Juin 2004

Ishii H, Nishijima M, Abe T et al (2004) Characterization of degradation process of cyanobacterial hepatotoxins by a Gram-negative aerobic bacterium. Water Res 38:2667–2676

8 Trends in Biological Degradation of Cyanobacteria and Toxins

Ishikawa K, et al (2002) Transport and accumulation of bloom-forming cyanobacteria in a large, mid-latitude lake: the gyre-Microcystis hypothesis. Limnology 3:87–96

Jochimsen EM, Carmichael WW, An JS, Cardo DM, Cookson ST, Holmes CEM, Antunes MB, de Melo-Filho DA, Lyra TM, Barreto VST, Azevedo SMFO, Jarvis WR et al (1998) Liver failure and death after exposure to microcystins at a hemodialysis center in Brazil. New Engl J Med 338:873–878

Jones GJ, Bourne DG, Blakeley RL, Doelle H et al (1994) Degradation of the cyanobacterial hepatotoxin microcystin by aquatic bacteria. Nat Toxins 2:228–235

Jones GJ, Orr PT (1994) Release and degradation of microcystin following algicide treatment of a *Microcystis aeruginosa* bloom in a recreational lake, as determined by HPLC and protein phosphatase inhibition assay. Water Res 8:871–876

Kato H, Imanishia SY, Tsuji K, Harada K et al (2007) Microbial degradation of cyanobacterial cyclic peptides. Water Res 41:1754–1762

Kaya K, Watanabe MM (1990) Microcystin composition of an axenic clonal strain of *Microcystis viridis* and *Microcystis viridis*–containing waterblooms in Japanese Freshwaters. J App Phycol 2:173–178

Klitzke S, Apelt S, Weiler C, Fastner J, Chorus I et al (2010) Retention and degradation of the cyanobacterial toxin cylindrospermopsin in sediments—the role of sediment preconditioning and DOM composition. Toxicon 55:999–1007

Krishnamurthy T, Szafraniec L, Hunt DF, Shabanowitz J, Yates JR, Hauer CR, Carmichael WW, Skulberg O, Codd GA, Issler S et al (1989) Structural characterization of toxic cyclic peptides from blue-green algae by tandem mass spectrometry. Proc Nat Acad Sci USA 86:770–774

Krishnamyrthy T, Szafraniec L, Hunt DF, Shabanowitz J, Yates JR, Hauer CR, Carmichael WW, Skulberg O, Codd GA, Missler S et al (1989) Structural characterization of toxic cyclic peptides from blue-green algae by tandem mass spectrometry. Proc Nat Acad Sci USA 86:770–774

Kusumi T, Ooi T, Watanabe MM, Takahashi H, Kakisawa H et al (1987) Cyanoviridin RR, a toxin from the cyanobacterium (blue-green alga) Microcystis viridis. Tetrahedron Let 28:4695–4698

Lavoie I, Laurion I, Vincent W et al (2007) Les fleurs d'eau de cyanobactéries, document d'information vulgarisée. Québec, INRS Eau, Terre et Environnement (INRS Eau, Terre et Environnement, Rapport de recherche; 917), 25 p. [http://www.ete.inrs.ca/pub/R919_2007.pdf]

Laybourn-Parry J, Jones K, Holdich JP et al (1987) Grazing by *Mayorella* sp. (protozoa: Sarcodina) on cyanobacteria. Funct Ecol 1:99–104

Leão PN, Pereira AR, Liuc WT, Ng J, Pevzner PA, Dorrestein PC, Königf GM, Vasconcelosa VM, Gerwick WH et al (2010) Synergistic allelochemicals from a freshwater cyanobacterium. Proc Natl Acad Sci USA 107:11183–11188

Lemes GAF, Kersanach R, Pinto LS, Dellagostin OA, Yunes JS, Matthiensen A et al (2008) Biodegradation of microcystins by aquatic *Burkholderia sp.* from a South Brazilian coastal lagoon. Ecotoxicol Environ Saf 69:358–365

Li C, Li XZ, Graham N, Gao NY (2008) The aqueous degradation of bisphenol A and steroid estrogens by ferrate. Water Res 42:109–120

Li D, Hayes PK (2008) Evidence for cyanophages active against bloom-forming freshwater cyanobacteria. Freshwater Biol 53:1240–1252

Luukkainen R, Sivonen K, Namikoshi M, Färdig M, Rinehart KL, Niemelä SI et al (1993) Isolation and identification of eight microcystins from 13 *Oscillatoria agardhii* strains: structure of a new microcystin. Appl Environ Microbiol 59:2204–2209

MacKintosh C, Beattie KA, Klumpps S, Cohen P, Codd G et al (1990) Cyanobacteria microcystin-LR is a potent and specific inhibitor of protein phosphatases 1 and 2A from both mammals and higher plants. FEBS Lett 264:187–192

Mazur-Marzec H, Meriluoto J, Plinski M et al (2006) The degradation of the cyanobacterial hepatotoxin nodularin (NOD) by UV radiation. Chemosphere 65:1388–1395

Meriluoto JAO, Sandström A, Eriksson JE, Remaud G, Grey A, Craig J et al (1989) Chattopadhyaya, Structure and toxicity of a peptide hepatotoxin from the cyanobacterium *Oscillatoria agardhii*. Toxicon 27:1021–1034

Michael Hogan C (2008) Makgadikgadi, The Megalithic Portal, ed. A. Burnham

Namikoshi M, Murakami T, Watanabe MF, Oda T, Yamada J, Tsujimura S et al (2003) Simultaneous production of homoanatoxin-a, anatoxin-a, and a new nontoxic 4-hydroxyhomoanatoxin-a by the cyanobacterium *Raphidiopsis mediterranea* Skuja. Toxicon 42:533–538

Ohki K, Fujita Y (1996) Occurrence of a temperate cyanophage lysogenising the marine cyanophyte *Phormidium persicum*. J Phycol 32:365–370

Oliver RL, Ganf GG (2000) Freshwater blooms. In: Whitton BA, Potts M (eds) The ecology of cyanobacteria. Their diversity in time and space. Kluwer Academic, Dordrecht, pp 149–194

Orr PT, Jones GJ, Hamilton GR et al (2004) Removal of saxitoxins from drinking water by granular activated carbon, ozone and hydrogen peroxide-implications for compliance with the Australian drinking water guidelines. Water Res 38:4455–4461

Padan E, Shilo M (1973) Cyanophages-viruses attacking bluegreen algae. Bacteriol Rev 37:343–370

Padan E, Shilo M, Kislev N et al (1967) Isolation of 'Cyanophages' from freshwater ponds and their interaction with *Plectonema boryanum*. Virology 32:234–246

Painuly P, Perez R, Fukai T, Shimizu Y et al (1988) The structure of a cyclic peptide toxin, cyanogenosin-RR from *Microcystis aeruginosa*. Tetrahedron Lett 29:11–14

Paldaviciene A, Mazur-Marzec H, Razinkovas H et al (2009) Toxic cyanobacteria blooms in the Lithuanian part of the Curonian Lagoon. Oceanologia 51:203–216

Park HD, Sasaki Y, Maruyama T, Yanagisawa E, Hirashi A, Kato K et al (2001) Degradation of the cyanobacterial hepatotoxin microcystin by a new bacterium isolated from a hyper-trophic lake. Environ Toxicol 16:337–343

Park HD, Watanabe MF, Harda K, Nagai H, Suzuki M, Watanabe M et al (1993) Hepatotoxin (microcystin) and neurotoxin (anatoxin-a) contained in natural blooms and strains of cyanobacteria from Japanese freshwaters. Nat Toxins 1:353–360

Pereira GC, Granato A, Figueiredo AR, Ebecken NFF et al (2009) Virioplankton abundance in trophic gradients of an upwelling field. Braz J Microbiol 40:857–865

Pitois S, Jackson MH, Wood BJB et al (2000) Problems associated with the presence of cyanobacteria in recreational and drinking waters. Internat J Environ Health Res 10:203–218

Redhead K, Wright SJ (1978) Isolation and properties of fungi that lyse blue-green algae. Appl Environ Microbiol 35:962–969

Redhead K, Wright SJ (1980) Lysis of the cyanobacterium Anabaena flos-aquae by antibiotic producing fungi. J Microbiol 119:95–101

Reim RL, Shane MS, Cannon RE et al (1974) The characterisation of a *Bacillus* capable of bluegreen bactericidal activity. Can J Microbiol 20:981–986

Reynolds CS, Walsby AE (1975) Water-blooms. Biol Rev 50:437–481

Rinehart KL, Harada KI, Namikoshi M, Chen C, Harvis CA, Munro MHG, Blunt JW, Mulligan PE, Beasley VR, Dahlem AM, Carmichael WW et al (1988) Nodularin, microcystin, and the configuration of Adda. J Am Chem Soc 110:8557–8558

Robarts RD, Zohary T (1987) Temperature effects on photosynthetic capacity, respiration, and growth rates of bloom-forming cyanobacteria. N Z J Mar Freshw Res 21:391–399

Rodriguez E, Majado ME, Meriluoto J, Acero JL et al (2007) Oxidation of microcystins by permanganate:Reaction kinetics and implications for water treatment. Water Res 41:102–110

Rook JJ (1974) Formation of haloforms during chlorination of natural waters. J Water Treat Exam 23:234–243

Safferman RS, Morris M (1962) Evaluation of natural products for algicidal properties. Appl Microbiol 10:289–292

Safferman RS, Morris M (1963) The antagonistic effects of actinomycetes on algae found in waste stabilisation ponds. Bacteriol Proc 14:56–56

Saitou T, Sugiura N, Itayama T, Inamori Y, Matsumura M (2002) Degradation of microcystin by biofilm in practical treatment facility. Water Sci Technol 46:237–244

Saitou T, Sugiura N, Itayama T, Inamoti Y, Matsumura M et al (2003) Degradation characteristics of microcystins by isolated bacteria from Lake Kasumigaura. J Water Supply Res T 52:13–18

Sallal AK (1994) Lysis of cyanobacteria with *Flexibacter* spp isolated from domestic sewage. Microbios 77:57–67

Santé Canada (1998) Les toxines-cyanobactériennes: Les microcystines dans l'eau potable, [Ottawa]: Santé Canada p.32. [http://www.ffck.org/renseigner/savoir/sante/pdf/microcysf.pdf]

8 Trends in Biological Degradation of Cyanobacteria and Toxins

Santé Canada (2002) Les toxines cyanobactériennes—Les microcystines-LR, [Ottawa]: Santé Canada, p 24. [http://tinyurl.com/3u7llj]

Shephard GS, Stockenström S, Villiers DD, Engelbrecht WJ, Wessels GFS et al (2002) Degradation of microcystin toxins in a falling film photocatalytic reactor with immobilized titanium dioxide catalyst. Water Res 36:140–146

Shi M, Zou L, Liu X, Gao Y, Zhang Z, Wu W, Wen D, Chen Z, An C (2006) A novel bacterium *Saprospira* sp. strain PdY3 forms bundles and lyses cyanobacteria. Frontiers in Bioscience 11:1916–1923

Shilo M (1970) Lysis of blue-green algae by *Myxobacter*. J Bacteriol 104:453–461

Sigee DC, Glenn R, Andrews MJ, Bellinger EG, Butler RD, Epton HAS, Hendry RD et al (1999) Biological control of cyanobacteria: principles and possibilities. Hydrobiologia 395/396:161–172

Sivonen K, Jones G (1999) Cyanobacterial toxins. In: Chorus I, Bartram J (eds) Toxic cyanobacteria in water: a guide to their public health consequences, monitoring, and management. E & FN Spon, NewYork, pp 41–111

Sivonen K, Namikoshi M, Evans WR, Carmichael WW, Sun F, Rouhiainen L, Luukkainen R, Rinehart KL (1992a) Isolation and characterization of a variety of microcystins from seven strains of the cyanobacterial genus *Anabaena*. App Environ Microbiol 58:2495–2500

Skulberg OM, Carmichael WW, Codd GA, Skulberg R et al (1993) Taxonomy of toxic cyanophyceae (cyanobacteria). In: Falconer IR (ed) Algal toxins in seafood and drinking water. Academic Press, San Diego, CA, pp 145–164

Smayda TJ (1997) What is a bloom? A commentary. Limnol Oceanogr 42:1132–1136

Smith JL, Boyer GL, Zimba PV (2008) A review of cyanobacterial odorous and bioactive metabolites: Impacts and management alternatives in aquaculture. Aquacult 280:5–20

Soppimath KS, Aminabhavi TM, Kulkarni AR, Rudzinski WE et al (2001) Biodegradable polymeric nanoparticles as drug delivery devices. J Contr Release 70:1–20

Stackebrandt E, Rainey FA, Ward-Rainey NL et al (1997) Proposal for a new hierarchic classification system, *Actinobacteria classis* nov. Int J Syst Bacteriol 47:479–491

Svrcek C, Smith DW (2004) Cyanobacteria toxins and the current state of knowledge on water treatment options: a review. J Environ Eng Sci 3:155–185

Tsuji RK, Asakawa M, Anzai Y, Sumino T, Harada K (2006). Degradation of microcystins using immobilized microorganism isolated in an eutrophic lake. Chemosphere 65:117–124

Ueno Y, Nagata S, Tsutsumi T, Hasegawa A, Watanabe MF, Park HD, Chen GC, Chen G, Yu SZ,(1996) Detection of microcystins, a blue–green algal hepatotoxin, in drinking water sampled in Haimen and Fusui, endemic areas of primary liver cancer in China by highly sensitive immunoassay. Carcinogen 17:1317–1321

Vauthier C, Dubernet C, Fattal E, Pinto-Alphandary H, Couvreur P (2003) Poly (alkylcyanoacrylates) as biodegradable materials for biomedical applications. Adv Drug Deliv Rev 55:519–548

Venter JC, Remington K, Heidelberg J, Halpern AL, Rusch D, Eisen JA (2004) Environmental genome shotgun sequencing of the Sargasso Sea. Science 304:66–74

Viaggiu E, Melchiorre S, Volpi F, Di A, Corcia, R. Mancini, L. Garibaldi et al (2004) Anatoxin a toxin in the cyanobacterium *Planktothrix rubescens* from a fishing pond in northern Italy. Environ Toxicol 19:191–197

Vincent WF, Quesada A (1993) Cyanobacterial responses to UV radiation: implications for antarctic microbial ecosystems. In: Weiler S, Penhale PA (eds) Ultraviolet radiation in antarctica: measurement and biological effects. American Geophysical Union, Washington, pp 111–124

Walsby AE, Mccallister GK et al (1987) Buoyancy regulation by Microcystis, Lake Okaro. N. Z J Mar Freshw Res 21:521–524

Watanabe MF, Oishi S, Harada KI, Matsuura K, Kawai H, Suzuki M et al (1988) Toxins contained in *Microcystis* species of cyanobacteria (blue-green algae). Toxicon 26:1017–1025

Watanabe MF, Tsuji K, Watanabe Y, Harada KI, Suzuki M,(1992) Release of heptapeptide toxin (microcystin) during the decomposition process of *Microcystis aeruginosa*. Nat Toxins 1:48–53

Whitton BA, Potts M (2000) Introduction to the cyanobacteria. In: Whitton BA, Potts M (eds) The ecology of cyanobacteria: their diversity in time and space.Kluwer Academic, Boston, pp 1–11

Wissing SA, Kayser O, Muller RH (2004) Solid lipid nanoparticles for parenteral drug delivery. Adv Drug Deliv Rev 56:1257–1272

World Health Organisation (WHO) (1999) Toxic cyanobacteria in water: a guide to their public health consequences, monitoring and management, 1ère edition

Wright SJ, Linton CJ, Edwards RA, Drury E (1991) Isoamyl alcohol (3-methyl-1-butanol), a volatile anticyanobacterial and phytotoxic product of some *Bacillus* spp. Lett Appl Microbiol 13:130–132

Wright SJ, Thompson RJ (1985) *Bacillus* volatiles antagonise cyanobacteria. FEMS Microbiol Lett 30:263–267

Yamamoto Y (1981) Observations on the occurrence of microbial agents which cause lysis of blue-green algae in Lake Kasumigaura. Jpn J Limnol 42:20–27

Yu SZ (1989) Drinking water and primary liver cancer. In: Tang ZY, Wu MC, Xia SS (eds) Primary liver cancer. Springer and China Acad. Publ, Berlin, Beijing, China, pp 30–37

Yu SZ (1995) Primary prevention of hepatocellular carcinoma. J Gastroenterol Hepatol 10:674–682

Chapter 9
Bioremediation of Pesticides from Soil and Wastewater

Reshma Anjum, Mashihur Rahman, Farhana Masood and Abdul Malik

Contents

9.1	Introduction	296
9.2	Pesticides and the Environment	299
	9.2.1 Pesticide Fate	299
	9.2.2 Pesticides in Groundwater	300
9.3	Beneficial Effects of Pesticides	301
	9.3.1 Improving Productivity	301
	9.3.2 Protect Crop Losses/Yield Reduction	302
	9.3.3 Improving Quality of Food	303
	9.3.4 Controlling Disease Vectors	303
	9.3.5 Other Benefits	304
9.4	Hazards of Pesticides	304
	9.4.1 Impact on Human Beings	304
	9.4.2 Impact on Aquatic Ecosystem	304
	9.4.3 Impact on Soil Microorganisms and Plants	305
9.5	Factors Influencing Pesticides Degradation in Soil and Wastewater	306
	9.5.1 Pesticide Structure	306
	9.5.2 Pesticide Concentration	306
	9.5.3 Soil Types	307
	9.5.4 Soil Moisture	308
	9.5.5 Temperature	308
	9.5.6 Soil pH	309
	9.5.7 Soil Salinity	309
	9.5.8 Soil Organic Matter	309
	9.5.9 Soil Biotic Components	310
9.6	Bioremediation	310
	9.6.1 Definition and History	312
	9.6.2 Bioremediation Types	312
	9.6.3 Bioremediation Techniques: Development and Application	313
9.7	Bioremediation of Pesticides	313
	9.7.1 Microbial Degradation	313

A. Malik (✉)
Department of Agricultural Microbiology, Faculty of Agricultural Sciences, Aligarh Muslim
University, Aligarh 202002, India
e-mail: ab_malik30@yahoo.com

A. Malik, E. Grohmann (eds.), *Environmental Protection Strategies for Sustainable
Development,* Strategies for Sustainability,
DOI 10.1007/978-94-007-1591-2_9, © Springer Science+Business Media B.V. 2012

9.7.2	Chemical Degradation	314
9.7.3	Photodegradation	314
9.7.4	Phytoremediation	314
9.7.5	Fungal Bioremediation	315
9.7.6	Bioremediation of Chlorinated Pesticide Contamination from Soil	316
9.7.7	Bioremediation of Pesticides from Wastewater	317
9.7.8	Biological Treatment Bioreactor and Other Ex Situ Bioremediation Methods	317
9.8	Sustainability	318
9.9	Economic Aspect	320
9.10	Conclusion and Perspectives	321
	References	321

Abstract The rapid increase in demand and development of industrial chemicals, fertilizers, pesticides and pharmaceuticals to sustain and improve quality of life worldwide have resulted in the contamination and high prevalence of these chemicals in air, water and soils, posing a potential threat to the environment. Pesticides are a common hazard around the world, as these chemicals are leaching into soils, groundwater and surface water and creating health concerns in many communities. The persistence of pesticides makes their removal and detoxification a more urgent undertaking. The toxicity or the contamination of pesticides can be reduced by the bioremediation process which involves the use of microbes or plants. Bioremediation technologies have been successfully employed in the field and are gaining more and more importance with increased acceptance of eco-friendly remediation solutions. Owing to complex nature of pesticides, more versatile and robust techniques need to be developed which can produce the desired result in a very cost-effective manner.

Keywords Bioremediation • Pesticides • Soil • Wastewater • Environmental pollution

9.1 Introduction

The rapid increase in populations worldwide has resulted in the need for greater fuel demand and development of industrial chemicals, fertilizers, pesticides and pharmaceuticals to sustain and improve quality of life (Chakrabarty et al. 1988). Although many of these chemicals are utilized or destroyed, a high percentage is released into the air, water and soil, representing a potential environmental hazard (Alexander 1995; Anwar et al. 2009). Consequently, earth's natural resources are not only being depleted, but are also becoming polluted and unfit for human use. As a result, many of the activities that we used to take for granted are now being carefully examined for potential damage to the environment. The use of pesticides has benefitted modern society by improving the quantity and quality of the world's production while keeping the cost of that food supply reasonable. Unsurprisingly, pesticide use has become an integral part of modern agricultural systems. Because of continuous pest problems, their usage possibly cannot be discontinued in the near future. Extensive and improper use of these chemicals had already caused considerable environmental pollution and leads to greater health risk to plants, animals and human population which have been reviewed from time to time by several workers (Kumar et al. 1996; Murugesan et al. 2010). Unfortunately, it is not possible within a short time to replace all the industrial processes generating polluting waste streams with clean alternatives. Therefore, treatment both at source and after release, whether accidental or not, must be considered as alternatives in many cases (Betts 1991). Current legislation and recent waste management strategies have placed significant emphasis on waste minimization, recycling and remediation rather than disposal (Colleran 1997). The persistence of organo-xenobiotics in the environment is a matter of significant public, scientific and regulatory concern because of the potential toxicity, mutagenicity, carcinogenicity and ability to bioconcentrate up the trophic ladder. These concerns continue to drive the need for the development and application of remediation techniques (Colleran 1997).

Chemical pesticides have contributed greatly to the increase of yields in agriculture by controlling pests and diseases and also towards checking the insect-borne diseases (malaria, dengue, encephalitis, filariasis, etc.) in the human health sector (Abhilash and Singh 2009). Owing to their efficiency; these compounds were considered a boon to the fields of agriculture and medical entomology. Organochlorine insecticides are more toxic to insects and less toxic to non-target organisms, but these chemicals can damage a wide variety of beneficial as well as harmful organisms due to their persistence in the environment. Therefore, organochlorine insecticides have important ecological effects in addition to those usually intended. Among these, the interaction of pesticides with microorganisms is important, since microorganisms are involved in many basic ecological processes, such as biogeochemical cycles, decomposition processes, energy transfer through trophic levels, and numerous microbe-microbe, microbe-plant, and microbe-animal interactions.

Millions of tons of pesticides are applied annually; however, less than 5% of these products are estimated to reach the target organism, with the remainder being deposited on the soil and non-target organisms, as well as moving into the atmosphere and water (Pimental and Levitan 1986). Microbes and plants are among the most important biological agents that remove and degrade waste materials to enable their recycling in the environment. Soil microflora, mainly bacteria, fungi, algae and protozoa make a valuable contribution in making the soil fertile through their primary catabolic role in the degradation of plants and animal residues in the cycling of the organic, inorganic nutrients content of soil. Pesticides that disrupt the activities of the soil microorganisms could be expected to affect the nutritional quality of soils and would therefore have serious ecological consequences (Handa et al. 1999). Concern for pesticide contamination in the environment in the current context of pesticide use has assumed great importance (Zhu et al. 2004). The fate of the pesticides in the soil environment in respect of pest control efficacy; non-target organism exposure and offsite mobility has become a matter of environmental concern (Hafez and Thiemann 2003) potentially because of the adverse effects of pesticidal chemicals on soil microorganisms (Araújo et al. 2003) which may affect soil fertility (Schuster and Schröder 1990). An ideal pesticide should be toxic only to the target organism, be biodegradable and undesirable residues should not affect non-target surfaces. However, the use of pesticide has been minimized or terminated in technologically advanced countries due to their persistence in nature, susceptibility to biomagnifications, and toxicity to higher animals. But the ever-increasing world population and poor health conditions, especially in developing countries, may outweigh the disadvantages caused by the extensive use of these insecticides.

Bioremediation in its formal sense, meaning any use of living organisms to degrade wastes, has been practiced since humans first populated the world and had to dispose of their trash. Without knowing about the microorganisms in soil and water, people relied on them to destroy waste products from human domestic, agricultural, and industrial activities by converting them to carbon dioxide, water, and additional microbial biomass. Other than this bioremediation is a technology that utilizes the metabolic potential of microorganisms to clean up contaminated environments. One important characteristic of bioremediation is that it is carried out in non-sterile open environments that contain a variety of organisms. Of these, bacteria, such as those capable of degrading pollutants, usually have central roles in bioremediation, whereas other organisms (e.g. fungi and grazing protozoa) also affect the process. A deeper understanding of the microbial ecology of contaminated sites is therefore necessary to further improve bioremediation processes. In the past two decades, molecular tools, exemplified by rRNA approaches, have been introduced into microbial ecology; these tools have facilitated the analysis of natural microbial populations without cultivation. Molecular ecological information is thought to be useful for the development of strategies to improve bioremediation and for evaluating its consequences and also risk assessment. Molecular tools are especially useful in bioaugmentation, in which exogenous microorganisms that are introduced to accelerate pollutant biodegradation need to be monitored. With the advent of wastewater treatment plants in the late 19th century (Frankland and Frankland 1894) biotreatment became a more formalized, better-engineered process, although it still was not called bioremediation. Direct land treatment of residues from wastewater treatment plants, refinery sludges, and municipal wastes, as well as composting, has been practiced widely for several decades. Microbiologists have now realized that natural microbial populations are much more diverse than those expected from the catalogue of isolated microorganisms. This is also the case for pollutant-degrading microorganisms, implying that the natural environment harbours a wide range of unidentified pollutant-degrading microorganisms that have crucial roles in bioremediation.

Raymond et al. (1975) have reported that by adding nutrients to subsurface soil, they could increase the numbers of bacteria that degrade hydrocarbons derived from petroleum and thereby boost the rate of removal of the contaminants. This was the origin of the process that is now called accelerated or enhanced *in situ* bioremediation which can be applied to subsurface soils or aquifers as well as to reservoirs, lakes, and other bodies of water. In the present scenario any transformation or removal of contaminants from the environment by organisms is considered to be bioremediation. Several types can be distinguished. Natural bioremediation is the simplest in which nothing is added to surficial soils or to the subsurface, but monitoring is performed to ensure that the contaminants are disappearing as a result of microbial action and not because of dilution or migration of the contaminant. Another type of bioremediation uses aboveground treatment and is the *ex situ* process. Excavated soil is treated directly in constructed containers or in a controlled environment, that is, in bioreactors. Bioreactors have the advantage in that they can be used to

9 Bioremediation of Pesticides from Soil and Wastewater

treat solid, liquid, or gaseous contaminants. Researchers who had been working on microbial processes for the degradation of various other organic compounds saw wider potential for bioremediation, however. There was/is particular interest in its use for dealing with chlorinated hydrocarbons, because these compounds have been widely released into the environment, have potential carcinogenic effects, and may contribute to ozone depletion. Soon efforts were under way to use bioremediation, either aerobically or anaerobically, to destroy pentachlorophenol, pesticides, and gaseous products such as styrene. The introduction of anaerobic biodegradation changed the field dramatically, because anaerobic bacteria can dechlorinate compounds more readily than can aerobic bacteria

9.2 Pesticides and the Environment

Once a pesticide is introduced into the environment, whether through an application or an accident, it is influenced by many processes. These processes determine a pesticide's persistence, movement, if any, and its ultimate fate.

9.2.1 Pesticide Fate

It is important to understand the behaviour and fate processes of pesticides after they have been applied as it can help every pesticide applicator ensure that applications are not only effective, but are also environmentally safe. Some of the processes that change or influence the availability, effectiveness, structure or physical identity of chemicals used as pesticides are discussed. The fate processes of a pesticide can be beneficial. They can move a pesticide to the target area or destroy its potentially harmful residues. However, some processes can be detrimental. Runoff can move a pesticide away from target sites and pests. As a result, the chemical is wasted, control is reduced, and the chance of damage to non-target plants, hazard to human health, and pollution of nearby soil and water increase.

9.2.1.1 Adsorption

The adsorption process binds pesticides to soil particles. The amount and the persistence of pesticide adsorption vary with pesticide properties, soil moisture content, soil acidity and soil texture. Soils high in organic matter or clay are more adsorptive than coarse, sandy soils, because of the availability of more particle surface area or sites onto which pesticides can bind. A soil-adsorbed pesticide is less likely to volatilize, leach or to be degraded by microorganisms. When pesticides are tightly held by soil particles, they are less available for absorption by plants or animals and other processes to affect them.

9.2.1.2 Absorption

Absorption is the process by which plants, animals, humans or microorganisms take up chemicals. Absorption of pesticides by target and nontarget organisms is influenced by environmental conditions and by the physiochemical properties of the pesticide and the soil.

9.2.1.3 Volatilization

Volatilization is the transformation of a solid or liquid into a gas. Vapour drift which is the movement of the pesticide vapours in the atmosphere can cause the transfer of a pesticide in a gaseous state from a treated area by air currents. Furthermore, vapour drift is invisible, unlike the drift of sprays and dusts that can sometimes be seen during an application.

9.2.1.4 Runoff

Runoff is the water movement over a sloping surface. The process carries pesticides either mixed in the water or bound to eroding soil. Factors like slope quality of the area; erodibility, texture and moisture content of the soil; and the amount and timing of rainfall and irrigation governs the severity of pesticide runoff. Certain physical and chemical properties of the pesticide like absorption by plant tissues or adsorption to soil also influence the process.

9.2.1.5 Leaching

Leaching is another process that moves pesticides in water. Contrary to runoff, which occurs as water moves on the surface of the soil, leaching occurs as water moves downward through the soil. Several factors influence the leaching of pesticides. These include the water solubility of the pesticide. A pesticide that dissolves in water can move readily with the water as it seeps through the soil. Soil structure and texture influence soil permeability as well as the amount and persistence of pesticide adsorption to soil particles. Adsorption is probably the most important factor influencing leaching of pesticides. If a pesticide is strongly adsorbed to soil particles, it is less likely to leach.

9.2.2 Pesticides in Groundwater

The pesticide fate and the numerous transfer and breakdown processes that occur in the environment determine whether pesticides reach groundwater or are degraded before reaching these underground waters. Geological characteristics, such as the depth of the water table and the presence of sinkholes, are also critical. If the water table is close to the soil surface, there may be few opportunities for adsorption and degradation to occur. On the soil surface and within the first few inches of soil, pesticides can be volatilized, adsorbed to soil particles, taken up by plants or broken down by sunlight, soil microorganisms and chemical reactions. The extent of pesticide leaching is affected by both pesticide and soil properties. Weather conditions and management practices also affect leaching of pesticides through the soil. Too much rain or irrigation water can leach pesticides or cause runoff beyond the treatment area. A pesticide that is not volatilized, absorbed by plants, bound to soil or broken down can potentially move through the soil to groundwater.

9.3 Beneficial Effects of Pesticides

9.3.1 *Improving Productivity*

Pesticides are a mixture of substances applied for the preventing, destroying, repelling or mitigating any pest and have been classified as shown in Fig. 9.1. Pesticides have had a key role in improving productivity to such an extent that India, a former country of famine has quadrupled grain production since 1951 (Jha and Chand 1999) and now not only feeds itself but exports produce (100 m tonnes in 2003—Indian export stats). Similarly outputs and productivity have increased dramatically in most countries, for example, wheat yields in the United Kingdom rose from

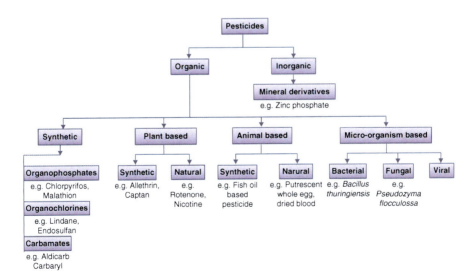

Fig. 9.1 Classification of pesticides

2.5 t/ha in 1948 to 7.5 t/ha in 1997 (Austin 1998). Corn yields in the USA went from 30 bushels per acre to over a hundred per acre over the period from 1920 to 1980 (Purdue 2006). Increases in productivity have been due to several factors including use of fertiliser, better varieties and use of machinery. However pesticides have been an integral part of the process by reducing losses from the weeds, diseases and pests that can markedly reduce the amount of harvestable produce.

Webster et al. (1999) stated that "considerable economic losses" would be suffered without pesticide use and quantified the yield increases and 50% increase in gross margin that result from pesticide use in British wheat production. Schmitz (2003) explored the potential effect on European crop productivity of possible EU legislation to reduce pesticide usage. He concluded that the evidence for the benefits of a reduction strategy were unconvincing and an imposed cut of 75% would reduce productivity to catastrophically low levels, that is, by between one third and two thirds of the original production levels.

These increased yields bring important secondary effects of increased revenue and rural nutrition, along with a diminished pressure to cultivate uncropped land and wild areas that provide habitats for endangered species (WWF undated) and attract national or international tourists.

9.3.2 Protect Crop Losses/Yield Reduction

Pesticides reduce yield losses that are caused by insect pests, diseases and weeds. Insecticides are used to control insect pests that feed on crops or carry plant diseases and thus prevent huge losses. Fungi are one of several causative agents of diseases of crops that can reduce yield and quality of food crops (Agrios 2002) and they sometimes produce toxic compounds making the produce unfit to eat. Fungicides now routinely protect many crops throughout the world from various diseases like late blight, powdery and downy mildew, leaf spot, scab, canker, rust, Botrytis grey mould. Weed competition is the major constraint that limits yield in many crops. Herbicides, i.e., chemical substances used to kill or control unwanted vegetation, are the most widely used type of pesticide and comprise around 50% of all crop protection chemicals used throughout the world, compared with insecticides and fungicides that are around 17% each (CropLife 2004). Timely weed control was found to be essential for better crop establishment, tillering, vigorous growth, higher fertilizer efficiency and increased yield. There would be estimated US $13.3 billion loss in farm income in the US without timely application of herbicides (Anon 2003). Yancy and Cecil (2005) put the figure higher for benefits of herbicide use at $21 billion annually, against a cost of $6.6 billion for the product and application, thereby reducing losses to weeds by 23% and reducing loss of farm income valued at $8 billion. Chemical weed control reduced soil erosion by 400% (40 tonnes/ha) and provided an average production increase of 16% due to increased soil moisture and reduced root damage compared with tilled systems (Pastor and Castro 1995).

Oerke et al. examined the role of pest management in meeting rapidly increasing global food needs and concluded that "pesticide-free [food] production would be a disaster".

9.3.3 Improving Quality of Food

In the developed countries, it is now observed that a diet containing fresh fruit and vegetables far outweighs potential risks from eating very low residues of pesticides in crops (Brown 2004). Increasing evidence (Dietary Guidelines 2005) shows that eating fruit and vegetables regularly reduces the risk of many cancers, high blood pressure, heart disease, diabetes, stroke, and other chronic diseases.

Gianessi (1999) largely attributed all year round availability of inexpensive and good quality fresh fruit and vegetables to the use of pesticides. Lewis et al. (2004, 2005) discussed the nutritional properties of apples and blueberries in the US diet and concluded that their high concentrations of antioxidants act as protectants against cancer, heart disease and other chronic diseases associated with oxidative stress and ageing. Lewis et al. (2005) attributed doubling in wild blueberry production and subsequent increases in consumption chiefly to herbicide use that improved weed control.

9.3.4 Controlling Disease Vectors

In public health, insecticides are used to control the insects that spread deadly diseases such as malaria (Delacollette 2004) that results in an estimated 5,000 deaths each day (Ross 2005). Gratz (1994) and Maroli (2004) had reached the same conclusions for control of dengue, yellow fever, Japanese encephalitis, malaria, leishmaniasis, and filariasis i.e., that control of the vectors is the most effective approach and insecticides are the basis for nearly all control campaigns. Bhatia et al. (2004) wrote that malaria is one of the leading causes of morbidity and mortality in the developing world and a major public health problem in India. The use of insecticide-treated nets has proved to be effective in reducing malaria mortality and morbidity in various epidemiological settings.

Disease control strategies are crucially important for livestock, too. Schukken et al. (2004) agreed that both insecticide treated target and insecticide pour-on control programs were associated with lower trypanosomiasis infection incidence in cattle compared with previous time periods without tsetse control. Cattle treated with insecticide benefit because parasite attack is reduced. Pesticides are also used by poultry farmers to control poultry red mite (Anon 2005), as well as many other ectoparasites.

9.3.5 Other Benefits

The transport sector makes extensive use of pesticides, particularly herbicides. Herbicides and insecticides are used to maintain the turf on sports pitches, cricket grounds and golf courses. Insecticides also protect buildings and other wooden structures from damage by termites and wood boring insects, thereby decreasing maintenance costs and increasing longevity of buildings and their safety.

9.4 Hazards of Pesticides

9.4.1 Impact on Human Beings

Pesticides have resulted in serious health implications to man and his environment. The high risk groups exposed to pesticides include the production workers, formulators, sprayers, mixers, loaders and agricultural farm workers. During manufacture and formulation, the possibility of health hazards may be higher because the processes involved are not risk free. In industrial settings, the workers are at increased risk since they handle various toxic chemicals including pesticides, raw materials, toxic solvents and inert carrier.

Studies suggest that pesticides may be related to various diseases, including cancers, as well as having neurological, mental and reproductive effects. Exposure to the pesticides can also lead to immune system disorders in humans. Children may be more susceptible to the effects of pesticides due to increased exposure via food and breast milk, underdeveloped detoxification pathways, and longer life expectancy in which to develop diseases with long latency periods (Cohen 2007).

Contamination of foods and crops with pesticides and consumption of contaminated foods and crops causes serious diseases in humans and animals.

9.4.2 Impact on Aquatic Ecosystem

Direct or indirect exposure of pesticide on the organisms in aquatic ecosystems may be acute and chronic. Acute effects, for example death of organisms, are often easily detected in toxicity tests during assessment of the pesticide. On the other hand, chronic effects, such as decreased reproduction success, disturbances in behaviour and change in community structure are far more difficult to detect and are often promoted by long-term exposure of pesticide at low concentrations. Therefore, chronic and acute effects may alter the niche which may have serious implications for the ecosystem. Mortality and disappearance of zooplankton following pesticide exposure is an example of direct effect of pesticides on aquatic organisms. Hence, due to a decreased grazing pressure, it leads to an indirect positive effect on phytoplankton, which may indirectly decrease the growth of macrophytes that experience increased shading (Wendt-Rasch et al. 2003). They also concluded that the effects of pesticides exposure on aquatic ecosystems are often diffuse and confusingly similar to eutrophication effects and that these two types of stressors may interact. Thus, the complex biological interactions that exist in an aquatic ecosystem make it very difficult to predict the ecological consequences that could be caused by pesticide exposure.

Most of the pesticides used are organic molecules with hydrophobic properties resulting in rapid sorption of the pesticide molecules to the soil particles that can be washed into the water. After the pesticides reach the aquatic ecosystem in dissolved form, they associate with organic matter in suspension or sediment. Sorbed pesticide molecules tend to be less degradable than their dissolved forms; since they are less accessible to the degrading action of microorganisms, UV-light and dissolved oxidative chemicals (Schwarzenbach et al. 1993; Ying and Williams 2000). Therefore, extensive use of pesticides causes accumulation of these compounds in sediments of freshwater and estuaries following their sorption and sedimentation.

9.4.3 Impact on Soil Microorganisms and Plants

In modern agriculture practices, application of insecticides which belong to diverse chemical groups (Table 9.1) has become a common practice to fight against insect pests for the treatment of soil and seed. The application of such insecticides causes their accumulation in soils and affects directly or indirectly the soil enzyme activities and physiological activities of non-target soil microbiota leading thereby to losses in fertility of soils. For the fertility and plant growth, soil microorganisms

Table 9.1 Examples of insecticides and their mode of action

Mode of action groups	Chemistry	Examples
Acetylcholinesterase inhibitors	Carbamates	aldicarb, carbaryl, carbofuran, propoxur, carbosulfan
	Organophosphates	phorate, chlorpyrifos, omethoate, parathion, methmidophos, malathion, diazinon
GABA-gated chloride channel antagonists	Cyclodienes and other organochlorines (OC)	lindane, aldrin, endosulfan
	Phenylpyrazoles (fiproles)	fipronil
Sodium channel modulators	OC	DDT
Acetylcholine receptor agonists	Neonicotinoids	imidacloprid, thiamethoxam
Acetylcholine receptor agonists allosteric	Spinosyns	spinosad
Voltage dependent sodium channel blocker	Oxadiazine	indoxacarb

Adapted from http://www.irac-online.org/

are an important and diverse community that catalyses many processes. There are many processes that are important for cycling of the nutrients from the soil and fertilizers and transfer of nutrients directly to the crops by microbes Therefore, these microbial soil communities are greatly influenced by various factors including the agrochemicals. In modern agricultural trends, agronomists recommend pesticides in order to augment the productivity of various crops. According to Zahran (1999) and Srinivas et al. (2008) continuous and abundant use of synthetic pesticides has become a major threat to beneficial soil microbes and in turn affects the sustainability of agricultural crops. It is globally, a greater concern that, how to minimize or reduce the effects of pesticides so that the consequential impact of these chemicals on the microorganisms involved in nutrient cycling, vis-a-vis the productivity of crops could be saved. Rajagopal et al. (1984) have reported both innocuous and inhibitory effects of certain pesticides on soil bacteria depending on the concentration used. For example the growth of *Azotobacter chroococcum* in nitrogen-containing culture medium was not affected at 0.5 and 5 ppm concentrations, but the growth of the same isolates was inhibited by the pesticides at higher concentrations.

9.5 Factors Influencing Pesticides Degradation in Soil and Wastewater

9.5.1 Pesticide Structure

The biodegradability of a pesticide is governed by its structure which determines its physical and chemical properties. Degradation is influenced by the substituents on the phenyl ring and the introduction of polar groups viz; OH, COOH, and NH_2 make the compound susceptible to microbial attack. Halogen or alkyl substituents tend to make the molecule more resistant to biodegradation (Cork and Krueger 1991). Chlorinated hydrocarbons such as DDT, pentalene and dieldrin are insoluble in water, sorb tightly to soil and are thus relatively unavailable for biodegradation whereas the insecticide carbofuran and the herbicide 2, 4-D, which are of different molecular structure, are readily degraded in few days in field soils. Some common pesticides structures are shown in Fig. 9.2.

9.5.2 Pesticide Concentration

The rate of pesticide biodegradation is determined by its concentration. As the degradation kinetics of many pesticides approaches first order, the rate of degradation decreases roughly in proportion with the residual pesticide concentration (Topp et al. 1997). Gupta and Gajbhiye (2002) reported that the half life of flufenacet in three Indian soils, viz., inseptisol, vertisol and ultisol, varied from 10.1 to 31.0 days

9 Bioremediation of Pesticides from Soil and Wastewater

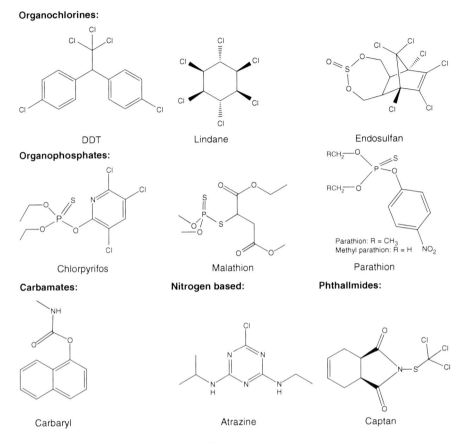

Fig. 9.2 Stuctures of some common pesticides

at low rate (1.0 μg/g soil) compared to 13.0–29.2 days at high rate (10.0 μg/g soil) of application. Prakash and Suseela Devi (2000) reported the reduced degradation rate of butachlor at higher initial concentrations, which could be attributed to limitation in the number of reaction sites in soils and toxic effect on microorganisms or enzyme inhibition.

9.5.3 Soil Types

Soil properties like organic matter, clay content, pH, etc. affect the degradation of pesticides in soil. Therefore, it is important to study the effect of soil types on pesticide degradation. Gold et al. (1996) reported that soil, pH and clay content greatly affect the persistence of bifenthrin, chlorpyrifos, cypermethrine, fenvelerate, permethrin and isofenphos under field conditions. Jones and Ananyeva (2001) reported that the degradation of metalaxyl and propachlor occurred at different rates in different soils. The half-lives in pasture, arable and pine forest soil were 10, 19 and 36 days respectively for metalaxyl and 2.6, 6.1 and 8.2 days for propachlor. The presence of organic matter and clay content might have posed synergistic effect in fluchloralin dissipation.

9.5.4 Soil Moisture

Soil moisture is essential for microbial functioning and hence pesticide degradation is slow in dry soils. Water acts as solvent for pesticide movement and diffusion and facilitates pesticide degradation. The rate of pesticide transformation generally increases with water content. Schroll et al. (2006) quantified the effect of soil moisture on the aerobic microbial mineralization of selected pesticides (isoproturon, benzolin-ethyl, and glyphosphate) in different soils. They found a linear correlation ($p < 0.0001$) between increasing soil moisture (within a soil water potential range of -20 and -0.015 MPa) and increased relative pesticide mineralization. Optimum pesticide mineralization was obtained at a soil water potential of -0.015 MPa. However the rate of diffusion of atmospheric oxygen is limited and anaerobic pesticide transformation can prevail over aerobic degradation in paddy soil. Phorate was found to be more persistent in flooded soil than in non flooded soil (Walter-Echols and Lichtenstein 1978). The herbicides atrazine and trifluraline disappeared more rapidly under anaerobic conditions than under aerobic conditions. DDT is fairly stable in aerobic soils, but is degraded rapidly to DDD in submerged soils (Topp et al. 1997). Thus, the transformation of pesticides in the submerged soils is different from that of the soils in field moist state. Phillips et al. (2006) reviewed the degradation of HCH in soil and reported many research findings that showed a degradation of HCH isomers under anoxic environment.

9.5.5 Temperature

The effect of temperature on pesticide degradation depends on the molecular structure of the pesticide. Temperature affects adsorption by altering the solubility and hydrolysis of pesticides in soil (Burns 1975; Racke et al. 1997). As adsorption processes are exothermic and desorption processes are endothermic, it is expected that adsorption will reduce with increase in temperature with a corresponding increase in pesticide solubility. Microbial activity is stimulated by increase in temperature and some ecological groups tend to dominate within certain temperature ranges. The maximum growth and activity of microorganisms in soils occur at 25–35°C (Alexander 1977) and the pesticide degradation is optimal in the mesophilic temperature range of around 25–40°C (Topp et al. 1997). A temperature range of 15 to 40°C was considered favourable for the degradation of pesticides by isolated pesticide-degrading bacteria (Singh et al. 2006; Hong et al. 2007). The bacterial isolates were able to rapidly degrade fenamiphos and chlorpyrifos between 15 and 35°C, but their degradation ability was sharply reduced at 5 or 50°C (Singh et al. 2006). Similar results were reported by Siddique et al. (2002), who studied biodegradation of HCH isomers in a soil slurry. They observed that an incubation temperature of 30°C was optimum for effective degradation of α- and γ-HCH isomers.

9.5.6 Soil pH

Soil pH is one of the major factors affecting the biodegradation of pesticides in soil (Arshad et al. 2007). The biodegradation of a compound is dependent on specific enzymes secreted by microorganisms. These enzymes are largely pH-dependent and bacteria tend to have optimum pH between 6.5 and 7.5, which equals their intracellular pH (Subhas and Irvine 1998). Soil pH may influence pesticide adsorption, abiotic and biotic degradation processes (Burns 1975). It affects the sorptive behaviour of pesticide molecules on clay and organic surfaces and thus, the chemical speciation, mobility and bioavailability (Hicks et al. 1990). For instance, the sorption of prometryn to clay montmorillonite is higher/more pronounced at pH 3 than at pH 7 (Topp et al. 1997). The effect of soil pH on degradation of a given pesticide depends greatly on whether a compound is susceptible to alkaline or acid catalyzed hydrolysis (Racke et al. 1997). Singh et al. (2006) and Hong et al. (2007) reported slower degradation rate of organophospate pesticides in lower pH soils in comparison with neutral and alkaline soils.

9.5.7 Soil Salinity

Degradation of pesticides in saline soils is slow and salinity is a severe problem in many arid, semiarid and coastal regions. Parathion was degraded faster in non saline soil than in saline soils and its stability increased with increasing electrical conductivity (Reddy and Sethunathan 1985). However, reports on the stability of pesticides in estuarine and seawater of varying degrees of salinity are available A high salt content in seawater may be innocuous or inhibitory to degradation.

9.5.8 Soil Organic Matter

Soil organic matter also affects biodegradation of pesticides in soil by providing nutrients for cell growth and controlling pesticide movement by adsorption/desorption processes (Spark and Swift 2002; Briceno et al. 2007). Soil organic matter plays a mixed role in pesticide degradation. It can either decrease the microbially mediated pesticide degradation by stimulating pesticide adsorption processes or enhance microbial activity (Perucci et al. 2000) by co-metabolism (Walker 1975; Thom et al. 1997). The addition of organic materials to flooded soils enhanced the bacterial degradation of some organochlorine insecticides such as BHC, DDT, methoxychlor and heptachlor (Yoshida 1978). Microbial degradation of linuron in nonsterilized soils was stimulated by organic matter amendment (Hicks et al. 1990). A certain minimum level of organic matter (probably greater than 1.0%) is essential to ensure the presence of an active autochthonous (the indigenous flora and fauna of a region) microbial population that can degrade pesticides (Burns 1975). The application of wheat residue-derived char released nutrients and stimulated microbial growth and degradation of pesticide (benzonitrile) in soil (Zhang et al. 2005). Phosphorus was primarily responsible for stimulation and degradation.

9.5.9 Soil Biotic Components

The role of microbes in pesticide degradation is evident from the fact that pesticide degradation under nonsterilized conditions is faster than under sterilized condition. Many articles reported microbial degradation of pesticides in soil (Adhya et al. 1987; Banerjee et al. 1999; Karpouzas et al. 1999; Sukul and Spiteller 2001; Hafez and Thiemann 2003). Degradation of phorate (Bailey and Coffey 1985), metalaxyl (Bailey and Coffey 1985; Droby and Coffey 1991) and fipronil (Zhu et al. 2004) proceeded more rapidly in nonsterilized than in sterile soils. The breakdown of pesticides in soils is brought about by a variety of biotic mechanisms. The principal route involves the use of pesticides as carbon, energy and nitrogen sources. Microorganisms can also degrade pesticides co-metabolically (Burns and Edwards 1980).

9.6 Bioremediation

Bioremediation consists of using living organisms (usually bacteria, fungi, actinomycetes, cyanobacteria and to a lesser extent, plants) to reduce or eliminate toxic pollutants. These organisms may be naturally occurring or laboratory cultivated. These organisms either eat up the contaminants or assimilate within them all harmful compounds from the surrounding area, thereby, rendering the region virtually contaminant-free. Generally, the substances that are eaten up are organic compounds, while those, which are assimilated within the organism, are heavy metals and pesticides. Bioremediation harnesses this natural process by promoting the growth and/or rapid multiplication of these organisms that can effectively degrade specific contaminants and convert them to nontoxic by-products. Importantly, bioremediation can also be used in conjunction with a wide range of traditional physical and chemical technologies to enhance their efficacy. The data derived from EPA REACH IT (2004) show that out of twelve technologies bioremediation is the most employed one (Fig. 9.3).

9 Bioremediation of Pesticides from Soil and Wastewater

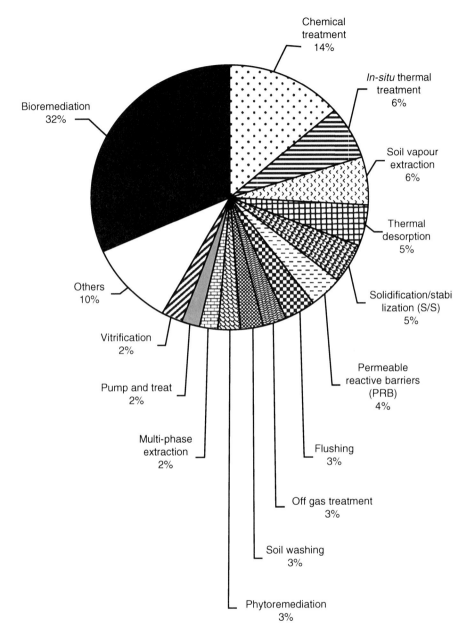

Fig. 9.3 Different technologies for decontamination of hazardous wastes (data derived from EPA REACH IT 2004)

9.6.1 Definition and History

The term "bioremediation" describes the process of contaminant degradation in the environment by biological methods using the metabolic potential of microorganisms to degrade a wide variety of organic compounds (Scragg 2005; Perelo 2010). For centuries, civilizations have used natural bioremediation in wastewater treatment, but intentional use for the reduction of hazardous wastes is a more recent development.

Modern bioremediation and the use of microbes to consume pollutants are credited, in part, to George Robinson (US Microbics 2003). He used microbes to consume an oil spill along the coast of Santa Barbara, California in the late 1960s. Since the 1980s, and bioremediation of oil spills and other environmental catastrophes, biological degradation of hazardous wastes has received more consideration (Shannon and Unterman 1993).

9.6.2 Bioremediation Types

9.6.2.1 Biostimulation

This refers to addition of specialized nutrients and suitable physiological conditions for the growth of the indigenous microbial populations. This promotes increased metabolic activity, which then degrades the contaminants (Trindade et al. 2005). Biostimulation has been used to remove pesticides contamination in the environment. This treatment stimulates the activity of the indigenous microorganisms by adding the organic and/or inorganic additives such as N or P, etc. The amendments added would be used by the indigenous microorganisms for cell growth resulting in an increase in cell number as well as their activities to degrade the pesticides. In addition, the amendments could be necessary as the enzyme-inducers and/or the co-metabolic substrates in the pesticide degradation pathways (Robles-Gonzalez IV et al. 2008; Plangklang and Reungsang 2010).

9.6.2.2 Bioaugmentation

Bioaugmentation is the introduction of exogenous microorganisms with specific catabolic abilities into the contaminated environment or into a bioreactor to initiate the bioremediation process. This can be an *in situ* or *ex situ* treatment process in which naturally occurring microbes are added to contaminated sites to eliminate toxic contaminants (Head and Oleszkiewicz 2004; Perelo 2010). The technique is reported to be an effective bioremediation approach for improving pesticide degradation in contaminated soils and water that lack indigenous microbial activity (Dams et al. 2007; Plangklang and Reungsang 2010). It has been employed to degrade a wide range of chemical contaminants such as ammonia, hydrogen sulphide, insecticides, petroleum compounds, and a growing number of toxic organic chemicals present in soil and water (Hwang and Cutright 2002; Jianlong et al. 2002; Quan et al. 2004).

Both biostimulation and bioaugmentation possess the ultimate goal of decontamination through biotransformation. The toxic substance is generally converted into other nontoxic forms (transformed) by a chemical reaction using microbes, aerobically or anaerobically. Anaerobic biotransformation is essentially used for degradation of organic compounds such as chlorinated hydrocarbons, polychlorinated phenols, and nitroaromatics which are recalcitrant to aerobic treatment processes (Ferguson and Pietari 2000; Marschner et al. 2001; Somsamak et al. 2001; Bhushan et al. 2003). Aerobic biotransformation finds use in degradation of thiocyanates, cyanates, aromatic hydrocarbons, gasoline monoaromatics, and methyl *tert*-butyl ether (Deeb and Cohen 2000; Shen et al. 2000).

9.6.3 Bioremediation Techniques: Development and Application

The process of developing bioremediation techniques may involve the following steps:

1. Isolating and characterizing naturally-occurring microorganisms with bioremediation potential.
2. Laboratory cultivation to develop viable populations.
3. Studying the catabolic activity of these microorganisms in contaminated material through bench monitoring and measuring the progress of bioremediation through chemical analysis and toxicity testing in chemically-contaminated media.

Field applications of bioremediation techniques using either/both steps:

1. *In-situ* stimulation of microbial activity by the addition of microorganisms and nutrients and the optimization of environmental factors at the contaminated site itself
2. *Ex-situ* restoration of contaminated material in specifically designated areas by land-farming and composting methods.

9.7 Bioremediation of Pesticides

9.7.1 Microbial Degradation

Microbial degradation occurs when microorganisms such as fungi and bacteria use pesticides as food sources. It is estimated that 1 g of soil contains more than one hundred million bacteria (5,000–7,000 different species) and more than ten thousand fungal colonies (Dindal 1990; Melling 1993). The use of microbial metabolic potential for eliminating soil pollutants provides a safe and economic alternative to other commonly used physico-chemical strategies (Vidali 2001). Indigenous microorganisms (natural attenuation) can be used for detoxification of contaminants in the environment (Siddique et al. 2003; Karpouzas et al. 2005; Kumar and Philip 2006). The application of *in situ* bioremediation with naturally occurring microorganism has been revealed in scientific reports (Swannell et al. 1996; Bhupathiraju et al. 2002; Moretti 2005). Microbial degradation can be rapid and thorough under soil conditions favouring microbial growth. Those conditions include warm temperatures, favourable pH levels, adequate soil moisture, aeration (oxygen) and fertility. The amount of adsorption also influences microbial degradation. Adsorbed pesticides, because they are less available to some microorganisms, are more slowly degraded.

9.7.2 Chemical Degradation

Chemical degradation is the breakdown of a pesticide by processes not involving a living organism. The adsorption of pesticides to the soil, soil pH levels, soil temperature and moisture all influence the rate and type of chemical reactions that occur. Many pesticides, especially the organophosphate insecticides, are susceptible to degradation by hydrolysis in high pH (alkaline) soils or spray mixes.

9.7.3 Photodegradation

Photodegradation is the breakdown of pesticides by the action of sunlight. Pesticides applied to foliage, the soil surface or structures vary considerably in their stability when exposed to natural light. Like other degradation processes, photodegradation reduces the amount of chemicals present and can reduce the level of pest control. Soil incorporation by mechanical methods during or after application or by irrigation of water or rainfall following application can reduce pesticide exposure to sunlight.

9.7.4 Phytoremediation

Phytoremediation technology comprises growing plants on contaminated sites so that polluting components percolate through the radical system of the plants and accumulate in various parts (roots, stems, leaves, etc.). The main advantages of phytoremediation are that: (1) it is far less disruptive for the environment (2) it has better public acceptance and (3) it avoids the need for excavation and heavy traffic (Matsumoto et al. 2009). Plants have a natural capacity to accumulate essential heavy metals (Fe, Mn, Zn, Cu, Mg, Mo, and Ni) and pesticides from soil or water for their growth and development. Plants have been shown to possess useful enzymatic mechanisms to degrade most pesticides (Hance 1973). There are other constraints to the use of plants alone in remediation; plant growth is dependent on a number of environmental factors, such as availability of nutrients and water, soil type and pH, etc. The maximum benefits of phytoremediation may therefore, be achieved in long-term applications, or when used in conjunction with other immediate remedial actions. Despite such limitations, plants are known to absorb a wide range of air-borne chemicals through the foliage surface. The *in situ* (on-site treatment) phytoremediation has extensive commercialization in Europe and the United States (Meharg and Cairney 2000; Gaur and Adholeya 2004). This technology is limited as contamination of soil should not exceed a certain depth so that the roots of the plant are in contact with the pollutants. It often takes a longer time period to decontaminate a site due to the limited growth rate of a selected plant species and confinement to the area covered by roots. It could also be necessary to proceed through several cycles of culture and harvest to restore a site completely. Lastly, once contaminated, vegetation must be disposed of in an appropriate manner (Mulligan et al. 2001).

9.7.5 Fungal Bioremediation

The ability of fungi to transform a wide variety of hazardous chemicals has aroused interest in using them in bioremediation (Alexander 1994). The fungi are unique among microorganisms in that they secrete a variety of extracellular enzymes. While most soil bacteria are ubiquitous and occur in a variety of moist soils, fungal species possess higher efficiency of pesticide degradation even in arid and semi-arid soil conditions (Baarschers and Heitland 1986; Bumpus et al. 1993; Twigg and Socha 2001). The soil fungi (e.g., *Fusarium, Penicillium*) are now known to degrade pesticides with greater efficacy (Twigg and Socha 2001). Highly recalcitrant pesticides like the chlorinated triazine herbicide 2-chloro-4- ethylamine-6-isopropylamino-1,3,4-triazine (atrazine) have been transformed by the white-rot fungi *Phanerochaete chrysosporium* and *Pleurotus pulmonarius*, yielding hydroxylated and *N*-dealkylated metabolites (Masaphy et al. 1993; Mougin et al. 1994; Beaudette et al. 1998; Van Acken et al. 1999). *Phanerochaete chrysosporium* has been shown to degrade a number of toxic xenobiotics such as aromatic hydrocarbons (Benzo alpha pyrene, Phenanthrene, Pyrene), chlorinated organics (Alkyl halide insecticides, Chloroanilines, DDT, Pentachlorophenols, Trichlorophenol, Polychlorinated biphenyls, Trichlorophenoxyacetic acid), nitrogen aromatics (2,4-Dinitrotoluene, 2,4,6-Trinitrotoluene-TNT) and several miscellaneous compounds such as sulfonated azodyes. Several enzymes that are released such as laccases, polyphenol oxidases, lignin peroxidases etc. play a role in the degradative process. In addition, a variety of intracellular enzymes such as reductases, methyl transferases and cytochrome oxygenases are known to play a role in xenobiotic degradation (Barr and

Aust 1994). Among the fungal systems, *Phanerochaete chrysosporium* is emerging as the model system for bioremediation. Oxidative enzymes play a major role in biodegradation. The white-rot fungi come equipped with a panoply of enzymes, particularly peroxidases, giving them an edge over bacteria, which require preconditioning/acclimatization to grow in any recalcitrant medium (Barr and Aust 1994).

The branching, filamentous mode of fungal growth allows for more efficient colonization and exploration of contaminated soil (Aust et al. 2004). White-rot fungi are filamentous organisms and offer advantages over bacteria in the diversity of compounds they are able to oxidise (Pointing 2001). In addition, they are robust organisms and are generally more tolerant to high concentrations of polluting chemicals than bacteria (Evans and Hedger 2001) and represent a powerful prospective tool in soil bioremediation and some species have already been patented (Sasek 2003).

Other fungi that can be used in bioremediation are obviously the members of Zygomycetes e.g., the mucoraceous fungi and the arbuscular mycorrhizal fungi. Aquatic fungi and anaerobic fungi are the other candidates for bioremediation. Among other fungi used in bioremediation, the yeasts, e.g., *Candida tropicalis, Saccharomyces cerevisiae, S. carlbergensis* and *Candida utilis* are important in clearing industrial effluents of unwanted chemicals.

9.7.6 Bioremediation of Chlorinated Pesticide Contamination from Soil

Despite decades passing since banning chlorinated pesticide use in most countries, these chemicals are still present worldwide. Considerable quantities could be found at former production sites and in obsolete pesticide stocks. States of the former communist block in Eastern Europe and Asia are especially affected by this problem. Over-production and central distribution of pesticides led in turn to accumulation of huge amounts of these chemicals. Poorly secured stocks, with a large contribution of chlorinated compounds, are now posing a serious threat to humans and the environment (Vijgen 2005).

Recent research has shown (Baczynski et al. 2010) that it is possible to remove chlorinated pesticides γ-HCH (γ isomer of HCH, known also as lindane), DDT and methoxychlor from a field-contaminated soil, using methanogenic granular sludge as an inoculum. However, final effectiveness of the process is limited by persistence of residuals, being desorption-resistant fractions resulting from pollution aging processes. Moreover, DDD, a product of DDT degradation, is only partially removed. To overcome these shortcomings the application of surfactant was proposed. Surfactant influence on bioremediation of soil contaminated with chlorinated pesticide has already been presented in a few reports (You et al. 1996; Walters and Aitkin 2001; Quintero et al. 2005). However, most of these studies used laboratory–spiked soil with very high concentrations of selected contaminant(s), where the effect of pollution aging was limited or neglected during experiments.

According to Baczynski et al. (2010) methanogenic granular sludge and wastewater fermented sludge were used as inoculum for batch tests of anaerobic bioremediation of chlorinated pesticide contaminated soil. Results obtained for both types of biomass were similar: 80 to over 90% of γ-hexachlorocyclohexane (γ-HCH), 1,1,1-trichloro-2,2-bis-(4-methoxyphenyl)ethane (methoxychlor) and 1,1,1-trichloro-2,2-bis-(4-chlorophenyl)ethane (DDT) removed in 4–6 weeks and residual fractions of these pesticides persisted till the end of the 16-week experiment. DDT was degraded through 1,1-dichloro-2,2-bis-(4-chlorophenyl)ethane (DDD). Both methanogenic granular sludge and fermented sludge proved to be good inocula for anaerobic bioremediation of chlorinated pesticide contaminated soil, capable of high removal of γ-HCH, DDT and methoxychlor. The removal efficiency was limited only by the persistence of residual concentrations of these compounds, manifesting after initial periods of fast and effective removal.

9.7.7 Bioremediation of Pesticides from Wastewater

Effluent discharge from pesticide manufacturing or formulating industry and agricultural runoff are major sources of pesticides in the environment and these pesticide industry wastewaters pose a high threat for aquatic systems (Rajeswari and Kanmani 2009). Bioremediation of pesticides from wastewater can be enhanced by the use of lignolytic enzymes such as laccases. Laccases oxidize, polymerise or transform phenolic or anthropogenic compounds to less toxic derivatives. Removal of phenolic or xenobiotic pollutants from wastewater using laccase can be accomplished in several ways: (1) using purified free enzyme; (2) using purified immobilised enzyme; (3) using enzyme obtained directly from culture broths; and (4) bioremediation in reactors with immobilized or free cells (Majeau et al. 2010).

Advanced oxidation processes (AOPs) in which hydroxyl radical (OH^-) is produced by means of chemical, photochemical, photocatalytic or electrochemical reactions, are very effective for treatment of wastewaters containing highly recalcitrant organic pollutants (Gogate and Pandit 2004). Rajeswari and Kanmani (2009) used TiO_2 for photocatalytic oxidation of pesticide wastewater. The degradation revealed 99% removal of active ingredient of pesticide and 76% mineralization. Complete mineralization could be achieved by further increasing the reaction time or by adding some additives, which would speed up reaction.

9.7.8 Biological Treatment Bioreactor and Other Ex Situ Bioremediation Methods

Biological treatment in bioreactors offers the benefits of degradation of wastes under controlled parameters with a continuous monitoring system and is an *ex situ* method of bioremediation. Bioreactor technology can be specially designed in a variety of configurations to maximize microbial activity (Plangklang and Reung-sang 2010). Several types of bioreactor are available worldwide: batch, continuous, sequential batch, membrane, fluidized bed, biofilm, and airlift bioreactors, etc. (Gander et al. 2000; Vischetti et al. 2004). They are employed in treatment of a wide array of organics (Sajc and Novakovic 2000; Zwiener and Frimmel 2003). Despite the advantage of a controlled environment for treatment, they suffer the limitations of high capital and operation costs and also excavation of contaminated sites.

Other *ex situ* bioremediation methods include landfarming (lately considered as a disposal alternative), composting, and biopiles (engineered systems—a combination of landfarming and composting). These methods suffer various disadvantages—large space requirements, extended treatment time, mass transfer problems, and restricted bioavailability of contaminants (Vidali 2001).

9.8 Sustainability

Various approaches to soil bioremediation have been developed and implemented, ranging from *in-situ* subsurface (unexcavated) processes, to land-farming and engineered soil pile approaches, to use of completely mixed-soil slurry reactor systems for treatment of excavated soils. The main and common objective in the various processes is to create the necessary environment to facilitate growth and contaminant degradation by the appropriate biological organisms. Bioremediation has now successfully been used to remediate sites contaminated with hydrocarbons.

The following are the advantages of bioremediation approaches:

1. They are generally the least expensive remediation alternatives (Grommen and Verstraete 2002).
2. The processes are flexible and adaptable to variable environmental conditions and, over time, microorganisms evolve that can degrade novel synthetic chemical structures (Mandelbaum et al. 1995).
3. The processes are perceived as being environmentally benign whereas incineration and more energy and equipment intensive processes are perceived as being more polluting.
4. The processes are implementable on site, indeed often *in situ*, and with dilute or widely diffused contaminants (Iwamoto and Nasu 2001).

On the negative side, there have also been many occasions in which bioremediation failed to reduce contaminant levels to defined concentration criteria, and the methods/practices are also often criticized as being too slow. Due to its history of failures owing to the presentation of "quick-fix" technologies consumers have been reluctant to use bioremediation technology (Srinivasan 2003). There can be many factors behind the slow bioremediation rates and failures; primarily, that the environmental conditions present are suboptimal for selection and growth promotion of the degrading strains. Besides, the kinetics of microbial growth and biodegradation are such that, as contaminant concentrations decline, so do the rates of their further degradation. Factors affecting the key rates and extents of contaminant degradation relate to the nature of the contaminant(s) (structure, water solubility, bioavailability, biodegradability, co-metabolism potential, substrate/metabolite concentration, and toxicity), the properties of the soil and the nature of the process (homo- or heterogeneous environment; contents of water, nutrients, and oxygen; presence of bioavailability enhancing agents), temperature, pH and, the size and make-up of the microbial population. Having more limited microbial intervention bioremediation processes tend to be more prolonged and unreliable. The cost of sampling and analyses are substantially increased where the process environment is non-homogeneous, and may become the dominant cost component in the project. Increased microbial technology intervention can lead to more accelerated processes, greater process reliability, and lower end-points (Ward et al. 2003). The durations of processes may range from 5 to 25 years for natural attenuation processes, 0.5–3 years for *in-situ* subsurface processes, 1–18 months for soil pile/composting processes, 1–12 months for land-farming and slurry phase systems, and 15 days for accelerated slurry phase systems (Ward and Singh 2004). Average daily rates of contaminant degradation can range from 5 to 10,000 ppm for natural attenuation processes to accelerated slurry phase systems. It has also been proved that prolonged bioremediation processes are typically disadvantageous with a diverse mixture of contaminants present, because significant amounts of contaminants may be removed by non-biological mechanisms and the extents of contaminant degradation achieved are often not adequate. Hence, in prolonged bioremediation processes the early loss of volatiles and/or the metabolism of low molecular-weight compounds can reduce the bioavailability and co-metabolic biodegradation potential of high-molecular-weight compounds. The resultant removal of the carbon and energy for microbial growth will lead to a decline in the hydrocarbon-degrading and general microbial population. Many authors have provided guidance for determining the suitability of bioremediation as a clean-up option and questions to be addressed related to the nature of the contaminants such as: (a) What is the impact of the duration of contamination of site on removal of the easily degradable compounds such that the persistent compounds may still require remediation? (b) What is the ability of known microbial systems and/or the microbial population at the site to degrade the contaminants? (c) What are the factors limiting population growth and contaminant degradation and the potential to achieve clean-up criteria? With regard to the selection of bioremediation configuration for treatment of different classes of chemicals, natural attenuation and electron donor delivery were considered to be the alternatives for treatment of chlorinated solvents, while biostimulation was an option for remediation of chlorinated solvents and phenols (Hughes et al. 2000). Bioventing (which uses low airflow rates to provide only enough oxygen to sustain microbial activity and prevent contaminant volatilization) was an option for treating polycyclic aromatic hydrocarbons (PAHs). Land treatment or composting was for nitroaromatics, phenols, and PAHs; and bioslurry processes for all of the above mentioned chemicals. Except electron-donor delivery, all treatment methods were potential approaches to monoaromatic hydrocarbon bioremediation. While all of this guidance is instructive, one cannot help getting the impression that there are a lot of barriers or pit falls to be aware of when embarking on a bioremediation project—just how reliable or robust is the technology? In the early 1990s, the perceived advantages of bioremediation processes resulted in significant research and commercial interest in bioremediation technologies, and investors, technologists, and entrepreneurs responded through creation of a substantial number of bioremediation companies whose missions were to develop and implement bioremediation technologies. Suffice it to say that these companies struggled at best and few have survived by sticking to their original missions. So, given that soil remediation opportunities exist widely, we are yet awaiting the development of a strong bioremediation-based industrial sector. One dimension of the problem is that bioremediation processes are perceived to be project-specific, requiring a lot of customization, which does not endear the technology to investors, who like more widely distributable technology. Case-by-case customization and technology implementation failures have retarded the development of environmental biotechnology enterprises, and more rigorous approaches to technology selection and its strategic development and commercialization are required. More versatile and robust processes that do not require research and development for each project need to be developed. Wider use of more controlled reactor-based accelerated bioremediation processes ought to be considered. The market, legislative decisions, and government funding initiatives all appear to favour pursuit of enhanced bioremediation approaches. Much of the activity in bioremediation technology is at the research and development level, and there is a need to develop strategies to successfully convert more of the new research findings into reliable processes.

9.9 Economic Aspect

Bioremediation is considered to be far more cost efficient than traditional cleaning technologies, with possible savings of 65–85%. Waste incineration, for instance, costs US$250–US$500 per ton, whereas biological treatment costs US$40–US$70 per ton (D. Brauer, pers. communication). It has been estimated that bioremediation of polluted soil is, at least, one third cheaper (Annon 1994); for example, the biotreatment costs are between US$50 and US$130 m^{-3}, compared with conventional costs of US$300–US$1000 m^{-3} for incineration and US$200–US$300 m^{-3} for disposal in landfills (Crawford and Crawford 1996). The market for soil decontamination is increasing and it has been predicted that the bioremediation market may increase more rapidly. Despite justified optimism for future growth, some problems are impeding further progress:

- Each waste site has unique characteristics, thus requiring costly tailor-made applications;
- Many industrial pollutants still cannot be degraded satisfactorily under natural conditions;
- *In situ* applications of altered microbial strains might pose significant ecological risks; and
- The technique is often time consuming.

It should be noted that adding exogenous microorganisms might not always be the best solution where indigenous populations are already *in situ* and could do well, if supplied with sufficient nutrients and oxygen (Caplan 1993).

9.10 Conclusion and Perspectives

Bioremediation is a rapidly developing technology having potential to provide safe, efficient and economic means of removing organic pollutants either from contaminated soils or wastewater without transferring them to another medium (Sarma et al. 2006). Bioremediation is far less expensive than other technologies that are often used to clean up hazardous waste. Bioremediation can often be accomplished where the problem is located ("*in-situ*"). This eliminates the need to transfer large quantities of contaminated waste off-site, and the potential threats to human health and the environment that can arise during such transportation. Bioremediation offers a viable alternative to the regular use of physicochemical methods of decontamination, which are not generally cost effective. The bioremediation process is influenced by various factors—existence of a specific microbial population, bioavailability of contaminants, and environmental factors (soil type, temperature, pH, nutrients, and presence of oxygen or other electron acceptors). Although bioremediation may not completely detoxify inorganic pollutants (metals and radionuclides), yet it can alter the oxidation state, aiding in adsorption, uptake, accumulation, and concentration in micro- or macroorganisms. A good bioremediation approach will involve strategic use of all native microbes in an engineered way to achieve the best possible detoxification levels. Bioremediation technologies have been successfully employed in the field and are gaining more and more importance with increased acceptance of eco-friendly remediation solutions. In summary, although bioremediation appears to be a promising alternative for the remediation of pesticide-contaminated sites, it is still in the developmental phase. Many bench-scale projects are being conducted to optimize bioremediation protocols and to expand the number of compounds for which bioremediation is feasible.

References

Abhilash PC, Singh N (2009) Pesticide use and application: An Indian Scenario. J Hazard Mater 165:1–12

Adhya TK, Wahid PA, Sethunathan N (1987) Persistence and biodegradation of selected organophosphorus insecticides in flooded versus non-flood soils. Biol Fertil Soils 4:36–40

Agrios GN (2002) Plant Pathology, IV Edition. ISBN 0120445646, Academic Press, London

Alexander M (1977) Introduction to soil microbiology, 2nd edn. Wiley Eastern Ltd, New Delhi

Alexander M (1994) Biodegradation and Bioremediation. Academic, New York

Alexander M (1995) How toxic are chemicals in soil? Environ Sci Technol 29:2713–2717

Ang EL, Zhao H, Obbard JP (2005) Recent advances in the bioremediation of persistent organic pollutants via biomolecular engineering. Enj Microbial Technol 37:487–496

Anon (1994) Biotechnology for a clean environment: prevention, Detection, Remediation, Organization for Economic Cooperation and Development, Paris, France

Anon (2003) Herbicide Use Essential to Crop Production, Chemical market reporter, vol 263, No 18, p4

Anon (2005) Ingenuity thrives in battle to beat pests and rodents. Poultry World 159(7):28

Anwar S, Liaquat F, Khan QM, Khalid ZM, Iqbal S (2009) Biodegradation of chlorpyrifos and its hydrolysis product 3,5,6-trichloro-2-pyridinol by *Bacillus pumilus* strain C2A1. J Hazard Mater 168:400–405

Araújo ASF, Monterio RTR, Abarkeli RB (2003) Effect of glyphosate on the microbial activity of two Brazilian soils. Chemosphere 52:799–804

Arshad M, Hussain S, Saleem M (2007) Optimization of environmental parameters for biodegradation of alpha and beta endosulfan in soil slurry by *Pseudomonas aeruginosa*. J Appl Microbiol 104:364–370

Aust SD, Swaner PR, Stahl JD (2004) Detoxification and metabolism of chemicals by white-rot fungi. In: Zhu JJPC, Aust SD, Lemley Gan AT (eds) Pesticide decontamination and detoxification, Oxford, Washington DC, pp 3–14

Austin RB (1998) Yield of wheat in the UK: Recent advances and prospects. Annual Meeting of the crop Science Society of America

Baczynski TP, Pleissner D, Groten Huis T (2010) Anaerobic biodegradation of organochlorine pesticides in contaminated soil–significance of temperature and availability. Chemosphere 78(1):22–28

Bailey AM, Coffey MD (1985) Biodegradation of Metalaxyl in avocado soils. Phytopathology 74:135–137

Banerjee A, Padhi S, Adhya TK (1999) Persistence and biodegradation of vinclozolin in tropical rice soils. Pestic Sci 55:1177–1181

Baarschers WH, Heitland HS (1986) Biodegradation of fenitrothion and fenitrooxon by the fungus Trichoderma viride. J Ag Food Chem 34(4):707–709

Barr DP, Aust SD (1994) Mechanisms the white-rot fungi use to degrade pollutants. Env Sci Technol 28:79A–87A

Beaudette LA, Davies S, Fedorak PM, Ward O P, Pickard MA (1998) Comparison of gas chromatography and mineralization experiments for measuring loss of selected polychlorinated biphenyl congeners in cultures of white-rot fungi. Appl Environ Microbiol 64:2020–2025

Betts WD (ed) (1991) Biodegradation: natural and synthetic materials. Springer, Germany

Bhatia MR, Fox-Rushby J, Mills A (2004) Cost-effectiveness of malaria control interventions when malaria mortality is low: insecticide-treated nets versus in-house residual spraying in India. Soc Sci Med 59:525–539

Bhupathiraju VK, Krauter P, Holman H-YN, Conrad ME, Daley PF, Templeton AS, Hunt JR, Hernandez M, Alvarez-Cohen L (2002) Assessment of *in-situ* bioremediation at a refinery waste-contaminated site and an aviationgasoline contaminated site. Biodegradation 13:79–90

Briceno G, Palma G, Duran N (2007) Influence of organic amendment on the biodegradation and movement of pesticides. Critic Rev Environ Sci Technol 37:233–271

Bhushan B, Paquet L, Halasz A, Spain JC, Hawari J (2003) Mechanism of xanthine oxidase catalyzed biotransformation of HMX under anaerobic conditions. Biochem Biophys Res Commun 306:509–515

Brown I (2004) UK pesticides residue committee report 2004: http://www.pesticides.gov.uk/uploadedfiles/Web_Assets/PRC/PRC

Bumpus JA, Kakar SN, Coleman RD (1993) Fungal degradation of organophosphorous insecticides. Appl Biochem Biotechnol 39:715–726

Burns RG (1975) Factors affecting pesticides loss from soil. In: Paul EA, McLaren AD (eds) Soil biochemistry, vol 4. Marcel Dekker, New York, pp 103–141

Burns RG, Edwards JA (1980) Pesticide breakdown by soil enzymes. Pest Sci 11:506–512

Caplan JA (1993) Trends Biotechnol 11:320–323

Chakrabarty T, Subrahmanyam PVR, Sundaresan BB (1988) Biodegradation of recalcitrant industrial wastes. In: Wise D (ed) Bio- treatment Systems, vol 2. CRC Press, Boca Raton. pp 172–234

9 Bioremediation of Pesticides from Soil and Wastewater

Cohen M (2007) Environmental toxins and health–the health impact of pesticides. Aus Fam Phy 36(12):1002–1004

Colleran E (1997) Uses of bacteria in bioremediation. In: Sheehan D (ed) Methods in Biotechnology, vol 2, Bioremediation Protocols. Humana Press, New Jersey, pp 3–22

Cork DJ, Krueger JP (1991) Mirobial transformation of herbicides and pesticides. Adv Appl Microbiol 36:1–66

Croplife (2004) International at: http://www.croplife.org/default.aspx

Crawford RL, Crawford DL (1996) Bioremediation principles and applications. Cambridge, UK

Dams RI, Paton GI, Killham K (2007) Rhizoremediation of pentachlorophenol by *Sphingobium chlorophenolicum* ATCC 39723. Chemosphere 68:864–70

Das AC, Mukherjee D (2000) Influence of insecticides on microbial transformation of nitrogen and phosphorus in typic orchragualf soil. Agric Food Chem 48:3728–3732

Das AC, Chakravarty A, Sen G, Sukul P, Mukherjee D (2005) A comparative study on the dissipation and microbial metabolism of organophosphate and carbamate insecticides in orchaqualf and fluvaquent soils of West Bengal. Chemosphere 58:579–584

Deeb RA, Cohen L (2000) Aerobic biotransformation of gasoline aromatics in multicomponent mixtures. Bioremediation J 4(1):1–9

Delacollette C (ed) (2004) Global strategic framework for integrated vector management. World Health Report

Dietary guidelines for Americans (2005) US Department of Health and Human Services, US Department of Agriculture

Dindal DL (1990) Soil biology guide. Wiley, New York

Droby S, Coffey MD (1991) Biodegradation processes and the nature of metabolism of metalaxyl in soil. Ann Appl Biol 118:543–553

Gianessi L (1999) Beneficial impacts of pesticide use for consumers in Ragsdale, Nancy, Seiber J (eds), Pesticides: Managing Risks and Optimizing Benefits, American Chemical Society Symposium Series #734, American Chemical Society, Washington DC, USA, p207

EPA REACH IT (2004) "Remediation and characterization innovative technologies." Information snapshots: Technologies by type

Evans C, Hedger J (2001) Degradation of cell wall polymers. In: Gadd G (ed) Fungi in Bioremediation. Cambridge, UK

Ferguson JF, Pietari JMH (2000) Anaerobic transformations and bioremediation of chlorinated solvents. Environ Pollut 107:209–215

Frankland P, Frankland MP (1894) Micro-organisms in Water. Longmans Green, London

Gander M, Jefferson B, Judd S (2000) Aerobic MBRs for domestic waste water treatment: a review with cost considerations. Sep Purif Technol 18:119–130

Gaur A, Adholeya A (2004) Prospects of arbuscular mycorrhizal fungi in phytoremediation of heavy metal contaminated soils. Curr Sci 86(4):528–534

Gold RE, Howell HN, Pawson BM, Wright MS, Lutz JL (1996) Persistence and bioavailability of termicides to subterranean termites from five soils types and location in Texas. Sociobiol 28:337–363

Gogate PR, Pandit AR (2004) A review of imperative technologies for wastewater treatment I: oxidation technologies at ambient conditions. Adv Environ Res 8:501–551

Gratz NG (1994) What role for insecticides in vector control programs? Am J Trop Med Hyg 50(6):11–20

Grommen R, Verstraete W (2002) Environmental biotechnology: the ongoing quest. J Biotechnol 98:113–123

Gupta S, Gajbhiye VT (2002) Effect of concentration, moisture and soil type on the dissipation of flufenacet from soil. Chemosphere 47:901–906

Hafez HFH, Theimann WHP (2003) Persistence and biodegradation of iazinone and imidacloprid in soil. Proc XII Symp Pest Chem, Congress Centre Universita Cattolica, Via Emilia Parmense 84, Piacenza, pp 35–42

Hance RJ (1973) The effects of nutrients on the decomposition of the herbicides atrazine and linuron incubated with soil. Pestic Sci 4:817–822

Handa SK, Agnihotri NP, Kulshreshtha G (1999) Effect of pesticide on soil fertility. In Pesticide residues; Significance, Management and analysis, pp 184–198

Head MA, Oleszkiewicz JA (2004) Bioaugmentation for nitrification at cold temperatures. Water Res 38:523–530

Hicks RJ, Stotzky G, Voris PV (1990) Review and evaluation of the effects of xenobiotic chemicals on microorganisms in soil. Adv Appl Microbiol 35:195–253

Hong, Q, Zhang ZH, Hong YF, Li SP (2007) A microcosm study on bioremediation of fenitrothion-contaminated soil using *Burkholderia* sp FDS-1. Int Biodeter Biodeg 59:55–61

Hughes JB, Neale CN, Ward CH (2000) Bioremediation. Enclyclopedia of microbiology, 2nd edn. Academic, New York, pp 587–610

Hwang S, Cutright TJ (2002) Biodegradability of aged pyrene and phenenthrene in a natural soil. Chemosphere 47:891–899

Iwamoto T, Nasu M (2001) Current bioremediation practice and perspective. J Biosci Bioeng 92:1–8

Jha D, Chand R (1999) National Centre for Agricultural Economics and Policy Research (ICAR), New Delhi, India from Agro-chemicals News in Brief Special Issue

Jianlong W, Xiangchun Q, Libo W, Yi Q, Hegemann W (2002) Bioaugmentation as a tool to enhance the removal of refractory compound in coke plant wastewater. Process Biochem, Oxford, UK, 38:777–781

Jitender K, Kumar J, Prakash J (1993) Persistence of thiobencarb and butachlor in soil incubated at different temperatures. In: Integrated weed management for sustainable agriculture. Proc. Indian Soc. Weed. Sci. Int. Seminar. Hisar, India, pp 123–124

Jones WJ, Ananyeva ND (2001) Correlations between pesticide transformation rate and microbial respiration activity in soil of different ecosystems. Biol Fertil Soils 33:477–483

Karpouzas DG, Fotopoulou A, Menkissoglu-Spiroudi U, Singh BK (2005) Non-specific biodegradation of the organophosphorus pesticides, cadusafos and ethoprophos, by two bacterial isolates. FEMS Microbiol Ecol 53:369–378

Karpouzas DG, Walker A, Williams RJF, Drennan DS (1999) Evidence for the enhanced biodegradation of ethoprophos and carbofuran in soils from Greece and the UK. Pest Sci 55:301–311

Kumar M, Philip L (2006) Enrichment and isolation of a mixed bacterial culture for complete mineralization of endosulfan. J Environ Sci Health 41:81–96

Kumar S, Mukerji KG, Lal R (1996) Molecular aspects of pesticide degradation by microorganisms. Critical Rev Microbiol 22(1):1–26

Lal R, Dadhwal M, Kumari K, Sharma P, Singh A, Kumari H, Jit S, Gupta SK, Nigam A, Lal D, Verma M, Kaur J, Bala K, Jindal S (2007) *Pseudomonas* sp to *Sphingobium indicum*: a journey of microbial degradation and bioremediation of Hexachlorocyclohexane. Indian J Microbiol 48:3–18

Lewis, Nancy M, Ruud J (2004) Apples in the American Diet. Nutrition in Clinical Care 7(2):82

Lewis, Nancy M, Ruud J (2005) Blueberries in the American Diet. Nutrition Today 40(2):92

Majeau J-A, Satinder KB, Tyagi RD (2010) Laccases for removal of recalcitrant and emerging pollutants. Biresour Technol 101:2331–2350

Mandelbaum RT, Allan DL, Wackett LP (1995) Isolation and characterization of a *Pseudomonas* sp that mineralizes the s-triazine herbicide atrazine. Appl Environ Microbiol 61:1451–1457

Maroli M (2004) Prevenzione e controllo dei vettori di leishmaniosi: attuali metodologie. Parassitologia 46:211–215

Marschner P, Yang CH, Lieberei R, Crowley DE (2001) Soil and plant specific effects on bacteria community composition in the rhizosphere. Soil Biol Biochem 33:1437–1445

Martikainen E, Haimi J, Ahtiainen J (1998) Effects of dimethoate and benomyl on soil organisms and soil processes: A microcosm study. Appl Soil Ecol 9:381–387

Masaphy S, Levanon D, Vaya J, Henis Y (1993). Isolation and characterization of a novel atrazine metabolite produced by the fungus *Pleurotus pulmonarius*, 2-chloro-4-ethylamino-6-(1-hydroxyisopropyl)amino-1,3,5-triazine. Appl Environ Microbiol 59:4342–4346

Matsumoto E, Kawanaka Y, Yun S, Oyaizu H (2009) Bioremediation of the organochlorine pesticides, dieldrin and endrin, and their occurrence in the environment. Appl Microbiol Biotechnol 84:205–216

9 Bioremediation of Pesticides from Soil and Wastewater 325

Megharaj M, Kantachote D, Singleton I, Naidu R (2000) Effects of long-term contamination of DDT on soil microflora with special reference to soil algae and algal transformations of DDT. Environ Pollution 109:35–42

Meharg AA, Cairney JWG (2000) Ectomycorrhizae—extending the capabilities of rhizosphere remediation? Soil Biol Biochem 32:1475–1484

Melling FB Jr (1993) Soil microbial ecology: applications in agricultural and environmental management. Marcel Dekker, New York

Moretti L (2005) *In situ* bioremediation of DNAPL source zones Washington, DC: US EPA, Office of Solid Waste and Emergency Response Technology Innovation and Field Services Division. Available at: http://wwwclu-inorg/download/studentpapers/moretti dnaplbioremediationpdf (available October 7, 2009)

Mougin C, Laugero C, Asther M, Dubroca J, Frasse P (1994) Biotransformation of the herbicide atrazine by the white-rot fungus *Phanerochaete chrysosporin*. Appl Environ Microbiol 60:705–708

Mulligan CN, Yong RN, Gibbs BF (2001) Remediation alternative treatment option for heavy metal bearing wastewaters: a review. Bioresour Technol 53:195–206

Murugesan AG, Jeyasanthi T, Maheswari S (2010) Isolation and characterization of cypermethrin utilizing bacteria from brinjal cultivated soil. Afr J Microbiol Res 4(1):10–13

Panda S, Sahu SK (1999) Effects of malathion on the growth and reproduction of Drawida willsi (Oligochaete) under laboratory conditions. Soil Biol Biochem 31:363–366

Panda S, Sahu SK (2004) Recovery of acetylcholine esterase activity of Drawida willsi (Oligochaete) following application of three pesticides to soil. Chemosphere 55:283–290

Pandey S, Singh DK (2004) Total bacterial and fungal population after chlorpyrifos and quinalphos treatments in groundnut (Arachis hypogaea L.) soils. Chemosphere 55:197–205

Pastor M, Castro J (1995) Soil management systems and erosion. Olivae 59:64–74

Perelo LW (2010) Review: *in situ* and bioremediation of organic pollutants in aquatic sediments. J Hazard Mater 177:81–89

Perucci P, Dumontet S, Bufo SA, Mazzatura A, Casucci C (2000) Effects of organic amendments and herbicide treatment on soil microbial biomass. Biol Fertil Soils 32: Texas Sociobiol 28:337–363

Phillips TM, Lee H, Trevors JT, Seech AG (2006) Full-scale *in situ* bioremediation of hexachlorocyclohexane-contaminated soil. J Chem Technol Biotechnol 81:289–298

Pimental D, Levitan L (1986) Pesticides: amounts applied and amounts reaching pests. Biosciences 36:86–91

Pointing SB (2001) Feasibility of bioremediation by white-rot fungi. App Microbiol Biotechnol 57:20–33

Prakash NB, Suseela Devi L (2000) Persistence of butachlor in soils under different moisture regime. J Ind Soc Soil Sci 48:249–256

Purdue (2006) University at: http://www.hort.purdue.edu/rhodcv/hort640c/nuse/nu00003.htm

Quan X, Shi H, Liu H, Wang J, Qian Y (2004) Removal of 2,4-dichlorophenol in a conventional activated sludge system through bioaugmentation. Process Biochem, Oxford, UK, 39:1701–1707

Quintero JC, Moreira MT, Feijoo G, Lema JM (2005) Effects of surfactants on the soil desorption of hexachlorocyclohexane (HCH) isomers and their anaerobic biodegradation. J Chem Technol Biotechnol 80(9):1005–1015

Plangklang P, Reungsang A (2010) Bioaugmentation of carbofuran by Burkholderia cepacia PCL3 in a bioslurry phase sequencing batch reactor. Process Biochemistry 45:230–238

Racke KD, Skidmore MW, Hamilton DJ, Unsworth JB, Miyamoto J, Cohen SZ (1997) Pesticide fate in tropical soils. Pure Appl Chem 69:1349–1371

Rajagopal BS, Rao VR, Nagendrappa G, Sethunathan N (1984) Metabolism of carbaryl and carbofuran by soil –enrichment and bacterial cultures. Can J Microbiol 30:1458–1466

Rajeswari R, Kanmani S (2009) A study on degradation of pesticide wastewater by TiO2 catalysis. J Sci Indus Res 68:1063–1067

Raymond RL, Jamison VW, Hudson JO Jr (1975) Final report on beneficial stimulation of bacterial activity in groundwater containing petroleum products. American petroleum institute, Washington

Reddy BR, Sethunathan N (1985) Salinity and the persistence of parathion in flooded soil. Soil Biol Biochem 17:235–239

Robles-Gonzalez IV, Fava F, Poggi-Varaldo HM (2008) A review on slurry bioreactors for bioremediation of soils and sediments. Microb Cell Fact 7:1–16

Ross G (2005) Risks and benefits of DDT. Lancet 366:1771–1772

Sahu SK, Patnaik KK, Sharmila M, Sethunnathan N (1990) Degradation of alpha-, beta- and gamma- Hexachlorocyclohexane by a soil bacterium under aerobic conditions. Appl Environ Microbiol 56:3620–3622

Sajc L, Novakovic GV (2000) Extractive bioconversion in a four phase external-loop airlift bioreactor. AIChE J 46(7):1368–1375

Sarma PN, Venkat Mohan S, Rama Krishna M, Shailaja S (2006) Bioremediation of pendimethalin contaminated soil by augmented bioslurry phase reactor operated in sequential batch (SBR) mode: Effect of substrate concentration. Indian J Biotechnol 5:169–174

Sasek (2003) Why mycoremediations have not yet come to practice. In: Sasek V et al (eds) The utilization of bioremediation to reduce soil contamination: problems and solutions, Kluwer Academis Publishers, New York, pp 247–276

Schmitz P (2003) Michael economic effects of chemical use reduction in European Agriculture, Institute of Agribusiness, University of GieBen, Germany

Schroll R, Becher HH, Dorfler U, Gayler S, Grundmann S, Hartmann HP, Ruoss J (2006) Quantifying the effect of soil moisture on the aerobic microbial mineralization of selected pesticides in different soils. Environ Sci Technol 40:3305–3312

Schukken YH, van Schaik G, McDermott JJ, et al (2004) Transition models to assess risk factors for new and persistent trypanosome infections in cattle-analysis of longitudinal data from the Ghibe Valley, southwest Ethiopia. J Parasit 90:1279–1287

Schuster E, Schröder D (1990) Side-effects of sequentially applied pesticides on non-target soil microorganisms: field experiments. Soil Biol Biochem 22:367–373

Schwarzenbach RP, Gschwend PM, Imboden DM (1993) Environmental organic chemistry. Wiley, New York

Scragg A (2005) Bioremediation. Environ Biotechnol 173–229

Shannon MJ, Unterman R (1993) Evaluating bioremediation: distinguishing fact from fiction. Ann Rev Microbiol v47(Annual 1993): pp 715(24)

Shen Y, West C, Hutchins SR (2000) In vitro cytotoxicity of aromatic aerobic biotransformation products in bluegill sunfish BF-2 cells. Ecotoxicol Environ Saf 45:27–32

Siddique T, Okeke BC, Arshad M, Frankenberger WT Jr (2002) Temperature and pH effects on biodegradation of hexachlorocyclohexane isomers in water and soil slurry. J Agric Food Chem 50:5070–5076

Siddique T, Okeke BC, Arshad M, Frankenberger WT (2003) Enrichment and isolation of endosulfan-degrading microorganisms. J Environ Qual 32:47–54

Simonich SL, Hites RA (1995) Organic pollutant accumulation in vegetation. Environ Sci Tech 29:2905–2914

Singh BK, Walker A, Wright DJ (2006) Bioremedial potential of fenamiphos and chlorpyrifos degrading isolates: Influence of different environmental conditions. Soil Biol Biochem 38:2682–2693

Somsamak P, Cowan RM, Haggblom MM (2001) Anaerobic biotransformation of fuel oxygenates under sulfate-reducing conditions. FEMS Microbiol Ecol 37:259–264

Spark KM, Swift RS (2002) Effect of soil composition and dissolved organic matter on pesticide sorption. Sci Total Environ 298(1–3):17–61

Srinivas T, Sridevi M, Mallaiah KV (2008) Effect of pesticides on Rhizobium and nodulation of green gram Vigna Radita (L) Wilczek ICFAI. J Life Sci 2:36–44

Srinivasan U (2003) US bioremediation markets. Frost and Sullivan research report, no. 7857. Frost and Sullivan, Palo Alto, California, pp 1–300

9 Bioremediation of Pesticides from Soil and Wastewater

Subhas KS, Irvine RL (1998) Bioremediation: Fundamentals and Applications, vol 1, Technomic Publishing

Sukul P, Spiteller M (2001) Influence of biotic and abiotic factors on dissipating metalaxyl in soil. Chemosphere 45:941–947

Swannell RPJ, Lee K, McDonagh M (1996) Field evaluations of marine oil spills bioremediation. Microbiol Rev 60:342–365

Thom E, Ottow JCG, Benckiser G (1997) Degradation of the fungicide difenoconazole in a silt loam soil as affected by pretreatment and organic amendement. Environ Poll 96:409–414

Timmis KN, Pieper DH (1999) Bacteria designed for bioremediation. Trends Biotechnol 17:201–204

Topp E, Vallayes T, Soulas G (1997) Pesticides: Microbial degradation and effects on microorganisms. In: Van Elsas JD, Trevors JT, Wellington EMH (eds) Modern soil microbiology. Mercel Dekker, New York, pp 547–575

Trindade PVO, Sobral LG, Rizzo AC L, Leite SGF, Soriano AU (2005) Bioremediation of a weathered and recently oil-contaminated soils from Brazil: A comparison study. Chemosphere 58:515–522

Twigg LE, Socha LV (2001) Defluorination of sodium monofluoroacetate by soil microorganisms from central Australia. Soil Biol Biochem 33:227–234

US Microbics. (2003) Annual Report FY-2003. http://www.bugsatwork.com/USMX/BUGS%20Report%20PRINT%20(07-13-04)%20Hawaii%20(paginate%201-8)

Van Acken B, Godefroid LM, Peres CM, Naveau H, Agathos SN(1999) Mineralization of 14C-U ring labeled 4-hydroxylamino-2,6-dinitrotoluene by manganese-dependent peroxidase of the white-rot basidiomycete *Phlebia radiata*. J Biotechnol 68:159–169

Vasileva V, Ilieva A (2007) Effect of presowing treatment of seeds with insecticides on nodulating ability, nitrate reductase activity and plastid pigments content of lucerne (*Medicago sativa* L.). Agron Res 5:87–92

Vidali M (2001) Bioremediation: An overview. Pure Appl Chem 73(7):1163–1172

Vijgen J (2005) Obsolete pesticides: how to solve a worldwide society problem? In: Lens P, Grotenhuis T, Malina G, Tabak H (eds) Soil and sediment remediation: mechanisms technologies and application. IWA Publishing, London, pp 331–340

Vischetti C, Capri E, Trevisan M, Casucci C, Perucci P (2004) Biomass bed: a biological system to reduce pesticide point contaminationat farm level. Chemosphere 55:823–828

Walker N (1975) Microbial degradation of plant protection chemicals. In: Walker N (ed) Soil microbiology. Butterwoths, London, pp 181–194

Walter-Echols G, Lichtenstein EP (1978) Movement and metabolism of 14C- phorate in a flooded soil system. J Agri Food Chem 26:599–604

Walters GW, Aitken MD (2001) Surfactant enhanced solubilization and anaerobic biodegradation of 1,1,1-trichloro-2,2-bis-(p-chlorophenyl)-ethane (DDT). Water Environ Res 73:15–23

Ward OP, Singh A (2004) Evaluation of current soil bioremediation technologies. In: Singh A, Ward OP (eds) Applied bioremediation and phytoremediation. Soil biology series, Vol 1. Springer, Berlin Heidelberg, New York, Texas. Sociobiol 28:337–363

Ward OP, Singh A, VanHamme J (2003) Accelerated bioremediation of petroleum hydrocarbon waste. J Ind Microbiol Biotechnol 30:260–270

Webster JPG, Bowles RG, Williams NT (1999) Estimating the economic benefits of alternative pesticide usage scenarios: wheat production in the United Kingdom. Crop Protection 18:83–89

Wendt-Rasch L, Pirzadeh P, Woin P (2003) Effects of metsulfuron methyl and cypermethrin exposure on freshwater model ecosystems. Aq Toxicol 63(3):243–256

WWF website (undated) at: http://www.panda.org/about_wwf/where_we_work/africa/problems/environments/index.cfm

Yancy J, Cecil H (2005) Study touts herbicide benefits, Southeast farm press 32(11):16

Ying GG, Williams B (2000) Dissipation of herbicides in soil and grapes in a South Australian vineyard. Ag Ecosys Environ 78(3):283–289

Yoshida T (1978) Microbial metabolism in rice soils. In: Soils and Rice, International Rice Research Institute, Philippines, pp 445–463

You G, Sayles GD, Kupferle MJ, Kim IS, Bishop PL (1996) Anaerobic DDT transformation: enhancement of surfactants and low oxidation reduction potential. Chemosphere 32(11):2269–2284

Zahran HH (1999) Rhizobium-legume symbiosis and nitrogen fixation under severe conditions and in an arid climate. Microbiol Mol Biol Rev 63:968–989

Zhang P, Sheng GY, Feng YC, Miller DM (2005) Role of wheat residue-derived char in the biodegradation of benzonitrile in soil: Nutritional stimulation versus adsorptive inhibition. Environ Sci Technol 39:5442–5448

Zhu G, Wu H, Guo J, Kimaro FME (2004) Microbial degradation of fipronil in clay loam soil. Water Air Soil Poll 153:35–44

Zwiener C, Frimmel FH (2003) Short term tests with a pilot sewage plant and biofilm reactors for the biological degradation of the pharmaceutical compounds clofibric acid, ibuprofen and diclofenac. Sci Total Environ 309:201–211

Chapter 10
Isolation and Characterization of Rhizobacteria Antagonistic to *Macrophomina phaseolina* (Tassi) Goid., Causal Agent of Alfalfa Damping-Off

L. B. Guiñazú, J. A. Andrés, M. Rovera and S. B. Rosas

Contents

10.1 Alfalfa Cultivation 330
    10.1.1 Alfalfa-Bacteria Interactions 331
    10.1.2 Alfalfa and Its Diseases 332
10.2 Biological Control: Application Perspectives for Suppression of *M. Phaseolina*
    in Alfalfa 333
    10.2.1 Growth Parameters 336
10.3 Conclusions 337
References 337

Abstract Alfalfa (*Medicago sativa* L.) is affected by several pathogens; however, those that attack crown and roots, such as *Macrophomina phaseolina*, directly define the longevity or productive period of the plant. Use of microorganisms with biological control capacity constitutes an alternative to chemical products. In this chapter, we have gathered some studies on this subject; in addition, we contribute data from our own researches on *in vitro* and *in vivo* antifungal activity of rhizobacteria isolated from alfalfa rhizosphere against the causal agent of damping-off, *M. phaseolina*. *In vitro* tests consisted in culturing a bacterium and the pathogen on Potato Dextrose Agar (PDA) medium. *In vivo* tests were carried out in pots containing soil infested with *M. phaseolina* sclerotia. Alfalfa seeds were inoculated with each selected isolate (single inoculation treatments) or co-inoculated with *Sinorhizobium meliloti* strain B399 (mixed inoculation treatment); also, non-inoculated seeds were sown in infested soil (control treatment). Experiments were conducted in a growth chamber for 30 days and the evaluated parameters were: germination percentage, damping-off at 5 and 8 days, number of surviving plants at 15 and 30 days, and number of nodules.

L. B. Guiñazú (✉)
Laboratorio de Interacción Planta-Microorganismo, Facultad de Ciencias Exactas,
Físico-Químicas y Naturales, Universidad Nacional de Río Cuarto, Ruta 36 Km 601, Río Cuarto,
Argentina
e-mail: lguinazu@exa.unrc.edu.ar

A. Malik, E. Grohmann (eds.), *Environmental Protection Strategies for Sustainable Development,* Strategies for Sustainability,
DOI 10.1007/978-94-007-1591-2_10, © Springer Science+Business Media B.V. 2012

The pathogen incidence was high in the control treatment, with a low germination rate, high occurrence of damping-off and death of all of the plants at the end of the experiment. *Pseudomonas* sp. Ch2 performed the best, since singly inoculated seeds reached a germination percentage of 83.3%; in addition, there was no damping-off incidence after 5 and 8 days neither in single inoculation nor in mixed inoculation treatments. The number of surviving plants at 15 days was 9/18 for the *Pseudomonas* sp. Ch2 treatment and 15/18 for the *Pseudomonas* sp. Ch2-*S. meliloti* B399 treatment. At 30 days, the number of surviving plants was 9/18 for both treatments. *Pseudomonas* sp. Ch2 showed antifungal activity against the alfalfa pathogen *M. phaseolina* in the *in vitro* as well as in the *in vivo* assays.

Keywords Alfalfa • Damping-off • Biological control • *Macrophomina phaseolina* • Mixed inoculants

10.1   Alfalfa Cultivation

The *Leguminosae* (*Fabaceae*) Family is constituted of approximately 18.000 species. There are three sub-families, *Caesalpinioideae*, *Mimosoideae* and *Papilionoideae*. Plants exhibit a great vegetative and floral diversity.

Alfalfa or lucerne (*Medicago sativa* L.), native from the Asian Southwest, is a legume widely used as forage (Smith 1981; Park et al. 2005) because of its worth in proteins, minerals and vitamins. Historical references mention that it was cultivated for the first time in Persia, from where it passed to Greece and later on to Spain. The arrival to the new world took place in 1519, in Mexico. Later on, through the Pacific route, it was transferred to Peru and Chile, and from these countries, by land, to Argentina (Tomes 1947). At the present time, its cultivation is widely extended in Asia, Europe and the Americas. The term "alfalfa" is of Arabic origin and it means the "best food".

Because the root can introduce itself up to 9 m deep in the ground, it can easily reach any available reserves of nutrients, but it is sensitive to the lack of O_2. It also adapts very well to varying climatic conditions, requiring certain conditions and an appropriate cultivation system. It presents a germination period of about 6 days, reaching a good foliate expansion at 30 days.

Its nutritious quality, forage production, growth habit, perpetuation, plasticity and symbiotic fixation capacity of atmospheric nitrogen, transform it into an essential species for many agricultural production systems, from intensive corral methods that include it in animal diet as harvested and processed forage (Zubizarreta 1992), to the pastoral methods that use it in direct shepherding (Wilberger 1984; Roberto and Viglizo 1993).

Argentina is the second country of importance in the world for alfalfa cultivation, with several millions of cultivated hectares. It has traditionally been the base for livestock production and, together with wheat and linen, one of the founding crops of the "Argentinean Pampas" agriculture. In most of the Argentinean regions,

10 Isolation and Characterization of Rhizobacteria Antagonistic

where milk and meat production is important, this forage species is basic in the feeding (Viglizzo 1982; Chimicz 1988; Spil and Salgado 1992). Nevertheless, the real dimension of its value arises when its fundamental role in the maintenance of both the structure and the nitrogen fertility of the soils in which it grows, correctly associated to specific rhizobia strains, is considered (Musiera Pardo and Ratera García 1984).

When compared with other species, alfalfa forage has a higher protein content; consequently, it has high requirements for nitrogen (Bickoff 1979; INTA-FAO 1986; Howarth 1988). A production of 15 Tn/ha of forage consumes around 450 kg of this element, which can cause a quick and constant loss of nitrogen fertility in the soil if a partial recycling method of nitrogen, together with an efficient biological nitrogen fixation system, does not exist.

10.1.1 Alfalfa-Bacteria Interactions

There are numerous reports that indicate that the interaction among rhizobia (intracellular-Plant Growth Promoting Rhizobacteria (PGPR)) and other rhizobacteria (extracellular-PGPR) can be favorable for cultivation (Gray and Smith 2005). Rovera et al. (2008) showed that *Pseudomonas aurantiaca* strain SR1 stimulates alfalfa growth and increases the number and size of nodules. Co-inoculation of alfalfa with *Pseudomonas* spp. and *Sinorhizobium meliloti* produced increase in nodulation, nitrogen fixation, plant biomass and grain yield (Knight and Langston-Unkefer 1988). Similar results have been observed in soybean when co-inoculating it with *Bradyrhizobium japonicum* and PGPR (Singh and Subba Rao 1979; Polonenko et al. 1987; Nishijima et al. 1988; Dashti et al. 1998; Rovera et al. 2008). Guiñazú et al. (2010) observed a beneficial effect of two strains isolated from alfalfa. *Pseudomonas* sp. FM7d caused a significant increase in shoot and root dry weight, length and surface area of roots, and number and symbiotic properties of alfalfa plants. Similarly, plants co-inoculated with *Sinorhizobium meliloti* strain B399 and *Bacillus* sp. strain M7c showed significant increases in the measured parameters.

Numerous authors conclude that mixed inoculants provide an enhanced nutritional balance to plants, and that improvements in N and P uptake would be the main mechanisms involved (Rodríguez and Fraga 1999). It has also been observed that some PGPR strains are able to additionally stimulate infection sites which are eventually occupied by rhizobia (Plazinski and Rolfe 1985). Besides the beneficial effect of rhizobacteria on alfalfa growth, their role as potential diseases control agents has also been studied. For instance, antibiotic-producing *Streptomyces* strains, isolated from Minnesota, Nebraska and Washington soils, were evaluated for their ability to inhibit plant pathogenic *Phytophthora medicaginis* and *Phytophthora sojae*. On alfalfa, isolates varied in their effect on plant disease severity, percentage of dead plants, and plant biomass in the presence of the pathogen (Kun et al. 2002).

10.1.2 Alfalfa and Its Diseases

Plagues and diseases can considerably reduce the quality, persistence and nutritious value of forage (Reed et al. 1994). Alfalfa is affected by several diseases that attack leaves, stems, crown and roots. Foliate pathogens cause intense defoliations during certain seasons; however, those that attack crown and roots define the longevity or productive period of alfalfa in a direct form. The main sanitary problems are: declination of plants with a scarce bud number, areas without plants and invaded by underbrushes, and low productivity. This allows the survival of pathogens through the whole annual cycle, which produces permanent re-infections with a high impact on forage quantity and quality.

Some of the most common pathogens affecting alfalfa are:

Alfalfa Mosaic Virus: Infected plants present a green yellowish coloration in their leaves. This disease can cause death of the plants. However, many of the infected plants never show recognizable signs of such disease (Hirnyck and Downey 2004).

Colletotrichum trifolii (Anthracnosis): Is the fungal pathogen that causes alfalfa anthracnosis (Barnes et al. 1969), which is one of the most economically important fungal diseases of alfalfa, particularly in hot and humid areas (Churchill et al. 1988). *C. trifolii* attacks leaves, stems, and crowns of susceptible alfalfa cultivars.

Sclerotinia trifoliorum (Sclerotinia/White Mold): Although it is only a minor problem in alfalfa seed production, *Sclerotinia* is a common disease issue in crimson clover seed. Spores of this fungus are wind blown, and plants of all ages are susceptible to colonization, though the disease is more common earlier in the season. Diseased leaves fall and are covered with white fungus growth, allowing the disease to spread to crown and roots. The majority of the damage is done during vegetative growth, but cultural control measures can be taken prior to planting (Pratt and Rowe 1991).

Sclerotium rolfsii (Southern blight): The fungus produces a cottony white growth on the stem or crown, near the soil surface. It forms sclerotia on stem and crown and on dead parts of the plant. Plants lose their color and die.

Macrophomina phaseolina (Tassi) Goid.: In varying climates, from arid to tropical, is the causal agent of plant blight, root and stem charcoal rot, and damping-off of several hosts. It is considered worldwide as one of the pathogens that cause higher economic losses (Sinclair 1982; Jiménez Díaz et al. 1983; Abawi and Pastor-Corrales 1990).

10.1.2.1 *Macrophomina phaseolina* (Tassi) Goid

It belongs to the *Deuteromycotina* sub-division, *Coelomycetes* class. Is a major soil-borne fungal pathogen that infects many agronomic, horticultural, and ornamental crops (Young 1949; Wyllie 1989; Mihail 1982). The diseases caused by this fungus have been described in more than 400 plant species and the wide host range includes important crops such as soybean, sorghum, cotton, peanut, alfalfa and several other plants used by man (Dhingra and Sinclair 1977; Diaz Polanco and Salas de Diaz 1980; Partridge 2000).

It is unusual that a wide variety of both monocots and dicots are parasitized. The pathogen infects young as well as mature plants, producing black lesions at first and, later on, extensive stains of a gray whitish color, in which the reproductive structures of the pathogen are located (sclerotia and pycnidia). Diseases caused by *M. phaseolina* are sometimes referred to as "charcoal rot" on account of small, black, macroscopically visible sclerotia that form in shredded, parasitized host tissue and cause an appearance of charcoal (Young 1949). Primary infection by *M. phaseolina* usually occurs in roots, but pathogenesis and sclerotia formation may extend above ground. Collapse and death of infected plants usually is favored by hot and dry conditions, but numerous sclerotia also may form in live host tissues without visible disease symptoms (Kendig et al. 2000). Sclerotia are the resting structures of this pathogen in soil and infested plant debris (Bhattacharya and Samaddar 1976; Papavizas 1977; Short et al. 1980; Wyllie 1989; Mihail 1982; Kendig et al. 2000) and their maximum size is of 80–100 μm of diameter, presenting a solid appearance, due to the zipping and collapse of external cells of the strongly pigmented peripheral hyphae, and an even cross-linking throughout all of their structure, without formation of special tissues or rings (Gunta Smits and Noguera 1988).

Implantation diseases caused by *M. phaseolina* are one of the factors affecting pastures yield the most. Pre-emergence "damping-off" is characterized by rotting of seeds, which are soften and filled with water; consequently, roots do not emerge. In post-emergence "damping-off", the pathogen attacks young stems causing the fall and death of the whole plant. Chemical control of this pathogen is expensive and not very feasible, since it can be found in soil as well as on seeds; for this, biological control offers an interesting alternative for its handling.

10.2 Biological Control: Application Perspectives for Suppression of *M. Phaseolina* in Alfalfa

Various disease management methods have been implemented to combat and eradicate pathogenic fungi. These include cultural, regulatory, physical, chemical and biological methods. All these methods are effective only when employed well in advance as precautionary measures (Sharma 1996; Kata 2000). Once a disease has appeared, these methods become impractical/ineffective. In that situation, chemical control offers a good option for growers to control diseases. Chemical pesticides have been in use since long and they provide quick, effective and economic management of plant diseases. Nevertheless, in recent past, it has been understood that the use of chemicals in agriculture is not as beneficial as it was visualised. Chemicals pose serious health hazards to an applicator as well as to a consumer of the treated material (Ramezani 2008). In addition to target organisms, pesticides

334   L. B. Guiñazú et al.

also kill various beneficial organisms and their toxic forms persist in soil and contaminate the whole environment (Hayes and Laws 1991). The understanding of the adverse effects caused by the indiscriminate use of chemical pesticides propitiated the worldwide resurgence of research on the use of formulations with microorganisms able to control pathogens and to improve plant growth. A biological control agent must be distributed in the root, multiply and survive during several weeks, in competition with other microorganisms of the indigenous microbiota, in order to suppress a plant disease (Duffy and Défago 1999). Natural organisms (rhizobacteria) are used to reduce the effects of undesirable organisms (pathogens); thus, crop production is favored. Among the populations that constitute soil biodiversity, several groups of beneficial microorganisms such as the genera *Bacillus*, *Burkholderia*, *Enterobacter*, *Pseudomonas*, *Paenibacillus* and *Serratia* are considered important for the control of plant diseases (Whipps 2001).

Another important aspect to keep in mind for using these biological control agents is their ecological relationship with other beneficial bacteria that also live on plant roots, and especially with rhizobia. This makes possible the development of mixed inoculants, useful to directly promote the growth of leguminous forages by supplying nitrogen and by protecting plants from fungal diseases (De La Fuente et al. 2002). Numerous species of *Pseudomonas* (*P. fluorescens*, *P. putida* and *P. aeruginosa*), isolated from plant tissues or soil, have been reported as potential biological control agents of phytopathogenic fungi (Weller 1988; Thomashow et al. 1990; Bakker et al. 1991; Duffy et al. 1996).

With this perspective, our research group assessed the *in vitro* and *in vivo* antifungal activity of rhizobacteria isolated from alfalfa rhizosphere against *M. phaseolina*, causal agent of alfalfa damping-off.

The phytopathogenic fungus *M. phaseolina* (Tassi) Goid. was isolated from infected plant tissue and is now a part of our culture collection. The bacterial strains used were: *Pseudomonas aurantiaca* SR1 and alfalfa rhizosphere soil isolates. Assays of *in vitro* antagonism were carried out by using surface inoculation methods, both simultaneously and with an offset. In addition, *in vivo* assays were performed with those strains that showed higher antagonistic activity in the *in vitro* biocontrol tests: *Bacillus* sp. Ball, *Pseudomonas* sp. Manf, *Pseudomonas* sp. UI, *Pseudomonas* sp. Ch2 (GenBank accession numbers: GQ853342, GQ853344, GQ853345 and GQ853346, respectively), *Sinorhizobium meliloti* B399 and *P. aurantiaca* SR1.

In the *in vivo* assays, in absence of the pathogen, the germination percentages were superior to 88% in both single and mixed inoculation treatments. Non-inoculated alfalfa seeds that were planted in infested soil showed a low germination percentage. In contrast, the germination percentage was considerably increased when planting inoculated seeds in infested soil (Table 10.1). A similar result was observed by Arora et al. 2001. It has been proven that inoculation of seeds with a potential antagonistic agent produces beneficial effects on plant growth. The inoculated population on seeds would not allow the invasion of the pathogenic microorganism that arrives to the plant to progress (Rovera 2004).

The incidence of "damping off" was higher at the 8th day than at the 5th day, since the number of fallen plants increased in most of the treatments with infested

Table 10.1 Effect of *M. phaseolina* and selected strains on germination of seeds, plant survival and "damping off" incidence

Biological control						
Treatment	Germinated seeds	Germination %	Damping off at the 5th day	Damping off at the 8th day	N° of surviving plants at 15 days	N° of surviving plants at 30 days
Alfalfa control	18	100	0	0	18	18
Pathogen control	5	27.8	2	3	0	0
S. meliloti B399 + *M. phaseolina*	9	50	2	3	3	3
P. aurantiaca + *M. phaseolina*	14	77.8	0	1	6	3
Bacillus sp. Bal1 + *M. phaseolina*	9	50	1	5	5	0
Pseudomonas spp Ch2 + *M. phaseolina*	15	83.3	0	0	9	9
Pseudomonas sp. Ul + *M. phaseolina*	2	11.1	0	1	1	0
Pseudomonas sp. Manf + *M. phaseolina*	10	55.6	1	2	4	2
S. meliloti B399 + *P. aurantiaca* + *M. phaseolina*	13	72.2	0	3	3	1
S. meliloti B399 + *Bacillus* sp. Bal1 + *M. phaseolina*	1	5.6	0	1	0	0
S. meliloti B399 + *Pseudomonas* sp. Ch2 + *M. phaseolina*	14	77.8	0	0	12	9
S. meliloti B399 + *Pseudomonas* sp. Manf + *M. phaseolina*	8	44.4	2	2	2	0
S. meliloti B399 + *Pseudomonas* sp. Ul + *M. phaseolina*	3	16.7	0	2	0	0

soil. In most of the treatments with infested soil, the number of surviving plants drastically decreased at 15 days post-planting; moreover, there was no surviving plants in some of the treatments. The affected plants showed necrosis symptoms at the crown and root apex level. In most of the treatments, the number of surviving plants at 30 days was very low (Table 10.1).

Seed inoculation with isolate *Pseudomonas* sp. Ch2 favored plant growth, since no disease symptoms were observed during the different stages of plant growth and development. The strain SR1 inoculation treatment in infested soil showed a high germination percentage, whereas the *Pseudomonas* sp. Ch2 and *S. meliloti/ Pseudomonas* sp. Ch2 inoculation treatments, in infested soil, showed the highest germination percentages; in addition, these treatments, both with a high survival at 30 days, did not show "damping off" incidence neither at the 5th nor at the 8th day.

10.2.1 Growth Parameters

Determination of root length showed that single inoculation with strain SR1, in infested soil, caused a clear growth-promoting effect. In absence of the pathogen, the *S. meliloti* B399/*P. aurantiaca* SR1 co-inoculation treatment stood out. Similar results were observed in shoot length of plants co-inoculated with the above mentioned strains. When analyzing root fresh weight, the *Pseudomonas* sp. Ch2 and *Pseudomonas* sp. Manf treatments showed mean values superior to those of the absolute control (non-infested soil) in some of the variables; as a consequence, we can infer that these strains not only protected seeds against the pathogen attack but also protected fully developed plants (data not shown).

Regarding the number and dry weight of nodules, no significant differences were observed between co-inoculation treatments in absence of the pathogen. In the treatments in infested soil, the number and fresh and dry weight of nodules were notoriously lower.

Co-inoculation treatments in infested soil allowed us to prove that seed inoculation with biocontrolling PGPR does not influence rhizobia growth-promoting activity in alfalfa. Similar results were observed by De la Fuente et al. (2002) when working with alfalfa, lotus and white clover plants.

Disease incidence decreased and plant biomass increased in an assay carried out in peanut, in soil infested with *M. phaseolina*. Indeed, plants that were inoculated with PGPR strains stayed healthy and did not show disease symptoms. Nodulation was low and there was a decrease in fresh weight of nodules, when comparing with control plants. The increase in biomass was higher in treatments planted in non-infested soil (Arora et al. 2001).

In our study, a decrease in *M. phaseolina* diseases incidence (damping-off and crown and root apex nechrosis) was observed after inoculation with antagonistic strains, since non-inoculated alfalfa seeds, planted in infested soil, had a low germination percentage (27.8%), presented damping-off symptoms and they did not survive until harvest (30 days post-sowing).

Assays carried out with electronic photomicrography in a dual culture of *M. phaseolina* and fluorescent *Pseudomonas* showed loss of sclerotia structural integrity through the interaction area, as well as hyphae looping and mycelium and sclerotia deformations; hence, the mentioned structures were lysed (Gupta et al. 2001).

10.3 Conclusions

At the present time, biological control occupies an important place within sustainable handling practices of plant disease caused by soil-borne fungal pathogens.

One of the greatest advantages of biological control is that biocontrol agents are frequently endowed with properties similar to those of the pathogen, such as multiplication and dispersion. They can act at several levels in the disease cycle, interfering in survival of the pathogen in the external environment, in its development on the host surface and entrance to the host, and later transmission among hosts; furthermore, the antagonist can compete with the pathogen inside the host´s tissues. In fact, it is an exploitation of certain ecological principles such as competition and antagonism that govern the interaction processes among living beings.

Biological control contemplates direct inoculation with selected antagonists or the creation of favorable conditions for the development of beneficial native microorganisms.

In summary, the tested rhizobacteria represent potential biological control agents of alfalfa fungal pathogens, as well as growth-promoting agents. The later observation was reflected by the increase in certain growth parameters in absence of the pathogen.

Acknowledgements This research was supported by Secretaría de Ciencia y Técnica of Universidad Nacional de Río Cuarto, Agencia Nacional de Promoción Científica y Tecnológica (ANPCyT) and Consejo Nacional de Investigaciones Científicas y Técnicas (CONICET), Argentina.

References

Abawi GS, Pastor MAC (1990) Root rots of beans in Latin America and Africa: Diagnosis, research methodologies, and management strategies. Cali, Colombia. Centro Internacional de Agricultura Tropical (CIAT). pp 114

Arora NK, Kang SC, Maheshwari DK (2001) Isolation of siderophore-producing strains of *Rhizobium meliloti* and their biocontrol potential against *Macrophomina phaseolina* that causes charcoal rot of groundnut. Curr Sci 81:673–677

Bakker PAHM, Van Peer R, Schippers B,(1991) Suppression of soil-borne plant pathogens by fluorescent *Pseudomonas*: mechanisms and prospects. In: Beemster, ABR, Bollen M, Gerirch M, Ruissen MA, Schippers B, Tempel A (eds) Biotic interactions and soil-borne diseases. Elsevier, Amsterdam, pp 221–230

Barnes DK, Ostazeski SA, Schillinger JA, Hanson CH (1969) Effects of anthracnose (*Colletotrichum trifolii*) infection on yield, stand, and vigor of alfalfa. Crop Sci 9:344–346

Bhattacharya M, Samaddar KR (1976) Epidemiological studies on jute diseases, survival of *Macrophomina phaseoli* (Maubl.) Ashby in soil. Plant Soil 44:27–36

338   L. B. Guiñazú et al.

Bickoff EM (1979) En Alfalfa Science and Technology. Agron No 15. Cap.12

Chimicz J (1988) Los sistemas de producción de leche en Argentina Revista. Argentina de Producción Animal 8:155–168

Churchill A, Baker C, O'Neill N, Elgin J (1988) Development of *Colletotrichum trifolii* race 1 and 2 on alfalfa clones resistant and susceptible to anthracnose. Can J Bot 66:75–81

Dashti N, Zhang F, Hynes R, Smith DL (1998) Plant growth promoting rhizobacteria accelerate nodulation and increase nitrogen fixation activity by field grown soybean [*Glycine max* (L.) Merr.] under short season conditions. Plant Soil 200:205–213

De La Fuente L, Quagliotto L, Bajsa N, Fabiano E, Altier N, Arias A (2002) Inoculation with *Pseudomonas fluorescens* biocontrol strains does not affect the symbiosis between rhizobia and forage legumes. Soil Biol Biochem 34:545–548

Dhingra OD, Sinclair JB (1977) An annotated bibliography of *Macrophomina phaseolina*, 1905–1975. Universidade Federal de Viçosa, M.G. Brasil, p 277

Díaz Polanco C, Salas de Díaz G (1980) Lista de patógenos de las plantas cultivadas en Venezuela. 2da. Edición. Boletín Técnico No. 20. CIARCO, p 62

Duffy B, Défago G (1999) Environmental factors modulating antibiotic and siderophore biosynthesis by *Pseudomonas fluorescens* Biocontrol strains. Appl Environ Microbiol 65:2429–2438

Duffy BK, Simon A, Weller DM (1996) Combination of *Trichoderma koningii* with fluorescent pseudomonads for control of take-all on wheat. Phytopathology 75:774–777

Gray EJ, Smith DL (2005) Intracellular and extracellular PGPR: commonalities and distinctions in the plant-bacterium signaling processes. Soil Biol Biochem 37:395–412

Guiñazú LB, Andrés JA, Del Papa MF, Pistorio M, Rosas SB (2010) Response of alfalfa (*Medicago sativa* L.) to single and mixed inoculation with phosphate-solubilizing bacteria and *Sinorhizobium meliloti*. Biol Fertil Soil 46:185–190

Gunta Smits B, Noguera R (1988) Ontogenia y morfogenesis de esclerocios y picnidios de *Macrophomina phaseolina*. Agronomía Tropical 38:69–78

Gupta CP, Dubey RC, Kang SC, Maheshwari DK (2001) Antibiosis-mediated necrotrophic effect of *Pseudomonas* GRC2 against two fungal plant pathogens. Curr Sci 81: 91–94

Hayes WJ, Laws ER (1991) Handbook of pesticide toxicology, vol 1. Academic Press Inc., New Delhi

Hirnyck R, Downey L (2004) Pre-Plant/Establishment. In: O'Neal Coates S (ed) Summary of pest management strategic plan for western u.s. alfalfa and clover seed production

Howarth RE (1988) En alfalfa and alfalfa improvement. Agronomy No 29, Cap.15

INTA-FAO (1986) Principios de manejo de praderas naturales. Técnicas para medir vegetación. Edit. INTA. pp 151–162

Jimenez Díaz RM, Blanco López MA, Sackston WE (1983) Incidence and distribution of charcoal rot of sunflower caused by *Macrophomina phaseoli* in Spain. Plant Dis 67:1033–1036

Kata J (2000) Physical and cultural methods for the management of soil borne pathogens. Crop Prot 19:725–731

Kendig SR, Rupe JC, Scott HD (2000) Effect of irrigation and soil water stress on densities of *Macrophomina phaseolina* in soil and roots of two soybean cultivars. Plant Dis 84:895–900

Knight TJ, Langston-Unkefer PJ (1988) Enhancement of symbiotic dinitrogen fixation by a toxin-releasing plan pathogen. Sci 241:951–954

Kun X, Kinkel L, Samac DA (2002) Biological Control of *Phytophthora* Root Rots on Alfalfa and Soybean with *Streptomyces*. Biol Control 23:285–295

Mihail JD (1982) Methods for research on soilborne phytopathogenic fungi. In: Singleton LL, Mihail JD, Rush CM (eds) Macrophomina. American Phytopathological Society, St. Paul, pp 34–136

Musiera Pardo E, Ratera García C (1984) La alfalfa. En: Praderas y forrajes Ed. Mundiprensa. Madrid España, pp 625–694

Nishijima F, Evans WR, Vesper SJ (1988) Enhanced nodulation of soybean by *Bradyrhizobium* in the presence of *Pseudomonas fluorescens*. Plant Soil 111:149–150

Papavizas GC (1977) Some factors affecting survival of sclerotia of *Macrophomina phaseolina* in soil. Soil Biochem 9:337–341

10 Isolation and Characterization of Rhizobacteria Antagonistic   339

Park BH, Kim SA, Kim TH et al (2005) Tchosaryo Zaweonhag (Science of Forage Crop Resources). Hyangmun Publishing, Seoul, 17–257 (in Korean.)
Partridge D (2000) *Macrophomina phaseolina.* NC State university. Department of plant pathology. Cited 24 June 2010 http://www.cals.ncsu.edu/course/pp728/Macrophomina/macrophomina_phaseolinia.HTM
Plazinski J, Rolfe BG (1985) Influence of *Azospirillum* strains on the nodulation of clovers by *Rhizobium* strains. Appl Environ Microbiol 49:984–989
Polonenko DR, Scher FM, Kloepper JW, Singleton CA, Laliberte M, Zaleska I (1987) Effects of root colonizing bacteria on nodulation of soybean roots by *Bradyrhizobium japonicum.* Can J Microbiol 33:498–503
Pratt RG, Rowe DE (1991) Differential responses of alfalfa genotypes to stem inoculations with *Sclerotinia sclerotiorum* and *S. trifoliorum.* Plant Dis 75:188–191
Ramezani H (2008) Biological Control of Root-Rot of Eggplant Caused by *Macrophomina phaseolina.* American-Eurasian J Agric Environ Sci 4:218–220
Reed R, Sanderson M, Read J (1994) Harvest management of switchgrass grown for biomass. In Forage Res. In TX. TAES P.R
Roberto ZE, Viglizzo EF (1993) Análisis del impacto de los recursos forrajeros en agrosistemas de la pampa semiárida. Revista Argentina de Producción Animal 10:47–54
Rodríguez H, Fraga R (1999) Phosphate solubilizing bacteria and their role in plant growth promotion. Biotechnol Advan 17:319–339
Rovera M (2004) Caracterización del pigmento antifúngico de *Pseudomonas aurantiaca* utilizada como agente de biocontrol. Tesis Doctoral. UNRC
Rovera M, Andrés J, Carlier E, Pasluosta C, Rosas S (2008) *Pseudomonas aurantiaca*: Plant growth promoting traits, secondary metabolites and inoculation response. In: Ahmad I, Pichtel J, Hayat S (eds) Plant-Bacteria Interactions. Strategies and Techniques to promote Plant growth. Wiley-VCH, Germany, pp 255–264. (ISBN: 978-3-527-31901-5. Chapter 8)
Sharma PD (1996) Plant pathology. Rastogi Publication Meerut, India
Short GE, Wyllie TD, Bristow PR (1980) Survival of *Macrophomina phaseolina* in soil in residue of soybean. Phytopatol 70:13–17
Sinclair JB (1982) Compendium of soybean diseases. American Phytopathology Society. 2nd edn, p 104
Singh CS, Subba Rao NS (1979) Associative effect of *Azospirillum brasilense* with *Rhizobium japonicum* on nodulation and yield of soybean (*Glycine max*). Plant Soil 53:387–392
Smith D (1981) Forage Management in the North, 4th edn. Kendall/Hunt Publishing, Dubuque, pp 5–99
Spil G, Salgado L (1992) Forrajeras en el oeste arenoso Revista CREA No 153, pp 26–31
Thomashow LS, Weller DM, Bonsall RF, Pierson IS (1990) Production of the antibiotic phenazine-1-carboxylic acid by fluorescent *Pseudomonas* species in the rhizosphere of wheat. Appl Environ Microbiol 56:908–912
Tomes G A (1947) La alfalfa en al Argentina. Anales Soc Rural Arg 81:82–90
Viglizzo EF (1982) Los potenciales de producción de carne en la región pampeana semiárida. Actas de las Primeras Jornadas de Producción Animal en la Región Pampeana Semiárida, pp 223–269
Weller DM (1988) Biological control of soil borne plant pathogens in the rhizosphere with bacteria. Ann Rev Phytopathol 26:379–407
Whipps JM (2001) Microbial interactions and biocontrol in the rhizosfere. J Experimen Bot 52:487–511
Wilberger JJ (1984) Tecnología disponible para mejorar la producción lechera en la región oeste de Buenos Aires y La Pampa. I simposio sobre la Integración Producción- Industria en la lechería Argentina. Santa Rosa La Pampa Arg
Wyllie TD (1989) Charcoal rot. In: Sinclair JB, Backman PA (eds) Compendium of Soybean Diseases, 3rd edn. American Phytopathological Society, St Paul, pp 30–33
Young PA (1949) Charcoal rot of plants in east Texas. Bull. 712, Texas Agr Exp Sta 33 p 4
Zubizarreta J (1992) Producción lechera en EE. UU. revista CREA No 156, pp 80–84

Chapter 11
Biofilm Formation by Environmental Bacteria

Mohd Ikram Ansari, Katarzyna Schiwon, Abdul Malik and Elisabeth Grohmann

Contents

11.1	Introduction	342
11.2	Biofilm Formation by Bacteria	343
11.3	Development of Resistance in Biofilm	344
11.4	Process of Biofilm Formation	345
	11.4.1 Attachment	345
	11.4.2 Microcolony Formation	345
	11.4.3 Formation of Three-Dimensional Structure and Maturation	346
	11.4.4 Detachment	346
11.5	Bacterial Extracellular Polysaccharides and Biofilm Formation	346
11.6	Factors Involved in Regulation of Biofilm Formation	348
	11.6.1 Surface	349
	11.6.2 Nutrients	349
	11.6.3 Environmental Cues	350
	11.6.4 Gene Regulation	350
11.7	Detection Methods of Biofilm Formation	351
	11.7.1 Screening of Bacterial Isolates for Biofilm Formation	351
	11.7.2 Detection of Quorum Sensing Molecules	352
	11.7.3 Biofilm Formation in a Bioreactor	353
	11.7.4 Fluorescence Tools to Visualize Biofilms	354
	11.7.5 Biofilm Imaging Techniques	355
11.8	Antibiotic Resistance of Bacterial Biofilms	355
	11.8.1 Occurrence and Architecture of Bacterial Biofilms	355
	11.8.2 Slow Growth and Low Oxygen Concentration	356
	11.8.3 Mutators	356
	11.8.4 Tolerance to Antibiotics and Efflux Pumps	356
	11.8.5 Mechanisms of Antimicrobial Resistance	357
11.9	Inhibition of Biofilm Formation	358
	11.9.1 Target- Based Screening	358
	11.9.2 Activity-Based Screening for QS Inhibitors	358
	11.9.3 Structure-Based Screening for QS Inhibitors	359
	11.9.4 Nucleotide Biosynthesis and DNA Replication Inhibitors	360

E. Grohmann (✉)
Department of Infectious Diseases, University Hospital Freiburg, Hugstetter Strasse 55, 79106 Freiburg, Germany
e-mail: elisabeth.grohmann@uniklinik-freiburg.de

A. Malik, E. Grohmann (eds.), *Environmental Protection Strategies for Sustainable Development*, Strategies for Sustainability,
DOI 10.1007/978-94-007-1591-2_11, © Springer Science+Business Media B.V. 2012

11.10	Incidence of Biofilm in the Environment	360
11.11	Adaptation of Biofilm Structure for Survival in Varying Environments	362
11.12	Role of Macromolecules in Biofilm Formation	363
11.13	Biofilm and Its Environmental Significance	363
11.14	Control of Biofilm Growth in the Environment	365
11.15	Applications of Quorum Sensing in Biotechnology	366
11.16	Conclusions and Perspectives	367
References		367

Abstract The majority of bacteria in the environment live associated with surfaces, in so called biofilms. Bacterial cells embedded in a biofilm can better withstand environmental stress, such as nutrient deprivation, unphysiological temperatures and pH changes. Within the biofilm they become more resistant to detachment, oxygen radicals, disinfectants, and antibiotics than the individual planktonic cells. In this chapter, the current status of biofilm research is summarized, with focus on the mechanims involved in formation of biofilms, characteristics of bacteria living in biofilms, e.g. the production of extracelluar polymeric substances (EPS) and the intercellular communication via quorum sensing. Detrimental and beneficial effects of microbial biofilms are described, as well as their application in modern biotechnology. An overview about state of the art techniques to analyse complex biofilms is given, as well as a summary on existing and emerging biofilm inhibitors. We developed a continuous upflow biofilm reactor system where mixed species environmental biofilms can form attached to glass beads. Studies on these biofilms by lectin-binding analysis and fluorescence microscopy are described. Experimental systems developed to visualize biofilms by fluorescent labels using confocal laser scanning microscopy (CLSM) and the current strategies in removing or controlling the biofilm are dicussed. The chapter ends with perspectives on the development of new emerging biofilm inhibitors and with an outlook on new promising techniques that will enable analysis of the composition as well as the structure of biofilms in even more detail.

Keywords Biofilm • Bacteria • Quorum sensing • Antibiotic resistance • Lectin

11.1 Introduction

Biofilms are surface-associated, three-dimensional multicellular structures whose integrity depends upon the extracellular matrix produced by their constituent bacterial cells. Biofilm formation occurs as a result of a sequence of events: adhesion of individual microbial cells to a surface, cell proliferation and aggregation into microcolonies, matrix production, and cell detachment. Initiation of biofilm formation is characterized by the interaction of bacterial cells with a surface and with each other. The biofilm matures through the production of extracellular matrix which is mainly composed of sticky extracellular polysaccharides (EPS) and proteins (Jefferson 2004; Branda et al. 2005). For a recent review on biofilms of the well-studied model bacteria, *Escherichia coli, Pseudomonas aeruginosa, Bacillus subtilis*, and *Staphylococcus aureus*, and the mechanisms by which extracellular signals trigger biofilm formation, reading of the review article of Lopez et al. (2010) is recommended.

Biofilms may constitute up to 80% of the total microbial population on plant surfaces (Lindow & Brandl 2003). Morris et al. (1998) reported that bacteria in biofilms constituted 10 and 40% of the total bacterial population on parsley and endive, respectively. Similarly, Fett (2000) observed that, more than half of the total mesophilic counts on alfalfa sprouts were present in biofilms. Donlan and Costerton (2002) and Adnan et al. (2010) have recently realized that more than 99% of all bacteria exist as biofilms, which are defined as a collection of microorganisms that are attached to a surface and enclosed in an extracellular matrix allowing growth and survival in sessile environment (Adnan et al. 2010).

Bacterial cells embedded within a biofilm can better withstand nutrient deprivation and pH changes. They become more resistant to detachment, oxygen radicals, disinfectants, and antibiotics than the individual cells (Jefferson 2004). The first mention of biocomplexity in the form of a biofilm was in the dental plaque and visualized at the onset of microbiology by Leeuwenhoeck (1683) following the development of the first microscope. He described them as different forms of "animalculi" adhering to his teeth. Biofilms are a complex society of interacting microbial communities that attach to various materials at the solid-liquid interface and also at the liquid-gas interface (Davey and O'Toole 2000; Branda et al. 2005). Remarkably, biofilms can have enhanced resistance to solvents and toxins when compared to their suspension counterparts (O'Toole et al. 2000; Hall-Stoodley et al. 2004; Anderson and O'Toole 2008). This distinction is observed when pathogenic microbes forming microbial biofilms exhibit enhanced resistance to antimicrobial agents and cause chronic infections and persistent disease (Costerton et al. 1999; O'Toole et al. 2000; Hall-Stoodley et al. 2004; Fux et al. 2005; Brady et al. 2008; Bryers 2008; Spormann 2008).

Apart from their natural environments, where biofilms are widely distributed, they have found an application in biotechnology as an immobilization method. Today, natural immobilization of bacteria as an industrial application is widely used for the treatment of wastewater, desulfurization of gas, and food production (Lazarova et al. 2000; Majumder and Gupta 2003; Kornaros and Lyberatos 2006), as well as for converting agriculturally derived materials into alcohols and organic acids such as acetic acid, ethanol, and butanol (Crueger and Crueger 1990; Demirci et al. 1997; Ho et al. 1997; Qureshi et al. 2004, 2005; El-Mansi and Ward 2007; Wang and Chen 2009).

11.2 Biofilm Formation by Bacteria

At present, there are several hypotheses of biofilm formation and surface association that make it such a widespread phenomenon. First, surfaces provide a space to be occupied and provide a degree of stability in the growth environment and might have catalytic functions through localizing cells in close proximity. Second, biofilm formation affords protection from a wide range of environmental challenges, such as UV exposure (Espeland and Wetzel 2001), metal toxicity (Teitzel and Parsek 2003), acid exposure (McNeill and Hamilton 2003), dehydration and salinity (Le Magrex-Debar et al. 2000), phagocytosis (Leid et al. 2002) and several antibiotics and antimicrobial agents (Mah and O'Toole 2001; Stewart and Costerton 2001; Gilbert et al. 2002). Bacterial biofilm is a structured community of bacterial cells enclosed in a self-produced polymeric matrix and adherent to an inert or living surface, which constitutes a protected mode of growth that allows survival in hostile environment. The biofilm-forming microorganisms have been shown to elicit specific mechanisms for initial attachment to a surface, formation of microcolonies leading to the development of the three-dimensional structure of mature biofilm. They differ from their free-living counterparts in their growth rate, composition of EPS and increased resistance to biocides, antibiotics and antibodies by virtue of up regulation and/or down regulation of approximately 40% of their genes (Prakash et al. 2003). This makes them highly difficult to eradicate with therapeutic doses of antimicrobial agents. A greater understanding of the mechanism of biofilm formation and survival under sessile growth conditions may help in devising control strategies.

11.3 Development of Resistance in Biofilm

Three mechanisms have been proposed to explain the general resistance of biofilms to biocidal agents. The first mechanism is due to the barrier properties of the slime matrix. This mechanism might be more relevant for reactive (chlorine bleach or superoxides), charged (metals) or large (immunoglobulin) antimicrobial agents that are neutralized or bound by the EPS and are effectively 'diluted' to sublethal concentrations before they can reach all of the individual bacterial cells within the biofilm. The barrier properties of the EPS hydrogel might also protect against UV light and dehydration, and might localize (may prevent enzymatic activity in other areas) enzymatic activity. For example, extracellular β-lactamase enzymatic activity in $P.$ $aeruginosa$ occurs within the matrix (Dibdin et al. 1996).

The second protective mechanism could involve the physiological state of biofilm organisms. Although many antibiotics can freely penetrate the EPS, cells within the biofilm are often still protected. The creation of starved, stationary phase dormant zones in biofilms seems to be a significant factor in the resistance of biofilm populations to antimicrobials (Spoering and Lewis 2001; Anderl et al. 2003; Walters et al. 2003) particularly against antibiotics such as β–lactams, which are effective against rapidly dividing Gram-positive bacteria by interruption of cell-wall synthesis. However, arguably all antibiotics require at least some degree of cellular activity to be effective, because the mechanism of action of most antibiotics involves disruption of a microbial process. Therefore, pockets of cells (group of cells) in a biofilm in stationary phase dormancy might represent a general mechanism of antibiotic resistance (Hall-Stoodley et al. 2004).

A third mechanism of protection could be the existence of subpopulations of resistant phenotypes in the biofilm (Suci and Tyler 2003), which have been referred to as 'persisters' (Spoering and Lewis 2001). Persisters comprise a small fraction of the entire biomass, whether in planktonic or biofilm culture, but as distinct phenotypes have yet to be cultured, it remains unclear if these organisms do indeed represent a distinct phenotype or are simply the most resistant cells within a population distribution. Roberts and Stewart (2005) reported when antibiotic treatment was simulated, bacteria near the biofilm surface were killed, but persisters in the depth of the biofilm were able to resist the antibiotic. When antibiotic treatment ceased, surviving persister cells quickly reverted and allowed the biofilm to regrow. The presence of persister cells in biofilms plays an important role in antibiotic resistance (Singh et al. 2009; Qu et al. 2010). Although the relative contribution of each of these mechanisms (and possibly others) varies with the type of biofilm and the nature of the environmental stress, the result is one of general protection (Hall-Stoodley et al. 2004).

11.4   Process of Biofilm Formation

Biofilm-forming microorganisms have been shown to elicit specific mechanisms for initial attachment to a surface, microcolony formation, development of a three-dimensional community structure and maturation, and detachment (Prakash et al. 2003).

11.4.1   Attachment

The bacterium approaches the surface so closely that its motility is slowed and it forms a transient association with the surface and/or other microbes previously attached to the surface. The solid–liquid interface between a surface and an aqueous medium (e.g. water, blood) provides an ideal environment for the attachment and growth of microorganisms (Costerton et al. 1999). In general, attachment will occur most readily on surfaces that are rougher, more hydrophobic and coated by surface 'conditioning' films. An increase in flow velocity, water temperature or nutrient concentration may also equate to increased attachment, if these factors do not exceed critical levels. Properties of the cell surface, in particular the presence of fimbriae, flagella and surface-associated polysaccharides or proteins, are also important and may possibly provide a competitive advantage for one organism when a mixed community is involved (Donlan 2002).

11.4.2   Microcolony Formation

After the bacteria adhere to the inert surface/living tissue, the association becomes stable for microcolony formation. The bacteria begin to multiply while emitting chemical signals that 'intercommunicate' among the bacterial cells. Once the signal intensity exceeds a certain threshold level, the genetic mechanisms underlying exopolysaccharide production are activated (Costerton et al. 1999). In this way, the bacteria multiply within the embedded exopolysaccharide matrix, thus giving rise to the formation of a microcolony (McKenney et al. 1998).

11.4.3 Formation of Three-Dimensional Structure and Maturation

During the attachment phase of biofilm development, after microcolony formation the transcription of specific genes takes place. These are required for the synthesis of EPS. Attachment itself can initiate synthesis of the extracellular matrix in which the sessile bacteria are embedded, followed by formation of water-filled channels. It has been proposed that these channels constitute primitive circulatory systems, delivering nutrients to and removing waste products from the communities of cells in the microcolonies (Prakash et al. 2005).

11.4.4 Detachment

Occasionally, for purely mechanical reasons, some bacteria are shed from the colony or (more frequently) some bacteria stop producing EPS and are thus 'released' into the surrounding environment. Biofilm cells may be dispersed by either shedding of daughter cells from actively growing cells, or detachment because of decreasing nutrient levels or quorum-sensing, or shearing of biofilm aggregates because of flow effects (Baselga et al. 1994). As the thickness of the EPS increases, anaerobic conditions develop within the biofilm with loci of the biofilm consisting of anaerobic bacteria. Because of film thickness and the activity of anaerobic species, the film detaches and sloughs-off from the surface of the substrate (Howell and Atkinson 1976). Polysaccharidase enzymes specific for the EPS of different organisms also contribute to detachment (Boyd and Chakrabarty 1994; Donlan 2002; Lequette et al. 2010).

11.5 Bacterial Extracellular Polysaccharides and Biofilm Formation

EPS produced by microorganisms are a complex mixture of biopolymers primarily consisting of polysaccharides, as well as proteins, nucleic acids, lipids and humic substances. EPS make up the intercellular space of microbial aggregates and form the structure and architecture of the biofilm matrix. The key functions of EPS comprise the mediation of the initial attachment of cells to different substrata and protection against environmental stress and dehydration (Vu et al. 2009).

The formation of biofilms is a prerequisite for the existence of all microbial aggregates (Flemming and Wingender 2001b; Sutherland 2001) as an essential step in the survival of bacterial populations (van Hullebusch et al. 2003). The proportion of EPS in biofilms can comprise between approximately 50–90% of the total organic matter (Flemming and Wingender 2001a; Donlan 2002). In Gram-negative bacteria, some of the polysaccharides are neutral or polyanionic. The presence of uronic acids or ketal-linked pyruvates enhances their anionic properties, thus allowing the association of divalent cations such as calcium and magnesium to increase the binding force in a developed biofilm. In some Gram-positive bacteria, the chemical composition of their EPS could be slightly different from that of Gram-negative bacteria due to their primarily cationic nature (Flemming and Wingender 2001a; Sutherland 2001; Vu et al. 2009). Lerner et al. (2009) reported that extracellular polysaccharides of the bacterium *Azospirillum brasilense* play an important role in its interactions with plant roots. The pRhico plasmid of *A. brasilense* Sp7, also named p90, carries several genes involved in synthesis and export of cell surface polysaccharides. Lerner and coworkers generated two Sp7 mutants impaired in two pRhico-located genes, *noeJ* and *noeL*, encoding mannose-6-phosphate isomerase and GDP-mannose 4,6-dehydratase, respectively. The results demonstrated that in *A. brasilense* Sp7, *noeJ* and *noeL* are involved in lipopolysaccharide and exopolysaccharide synthesis. *noeJ* and *noeL* mutant strains were significantly altered in their outer membrane and cytoplasmic/periplasmic protein profiles relative to the wild-type strain. Moreover, both *noeJ* and *noeL* mutations significantly affected the bacterial responses to several stresses and antimicrobial compounds. Disruption of *noeL*, but not *noeJ*, affected the ability of *A. brasilense* Sp7 to form biofilms. The pleiotropic alterations observed in the mutants could be due, at least partially, to their altered lipopolysaccharides and exopolysaccharides relative to the wild-type (Lerner et al. 2009). *Sinorhizobium meliloti* is a soil bacterium that elicits the formation of root organs called nodules on its host plant, *Medicago sativa*. Inside these structures, the bacteria are able to convert atmospheric nitrogen into ammonia, which is then used by the plant as a nitrogen source. The synthesis by *S. meliloti* of at least one exopolysaccharide, succinoglycan or EPS II, is essential for a successful symbiosis. While exopolysaccharide-deficient mutants induce the formation of nodules, they fail to invade them, and as a result, no nitrogen fixation occurs (Rinaudi and González 2009). Interestingly, the low-molecular-weight fractions of these exopolysaccharides are the symbiotically active forms, and it has been suggested that they act as signals to the host plant to initiate infection thread formation. The *ExpR/Sin* quorum-sensing system controls biofilm formation in *S. meliloti* through the production of EPS II, which provides the matrix for the development of structured and highly organized biofilms. Moreover, the presence of the low-molecular-weight fraction of EPS II is vital for biofilm formation, both in vitro and in vivo (Rinaudi and González 2009).

QS is known as one of the regulatory pathways for EPS production and biofilm formation in bacteria (Miller and Bassler 2001; Hall-Stoodley and Stoodley 2002; Rivas et al. 2005; Hooshangi and Bentley 2008; Ruiz et al. 2008). Also, phosphate and polyphosphate metabolism have been associated with biofilm development and the QS regulatory pathway (Rashid et al. 2000; Farah et al. 2005). However, in the QS regulatory system, biofilm formation and maintenance mechanisms are diverse among different bacterial species, so the role of QS in biofilm formation cannot be described in general terms (Hooshangi and Bentley 2008). For example, in *Pseudomonas aeruginosa*, QS is essential for adhesion, proper biofilm formation and virulence factors (Waters and Bassler 2005; Nakamura et al. 2008). Mutant *P. aeruginosa* cells that did not produce any QS signals were found to be more densely populated with a thinner biofilm than the wild type. In addition, mutation of the *LasI* gene resulted in an abnormal and undifferentiated biofilm formation process (Davies et al. 1998). In *E. coli*, cellular functions are controlled by the QS *LsrR/LsrK* system. Biofilm formation and architecture were found to be significantly altered in *lsrR* and *lsrK* mutants. There were differences observed in the cell fimbriae and matrix structure and in the thickness of the biofilm of the mutants compared to the wild type (Li et al. 2007). Lastly, a QS system, *AfeI/AfeR*, has recently been identified in *Acidithiobacillus ferrooxidans*, which is similar to the LuxI/LuxR proteins (Rivas et al. 2005; Ruiz et al. 2008).

The EPS produced by *A. ferrooxidans* consist of neutral sugars, predominantly rhamnose, fructose and glucose, and lipids (Harneit et al. 2006). The chemical constituents of the EPS vary depending on the type of substrate upon which the cells are grown. The mode of attachment also differs as a function of substrate, and hence results in the expression of different EPS genes (Gehrke et al. 1998). The majority of research investigating the nature of the EPS produced by *A. ferrooxidans* has taken place in studies involving pyrite, sulfur and ferrous sulfate substrates; mainly neutral sugars and lipids were found (Gehrke et al. 1998; Sand and Gehrke 2006).

11.6   Factors Involved in Regulation of Biofilm Formation

The growth of a biofilm is the result of a complex process that involves the transport of organic and inorganic molecules and microbial cells to the surface, a subsequent adsorption to the surface and finally attachment to the surface aided by the production of EPS (Beech 2004). Due to its complexity, the formation of biofilms is regulated at different stages via diverse mechanisms (Waters and Bassler 2005; Ruiz et al. 2008). The most studied regulatory mechanism that has been found to control the production of EPS, biofilm formation and differentiation is QS regulation (von Bodman et al. 1998; Davies et al. 1998; Donlan 2002; Rivas et al. 2005; Waters and Bassler 2005; Ruiz et al. 2008). QS allows bacteria to maintain cell-cell communication and also regulate the expression of specific genes in response to changes in cell population density (Davies et al. 1998; Rivas et al. 2005). In general, the QS process involves the production, release and detection of chemical signaling molecules, thus allowing microbial cells to regulate gene expression in a cell-density-dependent manner (Hooshangi and Bentley 2008). At a given population density, the genes involved in biofilm differentiation and maturation are activated (Donlan 2002; Waters and Bassler 2005; Ruiz et al. 2008).

The autoinducer-1 (AI-1) type is mainly involved in intra-species communication and the AI-2 type is associated with inter-species interaction (Farah et al. 2005). Gram-negative bacteria produce and release AI molecules, which are generally *N-acyl homoserine lactone* (AHL) molecules that serve as a function of controlling the cell-population density. Bacteria detect the accumulation of AHL signals. Above a certain threshold concentration, these signals are present in sufficient quantity to enable transcriptional effectors to activate silent genes. This alters their cell-density dependent gene expression and therefore their behavior (von Bodman et al. 1998; Ruiz et al. 2008). In Gram-positive bacteria, communication is carried out with modified oligopeptides generating the signals and membrane-bound sensor histidine kinases acting as receptors. Signaling is mediated by many phosphorylation steps, which control the activity of a response regulator. However, peptide signals are not diffusible across the membrane and therefore the signal release is mediated by oligopeptide exporters. Normally, signal release occurs concurrently with signal processing and modification (Miller and Bassler 2001; Waters and Bassler 2005). Factors such as availability of surface, nutrients and environmental cues also regulate biofilm formation.

11.6.1 Surface

The surface can be a dead or living tissue, or any inert surface. The attachment of microorganisms to the surface is a complex process, with many variables affecting the outcome (Pratt and Kolter 1998). Further, growth requires a complex developmental pathway involving a series of events that are regulated in response to environmental- and bacterial-derived signals. The surface may have several characteristics that are important in the attachment process. Microbial colonization appears to increase as the surface roughness increases (Characklis et al. 1990). This is because shear forces are less and surface area is larger on rougher surfaces. Most investigators have found that microorganisms attach more rapidly to hydrophobic, nonpolar surfaces such as teflon and other plastics than to hydrophilic materials such as glass or metals (Fletcher and Loeb 1979; Pringle and Fletcher 1983; Bendinger et al. 1993; Cross et al. 2007).

11.6.2 Nutrients

Increase in nutrient concentration correlated with an increase in the number of attached bacterial cells (Cowan et al. 1991). Biofilm bacteria acquire nutrients by concentrating trace organics on surfaces by the extracellular polymers, using the waste products from their neighbours and secondary colonizers, and by using different enzymes to break down food supplies. Because the biofilm matrix is often negatively charged, many nutrients (particulary cations) are attracted to the biofilm surface. Besides, nutrients with negative charge can exchange with ions on the surface. This provides bacterial cells within the biofilm with plenty of food compared to the surrounding (Prakash et al. 2003; Brar et al. 2010).

11.6.3 Environmental Cues

Other characteristics of the aqueous medium, such as pH, nutrient levels, iron, oxygen, ionic strength and temperature, may also play a role in the rate of microbial attachment to a substratum. Several studies have shown a seasonal effect on bacterial attachment and biofilm formation in different aqueous systems (Fera et al. 1989; Donlan et al. 1994). This effect may be due to water temperature or other unmeasured, seasonally affected parameters. Fletcher (1988) found that an increase in the concentration of several cations (sodium, calcium, lanthanum, ferric ion) affected the attachment of *P. fluorescens* to glass surfaces, presumably by reducing the repulsive forces between the negatively-charged bacterial cells and the glass surfaces (Caiazza et al. 2007; Gaddy and Actis 2009).

11.6.4 Gene Regulation

There is mounting evidence to show that both up- and down-regulation of a number of genes occurs in the attaching cells upon initial interaction with the substratum. Combaret et al. (1999) found that 22% of the genes were up-regulated and 16% down-regulated in biofilm-forming *P. aeruginosa*. Davies and Geesey (1995) demonstrated *algC* up-regulation within minutes of attachment to a surface in a flow cell system. Genes encoding for enzymes involved in glycolysis or fermentation (phosphoglycerate mutase, triosephosphate isomerase, and alcohol dehydrogenase) are up-regulated in biofilm-forming *Staphylococcus aureus* (Becker et al. 2001). Becker et al. (2001) surmised that the up-regulation of these genes could be due to oxygen limitation in the developed biofilm, favouring fermentation. A recent study by Pulcini (2001) also showed that *algD*, *algU*, *rpoS* and genes controlling polyphosphokinase synthesis were up-regulated in biofilm formation of *P. aeruginosa*. In *Escherichia coli*, two of the possibly important genes for biofilm growth are the *rpoS* and *bolA* gene. *RpoS* is also denominated a master regulator of general stress response. Even though many studies have revealed the importance of *rpoS* in planktonic cells, little is known about the functions of *rpoS* in biofilms. In contrast, *bolA*, which is a morphogene in *E. coli*, is overexpressed

11 Biofilm Formation by Environmental Bacteria

in cells experiencing stress resulting in round cell morphology. The morphogene *bolA* is mostly expressed under stress conditions or in stationary phase, suggesting that *bolA* could be implicated in biofilm development. Overexpression of *bolA* induces biofilm development, while *bolA* deletion decreases biofilms (Adnan et al. 2010).

11.7 Detection Methods of Biofilm Formation

A single standard method for the study of biofilm detection is not available, and this is certainly impeding progress in the field. It is very difficult if not impossible to compare the results obtained with biofilms of even the same species cultured and assayed under vastly different conditions. For now, several methods are available, and these will be briefly reviewed.

11.7.1 Screening of Bacterial Isolates for Biofilm Formation

There are many methods used to screen biofilm formation in bacterial isolates, here we describe the three most common methods used to screen biofilm formation.

To screen bacteria for biofilm formation, the most commonly used method is of Klingenberg et al. (2005). The method is performed in wells of a microtiter plate. Overnight culture of bacteria to be tested is grown in a polystyrene tissue culture plate. The culture supernatant is aspirated and the plates are washed with PBS. Culture is heat fixed and crystal violet is added. The plates are washed with tap water, with ethanol/acetone to remove the dye and the OD_{570} of adherent biofilm is measured in an ELISA reader. Bacteria showing OD_{570} comparable with those of control strains are considered as biofilm former.

For the biofilm formation assay, overnight cultures are prepared in brain heart infusion (BHI) broth, added to the wells of sterile flat-bottom polystyrene tissue culture plates and incubated in a closed humidified plastic container. Biofilms are fixed and stained with crystal violet. Quantitation of the total biofilm biomass is done spectrophotometrically. The extent of biofilm is determined by measuring the absorbance (A_{620}) of the resolubilized dye with a Microplate Reader.

In another assay, biofilm formation is tested in AB medium (Clark and Maaloe 1967) supplemented with glucose (Kjaergaard et al. 2000). Cultures are grown over night in a microplate and diluted in a new microplate and incubated for 48 h. Culture fluid with unbound cells is removed by tapping the inverted microplate on absorbent paper, and adherent cells are stained with crystal violet. Wells are washed thoroughly with water and dried overnight. Retained crystal violet is dissolved with ethanol-acetone, and absorbance (A) is measured at 600 nm.

11.7.2 Detection of Quorum Sensing Molecules

11.7.2.1 Acyl-Homoserine Lactone (AHL) Bio-Assays

The different bioassays all make use of reporter bacteria in which the proper production of AHL molecules has been blocked by mutation but which contain an AHL-responsive reporter gene. Expression of the reporter gene is therefore only possible in the presence of exogenous AHLs. Each reporter strain displays specificity towards different AHL molecules, and the use of multiple reporters therefore allows the detection of a wide range of AHLs and allows differentiating between different AHL production patterns. (1) Reporter strain *Agrobacterium tumefaciens* NTL4 (pCF218) (pCF372) is cross-streaked against the strain to be tested (Tempe´ et al. 1977). Production of specific AHLs by the tested strain induces β-galactosidase production in the reporter strain, and results in blue colonies due to X-Gal hydrolysis. (2) Reporter strain *Chromobacterium violaceum* CV026 is also used in a cross-streak test but on LB agar. In the presence of specific AHLs, this strain will produce purple colonies due to violacein production (McClean et al. 1997). (3) Reporter strain *Serratia liquefaciens* MG44 is point inoculated on the medium containing casamino acids and cell-free culture supernatant from the strain to be tested. In this reporter, specific AHLs induce swarming, which becomes apparent from the spreading of colonies over the surface of this medium (Eberl et al. 1996). (4) Finally, reporter strain *E. coli* MT102 (pJBA132) and cell-free culture supernatant are incubated and the production of green fluorescent protein is measured by fluorescence at 520 nm upon excitation at 485 nm in a fluorospectrophotometer.

11.7.2.2 Autoinducer-2 Bioassay

Autoinducer accumulation enables the cell to sense that a sufficient local concentration of bacteria (a quorum) has been reached, in order to initiate concerted population responses, including biofilm formation (Landini et al. 2010). Surette and Bassler (1998) described the assay in detail. Reporter strain *V. harveyi* BB170 grown in Autoinducer Bio-assay Vibrio (ABV) medium (Surette and Bassler 1998), and the diluted cell suspensions are dispensed into microplate wells containing the cell-free culture supernatants from the isolates to be tested. The microplate is incubated and the bioluminescence is measured in a fluorometer.

11.7.2.3 PQS Bioassay

Other quorum signaling molecules that have been identified in Gram-negative bacteria are 2-heptyl-3-hydroxy-4-quinolone (PQS) and diketopiperazines in *Pseudomonas aeruginosa* and other bacteria (Holden et al. 1999; Pesci et al. 1999). Overnight culture of reporter strain *P. aeruginosa* PAO1-R1 (pTS400) is mixed in a microplate with cell-free culture supernatant from the isolate to be tested, and incubated. Production of reporter enzyme β-galactosidase is measured. Cells are collected in a microcentrifuge and resuspended in Tris–HCl buffer and absorbance is measured at 600 nm. Permeabilization reagent as described by Schupp et al. (1995) is added and absorbance is measured at 420 nm. Then ONPG (o-nitrophenyl-β-galactopyranoside) is added and absorbance at 420 nm is measured at different times.

11.7.3 Biofilm Formation in a Bioreactor

Biofilm formation can also be investigated in a biofilm reactor as shown in Fig. 11.1. As substratum for bacteria attachment glass beads, polystyrene beads or "silicon particles" are provided. The reactor is continuously supplied with fresh medium. Additionally, internal circulation of the medium in the biofilm reactor is provided (Schiwon, K., Grohmann, E., unpublished data). Biofilm formation can be controlled visually or by Lectin-binding analysis.

11.7.3.1 Lectin-Binding-Analysis

The applicability of the method has been evaluated for single, dual and triple staining with a panel of fluor-conjugated lectins. It was shown that lectin-binding analysis was able to stain glycoconjugates within biofilm communities (Neu et al. 2001).

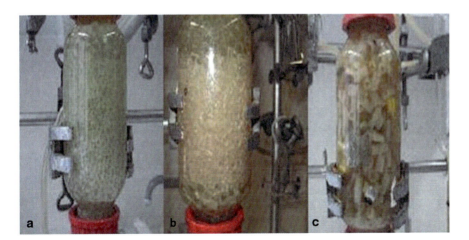

Fig. 11.1 Biofilm formation in a biofilm reactor by a *Bacillus* sp. isolate from tannery effluent contaminated soil in Kanpur (**a**) Glass beads showing biofilm, (**b**) Polystyrene beads showing no biofilm, and (**c**) Silicon particles as carrier material in the bioreactor showing no biofilm formation

Lectins represent useful probes for *in situ* techniques to examine the distribution of glycoconjugates in fully hydrated environmental biofilms in a three dimensional way (Neu and Lawrence 1999). Fluorescent lectins with fluorescein isothiocyanate (FITC) or tetramethyl rhodamine isothiocyanate (TRITC) and Cyanine (CY5) are employed alone or in combination for double and triple staining. The lectin-binding analysis can be combined with general nucleic acid stains to collect both nucleic acid and glycoconjugate signals (Neu et al. 2001). Lectins have been used as probes in a few studies of environmental (marine and freshwater) biofilm systems (Michael and Smith 1995; Neu and Lawrence 1997; Wolfaardt et al. 1998; Neu 2000). It has been suggested that lectins could be applied as a specific probe at the polymer level similar to specific rRNA targeted probes at the cellular level (Neu 2000). These studies indicated the utility of lectin-binding analysis; nevertheless, questions remain regarding the identity of the binding sites, the influence of fluorescent conjugates on the behaviour of the lectin and interactions between lectins in multiple-lectin staining.

11.7.4 Fluorescence Tools to Visualize Biofilms

Live-cell imaging techniques are important to obtain a better understanding of microbial functioning in biofilms. Autofluorescent proteins, such as the green fluorescent protein (GFP) and the red fluorescent protein (DsRed), are valuable tools for studying microbial communities in natural environments. Because of the functional limitations of DsRed, new and improved variants were generated such as mCherry. Lagendijk et al. (2010) developed new genetic tools on basis of mCherry for labeling Gram-negative bacteria to visualize them in their natural environment. The applicability of the new mCherry tools under the constitutive expression of the *tac* promoter was shown for *E. coli*, various *Pseudomonas* spp. and *Edwardsiella* sp. The expression of mCherry was analyzed by fluorescence microscopy and quantified by fluorometry. The suitability of the new constructs for the visualization of microbial communities was shown for biofilms formed on tomato roots. Lagendijk et al. (2010) showed that mCherry in combination with GFP is a suitable marker for studying mixed microbial communities.

Clarke et al. (2010) synthesized fluorescent nanoparticle quantum dot (QD) conjugates to target microbial species, including difficult to label Gram-negative bacteria. The QD conjugates provide contrast for both environmental scanning electron microscopy and fluorescence microscopy, permitting visualization of living and fixed bacteria and biofilms. These probes were applied for studying biofilms extracted from cold springs in the Canadian High Artic. In these biofilms, sulfur-metabolizing bacteria live in close association with unusal sulfur mineral formations. By simple labeling with the QD conjugates, Clarke et al. (2010) could image the sulfur-metabolizing bacteria in fully-hydrated samples and visualize their relationship to the sulfur minerals by environmental scanning electron microscopy and fluorescence microscopy.

11.7.5 Biofilm Imaging Techniques

Recently many techniques have been developed to visualize the biofilms directly in their environment. Apart from traditional microscopic techniques such as light and electron microscopy, new advanced techniques have been established including laser scanning microscopy (LSM), magnetic resonance imaging (MRI), scanning transmission X-ray microscopy (STXM), Raman microscopy (RM), surface-enhanced Raman scattering (SERS) and atomic force microscopy (AFM) (Neu et al. 2010; Ivleva et al. 2010; Wright et al. 2010). These new techniques allow in situ analysis of the structure, composition, processes and dynamics of microbial communities. The three techniques (LSM, MRI, STXM) open up quantitatively analytical imaging possibilities that were, until a few years ago, impossible. The microscopic techniques represent powerful tools for examination of mixed environmental microbial communities usually encountered in the form of aggregates and films. Therefore, LSM, MRI and STXM are being used in order to study complex microbial biofilm systems. These techniques are useful in visualizing the biofilm in living and hydrated states, and help in visualizing the three dimensional structure (Neu et al. 2010). Raman microscopy (RM) provides whole-organism fingerprints for biological samples with spatial resolution in the microm range and enables correlations between optical and chemical images to be made. Low water background makes RM beneficial for in situ studies of biofilms, since water is the major component of the biofilm matrix. Ivleva et al. (2010) discussed the feasibility of RM for chemical characterization of different structures in a multispecies biofilm matrix, including microbial constituents and EPS. They showed that by improving the sensitivity of RM with surface-enhanced Raman scattering (SERS) one can perform rapid biofilm analysis. In particular, by choosing appropriate SERS substrates and solving the problem of SERS measurement reproducibility one can carry out in situ studies of different components in the complex biofilm matrix (Ivleva et al. 2010). Atomic Force Microscopy (AFM), another technique has proven itself over recent years as an essential tool for the analysis of microbial systems. This technique also helps in visualization of biofilm in its hydrated and living form (Wright et al. 2010).

11.8 Antibiotic Resistance of Bacterial Biofilms

11.8.1 Occurrence and Architecture of Bacterial Biofilms

The microbes in biofilms are kept together and the bacterial consortium can consist of one or more species living in a sociomicrobiological way (Wingender et al. 2001; Whitchurch et al. 2002; Costerton et al. 2003; Bjarnsholt et al. 2009). The matrix is important since it provides structural stability and protection to the biofilm. Development of bacterial biofilms over time has been intensively studied in vitro by confocal scanning laser microscopy employing green fluorescent protein (GFP)-tagged bacteria. This technique has been combined with advanced in silico image analysis to produce three-dimensional images of the biofilm (Heydorn et al. 2002; Klausen et al. 2003). Development of an in vitro biofilm is initiated by planktonic (freely moving) bacteria that reversibly attach to a surface, which may be covered by a layer of proteins (a pellicle) (Marshall 1992; Kolenbrander and Palmer 2004). At this stage, the bacteria are still susceptible to antibiotics. The next step is irreversible binding to the surface within the next few hours and multiplication of the bacteria, which form microcolonies on the surface and begin to produce a polymer matrix around the microcolonies (Marshall 1992). The biofilm grows in thickness (up to 50 µm) and under in vitro conditions, mushroom-like or tower-like structures are often observed in the mature biofilm. At that stage, the biofilm shows maximum tolerance (= resistance) to antibiotics (Folkesson et al. 2008; Høiby et al. 2010).

11.8.2 *Slow Growth and Low Oxygen Concentration*

Inspection of environmental as well as in vitro biofilms has revealed that the oxygen concentration may be high at the surface but low in the centre of the biofilm where anaerobic conditions may be present (Costerton et al. 1995). Likewise, growth, protein synthesis and metabolic activity is stratified in biofilms, i.e. a high level of activity at the surface and a low level and slow or no growth in the centre, and this is one of the explanations for the reduced susceptibility of biofilms to antibiotics (Werner et al. 2004; Keren et al. 2004).

11.8.3 *Mutators*

The mutation frequency of biofilm-growing bacteria is significantly increased compared with planktonically growing isogenic bacteria (Driffield et al. 2008) and there is increased horizontal gene transmission in biofilms (Molin and Tolker-Nielsen 2003). These physiological conditions may explain why biofilm-growing bacteria easily become multidrug resistant by means of traditional resistance mechanisms against β-lactam antibiotics, aminoglycosides and fluoroquinolones, which are detected by routine susceptibility testing in the microbiology laboratory where planktonic bacterial growth is investigated. Thus, bacterial cells in biofilms may simultaneously produce enzymes that degrade antibiotics, have antibiotic targets of low affinity and overexpress efflux pumps that have a broad range of substrates.

11.8.4 *Tolerance to Antibiotics and Efflux Pumps*

The tolerance of bacterial biofilms to antibiotics is a consequence of the enhanced efflux pump activity in biofilms. Tolerance of biofilms to tobramycin is also mediated by low metabolic activity, and the high cell density that results in accumulation of extracellular signaling molecules is probably important, as it has been shown that tolerance to tobramycin of *P. aeruginosa* strain PAO1 biofilm is QS-mediated (Bjarnsholt et al. 2005). In addition, a non-specific mechanism for the tolerance of the metabolically active cells to colistin was shown to be up-regulation of the MexAB-OprM efflux (Pamp et al. 2008). Furthermore, increased efflux pump activity due to mutations has been shown to be a major resistance mechanism against aminoglycoside antibiotics and fluoroquinolones in *P. aeruginosa* (Jalal et al. 2000; Islam et al. 2009).

11.8.5 Mechanisms of Antimicrobial Resistance

There are several mechanisms for the resistance of bacteria to antibiotics in biofilm. First, the EPS secreted by biofilm bacteria, act as a physical/chemical barrier, thus preventing penetration of antibodies or many antibiotics (Costerton et al. 1995; Lewis 2001; Thien and O'toole 2001).

Second, embedded biofilm bacteria are generally not actively engaged in cell division, are smaller in size and less permeable to antibiotics. Virtually all antimicrobials are more effective in killing rapidly-growing cells. Further, transition from exponential to slow/no growth is generally accompanied by expression of antibiotic-resistance factors (Brown et al. 1988; Wentland et al. 1996; Thien and O'Toole 2001). Slow growth activates the RelA-dependent synthesis of ppGpp, which inhibits anabolic processes in bacterial cells. Interestingly, ppGpp suppressed the activity of a major *E. coli* autolysin, SLT57, which would make the cells more resistant to autolysis and could explain the mechanism of tolerance to antibiotics in slowly growing cells. ppGpp inhibits peptidoglycan synthesis, which would explain the decreased levels of activity of cell-wall synthesis inhibitors under starvation conditions (Lewis 2001; Wu et al. 2010).

Third, antibiotic degrading enzymes such as β-lactamases can effectively inactivate the incoming antibiotic molecules. It is interesting to note that biofilm cells of *P. aeruginosa* have been shown to produce 32-fold more β-lactamase than cells of the same strain grown planktonically (Tuomanen et al. 1986; Potera 1999).

Fourth, up to 40% of the cell-wall protein composition of bacteria in biofilms is altered from that of its planktonic community (Potera 1999; O'Toole et al. 2000). The membranes of biofilm bacteria might be better equipped to pump out antibiotics before they can cause damage, or even the targets of the antibiotics may disappear (Prakash et al. 2003).

Fifth, the antimicrobial agent is deactivated in the outer layers of the biofilm, faster than it diffuses. This is true for reactive oxidants such as hypochlorite and H_2O_2 (De Beer 1994; Chen and Stewart 1996; Xu et al. 1996; Thien and O'Toole 2001).

Biofilms also provide an ideal niche for the exchange of extrachromosomal DNA responsible for antibiotic resistance, virulence factors and environmental survival capabilities at accelerated rates, making it a perfect milieu for emergence of drug resistant pathogens (Hausner and Wuertz 1999; Ghigo 2001; Donlan 2002; Buchholz et al. 2010). Since plasmids may encode for resistance to multiple antimicrobial agents, biofilm association also provides a mechanism for selecting and promoting the spread of bacterial resistance to antimicrobial agents.

11.9 Inhibition of Biofilm Formation

11.9.1 Target- Based Screening

A basic strategy for the discovery of biofilm inhibitors is the direct screening of chemical compounds in biofilm formation assays (Junker and Clardy 2007; Richards et al. 2008; Rivardo et al. 2009). However, such a direct approach also selects for non-specific biofilm inhibitors, such as detergents or biosurfactants. Although these classes of molecules can display significant anti-biofilm activity under laboratory conditions, they often show limited activity, or lack of selective toxicity towards bacteria, if used in vivo. In recent years, the improvement in our understanding of the cellular processes controlling bacterial biofilms has allowed the development of target-oriented approaches for the discovery of biofilm inhibitors (Sivakumar et al. 2010). Development of target-based screening constitutes a rational and effective strategy for the discovery of biofilm inhibitors. Characterization of QS as an important regulatory mechanism in biofilm formation, and thus, as a potential target for antimicrobials (Smith and Iglewski 2003; Njoroge and Sperandio 2009), has led to the development of screening strategies for QS inhibitors. In turn, identification of biofilm inhibitors through a target-based approach has contributed to the elucidation of cellular processes controlling bacterial biofilms. The discovery that several compounds with anti-biofilm activity (e.g., halogenated furanones) are QS inhibitors (Hentzer et al. 2002; Manefield et al. 2002; Bjarnsholt et al. 2005; Persson et al. 2005; Rasmussen et al. 2005) confirmed the importance of this signaling system in biofilm formation. More recently, the search for novel biofilm inhibitors has selected targets other than QS, such as nucleotide biosynthesis (Attila et al. 2009; Ueda et al. 2009) and production of the signal molecule cyclic di-guanosine monophosphate (cdi- GMP; Antoniani et al. 2010).

11.9.2 Activity-Based Screening for QS Inhibitors

QS is a complex regulatory process dependent on bacterial cell density (Miller and Bassler 2001; Karatan and Watnick 2009) and is typically involved in the regulation of genes involved in biofilm maturation and maintenance (Hammer and Bassler 2003; Marketon et al. 2003; Vuong et al. 2003; Ueda and Wood 2009). Inhibitors of

QS, in addition to possessing anti-biofilm activity, could also counteract bacterial pathogenicity. Species-specific QS systems make use of other autoinducers, such as quinolonones in *P. aeruginosa* (McKnight et al. 2000), or the diffusible signal factor (DSF), a fatty acid (cis-11-methyl-dodecenoic acid) used as signal molecule by the plant pathogen *Xanthomonas campestris* (Barber et al. 1997). AHL autoinducers are synthesized by enzymes of the LuxI family and can bind transcription regulators of the LuxR family. AHL binding to LuxR activates the transcription of QS-dependent genes.

Binding of signaling peptides to sensor proteins in the cell membrane triggers a signal transduction cascade, which leads to phosphorylation of a response regulator and triggers QS-dependent gene expression. A model of QS systems in Gram-positive bacteria is the agr (accessory gene regulation) system of *S. aureus*, where autoinducer-dependent phosphorylation of the AgrA regulator, triggered by biofilm growth, leads to transcription activation of genes encoding virulence factors (Novick et al. 1993; Balaban and Novick 1995). The different chemical nature of signal molecules and the different molecular mechanisms involved in QS would suggest that QS inhibitors can only be directed against either Gram-positive or Gram-negative bacteria. However, furanones, an important class of inhibitors of QS in Gram-negative bacteria, also show killing activity against Gram-positive bacteria and even protozoa (Zhu et al. 2008; Lönn-Stensrud et al. 2009), suggesting that they might target cellular processes other than QS. Indeed, exposure of the Gram-positive bacterium *Bacillus subtilis* to furanones triggers induction of stress response genes in a QS-independent manner (Ren et al. 2004).

11.9.3 Structure-Based Screening for QS Inhibitors

In addition to activity-based assays, an alternative strategy for target-oriented discovery of QS inhibitors is represented by structure-based screening of chemical compounds. This strategy relies on the availability of a growing number of three-dimensional protein structures either predicted by computational biology methods or characterized through biochemical structural analysis. Using molecular modeling programs, it is possible to select potential inhibitors targeting catalytic domains or key amino acid residues for protein activity using virtual screening of small molecules with known structures and chemical properties (Li et al. 2008; Kiran et al. 2008; Zeng et al. 2008; Yang et al. 2009). This structure-based approach constitutes a primary virtual screening followed by a secondary activity-based assay using reporter genes controlled by QS-dependent promoters. Proteins involved in QS of Gram-negative bacteria, in particular the LasR transcriptional regulator of *P. aeruginosa*, have been used as a target in structure-based screening for biofilm inhibitors. This approach has led to the identification of several compounds showing significant inhibition of QS in *P. aeruginosa* (Smith et al. 2003; Müh et al. 2006; Geske et al. 2007; Amara et al. 2009); however, the number of inhibitors displaying broad anti-biofilm activity remains low, possibly due to yet not identified resistance mechanisms or to the inability of QS inhibitors to reach their target in biofilms. In Gram-positive bacteria, QS directly regulates biofilm maintenance and dispersal, rather than being a factor in its initial formation (Pratten et al. 2001; Yarwood et al. 2004).

11.9.4 Nucleotide Biosynthesis and DNA Replication Inhibitors

Over the last few years, it has become increasingly clear that modified nucleotides, such as c-di-GMP, play a pivotal role as signal molecules for biofilm regulation. Accumulation of c-di-GMP stimulates production of adhesion factors via a variety of different mechanisms, i.e., allosteric activation of protein activity, protein stabilization, or regulation of gene expression at the transcriptional and translational levels (Weinhouse et al. 1997; Simm et al. 2004; Kulasakara et al. 2006; Weber et al. 2006; Sudarsan et al. 2008). Intracellular levels of c-di-GMP are determined by two classes of enzymes with opposite activities: diguanylate cyclases (DGCs), which synthesize c-di-GMP, and cdi-GMP phosphodiesterases (PDEs) that hydrolyze it into the inactive diguanylate phosphate (pGpG) form (reviewed in Tamayo et al. 2007). Genes involved in c-di-GMP biosynthesis and turnover are conserved in all *Eubacteria*, while absent in animal species (Galperin 2004), thus suggesting that enzymes involved in c-di-GMP biosynthesis might be an interesting target for anti-biofilm agents.

However, while genes encoding DGCs and PDEs are present in remarkably high numbers in Gram-negative bacteria, they are much less abundant in Gram-positives (Galperin 2004). Consistent with this large discrepancy, the role of c-di-GMP in biofilm formation and maintenance has been well established in Gram-negative bacteria, while its importance in Gram-positive bacteria remains questionable (Holland et al. 2008). Thus, as observed for QS inhibitors, it appears that promising targets for biofilm control might follow a strict divide between Gram-positive and Gram-negative bacteria.

11.10 Incidence of Biofilm in the Environment

Biofilm microcolonies have been identified by morphology in the 3.3–3.4-billion-year-old South African Kornberg formation and filamentous biofilms have been identified in the 3.2-billion-year old deep-sea hydrothermal rocks of the Pilbara Craton, Australia (Rasmussen 2000). Similar biofilm structures can be found in modern hydrothermal environments, such as hot springs (Reysenbach, and Cady 2001) and deep-sea vents (Taylor et al. 1999). Interestingly, biofilm formation is also a characteristic of prokaryotic 'living fossils' in the most ancient lineages of the phylogenetic tree in both the *Archaea* and *Bacteria*—the *Korarchaeota* and *Aquificales,* respectively (Reysenbach et al. 2000; Jahnke et al. 2001).

11 Biofilm Formation by Environmental Bacteria

There are many recent reports on the characterization of natural or 'semi-natural' biofilms in aquatic systems reaching from drinking water biofilms to acid mine drainage solutions available. Some examples will be presented here. Jiao et al. (2010) examined the chemical composition of EPS extracted from two natural microbial pellicle biofilms growing on acid mine drainage solutions. The EPS from a mid-developmental-stage biofilm and a mature biofilm were qualitatively and quantitatively compared. EPS from mature biofilm contained higher concentrations of metals and carbohydrates than EPS from mid-developmental-stage biofilm. Fe was the most abundant metal in both samples, accounting for about 73% of the total metal content. The data of Jiao et al. (2010) demonstrated that the biochemical composition of the EPS from these acidic biofilms is dependent on maturity and is controlled by the microbial communities, as well as the local geochemical environment.

Shikuma and Hadfield (2010) investigated the abundance of *E. coli* and *Vibrio cholerae* in biofilms and water-column samples from three harbours in Hawai´i, which differ in their local and international ship traffic. *E. coli* and, in some cases *V. cholerae*, occurred in relatively high abundance in marine biofilms formed on abiotic surfaces, including the exterior hulls of ships. The community compositions of biofilms from different locations were more similar to each other than to water-column communities from the same locations. The data of Shikuma and Hadfield (2010) suggest that biofilms in harbours are an overlooked reservoir and a source of potential dissemination for *E. coli* and *V. cholerae*.

Balzer et al. (2010) studied the occurrence of faecal indicators (total coliform bacteria, *E. coli* and intestinal enterococci) in epilithic and sediment biofilms of German rivers. All of the biofilms displayed significant concentrations of these bacteria, which were several logs lower compared with the total cell number and the number of culturable heterotrophic plate counts indicating that faecal indicators represented a minor fraction of the biofilm communities. The biofilms contained approximately two orders of magnitude higher concentrations of faecal bacteria compared with the overlying water. The data of Balzer et al. (2010) showed that faecal indicator bacteria can survive in the presence of high concentrations of the authochthonous microflora in epilithic biofilms and sediments, suggesting that these biofilms might act as reservoirs for bacterial pathogens in polluted rivers.

To investigate dispersal and colonization patterns of fluorescently labeled bacterial cells in nascent and mature biofilms, Augspurger et al. (2010) grew complex biofilms in microcosms from natural surface waters in laminar and turbulent flow. Settling of the cells occurred in nonrandom spatial patterns governed by the interplay of local flow patterns and biofilm topography. Settling was higher in microcosms with nascent biofilms, with fewer cells remaining in the water column than in microcosms with mature biofilms. The flow regime had no effect on settling velocity. Small-scale variations in the flow pattern seemed to be more important than the overall flow regime. The results of Augspurger et al. (2010) showed that colonization of biofilms in a model stream environment is a heterogeneous process differently affected by biological and physical factors.

Cuzman et al. (2010) studied the microbial composition of biofilms on four fountains, two from Italy and two from Spain, by traditional and molecular techniques. The results indicated many similarities with regard to the phototrophic biodiversity for all the investigated fountains. Automated ribosomal RNA intergenic spacer analysis (ARISA) was used to examine the eubacterial and cyanobacterial community of two of the investigated fountains. The biodeteriogenic activity of microorganisms on stone cultural assets is well known, and monumental fountains are exposed to a high risk due to water availability which favors the development of a thick biofilm patina (Cuzman et al. 2010). The understanding of the microbial ecology of fountains will be the starting point for a scientific approach to develop suitable, effective, and long-lasting biofilm control methods (Cuzman et al. 2010).

Ramirez et al. (2010) investigated the structure and species composition of biofilms from the walls of one of the buildings at the archaeological site of Palenque, Mexico. The distribution of photosynthetic microorganisms in the biofilms, their relationship with the colonized substratum, and the three-dimensional structure of the biofilms were studied by image analysis. Ramirez et al. (2010) described the differences between local seasonal microenvironments at the Palenque site, the bioreceptivity of stone and the relationship between biofilms and their substrata. Knowledge on how different biofilms contribute to biodegradation or bioprotection of the monumental walls will help develop maintenance and conservation protocols for cultural heritage.

11.11 Adaptation of Biofilm Structure for Survival in Varying Environments

Intriguingly, the visual characteristics of biofilms growing in diverse environments are strikingly similar, indicating there are important convergent survival strategies that are conferred in part by structural specialization. Biofilms growing in fast-moving water tend to form filamentous streamers regardless of whether they occur in the drainage run-off from acid mines (Edwards et al. 2000), in hydrothermal photosynthetic mats (algal or bacterial) (Reysenbach and Cady 2001) or as periphyton in rivers. In quiescent waters, biofilms tend to form mushroom or mound-like structures that are similar to those of stromatolites. The overall patterns are isotropic with no obvious indication of flow direction. The structure of biofilms also changes with nutrient conditions (Stoodley et al. 1999; Klausen et al. 2003). The ability of prokaryotes to adopt different biofilm structures in response to environmental conditions—owing to genetic regulation (Sauer et al. 2002), selection (Klausen et al. 2003), or both (Ghigo 2003), or to localized growth patterns determined by mass transfer—gives them the flexibility to rapidly adapt to an extent that is not possible in multicellular eukaryotic organisms. The proclivity of bacteria to adhere to surfaces and form biofilms in so many environments is undoubtedly related to the selective advantage that surface association offers (Adnan et al. 2010; Andrews et al. 2010).

11.12 Role of Macromolecules in Biofilm Formation

The processes that govern the attachment of microbial cells to solid substrata and their subsequent development as biofilms are of great importance in natural ecosystems. However, while many studies have focussed on disease causing organisms or reducing contamination of medical implants, e.g. (Kierek and Watnick 2003; Morrow et al. 2005; Lichter et al. 2008; Matl et al. 2008; Ojha et al. 2008), less is known about the attachment processes of microbial cells found in polluted subsurface environments such as contaminated groundwater aquifers. The variable surface properties of the diverse minerals present, and changes to these surfaces by adsorbed molecules such as humic substances, produce distinctive and highly selective pressure on the indigenous community. The formation of biofilms has been shown to be important in influencing colloid mobility (Kim and Fogler 2000; Leon Morales et al. 2007), the treatment of contaminated groundwater (Geller et al. 2000; Komlos et al. 2004; Reardon et al. 2004; Schneider et al. 2006; Sublette et al. 2006; Sani et al. 2008) and the performance of wastewater treatment facilities (Hiley 2003; Larsen and Greenway 2004; Puigagut et al. 2007). EPS including polysaccharides and proteins (Sutherland 2001a, b), nucleic acids (Vlassov et al. 2007) and lipids (Lang and Philp 1998) alter the surface properties of the bacteria themselves to either promote or prevent initial attachment to a surface (Tsuneda et al. 2003) or cell aggregation (Eboigbodin et al. 2007). As biofilms develop, these molecules can provide cohesive forces between cells, thereby forming a matrix or rigid scaffold in which cells are embedded. Microscopic and spectroscopic techniques showed that extracellular macromolecules impacting cell attachment to surfaces, biofilm formation and structure were genus-specific for the strains tested; for *Rhodococcus* lipids were shown to play an important role, for *Pseudomonas* strains nucleic acids were important, and there are indications that proteins influenced differential attachment in *Sphingomonas* strains (Andrews et al. 2010). Within the three genera, the manner in which cells associated with EPS also influenced attached growth behaviour. In *Pseudomonas* sp. Pse1, DNA has been shown by both microscopic and spectroscopic techniques to play an important role in cell attachment to surfaces (Andrews et al. 2010).

11.13 Biofilm and Its Environmental Significance

The ability of many bacteria to adhere to surfaces and to form biofilms has major implications in a variety of industries where biofilms create a persistent source of contamination. The formation of a biofilm is determined not only by the nature of the attachment surface, but also by the characteristics of the bacterial cell and by environmental factors (Van Houdt and Michiels 2010). Biofilm formation by *B. subtilis* and related species permits the control of infections caused by plant pathogens, the reduction of mild steel corrosion, and the exploration of novel compounds. Although it is obviously important to control harmful biofilm formation, the exploitation of beneficial biofilms formed by such industrial bacteria may lead to a new biotechnology (Morikawa 2006). *Rhizobia* are non-spore-forming soil bacteria that fix atmospheric nitrogen into ammonia in a symbiosis with legume roots. However, in the absence of a legume host, rhizobia manage to survive and hence must have evolved strategies to adapt to diverse environmental conditions. The capacity to respond to variations in nutrient availability enables the persistence of rhizobial species in soil, and consequently improves their ability to colonize and to survive in the host plant. *Rhizobia,* like many other soil bacteria, persist in nature most likely in sessile communities known as biofilms, which are most often composed of multiple microbial species (Rinaudi et al. 2006). Rinaudi and coworkers (2006) employed in vitro assays to study environmental parameters that might influence biofilm formation in the *Medicago* symbiont *Sinorhizobium meliloti.* These parameters include carbon source, amount of nitrate, phosphate, calcium and magnesium as well as the effects of osmolarity and pH. The microtiter plate assay facilitates the detection of subtle differences in rhizobial biofilms in response to these parameters, thereby providing insight into how environmental stress or nutritional status influences rhizobial survival. Nutrients such as sucrose, phosphate and calcium enhance biofilm formation as their concentrations increase, whereas extreme temperatures and pH negatively affect biofilm formation (Rinaudi et al. 2006). Biofilm-associated microorganisms play crucial roles in terrestrial and aquatic nutrient cycling and in the biodegradation of environmental pollutants. Meng-Ying et al. (2009) determined biofilm formation for a total of 18 bacterial isolates obtained from the biofilms of wastewater treatment systems and soil. Among these isolates, seven showed strong biofilm-forming capacity. The phylogenetic affiliation of the isolates showing high biofilm formation capacity was determined through 16S rDNA sequencing and the isolates were grouped into seven bacterial species including *Pseudomonas* sp., *Pseudomonas putida, Aeromonas caviae, Bacillus cereus, Pseudomonas plecoglossicida, Aeromonas hydrophila*, and *Comamonas testosteroni.* The biofilm-forming capacity was closely related with presence of flagella, exopolysaccharide, and extracellular protein (Meng-Ying et al. 2009). According to the coefficient of determination, the relative importance of the five biological characteristics to biofilm formation was, in order from greatest to least, exopolysaccharide > flagella > N-acyl-homoserine lactones (AHLs) signaling molecules > extracellular protein > swarming motility (Meng-Ying et al. 2009).

Bacteria in marine environments are often under extreme conditions of e.g., pressure, temperature, salinity, and depletion of micronutrients, with survival and proliferation often depending on the ability to produce biologically active compounds. Some marine bacteria produce biosurfactants, which help transport hydrophobic low water soluble substrates by increasing their bioavailability. However, other functions related to heavy metal binding, QS and biofilm formation have been described. In the case of metal ions, bacteria developed a strategy involving the release of binding agents to increase their bioavailability. In the particular case of the $Fe(3+)$ ion, which is almost insoluble in water, bacteria secrete siderophores that form soluble complexes with the ion, allowing the cells to uptake the iron required for cell functioning (de Carvalho and Fernandes 2010). Adaptive changes in the

11 Biofilm Formation by Environmental Bacteria

lipid composition of marine bacteria have been observed in response to environmental variations in pressure, temperature and salinity. Some fatty acids, including docosahexaenoic and eicosapentaenoic acids, have only been reported in prokaryotes in deep-sea bacteria (de Carvalho and Fernandes 2010). Cell membrane permeability can also be adapted to extreme environmental conditions by the production of hopanoids, which are pentacyclic triterpenoids that have a function similar to cholesterol in eukaryotes. Bacteria can also produce molecules that prevent the attachment, growth and/or survival of challenging organisms in competitive environments. The production of these compounds is particularly important in surface attached strains and in those inside biofilms (de Carvalho and Fernandes 2010). The wide array of compounds produced by marine bacteria as an adaptive response to demanding conditions makes them suitable candidates for screening of compounds with commercially interesting biological functions. Biosurfactants produced by marine bacteria may be helpful to increase mass transfer in different industrial processes and in the bioremediation of hydrocarbon-contaminated sites (de Carvalho and Fernandes 2010).

Flexibility of gene expression in bacteria permits their survival in varied environments. The genetic adaptation of bacteria through systematized gene expression is not only important (for biofilm formation), but also relevant in their ability to grow biofilms in stress environments. Stress responses enable their survival under more severe conditions, enhanced resistance and/or virulence. *Listeria monocytogenes* is a food-borne facultative intracellular pathogen. It is widespread in the environment and has several distinct lifestyles. The key transcriptional activator PrfA positively regulates *L. monocytogenes* virulence genes to mediate the transition from extracellular, flagella-propelled cell to intracellular pathogen. Here PrfA has a significant positive impact on extracellular biofilm formation. Mutants lacking *prfA* were defective in surface-adhered biofilm formation. The Δ*prfA* mutant exhibited wild-type flagellar motility and its biofilm defect occurred after initial surface adhesion. Thus, PrfA positively regulates biofilm formation and suggests that PrfA has a global role in modulating *L. monocytogenes* lifestyle. The requirement of PrfA for optimal biofilm formation may provide selective pressure to maintain this critical virulence regulator when *L. monocytogenes* is outside of host cells in the environment (Lemon et al. 2010).

11.14 Control of Biofilm Growth in the Environment

Biofouling is the growth of bacteria, algae and animals, such as protozoans and crustaceans, on surfaces that experience prolonged contact with water. This undesired accumulation of living organisms and their secretions leads to contamination, colonization and corrosion of e.g., ship hulls, pipes, tanks, membrane bioreactors and decreases their efficiency (Choudhary and Schmidt-Dannert 2010).

Spettmann et al. (2007) developed an experimental system to simultaneously visualize biofouling, organic and inorganic particle fouling on separation membranes in water treatment systems. Fluorescently labeled model foulants were used: (1) drinking water bacteria stained with nucleic acid-specific dyes (biofouling), (2) synthetic clay minerals stained with rhodamine 6G (inorganic particle fouling), and (3) fluorescently labeled polystyrene microspheres (organic particle fouling). Polycarbonate and polyethersulfonate membranes were challenged with these foulants. On the basis of different fluorescent labels, the single foulants in these mixed deposits were visualized separately by confocal laser scanning microscopy, which, in combination with image analysis, allowed the generation of three-dimensional views of the complete deposits. The method of Spettmann et al. (2007) offers the possibility to estimate the quantitative surface coverage by foulants as well as the efficacy of cleaning measures with respect to the removal of different foulants.

Roeder et al. (2010) studied the long-term effect of disinfection methods on biofilm communities in drinking water systems. Old drinking water biofilms grown in silicone tubes were exposed to different preparations of disinfectants (free chlorine, chlorine dioxide, hydrogen peroxide combined with fruit acid, silver and silver with peracetic acid, respectively) and subsequently further exposed in the original drinking water. Comparison of the treated and regrown biofilm populations with untreated ones by denaturing gradient gel electrophoresis (DGGE) showed a considerable population shift caused by the disinfectants (Roeder et al. 2010). The disinfection methods induced a selection pressure on the biofilm populations depending on the compositions and concentrations. The similarities between the treated and untreated biofilms were generally low. Roeder et al. (2010) concluded that the disinfectants have a major impact on the drinking water biofilm communities and that the intervention possibly selects persisters and microorganisms, which can live on the residuals of the dead biofilm cells. To evaluate the efficiency of disinfection methods in drinking water installations, it is necessary not only to consider reduction of certain indicator bacteria but also to pay attention to the microbial biofilm community.

Shih and Lin (2010) studied the efficacy of copper-silver ionization against the growth of *Pseudomonas aeruginosa*, *Stenotrophomonas maltophilia*, and *Acinetobacter baumannii* in biofilms and planktonic phases from a model plumbing system. The authors concluded that at concentrations below the EPA limits, copper-silver ionization has the potential to control the three waterborne pathogens, in addition to *Legionella* and highlighted its potential application in hospital water systems for nosocomial infection control.

11.15 Applications of Quorum Sensing in Biotechnology

Choudhary and Schmidt-Dannert (2010) published a very interesting review article on applications of QS in biotechnology. Engineered QS-based circuits have a wide range of applications such as production of biochemicals, tissue engineering, and mixed species fermentations. They are also highly useful in designing whole-cell microbial biosensors to identify bacterial species in the environment.

11 Biofilm Formation by Environmental Bacteria   367

QS-control can be applied in the rhizosphere to protect plants from their major bacterial pathogens. Mae et al. (2001) demonstrated that transgenic tobacco plants that produced an AHL essential for virulence of the soft rot pathogen *Erwinia carotovora* displayed increased resistance to *E. carotovora* infection. Dong et al. (2000) successfully applied quorum-quenching enzymes to reduce bacterial virulence against plants.

Prevention of biofouling might be a further promising application of QS-control. Natural and synthetic QS inhibitors may be added to anti-fouling coatings. Alternatively, immobilized quorum-quenching enzymes or marine bacteria that have been engineered to secrete QS inhibitors may be incorporated into anti-fouling treatments (Choudhary and Schmidt-Dannert 2010).

11.16 Conclusions and Perspectives

Biofilms are found in most environments including sea water, fresh water, rocks acid mine drainage solution, in soils, stone monuments, ship hulls etc. Bacteria growing in biofilms are more resistant to antibiotics than their planktonic counterparts. How this transition occurs is unclear, but it is likely that there are multiple mechanisms of resistance that act together in order to provide an increased overall level of resistance to the biofilm. Several factors, such as the extracellular polysaccharide matrix, high cell density, and slow growth, have been implicated in the increase in resistance that biofilm cells exhibit over their planktonic counterparts. EPS play an important role in biofilm formation. The major factors involved in regulation of biofilm formation are the surface of the substrate, low nutrient availability and chromosomally or plasmid-encoded genes involved in the initiation of biofilm formation, maturation and persistence of biofilms. For many bacteria, biofilm dispersal plays an important role in the transmission of bacteria from environmental reservoirs to human hosts, in horizontal and vertical cross-host transmission. The molecular mechanisms of bacterial biofilm dispersal need to be elucidated in more detail. Biofilm dispersal is a promising area of research that may lead to the development of novel agents that inhibit biofilm formation or promote detachment of cells from biofilms. Such agents may be useful for the prevention and elimination/destruction of biofilms in a variety of industrial and environmental areas. QS molecules provide an important way to detect the biofilm formation. Using QS inhibitors biofouling can be controlled and biofilms can be eradicted.

References

Adnan M, Morton G, Singh J et al (2010) Contribution of rpoS and bolA genes in biofilm formation in *Escherichia coli* K-12 MG1655. Mol Cell Biochem. doi: 10.1007/s11010-010-0485-7
Amara N, Mashiach R, Amar D et al (2009) Covalent inhibition of bacterial quorum sensing. J Am Chem Soc 131:10610–10619

Anderl JN, Zahller J, Roe F et al (2003) Role of nutrient limitation and stationary-phase existence in *Klebsiella pneumoniae* biofilm resistance to ampicillin and ciprofloxacin. Antimicrob Agents Chemother 47:1251–1256

Anderson GG, O'toole GA (2008) Innate and induced resistance mechanisms of bacterial biofilms. Curr Top Microbiol Immunol 322:85–105

Andrews JS, Rolfe SA, Huang WE et al (2010) Biofilm formation in environmental bacteria is influenced by different macromolecules depending on genus and species. Environ Microbiol. doi:10.1111/j.1462-2920.2010.02223.x

Antoniani D, Bocci P, Maciag A et al (2010) Monitoring of di-guanylate cyclase activity and of cyclic-di-GMP biosynthesis by whole-cell assays suitable for high-throughput screening of biofilm inhibitors. Appl Microbiol Biotechnol 85:1095–1104

Attila C, Ueda A, Wood TK (2009) 5-Fluorouracil reduces biofilm formation in Escherichia coli K-12 through global regulator AriR as an antivirulence compound. Appl Microbiol Biotechnol 82:525–533

Augspurger C, Karwautz C, Mussmann M et al (2010) Drivers of bacterial colonization patterns in stream biofilms. FEMS Microbiol Ecol 72:47–57

Balaban N, Novick RP (1995) Autocrine regulation of toxin synthesis by Staphylococcus aureus. Proc Natl Acad Sci USA 92:1619–1623

Balzer M, Witt N, Flemming HC et al (2010) Faecal indicator bacteria in river biofilms. Water Sci Technol 61:1105–1111

Barber CE, Tang JL, Feng JX et al (1997) A novel regulatory system required for pathogenicity of Xanthomonas campestris is mediated by a small diffusible signal molecule. Mol Microbiol 24:555–566

Baselga R, Albizu I, Amorena B (1994) Staphylococcus aureus capsule and slime as virulence factors in ruminant mastits. A review. Vet Microbiol 39:195–204

Becker P, Hufnagle W, Peters G et al (2001) Detection of Differential Gene Expression in Biofilm-Forming versus Planktonic Populations of Staphylococcus aureus Using Micro-Representational-Difference Analysis. Appl Environ Microbiol 67:2958–2965

Beech IB (2004) Corrosion of technical materials in the presence of biofilms current understanding and state of the art methods of study. Int Biodeterior Biodegradation 53:177–183

Bendinger B, Rijnaarts HHM, Altendorf K et al (1993) Physicochemical cell surface and adhesive properties of coryneform bacteria related to the presence and chain length of mycolic acids. Appl Environ Microbiol 59:3973–3977

Bjarnsholt T, Jensen P-Ø, Burmølle M et al (2005) Pseudomonas aeruginosa tolerance to tobramycin, hydrogen peroxide and polymorphonuclear leukocytes is quorum-sensing dependent. Microbiol 151:373–83

Bjarnsholt T, Jensen PØ, Fiandaca MJ et al (2009) Pseudomonas aeruginosa biofilms in the respiratory tract of cystic fibrosis patients. Pediatr Pulmonol 44:547–58

Boyd A, Chakrabarty AM (1994) Role of alginate lyase in cell detachment of *Pseudomonas aeruginosa*. Appl Environ Microbiol 60:2355–2359

Brady RA, Leid JG, Calhoun JH, et al (2008) Osteomyelitis and the role of biofilms in chronic infection. FEMS Immunol Med Microbiol 52:13–22

Branda SS, Vik A, Friedman L et al (2005) Biofilms: the matrix revisited. Trends Microbiol 13:20–26

Brar SK, Verma M, Tyagi RD et al (2010) Engineered nanoparticles in wastewater and wastewater sludge–Evidence and impacts. Waste Manage 30(3):504–520

Brown MR, Allison DG, Gilbert PJ (1988) Resistance of bacterial biofilms to antibiotics: a growth-rate related effect? Antimicrob Chemother 22 777–780

Bryers JD (2008) Medical biofilms. Biotechnol Bioeng 100:1–18

Buchholz F, Wolf A, Lerchner J, Mertens F, Harms H, Maskow T (2010) Chip calorimetry for fast and reliable evaluation of bactericidal and bacteriostatic treatments of biofilms. Antimicrob Agents Chemother 54:1312–319

Caiazza NC, Merritt JH, Brothers KM et al (2007) Inverse Regulation of Biofilm Formation and Swarming Motility by *Pseudomonas aeruginosa* PA14. J Bacteriol 3603–3612

Characklis WG, McFeters GA, Marshall KC (1990) In: Characklis WG, Marshall KC (eds) Biofilms. Wiley, New York, pp 341–394

Chen X, Stewart PS (1996) Chlorine penetration into artificial biofilm is limited by a reaction diffusion interaction. Environ Sci Technol 30:2078–2083

Choudhary S, Schmidt-Dannert C (2010) Applications of quorum sensing in biotechnology. Appl Microbiol Biotechnol 86:1267–1279

Clark JD, Maaloe O (1967) DNA replication and the division cycle in Escherichia coli. J Mol Biol 23:99–112

Clarke S, Mielke RE, Neal A et al (2010) Bacterial and mineral elements in an arctic biofilm: a correlative study using fluorescence and electron microscopy. Microsc Microanal 16:153–165

Combaret PC, Vidal O, Dorel C et al (1999) Abiotic surface sensing and biofilm-dependent regulation of gene expression in *Escherichia coli*. J Bacteriol 181:5993–6002

Costerton JW, Lewandowski Z, Caldwell DE et al (1995) Microbial biofilms. Annu Rev Microbiol 49:711–745

Costerton JW, Stewart PS, Greenberg EP (1999) Bacterial biofilms: a common cause of persistent infections. Science 284:1318–1322

Costerton W, Veeh R, Shirtliff M et al (2003) The application of biofilm science to the study and control of chronic bacterial infections. J Clin Invest 112:1466–77

Cowan MM, Warren TM, Fletcher M (1991) Mixed- species colonization of solid surfaces in laboratory biofilms. Biofouling 3:23–34

Cross JL, Ramadan HH, Thomas JG (2007) The impact of a cation channel blocker (furosemide) on *Pseudomonas aeruginosa* PAO1 biofilm architecture Otolaryngology. Head Neck Surg 137(1):21–26

Crueger W, Crueger A (1990) Acetic acids. In: Brock TD (ed) Biotechnology: a textbook of industrial microbiology, pp 134–147. Science Tech, Sinauer Associates, Sunderland, MA

Cuzman OA, Ventura S, Sili C et al (2010) Biodiversity of phototrophic biofilms dwelling on monumental fountains. Microb Ecol 60(1):81–95

Davey ME, O'Toole GA (2000) Microbial biofilms: from ecology to molecular genetics. Microbiol Mol Biol Rev 64:847–67

Davies DG and Geesey GG (1995) Regulation of the alginate biosynthesis gene algC in Pseudomonas aeruginosa during biofilm development in continuous culture. Appl Environ Microbiol 61:860–867

Davies DG, Parsek MR, Pearson JP et al (1998) The involvement of cell-to-cell signals in the development of a bacterial biofilm. Science 280:295–298

De Beer D, Srinivasan R, Stewart PS (1994) Direct measurement of chlorine penetration into biofilms during disinfection. Appl Environ Microbiol 60:4339–4344

de Carvalho CC, Fernandes P (2010) Production of metabolites as bacterial responses to the marine environment. Mar Drugs 8(3):705–727

Demirci A, Pometto AL, Ho KL (1997) Ethanol production by *Saccharomyces cerevisiae* in biofilm reactors. J Ind Microbiol Biotechnol 19:299–304

Dibdin GH, Assinder SJ, Nichols WW et al (1996) Mathematical model of β-lactam penetration into a biofilm of Pseudomonas aeruginosa while undergoing simultaneous inactivation by released -lactamases. J Antimicrob Chemother 38:757–769

Dong YH, Xu JL, Li XZ et al (2000) AiiA, an enzyme that inactivates the acylhomoserine lactone quorum-sensing signal and attenuates the virulence of *Erwinia carotovora*. Proc Natl Acad Sci USA 97:3526–3531

Donlan RM (2002) Biofilms: Microbial Life on Surfaces. Emerg Infect Dis. 8(9):881–890

Donlan RM, Costerton JW (2002) Biofilms: survival mechanisms of clinically relevant microorganisms. Clinical Microbiol Rev 15:167–193

Donlan RM, Pipes WO, Yohe TL (1994) Biofilm formation on cast iron substrata in water distribution systems. Water Res 28:1497–1503

Driffield K, Miller K, Bostock M et al (2008) Increased mutability of *Pseudomonas aeruginosa* in biofilms. J Antimicrob Chemother 61:1053–1056

Eberl L, Winson MK, Sternberg C et al (1996) Involvement of *N*-acyl-L-homoserine lactone autoinducers in controlling the multicellular behaviour of *Serratia liquefaciens*. Mol Microbiol 20:127–136

Eboigbodin KE, Ojeda JJ, Biggs CA (2007) Investigating the surface properties of *Escherichia coli* under glucose controlled conditions and its effect on aggregation. Langmuir 23:6691–6697

Edwards KJ, Bond PL, Gihring TM et al (2000) An archaeal iron-oxidizing extreme acidophile important inacid mine drainage. Science 287:1731–1732

El-Mansi EMT, Ward FB (2007) Microbiology of industrial fermentation. In: El-Mansi EMT (ed) Fermentation, microbiology and biotechnology, pp 11–46. Taylor and Francis, Boca Raton

Espeland EM, Wetzel RG (2001) Complexation, stabilization, and UV photolysis of extracellular and surface-bound glucosidase and alkaline phosphatase: implications for biofilm microbiota. Microb Ecol 42:572–585

Farah C, Vera M, Morin D et al (2005) Evidence for a functional quorum-sensing type AI-1 system in the extremophilic bacterium *Acidithiobacillus ferrooxidans*. Appl Environ Microbiol 71:7033–7040

Fera P, Siebel MA, Characklis WG, Prieur D (1989) Seasonal variations in bacterial colonization of stainless steel, aluminum, and polycarbonate surfaces in seawater flow system. Biofouling 1:251–261

Fett WF (2000) Naturally occuring biofilms on alfalfa and other types of sprouts. J Food Prot 63:625–632

Flemming HC and Wingender J (2001a) Relevance of microbial extracellular polymeric substances (EPSs)—part I: structural and ecological aspects. Water Sci Technol 43:1–8

Flemming HC and Wingender J (2001b) Relevance of microbial extracellular polymeric substances (EPSs)—Part II: technical aspects. Water Sci Technol 43:9–16

Fletcher M (1988) Attachment of *Pseudomonas fluorescens* to glass and influence of electrolytes on bacterium-substratum separation distance. J Bacteriol 170:2027–2030

Fletcher M and Loeb GI (1979) Influence of substratum characteristics on the attachment of a marine pseudomonad to solid surfaces. Appl Environ Microbiol 37:67–72

Folkesson A, Haagensen JAJ, Zampaloni C et al (2008) Biofilm induced tolerance towards antimicrobial peptides. PLoS ONE 3(4):e1891. doi:10.1371/journal.pone.0001891

Fux CA, Costerton JW, Stewart PS et al (2005) Survival strategies of infectious biofilms. Trends Microbiol 13:34–40

Gaddy JA, Actis LA (2009) Regulation of *Acinetobacter baumannii* biofilm formation. Future Microbiol 4:273–278

Galperin MY (2004) Bacterial signal transduction network in a genomic perspective. Environ Microbiol 6:552–567

Gehrke T, Telegdi J, Thierry D et al (1998) Importance of extracellular polymeric substances from *Thiobacillus ferrooxidans* for bioleaching. Appl Environ Microbiol 64:2743–2747

Geller JT, Holman HY, Su G et al (2000) Flow dynamics and potential for biodegradation of organic contaminants in fractured rock vadose zones. J Contam Hydrol 43:63–90

Geske GD, O'Neill JC, Miller DM et al (2007) Modulation of bacterial quorum sensing with synthetic ligands: systematic evaluation of N-acylated homoserine lactones in multiple species and new insights into their mechanisms of action. J Am Chem Soc 129:13613–13625

Ghigo J M (2001) Natural conjugative plasmids induce bacterial biofilm development. Nature 412: 442–445

Ghigo JM (2003) Are there biofilm-specific physiological pathways beyond a reasonable doubt? Res Microbiol 154:1–8

Gilbert P, Allison DG, McBain AJ (2002) Biofilms in vitro and in vivo: do singular mechanisms imply cross-resistance? J Appl Microbiol 92:S98–S110

Hall-Stoodley L and Stoodley P (2002) Developmental regulation of microbial biofilms. Curr Opin Biotechnol 13:228–233

Hall-Stoodley L, Costerton JW, Stoodley P (2004) Bacterial biofilms: from the natural environment to infectious diseases. Nat Rev Microbiol 2:95–108

Hammer BK, Bassler BL (2003) Quorum sensing controls biofilm formation in Vibrio cholerae. Mol Microbiol 50:101–104

Harneit K, Goksel A, Kock D et al (2006) Adhesion to metal sulfide surfaces by cells of *Acidithiobacillus ferrooxidans*, *Acidithiobacillus thiooxidans* and *Leptospirillum ferrooxidans*. Hydrometallurgy 83:245–254

Hausner M, Wuertz S (1999) High rates of conjugation in bacterial biofilms as determined by quantitative in situ analysis. Appl Environ Microbiol 65:3710–3713

Hentzer M, Riedel K, Rasmussen TB et al (2002) Inhibition of quorum sensing in Pseudomonas aeruginosa biofilm bacteria by a halogenated furanone compound. Microbiology 148:87–102

Heydorn A, Ersboll B, Kato J et al (2002) Statistical analysis of *Pseudomonas aeruginosa* biofilm development: impact of mutations in genes involved in twitching motility, cell-to-cell signaling, and stationary-phase sigma factor expression. Appl Environ Microbiol 68:2008–2017

Hiley P (2003) Performance of wastewater treatment and nutrient removal wetlands (reedbeds) in cold temperate climates. In: Mander Ü, Jenssen PD (eds) Constructed wetlands for wastewater treatment in cold climates, pp 1–18. WIT Press, Southampton, UK

Ho KG, Pometto AI, Hinz PN et al (1997) Nutrient leaching and end product accumulation in plastic composite supports for L-(þ)-lactic acid biofilm fermentation. Appl Environ Microbiol 63:2524–2532

Høiby N, Bjarnsholt T, Givskov M et al (2010) Antibiotic resistance of bacterial biofilms. Internation J Antimicrob Agents 35(4):322–332

Holden MTG, Chhabra SR, de Nys R et al (1999) Quorum-sensing cross talk: isolation and chemical characterization of cyclic dipeptides from *Pseudomonas aeruginosa* and other gram-negative bacteria. Mol Microbiol 33:1254–1266

Holland LM, O'Donnell ST, Ryjenkov DA et al (2008) A staphylococcal GGDEF domain protein regulates biofilm formation independently of cyclic dimeric GMP. J Bacteriol 190:5178–5189

Hooshangi S and Bentley WE (2008) From unicellular properties to multicellular behavior: bacteria quorum sensing circuitry and applications. Curr Opin Biotechnol 19:550–555

Howell JA and Atkinson B (1976) Water Res 18:307–315

Islam S, Oh H, Jalal S et al (2009) Chromosomal mechanisms of aminoglycoside resistance in Pseudomonas aeruginosa isolates from cystic fibrosis patients. Clin Microbiol Infect 15:60–66

Ivleva NP, Wagner M, Horn H, Niessner R, Haisch C (2010) Raman microscopy and surface-enhanced Raman scattering (SERS) for in situ analysis of biofilms. J Biophotonics 3(8–9):548–556

Jahnke LL et al (2001) Signature lipids and stable carbon isotope analyses of octopus spring hyperthermophilic communities compared with those of aquificales representatives. Appl Environ Microbiol 67:5179–5189

Jalal S, Ciofu O, Høiby N et al (2000) Molecular mechanisms of fluoroquinolone resistance in Pseudomonas aeruginosa isolates from cystic fibrosis patients. Antimicrob Agents Chemother 44:710–712

Jefferson KK (2004) What drives bacteria to produce a biofilm? FEMS Microbiol Lett 236:163–173

Jiao Y, Cody GD, Harding AK et al (2010) Characterization of extracellular polymeric substances from acidophilic microbial biofilms. Appl Environ Microbiol 76:2916–2922

Junker LM, Clardy J (2007) High-throughput screens for small molecule inhibitors of *Pseudomonas aeruginosa* biofilm development. Antimicrob Agents Chemother 51:3582–3590

Karatan E, Watnick PI (2009) Signals, regulatory networks, and materials that build and break bacterial biofilms. Microbiol Mol Biol Rev 73:310–347

Keren I, Kaldalu N, Spoering A et al (2004) Persister cells and tolerance to antimicrobials. FEMS Microbiol Lett 230:13–18

Kierek K, Watnick PI (2003) Environmental determinants of *Vibrio cholerae* biofilm development. Appl Environ Microbiol 69:5079–5088

Kim DS, Fogler HS (2000) Biomass evolution in porous media and its effects on permeability under starvation conditions. Biotechnol Bioeng 69:47–56

Kiran MD, Adikesavan NV, Cirioni O et al (2008) Discovery of a quorum-sensing inhibitor of drug-resistant staphylococcal infections by structure-based virtual screening. Mol Pharmacol 73:1578–1586

Kjaergaard K, Schembri MA, Ramos C et al (2000) Antigen 43 facilitates formation of multispecies biofilms. Environ Microbiol 2:695–702

Klausen M, Heydorn A, Ragas P et al (2003) Biofilm formation by Pseudomonas aeruginosa wild type, flagella and type IV pili mutants. Mol Microbiol 48:1511–1524

Klingenberg C, Aarag E, Ronnestad A et al (2005) coagulase negative Staphylococcal sepsi in neonates: Association between antibiotic resistance, biofilm formation and the host inflammatory response. Ped inf dis J 24:817–822

Kolenbrander PE, Palmer Jr RJ (2004) Human oral bacterial biofilms. In: Ghannoum MA, O'Toole GA (eds) Microbial biofilms. ASM Press, Washington, DC

Komlos J, Cunningham AB, Camper AK et al (2004) Biofilm barriers to contain and degrade dissolved trichloroethylene. Environ Prog 23:69–77

Kornaros M, Lyberatos G (2006) Biological treatment of wastewaters from a dye manufacturing company using a trickling filter. J Hazard Mater 136:95–102

Kulasakara H, Lee V, Brencic A et al (2006) Analysis of Pseudomonas aeruginosa diguanylate cyclases and phosphodiesterases reveals a role for bis-(3'-5')-cyclic-GMP in virulence. Proc Natl Acad Sci USA 103:2839–2844

Lagendijk EL, Validov S, Lamers GE et al (2010) Genetic tools for tagging Gram-negative bacteria with mCherry for visualization in vitro and in natural habitats, biofilm and pathogenicity studies. FEMS Microbiol Lett 305:81–90

Landini P, Antoniani D, Burgess JG, Nijland R (2010) Molecular mechanisms of compounds affecting bacterial biofilm formation and dispersal. App Microbiol Biotechnol 86(3):813–823

Lang S, Philp JC (1998) Surface-active lipids in rhodococci. Anton Van Leeuwenhoek 74:59–70

Larsen E, Greenway M (2004) Quantification of biofilms in a sub-surface flow wetland and their role in nutrient removal. Water Sci Technol 49:115–122

Lazarova V, Perera J, Bowen M et al (2000) Application of aerated biofilters for production of high quality water for industrial reuse in West Basin. Water Sci Technol 41:417–424

Le Magrex-Debar E, Lemoine J, Gelle MP et al (2000) Evaluation of biohazards in dehydrated biofilms on foodstuff packaging. Int J Food Microbiol 55:239–234

Leewenhoeck A (1683) An abstract of a Letter from Mr. Anthony Leewenhoeck at Delft, Dated Sep. 17. 1683. Containing some microscopical observations, about animals in the serf of the teeth, the substance call'd worms in the nose, the cuticula consisting of scales. Philosop Trans 14:568–574

Leid JG, Shirtliff ME, Costerton JW et al (2002) Human leukocytes adhere, penetrate, and respond to *Staphylococcus aureus* biofilms. Infect Immun 70:6339–6345

Lemon KP, Freitag NE, Kolter R (2010) The virulence regulator PrfA promotes biofilm formation by Listeria monocytogenes. J Bacteriol 192(15):3969–3976

Leon Morales CF, Strathmann M et al (2007) Influence of biofilms on the movement of colloids in porous media. Implications for colloid facilitated transport in subsurface environments. Water Res 41:2059–2068

Lequette Y, Boels G, Clarisse M et al (2010) Using enzymes to remove biofilms of bacterial isolates sampled in the food-industry. Biofoul 26(4):421–431

Lerner A, Castro-Sowinski S, Valverde A et al (2009) The Azospirillum brasilense Sp7 noeJ and noeL genes are involved in extracellular polysaccharide biosynthesis. Microbiol 155(12):4058–4068

Lewis K (2001) Riddle of Biofilm Resistance. Int J Antimicrob Agents Chemother 45(4):999–1007

Li J, Attila C, Wang L et al (2007) Quorum sensing in *Escherichia coli* is signaled by AI-2/LsrR: effects on small RNA and biofilm architecture. J Bacteriol 189:6011–6020

Li M, Ni N, Chou HT et al (2008) Structurebased discovery and experimental verification of novel AI-2 quorum sensing inhibitors against Vibrio harveyi. ChemMed Chem 3:1242–1249

Lichter JA, Thompson MT, Delgadillo M et al (2008) Substrata mechanical stiffness can regulate adhesion of viable bacteria. Biomacromolecules 9:1571–1578

11 Biofilm Formation by Environmental Bacteria 373

Lindow SE, Brandl MT (2003) Microbiology of phyllosphere. App Environ Microbiol 69:1875–1883

Lönn-Stensrud J, Landin MA, Benneche T et al (2009) Furanones, potential agents for preventing Staphylococcus epidermidis biofilm infections? J Antimicrob Chemother 63:309–316

López D, Vlamakis H, Kolter R (2010) Biofilms. Cold Spring Harb Perspect Biol 2:a000398

Mäe A, Montesano M, Koiv V et al (2001) Transgenic plants producing the bacterial pheromone N-acyl-homoserine lactone exhibit enhanced resistance to the bacterial phytopathogen Erwinia carotovora. Mol Plant Microbe Interact 14:1035–1042

Mah TF O'Toole GA (2001) Mechanisms of biofilm resistance to antimicrobial agents. Trends Microbiol 9:34–39

Majumder PS, Gupta SK (2003) Hybrid reactor for priority pollutant nitrobenzene removal. Water Res 37:4331–4336

Manefield M, Rasmussen TB, Henzter M et al (2002) Halogenated furanones inhibit quorum sensing through accelerated LuxR turnover. Microbiol 118:1119–1127

Marketon MM, Glenn SA, Eberhard A et al (2003) Quorum sensing controls exopolysaccharide production in *Sinorhizobium meliloti*. J Bacteriol 185:325–331

Marshall KC (1992) Biofilms: an overview of bacterial adhesion, activity, and control at surfaces. ASM News 58:202–207

Matl FD, Obermeier A, Repmann S et al (2008) New anti-infective coatings of medical implants. Antimicrob Agents Chemother 52:1957–1963

McClean KH, Winson MK, Fish L et al (1997) Quorum sensing and Chromobacterium violaceum: exploitation of violacein production and inhibition for the detection of N-acylhomoserine lactones. Microbiol 143:3703–3711

McKenney D, Hubner J, Muller E et al (1998) The ica locus of *Staphylococcus epidermidis* encodes production of the capsular polysaccharide/adhesin. Infect Immunol 66:4711–4720

McKnight SL, Iglewski BH, Pesci EC (2000) The Pseudomonas quinolone signal regulates rhl quorum sensing in Pseudomonas aeruginosa. J Bacteriol 182:2702–2708

McNeill K, Hamilton IR (2003) Acid tolerance response of biofilm cells of *Streptococcus mutans*. FEMS Microbiol Lett 221:25–30

Meng-Ying LI, Ji Z, Peng LU et al (2009) Evaluation of Biological Characteristics of Bacteria Contributing to Biofilm Formation. Pedosphere 19(5):554–561

Michael T, Smith CM (1995) Lectins probe molecular films in biofouling : characterization of early films on non-living and living surfaces. Mar Ecol Progr Ser 119:229–236

Miller MB, Bassler BL (2001) Quorum sensing in bacteria. Annu Rev Microbiol 55:165–199

Molin S, Tolker-Nielsen T (2003) Gene transfer occurs with enhanced efficiency in biofilms and induces enhanced stabilisation of the biofilm structure. Curr Opin Biotechnol 14:255–261

Morikawa M (2006) Beneficial biofilm formation by industrial bacteria Bacillus subtilis and related species. J Biosci Bioeng 101(1):1–8

Morris CE, Monier JM, Jacques MA (1998) A technique to quantify the population size and composition of the biofilm component in communities of bacteria in the phyllosphere. App Environ Microbiol 64:4789–4795

Morrow JB, Stratton R, Yang HH et al (2005) Macro- and nanoscale observations of adhesive behavior for several *E. coli* strains (O157:H7 and environmental isolates) on mineral surfaces. Environ Sci Technol 39:6395–6404

Müh U, Schuster M, Heim R et al (2006) Novel Pseudomonas aeruginosa quorum-sensing inhibitors identified in an ultra-high-throughput screen. Antimicrob Agents Chemother 50:3674–3679

Nakamura S, Higashiyama Y, Izumikawa K et al (2008) The roles of quorum-sensing system in the release of extracellular DNA, lipopolysaccharide, and membrane vesicles from *Pseudomonas aeruginosa*. Jpn J Infect Dis 61:375–378

Neu TR (2000) *In situ* cell and glycoconjugate distribution of river snow as studied by confocal laser scanning microscopy. Aquat Microb Ecol 21:85–95

Neu TR, Lawrence JR (1997) Development and structure of microbial biofilms in river water studied by confocal laser scanning microscopy. FEMS Microbiol Ecol 24:11–25

Neu TR, Lawrence JR (1999) Lectin-binding analysis in biofilm systems. Methods Enzymol 310:145–152

Neu TR, Manz B, Volke F, Dynes JJ, Hitchcock AP, Lawrence JR (2010) Advanced imaging techniques forassessmentof structure, compositionand function in bioȼlm systems. FEMS Microbiol Ecol 72:1–21

Neu TR, Swerhone GDW, Lawrence JR (2001) Assessment of lectin-binding analysis for *in situ* detection of glycoconjugates in biofilm systems. Microbiol 147:299–313

Njoroge J, Sperandio J (2009) Jamming bacterial communication: new approaches for the treatment of infectious diseases. EMBO Molec Med 1:201–210

Novick RP, Ross HF, Projan SJ et al (1993) Synthesis of staphylococcal virulence factors is controlled by a regulatory RNA molecule. EMBO J 12:3967–3975

O'Toole GA, Kaplan HB, Kolter R (2000) Biofilm formation as microbial development. Annu Rev Microbiol 54:49–79

Ojha AK, Baughn AD, Sambandan D et al. (2008) Growth of *Mycobacterium tuberculosis* biofilms containing free mycolic acids and harbouring drug tolerant bacteria. Mol Microbiol 69:164–174

Pamp SJ, Gjermansen M, Johansen HK (2008) Tolerance to the antimicrobial peptide colistin in Pseudomonas aeruginosa biofilms is linked to metabolically active cells, and depends on the prm and mexAB-oprM genes. Mol Microbiol 68:223–240

Persson T, Hansen TH, Rasmussen TB et al (2005) Rational design and synthesis of new quorum-sensing inhibitors derived from acylated homoserine lactones and natural products from garlic. Org Biomol Chem 3:253–262

Pesci EC, Bilbank JBJ, Pearson JP et al (1999) Quinolone signaling in the cell-to-cell communication system of Pseudomonas aeruginosa. Proc National Acad Sci USA 96:11229–11234

Potera C (1999) Forging a link between biofilms and disease. Sci 283:1837–1839

Prakash B, Krishnappa G, Muniyappa L, Reddy MK (2005) *In vitro* phase variation studies of *Salmonella* Gallinarum in biofilm formation. Curr Sci 89(4):657–659

Prakash B, Veeregowda BM, Krishnappa G (2003) Biofilms: A survival strategy of bacteria Curr Sci 85(9):1299–1307

Pratt LA Kolter R (1998) Genetic analysis of *Escherichia coli* biofilm formation: defining the roles of flagella, motility, chemotaxis and type I pili. Mol Microbiol 30:285–294

Pratten J, Foster SJ, Chan PF et al (2001) Staphylococcus aureus accessory regulators: expression within biofilms and effect on adhesion. Microbes Infect 3:633–637

Pringle JH Fletcher M (1983) Influence of substratum wettability on attachment of freshwater bacteria to solid surfaces. Appl Environ Microbiol 45:811–817

Puigagut J, Salvado H, Garcia D et al (2007) Comparison of microfauna communities in full scale subsurface flow constructed wetlands used as secondary and tertiary treatment. Water Res 41:1645–1652

Pulcini E (2001) The effects of initial adhesion events on the physiology of *Pseudomonas aeruginosa*, Ph D dissertation, Montana State University, Bozeman

Qu Y, Daley AJ, Istivan TS et al (2010) Densely adherent growth mode, rather than extracellular polymer substance matrix build-up ability, contributes to high resistance of Staphylococcus epidermidis biofilms to antibiotics. J Antimicrob Chemother 65(7):1405–1411

Qureshi N, Annous BA, Ezeji TC et al (2005) Biofilm reactors for industrial bioconversion processes: Employing potential of enhanced reaction rates. Microb Cell Fact 4:24. doi:10.1186/1475-2859-4-24

Qureshi N, Karcher P, Cotta M et al (2004) High-productivity continuous biofilm reactor for butanol production: Effect of acetate, butyrate, and corn steep liquor on bioreactor performance. Appl Biochem Biotechnol 114:713–721

Ramirez M, Hernandez-Marine M, Novelo E et al (2010) Cyanobacteria-containing biofilms from a Mayan monument in Palenque, Mexico. Biofouling 26:399–409

Rashid MH, Rumbaugh K, Passador L et al (2000) Polyphosphate kinase is essential for biofilm development, quorum sensing, and virulence of *Pseudomonas aeruginosa*. Proc Natl Acad Sci USA 97:9636–9641

11 Biofilm Formation by Environmental Bacteria 375

Rasmussen TB, Skindersoe ME, Bjarnsholt T et al (2005) Identity and effects of quorum-sensing inhibitors produced by Penicillium species. Microbiol 151:1325–1340

Rasmussen, B (2000) Filamentous microfossils in a 3,235-millionyear-old volcanogenic massive sulphide deposit. Nature 405:676–679

Reardon CL, Cummings DE, Petzke LM et al (2004) Composition and diversity of microbial communities recovered from surrogate minerals incubated in an acidic uranium-contaminated aquifer. Appl Environ Microbiol 70:6037–6046

Ren D, Bedzyk LA, Setlow P et al (2004) Differential gene expression to investigate the effect of (5Z)-4-bromo-5-(bromomethylene)-3- butyl-2(5H)-furanone on *Bacillus subtilis*. Appl Environ Microbiol 70:4941–4949

Reysenbach AL Cady SL (2001) Microbiology of ancient and modern hydrothermal systems Trends Microbiol 9:79–86

Reysenbach AL, Ehringer M Hershberger K (2000) Microbial diversity at 83 degrees C in Calcite Springs, Yellowstone National Park: another environment where the Aquificales and 'Korarchaeota' coexist. Extremophiles 4:61–67

Richards JJ, Ballard TE, Huigens RW et al (2008) Synthesis and screening of an oroidin library against Pseudomonas aeruginosa biofilms. Chem Biochem 9:1267–1279

Rinaudi LV, González JE (2009) The low-molecular-weight fraction of exopolysaccharide II from Sinorhizobium meliloti is a crucial determinant of biofilm formation. J Bacteriol 191(23):7216–7224

Rinaudi L, Fujishige NA, Hirsch AM, Banchio E, Zorreguieta A, Giordano W (2006) Effects of nutritional and environmental conditions on *Sinorhizobium meliloti* biofilm formation. Res Microbiol 157(9):867–875

Rivardo F, Turner RJ, Allegrone G et al (2009) Anti-adhesion activity of two biosurfactants produced by Bacillus spp. prevents biofilm formation of human bacterial pathogens. Appl Microbiol Biotechnol 83:541–553

Rivas M, Seeger M, Holmes DS et al (2005) A Lux-like quorum sensing system in the extreme acidophile *Acidithiobacillus ferrooxidans*. Biol Res 38:283–297

Roberts ME, Stewart PS (2005) Modelling protection from antimicrobial agents in biofilms through the formation of persister cells. Microbiol 151:75–80

Roeder RS, Lenz J, Tarne P et al (2010) Long-term effects of disinfectants on the community composition of drinking water biofilms. Int J Hyg Environ Health 213:183–189

Ruiz LM, Valenzuela S, Castro M et al (2008) AHL communication is a widespread phenomenon in biomining bacteria and seems to be involved in mineral-adhesion efficiency. Hydrometallurgy 94:133–137

Sand W, Gehrke T (2006) Extracellular polymeric substances mediate bioleaching/biocorrosion via interfacial processes involving iron(III) ions and acidophilic bacteria. Res Microbiol 157:49–56

Sani RK, Peyton BM, Dohnalkova A (2008) Comparison of uranium(VI) removal by *Shewanella oneidensis* MR-1 in flow and batch reactors. Water Res 42:2993–3002

Sauer K, Camper AK, Ehrlich GD et al (2002) *Pseudomonas aeruginosa* displays multiple phenotypes during development as a biofilm. J Bacteriol 184:1140–1154

Schneider RP, Morano SC, Gigena MAC et al. (2006) Contamination levels and preliminary assessment of the technical feasibility of employing natural attenuation in 5 priority areas of Presidente Bernardes Refinery in Cubatao, Sao Paulo, Brazil. Environ Monit Assess 116:21–52

Schupp JM, Travis SE, Price LB et al (1995) Rapid bacterial permeabilization reagent useful for enzyme assays. Biotechniques 19:18–20

Shih HY, Lin YE (2010) Efficacy of copper-silver ionization in controlling biofilm- and plankton-associated waterborne pathogens. Appl Environ Microbiol 76:2032–2035

Shikuma NJ, Hadfield MG (2010) Marine biofilms on submerged surfaces are a reservoir for Escherichia coli and *Vibrio cholerae*. Biofouling 26:39–46

Simm R, Morr M, Kader A et al (2004) GGDEF and EAL domains inversely regulate cyclic di-GMP levels and transition from sessility to motility. Mol Microbiol 53:1123–1134

Singh R, Ray P, Das A et al (2009) Role of persisters and small-colony variants in antibiotic resistance of planktonic and biofilm-associated Staphylococcus aureus: an in vitro study. J Med Microbiol 58(8):1067–1073

Sivakumar PM, Prabhawathi V, Doble M (2010) Antibacterial activity and QSAR of chalcones against biofilm-producing bacteria isolated from marine waters. SAR QSAR Environ Res 21(3):247–263

Smith KM, Bu Y, Suga H (2003) Induction and inhibition of Pseudomonas aeruginosa quorum sensing by synthetic autoinducer analogs. Chem Biol 10:81–89

Smith RS, Iglewski BH (2003) *Pseudomonas aeruginosa* quorum sensing as a potential antimicrobial target. J Clin Invest 112:1460–1465

Spettmann D, Eppmann S, Flemming HC et al (2007) Simultaneous visualisation of biofouling, organic and inorganic particle fouling on separation membranes. Water Sci Technol 55:207–210

Spoering AL, Lewis K (2001) Biofilms and planktonic cells of *Pseudomonas aeruginosa* have similar resistance to killing by antimicrobials. J Bacteriol 183:6746–6751

Spormann AM (2008) Physiology of microbes in biofilms. Curr Top Microbiol Immunol 322:17–36

Stewart PS, Costerton JW (2001) Antibiotic resistance of bacteria in biofilms. Lancet 358:135–138

Stoodley P, Dodds I, Boyle JD et al (1999) Influence of hydrodynamics and nutrients on biofilm structure. J Appl Microbiol 85:19S–28S

Sublette K, Peacock A, White D et al (2006) Monitoring subsurface microbial ecology in a sulfate-amended, gasoline-contaminated aquifer. Ground Water Monitor Remed 26:70–78

Suci PA, Tyler BJ (2003) A method for discrimination of subpopulations of *Candida albicans* biofilm cells that exhibit relative levels of phenotypic resistance to chlorhexidine. J Microbiol Methods 53:313–325

Sudarsan N, Lee ER, Weinberg Z et al (2008) Riboswitches in eubacteria sense the second messenger cyclic di-GMP. Sci 321:411–413

Surette MG, Bassler BL (1998) Quorum sensing in *Escherichia coli* and *Salmonella typhimurium*. Proc Natl Acad Sci USA 95:7046–7050

Sutherland IW (2001a) Biofilm exopolysaccharides: a strong and sticky framework. Microbiol 147:3–9

Sutherland IW (2001b) The biofilm matrix—an immobilized but dynamic microbial environment. Trends Microbiol 9:222–227

Tamayo R, Pratt JT, Camilli A (2007) Roles of cyclic diguanylate in the regulation of bacterial pathogenesis. Annu Rev Microbiol 61:131–148

Taylor CD, Wirsen CO, Gaill F (1999) Rapid microbial production of filamentous sulfur mats at hydrothermal vents. Appl Environ Microbiol 65:2253–2255

Teitzel GM Parsek MR (2003) Heavy metal resistance of biofilm and planktonic *Pseudomonas aeruginosa*. Appl Environ Microbiol 69:2313–2320

Tempe J, Petit A, Holsters M et al (1977) Thermosensitive step associated with transfer of Ti plasmid during conjugation: possible relation to transformation in crown gall. Proc Natl Acad Sci USA 74:2848–2849

Thien FCM, O'toole GA (2001) Mechanisms of biofilm resistance to antimicrobial agents. Trends Microbiol 9:34–39

Tsuneda S, Aikawa H, Hayashi H et al (2003) Extracellular polymeric substances responsible for bacterial adhesion onto solid surface. FEMS Microbiol Lett 223:287–292

Tuomanen E, Durack DT, Tomasz A (1986) Antibiotic tolerance among clinical isolates of bacteria. Antimicrob Agents Chemother 30:521–527

Ueda A, Attila C, Whiteley M et al (2009) Uracil influences quorum sensing and biofilm formation in *Pseudomonas aeruginosa* and fluorouracilnis an antagonist. Microb Biotechnol 2:62–74

Ueda A, Wood TK (2009) Connecting quorum sensing, c-di-GMP, pel polysaccharide, and biofilm formation in *Pseudomonas aeruginosa* through tyrosine phosphatase TpbA (PA3885). PLoS Pathog 5:e1000483

Van Houdt R, Michiels CW (2010) Biofilm formation and the food industry, a focus on the bacterial outer surface. J Appl Microbiol 109(4):1117–1131

11 Biofilm Formation by Environmental Bacteria

van Hullebusch ED, Zandvoort MH, Lens PNL (2003) Metal immobilisation by biofilms: mechanisms and analytical tools. Rev Environ Sci Biotechnol 2:9–33

Vlassov VV, Laktionov PP, Rykova EY (2007) Extracellular nucleic acids. Bioessays 29:654–667

von Bodman SB, Majerczak DR, Coplin DL (1998) A negative regulator mediates quorum-sensing control of exopolysaccharide production in *Pantoea stewartii* subsp. *stewartii*. Proc Natl Acad Sci USA 95:7687–7692

Vu B, Chen M, Crawford RJ et al (2009) Bacterial Extracellular Polysaccharides Involved in Biofilm Formation. Molecules 14:2535–2554

Vuong C, Gerke C, Somerville GA et al (2003) Quorum-sensing control of biofilm factors in Staphylococcus epidermidis. J Infect Dis 188:706–718

Walters MC, Roe F, Bugnicourt A et al (2003) Contributions of antibiotic penetration, oxygen limitation, and low metabolic activity to tolerance of *Pseudomonas aeruginosa* biofilms to ciprofloxacin and tobramycin. Antimicrob Agents Chemother 47:317–323

Wang ZW, Chen S (2009) Potential of biofilm-based biofuel production. Appl Microbiol Biotechnol 83:1–18

Waters CM, Bassler BL (2005) Quorum sensing: cell-to-cell communication in bacteria. Annu Rev Cell Dev Biol 21:319–346

Weber H, Pesavento C, Possling A et al (2006) Cyclic-di-GMP-mediated signalling within the sigma network of Escherichia coli. Mol Microbiol 62:1014–1034

Weinhouse H, Sapir S, Amikam D et al (1997) c-di-GMP-binding protein, a new factor regulating cellulose synthesis in Acetobacter xylinum. FEBS Lett 416:207–211

Wentland EJ Stewart PS, Huang CT et al (1996) Spatial variations in growth rate within *Klebsiella pneumonia* colonies and biofilms. Biotechnol Prog 12:316–321

Werner E, Roe F, Bugnicourt A et al (2004) Stratified growth in *Pseudomonas aeruginosa* biofilms. Appl Environ Microbiol 70:6188–6196

Whitchurch CB, Tolker-Nielsen T, Ragas PC et al (2002) Extracellular DNA required for bacterial biofilm formation. Science 295:1487

Wingender J, Strathmann M, Rode A et al (2001) Isolation and biochemical characterization of extracellular polymeric substances from *Pseudomonas aeruginosa*. Methods Enzymol 336:302–314

Wolfaardt GM, Lawrence JR, Robarts RD et al (1998) *In situ* characterization of biofilm exopolymers involved in the accumulation of chlorinated organics. Microb Ecol 35:213–223

Wright CJ, Shah MK, Powell LC, Armstrong I (2010) Application of AFM from microbial cell to biofilm. Scanning Vol 31:1–16

Wu J, Long Q, Xie J (2010) (p)ppGpp and drug resistance. J Cellular Physiol 224(2):300–304

Xu X, Stewart PS, Chen X (1996) Transport limitation of chlorine disinfection of *Pseudomonas aeruginosa* entrapped in alginate beads. Biotechnol Bioeng 49:93–100

Yang L, Rybtke MT, Jakobsen TH et al (2009) Computer-aided identification of recognized drugs as Pseudomonas aeruginosa quorum-sensing inhibitors. Antimicrob Agents Chemother 53:2432–2443

Yarwood JM, Bartels DJ, Volper EM et al (2004) Quorum sensing in Staphylococcus aureus biofilms. J Bacteriol 186:1838–1850

Zeng Z, Qian L, Cao L et al (2008) Virtual screening for novel quorum sensing inhibitors to eradicate biofilm formation of Pseudomonas aeruginosa. Appl Microbiol Biotechnol 79:119–126

Zhu H, Kumar A, Ozkan J et al (2008) Fimbrolide-coated antimicrobial lenses: their in vitro and in vivo effects. Optom Vis Sci 85:292–300

Chapter 12
Biochemical Processes of Rhizobacteria and their Application in Biotechnology

M. S. Dardanelli, D. B. Medeot, N. S. Paulucci, M. A. Bueno, J. C. Vicario, M. García, N. H. Bensi and A. M. Niebylski

Contents

12.1	Introduction	380
12.2	Microbial Diversity of Soil	381
12.3	Biochemicals in Soil: Rhizodeposition	382
12.4	Root–Rhizobacteria Communication and Signals in the Rhizosphere	383
12.5	The Rhizosphere in Action	386
12.6	Benefits from Legume Plants: Food, Biochemical and Pharmaceuticals	388
12.7	Perspectives and Conclusion	392
	References	393

Abstract The rhizosphere is a multiple interface between soils, plant roots, microbes and fauna, where different biological components interact strongly. Rhizosphere interactions are based on complex exchanges that take place around plant roots. Beneficial, detrimental and neutral relationships between plant roots and microorganisms are all regulated by complex molecular signalling. Plants exude a variety of organic compounds (e.g. carbohydrates, carboxylic acids, phenolics, amino acids, flavonoids) as well as inorganic ions (protons and other ions) into the rhizosphere which change the chemistry and biology of the root microenvironment. All chemical compounds secreted by plants are collectively named rhizodepositions. In the rhizosphere, bacteria that exert beneficial effects on plant development are referred to as plant growth-promoting rhizobacteria (PGPR) because their application is often associated with increased rates of plant growth. On the other hand, although many technologies have been used in the improvement of stress tolerance in plants, fewer reports have been published on how PGPR can exert tolerance to salt, drought or heavy metals. In addition, the industrial use and technological application of compounds from plants and rhizobacteria are required to be successful in attaining sustainable microbial-based agrotechnologies. Among crops, legumes

M. S. Dardanelli (✉)
Departamento de Biología Molecular, Facultad de Ciencias Exactas, Físico-Químicas y Naturales, Universidad Nacional de Río Cuarto, Ruta Nacional N° 36, Km. 601, CP X5804BYA Río Cuarto, Córdoba, Argentina
e-mail: mdardanelli@exa.unrc.edu.ar

A. Malik, E. Grohmann (eds.), *Environmental Protection Strategies for Sustainable Development*, Strategies for Sustainability,
DOI 10.1007/978-94-007-1591-2_12, © Springer Science+Business Media B.V. 2012

are a good source of starch, dietary fibre, protein and minerals. It has long been recognized that legumes are functional foods that promote good health and have therapeutic properties. This chapter shows the significance of some biochemical and biological compounds derived from legumes and rhizobacteria with potential in biotechnology.

Keywords Rhizosphere • Rhizodeposition • Rhizobacteria • PGPR • Biochemicals

12.1 Introduction

Soils are the product of the activities of plants, which supply organic matter and play a pivotal role in weathering rocks and minerals. In the numerous interactions between plants and the soil, microorganisms play a key role (Lambers et al. 2009). Problems such as nutrient deficiency, soil erosion, degradation and desertification, soil salinization and sodification, inefficient water use, drought and low-temperature stresses, occur frequently on low-productivity and degraded soils in different regions. In addition, the continued use of chemical fertilizers and manures for enhanced soil fertility and crop productivity often results in unexpected harmful environmental effects, including leaching of nitrate into ground water, surface run-off of phosphorus and nitrogen run-off, and eutrophication of the soil. Integrated nutrient management systems are needed to maintain agricultural productivity and to protect the environment.

The world's human population is still increasing and is expected to reach about 8.3 billion by 2025, before attaining a stable level later in the next century.

Food production has also increased, due to several factors such as plant breeding, irrigation, use of multiple cropping, crop protection with agrochemicals, and increased availability of plant nutrients. The most common nutrient affecting yield is nitrogen (N) (Bøckman 1997). Biological nitrogen fixation (BNF) and mineral fertilizers provide the principal input (i.e., N) to agricultural soils. The atmospheric deposition of nitrogen oxides and ammonia also contributes to the supply of N to the soil. These depositions originate mainly from pollution. On the global scale, BNF provides the largest input of N to the soil. Various estimates of this contribution have been published, ranging from 44 to 200 Tg N/year. In general, about 140 Tg N/year is the estimate that is mostly obtained (Søderlund and Rosswall 1982).

In addition, the agricultural productivity in large terrestrial areas of the world is severely affected by abiotic and biotic stresses which are harmful to plant growth. Typical environmental stresses faced by legume nodules and their symbiotic partners (rhizobia) include photosynthate deprivation, osmotic stress, salinity, soil nitrate, temperature, heavy metals, and biocides (Walsh 1995). The damaging effects of salt accumulation in agricultural soils have negatively influenced both ancient and modern civilizations (Rengasamy 2006). As with most cultivated crops, the response to salinity of legumes varies greatly and depends on factors such as climatic conditions, soil properties, and stage of growth. Based on the beneficial effects

12 Biochemical Processes of Rhizobacteria and their Application in Biotechnology 381

of several microorganisms, studies using inoculant mixtures are very promising. Therefore, the integration of microorganisms into plant production systems, biotechnological development of new symbiotic, antibiotic, and antagonistic relationships between plants and microorganisms, creates new possibilities for sustainable agriculture.

On the other hand, during the last two decades, various new biotechnologies have been adapted to agricultural practices and have opened new possibilities for plant utilization. This will be intensified in the next decade. Plant biotechnology is changing the growth and development control, protecting plants against the ever-increasing threats of abiotic and biotic stress, and expanding the horizons by producing specialty foods, biochemicals and pharmaceuticals (Altman 1999).

12.2 Microbial Diversity of Soil

Soil microbial populations are involved in a framework of interactions known to affect key environmental processes, like biogeochemical cycling of nutrients, plant health and soil quality (Barea et al. 2005; Giri et al. 2005). The diversity of the microbial (bacterial and fungal) communities in the soil is extraordinary (Roselló-Mora and Amann 2001). Most of the dynamic microbial interactions take place near the plant roots and in the root–soil interface, an area called the rhizosphere (Lynch 1987; Barea et al. 2005; Bais et al. 2006; Prithiviraj et al. 2007). Most rhizosphere organisms occur within 50 μm from the root surface and populations within 10 mm from the root surface may reach 1.2×10^8 cells cm^{-3} or 10^9–10^{12} microbial cells g^{-1} soil. Despite the large numbers of bacteria in the rhizosphere, only 7–15% of the total root surface is generally occupied by microbial cells (Foster et al. 1983; Pinton et al. 2001).

Bacteria able to colonize plant root systems and promote plant growth are referred to as plant growth-promoting rhizobacteria (PGPR) (Kloepper and Schroth 1978). PGPR can affect plant growth either directly, by providing plants with a compound synthesized by the bacterium or facilitating the uptake of certain nutrients from the environment, or indirectly, by lessening or preventing the deleterious effects of one or more phytopathogenic organisms (Glick 1995). PGPR activity has been reported for strains belonging to a group that includes different diazotrophic bacterial species belonging to genera such as *Azoarcus, Azospirillum, Azotobacter, Bacillus, Burkholderia, Enterobacter, Gluconacetobacter, Herbaspirillum, Klebsiella, Paenibacillus, Pseudomonas,* and *Serratia* (Spaepen et al. 2009). In the broadest sense, PGPR include the N$_2$-fixing rhizobacteria that colonize the rhizosphere and provide N to plants, including the rhizobia of the well-characterized legume–rhizobia symbiosis. Rhizobia are a group of Gram-negative soil bacteria that form root nodules on legume plants. Most of these bacterial species belong to the Rhizobiaceae family belonging to the alpha-proteobacteria, however recent research has shown that there are other rhizobial species in the beta-protobacteria order Burkholderiales (Sawada et al. 2003). The species listed are all of the most currently published names for rhizobia, which consist of 76 species belonging to 13 genera named: *Allorhizobium,*

Azorhizobium, Blastobacter, Bradyrhizobium, Burkholderia, Cupriavidus, Devosia, Ensifer (formerly *Sinorhizobium*), *Mesorhizobium, Methylobacterium, Ralstonia, Rhizobium,* and *Shinella.* These genera are grouped in six families (Rhizobiaceae, Phyllobacteriaceae, Bradyrhizobiaceae, Methylobacteriaceae, Hyphomicrobiaceae and Burkholderiaceae).

All PGPRs have indirect positive effects on plant health by inhibiting soilborne pathogens by means of competition and antibiosis (Raaijmakers et al. 2009). However, most experiments examining the mechanisms of PGPRs deal with only a single host–single PGPR interaction. In nature, the rhizosphere contains millions of microorganisms including PGPR and pathogens. Further studies are needed to unravel these multiplex interactions at a molecular level to enhance their use for agricultural benefits (Badri et al. 2009). PGPR have been shown to cause positive effects when matched correctly to the right plant and the right environmental situation. A clear definition of which bacterial traits are useful and necessary for different environmental conditions and plants, and of which are the optimal bacterial strains to be selected or constructed is needed. Also, it would be very useful to have a better understanding of how different bacterial strains work together in the synergistic promotion of plant growth, and better knowledge on novel inoculants delivery systems, and on the environmental persistence of PGPR in the soil.

12.3   Biochemicals in Soil: Rhizodeposition

The rhizosphere is the (soil) layer influenced by the root. This area have a few milimeters and wher multifaceted ecological and biological processes take place. Hiltner (1904) was the first to define the "rhizosphere effect" by observing that the number and activity of microorganisms increased in the vicinity of plant roots.

Plant roots release a wide range of organic compounds (e.g. carbohydrates, carboxylic acids, phenolics, amino acids, flavonoids) (Sommers et al. 2004; Dardanelli et al. 2008a, 2010) as well as inorganic ions (protons and other ions) into the rhizosphere to change the chemistry and biology of the root microenvironment. Most root products are regular plant compounds which become available as substrate of colonizing microbes, including specific compounds typical of the secondary metabolism of each plant species (Badri et al. 2009). All chemical compounds secreted by the plant are collectively named rhizodepositions, and are released from living roots to the soil via several mechanisms (Gregory 2006). These mechanisms include:

1. Exudation of low molecular weight, water-soluble compounds, such as glucose, which are lost passively without the involvement of plant metabolic activity.
2. Secretion of higher molecular weight compounds, such as polysaccharide mucilage and enzymes, involving root metabolic processes.
3. Lysates released from sloughed off root cells and, with time, whole roots.
4. Gases such as CO_2, ethylene and hydrogen cyanide.

12 Biochemical Processes of Rhizobacteria and their Application in Biotechnology

Uren (2001) has suggested that, except for chlorophyll and some compounds associated with the photosynthetic process, root products probably contain every type of compounds that exists in plants (Table 12.1).

12.4 Root–Rhizobacteria Communication and Signals in the Rhizosphere

A large array of microbes can inhabit the rhizosphere and it is widely accepted that members from all microbial groups perform important functions in the rhizosphere. The plant root-soil interface is an environment with high microbial inoculum, composed of both pathogenic and beneficial microbes (Rouatt and Katznelson 1960; Lugtenberg and Kamilova 2009; Adesemoye and Kloepper 2009). Thus, plant roots are constantly exposed to an array of microbes, and must interact and defend according to the type of biotic stress (Bais et al. 2004, 2006).

Interestingly, specific compounds identified in root exudates have been shown to play roles in root–microbe interactions. A chemotactic response towards root-secreted organic and amino acids is the first step in root colonization (Zheng and Sinclair 1996). Most root products are regular plant compounds which become available as substrates of colonizing microbes, including specific compounds typical of the secondary metabolism of each plant species (Badri et al. 2009). A good example is the molecular integration of legume flavonoid signals by compatible rhizobia during the initiation of nitrogen-fixing symbiosis.

Legume hosts exude flavonoids continuously, but concentrations in the rhizosphere increase significantly in the presence of compatible rhizobial strains (Schmidt et al. 1994; Zuanazzi et al. 1998). Rhizobia perceive flavonoids of their legume hosts by binding of the flavonoid to a protein called NodD, which then activates a suite of rhizobial nodulation genes usually encoded on symbiotic (Sym) plasmids. Successful infection thread development depends probably on rhizobial production of extracellular polysaccharides and proteins, the secretion of which may also be induced by flavonoid structures (Broughton et al. 2000). A wide variety of flavonoids (chalcones, flavanones, isoflavones, flavonols) have been shown to have the *nod* gene inducing activity in different legume-rhizobia interactions (Aoki et al. 2000). A significant role of the isoflavonoid class of flavonoids as phytoalexins and phytoanticipins in disease response, in particular in legumes, has been postulated because of their broad spectrum in vitro antimicrobial activity (Dixon et al. 2002). Simple isoflavone compounds, such as daidzein, glycitein and formononetin glycosides are accumulated constitutively by many legume species (Dakora and Phillips 1996) and the corresponding aglycones are inhibitory to growth of microbial pathogens (VanEtten 1976; Kramer et al. 1984) and can thus be classified as phytoanticipins. Figure 12.1 shows how plant roots can communicate with rhizobacteria and establish active rhizospheric interactions.

Several beneficial microorganisms that reside in the rhizosphere can inhibit the growth and activity of soilborne pathogens. The activity and effects of beneficial

Table 12.1 Organic compounds released from plant roots. (Source: Uren (2001), Sommers et al. (2004), Faure et al. (2009))

Molecule group	Compounds	Biological functions
Amino acids and amides	All 20 proteinogenic amino acids, γ-aminobutyric acid, cystathionine, cystine, homoserine, mugenic acid, ornithine, phytosiderophores, betaine, stachydrine	Inhibit nematodes and root growth; microbial growth stimulation; chemoattractants, osmoprotectants; iron scavengers
Enzymes and proteins	Amylase, hydrolase, invertase, peroxidase, phenolase phosphatase, polygalacturonase, proteases, lectin	Plant defence; Nod factor degradation
Fatty acids	Linoleic, linolenic, oleic, palmitic, stearic acid	Plant growth regulation
Hormones	Auxin, ethylene and its precursor 1-aminocyclopropane-1-carboxylic acid (ACC), putrescine, jasmonate, salicylic acid	Plant growth regulation
Purines/pyrimidines	Adenine, guanine, uridine, cytidine	Microbial metabolism
Organic acids	Acetic, acetonic, aconitic, aldonic, butyric, citric, erythronic, formic, fumaric, gluconic, glutaric, glycolic, isocitric, lactic, maleic, malic, malonic, oxalic, oxaloacetic, oxaloglutaric, piscidic, propionic, pyruvic, shikimic, succinic, tartaric, tetronic, valeric acid	Plant growth regulation; chemoattractants; microbial growth stimulation
Phenolics	Flavanol, flavones, flavanones, anthocyanins, isoflavonoids, acetosyringone	Plant growth regulation; allelopathic interactions; plant defence; phytoalexins; chemoattractants; initiate legume-rhizobia, arbuscular mycorrhizal and actinorhizal interactions; microbial growth stimulation; stimulate bacterial xenobiotic degradation
Sterols	Campestrol, cholesterol, sitosterol, stigmasterol	Plant growth regulation
Sugars and polysaccharides	Arabinose, desoxyribose, fructose, galactose, glucose, maltose, oligosaccharides, raffinose, rhamnose, ribose, sucrose, xylose, mannitol, complex polysaccharides	Lubrication; protection of plants against toxins; chemoattractants; microbial growth stimulation
Vitamins	p-aminobenzoic acid, biotin, choline, n-methionylnicotinic acid, niacin, panthothenate, pyridoxine, riboflavin, thiamine	Microbial growth stimulation
Miscellaneous	Unidentified acyl homoserine lactone mimics, saponin, scopoletin, reactive oxygen species, nucleotides, calystegine, trigonelline, xanthone, strigolactones	Quorum quenching; plant growth regulation; plant defence; microbial attachment; microbial growth stimulation; initiate arbuscular mycorrhizal interactions

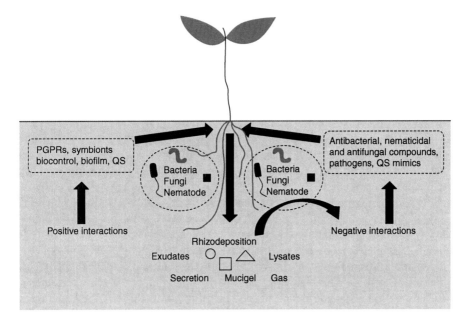

Fig. 12.1 Rhizodeposition and active rhizospheric interactions, (*QS* quorum sensing)

rhizosphere microorganisms on plant growth and health are well documented for bacteria like *Pseudomonas, Burkholderia* and fungi like *Trichoderma* and *Gliocladium*.

Similar to flavonoids in legume-rhizobia signalling, root secreted compounds modulate the interaction between plants and PGPRs. These interactions have been reviewed in several authors (Prithiviraj et al. 2007; Badri et al. 2009). Rudrappa et al. (2008) demonstrated that L-malic acid secreted from plant roots is involved in recruiting specifically *Bacillus subitilis* FB17 but not other *Bacillus* sp. This suggests that each beneficial rhizobacterium needs a specific signal to colonize the host. Recent evidence has shown that microbes can modulate plant root exudation of proteins. De-la-Pena et al. (2008) have clearly demonstrated that the compositions of the proteins present in the root exudates change upon the presence of a given microbial neighbour and that the exudation of proteins by a given bacterium is modulated by the presence of a specific plant neighbour. It was found that the interaction between *Medicago-Sinorhizobium meliloti* increases the secretion of several plant proteins such as hydrolases, peptidases and peroxidases, but that these proteins are not induced in the *Medicago-Pseudomonas syringae* interaction. Additionally and similarly to that found in *P. syringae*, these authors found that *S. meliloti* secretes different arrays of proteins in the presence of *Medicago* or *Arabidopsis*. These data provide concrete evidence that both plant root and bacterial protein secretion profiles change in response to the neighbour (De-la-Pena et al. 2008).

12.5 The Rhizosphere in Action

Optimal farming decisions should involve the analysis of a wide range of variables, including agronomic and environmental parameters. Agronomic parameters may include the crop species and cultivar, the crop rotation sequence, and the tillage practices. Environmental factors may include soil physicochemical characteristics as well as climatic variables such as rainfall, humidity, and temperature. Other parameters may include the use of agrochemicals such as pesticides, growth regulators, fertilizers etc. Microorganisms should also be considered. The better understanding of microbial genomes and proteomes gained in recent years, along with advances in recombinant technology, has significantly improved our ability to manipulate microorganisms for biotechnological applications. On the other hand, although many technologies have been used to improve stress tolerance in plants, fewer reports have been published on how PGPR can help improve plant tolerance to salt, drought, heavy metals, and other stress factors. In addition, the industrial use of compounds from plants and rhizobacteria and their technological application are required to be successful in attaining sustainable microbial-based agrotechnologies.

The fast industrialization and modernization all around the world leads to an unfortunate consequence: the production and release of considerable amounts of toxic wastes to the environment (Zhuang et al. 2007). Physical, chemical, and biological methods have been used to eliminate or control the pollutants in soils. Bioremediation is the application of biological processes for the cleanup of hazardous chemicals present in the environment (Gianfreda and Rao 2004). Heavy metals, which include cadmium, chromium, copper, lead, mercury, nickel and zinc, are the primary inorganic contaminants. Although PGPR were first used for the improvement of plant growth and for the biocontrol of plant diseases, much attention has been recently paid on bioremediation with PGPR (Huang et al. 2005). In contrast to inorganic compounds, microorganisms can degrade and even mineralize organic compounds in association with plants (Saleh et al. 2004).

Legumes, and their associated rhizobial bacteria, are important components of the biogeochemical cycles in agricultural and natural ecosystems. Thus, legumes and rhizobia are often desirable species during, and after, the remediation of arsenics (As)-contaminated land. Legumes have been identified as naturally occurring pioneer species on As-contaminated sites and free-living rhizobia are commonly found in soils with high As (Carrasco et al. 2005). Rhizobia tolerate high As concentrations in the growing medium by maintaining low internal As concentrations, i.e., absorbed As(V) is reduced to As(III) and then actively effluxed into the surrounding medium (Macur et al. 2001; Yang et al. 2005). Generally, As(III) is more toxic to plants than As(V) (Fitz and Wenzel 2002), and thus, rhizobial bacteria could potentially increase the sensitivity of legumes to As.

Nonrhizobial bacteria from genera such as *Pseudomonas*, *Bacillus* and *Flavobacterium* have shown promise for their growth-promoting impacts on plants used in the remediation of heavy metal contaminated sites (Belimov et al. 2005; Sheng and Xia 2006; Zaidi et al. 2006). However, the use of rhizobia as growth-promoting bacteria for the remediation of heavy metal contaminated sites has a potential advantage over other growth-promoting bacteria as rhizobia are already produced cheaply and in large quantities for the inoculation of legume crops and forage species. The potential use of rhizobia as growth-promoting bacteria for the remediation of heavy metal contaminated sites is an exciting new area of research (Reichman 2007). Inoculants of cadmium-resistant, rhizosphere-competent bacterial strains that had been isolated from metal-polluted soil substantially improved root elongation, root and shoot biomass production of *Brassica napus* grown on cadmium-polluted soil (Sheng and Xia 2006). Similarly, growth and biomass production of *Brassica juncea* grown on Pb–Zn mine tailings was improved substantially upon inoculation with a PGPR consortium consisting of N_2-fixing *Azotobacter chroococcum* HKN5, P-solubilising *Bacillus megaterium* HKP-1, and K-solubilising *Bacillus mucilaginosus* HKK-1 (Wu et al. 2006). The PGPR *Bacillus subtilis* strain SJ-101 capable of producing the phytohormone indole acetic acid and solubilising inorganic phosphates stimulated *Brassica juncea* growth on nickel polluted soil (Zaidi et al. 2006). Root growth and proliferation in polluted soils has also been shown to increase in the presence of ACC (1-aminocyclopropane-1-carboxylate) deaminase-producing bacteria (Arshad et al. 2007).

One of the most severe and widespread problems facing the agricultural industry is the degradation of soil quality due to desiccation and salinity. In fact, almost 40% of the world's land surface is affected by salinity-related problems (Zahran 1999). These two harsh environmental conditions can have a dramatic impact on the endogenous soil bacteria (Fierer et al. 2003; Griffiths et al. 2003). Of particular importance to the agricultural industry is the impact of these harsh environmental conditions on the PGPR. Desiccation produces many stress responses in the bacterial cell. In 1932, Fred and coworkers reported loss of viability in rhizobia used as seed inocula and suggested that the nature of the suspending medium, pH, and temperature are important factors in the survival of the inoculum in the dry state. This led to the recommendation for farmers to refrain from using rhizobia in dry form. Since these early observations, progress in understanding survival during desiccation stress of rhizobia has been slow and limited to testing the ability of selected strains to survive under controlled conditions (Vriezen et al. 2007). Rhizobia differ in their ability to respond to an increase in osmotic pressure and salt stress (Zahran 1999). Rhizobia accumulate potassium ions (Botsford and Lewis 1990), for which no new protein synthesis is required. This suggests that K uptake is regulated biochemically and used as a secondary messenger. Subsequently stressed cells accumulate certain compatible solutes, and uptake is preferred over synthesis. Compatible solutes include carbohydrates, disaccharides such as sucrose and trehalose, maltose, cellobiose, turanose, gentiobiose, palatinose, and amino acids, of which mainly glutamate and proline accumulate, though many genetic mechanisms involved in amino acid uptake are down-regulated (Vriezen et al. 2007). Furthermore, imino acids (e.g., pipecolate), ectoin, glycine betaine and stachydrin, *N*-acetylglutaminylglutamine amide, and dimethylsulfoniopropionate accumulate. Not all compounds are taken up from the medium, rather some are synthesized de novo, such as, sucrose and trehalose. It has been argued, however, that trehalose and glycine betaine are accumulated to prevent starvation rather than to function as osmotic stabilizers (Miller and Wood 1996; Zahran 1999; Vriezen et al. 2007).

Trehalose accumulates in stressed bacteria (Potts 1994) and especially in osmostressed rhizobia (Streeter 2003; Dardanelli et al. 2000, 2009a). Trehalose provides protection against desiccation by maintaining membrane integrity during drying and rewetting, and its presence may explain the increase in desiccation survival during the stationary phase and when the cell is exposed to NaCl (Vriezen et al. 2007). Different changes in biochemical and physiological functions may be triggered in rhizobia growing under stress conditions. In peanut rhizobia, alterations in membrane lipid composition in response to salinity and variation in the trehalose content have been detected in response to osmotic stress (Dardanelli et al. 2000, 2009a; Medeot et al. 2007, 2009, 2010). At 400 mM NaCl, the synthesis and accumulation of this compatible solute in the salt-tolerant strain *Bradyrhizobium* sp ATCC 10317 increased while in *Bradyrhizobium* TAL169 it diminished, indicating a strain dependent response. However, the growth of these peanut rhizobia under osmotic stress did not induce changes in their trehalose content (Dardanelli et al. 2009a). An explanation for this response can be the genetic variability of peanut symbionts or the use of other compatible solutes to modulate their response to stress.

The osmolality of rhizosphere soil water is expected to be elevated in relation to bulk-soil water osmolality as a result of the exclusion of solutes by plant roots during water uptake, the release of plant root exudates, and the production of exopolymers by plant roots and rhizobacteria. In contrast, the osmolality of water within highly hydrated bulk soil is low (less than 50 Osm/kg); thus the ability to adapt to elevated osmolality is likely to be important for successful rhizosphere colonization by rhizobacteria (Miller and Wood 1996). In order to survive and proliferate in the rhizosphere, rhizobacteria are therefore likely to possess mechanisms designed for adaptation to environments of high osmolality. Future research should be directed towards the identification and in-depth characterization of functional genes involved in the response to stress. A more detailed molecular and biochemical approach to NaCl mediated survival during desiccation can directly be applied in the understanding and manipulation of soilmicrobial communities, the directed identification of desiccation resistant microorganisms, and the production and development of dry seed inocula (Vriezen et al. 2007).

12.6 Benefits from Legume Plants: Food, Biochemical and Pharmaceuticals

Food production, for both quantity and quality, as well as for new plant commodities and products, in developed and developing countries around the globe, cannot rely solely on classical agriculture. Traditionally, agriculture was targeted to improving the production of plant-derived food, in terms of both quantity and quality. The new phase of plant biotechnology is now gradually being implemented: a shift

12 Biochemical Processes of Rhizobacteria and their Application in Biotechnology   389

from the production of low-priced food and bulk commodities to high-priced, specialized plant-derived products. This includes two major categories of biomaterials: (1) direct improvement and modification of specialized constituents of plant origin, and (2) the manufacture in plants of non-plant compounds (Altman 1999).

Biotechniques, mostly based on the engineering of metabolic pathways, are now available for modifying many plant constituents that are used in the food, chemical and energy industries. Moreover, the use of plants as "bioreactors" for the production of "foreign", non-plant compounds is gaining momentum and may eventually lead to alternative types of agriculture. This includes, for example, production of bioactive peptides, vaccines, antibodies and a range of enzymes mostly for the pharmaceutical industry. For the chemical industry, plants can be used to produce, e.g., polyhydroxybutyrate for the production of biodegradable thermoplastics, and cyclodextrins, which form inclusion complexes with hydrophobic substances (Altman 1999).

In addition to traditional food and forage uses, legumes can be milled into flour, used to make bread, doughnuts, tortillas, chips, and spreads, or used in liquid form to produce milks, yogurt, and infant formula (Garcia et al. 1998). Pop beans, licorice, and soybean candy provide novel uses for specific legumes (Graham and Vance 2003). These plants have been used industrially to prepare biodegradable plastics, oils, gums, dyes, and inks. Galactomannan gums derived from *Cyamopsis* spp. and *Sesbania* spp. are used in sizing textiles and paper, as a thickener, and in pill formulation.

Today the world appears to be increasingly interested in the health benefits of foods and have begun to look beyond the basic nutritional benefits of food stuffs to disease prevention and health enhancing ingredients of the same. Traditional systems of medicine owe their significance to the bioactive components that have their origin in plant sources and most of them were associated with routine food habits (Green 2004). Throughout history, plants have provided a rich source for the development of human medicines. Through empirical discovery, humans of many cultures have continually identified plants yielding beneficial health effects. The twentieth century brought further understanding of human health and the development of synthetic or semi-synthetic analogs of plant compounds that led to drugs with higher levels of potency. Over the past decade there has been an increased interest in phytochemicals for the purpose of human health and for benefits in the food industry (Green 2004).

The World Health Organization estimated that 80% of the earth's inhabitants rely on traditional medicine for their primary health care needs,and most of this therapy involves the use of plant extracts or their active components. Legumes were the primary source of protein and calcium before dairy products were added to the diet. Cereals and legumes, in general, play an important role in human nutrition. Recent studies have shown that cereals and beans contain constituents that have demonstrated health benefits for humans, such as antioxidants and anti-disease factors (Limsangouan and Isobe 2009).

Many legumes are currently used for nutraceutical and pharmaceutical purposes, whereas others contain medically important phytochemicals that are possibly effective or have been shown to have potential therapeutic effects. It is known that legumes can modify cardiovascular risk factor (CRF) (Morris 2004). These CRF include hypertension, dyslipidemia, oxidative stress, obesity and other. There is strong evidence that a diet high in whole grains is associated with lower body mass index, smaller waist circumference, and reduced risk of being overweight; that a diet high in whole grains and legumes can help reduce weight gain; and that significant weight loss is achievable with energy-controlled diets that are high in cereals and legumes (Williams et al. 2008).

The use of antioxidants to prevent disease is controversial (Meyers et al. 1996). In this sense, many studies have been conducted to evaluate the antioxidant capacity of different plants. Ethanolic extracts of peanut seed testa (EEPST) and its antioxidative component, ethyl protocatechuate (EP), showed activity on the inhibition of liposome peroxidation and they protect protein against oxidative damage. The antioxidant mechanism, for both EEPST and EP, could possibly be due to their scavenging effect on free radical and hydroxyl radical.

Lathyrus may represent an interesting source of phenolic compounds with high antioxidant activity that may be useful as natural antioxidants and contribute to revalorize the cultivation of these legumes. Antioxidant activity of seed phenolics was studied in several *Lathyrus* species. In general, non-cultivated *Lathyrus* species contained higher phenolic contents than cultivated ones. A negative correlation between seed size and phenolic contents was observed and was related to the higher proportion of hulls in the smaller seeds. This antioxidant activity was twice that observed in same amounts of extracted flours from commercial chickpea, lupin or soy. *Lathyrus* species are rich in phenolic compounds with higher antioxidant activity than phenolics of widely consumed legumes such as soy, chickpea or lupin (Pastor-Cavada et al. 2009).

Higher intake of legumes was associated with a decreased risk of several cancers including those of the upper aerodigestive tract, stomach, colorectum, and kidney, but not lung, breast, prostate or bladder (Aune et al. 2009). There is growing evidence to indicate that some anti-nutritional plant constituents inhibit the process of carcinogenesis and reduce the risk of developing certain cancers. Plant foods contain, in addition to their common macro nutrients and dietary fiber, a wide variety of biologically active micro components. Saponins and phytosterols are examples of the known phytochemicals that may favorably alter the likelihood of carcinogenesis. The proposed mechanisms for the anticarcinogenic properties of saponins include antioxidant effect, direct and select cytotoxicity of cancer cells, immune-modulation, acid and neutral sterol metabolism and regulation of cell proliferation. Phytosterols, on the other hand, are structurally related to cholesterol and have anticarcinogenic property. Alterations in the absorption and metabolism of acid and neutral sterols are considered to be the main mechanisms by which phytosterols afford their beneficial properties against cancer. Moreover, diosgenin, a naturally occurring steroid saponin found in legumes, is considered a chemopreventative/therapeutic agent againt several cancers. The anticancer mode of action of diosgenin has been demonstrated via modulation of multiple cell signaling events involving critical molecular candidates associated with growth, differentiation, apoptosis, and oncogenesis (Rao and Koratkar 1997; Raju and Mehta 2009).

In different herbs a wide variety of active phytochemicals, including the flavonoids, terpenoids, lignans, sulfides, polyphenolics, carotenoids, coumarins, saponins, plant sterols, curcumins, and phthalides have been identified. Several of these phytochemicals either inhibit nitrosation or the formation of DNA adducts or stimulate the activity of protective enzymes such as the Phase II enzyme glutathione transferase (Craig 1999).

A widely and safely used plant extract acts as a novel anti-inflammatory agent. Phenolic compounds have significant antiinflammatory effects, including inhibition of adhesion molecules, cytokine and chemokine gene expression; inhibition of platelet function; augmentation of endothelial nitric oxide release; suppression of smooth muscle activation; and other effects on proinflammatory factors such as endothelin and matrix metalloproteinases (Jiang and Dusting 2003). Other mechanisms involved in antiinflammatory activity are modulation of pro-inflammatory gene expression such as cyclooxygenase, lipoxygenase, nitric oxide synthases and several pivotal cytokines (Tuñón et al. 2009).

Flavonoids belong to a group of natural substances occurring normally in the diet that exhibit anti-inflammatory property. Several mechanisms of actions have been proposed to explain in vivo flavonoid anti-inflammatory actions, such as antioxidant activity, inhibition of eicosanoid generating enzymes or modulation of proinflammatory molecules production. Recent studies have also shown that some flavonoids are modulators of proinflammatory gene expression, thus leading to the attenuation of the inflammatory response (García-Lafuente et al. 2009). Peanut is a plant that synthesizes bioactive stilbenoids. Bioactivities of those stilbenoids have been meagrely investigated. Stilbenoids, mainly resveratrol and its derivatives, have exhibited potent antioxidant and anti-inflammatory activities (Djoko et al. 2007).

Isoflavones exhibit estrogenic, antiangiogenic, antioxidant, and anticancer activities (Dixon 1999), and are now popular as dietary supplements (Palevitz 2000). Major sources of isoflavones for humans are seed products of soybean (daidzein and genistein) and chickpea (biochanin A), and the health promoting activity of high-soy diets is believed to reside in their isoflavone components (Setchell and Cassidy 1999; Lamartiniere 2000). Epidemiological studies suggest a link between consumption of soy isoflavones and reduced risks of breast and prostate cancers in humans (Setchell and Cassidy 1999; Lamartiniere 2000). Isoflavones may possess other health-promoting activities, including chemoprevention of osteoporosis, and prevention of other postmenopausal disorders and cardiovascular diseases (Uesugi et al. 2001). Recent studies found that genistein exerts a novel non-genomic action by targeting on important signaling molecules in vascular endothelial cells (ECs). Genistein rapidly activates endothelial nitric oxide synthase and production of nitric oxide in ECs. This genistein effect is novel since it is independent of its known effects, but mediated by the cyclic adenosine monophosphate/protein kinase A (cAMP/PKA) cascade. Further studies demonstrated that genistein directly stimulates the plasma membrane-associated adenylate cyclases, leading to activation of the cAMP signaling pathway. In addition, genistein activates peroxisome proliferator-activated receptors, ligand-activated nuclear receptors important to normal vascular function. These new findings reveal the novel roles for genistein in the regulation of vascular function and provide a basis for further investigating its therapeutic potential for inflammatory-related vascular disease (Si and Liu 2007).

12.7 Perspectives and Conclusion

All human beings depend on agriculture that produces food of the appropriate quality at the required quantities. Legumes are agronomically and economically important in many cropping systems because of their ability to assimilate atmospheric N_2, and this importance is anticipated to increase with the need to develop sustainable practices.

During germination and growth, plants release many compounds into their rhizosphere, including amino acids, organic acids, sugars, aromatics and various other secondary metabolites. Therefore, the microbial community found in a particular rhizosphere is shaped by the abundance and diversity of the various components of plant-derived exudates. Such a situation can potentially lead to plant species-specific microflora. Rhizodeposition is a dynamic process that is developmentally regulated and varies with the plant species and cultivar; it is also altered upon biotic and abiotic stress. Because the concentration of fixed nitrogen is a limiting factor for growth, nitrogen fixers have a selective advantage that enables them to adapt to the most extreme conditions and to colonize diverse ecological niches. Indeed, nitrogen-fixing symbiotic microorganisms play an important role in the life of plants, ensuring not only their nutrition, but also their defense against pathogens and pests, and adaptation to various environmental stresses.

In addition, plants are used for other functions. Several cities now require that vehicles be powered in part by biodiesel fuel from soybean. Some states require that biodiesel be included at a fixed percentage in all diesel fuels (http://www.biodiesel.org). Plant-assisted bioremediation holds promise for in situ treatment of polluted soils. Enhancement of phytoremediation processes requires a sound understanding of the complex interactions in the rhizosphere. The role of rhizosphere processes in the phytoremediation of inorganic pollutants, in particular metals/metalloids, is much less investigated and only a few specific reviews are available on this topic.

In a world where population growth is outstripping food supply, agricultural advancements with special reference to plant-biotechnology and rhizospheric interactions, need to be swiftly implemented in all walks of life. Achievements today in plant biotechnology have already surpassed all previous expectations, and the future is even more promising. The full realization of the sustainable agriculture depends on both continued successful and innovative research and development activities and on a favourable regulatory climate and public acceptance. Learning more about beneficial plant-microbe interactions will yield a more complete picture of sustainable agriculture.

Acknowledgements This research was partially supported by the Secretaría de Ciencia y Técnica de la Universidad Nacional de Río Cuarto (SECyT-UNRC) and CONICET PIP 112-200801-00537. NP is fellow from CONICET. MSD is a member of the research career of CONICET, Argentina.

References

Adesemoye AO, Kloepper JW (2009) Plant microbes interactions in enhanced fertilizer-use efficiency. Appl Microbiol Biotechnol 85:1–12

Altman A (1999) Plant biotechnology in the 21st century: the challenges ahead. Electron J Biotechnol 2:51–55

Aoki T, Akashi T, Ayabe S (2000) Flavonoids of leguminous plants: structure, biological activity and biosynthesis. J Plant Res 113:475–488

Arshad M, Saleem M, Hussain S (2007) Perspectives of bacterial ACC deaminase in phytoremediation. Trends Biotechnol 25:356–362

Aune D, De Stefani E, Ronco A, Boffetta P, Deneo-Pellegrini H, Acosta G, Mendilaharsu M (2009) Legume intake and the risk of cancer: a multisite case-control study in Uruguay. Can Causes Cont 20:1605–1615

Badri DV, Weir TL, van der Lelie D, Vivanco JM (2009) Rhizosphere chemical dialogues: plant–microbe interactions. Curr Opin Biotech 20:1–9

Bais HP, Park SW, Weir TL, Callaway RM, Vivanco JM (2004) How plants communicate using the underground information superhighway. Trends Plant Sci 9:26–32

Bais HP, Weir TL, Perry LG, Gilroy S, Vivanco JM (2006) The role of root exudates in rhizosphere interactions with plants and other organisms. Ann Rev Plant Biol 57:233–266

Barea JM, Pozo MJ, Azcon R, Azcon-Aguilar C (2005) Microbial co-operation in the rhizosphere. J Exp Bot 56:1761–1778

Belimov AA, Hontzeas N, Safronova VI, Demchinskaya SV, Piluzza G, Bullitta S, Glick BR (2005) Cadmium-tolerant plant growth-promoting bacteria associated with the roots of Indian mustard (*Brassica juncea* L. Czern.). Soil Biol Biochem 37:241–250

Bøckman OC (1997) Fertilizers and biological nitrogen fixation as sources of plant nutrients: perspectives for future agriculture. Plant Soil 194:11–14

Botsford JL, Lewis TA (1990) Osmoregulation in *Rhizobium meliloti*: production of glutamic acid in response to osmotic stress. Appl Environ Microbiol 56:488–494

Broughton WJ, Jabbouri S, Perret X (2000) Keys to symbiotic harmony. J Bacteriol 182:5641–5652

Carrasco JA, Armario P, Pajuelo E, Burgos A, Caviedes MA, Lopez R, Chamber MA, Palomares AJ (2005) Isolation and characterisation of symbiotically effective *Rhizobium* resistant to arsenic and heavy metals after the toxic spill at the Aznalcollar pyrite mine. Soil Biol Biochem 37:1131–1140

Craig WJ (1999) Health-promoting properties of common herbs. Am J Clin Nut 70:491S–499S

Dakora FD, Phillips DA (1996) Diverse functions of isoflavonoids in legumes transcend antimicrobial definitions of phytoalexins. Physiol Mol Plant Pathol 49:1–20

Dardanelli MS, Fernández FJ, Espuny MR, Rodríguez MA, Soria ME, Gil Serrano AM, Okon Y, Megías M (2008a) Effect of *Azospirillum brasilense* coinoculated with *Rhizobium* on *Phaseolus vulgaris* flavonoids and Nod factor production under salt stress. Soil Biol Biochem 40:2713–2721

Dardanelli MS, González PS, Bueno MA, Ghittoni NE (2000) Synthesis, accumulation and hydrolysis of trehalose during growth of peanut rhizobia in hyperosmotic media. J B Microbiol 40:149–156

Dardanelli MS, González PS, Medeot DB, Paulucci NS, Bueno MS, Garcia MB (2009a) Effects of peanut rhizobia on the growth and symbiotic performance of *Arachis hypogaea* under abiotic stress. Symbiosis 47:175–180

Dardanelli MS, Manyani H, González-Barroso S, Rodríguez-Carvajal MA, Gil-Serrano AM, Espuny MR, López-Baena FJ, Bellogín RA, Megías M, Ollero FJ (2010) Effect of the presence of the plant growth promoting rhizobacterium (PGPR) *Chryseobacterium balustinum* Aur9 and salt stress in the pattern of flavonoids exuded by soybean roots. Plant Soil 328:483–493

De-la-Pena C, Lei Z, Watson BS, Sumner LW, Vivanco JM (2008) Root–microbe communication through protein secretion. J Biol Chem 283:25247–25255

Dixon RA (1999) Isoflavonoids: biochemistry, molecular biology and biological functions. In: Sankawa U (ed) Comprehensive natural products chemistry, vol 1. Elsevier, Oxford, pp 773–823

Dixon R, Achnine L, Kota P, Liu CJ, Reddy M, Wang L (2002) The phenylpropanoid pathway and plant defence a genomics perspective. Mol Plant Pathol 3:371–390

Djoko B, Chiou RY, Shee JJ, Liu YW (2007) Characterization of immunological activities of peanut stilbenoids, arachidin-1, piceatannol, and resveratrol on lipopolysaccharide-induced inflammation of RAW 264.7 macrophages. J Agric Food Chem 55:2376–2383

Faure D, Vereecke D, Leveau JHJ (2009) Molecular communication in the rhizosphere. Plant Soil 321:279–303

Fierer N, Schimel JP, Holden PA (2003) Influence of drying rewetting frequency on soil bacterial community structure. Microb Ecol 45:63–71

Fitz WJ, Wenzel WW (2002) Arsenic transformations in the soil–rhizosphere–plant system: fundamentals and potential application to phytoremediation. J Biotechnol 99:259–278

Foster RC, Rovira AD, Cock TW (1983) Ultrastructure of the root–soil interface. The American Phytopathological Society, St Paul, MN, p 157

Fred EB, Baldwin IL, McCoy E (1932) Some factors which influence the growth and longevity of the nodule bacteria. In: Baldwin IL, McCoy E, Fred EB (eds) Root nodule bacteria and leguminous plants, vol 5. University of Wisconsin, Madison, WI, pp 104–117

Garcia MC, Marina ML, Laborda F, Torre M (1998) Chemical characterization of commercial soybean products. Food Chem 62:325–331

García-Lafuente A, Guillamón E, Villares A, Rostagno MA, Martínez JA (2009) Flavonoids as anti-inflammatory agents: implications in cancer and cardiovascular disease. Inflamm Res 58:537–552

Glick BR (1995) The enhancement of plant growth by free-living bacteria. Can J Microbiol 41:109–117

Gianfreda L, Rao MA (2004) Potential of extra cellular enzymes in remediation of polluted soils: a review. Enzyme Microb Technol 35:339–354

Giri B, Giang PH, Kumari R, Prasad R, Varma A (2005) Microbial diversity in soils. In: Buscot F, Varma S (eds) Microorganisms in soils: roles in genesis and functions. Springer-Verlag, Heidelberg, Germany, pp 195–212

Graham PH, Vance CP (2003) Legumes: importance and constraints to greater use. Plant Physiol 131:872–877

Gregory P (2006) The rhizosphere. In: Gregory P (ed) Plant roots: growth, activity and interaction with soils. Blackwell Publishing, Iowa, pp 216–252

Green RJ (2004) Antioxidant activity of peanut plant tissues. Masters. Thesis. North Carolina State University

Griffiths RI, Whiteley AS, O'Donnell AG, Bailey MJ (2003) Physiological and community responses of established grassland bacterial populations to water stress. Appl Environ Microbiol 69:6961–6968

Hiltner L (1904) Über neuere Erfahrungen und Probleme auf dem Gebiete der Bodenbakteriologie unter besonderer Berücksichtigung der Gründüngung und Brache. Arbeiten der Deutschen Landwirtschaftlichen Gesellschaft 98:59–78

Huang XD, El-Alawi Y, Gurska J, Glick BR, Greenberg BM (2005) A multi-process phytoremediation system for decontamination of persistent total petroleum hydrocarbons (TPHs) from soils. Microchem J 81:139–147

Jiang F, Dusting GJ (2003) Natural phenolic compounds as cardiovascular therapeutics: potential role of their anti-inflammatory effects. Curr Vasc Pharmacol 1:135–156

Kramer R, Hindorf H, Jha H, Kallage J, Zilliken F (1984) Antifungal activity of soybean and chickpea isoflavones and their reduced derivatives. Phytochemistry 23:2203–2205

Kloepper JW, Schroth MN (1978) Plant growth-promoting rhizobacteria on radishes. Fourth International Conference on Plant Pathogen Bacteria. Angers, France, pp 879–882

Lamartiniere CA (2000) Protection against breast cancer with genistein: a component of soy. Am J Clin Nutr 71:1705S–1707S

12 Biochemical Processes of Rhizobacteria and their Application in Biotechnology 395

Lambers H, Mougel C, Jaillard B, Hinsinger P (2009) Plant-microbe-soil interactions in the rhizosphere: an evolutionary perspective. Plant Soil 321:83–115

Limsangouan N, Isobe S (2009) Effect of milling process on functional properties of legumes. Kasetsart J Nat Sci 43:745–751

Lugtenberg B, Kamilova F (2009) Plant-growth-promoting rhizobacteria. Annu Rev Microbiol 63:541–556

Lynch JM (ed) (1987) The Rhizosphere. Wiley Interscience, Chichester, UK

Macur RE, Wheeler JT, McDermott TR, Inskeep WP (2001) Microbial populations associated with the reduction and enhanced mobilization of arsenic in mine tailings. Environ Sci Technol 35:3676–3682

Medeot D, Bueno M, Dardanelli MS, García M (2007) Adaptational changes in lipids and fatty acids of *Bradyrhizobium* SEMIA 6144 nodulating peanut as a response to growth temperature and salinity. Curr Microbiol 54:31–35

Medeot DB, Paulucci NS, Albornoz A, Fumero MV, Bueno M, Garcia MB, Woelke M, Okon Y, Dardanelli MS (2009) Plant growth-promoting rhizobacteria-rhizobia and legume improvement. In: Saghir Md (ed) Microbes for legume improvement. Springer, New York

Medeot DB, Sohlenkamp C, Dardanelli MS, Geiger O, García de Lema M, López-Lara I (2010) Phosphatidylcholine levels of peanut-nodulating *Bradyrhizobium* sp. SEMIA 6144 affect cell size and motility. FEMS Microbiol Lett 303:123–131

Meyers DG, Maloley PA, Weeks D (1996) Safety of antioxidant vitamins. Arch Intern Med 156:925–935

Miller KJ, Wood JM (1996) Osmoadaptation by rhizosphere bacteria. Annu Rev Microbiol 50:101–136

Morris JB (2004) Legumes: nutraceutical and pharmaceutical uses. In: Goodman R (ed) Encyclopedia of Plant and Crop Science. Taylor and Francis, pp 651–655

Palevitz BA (2000) Soybeans hit main street. Scientist 14:8–9

Pastor-Cavada E, Juan R, Pastor JE, Alaiz M, Vioque J (2009) Antioxidant activity of seed polyphenols in fifteen wild *Lathyrus* species from South Spain. Food Sci Technol 42:705–709

Pinton R, Varanini Z, Nannipieri P (2001) The rhizosphere as a site of biochemical interactions among soil components, plants, and microorganisms. In: Pinton R, Varanini Z, Nannipieri P (eds) The rhizosphere. Mercel Dekker, Inc, New York, USA, pp 1–18

Potts M (1994) Desiccation tolerance of prokaryotes. Microbiol Rev 58:755–805

Prithiviraj B, Paschke MW, Vivanco JM (2007) Root communication: the role of root exudates. Encycl Plant Crop Sci 1:1–4

Raaijmakers JM, Paulitz TC, Steinberg C, Alabouvette C, Moenne-Loccoz Y (2009) The rhizosphere: a playground and battlefield for soilborne pathogens and beneficial microorganisms. Plant Soil 321:341–361

Raju J, Mehta R (2009) Cancer chemopreventive and therapeutic effects of diosgenin, a food saponin. Nutr Cancer 61:27–35

Rao AV, Koratkar R (1997) Anticarcinogenic effects of saponins and phytosterols. In: Shahidi F (ed) Antinutrients and phytochemicals in food, American Chemical Society, pp 313–324

Reichman SM (2007) The potential use of the legume–*Rhizobium* symbiosis for the remediation of arsenic contaminated sites. Soil Biol Biochem 39:2587–2593

Rengasamy P (2006) World salinization with emphasis on Australia. J Exp Bot 57:1017–1023

Roselló-Mora R, Amann R (2001) The species concept for prokaryotes. FEMS Microbiol Rev 25:39–67

Rouatt JW, Katznelson H (1960) Influence of light on bacterial flora of roots. Nature 186:659–660

Rudrappa T, Czymmek KJ, Pare PW, Bais HP (2008) Root-secreted malic acid recruits beneficial soil bacteria. Plant Physiol 148:1547–1556

Saleh S, Huang XD, Greenberg BM, Glick BR (2004) Phytoremediation of persistent organic contaminants in the environment. In: Singh A, Ward O (eds) Soil biology, vol 1. Applied Bioremediation and Phytoremediation. Springer, Berlin, pp 115–134

Sawada H, Kuykendall D, Young JM (2003) Changing concepts in the systematic of bacterial nitrogen-fixing legume symbionts. J Gen Appl Microbiol 49:155–179

Schmidt PE, Broughton WJ, Werner D (1994) Nod factors of *Bradyrhizobium Japonicum* and *Rhizobium* sp NGR234 induce flavonoid accumulation in soybean root exudate. Mol Plant Microbe Interac 7:384–390

Setchell KDR, Cassidy A (1999) Dietary isoflavones: biological effects and relevance to human health. J Nutr 129:758S–767S

Sheng XF, Xia JJ (2006) Improvement of rape (*Brassica napus*) plant growth and cadmium uptake by cadmium resistant bacteria. Chemosphere 64:1036–1042

Si H, Liu D (2007) Phytochemical genistein in the regulation of vascular function: new insights phytochemical genistein in the regulation of vascular function: new insights. Curr Med Chem 14:2581–2589

Sommers E, Vanderleyden J, Srinivasan M (2004) Rhizosphere bacterial signalling: a love parade beneath our feet. Crit Rev Microbiol 30:205–240

Spaepen S, Vanderleyden J, Okon Y (2009) Plant growth-promoting actions of rhizobacteria. In: van Loon LC (ed) Advances in botanical research, vol 51. Academic Press, Burlington, pp 283–320

Søderlund R, Rosswall T (1982) The nitrogen cycle. In: Hutzinger O (ed) The handbook of environmental chemistry, vol 1B. The natural environment and the biogeochemical cycles. Springer, Berlin, Heidelberg, pp 60–81

Streeter JG (2003) Effect of trehalose on survival of *Bradyrhizobium japonicum* during desiccation. J Appl Microbiol 95:484–491

Tuñón MJ, García-Mediavilla MV, Sánchez-Campos S, González-Gallego J (2009) Potential of flavonoids as anti-inflammatory agents: modulation of pro-inflammatory gene expression and signal transduction pathways. Curr Drug Metab 10:256–271

Uesugi T, Toda T, Tsuji K, Ishida H (2001) Comparative study on reduction of bone loss and lipid metabolism abnormality in ovariectomized rats by soy isoflavones, daidzin, genistin, and glycitin. Biol Pharm Bull 24:368–372

Uren NC (2001) Types, amounts, and possible functions of compounds released into the rhizosphere by soil-grown plants. In: Pinton R, Varanini Z, Nannipieri P (eds) The rhizosphere: biochemical and organics substances at the soil-plant interface. Marcel Dekker, New York, pp 19–39

VanEtten H (1976) Antifungal activity of pterocarpans and other selected isoflavonoids. Phytochemistry 15:655–659

Vriezen JAC, de Bruijn FJ, Nüsslein K (2007) Responses of rhizobia to Desiccation in relation to osmotic stress, oxygen, and temperature. Appl Environ Microbiol 73:3451–3459

Walsh KB (1995) Physiology of the legume nodule and its response to stress. Soil Biol Biochem 27:637–655

Williams PG, Grafenauer SJ, O'Shea JE (2008) Cereal grains, legumes, and weigth management: a comprehensive review of the scientific evidence. Nut Rev 66:171–182

Wu SC, Cheung KC, Luo YM, Wong MH (2006) Effects of inoculation of plant growth-promoting rhizobacteria on metal uptake by *Brassica juncea*. Environ Pollut 140:124–135

Yang HC, Cheng JJ, Finan TM, Rosen BP, Bhattacharjee H (2005) Novel pathway for arsenic detoxification in the legume symbiont *Sinorhizobium meliloti*. J Bacteriol 187:6991–6997

Zaidi S, Usmani S, Singh BR, Musarrat J (2006) Significance of *Bacillus subtilis* strain SJ-101 as a bioinoculant for concurrent plant growth promotion and nickel accumulation in *Brassica juncea*. Chemosphere 64:991–997

Zahran HH (1999) *Rhizobium*-legume symbiosis and nitrogen fixation under severe conditions and in arid climate. Microbiol Mol Biol Rev 63:968–989

Zheng XY, Sinclair JB (1996) Chemotactic response of *Bacillus megaterium* strain B153-2-2 to soybean root and seed exudates. Physiol Mol Plant Pathol 48:21–35

Zhuang X, Chen J, Shim H, Bai Z (2007) New advances in plant growth-promoting rhizobacteria for bioremediation Environ Int 33:406–413

Zuanazzi JAS, Clergeot PH, Quirion JC, Husson HP, Kondorosi A, Ratet P (1998) Production of *Sinorhizobium meliloti nod* gene activator and repressor flavonoids from *Medicago sativa* roots. Mol Plant-Microbe Interac 11:784–794

Chapter 13
Pulp and Paper Industry—Manufacturing Process, Wastewater Generation and Treatment

Saima Badar and Izharul Haq Farooqi

Contents

13.1	Introduction	398
13.2	Global Paper Industry	399
	13.2.1 Industry Structure	399
	13.2.2 Demand–Supply Scenario	400
13.3	Status of Pulp and Paper Industries in India	401
13.4	Classification of Pulp and Paper Industries in the country	401
	13.4.1 Based on Scale of Operation	401
	13.4.2 Based on Raw Material Usage	402
	13.4.3 Based on Products Manufactured	402
13.5	Raw Material Preparation	403
	13.5.1 Wood Based Mills—Selection of Chippers and Conveying Equipment	403
	13.5.2 Waste Paper based Mills	403
13.6	Manufacturing Process	403
	13.6.1 Wood Based Pulp and Paper Manufacturing Process	404
	13.6.2 Agro Based Pulp and Paper Manufacturing Process	410
	13.6.3 Waste Paper Pulping	413
13.7	Pollution Outputs in the Production Line	414
	13.7.1 Debarking	415
	13.7.2 Kraft Pulping	415
	13.7.3 Sulphite and Semichemical Pulping	416
	13.7.4 Chlorine Bleaching	416
	13.7.5 Mechanical and Chemithermomechanical Pulping	417
13.8	Adsorbable Organic Halides (AOX)	418
13.9	Cleaner Technologies for AOX Reduction	420
	13.9.1 Technologies for Kappa no. Reduction	420
	13.9.2 Extended Delignification	420
	13.9.3 Improved Pulp Washing	421
	13.9.4 Oxygen Delignification	421
	13.9.5 Chlorine Dioxide Substitution	421
	13.9.6 Oxidative Alkali Extraction Bleaching	421
13.10	Methods Employed for AOX Degradation	422

I. H. Farooqi (✉)
Department of Civil Engineering, Environmental Engineering Section, Aligarh Muslim University, Aligarh 202002, India
e-mail: farooqi_izhar@yahoo.com

A. Malik, E. Grohmann (eds.), *Environmental Protection Strategies for Sustainable Development,* Strategies for Sustainability,
DOI 10.1007/978-94-007-1591-2_13, © Springer Science+Business Media B.V. 2012

13.10.1	Physical, Chemical and Electrochemical Methods of AOX Removal	422
13.10.2	Biological Treatment	424
13.11	Water Conservation Options	429
13.11.1	Use of Better Pulp Washing Technology Instead of Obsolete Technologies like Potcher Washing	429
13.11.2	Optimum Use of Cooling Wastewater	430
13.11.3	Optimum Use of Paper Machine Clarified Wastewater in Sections Other than Paper Machine	431
13.11.4	Recycling of Treated Effluent for Use Within the Mill for Non-Process	432
13.11.5	Tertiary Treatment of Wastewater for Recycling	432
References		433

Abstract Pulp and paper mills are categorized as a core sector industry and are the fifth largest contributor to industrial water pollution. Pulp and paper production has increased globally and will contribute to increase in the near future. For every tonne of paper produced, these mills generate 220–380 m^3 of highly coloured and potentially toxic wastewater. The pulp and paper mill is a major industrial sector utilizing a huge amount of lignocellulosic materials and water during the manufacturing process, and releases chlorinated lignosulphonic acids, chlorinated resin acids, chlorinated phenols and chlorinated hydrocarbons in the effluent. About 500 different chlorinated organic compounds have been identified including chloroform, chlorate, resin acids, chlorinated hydrocarbons, phenols, catechols, guaiacols, furans, dioxins, syringols, vanillins, etc. In wastewater these compounds are estimated collectively as "Adorbable Organic Halides" AOX. This paper is the state of the art review of the manufacturing process, treatability of the pulp and paper mill wastewater and performance of available treatment processes. A comparison of all treatment processes is presented emphasis being made on the treatability studies of AOX.

Keywords Adsorbable organic halides • Pulp and paper industry • Aerobic degradation • Anaerobic degradation • Wastewater treatment

13.1 Introduction

The rapid increase in population and the increased demand for industrial establishments to meet human requirements have created problems such as overexploitation of available resources, leading to pollution of the land, air and water environments. The pulp and paper industry is one of the most important industries of the North American economy and ranks as the fifth largest in the U.S. economy (Nemerow and Dasgupta 1991). The pulp and paper industry is considered as the third largest polluter in the United States (US). It has been estimated that the pulp and paper industry is responsible for 50% of all wastes dumped into Canada's waters (Sinclair 1990). There are 308 paper mills in India, producing a million tonne of paper

13 Pulp and Paper Industry—Manufacturing Process

using variety of wood materials (Pandey and Carney 1998). Out of these there are 34 large mills, which contribute more than 51% of the total paper production and rest 274 small paper mill produces 49% of the total paper produce in country. It is estimated that about 273–455 m^3 (60,000–1,00,000 gallons) of water is required per tonne of paper production and discharges more than 47,000–80,000 gallons of wastewater containing lignin and chlorophenols (CPs), which causes soil as well as aquatic pollution. Lignin is the polymer of phenolic compounds such as P. coumaryl alcohol, coniferyl alcohol, sinapyl alcohol and is main component of vascular plants, providing the plants strength and rigidity (Burnow 2001). Lignin is heterogenous polymer containing various biologically stable carbon-to-carbon linkages are interspersed with hemicelluloses. The color of effluent is mainly due to the presence of lignin and its derivatives. Color not only is aesthetically unacceptable but also leads to chain of adverse effect on the aquatic ecosystem. Similarly, pentachlorophenol (PCP) from category of CPs, one of the most hazardous classes of environmental pollutants, has been produced in thousands of tons annually by the pulp and paper and agrochemical industries. The combined wastewater coming out in each process is toxic to all trophic level in the aquatic ecosystem, affecting flora and fauna, both in water column and sediment (Sodergren 1993). PCP is expected to be recalcitrant to aerobic biodegradation because it is highly chlorinated and, in general, aromatic compounds with higher amounts of chlorine are more resistant to biodegradation (Anandarajah et al. 2000). Bleached chemical pulp mill effluents have been identified as toxic according to the Canadian Environmental Protection Act due to the presence of high quantity of CPs (Schnell et al. 2000).

13.2 Global Paper Industry

13.2.1 Industry Structure

The global industry is configured for volume driven operations with distinct pulp manufacturers (which is essentially commodity trade) and paper makers (differentiated brand driven industry). Europe and Canada dominate the pulping industry while North America (specifically US), Western Europe and parts of Asia dominate paper manufacturing. The world paper industry can broadly be classified into:

1. Paper and Paperboard

   - Printing and Writing
   - Packing
   - Tissue and Sanitary

2. Newsprint

The global paper consumption in FY 2000 was approximately 325 mn tonnes. Writing & Printing segment accounted for 32% of the global paper consumption while Packaging, Tissue & Sanitary and Newsprint accounted for 50, 6 and 12% respectively.

13.2.2 Demand–Supply Scenario

Geographically, Asia accounts for around 32% of the global paper consumption while North America and Europe account for 31 and 28% respectively. Asian countries have experienced higher growth in demand due to higher economic growth, ranging from 7 to 10% per annum. As of 2000, the aggregate global capacity of paper and paper boards stood at 364 mn tpa resulting in excess supply to the extent of 40 mn tpa. However, due to geographical capacity inequalities (particularly capacity shortage in Asia which is the fastest growing region), global prices have remained steady. Global paper & board demand is recovering due to revival in demand in US and robust demand growth in China along other Asian economies. The growth in world paper demand and consumption pattern for the last 5 years is indicated in the chart. As per global industry estimates, paper and paper boards demand is expected to grow at a CAGR of 2% p.a. and is expected to touch 370 mn tpa by 2005.

The pulp and paper industry plays an important role in European economic cluster, generating an annual turnover of more than EUR 400 billion (Gebart 2006). In 2007, European pulp and paper industry produced a total of about 102 million tonnes of paper and board (FAO 2008). Most of the pulp and paper produced utilizes wood as the basic raw material (90%). The major components of wood are cellulose (40–50%), hemicelluloses (25–30%) and lignin (25–30%). There are two significant pulping technologies available that differ greatly in terms of process i.e. mechanical and chemical pulping. Approximately 30% of the total pulp production in European Union is from mechanical pulping while the rest is produced by means of chemical pulping (Swedish forest agency 2008). North America has major pulp and paper industry, about 21% of the total pulp produced is from mechanical pulping and rest is produced chemically, see Table 13.1.

Table 13.1 Pulp production in North America and European Union (EU)

Region	Pulp production[a] (million tonnes)	2004	2005	2006	2007
North America	Chemical wood pulp	59.6	59.1	57.3	55.6
	Mechanical wood pulp	16.3	16.2	15.3	14.4
	Total production	75.9	75.3	72.6	70.0
EU	Chemical wood pulp	26.8	25.9	27.5	27.3
	Mechanical wood pulp	11.5	11.2	12.4	12.1
	Total production	38.3	37.1	39.9	39.4

[a] Based on Food and Agriculture Organization (FAO) Database (FAO 2008).

13 Pulp and Paper Industry—Manufacturing Process

Table 13.2 Consumption pattern of paper and paper board products in India. (Source: Saxena and Rajat (2006))

Type of paper	Main varieties	Percent of total Consumptions (%)
Cultural paper	cream woven, maplitho, bond paper, Chromo paper	41
Industrial paper	kraft paper, paper board—paper board—single layer board, multilayer board, duplex board,	43
Specialty paper	Security paper, grease proof paper, electrical grades of paper	4
Newsprint	glazed, non-glazed	12

13.3 Status of Pulp and Paper Industries in India

Indian industries produce types/grades of paper for variety of uses. The paper and paperboard product segment constitutes of cultural paper, industrial paper and specialty paper. Cultural paper comprises of writing and printing paper, Art/Media paper, Bond paper, Copier paper, Cream wove, Maplitho, Ledger paper, etc. Industrial paper comprises of Duplex Board/Paper, Kraft Paper, and other Board/Paper.

The major types of paper that are produced in the country along with main varieties and their consumption pattern (demand indicator) are presented in the Table 13.2

13.4 Classification of Pulp and Paper Industries in the country

At present, around 309 pulp and paper manufacturing mills are operating in the country. Out of these, 198 mills are operating under large scale category with actual capacity of 5.2 million tonnes per annum and 111 mills are operating under small scale category with actual capacity of 0.3 million tonnes per annum.

13.4.1 Based on Scale of Operation

The pulp and paper mills based on the scale of operation are classified as those having an installed capacity of 25,000 t per year & above as large scale and less than 25,000 t but greater than 5,000 t per year as medium scale and up to 5000 t per year as small scale. The distribution of large/medium and small scale pulp and paper mills in the country is given below in Table 13.3.

Table 13.3 Distribution of large/medium and small scale pulp and paper mills. (Source: Saxena and Rajat (2006))

Sl. No	Scale of operation	No of Mills	Actual capacity, million tonnes per annum	
1	Large/Medium Scale	198	5.2	94.5%
2	Small Scale	111	0.3	5.5%

Table 13.4 Distribution of pulp and paper mills based on raw material used (Source: Saxena and Rajat (2006))

	Agro	Integrated	Wastepaper	Wood
No of Mills	66	3	219	21
% No. of Mills	21	1.0	71	7
Production, million TPA	1.0	0.14	2.8	1.6
% Production	18.0	2.5	50.5	28

13.4.2 Based on Raw Material Usage

The pulp and paper industry is segmented as wood/forest-based, agro-based and waste paper based with the former accounting for 21%, agro-based 71%, waste paper based 7%, wood based and integrated for 1% of the total actual production. The number of pulp and paper mills under each classification is given below in Table 13.4.

13.4.3 Based on Products Manufactured

The Indian paper industry is classified broadly into two categories based on product manufactured:

- Paper and Paper board products
- Newsprint

The number of pulp and paper mills producing Paper and Paper board and News-print along with actual production in the country is given in Table 13.5.

Table 13.5 Distribution of pulp and paper mills based on products manufactured. (Source: Saxena and Rajat (2006))

Category	No of Mills	%	Production, million TPA	%
Cultural—high grade	66	21.3	2.27	41
Cultural—low grade	20	6.5	0.57	10.2
Industrial	187	60.5	1.74	31.4
Newsprint	21	6.8	0.62	11.2
Pulp	3	1	0.22	4
Specialty	12	3.9	0.12	2.2

13.5 Raw Material Preparation

13.5.1 Wood Based Mills—Selection of Chippers and Conveying Equipment

Chippers are one of the major consumers of power in a wood based paper mill. It is suggested that energy efficient, high capacity, chippers be selected. The unitisation of equipment (select only one high capacity chipper) results in obtaining good operating efficiency. Another factor to be seen in achieving optimum efficiency at chipper house is the optimal feed rate. To achieve the optimal feed rate, mechanised conveying to chipper is to be planned at design stage. This will result in better capacity utilisation of the chippers. Conveying of chips also consumes considerable energy. The different methods of conveying chips are pneumatic conveying, screw conveying or belt conveying. The most energy efficient chips conveying is through belt conveyor. During design stage consideration should be given for layout to minimise transportation. This will result in power savings in chipper house. Planning of belt conveying systems for chips transport will also reduce maintenance cost (due to lower wear and tear compared to pneumatic conveying systems) and reduce atmospheric emissions from chippers. During design stage, the paper mills should plan to burn chip-dust in the paper mill's power boilers. This will reduce consumption of purchased fuels in the paper mill's power boilers and eliminate environmental problems associated with chip-dust disposal.

13.5.2 Waste Paper based Mills

In countries like India, where there is a shortage of conventional raw material, every effort should be made to encourage utilisation of waste paper. It can be either imported or indigenous.

13.6 Manufacturing Process

In general, paper is manufactured by applying a liquid suspension of cellulose fibres to a screen, which allows the water to drain, and leaves the fibrous particles behind in a sheet. The liquid fibrous substrate formed into paper sheets is called pulp.

Processes in the manufacture of paper and paperboard can, in general terms, be split into three steps: pulp making, pulp processing, and paper/paperboard production. Paperboard sheets are thicker than paper sheets; paperboard is normally thicker than 0.3 mm. Generally speaking, however, paper and paperboard production processes are identical. First, a stock pulp mixture is produced by digesting a material into its fibrous constituents via chemical, mechanical, or a combination of both. In the case of wood, the most common pulping material, chemical pulping actions

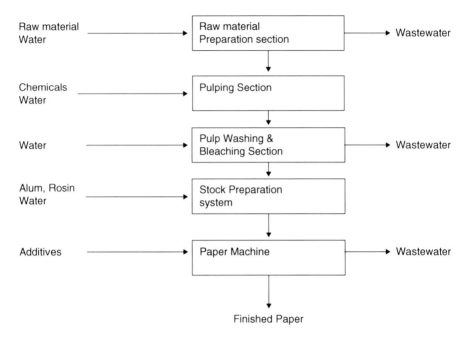

Fig. 13.1 A typical flow sheet for paper manufacture

release cellulose fibres by selectively destroying the chemical bonds in the glue-like substance (lignin) that binds the fibres together. After the fibres are separated and impurities have been removed, the pulp may be bleached to improve brightness and processed to a form suitable for paper-making equipment. Currently, one-fifth of all pulp and paper mills practice bleaching. At the paper-making stage, the pulp can be combined with dyes, strength building resins, or texture adding filler materials, depending on its intended end product. Afterwards, the mixture is dewatered, leaving the fibrous constituents and pulp additives on a wire or wire-mesh conveyor. Additional additives may be applied after the sheet making step. The fibres bond together as they are carried through a series of presses and heated rollers. The final paper product is usually spooled on large rolls for storage.

A typical flow sheet for the manufacture of paper is shown in Fig. 13.1 and a generalised process of Paper making from different types of raw material is shown in Fig. 13.2

13.6.1 Wood Based Pulp and Paper Manufacturing Process

13.6.1.1 Pulping

At the pulping stage, the processed furnish (wood or other fibre source) is digested into its fibrous constituents. The bonds between fibres may be broken chemically,

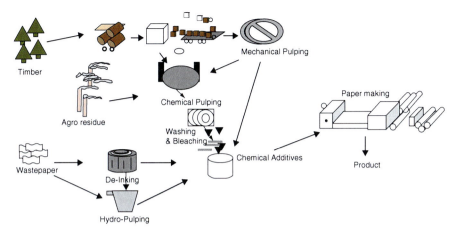

Fig. 13.2 Generalised process of paper making from different types of raw material

mechanically, or by a combination of the techniques called semi-chemical pulping. The choice of pulping technique is dependent on the type of furnish and the desired qualities of the finished product, but chemical pulping is the most prevalent. The three basic types of wood pulping processes (1) chemical pulping, (2) mechanical pulping and (3) semi-chemical pulping

Chemical Pulping Process

This process utilises significantly large amounts of chemicals to break down the wood in the presence of heat and pressure. The spent liquor is then either recycled or disposed of by burning for heat recovery. Chemical pulps are typically manufactured into products that have high-quality standards or require special properties. Chemical pulping degrades wood by dissolving the lignin bonds holding the cellulose fibres together. Generally, this process involves the cooking/digesting of wood chips in aqueous chemical solutions at elevated temperatures and pressures. There are two major types of chemical pulping (a) kraft/soda pulping and (b) sulphite pulping.

Kraft Pulping (or sulphate) processes: The Kraft process uses a sodium-based alkaline pulping solution (liquor) consisting of sodium sulphide (Na_2S) and sodium hydroxide (NaOH) in 10% solution. This liquor (white liquor) is mixed with the wood chips in a reaction vessel (digester). The output products are separated wood fibres (pulp) and a liquid that contains the dissolved lignin solids in a solution of reacted and unreacted pulping chemicals (black liquor). The black liquor undergoes a chemical recovery process to regenerate white liquor for the first pulping step. Overall, the Kraft process converts approximately 50% of input furnish into pulp. Figure 13.3 shows a Kraft pulping process with chemical recovery.

Sulphite Pulping Process: The sulphite pulping process relies on acid solutions of sulphurous acid (H_2SO_3) and bisulphite ion (HSO_3^-) typically from sodium sulphite

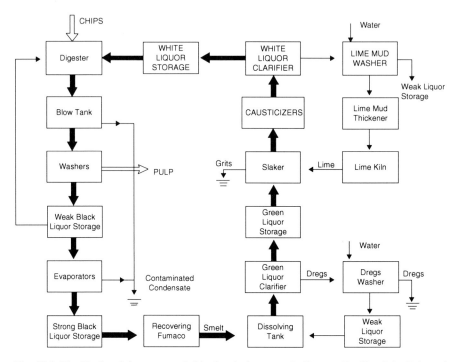

Fig. 13.3 The Kraft pulping process (with chemical recovery). (Source: Profile of the Pulp and Paper Industry (2002))

to degrade the lignin bonds between wood fibres. Softwood is the predominant furnish used in sulphite pulping processes. However, only non-resinous species are generally pulped. Sulphite pulps have fewer colours than Kraft pulps and can be bleached more easily, but are not as strong as Kraft pulps. The efficiency and effectiveness of the sulphite process is also dependent on the type of wood furnish and the absence of bark. For these reasons, the use of sulphite pulping has declined in comparison to Kraft pulping.

Mechanical Pulping Process

This process utilises mechanical forces in the presence of water. The process involves passing a block of wood, usually debarked, through a rotating grindstone where the fibres are stripped of and suspended in water. Mechanically produced pulp is of low strength and quality. Such pulps are used principally for newsprint and other non-permanent paper goods. Mechanical pulping relies on physical pressure instead of chemicals to separate furnish fibres. However, chemicals are sometimes added at the various stages of refining. Processes include: (1) stone ground wood, (2) refiner mechanical, (3) thermo-mechanical, (4) chemi_mechanical, and (5) chemi-thermo-mechanical. The stone ground wood process simply involves mechanical grinding of wood in several high-energy refining systems. The refiner mechanical process involves refining wood chips at atmospheric pressure while the thermo-mechanical process uses steam and pressure to soften the chips before mechanical refining. In the chemi-mechanical process, chemicals can be added throughout the process to aid the mechanical refining. The chemi-thermo-mechanical process involves the treatment of chips with chemicals for softening followed by mechanical pulping under heat and pressure. Mechanical pulping typically results in high pulp yields, up to 95% as compared to chemical pulping yields of 45- 50%, but the energy usage is also high. To offset its structural weakness, mechanical pulp is often blended with chemical pulp.

Semi-Chemical Pulping

The major process difference between chemical pulping and semi-chemical pulping is that semi-chemical pulping uses lower temperatures, more dilute cooking liquor or shorter cooking times, and mechanical disintegration for fiber separation. At most, the digestion step in the semi-chemical pulping process consists of heating pulp in sodium sulphite (Na_2SO_3) and sodium carbonate (Na_2CO_3) Other semi-chemical processes include the Permachem process and the two-stage vapour process. The yield of semi-chemical pulping ranges from 55 to 90%, depending on the process used, but pulp residual lignin content is also high so bleaching is more difficult.

13.6.1.2 Pulp Processing

After pulp production, pulp processing removes impurities, such as uncooked chips, and recycles any residual cooking liquor via the washing process. Pulps are processed in a wide variety of ways, depending on the method that generated them (e.g., chemical, semi-chemical). Some pulp processing steps that remove pulp impurities include screening, de fibering, and de knotting. Pulp may also be thickened by removing a portion of the water. At additional cost, pulp may be blended to ensure product uniformity. If pulp is to be stored for long periods of time, drying steps are necessary to prevent fungal or bacterial growth. Residual spent cooking liquor from chemical pulping is washed from the pulp using brown stock washers. Efficient washing is critical to maximize return of cooking liquor to chemical recovery and to minimize carryover of cooking liquor (known as brown stock washing loss) into the bleach plant, because excess cooking liquor increases consumption of bleaching chemicals. Specifically, the dissolved organic compounds (lignins and hemicelluloses) contained in the liquor will bind to bleaching chemicals and thus increases bleach chemical consumption. In addition, these organic compounds function as precursors to chlorinated organic compounds (e.g., dioxins, furans), increasing the probability of their formation. The most common washing technology is rotary vacuum washing, carried out sequentially in two or four washing units. Other washing technologies include diffusion washers, rotary pressure washers, horizontal belt filters, wash presses, and dilution/extraction washers.

Pulp screening, removes remaining oversized particles such as bark fragments, oversized chips, and uncooked chips. Centrifugal cleaning (also known as liquid cy-

408   S. Badar and I. H. Farooqi

clone, hydro cyclone, or centricleaning) is used after screening to remove relatively dense contaminants such as sand and dirt. Rejects from the screening process are either repulped or disposed of as solid waste.

13.6.1.3 Chemical Recovery

The chemical recovery system is a complex part of a chemical pulp and paper mill and is subject to a variety of environmental regulations. Chemical recovery is a crucial component of the chemical pulping process: it recovers process chemicals from the spent cooking liquor for reuse. The chemical recovery process has important financial and environmental benefits for pulp and paper mills. Economic benefits include savings on chemical purchase costs due to regeneration rates of process chemicals approaching 98%, and energy generation from pulp residue burned in a recovery furnace. Environmental benefits include the recycle of process chemicals and lack of resultant discharges to the environment.

13.6.1.4 Bleaching

Bleaching is defined as any process that chemically alters pulp to increase its brightness. Bleached pulps create papers that are whiter, brighter, softer, and more absorbent than unbleached pulps. Bleached pulps are used for products where high purity is required and yellowing (or colour reversion) is not desired (e.g. printing and wrapping papers, food contact papers). Unbleached pulp is typically used to produce boxboard, linerboard, and grocery bags.

Any type of pulp may be bleached, but the type(s) of fibre furnish and pulping processes used, as well as the desired qualities and end use of the final product, greatly affect the type and degree of pulp bleaching possible. Printing and writing papers comprise approximately 60% of bleached paper production. The lignin content of a pulp is the major determinant of its bleaching potential. Pulps with high lignin content (e.g., mechanical or semi-chemical) are difficult to bleach fully and require heavy chemical inputs. Excessive bleaching of mechanical and semi-chemical pulps results in loss of pulp yield due to fibre destruction. Chemical pulps can be bleached to a greater extent due to their low (10%) lignin content.

Chemical Pulp bleaching is done using elemental chlorine free (ECF) and total chlorine free (TCF). The difference between ECF and TCF is that ECF may include chlorine dioxide (ClO_2) and hypochlorite ($HClO$, $NaOCl$, and $Ca(OCl)_2$) based technologies. ECF technologies were used for about 95% of bleached pulp production, TCF technologies were used for about 1% of bleached pulp production, and elemental chlorine was used for about 4% of production. Chemical pulp is bleached in traditional bleach plants where the pulp is processed through three to five stages of chemical bleaching and water washing. The number of cycles is dependent on the whiteness desired, the brightness of initial stock pulp, and plant design.

13 Pulp and Paper Industry—Manufacturing Process

Semi-Chemical Pulps are typically bleached with hydrogen peroxide (H_2O_2) in a bleach tower.

Mechanical Pulps are bleached with hydrogen peroxide (H_2O_2) and/or sodium hydrosulfite (Na_2SO_3). Bleaching chemicals are either applied without separate equipment during the pulp processing stage (i.e., in-line bleaching), or in bleaching towers. Full bleaching of mechanical pulps is generally not practical due to bleaching chemical cost and the negative impact on pulp yield.

13.6.1.5 Different Types of Equipment/Technologies Used for Washing Bleached Pulp are Listed Below

Batch Process

Potcher washing is the oldest technology used in batch washing. Potcher consists of a series of beaters or engines used in washing and preparing pulp. This process consumes huge quantity of water.

Continuous Counter-Current Processes

Different continuous processes are highlighted below:

1. **Hydraulic Drum Washing**
   A hydraulic drum washer does not require barometric leg and works on hydraulic principle and therefore has minimum operating costs. A sketch of hydraulic drum is shown in Fig. 13.4.
2. **Vacuum Drum Washing**
   In vacuum drum washing, each stage consists of a rotating screen drum which has a partial vacuum applied to interior. The drum sits in a tank where pulp is diluted with wash water. The vacuum draws a pulp mat against the surface and wash- water through the mat. The drum rotation advances the washed pulp mat to the next dilution tank. Wash water discharged from this wash stage is sent to the previous washing stage.
3. **Pressure Washing**
   Pressure washing is similar to vacuum drum, but differs in such a way that water is prayed under pressure through the pulp mat as the drum rotates.
4. **Diffusion Washing**
   Diffusion washing is a counter flow process that takes place in one or more stages. Pulp flow is upward and is carried on a perforated plate. Water flows downward through a series of baffles.
5. **Chemi or Belt Washing**
   A Chemi or belt washer is perhaps the simplest washing system in terms of design. It offers excellent washing with reduced water usage. Belt washing is a counter flow

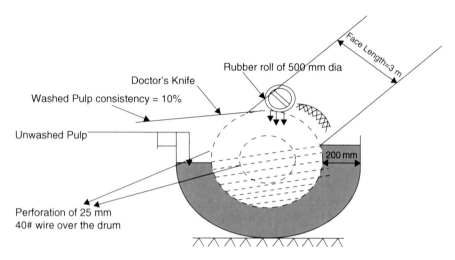

Fig. 13.4 Hydraulic Washer

process where pulp enters the washer area on a wire belt. Washing takes place under a series of showers. Clean water enters on the opposite end from the pulp and is sprayed vertically through the pulp. The used wash water is then collected and reapplied to the dirtier pulp by the next washing head. Counter current to the direction that the pulp moves. This process is continued through at least seven stages until the wash water is saturated with liquor after washing immediately coming pulp. The wash water is then sent to the recovery process. Diffusion washing and belt washing can reduce the amount of water used per tonne of pulp in brownstock washing by 50% or more according to published data.

6. **Twin Roll Press Washer**
   Twin wire roll press washer works on the general principle of dewatering, displacement and pressing. It consists of a twin-wire dewatering unit which allows controlled washing of the pulp. The two-sided dewatering and strong turbulence of the washer facilitate two to three times' higher capacities per unit of width than conventional washer technologies. An additional main feature of this technology is that the ash and fines removal can be controlled depending on the targeted levels. This technology can be used for high consistency pulp washing and results in reduced water consumption. Figure 13.5 shows a twin wire roll press washer installed in one of pulp and paper mill in India.

13.6.2 *Agro Based Pulp and Paper Manufacturing Process*

The process of pulp and paper making from agro-based residues is described below:

13 Pulp and Paper Industry—Manufacturing Process

Fig. 13.5 Twin wire press roll washer

The various production stages are as follows:

1. Raw Material Preparation: Dedusting, Depithing, Leaf Removal
2. Pulping Section: Cooking, beating, pulp washing, refining, bleaching, cleaning and, thickening
3. Stock preparation: Blending, pulp conditioning
4. Paper machine: Refining, Centricleaning, dewatering, drying of paper

13.6.2.1 Raw Material Preparation

The agro residue based raw material (RM) is procured by the mills from nearby farms. In some mills raw material is screened at the site itself. The dust from the screening section is disposed of as solid waste along with municipal waste. In very small mills, bagasse is not depithed. The prepared agro raw material is then conveyed to spherical digesters.

13.6.2.2 Pulping Section

Pulping comprises of cooking or digestion followed by washing, bleaching and centri-cleaning.

Cooking

There are two types of digestion processes employed similar to wood pulping;

Batch digestion carried out in spherical digesters and Continuous digestion process carried out in a Pandia type digester. Unlike wood pulping two different chemical pulping processes are employed, namely, Kraft process and soda process.

The agro residue is chemically digested in a digester at 150–160°C and 6–7 atm pressure for about six hours. Charging and discharging takes 1.5 and 0.5 h respectively. The cooking in small agro-based mills is done with caustic soda (NaOH) and steam. The quantity of NaOH charged, varies from 6 to 14% of raw material, depending on the type of agro residue. For every tonne of agro residue, about 1.5–2.0 t of steam are used, depending on the pulp required (hard cooked or soft cooked). During digestion, solid to liquid (bath ratio) is maintained in the range of 1:3–1:4

(a) **Blow Tank** After cooking, the content of the digester is discharged, under pressure, either into a blow tank where the pressure is released or directly into potchers. Water is added to reduce the pulp consistency from an inlet value of 12–14% to about 3–4%, so that it can be pumped to the washing and cleaning section.

(b) **Washing** The pulp is then pumped to the washers for washing with fresh water in the final stage and backwater in the previous stages. The washing operation takes about four to six hours. The wash water, known as black liquor, has total solids content of around 7–10% mainly because of residual alkali and lignin. This liquor is mostly discharged to drains as chemical recovery has so far been economically unviable.

(c) **Screening** The washed pulp contains sand and uncooked agro residue as impurities. The impurities are removed through screening and centricleaning. The rejects from the screening (Johnson and / or Hill screen) are normally drained out. After screening, which is carried out at 1% consistency, the pulp is thickened to about 4% for next operation, namely bleaching. The filtrate, called back water, generated during thickening operation, is generally collected and used for pulp washing (previous operation). The pulp for making unbleached kraft paper (for packaging purpose) is not bleached and is taken directly for stock preparation.

(d) **Bleaching** The bleaching in small mills is carried out using calcium hypochlorite (hypo), which is added in two stages in order to provide sufficient retention time for hypo and to minimize the fibre degradation. Fifty percent of the hypo is added in the screened pulp storage chest and the rest is added in the bleacher. A retention time of about two hours is provided in the storage chest. After bleaching, the pulp is washed, partly with fresh water and partly with white water (paper machine back water). The wash water from bleaching operation contains chloro-lignates and residual chlorine preventing the wash water from direct reuse. A typical vacuum drum bleach washer is shown in Fig. 13.6

(e) **Stock Preparation** The bleached pulp is mixed with the long fibre pulp, comprising mainly rags and wastepaper pulp. The mix depends upon the agro residue being processed and the type of paper to be manufactured. The mix pulp is

13 Pulp and Paper Industry—Manufacturing Process

Fig. 13.6 Conventional vacuum drum washer after bleaching

blended with additives and fillers in the blending chest. The chemicals added to the blending chest are rosin, alum, talc, dye (optional), optical whitener and high gum. The chemicals (additives, fillers etc) solutions are prepared and added manually in every batch.

13.6.2.3 Paper Machine

The blended pulp is again centricleaned to remove impurities and finally fed to the paper machine through a head box. From the dewatering and paper making angle, the machine has three principal stages:

- The gravitational and vacuum dewatering stage (wire part)
- The mechanical dewatering stage (press rolls part)
- The thermal drying stage (indirect steam dryers)

On the wire part of the paper machine, the dewatering of pulp takes place by gravity and vacuum. The water from the wire mesh is collected in a fan pump pit and is continuously recycled to dilute the pulp fed into the paper machine centricleaner. In some mills, the wire is continuously washed with fresh water showers. The water is collected and fibre is recovered through Krofta saveall. The clear water from save all is recycled back to different consumption points. Excess is discharged to the drain. After the wire part, the edge cutting operation is carried out to obtain paper of a definite width. The edge cuts of the pulp web falls in the couch pit and are recycled to the machine chest. Towards the end of the wire part of the machine, the consistency of pulp rises to about 20%. Further dewatering is carried out by press rolls to raise the consistency to about 55%. The paper is finally dried through an indirect steam dryer to about 94% solids and is collected in rolls as the final product.

13.6.3 *Waste Paper Pulping*

Secondary fibre recycling or wastepaper pulping is another important sector. The repulping of recycled paper and deinking is described as under. Recycled paper, newsprint and magazine are charged in a hydraulic pulper. Water is added till waste paper is converted into slurry form with high consistency pulp is formed. The hydro pulped pulp is cleaned in a high density cleaner followed by turbo separator for heavy weight and light weight impurities respectively. It is continuously forwarded to centricleaner after passing through screen. At centricleaner, the sand is separated due to centrifugal force. The pulp is then taken to Decker thickener where the wastewater is removed and pulp is thickened. The thickened pulp is processed to a chest through refiner by which the pulp is thickened. The thickened pulp is processed to a chest through refiner by which the pulp becomes finer as per process requirement. It is then transferred to machine chest where addition of dye, chemical takes place.

13.6.3.1 Deinking Process

Deinking is a recycling technique that can produce high quality recycled pulp from recovered papers. Ink detachment is an important step. Flotation method is commonly used for this purpose. Flotation deinking makes ink particles hydrophobic by means of a collector in a flotation cell. The air bubbles generated at the bottom of the cell carry the ink particles to the surface where they are confined in foam which is then removed.

13.6.3.2 Paper Making

Papermaking is common to all types of categories.

13.7 Pollution Outputs in the Production Line

Pulp and paper mills use and generate materials that may be harmful to air, water, and land: pulp and paper processes generate large volumes of wastewaters which might adversely affect freshwater or marine ecosystems, residual wastes from wastewater treatment processes may contribute to existing local and regional disposal problems, and air emissions from pulping processes and power generation facilities may release odours, particulates, or other pollutants. Major sources of pollutant released in pulp and paper manufacture are at the pulping and bleaching stages respectively. As such, non-integrated mills (i.e., those mills without pulping facilities on-site) are not significant environmental concerns as compared to integrated mills or pulp mills.

General water pollution concerns for pulp and paper mills are effluent solids, biochemical oxygen demand, and colour. Toxicity concerns historically occurred from the potential presence of chlorinated organic compounds such as dioxins, furans, and others (collectively referred to as adsorbable organic halides, or AOX) in wastewaters after the chlorination/extraction sequence. With the substitution of chlorine dioxide for chlorine, effluent loads of the chlorinated compounds decreased dramatically.

13 Pulp and Paper Industry—Manufacturing Process

Table 13.6 Potential water pollutants from pulp and paper processes. (Source: Profile of the Pulp and Paper Industry (2002))

Source	Effluent characteristics
Water used in wood handling/ debarking and chip washing	Solids, BOD, color
Chip digester and liquor evaporator condensate	Concentrated BOD, reduced sulfur compounds
"White waters" from pulp screening, thickening, and cleaning	Large volume of water with suspended solids, can have significant BOD
Bleach plant washer filtrates	BOD, color, chlorinated organic compounds
Paper machine water flows	Solids
Fiber and liquor spills	Solids, BOD, color

Pollution from the pulp and paper industry can be minimized by various internal process changes and management measures such as the Best Available Technology (BAT). Dube et al. (2000) reported a 60% reduction in effluent BOD due to an internal process change in Irving Pulp and Paper Limited, Canada. The However, the average water use for the pulp and paper mills in India was still 200–259 m^3/t of paper production (Gune 2000).

The major sources of effluent pollutants in a pulp and paper mill are presented below in Table 13.6.

13.7.1 Debarking

The first stage in the manufacture of pulp is wood debarking, usually wet drum debarking. In this process, water is used to thaw the frozen logs, remove the bark, and wash the logs after debarking. The contribution of wet debarking water to a Finnish chemical pulp mill total effluent is about 5% as BOD 7 or COD, and 16% as SS (Virkola and Honkanen 1985). The soluble COD, of approximately 1000 mg/L of various debarking effluents, consists of tannins (30–55%), monomeric phenols (10–20%), simple carbohydrates (30–40%) and resin compounds (5%), as reviewed by Field et al. (1988). Dry debarking, introduced especially in chemical pulp mills has decreased the BOD 7 load to 10% from that of conventional wet debarking.

13.7.2 Kraft Pulping

In the Kraft process, wood chips are cooked under pressure with a mixture of hot caustic soda and sodium sulphide to promote cleavage of various ether bonds in the lignin. The inorganic chemicals are recovered from the spent black liquor and the pulp washing liquor in the recovery boiler where the organic residue is corn busted to generate steam for the process. In the kraft process, approximately 55% of the total weight of the wood is dissolved in the pulping liquor which contains degradation products of its constituents; i.e. lignin, polysaccharides, and wood extractives pulp is washed to remove the spent black liquor from the fibres. In screening, partially cooked fibres, shives, and other materials, are removed. The washing and screening waters are recycled as much as possible to reduce the discharge load. The BOD load from washing is 6–10 kg/t pulp (Virkola and Honkanen 1985) which leaves the process through bleaching and partially open washing. The evaporation of spent black liquor produces condensates containing methanol and reduced-sulphur compounds as the main components (Blackwell et al. 1979). Resin and fatty acids, terpenes and ethanol are present in lower concentrations. Most of the BOD is concentrated in a small volume fraction of evaporator condensates. Digester and evaporator condensates represent about 5% of the mill effluent volume and account for 5–10% and 10–20% of the total BOD discharge of a mill producing bleached and unbleached sulphate pulp, respectively.

13.7.3 Sulphite and Semichemical Pulping

The active chemical in sulphite process cooking liquor is calcium, magnesium, ammonium or sodium sulphite which in acidic or neutral conditions solubilizes the wood lignin as lignosulfonic acids. Semichemical pulping involves the cooking of chips with a neutral or slightly alkaline sodium sulphite solution and mechanical separation of fibres. After chemical recovery, the remaining spent sulphite liquor flows with the pulp and finally leaves the process with bleaching and pulp-dewatering effluents. The wastewater sources for the sulphite process include spills from the digester area, digester relief and blow condensate, washing and screening, and from the recovery system as evaporator condensate. Spent sulphite liquor mainly consists of lignosulfonates and carbohydrates. In acid spent liquor, carbohydrates appear as monosaccharides (Pfister and Sjrstrrm 1977) and in bisulphite liquor as oligo or polysaccharides (Jurgensen and Patton 1979). Evaporator condensates may account for about 15% of sulphite pulp mill total effluent volume and 30–50% of total BOD load (Benjamin et al. 1984). Various evaporator condensates contain acetic acid (1.6–8.2 g/L), methanol (0.2–1.2 g/L), and furfural (0.2–1.0 g/L) as the major organic components with smaller concentrations of formaldehyde, formic acid, acetaldehyde, and methylglyoxal (Ruus 1964). The sulphur is mainly present as free SO_2, loosely bound SO_2 (hydroxysulfonic acids), and sulfonates including lignosulfonates and other organic sulphur (Rexfelt and Samuelson 1970).

13.7.4 Chlorine Bleaching

In chemical pulping about 5–10% of the lignin remains in the pulp which is subsequently depolymerised and removed in multistage bleaching (Rydholm 1965). The lignin is converted to alkaline-soluble compounds by treatment with oxidiz-

13 Pulp and Paper Industry—Manufacturing Process

ing agents (chlorine, chlorine dioxide and hypochlorite), and then washed out with sodium hydroxide. Recently, the use of small amounts of oxygen and/or hydrogen peroxide in the alkali stages has been introduced for partial substitution of other oxidizing agents. A commonly used multistage bleaching sequence for softwood is as follows: chlorine/chlorine dioxide-sodium hydroxide-chlorine dioxide-sodium hydroxide-chlorine dioxide. For example, the COD and AOX load from kraft pulp bleaching varies from 39 to 80 kg COD per tonne and from 2.8 to 7 kg AOX per tonne, respectively, depending on the delignification method applied (Gullichsen 1991).

Bleaching effluents mainly contain degradation products of lignin. Smaller amount of polysaccharide and wood-extractive degradation products are generated. Methanol and various hemicelluloses are dominant organic compounds (over 90%) in bleaching liquors. A vast variety of organochlorines are created as reviewed e.g. by Dellinger (1980), Voss et al. (1980), Kringstad and Lindstrom (1984) and McLeay (1987). About 70 and 95% of the organically bound chlorine has been reported to be as high-relative-molar-mass material (MW> 1000) in spent chlorination and alkali extraction liquors, respectively, and the major part of this material consists of cross-linked aliphatic compounds (Kringstad and Lindstrom 1984). Jokela and Salkinoja-Salonen (1992) showed that the molecular sizes of pulp bleaching organic halogens were smaller than generally assumed. They concluded that over 85% of these compounds were of low molecular size (<1000 g/mol). The low-molar-mass compounds containing organically bound chlorine consist of acidic, phenolic, and neutral compounds. Typical chlorophenolic bleaching waste constituents are presented in Fig. 13.7. About 10% of the low-molar-mass chlorinated compounds have been identified so far. The BOD load of a conventional bleaching accounts for 50–60% of the total BOD load of a pulp mill (Virkola and Honkanen 1985).

13.7.5 Mechanical and Chemithermomechanical Pulping

In mechanical pulping, the wood is converted into fibres by physical or mechanical grinding, aided in some processes by heat or chemicals. The main dissolved compounds in groundwood pulping effluents are carbohydrates (80–90%), extractives and acids (10–20%) (Jarvinen et al. 1980). Thermo Mechanical Pulping (TMP) involves heating wood chips under pressure prior to mechanical refining.

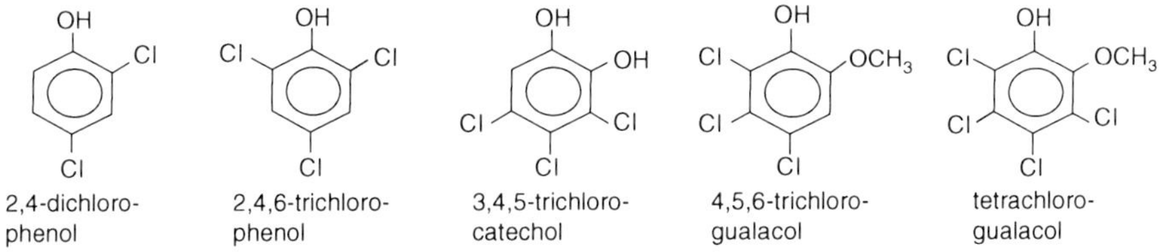

Fig. 13.7 Typical chlorophenolic compounds found in bleaching effluents

The organic compounds in TMP effluents consist of lignin (40%), carbohydrates (40%), and extractives (20%) (Jarvinen et al. 1980). In Chemi thermo mechanical pulping (CTMP), the wood chips are impregnated with sodium sulfite and steamed before refining. The CTMP effluents contain 1–5 g/L BOD or 2.5–13 g/L COD as reviewed by Cornacchio and Hall (1988). Polysaccharides (10–15%), organic acids (35–40%), and lignin (30–40%) are principal constituents of CTMP effluents (Pichon et al. 1988). Resin and fatty acid concentrations are high in CTMP effluents (Walden and Howard 1981; Cornacchio and Hall 1988). Sulfur is present mainly as sulfate and sulfite with minor amounts of lignosulfonates (Pichon et al. 1988). In dithionite bleaching of mechanical pulp the discharge load is low (2–8 kg BODv/ tonne pulp), whereas effluent dissolved compounds from alkaline peroxide bleaching (5–20 kg/t pulp) consist of carbohydrates (60%) and acetic acid and formic acid and methanol (40%) (Jarvinen et al. 1981). Typical COD/BOD ratios for TMP and CTMP effluents are between 2.2 and 3 (Cornacchio and Hall 1988).

The pulp and paper industry is a water intensive industry and ranks only third in the world, after the primary metals and the chemical industries, in terms of freshwater withdrawal. Even with the most modern and efficient operational techniques, about 60 m^3 of water is required to produce a ton of paper resulting in the generation of large volumes of wastewater. Of the different waste streams, bleach plant effluents are most toxic due to various chlorinated organic compounds generated during the bleaching of pulp. About 500 different chlorinated organic compounds have been identified including chloroform, chlorate, resin acids, chlorinated hydrocarbons, phenols, catechols, guaiacols, furans, dioxins, syringols, vanillins, etc. In wastewater, these compounds are estimated collectively as "adsorbable organic halides" (AOX). Amount of these compounds is directly proportional to consumption of chlorine.

13.8 Adsorbable Organic Halides (AOX)

Adsorbable organic halides (AOX) are generated in the pulp and paper industry during the bleaching process. These compounds are formed as a result of reaction between residual lignin from wood fibres and chlorine/chlorine compounds used for bleaching. Many of these compounds are recalcitrant and have long half-life periods. Some of them show a tendency to bio-accumulate while some are proven carcinogens and mutagens. Bleaching of pulp by chlorine based agents as a part of pulp processing for paper making industry is still practiced in India and many other developing countries. Several organochlorine compounds are generated during chlorine bleaching (Suntio et al. 1988), which are collectively termed as adsorbable organic halides (AOX). These compounds are usually biologically persistent, recalcitrant and highly toxic to the environment (Baig and Liechti 2001; Thomson et al. 2001). The toxic effects of AOX range from carcinogenicity, mutagenicity to acute and chronic toxicity (Savant et al. 2006). The toxic effects of AOX range from carcinogenecity, mutagenecity to very acute and chronic toxicity (Table 13.7).

13 Pulp and Paper Industry—Manufacturing Process

Table 13.7 Toxic effects of major AOX compounds. (Source: Savant et al. 2006)

Compound	Toxic effect
Chlorophenols	2,4-Dichlorophenol (DCP), 2,4,5-TCP and PCP are Group 2B carcinogens. PCP is the most toxic chlorophenol. Chronic exposition results in liver and kidney damage, loss in weight, general fatigue and low appetite. In fish, these compounds cause impaired function of liver, enzyme system, metabolic cycle, increase in the incidence of spinal deformities and reduced gonad development
Chlorocatechols	Strong mutagens
Chloroguaiacols	Tetra- and trichloroguaiacols are known to bioaccumulate in fish
Chlorobenzenes	Exposure to 60 ppm is known to cause drousiness, headache, eye irritation, sore throat. Chronic exposures are known to cause adverse effects on lungs, renal degeneration and porphyria. Hexachlorobenzene is carcinogenic in animal tests. Mono-chlorobenzenes is known to cause multiple effects on central nervous system—headache, dizziness, cyanosis, hyperesthesia and muscle spasms
Chlorinated dibenzodioxins and dibenzofurans	Highly toxic, teratogenic. Acute exposures cause severe skin rash, changes in skin colour, hyperpigmentation, polyneuropathies in arms and legs. Act as endocrine disrupting factors by interfering production, release, transport, metabolism, binding action or elimination of natural hormones in the body weight. They may cause reproductive and immune system disorders and abnormal fetal development. In fish, they decrease growth rate, increase egg mortality and produce histological changes in liver

Increased awareness of the harmful effects of these pollutants has resulted in stringent regulations on AOX discharge into the environment (Bajpai and Bajpai 1997). According to PARCOM (Paris Convention for Prevention of Marine Pollution for Land Based Sources and Rivers), twelve European countries have signed for a general AOX emission limit of 1 kg/t for bleached chemical pulp in 1995. The discharge limits were then lowered gradually up to 0.3–0.5 kg/t. In Japan, all Kraft mills now operate at an AOX discharge rate of less than 1.5 kg/t of pulp. Germany has banned production of pulp using chlorine containing chemicals and has also banned consumption of pulps other than total chlorine free (TCF) pulps. The USA has proposed cluster rule in 1992 to restrict the discharge limit at 2.5–5.0 ppb for polychlorinated phenolics. In India, the Central Pollution Control Board has recently implemented a discharge limit of 1.5 kg/t pulp within next two years and 1.0 kg/t pulp within next five years for large pulp and paper mills.

It is therefore, essential to treat the pulp and paper industry wastewater not only for the conventional parameters like COD, BOD, and color, but also for AOX, prior to its discharge into the environment. There are several modifications made to reduce the generation of chlorinated organic compounds from bleach plant effluents using one or more of the following strategies: (1) removing more lignin before starting the chlorination, i.e., reducing the kappa number of unbleached pulp, (2) modifying the conventional bleaching process to elemental chlorine free bleaching (ECF) and total chlorine free bleaching (TCF). These methods may be physicochemical or biochemical in nature or a combination thereof (Bajpai and Bajpai 1997). Though these modifications reduce chlorinated compounds, such processes are not considered economical in the developing countries; due to which bleaching of pulp using elemental chlorine is still preferred.

13.9 Cleaner Technologies for AOX Reduction

13.9.1 Technologies for Kappa no. Reduction

The Kappa no. is an index used by the pulp and paper industry to express the lignin content of a pulp. Lignin is responsible for the brown coloration of paper, and is removed by bleaching. Therefore, the lignin content must be well known, so that only a minimum amount of bleach is used. Higher the lignin content more is the kappa no. The pulp having high lignin content termed as hard cooked pulp and the pulp with low lignin content is termed as soft cooked pulp. The hard cooked pulp required more bleaching chemicals to attain particular brightness as compared to soft cooked pulp.

Pulp and paper industries have incorporated various measures to reduce the kappa no. and also to minimize the carryover of organic matter along with pulp as it governs the bleach chemical demand during the bleaching process. Some of these measures include oxygen delignification, extended delignification, improved pulp washing, substitution of elemental chlorine with chlorine dioxide, oxidative alkali extraction stage bleaching etc.

13.9.2 Extended Delignification

The pulp and paper industries normally use kraft process in batch or continuous digesters to remove the lignin as much as possible during pulping of wood based fibrous raw material but the process has limitation that the wood based fibrous raw material cannot be delignified to a low kappa number. Since the kappa number is the main factor which governs the demand of chemicals required for bleaching of the pulp, the process was modified to achieve maximum possible delignification during cooking of raw materials and now most of the industries in developed countries are employing RDH, modified continuous cooking, super batch process, etc., to reduce the kappa number of the unbleached pulp. Modified pulping processes are energy efficient, require fewer chemicals for cooking of raw materials and produce the pulp of low kappa number with better strength properties as compared to conventional pulping processes. However, the high capital investment and high level of operation restrict the adoption of these technologies in Indian pulp and paper industries. Agro based pulp and paper industries normally use soda pulping process.

13.9.3 Improved Pulp Washing

The pulp mill section of paper industries normally uses brown stock washers for extraction of black liquor and for washing of pulp. The washing efficiency of these washers depends on nature and quality of fibrous raw materials. Most of the small industries use brown stock washers for washing of pulp produced from agro residues but the efficiency of these washers are not satisfactory as high carryover of black liquor along with pulp is observed in agro based industries. Since the pulp from agro residues is difficult to dewater, the industries may use the modified washing systems such as belt filter press, double wire washer etc., to minimize the carryover of the black liquor with pulp entering the bleaching section.

13.9.4 Oxygen Delignification

Oxygen delignification is a well established technology and most of the pulp mills abroad are using this process to reduce the kappa number of pulp before bleaching stage. Single stage oxygen pre bleaching of the pulp reduces the pulp kappa number by 50–60% and two stage oxygen pre-bleaching reduces the pulp kappa number by 80%. The process is used in large pulp and paper industries in the developed countries. Indian paper industries have limitations in adopting the process due to high capital investments involved and low scale of their operation.

13.9.5 Chlorine Dioxide Substitution

The elemental chlorine is a major source in the generation of toxic chlorinated phenolics and dioxins compounds and contributes more than 70% of total AOX. The chlorine dioxide, because of its high oxidation potential, decreases the formation of chlorinated phenolics, colour, AOX, dioxins etc. in addition to improved quality of pulp. Most of the pulp mill in developed countries have substituted or replaced elemental chlorine with chlorine dioxide. The large pulp and paper industries in India have now started the use of chlorine dioxide also.

13.9.6 Oxidative Alkali Extraction Bleaching

The addition of small amount of oxygen or peroxide in alkali extraction stage improves the quality of bleach plant effluent by reducing colour and AOX. Most of the large paper industries in India have already started the use of oxygen or hydrogen peroxide in alkali extraction stage. The adoption of modified pulping and bleaching processes in pulp mill in developed Countries has resulted in an increased recycling or reuse of the wastewater to the internal process and efforts are being continued to achieve zero discharge. These industries are, however, required to operate the pulp mill under controlled conditions to reduce the kappa number and also to modify their pulp washing system to minimize the carryover of black liquor along with pulp in order to reduce the discharge of chlorinated phenolic compounds.

13.10 Methods Employed for AOX Degradation

Various methods have been developed for the degradation of AOX. They can be classified broadly as physical, chemical, electrochemical and biological methods.

13.10.1 Physical, Chemical and Electrochemical Methods of AOX Removal

13.10.1.1 Adsorption

It involves the use of activated carbon which is characterised by extremely large surface areas to unit weight ratios (450–1800 m^2/g). The large surface areas results in substantial adsorptive capacity. The rate of adsorption is a function of carbon particle size. Carbon pore size, particle size, colour component molecular weight and effluent pH are the significant factors influencing the performance efficiency of activated carbon. The high costs of process and its regeneration requirements are the major handicap in its wide scale application.

13.10.1.2 Membrane Filtration

Ultra filtration (UF) and reverse osmosis (RO) are two common examples of such process which are based on the principle of separation of higher and lower molecular fractions when an effluent passes through a semi permeable membrane. The driving force in such process is fluid pressure. With the development of new membrane polymers, dynamic membranes, nano filtration membranes and system with lower energy requirements the technique has undergone more acceptability abroad in recent times. However, due to its high cost this technique is not very common in India.

13.10.1.3 Electroflocculation

The process involves the application of an electric current to sacrificed metal electrodes having coagulating properties like Al, Fe, Mg generating metal ions and gas bubbles simultaneously. The metal ion released combines with pollutant and coagulates which are captured by the gas bubbles resulting in flotation of most of the pollutant on the surface.

Physical, chemical and electrochemical methods reported to remove AOX compounds are not economically viable. The major drawback in most of these technologies is the ultimate disposal of sludge or concentrates which is more difficult and costly than the initial removal and separation. Different types of aerobic, anaerobic and combined biological treatment processes have been developed for treatment of pulp and paper industry wastewater. Biological treatment often results in the complete mineralization of organics. Savant et. al. (2006) has presented an excellent review of the treatment technologies for AOX removal. The physic chemical treatment of AOX has been suarised as under:

Fly ash as an adsorbing medium is known to remove chlorinated organics and colour efficiently (Nancy et al. 1996). Shawwa et al. (2001) reported the use of delayed petroleum coke to bring about significant removal of AOX from bleaching wastewater at a concentration higher than 15,000 mg/L the resulting bleaching wastewater being more susceptible to biological treatment. Torrades et al. (2001) described a photocatalytic treatment, which removed the entire colour and most of the TOC, AOX and COD in a highly loaded D stage effluent within 20 h. Moiseev et al. (2004) have also shown that the use of photocatalytic oxidation as a pretreatment step enhances the biodegradability of wastewater containing recalcitrant or inhibitory compounds and is an alternative to a long and energy intensive total pollutant mineralization. Using ultrafiltration, 99% reduction in AOX was achieved by Yao et al. (1994). Reverse osmosis with pressures in the range of 3.5–5.5 MPa or higher can also be useful for AOX removal. Nano-filtration in combination with electrodialysis at a pilot scale is reported to remove over 95% of the contaminating toxic organic halides, salts and colourants leaving the treated effluents suitable for process reuse (Seiss et al. 2001). However, these techniques require pretreatment and are capital intensive. Membrane fouling is another problem associated with these techniques. Amongst chemical methods, Milstein et al. (1991) reported 75%, 59% and 80% removal of AOX, COD and colour, respectively, with the help of a mixture of polyethylene and modified starches. Milosevich and Hill (1991) could achieve 60–70% AOX removal in 1 h at 50 _C by neutralization of bleach plant effluent with lime mud, followed by addition of alkaline sulfide process liquor. Stephensen and Duff (1996) showed that toxicity of bleached kraft mill effluent could be markedly reduced with the use of chloride and sulfate salts of iron and aluminium. With chitosan as a coagulant, 90% and 70% reduction of colour and TOC is reported (Ganjidoust et al. 1996). Wingate et al. (2001) developed a technology that utilizes a novel class of iron III macrocyclic tetraamide complexes to catalytically activate H_2O_2 in oxidative reactions. Pretreatment with ozone or peroxide also enhances the biodegradability of kraft mill caustic extraction stage effluents (Hilleke 1993). Hostachy et al. (1997) reported complete detoxification and removal of residual COD of bleached kraft mill effluent at low ozone doses, for example, 0.5–1.0 kg/t air dried pulp. Mobius and Helble (2004) have described a combination of ozonation followed by biodegradation in a biofilm reactor to achieve a far reaching elimination of AOX, colour and other disturbing substances. The major problem associated with chemical methods is that they produce voluminous sludges which pose difficulty in disposal. Supercritical water oxidation is another novel method of AOX removal in which the waste stream is mixed with an oxidant and heated to a temperature above the critical point of water (about 374 _C) at a pressure of 250 kg/cm^2. Within a few seconds, almost 100% of the organic waste is oxidized. Inorganic compounds can be easily separated as they are insoluble at such high temperatures. The technology is suitable for high strength effluents. The technology has been tested on pulp mill sludges, mixtures of kraft bleaching effluent and sludge from a primary clarifier, achieving a removal of 99.99% of the organic waste. However, the cost of applying this technology is very high and the technology is still underdeveloped (Allen and Liu 1998). The chlorine and hypochlorite generated by electrolysis of chlorides can be used to oxidize organic compounds in the effluent (Springer et al. 1994). Although the electrochemical systems are effective, they are high in operating costs because much of electrochemical energy is consumed in undesirable side reactions. In general, physical, chemical and electrochemical treatment technologies for removal of AOX compounds are uneconomical when applied alone at field scale operations. This necessitates consideration of developing economical and eco-friendly methods for removal of AOX compounds.

13.10.2 Biological Treatment

Biological oxidation is the most widely used technique to remove BOD consisting complete oxidation of organic compound to CO_2 and water. Most commonly used aerobic reactors for the treatment of pulp and paper industry wastewater are aerated lagoon, aerated stabilization basin, activated sludge and sequential batch reactors. Wilson and Holloran (1992) reported 15–60% removal of AOX from bleach kraft mill effluent with an average removal of 30%. It was observed that LMW AOX was effectively removed in (43–63%) as compared to HMW AOX (4–31%) in aerated stabilization basins. (Bryant et al. 1992). Aerated stabilization basins do not require nutrient addition but their land requirement is high. Gergov et al. (1988) investigated pollutant removal efficiencies in mill scale biological treatment systems. They observed about 48–65% AOX removal in the activated sludge process. Sequential batch reactors have advantages of low operating costs, no sludge settler required, no recycling pumps required, good control over filamentous bulking, tolerance to shock loads and peak flow denitrification during anoxic fill and settle stage. However, the technology is not commonly used for treatment of AOX. Advanced aerobic reactors can bring about 80% AOX reduction. However, their energy requirement is high. Fulthorpe and Allen (1995) compared relative organochlorine removal from bleached kraft pulp and paper mill effluents by specific bacteria such as Pseudomonas P1, Ancylobacter aquaticus A7 and Methylobacterium CP13 and found that A. Aquaticus A7 exhibited broadest substrate range and highest AOX degradation in softwood effluents. Similar experiments with other micro-organisms showed that AOX removal efficiency can be increased with use of specific dechlorinating bacteria.

In pulp- and paper-industry wastewater treatment, anaerobic process is usually followed by aerobic post-treatment. The actual treatment systems depend on mill-specific conditions and especially on energy and sludge treatment and disposal costs. In many cases, only some of the waste streams of the pulp and paper integrate are treated in the anaerobic unit while the other streams are directly conducted into aerobic treatment, which often serves as post-treatment for the anaerobically treated streams. The role of the aerobic post-treatment is to degrade compounds which are non-degradable anaerobically and to reduce the effluent concentrations of the other compounds. For example, resin acids are aerobically degradable (Leach et al. 1978) while only low degradation has been achieved in the anaerobic process (Sierra-Alvarez et al. 1990a). Aerobic post-treatment of some anaerobically treated pulp- and paper-industry wastewater streams eliminates acute toxicity not removed in the anaerobic treatment (Schnell et al. 1990).

Pokhrel and Viraraghavan (2004) reviewed the methods used for overall treatment of pulp and paper mill wastewater including COD reduction, AOX and colour removal. These authors recommended a method consisting of anaerobic followed by aerobic treatment as a better option. They expressed certain reservations on utility of anaerobic processes for AOX degradation and hence recommended use of physico-chemical process. However, physico-chemical processes are uneconomical (Allen and Liu 1998; Savant et al. 2006). The biodegradation of chlorinated organic compounds under aerobic (Schmidt et al. 1983; Haggblom 1992; Jung-Hwa et al. 2002; Vasconcelos et al. 2006) and anaerobic (Mohn and Tiedje 1992; Boyd and Shelton 1994; Christiansen and Ahring 1996) conditions has been reported in the literature. Highly chlorinated compounds are degraded easily under anaerobic conditions rather than the aerobic one (Van Eekert and Schraa 2001). Savant et al. (2006) have extensively reviewed the methods employed for degradationor removal of AOX from PAP wastewater. These methods include physico-chemical, anaerobic and aerobic-biological as well as combination of these processes. On the basis of the comparison of process efficiencies these authors felt that anaerobic microbial degradation was the most economical and environment friendly process for the treatment of AOX containing wastewaters. Halorespiration and co-metabolisms are the common dechlorination mechanisms observed in anaerobic processes. Microbial reduction of organochlorine compounds essentially requires input of electrons to replace chlorine (Dolfing 1990). This reduction is accomplished through co-metabolism which is considered to be one of the preferred routes of degradation of priority pollutants (Richards and Shieh 1986). The electrons needed for reductive dehalogenation are generated from the oxidation of H2 (electron donor) which originates from the fermentation of organic compounds like glucose, acetate, butyrate etc. Supplementation of the electron donors provides thermodynamically most favourable conditions for microbial dechlorination reaction (Doong et al. 1996). Increase in microbial dechlorination of different chloroorganics such as chlorophenols, carbontetrachloride, chloroguaiacols, tetrachloroethylene etc. due to addition of electron donors is reported in the literature (Mohn and Tiedje 1992; Boyd and Shelton 1994; Christiansen and Ahring 1996; Gao et al. 1997; Van Eekert and Schraa 2001). Ali and Sreekrishnan (2000) had reported improvement in AOX degradation during the anaerobic treatment of bleach effluent when glucose was supplemented. This study, however, was limited to batch treatment. On the other hand Peijeie and Thomas (1994) had reported insignificant effect of electron donors on anaerobic AOX degradation in continuous mode of reactor operation. This suggests that more studies are essential to decide effect of addition of electron donor in AOX degradation in anaerobic treatment process. Savant et al. (2006) have traced the need to evolve new strategies for AOX degradation.

In general, the pulp- and paper-industry effluents are nutrient deficient. The addition of phosphorus and nitrogen is a common practice in activated-sludge treatment. A commonly used COD:N:P ratio in the activated-sludge plants is 100:5:1 while in the anaerobic plants a ratio of 350:5:1 has been used (Maat 1990). With chlorine-bleached effluents, anaerobic treatment could be advantageous. Reductive dechlorination has been shown to remove efficiently polychlorinated compounds (Woods et al. 1989). Furthermore, sulphate can be converted anaerobically to sulfide which can then be recovered as elemental sulphur by using either a chemical (Sarner 1990) or a biological (Buisman et al. 1988, 1991) method. The susceptibility of methanogens and acetogens to several inhibitors has been thought to limit the applicability of anaerobic systems for the pulp- and paper industry wastewaters (Lettinga et al. 1991). Both long-term and short-term toxicity problems have been associated with some failures in fullscale anaerobic-plant performances (Saslawsky 1988; Paasschens et al. 1991). Furthermore, technical problems such as clogging of the reactor feed inlets or the biomass support material have occurred during the introduction of the anaerobic system (Saslawsky et al. 1988; MacLean et al. 1990; Paasschens et al. 1991). Chandra (2001) reported efficient removal of color, BOD, COD, phenolics, and sulphide by microorganisms such as Pseudomonas putida,Citrobacter sp., and Enterobacter sp. in the activated sludge process.

However, it must be noted that the aerobic treatment systems used in the pulp and paper industries have also had operational problems, and, for example, large equalization basins have been built to insure constant wastewater characteristics. Both the investment and operation costs of treatment systems as well as their technical suitability greatly depend on each case of application. In general, anaerobic-aerobic systems have been reported to have lower operations costs than aerobic treatment because of the reduced aeration-energy equirements and excess sludge production, and because of the methane recovery (Rekunen et al. 1985; Minami et al. 1991; Garvie 1992). Peerbhoi (2000) investigated anaerobic treatability of black liquor by a UASB reactor in her study at the University of Roorkee, India. The author concluded that anaerobic biological treatment of black liquor was not feasible, as the pollutants were not readily degradable.

13.10.2.1 Up-flow Anaerobic Sludge Blanket (UASB) Reactor

UASB was developed at Wageningen Agricultural University, Netherlands. Lettinga et al. (1980) initiated the development of the first full scale installation of an up-flow anaerobic sludge blanket reactor (UASB) at the Central Sugar Manufacturing Plant in Netherlands. UASB employs anaerobic bacteria especially methanogens, which have capability to form self-immobilized granular structure with good settling properties inside the reactor. These anaerobic bacteria granules make a "blanket" through which the effluent flows up the reactor. The substrate present in the effluent diffuses into the sludge granule where it is degraded by

13 Pulp and Paper Industry—Manufacturing Process

Fig. 13.8 UASB reactor

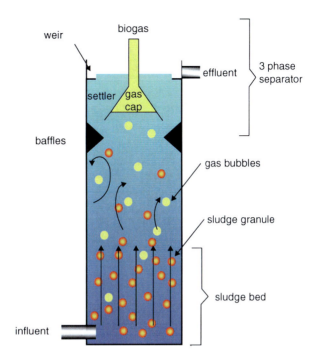

the anaerobic bacteria. Thus, these reactors due to their high biomass concentrations can achieve conversions several folds higher than that possible by the conventional anaerobic processes and tolerate fluctuations in the influent feed, temperature and pH. Anaerobic treatment resulted in higher AOX removal and better tolerance to shock loads. Studies showed that by using UASB reactor alone for AOX degradation removal efficiencies greater than 60% could be achieved but if it is used in combination with some aerobic process the removal efficiency up to 90% could be achieved. Figure 13.8 shows a typical diagram of a UASB reactor.

13.10.2.2 Sequencing Batch Reactors (SBR)

SBR systems have been successfully used for the treatment of both domestic and industrial wastewaters. It is aerobic process. SBR, as its name implies, is a batch reactor system, which treats wastewater in one vessel and accomplishes different events in a timed sequence.

In its most basic form, the SBR system is a set of tanks that operate on a fill-and draw basis. Each tank in the SBR system is filled during a discrete period of time and then operated as a batch reactor. After desired treatment, the mixed liquor is allowed to settle and the clarified supernatant is then drawn from the tank. The cycle for each tank in a typical SBR is divided into five discrete time periods: Fill, React, Settle,

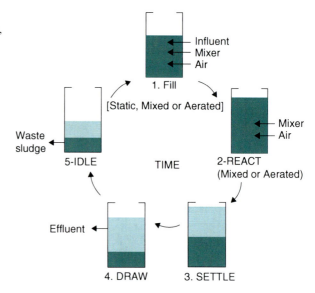

Fig. 13.9 SBR operation showing periods of fill, react, settle, draw, and idle

Draw and Idle as shown in Fig. 13.9. There are several types of Fill and React periods, which vary according to aeration and mixing procedures. Sludge wasting may take place near the end of React, or during Settle, Draw, or Idle. The SBR process is a flexible and high-performance treatment technology for wastewater treatment, especially for the pulp and paper mill wastewater. Sequential batch reactors have advantages of low operating costs, no sludge settler required, no recycling pumps required, good control over filamentous bulking and tolerance to shock loads. Studies shows that SBR was successfully applied for treating the paper mill effluent with high organic strength and the average removal efficiency greater than 90% could be achieved.

Tsang et al (2006) studied the effects of operating parameters, including mixed liquor suspended solid (MLSS) concentration, volumetric exchange rate (VER), aeration time, temperature and daily operation cycle on biological treatment of the pulp and paper mill effluent using four 4 l sequencing batch reactors (SBR). The results revealed that chemical oxygen demand (COD) removal efficiency was up to 93.1±0.3% and the volumetric loading reached 1.9 kgBODm^{-3}day^{-1} under optimal operating conditions. The treatment performance of organic removal by the SBR system remained stable during the operation. The effluent quality was satisfied with the discharge standard set by the local authority and the filamentous bulking problem was solved. At the same time, the sludge settleability, in terms of sludge volume index (SVI), was improved to the healthy level.

Qiu, Zet al (2006) demonstrated that the SBR technology could be successfully employed in the pulp and paper industry in China to treat wastewater. The two full scale SBR treatment plants described in this paper have been put on line for more than half a year. The performance data from one case study showed that SBR could handle higher than design load without compromising the effluent quality. In the second case, the flow fluctuation and the lack of the adequate sludge handling capability has impacted the performance of the SBR a little, but the overall effluent quality is good and most of the time is within discharge limits. The two case studies confirm that SBR technology treating pulp and paper wastewater is effective, that the operation of SBR basins is simple and flexible, and that the SBR is resilient to flow and load fluctuations.

Nair Indu C. et al (2007) carried out the treatment of the paper factory effluent with free and immobilized cells of a phenol degrading *Alcaligenes* sp. d2. The free cells could bring a maximum of 99% reduction in phenol and 40% reduction in chemical oxygen demand (COD) after 32 and 20 h of treatment, respectively. In the case of immobilized cells, a maximum of 99% phenol reduction and 70% COD reduction was attained after 20 h of treatment under batch process. In the continuous mode of operation using packed bed reactor, the strain was able to give 99% phenol removal and 92% COD reduction in 8 h of residence time.

Afzal, M et al. (2008) used pilot scale reactor based on combined biological–coagulation–filtration treatments and evaluated for the treatment of effluent from a paper and board mill. Biological treatment by fed batch reactor (FBR) followed by coagulation and sand filtration (SF) resulted in a total COD and BOD reduction of 93% and 96.5%, respectively. A significant reduction in both COD (90%) and BOD (92%) was also observed by sequencing batch reactor (SBR) process followed by coagulation and filtration. Untreated effluent was found to be toxic, whereas the treated effluents by either of the above two processes were found to be non-toxic when exposed to the fish for 72 h.

13.11 Water Conservation Options

The high water consumption in Indian pulp and paper industry is mainly due to obsolete process technology, poor water management practices and inadequate wastewater treatment. Water once used is generally thrown without any further use, even if the water is not much contaminated.

Segregation of wastewater from various processes into clean wastewater, (that can be reused) and contaminated water is therefore one of the very important step to be taken towards water conservation. This would avoid the uncontaminated water getting contaminated after mixing and is discharged as effluent. Another important step towards water conservation would be rainwater harvesting. This would help the industries to meet a substantial part of their annual water requirement even as demand on local sources is minimised. Water conservation options are as detailed below.

13.11.1 Use of Better Pulp Washing Technology Instead of Obsolete Technologies like Potcher Washing

Conventional potcher washing consumes huge quantity of water as it is a batch process. Compared with this continuous countercurrent processes consume less water. Usually three washings are applied in these processes. It is suggested to use fresh water only at the last stage (i.e. at third stage) and recycle the effluent in first and second washings. This system not only reduces water consumption but also allows recycle of wastewater to some extent leading to saving of bleaching chemicals.

Some of the continuous processes are: Hydraulic drum washing, Vacuum Drum washing, Pressure washing, Diffusion washing, Chemi or Belt washing, Twin roll press washer.

13.11.2 Optimum Use of Cooling Wastewater

In every mill, cooling with water is required at various sites like pump gland cooling/sealing, steam turbine cooling, compressor cooling, refiner gland cooling, rewinder brake cooling, etc. In most of the industries, fresh water is being used for pump jacket cooling on once through basis. This water is most of the time non-contaminated. There are different alternatives in which this water use can be optimized as described below:

1. Collection of once-through cooling water and reuse it in different process operation: Collection involves installation of several small sumps or tanks from which water is subsequently pumped to process water tanks. It is also possible to directly use the water for specific applications like shower systems. The higher temperature of cooling wastewater is advantageous to the shower system as it increases the water drainage property of the web.
2. Converting once-through system into a closed –loop system: This requires installation of cooling tower, a temperature controller and a cartridge filter to remove any suspended particles present in the waste cooling water. Further, periodic injection of fresh water is required as a make up for the evaporation loss. One such figure of a closed loop re-circulation system is shown in Fig. 13.10.

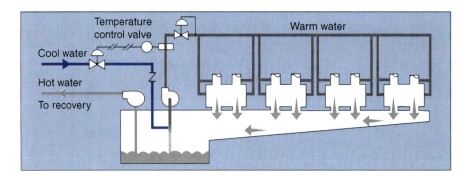

Fig. 13.10 Temperature controlled re-circulation system

13.11.3 Optimum Use of Paper Machine Clarified Wastewater in Sections Other than Paper Machine

In order to make maximum use of paper machine clarified wastewater, it is important to have a sufficient storage capacity. The requirement of backwater in other sections is not regular and is often intermittent. The various applications of this water are Decker thickener showers

- Vacuum washers
- Centri-cleaner reject dilution
- Pulp dilutions before bleaching stage, etc
- Johnson screen showers, etc.

Further, clarified water storage tank can be modified to enable further separation of fibres from the filtrate. One such design has been suggested by Arjo Wiggins called 'Stowford separator'. Figure 13.11 depicts stowford separator used as filtrate storage tank.

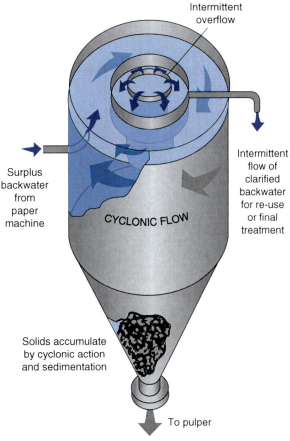

Fig. 13.11 The 'stowford' separator used as clarified wastewater storage tank

13.11.4 Recycling of Treated Effluent for Use Within the Mill for Non-Process

In most of the mills, fresh water is used for the following purposes:

- Plantation
- Gardening
- Floor washing and toilet flushing

Since quality of water is not essential for these activities, treated wastewater from ETP can be used. This will reduce fresh water consumption. The saving in fresh water consumption by this system is expected to be 50–100 m^3/d.

13.11.5 Tertiary Treatment of Wastewater for Recycling

In order to recycle the wastewater completely back to the system, biologically treated wastewater requires undergoing a number of treatment processes to achieve nearly inlet water quality. In a pulp and paper mill, colour removal is an important requirement to use the finally treated wastewater as process water. Figure 13.12 gives schematic diagram of treatment process. The wastewater after primary treatment involving physio-chemical separation is given secondary treatment. Secondary treatment uses microorganisms to accelerate the natural decomposition of organic waste. The two main methods used are aerated stabilisation and activated sludge treatment—both are known as aerobic treatments.

The efficiency of these two systems varies widely, depending on climate, influent quality, pulp type, fibre source and mill practice. In ideal conditions, activated sludge performs better at reducing BOD and suspended solids. Other methods include anaerobic processes like USAB processes followed by aerobic systems have been successfully been used in the pulp and paper industry.

The secondary treated wastewater is then sent to tertiary treatment plant. In tertiary treatment, Aluminum oxide, ferric oxide and polyelectrolytes may assist coagulation of waste in the effluents, which are then sand filtered.

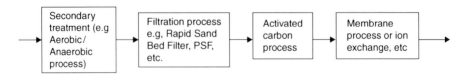

Fig. 13.12 Tertiary treatment options for pulp and paper waste treatment

References

Afzal M, Ghulam S, Irshad H, Zafar MK (2008) Paper and board mill effluent treatment with the combined biological–coagulation–filtration pilot scale reactor. Biores Tech 99:7383–7387

Ali M, Sreekrishnan, TR (2000) Anaerobic treatment of agricultural residue based pulp and paper mill effluents for AOX and COD reduction. Process Biochem 36:25–29

Allen DG, Liu HW (1998) Pulp mill effluent remediation. In: Meyers RA (ed) Encyclopedia of environmental analysis and remediation. Wiley, New York, pp 3871–3887

Anandarajah K, Kiefer, PM, Donohoe BS, Copley SD (2000) Recruitment of a double bond isomerase to serve as a reductive dehalogenase during biodegradation of pentachlorophenol. Biochemistry 39:5303–5311

Baig S, Liechti PA (2001) Ozone treatment for biorefractory COD removal. Sci Technol 43:197–204

Bajpai P, Bajpai PK (1997) Reduction of organochlorine compounds in bleach plant effluents. Adv Biochem Eng Biot 57:213–259

Benjamin MM, Woods SL, Ferguson JE (1984) Anaerobic toxicity and biodegradability of pulp mill waste Constituents. Wat Res 18:601–607

Blackwell BR, MacKay WB, Murray FE, Oldham WK (1979) Review of kraft foul condensates. Sources, quantities, chemical composition and environmental effects. TAPPI J 62:33–37

Boyd SA, Shelton DR (1994) Anaerobic biodegradation of chlorophenols in fresh and acclimated sludge. Appl Environ Microb 47:50–54

Bryant CW, Avenell JJ, Barkley WA, Thut RN (1992) The removal of chlorinated organics from conventional pulp and paper wastewater treatment systems. Water Sci Technol 26 (1–2):417–425

Buisman CJ, Witt B, Lettinga G (1988) A new biotechnological process for sulphide removal with sulphur production. In: Fifth international symposium on anaerobic digestion. Bologna, Italy, pp 19–22

Buisman CJN, Lettinga G, Paasschens CWM, Habets LHA (1991) Biotechnological sulphide removal from effluents. Wat Sci Tech 24:347–356

Burnow G (2001) Method to reveal the structure of lignin. Lignin humic substances and coal. In: Steinbuchel A, Hofrichter M, Steinbuchel A (eds) Biopolymers, vol 1. Wiley—VCH0, Germany, pp 89–116

Chandra R (2001) Microbial decolourisation of pulp mill effluent in presence of nitrogen and phosphorous by activated sludge process. J Environ Biol 22(1):23–7

Christiansen N, Ahring BK (1996) Desufitobacterium hafniense sp. nov. ananaerobic reductively dechlorinating bacterium. Int J Syst Bacteriol 46:442–448

Cornacchio LA, Hall ER (1988) Characteristics of CTMP wastewaters. Paper presented at Seminar on Treatment of Chemithermomechanical Pulping Effluents, 27 Oct 1988, Vancouver, British Columbia, Canada

Dellinger RW (1980) Development document for effluent limitations guidelines and standards for the pulp, paper and paperboard and the builders' paper and board mills.US Environmental Protection Agency Report EPA-440/025-b, Washington DC, USA

Dolfing J (1990) Reductive dechlorination of 3-chloro-benzoate is coupled to ATP production and growth in an anaerobic bacterium, DCB-1. Arch Microbiol 153:264–266

Doong RA, Chen TF, Wu YW (1996) Effects of electron donor and microbial concentration on the enhanced dechlorination of carbon tetrachloride by anaerobic consortia. Appl Microbiol Biot 46:183–186

Dube M, McLean R, MacLatchy D, Savage P (2000) Reverse osmosis treatment: effects on effluent quality. Pulp Pap Can 101(8):42–45

FAOSTAT—Forestry, Food and Agriculture Organization of the United Nations (2008) <http://faostat.fao.org/site/630/default.aspx>. Cited January 2008

Field JA, Leyendeckers MJH, Sierra-Alvarez R, Lettinga G, Habets LHA (1988) The methanogenic toxicity of bark tannins and the anaerobic biodegradability of water soluble bark matter. Wat Sci Technol 20:219–240

Fulthorpe RR, Allen DG (1995) A comparison of organochlorine removal from bleached Kraft pulp and paper-mill effluents by dehalogenating Pseudomonas, Ancylobacter and Methylobacterium strains. Appl Microbiol Biotechnol 42(5):782–789

Ganjidoust H, Tatsumi K, Yamagishi T, Gohlian R (1996) Effect of synthetic and natural coagulants on lignin removal from pulp and paper wastewater. In: Proceedings of 5th IAWQ Symposium on Industrial Wastewaters. Vancouver, Canada, p 305

Gao J, Skeen RS, Hooker BS, Quesenberry RD (1997) Effect of several electron donors on tetrachloroethylene dechlorination in anaerobic soil microcosms. Water Res 31:2479–2486

Garvie R (1992) Anaerobic/aerobic treatment of NSSC/ CTMP effluent and biogas utilization. Pulp & Paper Canada J 93:56–60

Gebart R (2006) The BLG2 program: BLG—enabling technology for renewable transportation fuels. Application for financial support to MISTRA

Gergov M, Priha M, Talka E, Valtilla O, Kanges A, Kukkonen K (1988) Chlorinated organic compounds in effluent treatment of kraft mill. Tappi J 71(12):175–184

Gullichsen J (1991) Process internal measures to reduce pulp mill pollution load. Wat Sci Technol 24:45–53

Gune NV (2000) Total water management in pulp and paper industry with focus on achieving 'zero effluent discharge' status. IPPTA J 12(4):137–42

Haggblom MM (1992) Microbial breakdown of halogenated aromatic pesticides and related compounds. FEMS Microbiol Rev 103:29–72

Hilleke J (1993) In: Springer AM (ed) Industrial environmental control—pulp and paper industry. Tappi press, Atlanta, GA, pp 304–343

Hostachy JC, Lenon G, Pisicchio JL, Coste G, Legay C (1997) Reduction of pulp and paper mill pollution by ozone treatment. Water Sci Technol 35(2–3):261–268

Jarvinen R, Vahtila M, Mannstr6m B, Sundholm J (1980) Reduce environmental load in TMP. Pulp & Paper Canada 81:39–43

Jarvinen R, Langi A, Anhava J, Savolainen E (1981) Reduction of the pollution load from an integrated mechanical pulp and paper mill. SITRA report TESI 1.1.,SITRA, Helsinki, Finland (in Finnish)

Jokela JK, Salkinoja-Salonen MS (1992) Molecular weight distributions of organic halogens in bleached kraft pulp mill effluents. Environ Sci Tech 26:1190–1197

Jung-Hwa K, Keun K, Sung Taik OL, Seung-Wook K, Suk H (2002) Biodegradation of phenol and chlorophenols with defined mixed culture in shake-flasks and a packed bed reactor. Process Biochem 37:1367–1373

Jurgensen ME, Patton JT (1979) Bioremoval of lignosulfonates from sulphite pulp mill effluents. Process Biochem 14:2–4

Kringstad KP, Lindstrm K (1984) Spent liquors from pulp bleaching. Environ Sci Technol J 18:236–248

Leach JM, Mueller JC, Walden CC (1978) Biological detoxification of pulp mill effluents. Process Biochem 13:18–21

Lettinga G, Field JA, Sierra-Alvarez R, van Lier JB, Rintala J (1991) Future perspectives for the anaerobic treatment of forest industry wastewaters. Wat Sci Tech 24:91–102

Maat DZ (1990) Anaerobic treatment of pulp and paper effluents. In: Proceedings TAPPI (1990) Environmental Conference, Atlanta, GA, p 757

MacLean B, de Vegt A, van Driel E (1990) Full-scale anaerobic/aerobic treatment of TMP/BCTMP effluent at Quesnel River Pulp. In: Proceedings TAPPI 1990 Environmental Conference, Atlanta, GA, p 647

McLeay DJ (1987) Aquatic toxicity of pulp and paper mill effluent: a review. Environment Canada Report EPS 4/PF/1. Ottawa, Ontario, Canada

Milosevich GM, Hill DA (1991) Reduction of AOX in bleach plant effluent by addition of mill process alkalis. In: Proceedings of 77th annual meeting of canadian pulp and paper association. Montreal, Canada, pp 309–318

13 Pulp and Paper Industry—Manufacturing Process 435

Milstein O, Haars A, Krause F, Huettermann A (1991) Decrease of pollutant level of bleaching effluents and winning valuable products by successive flocculation and microbial growth. Water Sci Technol 24(3/4):199–206

Minami K, Okamura K, Shigemichi O, Naritomi T (1991) Continuous anaerobic treatment of wastewater from a kraft pulp mill. Ferment Bioeng 74:270–274

Mobius CH, Helble A (2004) Combined ozonation and biofilm treatment for reuse of paper mill wastewaters. Water Sci Technol 49(4):319–323

Mohn WW, Tiedje JM (1992) Microbial reductive dehalogenation. Microbiol. Rev 56:482–507

Moiseev A, Schroeder H, Kotsaridou-Nagel M, Geissen SU, Vogelpohl A (2004) Photocatalytical polishing of paper-Mill effluents. Water Sci Technol 49(4):325–330

Nair Indu C, Jayachandran K, Shashidhar S (2007) Treatment of paper factory effluent using a phenol degrading Alcaligenes sp. under free and immobilized conditions Biores Tech 98:714–716

Nancy SJ, Norman JC, Vandenbusch MB (1996) Removing colour and chlorinated organics from pulp mill bleach plant effluent by use of Flyash. Resour Conserv Recycl 10(4):279–299

Nemerow NL, Dasgupta A (1991) Industrial and hazardous waste management. New York

Paasschens CWM, de Vegt AL, Habets LHA (1991) Five years full scale experiences with anaerobic treatment of recycled paper mill effluent at Industriewater Eerbeek in the Netherlands. In: Proceedings TAPPI 1991 Environmental Conference, Atlanta, GA, pp 879

Pandey GN, Carney GC (1998) Environment engineering. Tata McGraw-Hill Publishing Company, New Delhi, p 346

Peerbhoi Z (2000) Treatability studies of black liquor by UASBR-PhD thesis. University of Roorkee, India

Peijeie Y, Thomas W (1994) Anaerobic treatment of kraft bleaching plant effluent. Appl Microbio Biot 40:806–811

Pfister K, Sjstrm E (1977) The formation of monosaccharides and aldonic and uronic acids during sulphite cooking. Paperi ja Puu—Paper Timber 59:711–720

Pichon M, Rouger J, Junet E (1988) Anaerobic treatment of sulphur containing effluents. Wat Sci Tech 20:133–141

Pokhrel D, Viraraghavan T (2004) Treatment of pulp and paper mill wastewater—a review. Sci Total Environ 333:37–58

Profile of the Pulp and Paper Industry (2002) EPA Office of Compliance Sector Notebook Project, 2nd edn. U.S. Environmental Protection Agency

Qiu Z, McCarthy PJ, Brown GJ, Zhang Y, Wang H (2006) Treatment of Pulp and Paper Effluent in China with SBR Technology. Wat Pract Tech 1(3). doi:10.2166/WPT.2006067

Rekunen S, Kallio O, Nystrm T, Oivanen O (1985) The TAMAN anaerobic process for wastewater from mechanical pulp and paper production. Wat Sci Tech 17:133–144

Rexfelt J, Samuelson O (1970) The composition of condensates from the evaporation of sulfite spent liquor. Svensk Papperstidnin 73:689–695

Richards DJ, Shieh WK (1986) Biological fate of organic priority pollutants in the aquatic environment. Water Res. 9:1077–1090

Ruus L (1964) A study of wastewater from the forest products industry: 4. Composition and biochemical oxygen demand of condensate from spent Sulfite liquor evaporation. Svensk Papperstidning 67:221–225

Rydholm SA (1965) Pulping Processes, Interscience Publishers, New York

Sarner E (1990) Removal of sulphate and sulphite in an anaerobic trickling (ANTRIC) filter. Wat Sci Tech 22:395–404

Saslawsky J, Liziard Y, Chave E (1988) Anaerobic treatment of evaporator condensates in a sulfite pulp mill. In: Hall ER, Hobson PN (eds) Anaerobic digestion. Pergamon Press, New York, pp 499–505

Savant DV, Abdul-Rahman R, Ranade DR (2006) Anaerobic degradation of Adsorbable Organic Halides (AOX) from pulp and paper industry wastewater Biores Tech 97:1092–1104

Saxena AK, Rajat G, Development of guidelines for water conservation in pulp and paper sector (2006) Environment group, National productivity council, New Delhi

Schmidt E, Hellwig M, Knackmuss HJ (1983) Degradation of chlorophenols by a defined mixed microbial community. Appl Environ Microb 46:1038–1944

Schnell A, Dorica J, Ho C, Ashikawa M, Munnoch G, Hall ER (1990) Anaerobic and aerobic pilot-scale effluent detoxification studies at an integrated newsprint mill. Pulp Paper Canada 91:75–80

Schnell A, Stell P, Melcer H, Hudson PV, Carey JH (2000) Enhanced biological treatment of bleached kraft mill effluents removal of chlorinated organic compounds and toxicity. Water Res 34:493–500

Seiss M, Gahr A, Neissner R (2001) Improved AOX degradation in UV oxidative wastewater treatment by dialysis with nano-filtration membrane. Water Res 35(13):3242–3248

Shawwa AR, Smith DW, Sego DC (2001) Color and chlorinated organics removal from pulp mills wastewater using ctivated petroleum coke. Water Res 35(3):745–749

Sierra-Alvarez R (1990) The role of natural wood constituents on the anaerobic treatability of forest industry wastewaters. PhD thesis, Wageningen Agricultural University, Wageningen, The Netherlands

Sinclair WF (1990) Controlling pollution from Canadian pulp and paper manufactures: a federal perspective. Canadian Government Publishing Centre, Ottawa

Sodergren A (1993) Bleached pulp mill effluent; composition, fate and effects in the Baltic Sea: Swedish. Environmental Protection Agency Report, p 4047

Springer AM, Hand VC, Jarvis TS (1994) Electrochemical decolourization of bleached kraft effluents. In: Proceedings of 1994 Tappi international environmental conference. Portland, OR, pp 271–279

Stephensen RJ, Duff SJB (1996) Coagulation and precipitation of a mechanical pulping effluent— I. Removal of carbon, colour and turbidity. Water Res 30(4):781–792

Suntio LR, Shiu WY, Mackay D (1988) A review of the nature and properties of chemicals present in pulp mill effluents. Chemosphere 17:1249–1290

Swedish forest agency, Forestry statistics. Jönköping, Sweden (2008). www.svo.se Cited 15 Dec 2008

Thomson G, Swain J, Kay M, Forster CF (2001) The treatment of pulp and paper mill effluent: a review. Bioresour Technol 77:275–286

Torrades F, Peral J, Perez M, Domenech X, Garcia Hortal JA, Riva MC (2001) Removal of organic contaminants in bleached kraft effluents using heterogenous photocatalysis and ozone. Tappi J. 84(6):63–64

Tsang YF, Hua FL, Chua H, Sin SN, Wang YJ (2006) Optimization of biological treatment of paper mill effluent in a sequencing batch reactor. Biochem Engg J 34:193–199

Van Eekert MHA, Schraa G (2001) The potential of anaerobic bacteria to degrade chlorinated compounds. Water Sci Technol 44:49–56

Vasconcelos De LC, Santoyo D, Tepole F, Juarez Ramírez C, Ruiz-Ordaz N, Galíndez Mayer CJJ (2006) Cometabolic degradation of chlorophenols by a strain of Burkholderia in fed-batch culture. Enzyme Microb Technol 40:57–60

Virkola N-E, Honkanen K (1985) Wastewater characteristics. Wat Sci Technol 17:1–28

Voss RH, Wearing JT, Mortimer RD, Kovacs T, Wong A (1980) Chlorinated organics in kraft bleachery effluents. Paperi ja Puu—Paper and Timber 62:809–814

Walden CC, Howard TE (1981) Toxicity of pulp and paper mill effluents—a review. Pulp & Paper Canada 82:143–147

Water Wilson DG, Holloran MF (1992) Decrease of AOX with various external effluent treatments. Pulp Pap Canad 93:T372–378

Wingate KG, Robinson MJ, Stuthridge TR, Collins TJ, Wright IJ (2001) Colour and AOX removal from bleached kraft mill wastewaters using catalytic peroxide activators. In: Proceedings of 2001 International Environmental, Health and Safety Conference and Exhibit. Tappi Press, Atlanta, GA, USA

Woods SL, Ferguson JE, Benjamin MM (1989) Characterization of chlorophenol and chloromethoxybenzene biodegradation during anaerobic treatment. Environ Sci Tech 23:62–68

Yao WX, Kennedy KJ, Tam CM, Hazlett JD (1994) Pretreatment of kraft pulp bleach plant effluent by selected ultrafiltration membranes. Canad J Chem Eng 72(6):991–999

Chapter 14
A Review of Environmental Contamination and Remediation Strategies for Heavy Metals at Shooting Range Soils

Mahtab Ahmad, Sang Soo Lee, Deok Hyun Moon, Jae E. Yang
and Yong Sik Ok

Contents

14.1	Shooting Ranges: A Menace to the Environment	438
	14.1.1 Backgrounds	438
	14.1.2 Environmental Concerns	439
	14.1.3 Environmental Management Practices (EMPs)	441
14.2	Remediation of Shooting Ranges Soil	442
	14.2.1 Soil Characteristics and Remediation Technologies	442
14.3	Remediation of Shooting Range Soils Using Lime-Based Waste Materials	444
	14.3.1 Lead Immobilization Using Waste Materials	445
	14.3.2 Use of Lime-Based Waste Materials	446
14.4	Conclusion	446
References		447

Abstract Many shooting ranges are contaminated by heavy metals and the used bullets have been known as a primary source. Once the bullets perch on soils, toxic metals such as lead (Pb), copper (Cu), nickel (Ni), antimony (Sb), and zinc (Zn) can be released into the soils and further transformed into available forms threatening the surrounding environment. In this review, we evaluated different sources of waste materials as soil amendments to stabilize heavy metals in soils. Amendments such as red mud, sugar foam, poultry waste, and dolomitic residue have been used to stabilize Pb at shooting ranges. Among various amendments, lime-based waste materials such as oyster shell and eggshell can effectively immobilize heavy metals, thereby reducing their bioavailability in soils. The main mechanism of Pb immobilization is closely associated with sorption and precipitation at high soil pH. Calcium aluminate hydrate (CAH) and calcium silicate hydrate (CSH) also can be formed to retain the metals in hardened soils. Overall, the use of lime-based wastes is applicable to immobilize toxic metals at shooting range soils.

Y. S. Ok (✉)
Department of Biological Environment, Kangwon National University, Chuncheon 200-701, Korea
e-mail: soilok@kangwon.ac.kr

A. Malik, E. Grohmann (eds.), *Environmental Protection Strategies for Sustainable Development,* Strategies for Sustainability,
DOI 10.1007/978-94-007-1591-2_14, © Springer Science+Business Media B.V. 2012

437

Keywords Heavy metals • Environmental contamination • Shooting range • Remediation • Immobilization

14.1 Shooting Ranges: A Menace to the Environment

14.1.1 Backgrounds

Shooting ranges, which are also known as firing ranges, are designed to use firearms for purposes of military training or public recreation (Scheetz and Rimstidt 2009). To secure a sufficient distance or to reduce noise pollution, most shooting ranges are constructed as outdoor facilities and are located in desolate areas. In some cases, the indoor shooting ranges are situated at urban areas for public recreation purpose. All firearms such as rifles, pistols, shot guns, and machine guns must be used at properly designed shooting ranges as shown in Fig. 14.1 (Tardy et al. 2003; USEPA 2003). Depending on types of firearm and capacity of firing station, a sufficient distance and width in shooting ranges vary. Moveable or fixed target is made of paper, plastic or metal, and lead (Pb) is mostly used as a bullet material because of its versatility and performance. A backstop with heights from 5 to 8 m is distinguished by building materials that consist of earthen berms, sand traps, steel traps, or rubber traps behind a target to gather bullets and bullet fragments (ITRC 2005). However, absorbed bullets and their fragments can be weathered over time by process in metal transformation, thereby contaminating soils and surrounding environments.

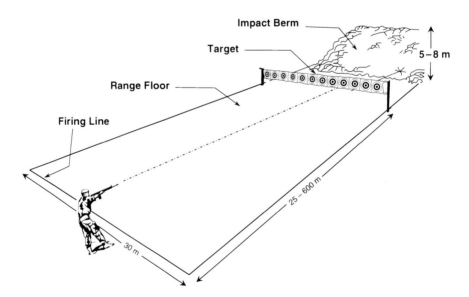

Fig. 14.1 A typical shooting range (modified from Tardy et al. 2003)

14.1.2 Environmental Concerns

14.1.2.1 Heavy Metal Contamination

Used bullets at shooting ranges are regarded as a source of soil pollution due to the release of metals and metalloids into surrounding environments (Cao et al. 2003a; Dermatas et al. 2006a; Sorvari et al. 2006). A bullet pellet typically consists of lead (Pb; >90%), antimony (Sb; 2–7%), arsenic (As; 0.5–2%), nickel (Ni; <0.5%), and traces of bismuth (Bi) and silver (Ag) (Chrastný et al. 2010; Dermatas et al. 2006a; Robinson et al. 2008; Sorvari et al. 2006), and a bullet jacket is mainly made of copper (Cu; 89–95%), and zinc (Zn; 5–10%) (Robinson et al. 2008; USEPA 2003). Once a bullet and its fragments perch on the soil, these materials can be dissolved by chemical reactions such as oxidation, carbonation, and hydration (Ma et al. 2007). Soils at military shooting ranges are also contaminated with energetic compounds such as trinitrotoluene (TNT; $C_6H_2[NO_2]_3CH_3$), dinitrotoluene (DNT; $C_6H_3[CH_3]$ $[NO_2]_2$), trinitrobenzene (TNB; $C_6H_3N_3O_6$), nitroglycerin (NG; $C_3H_5N_3O_9$), and nitrocellulose (NC; $C_6H_7[NO_2]_3O_5$) (Berthelot et al. 2008; Ragnvaldsson et al. 2007).

Shooting ranges are world widely distributed as shown in Table 14.1. About 60,000 MT of Pb is being deposited every year from ammunition in the United States (Ma et al. 2002). In addition, several studies reported that heavy metals such as Pb, Sb, and Cu are highly concentrated in shooting range soils as shown in Table 14.2. According to the USGS report in 2002, shooting ranges are the second largest source of Pb contamination in the United Sates. Generally, factors such as the firing period and frequency, a type of used ammunition, soil properties, climate, and management practices influence on a soil contamination level at shooting ranges. For example, at common shooting ranges, the concentrations of Pb reached serious levels ranging from 385 to 49,228 mg kg^{-1} in the United States, and from 8,684 to 29,200 mg kg^{-1} in East Asian countries. However, the importance of heavy metal contamination in the shooting range soils, particularly Pb, has been reported from only few countries such as the United States and Scandinavian Peninsula.

Table 14.1 Lead (Pb) deposition at shooting ranges

Country	Pb deposition (ton y^{-1})	Number of shooting ranges	Reference
Canada	10–30	NA	Scheuhammer and Norris (1996)
Denmark	800	NA	Lin (1996)
Finland	530	2,000–2,500	Hartikainen and Kerko (2009)
Korea	267	>690	KMND[a] (2002)
Norway	103	NA	Heier et al. (2009)
Sweden	500–600	NA	Lin (1996)
Switzerland	400–500	>2,000	Johnson et al. (2005)
United States	60,000	>10,000	Cao et al. (2008)

NA Not available

[a] The Korea Ministry of National Defense

440 M. Ahmad et al.

Table 14.2 Total contents of lead (Pb), antimony (Sb) and copper (Cu) at shooting ranges

Country	Pb (ton y^{-1})	Sb (ton y^{-1})	Cu (ton y^{-1})	Reference
Canada	100–14,600	3.5–314	330–835	Ragnvaldsson et al. (2007)
Finland	15,500–41,800	NA	NA	Hartikainen and Kerko (2009)
Germany	16,760	437	817	Spuller et al. (2007)
Korea	166	NA	161	Lee et al. (2002)
	8,684	NA	285	Moon et al. (2010)
Japan	29,200	NA	NA	Hashimoto et al. (2009)
Sweden	71–24,500	NA	NA	Lin (1996)
Switzerland	110–67,860	5–3,020	20–2,250	Vantelon et al. (2005)
	1,450–515,800	35–1,750	100–445	Johnson et al. (2005)
	12,533	675	149	Knechtenhofer et al. (2003)
United States	2,520–35,868	NA	NA	Cao et al. (2008)
	3,165	NA	NA	Dermatas et al. (2006a)
	12,710–48,400	NA	NA	Cao et al. (2003a)
	1,025–49,228	NA	NA	Dermatas et al. (2006b)
	385–12,400	NA	NA	Labare et al. (2004)

NA Not available

14.1.2.2 Heavy Metal Mobility

High concentration of Pb from shooting range soils is commonly observed due to increased solubility and bioavailability of Pb pellets. In the past, metallic Pb in bullets and bullet fragments has been assumed as a material that is naturally stable in the soil and it has no detrimental impact on the environment (Ma et al. 2002). However, recent studies found that metallic Pb in the soils can be transformed into active Pb species, thereby increasing the mobility of Pb in the surrounding environments (Astrup et al. 1999; Cao et al. 2003a, 2008; Ma et al. 2007).

Lead transformation mechanism has been well documented by Ma et al. (2007). Once, a bullet settles into soils, the weathering of Pb pellets is being processed with exposure to air and water. Oxidation of metallic Pb results in the formation of crust around bullet fragments consisting of secondary minerals such as massicot (PbO), cerussite ($PbCO_3$), hydrocerussite ($Pb_3[CO_3]_2[OH]_2$), and anglesite ($PbSO_4$). Dissolution of these secondary minerals has also occurred by the mobilization of Pb (Cao et al. 2003a; Chrastný et al. 2010; Ma et al. 2007). Weathering and transformation of Pb pellets are affected by factors such as soil particle size, water flow rate, soil pH, redox-potential, available anion, cation exchange capacity (CEC), soil organic matter (SOM), and carbon dioxide (CO_2) partial pressure in soil solution (Cao et al. 2003b; Dermatas et al. 2008; Heier et al. 2009; Manninen and Tanskanen 1993; Martin et al. 2008; Sorvari et al. 2006; Takamatsu et al. 2010). Generally, Pb in soils may be readily mobilized at the acidic condition compared to the neutral or alkali condition (Manninen and Tanskanen 1993; Ok et al. 2007b). Enriched soil with high SOM leads to increase the weathering rate of Pb pellets (Dermatas et al. 2006b; Heier et al. 2009). Mobility of Pb can also be significantly higher in the sandy textured-soil than the other textured-soils (Sorvari et al. 2006). Spuller et al. (2007) reported that iron hydroxides ($Fe[OH]_2$ and $Fe[OH]_3$) and secondary mineral

14   A Review of Environmental Contamination and Remediation Strategies   441

phases ($Ca[Sb(OH)_6]_2$ and $Pb[Sb(OH)_6]_2$) control the Sb mobility in shooting range soils. The high mobility of metal elements from the weathering of bullets and bullet fragments increases the bioavailability of metals (Ma et al. 2007).

14.1.2.3   Environmental Risks

Soluble phase of heavy metals induces fatal environmental risks due to its high mobility and bioavailability. Surface runoff and erosion may cause migration of a relatively less-soluble form of heavy metals in shooting range soils. The Pb is a toxic heavy metal that generates the most concern in shooting range soils. The bioavailability of Pb depends on its species in the soil. As mentioned above, metallic Pb which is contained in bullets and bullet fragments is being weathered and transformed into secondary minerals, especially $PbCO_3$ which is a highly bioavailable form (USEPA 2003). The bioavailability of Pb can be measured by the juvenile swine test (as mimic childhood Pb exposure), stable Pb isotope technique (Pb ingestion in adults), and physiological extraction test (PBET; *in vitro* test) (USEPA 2003). Investigation of metal leachability also determines an environmental risk (Cao et al. 2008). The United State Environmental Protection Agency (USEPA 1992, 1994) has developed several procedures to analyze the heavy metal leachability such as the toxicity characteristics leaching procedure (TCLP) and the synthetic precipitation leaching procedure (SPLP). The TCLP is often used to simulate specific conditions in a landfill, while the SPLP is used to reproduce the acid rain condition. An environmental risk at shooting ranges has been determined from contamination levels of groundwater and surface water (Heier et al. 2009; Sorvari et al. 2006), soil enzymatic activity (Lee et al. 2002), and accumulation of heavy metals into plant tissues (Labare et al. 2004; Ma et al. 2002; Mozafar et al. 2002; Rooney et al. 1999), regarding to a human being or other animals (Braun et al. 1997; Gulson et al. 2002; Migliorini et al. 2004; Wixson and Davies 1994). Nowadays, concerns about heavy metal contamination at shooting range soils have been rapidly raised. In 1993, Pb residues in shooting range soils were defined as hazardous materials by the United States Court of Appeals (USCA), the Resource Conservation and Recovery Act (RCRA) (USEPA 2001). However, no strict regulation has been implemented in most of the countries.

14.1.3   Environmental Management Practices (EMPs)

A potential risk from contaminated shooting range soils needs to be managed by range owners cooperated with a trained specialist under controls of regulatory authorities and environmental management practices (EMPs). The EMPs mainly focus on (i) recycling of used ammunition, (ii) prevention of groundwater and surface water contaminations by Pb, (iii) removal and recycling of Pb from a backstop, and (iv) remediation of shooting range soils which are contaminated with heavy metals (Cohen 2000; FDEP 2004). Investigation of shooting range soils should be periodi-

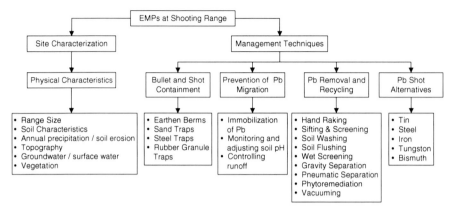

Fig. 14.2 Environmental management practices (EMPs) at outdoor shooting ranges (FDEP 2004)

cally implemented following Fig. 14.2 (FDEP 2004). In addition, the prevention of Pb migration into the soils by means of the immobilization or stabilization is the most essential part of remediation procedure (Ok et al. 2010a, 2011a). Physical removal of Pb from a backstop and exploration of its alternative such as steel (Fe compound), tungsten (W), and tin (Sn) have been suggested; however, these remediation managements require additional safety and preceded verification (FDEP 2004).

14.2 Remediation of Shooting Ranges Soil

Concentration of Pb in shooting range soils usually exceeds 1% due to the weathering of used Pb bullets (Levonmäki et al. 2006). Indeed, the analytical results commonly showed >20% Pb concentration in shooting range soils (Dermatas et al. 2006a; Lin 1996). The Pb may not be biologically degraded whereas it can readily be transformed into oxidation states or organic complexes (Garbisu and Alkorta 2001).

14.2.1 Soil Characteristics and Remediation Technologies

At a shooting range, soil physiochemical properties should be considered for the selection of a proper remediation technique. Soil pH is an important soil characteristic that can be mostly affecting the bioavailability of heavy metals (Jin et al. 2005; Ok et al. 2007a, c). Range of optimum soil pHs in shooting range soils is from 6.5 to 8.5 because Pb may be readily reacted and became a highly mobile material under the acidic pHs <6.5 (Cao et al. 2003a) or alkaline pHs >8 thereby inducing the rapid weathering of Pb in contaminated shooting range soils (USEPA 2001).

Some anionic ligands, especially phosphate (PO_4^{3-}) and carbonate (CO_3^{2-}), are very effective in controlling Pb solubility because of the formation of less soluble forms which are stimulating Pb stabilization. Soil texture and SOM content are also critical factors for heavy metal remediation. Soils which have high clay and silt contents may have difficulty to extract heavy metals because these can be easily absorbed by iron-manganese oxide (Fe-Mn oxide) at soil particle surface. Moreover, specific site conditions such as bedrock appearance, large boulder clay, soil moisture content, and oily patch should be considered for the selection of proper remediation practices (Evanko and Dzombak 1997).

Excavation is a physical removal of surface soil contaminated with heavy metals and reburial in special landfills (Conder et al. 2001; Jing et al. 2007; Lanphear et al. 2003). This ex-situ technique is often used for closed or inactive shooting ranges. Sorvari et al. (2006) reported that >90% of shooting range soils in Finland has been remediated by the excavation management practice. Contaminants from shooting range soils can be permanently removed; however, the excavation practice is too costly and produces the secondary environmental problems such as loss of biota or habitat, and dust pollution.

Soil washing is an ex-situ technique to remediate the soils contaminated with heavy metals (Khan et al. 2004; Peters 1999). Basically, the soil washing technique isolates contaminants from soils using water. A mixture of water and synthesized complex agents such as ethylendiaminetetraaceticacid (EDTA) and nitrilotriacetate (NTA) is occasionally used to enhance the removal ability of heavy metals from soils (Arwidsson et al. 2010; USEPA 1991). As a remediation technique at shooting ranges, advantages of soil washing technique have been reported by Fristad (2006) and Lin et al. (2001). Lin et al. (2001) found that >0.15 mm metallic Pb particles are effectively removed by gravity separation and the solution of sodium chloride (NaCl) with sodium hypochlorite (NaOCl) as an oxidant is an effective way to remove <0.15 mm metallic Pb particles from the soils. However, this technique may generate the secondary environmental pollution and is not effective on soils having high clay and SOM contents.

Electrokinetic remediation is an in-situ technique using a direct current with low voltage into the soil. Once applying electricity into contaminated soils, heavy metals migrate to electrode chamber by the electromigration reaction (Cang et al. 2009). Solution or water can be used to enhance the electromigration reaction (Smith et al. 1995). Braida et al. (2007) conducted a remediation experiment using a 30 V absolute voltage at contaminated shooting range soils. They found that heavy metals of W, Cu, and Pb in shooting range soils are effectively removed or stabilized; however, the effectiveness of electrokinetic remediation depends on soil bulk density, soil particle size, and heavy metal mobility in soils.

Phytoremediation is an in-situ technique using specific plant species to extract heavy metals from contaminated soils, which is also known as phytoextraction (Bennett et al. 2003; Ok and Kim 2007), or to physically stabilize contaminated soils adjacent to the aboveground root system, as also known as phytostabilization (Jing et al. 2007). Wilde et al. (2005) showed that the use of *Vetiver* grass combined with chelating agents (EDTA) is a well suited remediation technique for shooting ranges soils. Hashimoto et al. (2008) found that the use of proper plant species with poultry waste as an immobilizing agent reduces Pb leaching and enhances the stabilization of Pb in shooting range soils.

Stabilization/Solidification (S/S) technique chemically stabilizes and physically encapsulates heavy metal contaminants in the soils (Cao et al. 2008; Ok et al. 2010b). The heavy metals can be transformed into less soluble forms using the S/S technique, thereby reducing a potential health risk. The S/S technique induces physical heavy metal encapsulation in a solid matrix using a pozzolanic solidification system or using chemical additives (Cao et al. 2008). Applications of P, zeolite, lime, and Fe-Mn oxides have been considered to stabilize Pb in shooting range soils (Alpaslan and Yukselen 2002; Ma et al. 1995). Solidification of Pb using lime-based materials and the formation of crystalline phases (less soluble forms) which are involved in the stabilization of Pb can be effective ways as one of the S/S remediation techniques (Cao et al. 2008; Moon et al. 2010). For the formation of crystalline phases, the reaction of calcium hydroxide (CH; $Ca[OH]_2$) and silicic acid (SH; $[SiO_x(OH)_{4-2x}]_n$) is required to produce calcium silicate hydrate (CSH; $CaH_2SiO_2.2H_2O$) and calcium aluminate hydrate (CAH; $CaOAl_2O_3.10H_2O$) as solidification process. The Pb can also be precipitated as a metal hydroxide (OH^-) under the alkaline pH condition associated with cementitious lime materials, thereby reducing the Pb bioavailability (Moon and Dermatas 2006; Palomo and Palacios 2003). To ensure the effectiveness of S/S technique, sequential or ascertaining extraction tests are commonly used (Ruby et al. 1996; Ryan et al. 2001).

14.3 Remediation of Shooting Range Soils Using Lime-Based Waste Materials

Soils contaminated with heavy metals pose a potential threat to the environment. In recent, exploration of cost-effective remediation techniques has received increased attention. The S/S techniques were recommended as the Best Demonstrated Available Technology (BDAT) by the USEPA and are being widely used at contaminated soils with heavy metals (Singh and Pant 2006). To accomplish stabilization or immobilization, several soil amendments were commonly used to contaminated soils for its remediation, revitalization, or recycle. Soil amendments are achieved from various sources of natural materials, industrial or agricultural by-products, and artificial or synthetic materials. Recently, the use of a natural or waste material is highlighted as a cost-effective and friendly-environmental remediation method (Guo et al. 2006).

The biologically available fraction of metal can be taken up by organisms depending on metabolic activities and metal bioavailability (Geebelen et al. 2003). Soil amendment may reduce the bioavailability of metals in the soils by immobilization. Factors such as cation exchange, adsorption, precipitation, and surface complexation mainly influence on the metal immobilization in the soils (Chen et al.

2009; Guo et al. 2006; Kumpiene et al. 2008). However, different mobilities of metals and distinct organism speciation in the soils indicate difficulty to determine an appropriate soil amendment for simultaneously immobilizing various metals. Therefore, the exploration and selection of cost-effective and suitable soil amendments are critical to success.

14.3.1 Lead Immobilization Using Waste Materials

Most shooting range soils are contaminated by Pb which is contained in used bullets through its weathering and transformation. Toxicity of Pb has been well known as one of the adverse heavy metals on human health and environment. With this reason, several approaches using waste materials have been suggested to stimulate the Pb immobilization in contaminated soils. Different waste materials have been applied

Table 14.3 Different waste materials to immobilize lead (Pb) in contaminated soils

Waste material	Source	Reference
Red mud	Waste alumina industry	Lee et al. (2009)
		Garau et al. (2007)
		Brown et al. (2005)
		Lombi et al. (2002)
Beringite	Burning of coal refuse	Lombi et al. (2002)
Cyclonic ash	Burning of coal refuse	Brown et al. (2005)
		Geebelen et al. (2003)
Fly ash	Coal fired power stations	Ciccu et al. (2003)
	Flue gas desulfurizing product	Clark et al. (2001)
Furnace slag	Steel industry waste	Lee et al. (2009)
Poultry waste	Poultry farming industry	Hashimoto et al. (2008)
Dolomitic residue	Steel industry waste	Rodríguez-Jordá et al. (2010)
		Garrido et al. (2005)
Sugar foam	Sugar manufacturing industry	Rodríguez-Jordá et al. (2010)
		Garrido et al. (2005)
Waste oyster shells	Oyster farming industry	Ok et al. (2010a, b)
		Moon et al. (2010)
		Lim et al. (2009)
Waste eggshell	Food waste	Lim et al. (2009)
Steel shot	Steel-iron industry waste	Geebelen et al. (2003)
Water and wastewater	Treatment Residue Water and wastewater treatment plants	Spuller et al. (2007)
		Brown et al. (2005)
Red gypsum	TiO_2 industry waste	Rodríguez-Jordá et al. (2010)
		Garrido et al. (2005)
Phosphogypsum	Production of phosphoric acid from rock phosphate	Rodríguez-Jordá et al. (2010)
		Garrido et al. (2005)

to soils for the Pb immobilization as shown in Table 14.3. Natural alkaline materials such as red mud, beringite, cyclonic ash, fly ash, furnace slag, poultry waste, dolomitic residue, sugar foam, waste oyster shell, and eggshell generally increase soil pH; therefore, these materials reduce Pb solubility in the soils. Lindsay (1979) found that Pb can be precipitated as OH^- at pH 8 and became a less bioavailable form. Sorption of Pb to soil particles is also increasing at high soil pH because of an increase in a net negative charge of colloidal particles such as clay, SOM, and Fe-Al oxides (Lee et al. 2009). This review focuses on the use of waste materials or industrial by-products for stabilizing Pb in shooting range soils and evaluates their effectiveness.

14.3.2 Use of Lime-Based Waste Materials

The S/S technique using P has been considered as an effective way to remediate Pb in shooting range soils. However, the overuse of P may induce the secondary environmental pollution like eutrophication resulting from leaching of excess P and extremely slow reaction between Pb and P (Dermatas et al. 2008). Use of lime-based waste materials is an emerging remarkable way to remediate heavy-metal-contaminated soils with advantages of cost and time. Ok et al. (2010a) found that the use of oyster shell waste effectively remediates Cd and Pb in the soils, and used oyster shells as a source of $CaCO_3$ for immobilizing heavy metals as shown below:

$$CaCO_3 + H_2O \rightarrow Ca^{2+} + HCO_3^- + OH^-$$

$$M + OH^- \rightarrow M - OH$$

Ok et al. (2010b) also suggested that a form of calcium oxide (CaO) or quick lime which is obtained from the calcination of $CaCO_3$ at a high temperature (900°C) may increase the efficiency of waste oyster shell on heavy metal remediation in the soils:

$$CaCo_3 + heat \rightarrow Cao + Co_2$$

The CaO can effectively immobilize Pb and Cd in the soils because of pozzolanic reactions compared to $CaCO_3$. Formation of CAH or CSH by CaO in the soils results in hardening of soil particles, thereby reducing the bioavailability of metals (Kostarelos et al. 2006). Moon et al. (2010) applied both calcined and uncalcinated oyster shell to highly contaminated shooting range soils indicating 29,000 mg kg^{-1} Pb and found that the calcined oyster shell material has a higher Pb immobilization rate because of the formation of CAH or CSH at high soil pHs.

14.4 Conclusion

Lime-based waste materials such as eggshell, oyster shell and mussel shell are easily accessible in our surrounding. More than 100,000 tons of oyster shell and 50,000 tons of eggshell are being annually generated in Korea (Lee et al. 2005; Ok et al.

2011b; Palka 2002). Recycling or reuse of such waste materials promises to not only reduce environment pollution but also be a cost-effective way to remediate the shooting range soils which are highly contaminated with heavy metals. Furthermore, the exploration or evaluation of lime-based waste materials would be a valuable study to select an appropriate remediation material depending on heavy metal contaminants. Longevity of lime-based waste materials or their additional advantages of plant growth also needs to be addressed in the future.

References

Alpaslan B, Yukselen MA (2002) Remediation of lead contaminated soils by stabilization/solidification. Water Air Soil Pollut 133:253–263

Arwidsson Z, Elgh-Dalgren K, von Kronhelm T, Sjöberg R, Allard B, van Hees P (2010) Remediation of heavy metal contaminated soil washing residues with amino polycarboxylic acids. J Hazard Mater 173:697–704

Astrup T, Boddum JK, Christensen TH (1999) Lead distribution and mobility in a soil embankment used as a bullet stop at a shooting range. J Soil Contam 8(6):653–665

Bennett LE, Burkhead JL, Hale KL, Terry N, Pilon M, Pilon-Smits EAH (2003) Bioremediation and biodegradation—analysis of transgenic Indian mustard plants for phytoremediation of metal-contaminated mine tailings. J Environ Qual 32:432–440

Berthelot Y, Valton E, Auroy A, Trottier B, Robidoux PY (2008) Integration of toxicological and chemical tools to assess the bioavailability of metals and energetic compounds in contaminated soils. Chemosphere 74:166–177

Braida W, Christodoulatos C, Ogundipe A, Dermatas D, O'Connor G (2007) Electrokinetic treatment of firing ranges containing tungsten-contaminated soils. J Hazard Mater 149:562–567

Braun U, Pusterla N, Ossent P (1997) Lead poisoning of calves pastured in the target area of a military shooting range. Schweiz Arch Tierheilkd 139:403–407

Brown S, Christensen B, Lombi E, McLaughlin M, McGrath S, Colpaert J, Vangronsveld J (2005) An inter-laboratory study to test the ability of amendments to reduce the availability of Cd, Pb, and Zn in situ. Environ Pollut 138:34–45

Cang L, Zhou DM, Wang QY, Wu DY (2009) Effects of electrokinetic treatment of a heavy metal contaminated soil on soil enzyme activities. J Hazard Mater 172:1602–1607

Cao X, Dermatas D, Xu X, Shen G (2008) Immobilization of lead in shooting range soils by means of cement, quicklime, and phosphate amendments. Environ Sci Pollut Res Int 15:120–127

Cao X, Ma LQ, Chen M, Hardison DW Jr, Harris WG (2003a) Weathering of lead bullets and their environmental effects at outdoor shooting ranges. J Environ Qual 32:526–534

Cao X, Ma LQ, Chen M, Hardison DW Jr, Harris WG (2003b) Lead transformation and distribution in the soils of shooting ranges in Florida, USA. Sci Total Environ 307:179–189

Chen QY, Tyrer M, Hills CD, Yang XM, Carey P (2009) Immobilisation of heavy metal in cement-based solidification/stabilisation: A review. Waste Manage 29:390–403

Chrastný V, Komárek M, Hájek T (2010) Lead contamination of an agricultural soil in the vicinity of a shooting range. Environ Monit Assess 162:37–46

Ciccu R, Ghiani M, Serci A, Fadda S, Peretti R, Zucca A (2003) Heavy metal immobilization in the mining-contaminated soils using various industrial wastes. Miner Eng 16:187–192

Clark RB, Ritchey KD, Baligar VC (2001) Benefits and constraints for use of FGD products on agricultural land. Fuel 80:821–828

Cohen SZ (2000) Testing your outdoor range—using the right tools. http://www.nssf.org/ranges/resources/NSRS/04Policy Track/Testing Range.pdf. Accessed 11 February 2009

Conder JM, Lanno RP, Basta NT (2001) Assessment of metal availability in smelter soil using earthworms and chemical extractions. J Environ Qual 30:1231–1237

Dermatas D, Cao X, Tsaneva V, Shen G, Grubb DG (2006a) Fate and behavior of metal(loid) contaminants in an organic matter-rich shooting range soil: implications for remediation. Water Air Soil Pollut 6:143–155

Dermatas D, Shen G, Chrysochoou M, Grubb DG, Menounou N, Dutko P (2006b) Pb speciation versus TCLP release in army firing range soils. J Hazard Mater 136:34–46

Dermatas D, Chrysochoou M, Grubb DG, Xu X (2008) Phosphate treatment of firing range soils: lead fixation or phosphorous release. J Environ Qual 37:47–56

Evanko CR, Dzombak DA (1997) Remediation of metals-contaminated soils and groundwater. Technology evaluation report TW-97-01. http://www.clu-in.org/download/toolkit/metals.pdf. Accessed 28 March 2010

FDEP (2004) Best management practices for environmental stewardship of Florida shooting ranges. Florida Department of Environmental Protection. http://www.dep.state.fl.us/waste/quick_topics/publications/shw/hazardous/shootingrange/FloridaBMP-2004reducedsize.pdf. Accessed 11 February 2009

Fristad WE (2006) Case study: Using soil washing/leaching for the removal of heavy metal at the twin cities army ammunition plant. Remed J 5:61–72

Garau G, Castaldi P, Santona L, Deiana P, Melis P (2007) Influence of red mud, zeolite and lime on heavy metal immobilization, culturable heterotrophic microbial populations and enzyme activities in a contaminated soil. Geoderma 142:47–57

Garbisu C, Alkorta I (2001) Phytoextraction: a cost-effective plant-based technology for the removal of metals from the environment. Bioresour Technol 77:229–236

Garrido F, Illera V, García-González MT (2005) Effect of the addition of gypsum- and lime-rich industrial by-products on Cd, Cu and Pb availability and leachability in metal-spiked acid soils. Appl Geochem 20:397–408

Geebelen W, Adriano DC, van der Lelie D, Mench M, Carleer R, Clijsters H, Vangronsveld J (2003) Selected bioavailability assays to test the efficacy of amendment-induced immobilization of lead in soils. Plant Soil 249:217–228

Gulson BL, Palmer JM, Bryce A (2002) Changes in blood lead of a recreational shooter. Sci Total Environ 293:143–150

Guo G, Zhou Q, Ma LQ (2006) Availability and assessment of fixing additives for the in situ remediation of heavy metal contaminated soils: a review. Environ Monit Assess 116:513–528

Hartikainen H, Kerko E (2009) Lead in various chemical pools in soil depth profiles on two shooting ranges of different age. Boreal Environ Res 14(suppl. A):61–69

Hashimoto Y, Matsufuru H, Sato T (2008) Attenuation of lead leachability in shooting range soils using poultry waste amendments in combination with indigenous plant species. Chemosphere 73:643–649

Hashimoto Y, Taki T, Sato T (2009) Sorption of dissolved lead from shooting range soils using hydroxyapatite amendments synthesized from industrial byproducts as affected by varying pH conditions. J Environ Manage 90:1782–1789

Heier LS, Lien IB, Strømseng AE, Ljønes M, Rosseland BO, Tollefsen KE, Salbu B (2009) Speciation of lead, copper, zinc and antimony in water draining a shooting range—Time dependant metal accumulation and biomarker responses in brown trout (*Salmo trutta* L.). Sci Total Environ 407:4047–4055

ITRC (2005) Environmental management at operating outdoor small arms firing ranges. Interstate Technology and Regulatory Council, Small Arms Firing Range Team. Washington, DC. http://www.itrcweb.org/Documents/SMART-2.pdf. Accessed 25 April 2010

Jin CW, Zheng SJ, He YF, Zhou GD, Zhou ZX (2005) Lead contamination in tea garden soils and factors affecting its bioavailability. Chemosphere 59:1151–1159

Jing YD, He ZL, Yang XE (2007) Role of soil rhizobacteria in phytoremediation of heavy metal contaminated soils. J Zhejiang Univ Sci B 8:192–207

Johnson CA, Moench H, Wersin P, Kugler P, Wenger C (2005) Solubility of antimony and other elements in samples taken from shooting ranges. J Environ Qual 34:248–254

Khan FI, Husain T, Hejazi R (2004) An overview and analysis of site remediation technologies. J Environ Manag 71:95–122

14 A Review of Environmental Contamination and Remediation Strategies

Knechtenhofer LA, Xifra IO, Scheinost AC, Flühler H, Kretzschmar R (2003) Fate of heavy metals in a strongly acidic shooting-range soil: small-scale metal distribution and its relation to preferential water flow. J Plant Nutr Soil Sci 166:84–92

Kostarelos K, Reale D, Dermatas D, Rao E, Moon DH (2006) Optimum dose of lime and fly ash for treatment of hexavalent chromium-contaminated soil. Water Air Soil Pollut 6:171–189

Kumpiene J, Lagerkvist A, Maurice C (2008) Stabilization of As, Cr, Cu, Pb and Zn in soil using amendments—a review. Waste Manag 28:215–225

Labare MP, Butkus MA, Reigner D, Schommer N, Atkinson J (2004) Evaluation of lead movement from the abiotic to biotic at a small-arms firing range. Environ Geol 46:750–754

Lanphear BP, Succop P, Roda S, Henningsen G (2003) The effect of soil abatement on blood lead levels in children living near a former smelting and milling operation. Public Health Rep. 118:83–91

Lee IS, Kim OK, Chang YY, Bae B, Kim HH, Baek KH (2002) Heavy metal concentrations and enzyme activities in soil from a contaminated Korean shooting range. J Biosci Bioeng 94:406–411

Lee JY, Lee CH, Yoon YS, Ha BH, Jang BC, Lee KS, Lee DK, Kim PJ (2005) Effects of oystershell meal on improving spring Chinese cabbage productivity and soil properties. Korean J Soil Sci Fert 38:274–280

Lee SH, Lee JS, Choi YJ, Kim JG (2009) In situ stabilization of cadmium-, lead-, and zinc-contaminated soil using various amendments. Chemosphere 77:1069–1075

Levonmäki M, Hartikainen H, Kairesalo T (2006) Effect of organic amendment and plant root on the solubility and Immobilization of lead in soils at a shooting range. J Environ Qual 35:1026–1031

Lim JE, Moon DH, Kim D, Kwon OK, Yang JE, Ok YS (2009) Evaluation of the feasibility of oyster-shell and eggshell wastes for stabilization of arsenic-contaminated soil. Korean Soc Environ Eng 31:1095–1104

Lin Z (1996) Secondary mineral phases of metallic lead in soils of shooting ranges from Örebro County, Sweden. Environ Geol 27:370–375

Lin HK, Man XD, Walsh DE (2001) Lead removal via soil washing and leaching. J Miner Met Mater Soc 53:22–25

Lindsay W (1979) Chemical equilibria in soil. Wiley, New York

Lombi E, Zhao F, Zhang G, Sun B, Fitz W, Zhang H, McGrath SP (2002) In situ fixation of metals in soils using bauxite residue: chemical assessment. Environ Pollut 118:435–443

Ma LQ, Cao RX, Hardison D, Chen M, Harris WG, Sartain J (2002) Environmental impacts of lead pellets at shooting ranges and arsenical herbicides on golf courses in Florida. Report # 02-01. http://noflac.org/wp-content/uploads/2010/02/Environmental-Impacts-of-Lead-Pellets-at-Shooting-Ranges-%E2%80%93-FL-2000.pdf. Accessed 11 February 2009

Ma LQ, Hardison DW, Harris WG, Cao X, Zhou Q (2007) Effects of soil property and soil amendment on weathering of abraded metallic Pb in shooting ranges. Water Air Soil Pollut 178:297–307

Ma QY, Logan TJ, Traina SJ (1995) Lead immobilization from aqueous solutions and contaminated soils using phosphate rocks. Environ Sci Technol 29:1118–1126

Manninen S, Tanskanen N (1993) Transfer of lead from shotgun pellets to humus and three plant species in a Finnish shooting range. Arch Environ Contam Toxicol 24:410–414

Martin WA, Larson SL, Felt DR, Wright J, Griggs CS, Thompson M, Conca JL, Nestler CC (2008) The effect of organics on lead sorption onto Apatite II™. Appl Geochem 23:34–43

Migliorini M, Pigino G, Bianchi N, Bernini F, Leonzio C (2004) The effects of heavy metal contamination on the soil arthropod community of a shooting range. Environ Pollut 129:331–340

Moon DH, Dermatas D (2006) An evaluation of lead leachability from stabilized/solidified soils under modified semi-dynamic leaching conditions. Eng Geol 85:67–74

Moon DH, Cheong KH, Kim TS, Khim J, Choi SB, Ok YS, Moon OR (2010) Stabilization of Pb contaminated army firing range soil using calcinated waste oyster shells. Korean Soc Environ Eng 32:1353–1358

Mozafar A, Ruh R, Klingel P, Gamper H, Egli S, Frossard E (2002) Effect of heavy metal contaminated shooting range soils on mycorrhizal colonization of roots and metal uptake by leek. Environ Monit Assess 79:177–191

450 M. Ahmad et al.

Ok YS, Kim JG (2007) Enhancement of cadmium phytoextraction from contaminated soils with *Artemisia princeps* var. *orientalis*. J Appl Sci 7:263–268

Ok YS, Chang SX, Feng YS (2007a) Sensitivity to acidification of forest soils in two watersheds with contrasting hydrological regimes in the oil sands region of Alberta. Pedosphere 17:747–757

Ok YS, Chang SX, Feng Y (2007b) The role of atmospheric N deposition in soil acidification in forest ecosystems. In: Muñoz SI (ed) Ecology research progress. Nova Science Publishers, New York

Ok YS, Yang JE, Zhang YS, Kim SJ, Chung DY (2007c) Heavy metal adsorption by a formulated zeolite-Portland cement mixture. J Hazard Mater 147:91–96

Ok YS, Lim JE, Moon DH (2010a) Stabilization of Pb and Cd contaminated soils and soil quality improvements using waste oyster shells. Environ Geochem Health 33:83–91

Ok YS, Lim JE, Ahmad M, Hyun S, Kim KR, Moon DH, Lee SS, Lim KJ, Yang JE (2010b) Effects of natural and calcined oyster shells on Cd and Pb immobilization in contaminated soils. Environ Earth Sci 61:1301–1308

Ok YS, Kim SC, Kim DK, Skousen JG, Lee JS, Cheong YW, Kim SJ, Yang JE (2011a) Ameliorants to immobilize Cd in rice paddy soils contaminated by abandoned metal mines in Korea. Environ Geochem Health 33:23–30

Ok YS, Lee SS, Jeon WT, Oh SE, Usman ARA, Moon DH (2011b) Application of eggshell waste for the immobilization of cadmium and lead in a contaminated soil. Environ Geochem Health 33:31–39

Palka K (2002) Chemical composition and structure of foods. In: Sikorski ZE (ed) Chemical and functional properties of food components, 2nd edn. CRC press, Florida

Palomo A, Palacios M (2003) Alkali-activated cementitious materials: Alternative matrices for the immobilization of hazardous wastes: Part II. Stabilization of chromium and lead. Cement Concrete Res 33:289–295

Peters RW (1999) Chelant extraction of heavy metals from contaminated soils. J Hazard Mater 66:151–210

Ragnvaldsson D, Brochu S, Wingfors H (2007) Pressurized liquid extraction with water as a tool for chemical and toxicological screening of soil samples at army live-fire training ranges. J Hazard Mater 142:418–424

Robinson BH, Bischofberger S, Stoll A, Schroer D, Furrer G, Roulier S, Gruenwald A, Attinger W, Schulin R (2008) Plant uptake of trace elements on a Swiss military shooting range: Uptake pathways and land management implications. Environ Pollut 153:668–676

Rodríguez-Jordá MP, Garrido F, García-González MT (2010) Potential use of gypsum and lime rich industrial by-products for induced reduction of Pb, Zn and Ni leachability in an acid soil. J Hazard Mater 175:762–769

Rooney CP, McLaren RG, Cresswell RJ (1999) Distribution and phytoavailability of lead in a soil contaminated with lead shot. Water Air Soil Pollut 116:535–548

Ruby MV, Davis A, Schoof R, Eberle S, Sellstone CM (1996) Estimation of lead and arsenic bioavailability using a physiologically based extraction test. Environ Sci Technol 30:422–430

Ryan JA, Zhang P, Hesterberg D, Chou J, Sayers DE (2001) Formation of chloropyromorphite in a lead-contaminated soil amended with hydroxyapatite. Environ Sci Technol 35:3798–3803

Scheetz CD, Rimstidt JD (2009) Dissolution, transport, and fate of lead on a shooting range in the Jefferson National Forest near Blacksburg, VA, USA. Environ Geol 58:655–665

Scheuhammer AM, Norris SL (1996) The ecotoxicology of lead shot and lead fishing weights. Ecotoxicol 5:279–295

Singh TS, Pant KK (2006) Solidification/stabilization of arsenic containing solid wastes using portland cement, fly ash and polymeric materials. J Hazard Mater 131:29–36

Smith LA, Means JL, Chen A, Alleman, B, Chapman CC, Tixier JS Jr, Brauning SE, Gavaskar AR, Royer MD (1995) Remedial options for metals-contaminated sites. Lewis Publisher, Boca Raton

Sorvari J, Antikainen R, Pyy O (2006) Environmental contamination at Finnish shooting ranges—the scope of the problem and management options. Sci Total Environ 366:21–31

Spuller C, Weigand H, Marb C (2007) Trace metal stabilisation in a shooting range soil: Mobility and phytotoxicity. J Hazard Mater 141:378–387

14 A Review of Environmental Contamination and Remediation Strategies 451

Takamatsu T, Murata T, Koshikawa MK, Watanabe M (2010) Weathering and dissolution rates among Pb shot pellets of differing elemental compositions exposed to various aqueous and soil conditions. Arch Environ Contam Toxicol 59:91–99

Tardy BA, Bricka RM, Larson SL (2003) Chemical stabilization of lead in small arms firing range soils. http://el.erdc.usace.army.mil/elpubs/pdf/trel03-20.pdf. Accessed 1 July 2010

USEPA (1991) Innovative treatment technologies. Semi-annual status report. http://www.clu-in.org/download/remed/asr/asr3.pdf. Accessed 13 March 2010

USEPA (1992) Test methods for evaluating solid waste, physical/chemical methods. Method 1311. http://www.epa.gov/osw/hazard/testmethods/sw846/pdfs/1311.pdf. Accessed 11 February 2009

USEPA (1994) Test methods for evaluating solid waste, physical/chemical methods. Method 1312. http://www.epa.gov/osw/hazard/testmethods/sw846/pdfs/1311.pdf. Accessed 11 February 2009

USEPA (2001) Best management practices for lead at outdoor shooting ranges. http://www.epa.gov/region2/waste/leadshot/epa_bmp.pdf. Accessed 10 March 2010

USEPA (2003) TRW recommendations for performing human health risk analysis on small arm shooting ranges. http://www.epa.gov/superfund/lead/products/firing.pdf. Accessed 11 February 2009

USEPA (2004) Treatment technologies for site cleanup. http://www.epa.gov/tio/download/remed/asr/11/asr.pdf. Accessed 28 March 2010

Vantelon D, Lanzirotti A, Scheinost AC, Kretzscmar R (2005) Spatial distribution and speciation of lead around corroding bullets in a shooting range soil studied by micro-X-ray fluorescence and absorption spectroscopy. Environ Sci Technol 39:4808–4815

Wilde EW, Brigmon RL, Dunn DL, Heitkamp MA, Dagnan DC (2005) Phytoextraction of lead from firing range soil by Vetiver grass. Chemosphere 61:1451–1457

Wixson BG, Davies BE (1994) Guidelines for lead in soil. Proposal of the society for environmental geochemistry and health. Environ Sci Technol 28:26A–31A

Chapter 15
Peroxidases as a Potential Tool for the Decolorization and Removal of Synthetic Dyes from Polluted Water

Qayyum Husain and Maroof Husain

Contents

15.1	Dyes: A Colorful Environmental Concern		454
15.2	Toxicity of Dyestuffs		455
15.3	Classification of Dyes		456
	15.3.1	Classification Based on Constitution	457
	15.3.2	Classification Based on Application	457
15.4	Classical Methods for Dye Removal		457
15.5	Microbiological Decomposition of Synthetic Dyes		459
	15.5.1	Bacterial Treatment	459
	15.5.2	Fungal Treatment	460
	15.5.3	Algal Treatment	462
	15.5.4	Factors Affecting Color Removal by Microbiological Methods	462
	15.5.5	Demerits of Microbiological Treatment	463
15.6	Enzymatic Approach for Wastewater Treatment		463
	15.6.1	Microbial Peroxidases	465
	15.6.2	Plant Peroxidases	473
	15.6.3	Microperoxidase-11 (MP-11)	477
15.7	Dye Removal by Immobilized Peroxidases		477
15.8	Role of Redox Mediators in Peroxidase Catalyzed Decolorization of Dyes		481
15.9	Future Outlook		484
References			485

Abstract Environmental pollution by discharge of dye-containing effluents represents a serious ecological problem all over the world. Public demands for colour-free discharges to receiving waters have made decoloration of a variety of industrial wastewater a top priority. The current existing chemical, physical and biological techniques used for the removal of colored pollutants have several

---

Q. Husain (✉)
Faculty of Applied Medical Sciences, Jazan University, Jazan
Post Box 2092, Kingdom of Saudi Arabia
e-mail: qayyumbiochem@gmail.com

M. Husain
Department of Microbiology, School of Medicine, University of Colorado,
Health Sciences Center, Aurora, Colorado, USA

A. Malik, E. Grohmann (eds.), *Environmental Protection Strategies for Sustainable Development,* Strategies for Sustainability,
DOI 10.1007/978-94-007-1591-2_15, © Springer Science+Business Media B.V. 2012

drawbacks such as high cost, low efficiency, use of large amounts of chemicals and formation of side toxic products. It has necessitated for the search of alternative procedures such as those based on oxidative enzymes. This approach is believed to be a promising technology since it is cost-effective, environmentally friendly, highly specific and does not produce sludge. Enzymatic transformation of synthetic dyes can be described as the conversion of dye molecules by enzymes into simpler and generally colorless products. Peroxidases have attracted much attention from researchers in last decades due to their ability to oxidize both phenolic and nonphenolic compounds as well as highly recalcitrant environmental pollutants, which makes them a very useful tool for the removal of colored pollutants from industrial effluents. Detailed characterization of the metabolites produced during enzymatic transformation of synthetic dyes as well as ecotoxicity studies is of great significance to examine the effectiveness of the biodegradation processes. However, most reports on the bio-treatment of dyes mainly deal with decoloration and there are few reports on the reduction in toxicity or on the identification of the biodegradation products. In this work an effort has been made to review the literature based on the applications of peroxidases for the removal of colored compounds from wastewater. Peroxidases immobilized on numerous supports have been used to treat dyes in batch as well as in continuous processes. The role of various redox mediators in peroxidase catalyzed dye decolorization has also been discussed in this chapter.

Keywords Peroxidase • Decolorization • Synthetic dyes • Treatment

15.1 Dyes: A Colorful Environmental Concern

The production of synthetic compounds and their application is essential but there is an undesirable discharge of poorly biodegradable wastes from various manufacturing operations, e.g. coal conversion, petroleum refining, resin, dye and other organic compound manufacturing, dyeing and textile, mining, pulp and paper (Duran and Esposito 2000; Husain and Jan 2000). Synthetic dyes are extensively used as coloring material for textiles, paper, leather, hair, fur, plastics, wax, cosmetic bases and foodstuff (O'Neill et al. 2000). However, inefficiencies in dyeing processes resulted into loss of large amounts of dyestuff in effluent, which posed serious threat to the environment (Robinson et al. 2001a; Keharia and Madamwar 2003). Wastewater from textile and dyestuff industries contains synthetic and complex molecular compounds, which make them more stable and difficult to degrade (Padmesh et al. 2005; Palmieri et al. 2005; Al-Aseeri et al. 2007). Moreover, dye effluent usually contains chemicals that may be toxic, carcinogenic and mutagenic to various microbes, aquatic animals and human beings (Verma et al. 2003; Golob and Ojstrsek 2005; Beak et al. 2009). The seriousness of the problem is apparent from the magnitude of research that has been done in this field in the last decade (Robinson et al. 2001a, b; Jager et al. 2004; Pricelius et al. 2007; Ghodake et al. 2009).

15.2 Toxicity of Dyestuffs

Interest in the remediation of synthetic dyes has primarily been prompted by concern over their possible toxicity and carcinogenicity (Koyuncu 2002; Novotny et al. 2006; Kokol et al. 2007). Roughly 60–70% of the dyes used in textile industries are azo compounds, i.e. molecules with one or more azo (N=N) bridges (Stolz 2001). The azo dyes are divided in two groups according to their water solubility. Water soluble azo dyes are anionic (acidic dyes) or cationic (basic dyes) while the water insoluble dyes are non-ionic (neutral). Dyes are highly visible; some can be detected in concentration < 1 mg L^{-1} and are synthesized to be chemically and photolytically stable thus persist in natural environments (Nigam et al. 2000). However, most of the azo dyes are not toxic by themselves but after releasing into the aquatic environment, these compounds might be converted into potentially carcinogenic amines and other aromatic compounds (Stolz 2001; Neamtu et al. 2004; Umbuzeiro et al. 2005a, b; Ozturk and Abdullah 2006; Lima et al. 2007). Consequently, the release of potentially hazardous dyes in the environment can be an ecotoxic risk and can affect human beings through the food chain (Cicek et al. 2007; Khenifi et al. 2007; Mondal 2008). The risk, which dyes represent in wastewaters, depends on their chemical structure, physical properties, concentration and exposure time (Robinson et al. 2001a; Golob and Ojstrsek 2005). Research has been carried out to investigate the effects of dyestuffs and dye containing effluents on the activity of both aerobic and anaerobic bacteria in wastewater treatment systems. The acute toxicity of azo dyes is rather low. Algal growth and fish mortality is not affected by dye concentrations below 1 mg L^{-1}. The most acutely toxic dyes for fish and algae are acid and basic dyes, especially those with a triphenylmethane structure (Greene and Baughman 1996). Mortality tests with rats showed that only 1% out of 4461 commercial dyestuffs tested had LD_{50} values below 250 mg kg^{-1} body weight whereas a majority of dyes showed LD_{50} values between 250 and 2,000 mg kg^{-1} body weight. Therefore, the chances of human mortality due to acute dyestuff toxicity are probably very low. However, in humans some azo dyestuffs have been reported to cause allergic reactions, i.e. eczema or contact dermatitis (Giusti et al. 2002; Giusti and Seidenari 2003). The majority of sensitizing dyes present in clothes practically all belong to the group of disperse dyes (Pratt and Taraska 2000; Ryberg et al. 2009). Disperse dyes are the most heavily used textile dyes. These are structurally classified as mainly an azo and anthraquinone chromophoric system with small molecular size and low aqueous solubility (Golob and Ojstrsek 2005). Anthraquinone based dyes are most resistant to bacterial degradation due to their fused aromatic structures (Nigam et al. 2000). Reactive azo dyes are problematic due to their excessive consumption and high water solubility (Keharia and Madamwar 2003). Metal-based complex dyes such as chromium-based dyes can lead to the release of chromium, which is carcinogenic (Banat et al. 1996; Capar and Yetis 2006).

Chronic effects of dyestuffs, especially of azo dyes, have been studied for several decades. Some azo dyes in purified form showed mutagenic or carcinogenic property (Umbuzeiro et al. 2005b). Intestinal cancers are common in highly industrialized societies and possible connection between these tumours and the use of azo dyes has been investigated (Keharia and Madamwar 2003). Numerous dyes were found to cause cerebral and skeletal abnormalities in foetus (Murugesan and Kalaichelvan 2003). Textile dyeing, paper printing and leather finishing industry workers exposed to benzidine based dyes had a higher than normal incidence of urinary bladder cancer (Rothman et al. 1997; Vineis and Pirastu 1997; Pielesz et al. 2002). Benzidine based dyes when administered to various experimental animals undergo reduction of azo bonds with appearance of human bladder carcinogen, benzidine and benzidine metabolites in the urine (Platzek et al. 1999; Murugesan and Kalaichelvan 2003). The carcinogenicity mechanism probably includes the formation of acyloxy amines through N-hydroxylation and N-acetylation of the aromatic amines followed by O-acylation. These acyloxy amines can be converted to nitrenium and carbonium ions that bind to DNA and RNA thus inducing mutations and tumour formation (Hathway and Kolar 1980).

Generally stated, genotoxicity is associated with all aromatic amines containing benzidine moieties as well as with some aromatic amines with toluene, aniline and naphthalene moieties (Pavanello and Clonfero 2000). The toxicity and carcinogenicity of aromatic amines depends on the three-dimensional structure of the molecule and on the location of amino group(s). For instance, 2-naphthylamine is a carcinogen while 1-naphthylamine is much less toxic. Moreover, the nature and the position of other substituents could increase (nitro, methyl or methoxy) or decrease (carboxyl or sulphonate) the toxicity (Chung and Cerniglia 1992). Sulphonated aromatic amines in contrast to some of their unsulphonated analogs have either no or very low genotoxic and tumorigenic potential. Dyes are toxic to some aquatic life due to presence of metals and chlorides (Arslan 2001; Daneshvar et al. 2007; Husseiny 2008; Rezaee et al. 2008). These concerns have led to new and strict regulations concerning colored wastewater discharges, compelling the dye manufacturers and users to adopt "cleaner technology" approaches. Therefore, effective means for solving this problem by the development of new lines of ecologically safe dyeing auxiliaries and improvement of exhaustion of dyes on to fiber must be adopted in order to preserve the quality of life for future generations (Hao et al. 2000; Rott 2003; Hai et al. 2007).

15.3 Classification of Dyes

Aromatic compounds that absorb light/electromagnetic energy with wavelengths in the visible range (350–700 nm) are colored. Dyes contain *chromophores*, delocalized electron systems with conjugated double bonds and *auxochromes*, electron-withdrawing or electron donating substituents that intensify the color of the chromophore by altering the overall energy of the electron system. Usual chromophores are $-C=C-$, $-C=N-$, $-C=O$, $-N=N-$, $-NO_2$ and quinoid rings whereas usual auxochromes are $-NH_3$, $-COOH$, $-SO_3H$ and $-OH$ (van der Zee et al. 2003). The Color

Index (CI) number, developed by the Society of Dyers and Colorists, is used for dye classification (Kiernan 2001). Once the chemical structure of a dye is known, a five digit CI number is assigned to it. The first word is the dye classification and the second word is the hue or shade of the dye. For example, CI Acid Yellow 36 (CI 13065) is a yellow dye of the acid type. Further dyes are classified either according to their constitution or method of application.

15.3.1 *Classification Based on Constitution*

Chemical structures determine the colors, properties and uses of dyes, and provide the only rational basis for the classification of these compounds (Hao et al. 2000). There are numerous groups of dyes based on the presence of different structural units/chemical structures/chromophores. The most important group of dyes includes azo (monoazo, disazo, triazo, polyazo), anthraquinone, naphthoquinone, arylmethane (diarylmethane, triarylmethane) phthalocyanine and polymethine dyes. Other groups are indigoid, azine, oxazine, thiazine, xanthene, nitro, nitroso, thiazole, indamine, indophenol, lactone, aminoketone and hydroxyketone dyes and dyes of undetermined structures such as stilbene and sulphur dyes (Kiernan 2001).

15.3.2 *Classification Based on Application*

A vast array of dyes/colorants is classified depending on application characteristics. These include acid dyes, basic dyes, direct dyes, disperse dyes, fiber reactive dyes, insoluble azo dyes, vat dyes and mordant dyes (Hao et al. 2000). Various industrially important dyes, their classes, structures and applications have been summarized in Table 15.1.

15.4 Classical Methods for Dye Removal

Dyestuffs give colored wastewaters that have high chemical oxygen demand (COD) and total organic carbon (TOC) values and low biological oxygen demand (BOD) values (Pearce et al. 2003). Various physical, chemical and biological treatment techniques have been developed to remove colored pollutants from wastewaters. Choosing the most appropriate treatment methods or their combinations depends on the dyestuffs and the dyeing methods used in the textile production. Physico-chemical processes have been used to remove high molecular weight (M_r) organic compounds and their color, toxicity, suspended solids and

458   Q. Husain and M. Husain

Table 15.1 Industrially important dyes, their classes and applications

Class	Chemical types	Applications
Acid	Azo, anthraquinone, triarylmethane	Nylon, wool, silk, paper, inks, leather
Azoic	Azo	Cotton, rayon, cellulose acetate, polyester
Basic	Cyanine, azo, azine, triarylmethane, xanthene, acridine, oxazine, anthraquinone	Paper, polyacrylonitrile modified nylon, polyester, inks
Direct	Azo, phthalocyanine, stilbene, oxazine	Cotton, rayon, paper, leather, nylon
Disperse	Azo, anthraquinone, styryl, nitro	Polyester, polyamide, acrylic, and plastics
Fluorescent brighteners	Stilbene, pyrazoles, coumarin, naphthalimides	Soaps, detergents, all fibres, oils, paints, plastics
Food, drugs, cosmetics	Azo, anthraquinone, carotenoid, triarylmethane	Foods, drugs, cosmetics
Mordant	Azo, anthraquinone	Wool, leather, anodized aluminium
Natural	Anthraquinone, flavonols, flavones, indigoids, chroman	Food
Oxidation bases	Aniline black and indeterminate structures	Hair, fur, cotton
Pigments	Azo, basic, phthalocyanine, quinacridone, indigoid	Paints, inks, plastics, textiles
Reactive	Azo, anthraquinone, phthalocyanine, formazan, oxazine, basic	Cotton, wool, silk, nylon
Solvent	Azo, triphenylmethane, anthraquinone, phthalocyanine	Plastics, gasoline, varnish, lacquer, stains, inks, fats, oils, waxes
Sulfur	Indeterminate structures	Cotton, rayon
Vat	Anthraquinone (including polycyclic quinones), indigoids	Cotton, rayon, wool

COD but BOD and compounds of low M_r are not effectively removed. Currently available methods include chemical oxidation, reverse osmosis, adsorption, membrane filtration, coagulation/flocculation, sorption, electrolysis, advanced oxidation processes (chlorination, bleaching, ozonation, Fenton oxidation and photocatalytic oxidation) and chemical reduction. These methods suffer from certain drawbacks such as high cost and salt content in the effluent, problems related to disposal of concentrate and excessive use of chemicals and energy (Ghoreishi and Haghighi 2003; Mielgo et al. 2003; Anjaneyulu et al. 2005). Ozonation, flocculation, photocatalytic oxidation and electrochemical methods result in poor color removal and formation of useless and toxic products. Moreover, chemical coagulation, membrane techniques and reverse osmosis have been successfully used for the treatment of major portion of the colored pollutants but these processes were found highly expensive (O'Neill et al. 2000; Robinson et al. 2001a).

15.5 Microbiological Decomposition of Synthetic Dyes

Biological degradation is an important and simple method for the breakdown/removal of synthetic dyes (Forgacs et al. 2004; Park et al. 2007). It is an attractive and viable alternative to physico-chemical methods mainly because of relatively inexpensive and effective nature of the process. Furthermore, the end products of complete mineralization are not toxic which contribute to environmental benignity and publicly acceptable treatment technology (McMullan et al. 2001; Mohan et al. 2002; Chen et al. 2003). Biotechnological approaches suggested by recent research are of potential interest towards combating the pollution source in an ecoefficient manner involving the use of bacteria or fungi, often in combination with physico-chemical processes (Borchert and Libra 2001; McMullan et al. 2001; Zissi and Lyberatos 2001; Asses et al. 2009). In recent years, numerous types of microorganisms have been identified and isolated that are able to degrade recalcitrant dyes (Fu and Viraraghavan 2001; Hao et al. 2007; Shimokawa et al. 2008).

15.5.1 Bacterial Treatment

Chung et al. (1978) for the first time isolated bacterial cultures of *Bacillus subtilis* which were capable of degrading azo dyes. Later on numerous other dye degrading bacteria were discovered (Pandey et al. 2007; Khalid et al. 2008; Tan et al. 2009). *Pseudomonas* sp. was isolated from an anaerobic-aerobic dyeing house wastewater treatment facility as the most active azo dye degraders (Knapp and Newby 1999). Elisangela et al. (2009) have shown that a facultative *Staphylococcus arlettae* strain VN-11, is very effective in azo dye decolorization in a sequential micro-aerophilic/aerobic process. Recently, it was reported that a recombinant bacterial CotA-laccase from *Bacillus subtilis* was able to decolorize a variety of structurally different synthetic dyes in the absence of redox mediators at alkaline pH (Pereira et al. 2009).

The ability of whole bacterial cells to metabolize azo dyes has been extensively investigated (Pearce et al. 2003). For the degradation of dyes from wastewater, the use of whole cells rather than isolated enzymes is advantageous. This is due to the cost associated with enzyme purification is negated and the cell could also offer protection from the harsh process environment to the enzymes. Also, degradation is often carried out by a number of enzymes working sequentially. Under aerobic conditions, azo dyes are not readily metabolized (Robinson et al. 2001a). However, under anaerobic conditions many bacteria reduce the highly electrophilic azo bond in the dye molecule, reportedly by the activity of low specificity cytoplasmic azo reductases to colorless aromatic amines. These amines are resistant to further anaerobic mineralization and could be toxic or even carcinogenic/mutagenic to humans and animals. Fortunately, once the xenobiotic azo components of the dye molecule are removed, the resultant amino compounds are found to be good substrates for aerobic biodegradation. Lourenco et al. (2000) developed a sequential anaerobic-aerobic system for wastewater treatment.

The utilization of mixed cultures/microbial consortia offered considerable advantages over the use of pure cultures in the degradation of synthetic dyes. Many researchers have reported that a higher degree of biodegradation and mineralization could be expected when co-metabolic activities within a microbial community complement each other (Khelifi et al. 2009; Su et al. 2009). Junnarkar et al. (2006) have reported biodegradation of colored wastewater using mixed bacterial cultures. Optimal conditions for the microbial decolorization of dyes show a marked diversity both in anaerobic and aerobic as well as in mixed anaerobic/aerobic processes. However, mixed cultures only provide an average macroscopic view of what is happening in the system and results are not easily reproduced, making thorough and effective interpretation difficult. The use of a pure culture system ensured that the data were reproducible and the interpretation of experimental observations was easier. The quantitative analysis of the kinetics of azo-dye decolorization by a particular bacterial culture could be undertaken meaningfully and response of the system to changes in operational parameters could also be studied (Chang and Lin 2000).

15.5.2 Fungal Treatment

The majority of biological decolorization procedures have focused on fungal strains, white rot fungi (WRF). These strains are versatile and robust microorganisms (Murugesan et al. 2007a, b; Vivekanand et al. 2008; Li et al. 2009). These constitute a diverse ecophysiological group comprising mostly basidiomycetous and to a lesser extent litter-decomposing fungi capable of extensive aerobic lignin depolymerization and mineralization. This property is based on the WRF's capacity to produce one or more extracellular lignin-modifying enzymes (LME) such as laccase, lignin peroxidase (LiP), phenol oxidase, Mn dependent peroxidase and Mn independent peroxidase (Pocedic et al. 2009). These LMEs oxidize highly stable natural polymers; lignin, hemicellulose, cellulose, etc. (Moreira et al. 2000; Wesenberg et al. 2003). Because of their high biodegradation capacity, these fungi have shown their potential in biotechnological and other applications, particularly, in the decolorization of synthetic dyes present in wastewaters (Robinson et al. 2001b; Tekere et al. 2001; Pazarlioglu et al. 2005; Chandralata et al. 2008; Kaushik and Malik 2009). However, for successful implication of these enzymes at large scale, some molecular challenges need to be accomplished. These are (i) enhancement of operational stability, specifically to H_2O_2 in case of fungal peroxidases, (ii) increase of enzyme redox potential in order to widen the substrate range, (iii) development of heterologous expression and industrial production (Ayala et al. 2008).

Numerous WRF strains have been employed for the decolorization of distinct synthetic dyes and dye effluents (McMullan et al. 2001; Jarosz-Wilkolazka et al. 2002; Park et al. 2007). Earlier results on decolorization of colored wastewaters by fungi have been reviewed by several workers; *Aspergillus foetidus* (Sumathi

15 Peroxidases as a Potential Tool for the Decolorization and Removal of Synthetic Dyes   461

and Manju 2000), *Phanerochaete chrysosporium* (Mielgo et al. 2001; Soares et al. 2002), *Trametes versicolor* (Borchert and Libra 2001), *Trametes hirsuta* (Abadulla et al. 2000), *Coriolus versicolor* (Knapp and Newby 1999; Kapdan and Kargi 2002), *Cunninghamella polymorpha* (Sugimori et al. 1999), *Geotrichum candidum* and *Rhizopus arrhizus* (Aksu and Tezer 2000; Asses et al. 2009). Moreover, fungi other than WRF, such as *Umbelopsis isabellina, Penicillium* sp., *Aspergillus niger* and *Rhizopus oryzae* could also decolorize or biosorb diverse dyes (Assadi and Jahangiri 2001; Yang et al. 2003). In fungal decolorization of dye wastewater, these fungi can be classified into two kinds according to their life state: living cells to biodegrade and biosorb dyes and dead cells (fungal biomass) to adsorb dyes (Fu and Viraraghavan 2001).

The most widely studied WRF in regards to xenobiotic degradation is *Phanerochaete chrysosporium*. This fungus has been frequently employed for the biodegradation of synthetic dyes due to its high enzyme production (Enayatzamir et al. 2009). The decomposition of Indigo Carmine by the fungus has also been studied and it showed the involvement of ligninolytic enzymes in this process (Podgornik et al. 2001). The biodegradation of Amaranth, New Coccine, Orange G and tartrazine by *Phanerochaete chrysosporium* and *Pleurotus sajor-caju* was compared. It has been found that the addition of activators; Tween-80, veratryl alcohol (VA) and manganese (IV) oxide for the production of lignolytic enzymes from *Phanerochaete chrysosporium* increased the decomposition rate of the dyes, Poly R-478 and Poly-448 (Couto et al. 2000a, b). Manganese peroxidase (MnP) was the main enzyme involved in dye decolorization by *Phanerochaete chrysosporium* (Chagas and Durrant 2001). Although several workers have reported that LiP from *Phanerochaete chrysosporium* was the main decolorizing agent. A recent investigation of the degradation of selected phthalocyanine dyes and their degradation products has shown the involvement of laccase and MnP (Conneely et al. 2002).

Other WRF have been used for the decoloration of different dyes. *Trametes versicolor* decomposed anthraquinone, azo and indigo-based dyes (Wang and Yu 1998; Casas et al. 2009) and *Trametes hirsuta* was able to decompose triarylmethane, indigoid and anthraquinone dyes (Abadulla et al. 2000). It has been established that the higher degradation rate was achieved by *Irpex lacteus* and *Pleurotus ostreatus* (Novotny et al. 2001; Pocedic et al. 2009). The results indicated that the decomposition rate depended on both chemical structure of the dye and the character of the fungi. The decolorization of the phthalocyanine dyes, Reactive Blue 15 and 38, by *Bjerkandera adusta* was studied in detail and it was found that the main metabolites were sulfophthalimides (Heinfling-Weidtmann et al. 2001). The potential of some WRF to decolorize indigo dye has been compared by some workers (Balan and Monteiro 2001). The decomposition was highest by *Phellinus gilvus* followed by *Pleurotus sajor-caju, Pycnoporus sanguineus* and *Phanerochaete chrysosporium*. It was reported that *Phlebia tremellose* catalyzed decomposition of synthetic dyes but complete mineralization was not achieved (Kirby et al. 2000). *Trametes hispida* produced lignolytic enzymes at higher rate than was obtained by *Pleurotus ostreatus* in solid state cultures on whole oats (Rodriguez et al. 1999). The assays carried out on another set of WRF indicated that *Coriolus versicolor* exhibited highest decomposition capacity (Knapp and Newby 1999). An interesting combined method has been described for the decolorization of Acid Violet 7. Pellets have been prepared from the mycelium of *Trametes versicolor* and activated carbon powder and their decolorization rate has been shown to be higher than those of the individual components, pure mycelium or activated carbon (Zhang and Yu 2000). Eighteen fungal strains were used for the degradation of lignocellulosic material or lignin derivatives with azo dyes; Reactive Orange 96, Reactive Violet 5 and Reactive Black 5. *Bjerkandera adusta, Trametes versicolor* and *Phanerochaete chrysosporium* were able to decolorize all azo dyes (Heinfling et al. 1997). However, despite the fact that laccases from *Trametes versicolor, Polyporus pinisitus* and *Myceliophthora thermophila* were found to decolorize anthraquinone and indigoid-based dyes at high rates, an azo dye, Direct Red 29 (Congo Red) was a very poor substrate for laccases (Claus et al. 2002). Other fungi such as, *Hirschioporus larincinus, Inonotus hispidus, Phlebia tremellose, Coriolus versicolor* and *Daedaleopsis* sp. have also been employed to decolorize dye-containing effluent (Banat et al. 1996; Prasongsuk et al. 2009).

15.5.3 Algal Treatment

The biodegradation of azo dyes by algae; *Chlorella pyrenoidosa, Chlorella vulgaris* and *Oscillatoria tenuis* has been demonstrated. It has been described that the azo reductase of algae was the main enzyme which was responsible for the degradation of these dyes into aromatic amines by breaking azo linkages (Liu and Liu 1992). Algae have been used to degrade several aromatic amines and dyes, even sulphonated ones (Semple et al. 1999). Yan and Pan (2004) have reported that more than 30 azo compounds were decolorized and biodegraded into simpler aromatic amines by *Chlorella pyrenoidosa, Chlorella vulgaris* and *Oscillatoria tenuis*. The potential of *Cosmarium* sp., belonging to green algae, was investigated as a viable biomaterial for biological treatment of triphenylmethane dye, Malachite Green (Daneshvar et al. 2007).

15.5.4 Factors Affecting Color Removal
by Microbiological Methods

Several factors determine the technical and economic feasibility of each single dye removal technique. These include dye type, waste composition, dose and costs of required chemicals, operation costs, environmental fate and handling costs of generated waste products. The level of aeration, temperature, pH and redox potential of the system must be optimized to catalyze maximum dye reduction. The concentrations of electron donor and redox mediator must be balanced with the amount of biomass in the system and the quantity of dye present in wastewater. The composition of textile wastewater varies and includes organic compounds, nutrients, salts, sulphur compounds and toxicants. Any of these compounds may have an inhibitory effect on the dye reduction process. The effect of each of the factors on the color removal process must be investigated before the system can be used to treat industrial wastewater. For the environmental hazard assessment of chemicals, estimation of likely environmental concentration and comparison of predicted concentrations with experimentally determined toxic effect levels is essential.

15.5.5 *Demerits of Microbiological Treatment*

Dye effluents have been poorly decolorized by conventional biological treatments and might be toxic for the microorganisms present in the treated effluent plants. Sometimes, anaerobic digestion of nitrogen-containing dyes resulted into release of aromatic amines that were found more toxic and even sometimes mutagenic than the parent molecules (Gottlieb et al. 2003; Zouari-Mechichi et al. 2006). Bacterial, fungal and algal degradation of aromatic compounds is attributed to secondary metabolic pathways. Hence, appropriate growth conditions have to be accomplished by additional loads of chemicals. Moreover, the expression of enzymes involved in phenol, aromatic amine and dye degradation is not constant with time but dependent on the growth phase of the organisms and is influenced by inhibitors that might be present in the effluent (Wesenberg et al. 2003). Certain other limitations of using microbes for treating pollutants are high costs of production of microbial cultures, limited mobility and survival of cells in the soil, alternative carbon source, completeness of the indigenous populations, metabolic inhibition and slow decolorization/degradation of dyes requiring several days to months (Duran and Esposito 2000; Husain and Jan 2000; Nazari et al. 2007). Additionally, the sludge volume increases due to the generation of biomass.

Biological degradation of dyes includes properties such as water solubility, large M_r and fused aromatic ring structures, which inhibit permeation through biological cell membranes. Algae and higher plants exposed to dye containing effluents have been shown to accumulate high concentrations of certain disperse dyes and heavy metals. These techniques used for the treatment of organic pollutants discussed so far have their own limitations and suffer from some serious drawbacks, which render them ineffective for application in removing dyes and other aromatic pollutants at large scale. Therefore, there should be a need to find simple and efficient alternative treatment methods that are effective in removing pollutants from large volumes of effluents.

15.6 Enzymatic Approach for Wastewater Treatment

The variety of chemical transformations catalyzed by enzymes has made these catalysts a prime target for exploitation by the emerging biotechnological industries. Recent advances in this direction, through better isolation and purification procedures have allowed the production of cheaper and more readily available enzymes that can be used in many remediation processes to target specific persistent and recalcitrant pollutants present in wastewaters (Husain 2006; Xu and Salmon 2008; Hamid and Rehman 2009). Enzymes isolated from their parent organisms have been often preferred over intact organisms containing the enzyme because the isolated enzymes act with greater specificity, their activity could be better standardized, are easy to handle and store and enzyme concentration is not dependent on the growth rate of microorganisms (Wagner and Nicell 2003). Moreover, unlike chemical catalysts, the enzymatic systems have the potential of accomplishing complex chemical conversions under mild environmental conditions with high efficiency and reaction velocity (Husain and Husain 2008; Michniewicz et al. 2008). Due to their high specificity to individual species, enzymatic processes have been developed to specifically target selected compounds that cannot be treated effectively or reliably using traditional techniques (Ryan et al. 2003; Couto et al. 2005a; Husain et al. 2009). Alternatively, enzymatic treatment has been used as a pretreatment step to remove one or more compounds that interfere with subsequent downstream treatment processes. For example, if inhibitory or toxic compounds can be removed selectively, the bulk of the organic material could be treated biologically, thereby minimizing the cost of treatment (Gianfreda and Rao 2004). Thus, the potential advantages of enzymatic treatment as compared to conventional treatments can be summed up as; application to recalcitrant materials, operation at low and high pollutant concentrations over a wide pH, temperature and salinity range, absence of shock loading effects, delays associated with the acclimatization of biomass, reduction in sludge volume, the ease and simplicity of controlling the process, need of bio-acclimatization and remediation of various aromatic compounds under dilute conditions (Held et al. 2005; Husain 2006). Enzymes catalyze detoxification/decolorization of dyes quite fast, i.e. within minutes to few hours as compared to algae and fungi which take several days or even months to provide the same results (Sumathi and Manju 2000; Borchert and Libra 2001; Rodriguez 2009). Oxidoreductive enzymes; peroxidases and polyphenol oxidases have shown their potential in the remediation and conversion of aromatic pollutants such as phenols, aromatic amines, biphenyls, bisphenols and dyes to less toxic insoluble compounds, which could be easily removed out of wastewater (Torres et al. 2003; Husain 2006; Husain et al. 2009; Husain 2010).

Peroxidases (E.C.1.11.1.7.) are ubiquitous heme containing oxidoreductases having protoporphyrin IX with Fe^{+3} (Hiraga et al. 2001; Duarte-Vazquez et al. 2003). The iron ion is coordinated to 4 pyrrole nitrogens of the heme and nitrogen of an axial histidine. Peroxidases have broad substrate specificity; can use various organic and inorganic substrates which act as hydrogen donors in vitro in the presence of H_2O_2 (Vianello et al. 1997). These enzymes are present in animals, plants and microorganisms, having M_r from 30–150 kDa (Regalado et al. 2004). Based on their structural and catalytic properties, these are divided into three superfamilies: (i) peroxidases in animals (glutathione peroxidase, myeloperoxidase and lactoperoxidase); (ii) catalases in animals, plants, bacteria, fungi and yeast; (iii) peroxidases in plants, fungi, bacteria and yeast. The amino acid sequence among members of

15 Peroxidases as a Potential Tool for the Decolorization and Removal of Synthetic Dyes   465

plant peroxidase superfamily had been found to be highly variable, with less than 20% identity in the most divergent cases. Based on differences in primary structure, the plant peroxidase superfamily is further categorized into three classes (Welinder 1992). Class I, the intracellular peroxidases, includes yeast cytochrome c peroxidase, ascorbate peroxidase and bacterial peroxidases (Passardi et al. 2007). Class II consists of extracellular fungal peroxidases; LiP and MnP. These are monomeric glycoproteins involved in the degradation of lignin. The peroxidases most commonly studied for dye decolorization are fungal LiP and MnP. Class III consists of secretory plant peroxidases, which have multiple tissue specific functions: e.g. removal of H_2O_2 from chloroplasts and cytosol, oxidation of toxic compounds, biosynthesis of the cell wall, defence responses towards wounding, indole-3-acetic acid catabolism, ethylene biosynthesis, etc. Some of the well known peroxidases of this class are horseradish peroxidase (HRP), turnip peroxidase (TP), bitter gourd peroxidase (BGP), and soybean peroxidase (SBP). Class III peroxidases are also monomeric glycoproteins, having four conserved disulphide bridges and require calcium ions (Conesa et al. 2002). Peroxidases in particular have been extensively studied for the treatment of dyes as these enzymes showed many attractive properties such as wide specificity, high stability in solution and easy accessibility from plant materials and fungal sources (Magri et al. 2005; Biswas et al. 2007; Husseiny 2008; Husain et al. 2009).

15.6.1 *Microbial Peroxidases*

Table 15.2 demonstrates various microbial peroxidases and their role in decolorization of dyes and dye effluents.

Lignin peroxidase (LiP) also known as ligninase or diaryl propane oxygenase, was the first discovered enzyme that catalyzed the partial depolymerization of methylated lignin in vitro (Tien and Kirk 1983). LiP, a heme-containing glycoprotein has an unusually low pH-optimum and was responsible for the oxidation of non-phenolic aromatic lignin moiety and variety of similar compounds (Christian et al. 2005). The oxidation of such compounds by LiP resulted in cleavage of C_α–C_β bond, the aryl–C_α bond, aromatic ring opening, phenolic oxidation and demethoxylation. Due to its high redox potential and enlarged substrate range in the presence of specific mediators, LiP has been used to mineralize a variety of xenobiotic compounds including polycyclic aromatic hydrocarbons (PAH), polychlorinated phenols, nitro aromatics and azo dyes (Krcmar and Ulrich 1998; Abadulla et al. 2000; Husain 2006; Jadhav et al. 2009). Heinfling et al. (1998a) have described transformation of six industrial azo and phthalocyanine dyes by ligninolytic peroxidases from *Bjerkandera adusta* and other WRF. LiP from *Phanerochaete chrysosporium* has been shown to play a major role in decolorization of azo, triphenylmethane, heterocyclic and polymeric dyes (Pointing et al. 2000). Partially purified LiP from *Phanerochaete chrysosporium* grown on neem hull waste was demonstrated to decolorize more than 50% of Procion Brilliant Blue HGR, Ranocid Fast Blue, Acid Red 119

Table 15.2 Dye decolorization by microbial peroxidases. (Modified from Husain 2010)

Enzyme	Source	Dye/effluent	Experimental conditions	Reference
LiP	*Phanerochaete chrysosporium*	Navidol Fast Black, Procion Brilliant Blue HGR, Acid Red 119, Ranocid Fast Blue	Maximum decolorization was 80% for Porocion Brilliant Blue HGR, 83% for Ranocid Fast Blue, 70% for Acid Red 119 and 61% for Navidol Fast Black MSRL	Verma and Madamvar (2002b)
		Methylene Blue (MB)	The use of an MB:H_2O_2 molar ratio of 1:5 resulted in efficient removal of 90% color of MB (50 mg mL^{-1})	Ferriera-Leitao et al. (2007)
		Xylene Cyanol, Fuchsine, Rhodamine B	Degradation of these dyes by LiP coupled with glucose oxidase showed that present H_2O_2 supply strategy was very effective for improvement of the efficiency of the decolorization of dyes	Lan et al. (2006)
	Pseudomonas desmolyticum NCIM 2112	Reactive Brilliant Red K-2BP	LiP decolorized a dye concentration of 60 mg/L and below to no less than 85%	Yu et al. (2006)
	Acinetobacter calcoaceticus NCIM 2890	Ten textile dyes of	Most of dyes decolorized up to 90%. LiP activity was stabilized by tryptophan during decolorization of dyes	Ghodake et al. (2009)
		Methylene Blue	Decolorization process gave higher removal of 90% in agitation mode compared to the static mode with 65% in 60 min by LiP	Alam et al. (2009)
		Astrazon Red FBL	Maximum 87% decolorization of dye and 42% removal of COD, both occurred when only Tween80 (0.05%, w/v) was added to the effluent	Sedighi et al. (2009)
	Trametes versicolor	Amaranth	Decolorized mono-azo substituted napthalenic dye	Gavril et al. (2007)
MnP	*Phanerochaete sordida*	Azo and anthraquinone dyes	The dye (500 mg L^{-1}) was treated in a reaction mixture (5 mL) containing MnP (0.5 U), 0.5 mM MnSO$_4$, 1.0% Tween 80, and 50 mM malonate, pH 4.5 and the reaction mixtures were shaken at 150 rpm at 30°C. These dyes were decolorized to 90%	Harazono and Nakamura (2005)
		RB 5	MnP from these fungi was responsible for decolorization of RB 5	Yang et al. (2003)
	Debaryomyces polymorphus *Candida tropicalis* and *Umbelopsis isabellina* *Thelephora* sp.	Orange G (OG)	MnP (15 U mL^{-1}) decolorized a maximum of 10.8% of the dye	Selvam et al. (2003)

Table 15.2 (continued)

Enzyme	Source	Dye/effluent	Experimental conditions	Reference
MnP	*Irpex lacteus*	Dye effluent	A nearly total decolorization was possible in 10 min and at high dye concentration up to 1500 mg L^{-1}	Mielgo et al. (2003)
		RO 16	A significant increase in the decolorization by the agitated *I. lacteus* cultures was observed after adding 0.1% Tween 80, due to higher MnP production	Svobodova et al. (2006)
	Pleurotus ostreatus	Sulpho-naphthalein (SP) dyes	Complete Decolorization of SP dyes by MnP was achieved at pH 4.0. The higher K_m for m-cresol purple (40 μM) and lower K_m for o-cresol red (26 μM) for MnP activity explained the preference for the position of methyl group at ortho than meta on chromophore	Christian et al. (2003), Shrivastava et al. (2005)
	Pleurotus calyptratus	Orange G (OG), RBBR	This strain decolorized up to 91% of OG and 85% of RBBR in liquid culture and more than 50% of these dyes on agar plates within 14 days	Eichlerova et al. (2006)
	Collybia dryophila	Poly B-411, RB 5, RO 16 and RBBR	The decolorization of 100 mg L^{-1} dyes after 28 days ranged from 80–95% for RBBR, 60–95% for Poly B-411, 58–85% for RB 5 and 45–82% for RO 16	Baldrian and Snajdr (2006)
	Bjerkandera adusta	RB 5 dye bath	Comparing the fungal and enzymatic treatments of RB 5 dye bath for 120 h, the enzymatic procedure exhibited nearly 1.5 times greater color removal as compared to fungus treatment	Mohorcic et al. (2006)
	Schizophyllum sp.	Congo Red (CR), OG, Orange IV	Purified MnP effectively catalyzed decolorization of azo dyes such as CR, OG and Orange IV	Cheng et al. (2007)
	Ischnoderma resinosum	RB 15, Reactive Blue 19, Reactive Red 22, Reactive Yellow 15	The culture liquid from I. resinosum cultures was also able to decolorize all dyes as well as the synthetic dye baths in presence of VA and HOBT. The highest decolorization was found at acidic pH, 3.0–4.0	Kokol et al. (2007)
	Nematoloma frowardii	Flame Orange and Ruby Red	MnP N-demethylated these dyes and concomitantly polymerized them to some extent	Pricelius et al. (2007)

Table 15.2 (continued)

Enzyme	Source	Dye/effluent	Experimental conditions	Reference
MnP	Dichomitus squalens	Azo and anthraquinone dyes	The purified MnP was able to decolorize selected azo and anthraquinone dyes more rapidly than laccase. In vitro dye decolorization showed a synergistic cooperation of MnP and laccase	Xiao-Bin et al. (2007)
	Lentinula edodes	Brilliant Cresyl Blue, MB	The decolorization of dyes was remarkably decreased in the absence of manganic ions and H_2O_2. MnP was found to be the main enzyme which participated in dyes decolorization	Boer et al. (2004)
	Basidiomycete PV002	Ranocid Fast Blue and Acid Black 210	The degradation of azo dyes under different conditions was strongly correlated with the ligninolytic activity	Verma and Madamvar (2005)
	Basidiomycetous fungi	RBBR	RBBR-decolorization is a simple indicative method for a multienzymatic system for xenobiotic biodegradation	Machado et al. (2005)
DyP	Geotrichum candidum Dec 1	Synthetic dyes	The optimal temperature for DyP activity was 30°C	Kim and Shoda (1999b)
rDyP	Aspergillus oryzae	RBBR	rDyP immobilized on FSM-16 decolorized eight sequential batches of an anthraquinone dye and RBBR in repeated-batch decolorization at pH 4.0	Shakeri and Shoda (2008a)
			rDyP (4600 U) decolorized 5.07 g RBBR at the apparent decolorization rate of 17.7 mg L^{-1} min^{-1}	Shakeri and Shoda (2008b)
TcVP1/ DyP	Thanatephorus cucumeris Dec 1	RB 5	Coapplication of TcVP1 and DyP was capable to completely decolorize Reactive Blue 5 in vitro	Sugano et al. (2006)
VP	Bjerkandera sp.	RBBR	RBBR decolorizing specific activity was 11 U per mg protein at pH 5.0	Moreira et al. (2006)
	Bjerkandera adusta	27 industrial dyes	The presence of redox mediator, tryptophan enhanced decolorization of dyes	Tinoco et al. (2007)
VHP	Curvularia inaequalis	Chicago Sky Blue 6B	r-VHP catalyzed bleaching of Chicago Sky Blue 6B in the presence of H_2O_2	ten Brink et al. (2000)

and Navidol Fast Black (Verma and Madamwar 2002a). LiP was the principal enzyme involved in the decolorization of dyes by *Bjerkandera adusta* (Robinson et al. 2001a, b). Christian et al. (2005) have reported that LiP from *Trametes versicolor* decolorized Remazol Brilliant Blue R (RBBR) and that this dye can also be used as a substrate for the estimation of LiP activity.

An effort had been made to couple a H_2O_2 producing enzymatic reaction to the LiP catalyzed oxidation of dyes (Lan et al. 2006). H_2O_2 was produced by glucose oxidase and its substrate glucose, due to controlled release of H_2O_2, a sustainable constant activity of LiP was observed. Degradation of three dyes; Xylene Cyanol, Fuchsine and Rhodamine B by LiP coupled with glucose oxidase indicated that H_2O_2 was very effective for the improvement of efficiency of the decolorization of dyes. Ferreira-Leitao et al. (2007) compared the usefulness of fungal LiP with HRP in relation to the degradation of Methylene Blue and its demethylated derivatives. However, it was reported that both enzymes could oxidize Methylene Blue and its derivatives but HRP required higher H_2O_2 and showed a considerably lower reaction rate contrary to LiP. LiP catalyzed the cleavage of the aromatic ring of the investigated compounds while HRP was unable to achieve this. The oxidation potential of LiP was roughly twice of the HRP. Gavril et al. (2007) investigated the involvement of LiP from *Trametes versicolor* in decolorization of mono-azo substituted napthalenic dye Amaranth. The verification study confirmed that LiP has a direct influence on the initial decoloration rate and showed that another enzyme, which does not need H_2O_2 to function and is not a laccase, also plays a role during decolorization.

Recently it has been reported that LiP from different bacterial sources was involved in decolorization of dyes. *Pseudomonas desmolyticum* NCIM 2112 was able to degrade a diazo dye, Direct Blue 6 (100 mg L^{-1}) completely within 72 h of incubation with 88.95% reduction in COD in static anoxic condition. Decolorization of Direct Blue 6 in batch culture represented the role of oxidative enzymes; LiP, laccase and tyrosinase (Kalme et al. 2007). LiP from *Acinetobacter calcoaceticus* NCIM 2890 was able to oxidize a variety of substrates including Mn^{2+}, tryptophan, mimosine, L-dopa, hydroquinone, xylidine, *n*-propanol, VA and textile dyes. Four structurally different groups such as azo, thiazin, heterocyclic, and polymeric dyes were decolorized up to 90%. Tryptophan stabilized the LiP activity during decolorization of dyes (Ghodake et al. 2009). Alam et al. (2009) investigated standardization of decolorization of Methylene Blue dye by LiP produced by WRF *Phanerochaete chrysosporium* using sewage treatment plant sludge as a major substrate. Optimization by the one-factor-at-a-time and statistical approach was performed to understand the process conditions on optimum decolorization of the dye using LiP in static mode. This method exhibited that the optimum conditions for decolorization of dye removal: 14–40%, was at 55°C, pH 5.0 with 4.0 mM H_2O_2, 20 mg L^{-1} dye concentration and 0.487 U mL^{-1} of LiP activity. The addition of VA to the reaction mixtures did not contribute any further increases in decolorization. The initial concentration of the dye and the activity of LiP were further optimized using response surface methodology. The contour and surface plots suggested that the optimum initial concentration of the dye and LiP activity predicted were 15 mg L^{-1} and

0.687 U mL^{-1}, respectively, for the removal of 65%. The validation of the model showed that the decolorization process gave higher/more than 90% removal of dye in agitation mode as compared to the static mode with 65% by LiP in 60 min.

The reaction mechanism of MnP started with the oxidation of the enzyme by H_2O_2 to an intermediary oxidized state that, in turn, promoted the oxidation of Mn^{2+} to Mn^{3+} during the catalytic cycle (Hofrichter 2002). Afterwards, in a mechanism involving two successive electron transfer reactions, substrates such as azo dyes reduce the enzyme to its original form (Stolz 2001). Mn^{3+} was stabilized by organic acids such as oxalic acid and the resulting Mn^{3+}-organic acid complex acts as an active oxidant (Schlosser and Hofer 2002). Thus, MnP oxidized its natural substrate, i.e. lignin as well as textile dyes (Heinfling et al. 1998b; Stolz 2001). Chagas and Durrant (2001) have shown that MnP was the main enzyme involved in dye decolorization by *Phanerochaete chrysosporium*. Moreira et al. (2001) have demonstrated the role of MnP in degradation of highly recalcitrant polymeric dye, Poly R-478. MnP catalyzed not only the destruction of chromophoric groups but also a noticeable breakdown of chemical structure of the dye. MnP from *Clitocybula dusenii* participated in the breakdown of dyes in the real dye-containing effluent (Wesenberg et al. 2003). Several investigators have developed MnP based membrane reactors for the oxidation of azo dyes (Lopez et al. 2002, 2004, 2007; Mielgo et al. 2003). A novel dye-decolorizing strain of the bacterium *Serratia marcescens* efficiently decolorized two chemically different dyes; Ranocid Fast Blue and Procion Brilliant Blue HGR which belonged to the azo and anthraquinone groups, respectively. However, it was observed that MnP was involved in the decolorization of these dyes (Verma and Madamwar 2002b). MnP was detected during dye decolorization by culture of *Phlebia tremellose* (Kirby et al. 2000). Selvam et al. (2003) reported that an azo dye, Orange G was decolorized by 10.8% by 15 U mL^{-1} of MnP present in WRF *Thelephora* sp. The WRF, *Irpex lacteus* decolorized the textile industry wastewater efficiently without adding any chemical. The degree of decolorization of the dye effluent by shaking or stationary cultures on 8th day was 59 and 93%, respectively. Higher decolorization was related to the activity of MnP which was detected in higher amount in stationary cultures than in the shaking cultures (Shin 2004). High MnP activity but very low LiP and laccase activities were detected in the culture of the WRF, *Lentinula edodes*. These findings have shown that MnP was the main enzyme which was responsible for the capability of *Lentinula edodes* to decolorize synthetic dyes (Boer et al. 2004). Machado et al. (2005) used RBBR as substrate to evaluate ligninolytic activity in 125 basidiomycetous fungi isolated from tropical ecosystems. Extracellular extracts of 30 selected fungi grown on solid medium with sugar cane bagasse showed RBBR decolorization and peroxidase activity. Eight fungi produced MnP activity which had RBBR decolorization capacity. Basidiomycete PV002, a white-rot strain could efficiently decolorize 96% of Ranocid Fast Blue and 70% of Acid Black 210 on 5th and 9th day under static conditions (Verma and Madamwar 2005). The degradation of azo dyes under different conditions was strongly correlated with high MnP activity.

Christian et al. (2003) have investigated the decolorization of sulfonaphthalein (SP) dyes by MnP and almost all dyes were maximally decolorized at pH 4.0. The

15 Peroxidases as a Potential Tool for the Decolorization and Removal of Synthetic Dyes   471

order of preference for SP dyes as substrate for the MnP-catalyzed decolorization was Phenol Red> *o*-cresol Red>*m*-cresol Purple>Bromophenol Red>Bromocresol Purple>Bromophenol Blue (BPB)>Bromocresol Green (Shrivastava et al. 2005). Harazono et al. (2003) reported decolorization of an azo-reactive dye, Reactive Red 120, by a white-rot basidiomycete, *Phanerochaete sordida* strain YK-624. In liquid culture of *P. sordida* in a medium containing 3% malt extract and 200 mg L^{-1} of the dye, the dye was decolorized by 90.6% after 7 days. MnP activity was noticed during decolorization process. The dye could be decolorized by purified MnP of *P. sordida* in the presence of Mn(II) and Tween 80. The involvement of lipid per-oxidation during decolorization with MnP was considered. Further, *Phanerochaete sordida* was used to decolorize mixtures of four reactive textile dyes, including azo and anthraquinone dyes. These dye mixtures (200 mg L^{-1}) were decolorized by 90% within 48 h in nitrogen-limited glucose-ammonium media. MnP was the main en-zyme responsible for decolorization of dyes by *Phanerochaete sordida* (Harazono and Nakamura 2005).

Comparison of litter-decomposing basidiomycete fungi with WRF for lignino-lytic enzymes production and decolorization of synthetic dyes (Poly B-411, Reac-tive Black 5, Reactive Orange 16 and RBBR) revealed that the highest MnP activ-ity was observed from the culture of *Collybia dryophila* with the activity more than 30 U L^{-1}. Strains with high levels of MnP and laccase activities were able to perform the fastest degradation of Poly B-411 while the decolorization of other dyes did not depend so strictly on enzyme activity (Baldrian and Snajdr 2006). Kariminiaae-Hamedaani et al. (2007) evaluated decolorization of 12 different azo, diazo and anthraquinone dyes by a new isolate of WRF strain L-25 and 84.9–99.6% decolorization efficiency was achieved by cultivation for 14 days using an initial dye concentration of 40 mg L^{-1}. MnP produced by strain L-25 was used for the enzymatic decolorization of dyes, thus it confirmed the capability of the enzyme in this process.

A partially purified MnP from *Bjerkandera adusta* was used for the decol-orization of several artificial dye baths. The most effective decolorization was recorded in dye bath of anthraquinone dye; Reactive Blue 19, diazo dyes; Reac-tive Black 5 (RB5) and Acid Orange 7 (Mohorcic et al. 2006). MnP production by *Phanerochaete chrysosporium* and the level of decolorization of 13 dyes was performed using static, agitated batch and continuous cultures. The decoloriza-tion efficiency was over 90% for 100 mg L^{-1} of Acid Black 1, Reactive Black 5, Reactive Orange 16 and Acid Red 27. A significant increment in primary post-metabolism biomass was observed in batch cultures with Acid Black 1 and Re-active Black 5. It was possible to explore the response of the continuous system during 32–47 days for these dyes (25–400 mg L^{-1}), obtaining greater than 70% decolorization for 400 mg L^{-1} (Urra et al. 2006). It was reported that the azo dyes; Congo Red, Orange G and Orange IV were efficiently decolorized by MnP puri-fied from *Schizophyllum* sp. (Cheng et al. 2007). However, purified MnP from *Ischnoderma resinosum* decolorized Reactive Black 5, Reactive Blue 19, Reac-tive Red 22 and Reactive Yellow 15 and the highest decolorization was achieved at acidic pH (Kokol et al. 2007).

Park et al. (2007) reported decolorization of six commercial dyes by 10 fungal strains exhibiting extracellular laccase and MnP activities. The decolorization mechanisms by *Funalia trogii* ATCC 200800 involved a complex interaction of enzyme activity and biosorption. This study suggested that it was possible to decolorize high concentrations of commercial dyes, which could be a great advancement in the treatment of dye containing wastewater. The conversion of azo dyes; Flame Orange and Ruby Red into their N-demethylated form and their polymerization by different oxidoreductases has been investigated (Pricelius et al. 2007). Laccase from *Pycnoporus cinnabarinus*, MnP from *Nematoloma frowardii* and the novel *Agrocybe aegerita* peroxidase used a similar mechanism to decolorize/degrade azo dyes. Susla et al. (2008) purified CIM1 MnP from *Dichomitus squalens* and this enzyme preparation was able to decolorize selected azo and anthraquinone dyes more rapidly than laccase. In vitro dye decolorization showed a synergistic cooperation of MnP and laccase. In the case of CSB degradation MnP prevented from the production of a differently colored substance that could be produced after CSB degradation by laccase-HOBT system. Bibi et al. (2009) investigated decolorization of some direct textile dyes by WRF *Ganoderma lucidum* IBL-05. Maximum decolorization ($83.78 \pm 5\%$) was observed for Solar Golden Yellow R at pH 4 and 30°C after 6th day. MnP was the main enzyme (256 ± 5 U mL^{-1}) secreted by *G. lucidum* IBL-05 which was responsible for the decolorization of textile dyes.

DyP, a unique dye-decolorizing enzyme, is a member of a novel heme peroxidase family (DyP-type peroxidase family). DyP isolated from the fungus *Thanatephorus cucumeris* Dec 1 (formerly named *Geotrichum candidum* Dec 1) is a 58 kDa glycoprotein having one heme as a cofactor. This enzyme requires H$_2$O$_2$ for all enzymatic reactions indicating that it functions as a peroxidase (Sugano et al. 2000; Saijo et al. 2005). DyP degraded all typical peroxidase substrates, but it also degraded hydroxyl-free anthraquinone, which is not a substrate of other peroxidases (Kim and Shoda 1999a; Sugano et al. 2000, 2006). This is a very important characteristic because many synthetic dyes are derived from anthraquinone compounds. Thus, DyP is a promising enzyme for the treatment of dye-contaminated water because it degrades azo and anthraquinone dyes effectively (Sato et al. 2004; Shakeri et al. 2007; Shimokawa et al. 2008; Sugano 2009). The degradation and decolorization of many xenobiotic compounds such as synthetic dyes, food coloring agents, molasses, organic halogens, lignin and kraft pulp effluents was investigated in the presence of a newly isolated fungus, *Geotrichum candidum* Dec 1 peroxidase (Kim and Shoda 1999b). Sugano et al. (2000) have isolated a new DyP from *Trametes cucumeris* Dec 1, which decolorized more than 30 types of synthetic dyes. It has been reported that DyP from *Trametes cucumeris* Dec 1 was able to decolorize a representative of anthraquinone dye, Reactive Blue 5 to light red-brown compounds (Sugano et al. 2006). A crude recombinant dye decolorizing peroxidase (rDyP) obtained from *Aspergillus oryzae* was used for the decolorization of anthraquinone dye, RBBR. In batch culture, equimolar batch addition of H$_2$O$_2$ and RBBR produced complete decolorization of RBBR by rDyP, with a turnover capacity of 4.75. In stepwise fed-batch addition of H$_2$O$_2$ and enzyme, the turnover capacity increased to 5.76 and 14.3, respectively. When H$_2$O$_2$ was added in continuous fed-batch and

15 Peroxidases as a Potential Tool for the Decolorization and Removal of Synthetic Dyes   473

1.6 mM dye was added in stepwise fed-batch mode, 102 g of RBBR was decolorized by 5000 U of crude rDyP in 650 min increasing the turnover capacity to 20.4 (Shakeri and Shoda 2007, 2008a).

VP has been recently recognized as a new group of ligninolytic peroxidases, together with LiP and MnP obtained from *Phanerochaete chrysosporium* (Martinez 2002). VP from *Bjerkandera adusta* was reported to show a hybrid molecular architecture between LiP and MnP. This hybrid combines the catalytic properties of these two peroxidases, being able to oxidize typical LiP and MnP substrates. The catalytic mechanism of VP is similar to that of classical peroxidases; the substrate oxidation is carried out by two-electron multistep reactions at the expense of H_2O_2 (Pogni et al. 2005). VP oxidizes Mn^{2+} to Mn^{3+} at around pH 5.0 while also oxidizing aromatic compounds including dyes at around pH 3.0, regardless of the presence of Mn^{2+} (Ruiz-Duenas et al. 2001; Tinoco et al. 2007). VPs from various sources *Pleurotus pulmonarius* (Camarero et al. 1996), *Pleurotus ostreatus* (Cohen et al. 2002), *Bjerkandera adusta* (Heinfling et al. 1998a, b, c), *Pleurotus eryngii* (Gomez-Toribio et al. 2001) and *Bjerkandera* sp. (Palma et al. 2000; Moreira et al. 2005, 2006) have shown their potential in azo dye degradation. Sugano et al. (2006) purified and characterized a new VP from the dye decolorizing microbe, *Thanatephorus cucumeris* Dec 1 (TcVP1). TcVP1 exhibited particularly high decolorizing activity towards azo dyes. Furthermore, co-application of TcVP1 and the DyP from *Thanatephorus cucumeris* Dec 1 was able to completely decolorize Reactive Blue 5. DyP decolorized Reactive Blue 5 to light red-brown compounds followed by decolorization of these colored intermediates to colorless products by TcVP1. Sugano (2009) for the first time proposed a possible in vitro degradation pathway for Reactive Blue 5 catalyzed by VP and suggested that VP is a novel bifunctional enzyme.

15.6.2 Plant Peroxidases

Various plant peroxidases and their dye and dye effluent decolorization and removal applications are listed in Table 15.3.

HRP has been reported to be an effective enzyme for the degradation of industrially important azo dyes (Bhunia et al. 2001; Mohan et al. 2005; Maddhinni et al. 2006). These investigators have evaluated specificity of HRP toward different dyes, such as Remazol Blue, Cibacron Red, Acid Black 10 BX and Direct Yellow 10. The enzyme activity for Remazol Blue was found to be far better at pH 2.5 than at neutral pH whereas Remazol Blue acted as a strong competitive inhibitor of HRP at neutral pH (Bhunia et al. 2001). Mohan et al. (2005) described the significance of HRP catalyzed reaction in the treatment of an acid azo dye, Acid Black 10 BX. The performance of HRP catalyzed reaction was found to be dependent upon the aqueous phase pH, contact time, H_2O_2, dye and HRP concentrations. The effective degradation of Direct Yellow 12 by HRP in the presence of H_2O_2 was found a viable approach for the degradation of azo dyes from aqueous solutions (Maddhinni et al. 2006). Ulson de Souza et al. (2007) investigated the potential of HRP for the

Table 15.3 Applications of plant peroxidases in dye decolorization. (Adapted from Husain (2010) Rev. Environ. Sci. Biotechnol. 9:117–140)

Source	Dye/effluent	Experimental conditions	Reference
HRP, Horseradish	Remazol Blue, Cibacron Red	HRP activity was better for Remazol Blue at pH 2.5 as compared at neutral pH. HRP exhibited broad substrate specificity toward a variety of azo dyes.	Bhunia et al. (2001)
	Methylene Blue and Azure B	N-demethylation of both dyes was carried by HRP, but exhibited much slower reaction kinetics than LiP and required higher H_2O_2 concentrations.	Ferreira-Leitao et al. (2003)
	Remazol Turquoize Blue G, Lancet Blue 2 R	The decolorization of Remazol Turquoise Blue G 133% and Lanaset Blue 2R were 59% and 94%, respectively. For textile effluent, the decolorization was 52%. The toxicity of the dyes towards *Daphnia magna* was decreased after enzymatic treatment.	Ulson de Souza et al. (2007)
	Acid Black 10BX	The decolorization and removal of dye was dependent upon the reaction time, pH and on the concentrations of dye, enzyme and H_2O_2.	Mohan et al. (2005)
	Direct Yellow 12	Direct Yellow 12 was decreased by HRP treatment, the oxidation capacity increased at pH 4.0 with increasing concentrations of HRP and H_2O_2.	Maddhinni et al. (2006)
	Orange II	The electrolytic experiments were carried out with 0.42 U mL^{-1} HRP at 0.5 V for electro-enzymatic degradation of Orange II. The removal of Orange II was partial due to its adsorption on the graphite felt.	Kim et al. (2005)
	Orange II	More than 90% Orange II degradation efficiency was maintained for 36 h during continuous operation.	Shim et al. (2007)
	BPB, Methyl Orange	Modified HRP showed good decolorization of dye; 8 to 24 or 32 mol L^{-1} at 300 mol L^{-1} H_2O_2: Decolorization for BPB and Methyl Orange was 1.8% and 12.4%.	Liu et al. (2006)
TP, Turnip	Five acid dyes	The decolorization of dyes (40–170 mg L^{-1}) was 62–100% in the presence of 2.0 mM HOBT at pH 5.0 and 40°C.	Kulshrestha and Husain (2007)
	Direct dyes	The decolorization of direct dyes was maximal in the presence of 0.6 mM redox mediator at pH 5.5 and 30°C.	Matto and Husain (2007)
	Textile carpet effluent	Textile carpet effluent red and blue were decolorized to 75 and 80% by TP (0.423 U mL^{-1}) at pH 5.0 and 40°C	Husain and Kulshrestha (2010)
TMP, Tomato	Textile carpet effluent	TMP (0.705 U mL^{-1}) decolorized effluent red and blue to 69% and 59% at pH 6.0 and 40°C, respectively	Husain and Kulshrestha (2010)
TMP, Tomato	Direct dyes	Dyes were maximally decolorized at pH 6.0 and 40°C	Matto and Husain (2008)

Table 15.3 (continued)

Source	Dye/effluent	Experimental conditions	Reference
BGP, Bitter gourd	Twenty one textile and non-textile dyes and their mixture	Various dyes were decolorized from 7 to 100% in the presence of 1.0 mM HOBT, 0.6 mM H_2O_2 at pH 5.6 and 37°C in 1 h.	Akhtar et al. (2005b)
	Textile effluent	The maximum effluent decolorization was 70% in the presence of 1.0 mM HOBT at pH 5.0 and 40°C. Entrapped BGP retained 59% decolorization reusability after its tenth repeated use.	Matto and Husain (2009), Matto et al. (2010)
	Disperse Red 17 (DR 17), Disperse Brown 1 (DB 1)	Disperse dyes were decolorized to 90% and 65% in the presence of 1.0 mM HOBT in 0.1 M glycine HCl buffer, pH 3.0 at 37°C in 1 h.	Satar and Husain (2009a)
	DR 17 and DB 1	Maximum decolorization of these dyes in the presence of 0.75 mM H_2O_2 in 30 min at pH 3.0 and 40°C; was 60% for DR 17 in the presence of 0.2 mM phenol and for DB 1 was only 40% in the presence of 0.4 mM phenol.	Satar and Husain (2010)
	Tannery effluent	Tannery effluent was maximally decolorized by 0.6 mL (4500 U) of the enzyme at pH 5.0 and 40°C in 4 h.	Subramaniam et al. (2010)
HRP/SBP, Horseradish/ Soybean	Stilbene dye (Direct Yellow 11) & methine dye, Basazol 46 L	Both enzymes were quite effective in chromophore removal. When compared to laccase in combination with a mediator (ABTS), SBP was more effective at oxidative dye removal, especially for the methine dye.	Knutson et al. (2005)
WRP, White radish	Reactive Red 120 Reactive Blue 171	Maximum decolorization of the dyes was 56–81% at pH 5.0 and 40°C in 1 h in the presence of 1.0 mM HOBT	Satar and Husain (2009c)

decolorization of textile dyes and industrial effluents. The obtained decolorization of Remazol Turquoise Blue G 133%, Lanaset Blue 2R and the textile effluent was 59, 94 and 52%, respectively.

The potential of peroxidases from *Momordica charantia* in decolorizing industrially important dyes has been investigated. *Momordica charantia* peroxidase could decolorize 21 dyes used in textile and other important industries (Akhtar et al. 2005a, b). The greater fraction of the color was removed when the textile dyes were treated with increasing concentrations of enzyme but four out of eight reactive dyes were recalcitrant to decolorization by BGP. The rate of decolorization was enhanced when the dyes were incubated with a fixed quantity of enzyme for increasing times. Decolorization of non-textile dyes resulted in the degradation and removal of dyes from the solution without any precipitate formation. Thus it indicated that BGP was an effective biocatalyst for the treatment of effluents containing recalcitrant dyes from textiles, dye manufacturing and printing industries. Recently, it has been reported that BGP could be used in the detoxification and biotransformation of several aromatic amines, phenols and dyes present in wastewater/industrial effluents (Akhtar and Husain 2006; Kulshrestha and Husain 2007). Catalytic efficiency of BGP has been enhanced in presence of redox mediators. Complex mixtures of dyes were also significantly decolorized by BGP in the presence of 1.0 mM 1-hydroxybenzotriazole (HOBT) (Akhtar et al. 2005b; Matto and Husain 2009a). Recently some investigators have used BGP for the decolorization of tannery effluent. The maximum decolorization was found at pH 5.0, 40°C for 4 h with 4500 U of the enzyme (Subramaniam et al. 2010).

Multiple efforts have been directed towards optimization of processes in which peroxidases from cheap plant sources were used to remove dyes from polluted water (Husain 2006). Matto and Husain (2008) described the role of partially purified TMP in decolorizing direct dyes; Direct Red 23 and Direct Blue 80. These dyes were maximally decolorized by TMP at pH 6.0 and 40°C. The absorption spectra of the treated dyes exhibited marked differences in the absorbance at various wavelengths as compared to untreated dyes. Decolorization and decontamination of two textile carpet industrial effluents by TMP has been investigated (Husain and Kulshrestha 2010). Textile carpet effluent red and blue were decolorized by 0.705 U mL^{-1} TMP at pH 6.0 and 40°C to 69 and 59%, respectively. The TMP treated effluents exhibited significant loss of TOC.

Turnip roots, which are readily grown in several countries, are a rich source of peroxidase activity. Because of the kinetic and biochemical properties, TP has a high potential as an economic alternative to HRP (Duarte-Vazquez et al. 2002; Kulshrestha and Husain 2006a; Matto and Husain 2006; Husain and Kulshrestha 2010). Peroxidases from turnip roots were highly effective in decolorizing direct dyes with a wide spectrum of chemical groups (Matto and Husian 2007). Dye solutions, containing 40–170 mg L^{-1} dye in the presence of 2.0 mM HOBT were successfully treated by TP at pH 5.0 and 40°C. Complex mixtures of acid dyes were also significantly decolorized by TP in the presence of HOBT (Kulshrestha and Husain 2007). Wood shaving bound TP has successfully been employed for the

15 Peroxidases as a Potential Tool for the Decolorization and Removal of Synthetic Dyes   477

treatment of direct dyes in batch as well as in continuous reactors in the presence of 0.6 mM HOBT (Matto and Husain 2009b).

15.6.3 Microperoxidase-11 (MP-11)

Microperoxidases are small heme-peptides obtained by proteolytic digestion of cytochrome c, exhibiting peroxidase activity. These enzymes consisted of a short or medium length polypeptide chain, covalently linked to an iron protoporphyrin IX moiety via two thioether bonds involving Cys residues at the c-porphyrin A and B pyrrole rings (Tullio et al. 2005). The decolorization of water insoluble synthetic dyes by MP-11 in 90% methanol was attempted. MP-11 exhibited effective decolorization activity against azo and anthraquinone type of dyes. The degradation pathway for Solvent Orange 7 was investigated and it showed that MP-11 catalyzed the oxidative cleavage of azo linkage to generate 1,2-naphthoquinone and 2,4-dimethylphenol as key intermediates (Wariishi et al. 2002). MP-11 effectively catalyzed the oxidative decolorization of an azo dye, Solvent Yellow 7 and an anthraquinone dye, Solvent Blue 11 in hydrophobic organic solvent in the presence of H_2O_2 (Okazaki et al. 2002). Pisklak et al. (2006) reported the role of MP-11 in the oxidation of dyes, Amplex® Ultra Red and Methylene Blue.

15.7 Dye Removal by Immobilized Peroxidases

Table 15.4 shows various immobilized peroxidase preparations and their application in the removal and decolorization of various synthetic dyes and dye effluents.

Enzymatic detoxification and decolorization has been successfully employed for the removal of colored compounds from wastewater because the resulted polymeric products were insoluble and therefore such products were easily removed by filtration or centrifugation (Husain 2006; Yang et al. 2008). Nevertheless, the use of free enzymes also showed some drawbacks such as susceptibility to inactivation by the product, non-reusability, thermal instability, susceptibility to attack by proteases, high sensitivity to several denaturing agents and the impossibility of separating and reusing of free catalyst at the end of the reaction (Husain and Jan 2000; Husain 2006; Husain and Husain 2008). In order to overcome such constraints, enzyme immobilization is one of the best alternatives to exploit enzymes at industrial level.

The main objective of immobilization is to apply the benefits of heterogeneous catalysis to the soluble enzymes used routinely in academic biosciences and industrial biotechnology (Chen et al. 2009; Cho et al. 2009; Enayatzamir et al. 2009; Rotkova et. al. 2009). Benefits offered by immobilization can be summarized as the retention of high concentration of catalysts in the reactor, control of catalyst microenvironment, quantitative and rapid removal of the catalyst, enhanced stability, easier product recovery and purification, protection of enzymes against denaturing

Table 15.4 Use of immobilized peroxidases in dye removal. (Adapted from Husain and Ulber (2011) Crit. Rev. Environ. Sci. Technol. 41(8): 770–804)

Name of enzyme	Type of source	Name of support	Type of dye/s or effluent treated	Reference/s
Catalase-peroxidase	Bacillus SF	Various alumina based supports	Textile bleaching effluent	Fruhwirth et al. (2002)
Peroxidase	*Saccharum uvarum*	Modified polyethylene	Supranol Green, Procion Brilliant Blue H-7G, Procion Navy Blue HER, Procion & Green HE-4BD	Shaffiqu et al. (2002)
HRP	Horseradish	Graphite, Porous Celite R646	Orange II	Kim et al. (2005); Shim et al. (2007); Cho et al. (2009)
		Polyacrylamide gel	Acid Black 10BX	Mohan et al. (2005)
		Alginate, polyacrylamide	Polyacrylamide entrapped HRP oxidized greater concentration of Direct Yellow 12 as compared to alginate entrapped HRP & free HRP	Maddhinni et al. (2006)
		Cellulose, chitosan, and ethylene-vinyl alcohol copolymer	Orange II, Crystal Violet, Astrogen Red, & RBBR	Maki et al. (2006)
		Citraconic-anhydride	Bromophenol Blue and Methyl Orange	Liu et al. (2006)
		Modified PET fibers	Effectively removes azo dyes from aqueous solutions	Arslan (2010)
BGP	Bitter gourd	Con A-Sephadex	Reactive textile dyes	Akhtar et al. (2005b)
		Calcium alginate-starch beads	Industrial effluent	Matto and Husain (2009b); Matto et al. (2009)
TMP	Tomato	Con A-cellulose	Reactive dyes	Matto and Husain (2008)
TP	Turnip	Con A-wood shaving	Textile carpet effluents	Matto and Husain (2009a)
TP/TMP	Turnip/Tomato	Con A-cellulose	Reactive Red 120 & Reactive Blue 170	Husain and Kulshrestha (2010)
WRP	White radish	Celite 545	Decolorization of recalcitrant azo dyes	Satar and Husain (2009c)
MnP	*Phanerochaete chrysosporium*	Ca-alginate beads	Direct Violet 51, Reactive Black 5, Ponceau Xylidine and Bismark Brown R in successive batch cultures	Enayatzamir et al. (2009)
FSP	Feenugreek seed	Membrane reactor	Textile industrial effluent	Husain et al. (2010)
MP-11	Undecapeptide from horse heart cytochrome C	Silica gel	Water insoluble synthetic dyes	Wariishi et al. (2002)
		Hybrid periodic mesoporous organo-silica materials & a nano-crystal-line metal organic	Amplex-Ultra Red & Methylene Blue	Pisklak et al. (2006)
rDyP	*Aspergillus oryzae*	Silica-based mesoporous materials	RBBR, Direct dyes & their mixtures	Shakeri and Shoda (2008a)

15 Peroxidases as a Potential Tool for the Decolorization and Removal of Synthetic Dyes   479

factors, enzyme reusability, minimum effluent problems and material handling (Wang et al. 2001; Duran et al. 2002; Magri et al. 2005; Pedroche et al. 2007). However, the high cost and low yield of immobilized enzyme preparations are two main limitations in their applications at industrial level (Mateo et al. 2007). Thus, the success of enzyme immobilization technology clearly depends on the choice of carrier and the method of immobilization. Several techniques have been employed to immobilize enzymes on solid supports. They are mainly based on chemical and physical mechanisms (Khan et al. 2006; Kulshrestha and Husain 2006a, b; Matto and Husain 2006; Fatima and Husain 2007a; Gomez et al. 2007; Quintanilla-Guerrero et al. 2008; Yang et al. 2008; Matto and Husain 2009c). Immobilized peroxidases have been used in various fields but their recent use for the treatment of dyes present in wastewater has attracted the attention of enzymologists (Husain et al. 2009; Husain 2010).

An electroenzymatic process is an interesting approach that combines enzyme catalysis and electrode reactions. Kim et al. (2005) studied an electroenzymatic method which used an immobilized HRP to degrade Orange II (azo dye) within a two-compartment packed-bed flow reactor. The electroenzymatic degradation of Orange II was carried out by 0.42 U mL^{-1} of HRP at 0.5 V. The overall application of the electroenzymatic approach led to a greater degradation of dye than the use of electrolysis alone. Degradation of Orange II by an electroenzymatic method using HRP bound to an inexpensive and stable inorganic support Celite R-646 beads bound with 2% aqueous glutaraldehyde was studied in a continuous electrochemical reactor with in situ generation of H_2O_2 (Shim et al. 2007). Based on the parametric studies, over 90% of Orange II was degraded during continuous operation for 36 h. From the results of GC/MS analysis, degradation products were identified and a possible breakdown pathway for Orange II was proposed.

Acrylamide gel immobilized HRP exhibited effective performance compared to alginate entrapped and free HRP in the removal of Acid Black 10 BX. Alginate entrapped HRP showed inferior performance over the free enzyme due to the consequence of non-availability of the enzyme to the dye molecule (Mohan et al. 2005). Maddhinni et al. (2006) investigated the decolorization of Direct Yellow 12 by soluble and immobilized HRP under various experimental parameters; pH, H_2O_2, dye and enzyme concentrations. The efficiency of polyacrylamide entrapped HRP for the oxidation of Direct Yellow 12 was higher followed by alginate entrapped HRP as compared to free HRP. The alginate and polyacrylamide immobilized HRP preparations were further used 2–3 times for the removal of the same dye with lower efficiency, respectively. Methyl Orange and BPB removal capability of citraconic anhydride-modified HRP was compared with those of free HRP (Liu et al. 2006). Upon chemical modification, the decolorization efficiency was increased by 1.8 and 12.4% for BPB and Methyl Orange, respectively. Lower dose of citraconic anhydride-modified HRP was required than that of free enzyme for the decolorization of both dyes to obtain the similar decolorization efficiency. Citraconic anhydride-modified HRP exhibited a good decolorization of dye over a wide range of dye concentrations from 8 to 24 or 32 μmol L^{-1} at 300 μmol L^{-1} H_2O_2, which would meet industrial expectations.

MnP oxidizes a wide range of substrates, rendering it as an interesting enzyme for potential applications. The significant decolorization of azo dyes in static and shaky situation by gelatin-immobilized MnP was studied. There was no loss of immobilized enzyme activity after two repeated uses in batch process (Xiao-Bin et al. 2007). MP-11 entrapped in the system of bis (2-ethylhexyl) sulphosuccinate sodium salt (AOT)-reversed micelles exhibited peroxidase activity in the presence of H_2O_2. It effectively catalyzed the oxidative decolorization of an azo dye, Solvent Yellow 7 and an anthraquinone dye, Solvent Blue 11 in the hydrophobic organic solvent (Okazaki et al. 2002). MP-11 was immobilized in hybrid periodic mesoporous organosilica materials and in a nano-crystalline metal organic framework. The conversion of Amplex® Ultra Red and Methylene Blue to their respective oxidation products was catalyzed successfully in the presence of immobilized MP-11 (Pisklak et al. 2006). MnP secreted by *Phanerochaete chrysosporium* immobilized into Ca-alginate beads could decolorize recalcitrant azo dyes Direct Violet 51, Reactive Black 5, Ponceau Xylidine and Bismark Brown R in successive batch cultures (Enayatzamir et al. 2009).

Fruhwirth et al. (2002) used a catalase peroxidase from the newly isolated *Bacillus* SF to treat textile-bleaching effluents. Various alumina-based carriers of different shapes were used to immobilize the enzyme. Bleaching effluent was treated in a horizontal packed-bed reactor containing 10 kg of the immobilized catalase peroxidase at a textile-finishing company. The treated liquid (500 L) was reused within the company for dyeing fabrics with various dyes, resulting in acceptable color differences of below Delta E* = 1.0 for all dyes.

BGP immobilized on concanavalin A (Con A)-Sephadex bioaffinity support was used for the decolorization of industrially important dyes from polluted water (Akhtar et al. 2005a). Maximum decolorization was obtained at pH 3.0 and 40°C. This enzyme was repeatedly used for the decolorization of eight reactive textile dyes and after 10th repeated use the immobilized enzyme retained nearly 50% of its decolorization activity. The mixtures of dyes were also successfully decolorized by immobilized BGP. Decolorization and decontamination of two textile carpet industrial effluents was carried out by using Con A-cellulose bound turnip and tomato peroxidases (Husain and Kulshrestha 2010). Textile carpet effluent red and blue were decolorized to 78 and 84% by immobilized TP (0.423 U mL^{-1}), respectively. However, 0.705 U mL^{-1} of immobilized TMP decolorized effluent red to 73% and effluent blue to 74%. The immobilized TP treated effluent even exhibited significant loss of TOC from the solution. TMP immobilized on a bioaffinity support, Con A-cellulose was highly effective in decolorizing direct dyes as compared to free TMP (Matto and Husain 2008). More than 70% of Direct Red 23 and Direct Blue 80 were decolorized by TMP at pH 6.0 and 40°C. Immobilized TMP showed a lower Michaelis constant, K_m than the free enzyme for the direct dyes. Shakeri and Shoda (2008b) demonstrated the immobilization of rDyP produced from *Aspergillus oryzae* using silica-based mesoporous materials, FSM-16 and AlSBA-15. rDyP immobilized on FSM-16 at pH 4.0 decolorized eight sequential batches of an anthraquinone dye, RBBR in repeated-batch decolorization process.

15 Peroxidases as a Potential Tool for the Decolorization and Removal of Synthetic Dyes   481

Satar and Husain (2009c) reported a comparative decolorization of reactive dyes: Reactive Red 120 and Reactive Blue 171 by soluble and Celite immobilized WRP in the presence of different redox mediators; HOBT, syringaldehyde, VA and VLA. The investigated dyes were decolorized to different extents in the presence of tested redox mediators. However, the maximum decolorization of the dyes was observed at pH 5.0 and 40°C in 1 h. The operational stability performance of soluble and immobilized WRP in treatment of dyes was checked in the presence of various denaturing and inhibiting agents such as sodium azide, organic solvents and mercuric chloride. Immobilized peroxidase decolorized dyes more effectively in batch process. Efficiency of immobilized peroxidase was checked in a continuous reactor where the immobilized enzyme exhibited 73% decolorization of Reactive Red 120 even after 1 month of continuous operation of the reactor.

Activated fibers were used as a new support material for the immobilization of HRP. Poly(ethylene terephthalate) (PET) fibers were grafted with glycidyl methacrylate (GMA) using benzoyl peroxide (Bz_2O_2) as initiator. 1,6-diaminohexane (HMDA) was then covalently attached to this GMA grafted PET fibers. HMDA-GMA-g-PET fibers were activated with glutaraldehyde and HRP was successfully immobilized. Both free enzyme and immobilized enzyme were used in a batch process for the degradation of azo dye. About 98% of azo dye removal was observed with immobilized HRP, while 79% of azo dye removal was found with the free HRP. 45 min of the contact time is sufficient for the maximum azo dye removal. The HRP immobilized on modified PET fibers were very effective for removal of azo dye from aqueous solutions (Arslan 2010).

15.8   Role of Redox Mediators in Peroxidase Catalyzed Decolorization of Dyes

Redox mediators are compounds that speed up the reaction rate by shuttling electrons from the biological oxidation of primary electron donors or from bulk electron donors to the electron-accepting aromatic compounds (Fabbrini et al. 2002). Redox mediators provide high redox potentials (>900 mV) to attack recalcitrant structural analogs and are able to migrate into aromatic structure of the compounds and accelerate reactions by lowering the activation energy of the total reaction. In some cases, the presence of these mediators might even be a prerequisite for the reaction to take place (van der Zee and Cervantes 2009). It has been shown that these small molecules acted as electron transfer mediators and were able to oxidize non-phenolic compounds, thus expanding the range of compounds that can be oxidized by enzymes (Crestini et al. 2003). The catalytic effect of such organic molecules with redox mediating properties on the bio-transformation of a wide variety of organic and inorganic compounds has been extensively explored (Dos Santos et al. 2004; Guo et al. 2008; Husain and Husain 2008; Jing et al. 2009). The need and nature of redox mediator for the degradation of a specific dye depends on the source of enzyme (Xu et al. 2000).

The role and mechanism of action of laccase-mediator system is well characterized and can also be applied for other enzymes (Couto et al. 2005b). Laccase catalyzed its substrate into cation radical which acted as a redox mediator, it is a short-lived intermediates that co-oxidized non-substrates. These cation radicals were formed by two mechanisms; either the redox mediator performed one-electron oxidation of the substrate to a free radical cation (Xu et al. 2000, 2001), or it abstracted a proton from the substrate by converting it into a radical (Fabbrini et al. 2002). For example, 2,2-azino-bis-(3-ethylbenzthiazoline-6-sulfonic acid) (ABTS) acted by the first mechanism (Potthast et al. 2001) whereas HOBT followed the second mechanism (Fabbrini et al. 2002; Hirai et al. 2006). A correlation between the enzyme redox potential and its activity toward substrates has been described. The driving force for the redox reaction catalyzed by oxidoreductive enzymes is expected to be proportional to the difference between redox potentials of oxidizing enzyme and reducing substrate/dye (Zille et al. 2004; Sadhasivam et al. 2009). Among the mediators, those presenting the $>$N–OH moiety; HOBT, N-hydroxyphthalimide and violuric acid (VLA) proved very efficient towards benzylic substrates, through a radical H-abstraction route of oxidation involving an aminoxyl radical ($>$N–O$^{.}$) intermediate (d'Acunzo et al. 2006).

The presence of oxidizing mediators exhibited an enhancement in the decolorization of dyes (Astolfi et al. 2005; Tinoco et al. 2007). LiP from *Bjerkandera adusta* showed low activity with most of the azo and phthalocyanine dyes. However, the specific activity increased 8–100-fold when VA was added to the reaction mixture (Heinfling et al. 1998b). Maximum decolorization achieved by partially purified LiP from *Phanerochaete chrysosporum* was 80% for Procion Brilliant Blue HGR, 83% for Ranocid Fast Blue, 70% for Acid Red 119 and 61% for Navidol Fast Black MSRL (Verma and Madamwar 2002a). This decolorization efficiency was observed at 0.2 and 0.4 mM H_2O_2, 2.5 mM VA and pH 5.0 after 1 h. LiP produced by *Trametes versicolor* decolorized RBBR in the presence as well as in the absence of VA (Christian et al. 2005). Decolorization of Reactive Brilliant Red K-2BP by LiP with higher addition of H_2O_2 and VA was enhanced to 89%, whereas decolorization by MnP was optimized only with a suitable dose of H_2O_2 (0.1 mM) and decreased by the addition of Mn^{2+} (Yu et al. 2006). Purified MnP decolorized Reactive Black 5, Reactive Blue 19, Reactive Red 22 and Reactive Yellow 15 whereas laccase was ineffective to decolorize Reactive Black 5 and Reactive Red 22. However, all these dyes were decolorized after addition of redox mediators; VLA and HOBT (Kokol et al. 2007). *Phanerochaete chrysosporium* MnP catalyzed decolorization of azo dyes; Direct Blue 15, Direct Green 6, and Congo Red was enhanced on addition of Tween-80 (Urek and Pazarlioglu 2005).

Low molecular mass redox mediators like ABTS have been found necessary for laccase-catalyzed decolorization of most of the dyes (Lu et al. 2005, 2007). Decolorization of RBBR was only observed when redox mediators, VLA, HOBT and phenothiazine-10-propionic acid (PTPA) were added together with the laccase (Soares et al. 2001). However, RBBR was completely decolorized within 20 min in the presence of 5.7 mM VA while this dye was decolorized at about a twofold slower rate in the presence of 11.0 mM HOBT. HOBT was found to be the most effective mediator and it showed decolorization of Sella Solid Red to 88% in 10 min and Luganil Green to 49% in 20 min by laccase (Couto et al. 2005b). The addition of 2.0 mM HOBT improved the rate of laccase catalyzed decolorization of Direct Red 28, Reactive Black 5, Acid Blue 25 and Azure B by 17, 63, 12 and 56%, respectively (Claus et al. 2002). Decolorization of Acid Blue 225, Acid Violet 17 and Reactive Black 5 by laccase was increased to 2–6-fold in the presence of HOBT (Nyanhongo et al. 2002). Murugesan et al. (2007b) reported the presence of HOBT was essential for the decolorization of Reactive Black 5 by purified laccase from *Pleurotus sajor-caju*. The potential of crude and partially purified WL1 laccase from *Trichoderma harzianum* for the decolorization of synthetic dyes, Rhodamine 6G, Erioglaucine and Trypan Blue has been evaluated in the presence of HOBT (Sadhasivam et al. 2009). Moldes and Sanroman (2006) have studied the effect of redox mediators on dye decolorization by laccase isoenzymes from *Trametes versicolor*. All the tested redox mediators; HOBT, promazine, *p*-hydroxybenzoic acid and 1-nitroso-2-naphthol-3,6-disulfonic acid led to higher dye decolorization than that obtained without mediator addition. Promazine was the most effective mediator while *p*-hydroxybenzoic acid had no significant effect on the decolorization of dye.

Sequential decolorization of reactive dyes was carried out by a laccase mediator system using VA and HOBT as mediators (Tavares et al. 2008a). VA resulted in a high level of decolorization on the first and second cycles for Reactive Blue 114 ($\geq$95%), Reactive Yellow 15 and Reactive Red 239 ($\geq$80%) while for Reactive Black 5 a slightly lower value (70%) was observed on the second cycle. The degree of Reactive Blue 114 decolorization remained 90% after the third cycle and about 60% after seven cycles. When HOBT was used as mediator a slight decrease in decolorization efficiency was observed. Tavares et al. (2008b) performed a screening test on the degradation of six reactive textile dyes by several laccase mediators; ABTS, HOBT, N-hydroxyacetanilide, polioxometalates, VLA and 2, 2, 6, 6-tetramethylpiperidin-1-yloxy (TEMPO). ABTS was found to be the most effective mediator. The efficiency of ABTS depended on the type of dye, pH, temperature and dye concentration. Hu et al. (2009) reported that the decolorization capabilities of laccase/mediator system were related to the types of mediator, the dye structure and decolorization conditions. After screening 14 different compounds with Indigo Carmine (indigoid dye) as a substrate, phloroglucinol, thymol, and VLA were selected as laccase mediators resulting in 90–100% decolorization of this dye in 1 h. Thus, these three compounds were used as mediators for the decolorization of other four dyes. VLA was very effective in decolorizing RBBR (anthraquinoid dye), Coomassie Brilliant Blue G-250 (CBB, triphenylmethane dye), and Acid Red (diazo dye). Thymol was able to mediate decolorization of RBBR and Azure A (heterocyclic dye). Phloroglucinol had no mediating capability in decolorization of these dyes. Two recalcitrant dyes; stilbene dye, Direct Yellow 11 and methine dye, Basazol 46L were effectively decolorized by HRP, SBP and laccase in the presence of ABTS as a mediator (Knutson et al. 2005). The stilbene dye, Direct Yellow 11 responded to both SBP and laccase/ABTS. SBP was more effective in the oxidative removal of methine dye, Basazol 46L as compared to the other peroxidases.

The presence of redox mediators VA, acetosyringone or TEMPO as oxidizing mediators generally enhanced the rate of dye decolorization by VP from *Bjerkandera adusta* (Tinoco et al. 2007). Akhtar et al. (2005b) have demonstrated the decolorization of 21 different reactive textile and other industrially important dyes by BGP. The decolorization of dyes and their mixtures was drastically increased in the presence of 1.0 mM HOBT. Textile effluent was also significantly decolorized by BGP in the presence of 1.0 mM HOBT (Matto and Husain 2009a). The decolorization of acid dyes by TP was significantly enhanced in the presence of 2.0 mM HOBT (Kulshrestha and Husain 2007). Redox mediated decolorization of direct dyes by TP has been extensively studied (Matto and Husain 2007, 2009b). Dyes were recalcitrant to the action of enzyme without a redox mediator and were decolorized to different extents in the presence of all used redox mediators (2.0 mM); vanillin, L-histidine and VLA. However, 0.6 mM HOBT emerged as a potential redox mediator for TP catalyzed decolorization of direct dyes and their mixtures. A comparative study was performed for the decolorization and removal of two textile carpet industrial effluents by TP and TMP (Husain and Kulshrestha 2010). The decolorization of effluents was enhanced in the presence of 2.0 mM HOBT. Matto and Husain, (2008) investigated decolorization of Direct Red 23 and Direct Blue 80 by TMP in the presence of HOBT, phenol, vanillin, VLA, VA and syringaldehyde. HOBT decolorized the direct dyes significantly. Decolorization of Acid Red 27 by azoreductase was enhanced effectively by quinone redox mediators; lawsone and menadione (Liu et al. 2009).

Effect of nine different redox mediators; bromophenol, 2,4-dichlorophenol, guaiacol, HOBT m-cresol, quinol, syringaldehyde, violuric acid, and vanillin on decolorization of disperse dyes; Disperse Red 17 and Disperse Brown 1 by BGP has been investigated. Among these redox mediators, HOBT was the most effective mediator. Disperse Red 17 was maximally decolorized to 90% in the presence of 0.1 mM HOBT while Disperse Brown 1 decolorized 65% in the presence of 0.2 mM HOBT in 1 h at pH 3.0 and 40°C (Satar and Husain 2009a). Further same workers investigated decolorization of these disperse dyes by BGP in the presence of redox mediator, phenol. BGP (0.215 U mL^{-1}) could decolorize about 60% of Disperse Red 17 in the presence of 0.2 mM phenol, whereas Disperse Brown 1 was decolorized by only 40% in the presence of 0.4 mM phenol. Maximum decolorization of dyes was achieved in the presence of 0.75 mM H_2O_2 in a buffer of pH 3.0 and 40°C in 30 min. (Satar and Husain 2009). Husain et al. (2010) have investigated the role of six redox mediators for the decolorization of a textile effluent by fenugreek seeds peroxidase (FSP). However, the effluent was decolorized maximally in the presence of 1.0 mM HOBT. The decolorization of textile effluent in batch process by this enzyme was 85% in 5 h whereas the complete decolorization of textile effluent by membrane- entrapped peroxidase was observed within 11 h of its operation.

15.9 Future Outlook

The most important obstacles to commercial application of peroxidases are the lack of sufficient enzyme stocks and the cost of redox mediators. Marked progress has been made over the last few years to solve these problems and it is expected that peroxidases will be able to compete with other known effective processes. Thus, efforts have to be made in order to achieve cheap overproduction of this biocatalyst in heterologous hosts and also their modification by chemical means or protein engineering to obtain more robust and active enzymes. Enzymes obtained from cheap sources and immobilzed on inexpensive supports have great future for the treatment of colored effluents. The use of simple, inexpensive and nontoxic redox mediator will increase the range of treatment of aromatic colored pollutents by using peroxidases. On the other hand, the development of an effective and suitable combined system of immobilized peroxidase and redox mediator will be most effective for the remediation of such compounds.

Acknowledgments The authors are thankful to Dr. Faisal M. Al-Tobaigy, Dean, Faculty of Applied Medical Sciences, Jazan University, Jazan, Kingdom of Saudi Arabia for constant encouragement and support during writing of this book chapter.

References

Abadulla E, Tzanov T, Costa S, Robra KH, Cavaco-Paulo A, Gubitz GM (2000) Decolorization and detoxification of textile dyes with a laccase from *Trametes hirsuta*. Appl Environ Microb 66:3357–3362

Akhtar S, Husain Q (2006) Potential applications of immobilized bitter gourd (*Momordica charantia*) peroxidase in the removal of phenols from polluted water. Chemosphere 65:1228–1235

Akhtar S, Khan AA, Husain Q (2005a) Potential of immobilized bitter gourd (*Momordica charantia*) peroxidases in the decolorization and removal of textile dyes from polluted wastewater and dyeing effluent. Chemosphere 60:291–301

Akhtar S, Khan AA, Husain Q (2005b) Partially purified bitter gourd (*Momordica charantia*) peroxidase catalyzed decolorization of textile and other industrially important dyes. Biores Technol 96:1804–1811

Aksu Z, Tezer S (2000) Equilibrium and kinetic modelling of biosorption of Remazol Black B by *R. arrhizus* in a batch system: effect of temperature. Process Biochem 36:431–439

Alam MZ, Mansor MF, Jalal KCA (2009) Optimization of decolorization of methylene blue by lignin peroxidase enzyme produced from sewage sludge with *Phanerocheate chrysosporium*. J Hazard Mater 162(2–3):708–715

Al-Aseeri M, Bu-Ali Q, Haji S, Al-Bastaki N (2007) Removal of Acid Red and sodium chloride mixtures from aqueous solutions using nanofiltration. Desalination 206:407–413

Anjaneyulu Y, Chary NS, Raj, DSS (2005) Decolorization of industrial effluents-available methods and emerging technologies: a review. Rev Environ Sci Biotechnol 4:245–273

Arslan I. (2001) Treatability of a simulated disperse dye-bath by ferrous iron coagulation, ozonation, and ferrous iron-catalyzed ozonation. J Hazard Mat 85:229–241

Arslan M (2010) Immobilization horseradish peroxidase on amine-functionalized glycidyl methacrylate-*g*-poly(ethylene terephthalate) fibers for use in azo dye decolorization Polym. Bull. doi:1007/s00289-010-0316-8

Assadi MM, Jahangiri MR (2001) Textile waste water treatment by *Aspergillus niger*. Desalination 141:1–6

Asses N, Ayed L, Bouallagui H, Ben Rejeb I, Gargouri M, Hamdi M (2009) Use of *Geotrichum candidum* for olive mill wastewater treatment in submerged and static culture. Biores Technol 100:2182–2188

Astolfi P, Brandi P, Galli C, Gentili P, Gerini MA, Greci L, Lanzalunga O (2005) New mediators for the enzyme laccase: mechanistic features and selectivity in the oxidation of non-phenolic substrates. New J Chem 29:1308–1317

Ayala M, Pickard MA, Vazquez-Duhalt R (2008) Fungal enzymes for environmental purposes, a molecular biology challenge. J Mol Microb Biotechnol 15:172–180

Balan, DSL, Monteiro, RTR (2001) Decolorization of textile Indigo dye by ligninolytic fungi. J Biotechnol 89:141–145

Baldrian P, Snajdr J (2006) Production of ligninolytic enzymes by litter-decomposing fungi and their ability to decolorize synthetic dyes. Enzyme Microb Technol 39:1023–1029

Banat IM, Nigam P, Singh D, Marchant R (1996) Microbial decolorization of textile-dye-containing effluents: a review. Biores Technol 58:217–227

Beak MH, Ijagbemi CO, Kim DS (2009) Treatment of Malachite Green-containing wastewater using poultry feathers as adsorbent. J Environ Sci Health A 44:536–542

Bhunia A, Durani S, Wangikar PP (2001) Horseradish peroxidase catalyzed degradation of industrially important dyes. Biotechnol Bioeng 72:562–567

Bibi I, Bhatti HN, Asgher M (2009) Decolourisation of direct dyes with manganese peroxidase from white rot basidiomycete *Ganoderma lucidum*-IBL-5. Canad J Chem Eng 87(3):435–440

Biswas MM, Taylor KE, Bewtra JK, Jatinder, Biswas N (2007) Enzymatic treatment of sulfonated aromatic amines generated from reductive degradation of reactive azo dyes. Water Environ Res 79:351–356

Boer CG, Obici L, de Souza, CGM, Peralta RM (2004) Decolorization of synthetic dyes by solid state cultures of *Lentinula (Lentinus) edodes* producing manganese peroxidase as the main ligninolytic enzyme. Biores Technol 94:107–112

Borchert M, Libra JA (2001) Decolorization of reactive dyes by the white rot fungus *Trametes versicolor* in sequencing batch reactors. Biotechnol Bioeng 75:313–321

Bras R, Gomes A, Ferra, MIA, Pinheiro HM, Goncalves IC (2005) Monoazo and diazo dye decolorization studies in a methanogenic UASB reactor. J Biotechnol 115:57–66

Camarero S, Bockle B, Martinez MJ, Martinez AT (1996) Manganese-mediated lignin degradation by *Pleurotus pulmonarius*. Appl Environ Microb 62:1070–1072

Capar G, Yetis U (2006) Effect of color and surfactants on nanofiltration for the recovery of carpet printing wastewaters. Sep Sci Technol 41:2771–2784

Casas N, Parella T, Vicent T, Caminal G, Sarra M (2009) Metabolites from the biodegradation of triphenylmethane dyes by *Trametes versicolor* or laccase. Chemosphere 75:1344–1349

Chagas EP, Durrant LR (2001) Decolorization of azo dyes by *Phanerochaete chrysosporium* and *Pleurotus sajor-caju*. Enzyme Microb Technol 29:473–477

Chandralata R, Donna DT, Kumar, V.A. (2008) Treatment of colored effluents with lignin-degrading enzymes: an emerging role of marine-derived fungi. Crit Rev Microb 34:189–206

Chang JS, Lin YC (2000) Fed-batch bioreactor strategies for microbial decolorization of azo dye using a *Pseudomonas luteola* strain. Biotechnol Prog 16:979–985

Chen BY, Yen CY, Hsueh CC (2009) Cost-effective biostimulation strategy for wastewater decolorization using immobilized-cell systems. Biores Technol 100:2975–2981

Chen KC, Wu JY, Huang CC, Liang YM, Hwang, SCJ (2003) Decolorization of azo dye using PVA-immobilized micro-organisms. J Biotechnol 101:241–252

Cheng X, Jia R, Li P, Tu S, Zhu Q, Tang W, Li X (2007) Purification of a new manganese peroxidase of the white-rot fungus *Schizophyllum* sp. F17, and decolorization of azo dyes by the enzyme. Enzyme Microb Technol 41:258–264

Cho SH, Shim J, Moon. S.H. (2009) Detoxification of simulated textile wastewater using a membraneless electrochemical reactor with immobilized peroxidase. J Hazard Mat 162:1014–1018

Christian V, Shrivastava R, Novotny C, Vyas BR (2003) Decolorization of sulfonphthalein dyes by manganese peroxidase activity of the white-rot fungus *Phanerochaete chrysosporium*. Folia Microb 48:771–774

Christian V, Shrivastava R, Shukla D, Modi H, Rajiv B, Vyas M (2005) Mediator role of veratryl alcohol in the lignin peroxidase-catalyzed oxidative decolorization of Remazol Brilliant Blue R. Enzyme Microb Technol 36:426–431

15 Peroxidases as a Potential Tool for the Decolorization and Removal of Synthetic Dyes   487

Chung KT, Cerniglia CE (1992) Mutagenicity of azo dyes: structure-activity relationships. Mut Res 277:201–220

Chung KT, Fulk GE, Egan M (1978) Reduction of azo dyes by intestinal anaerobes. Appl Environ Microb 35:558–562

Cicek F, Ozer D, Ozer A. (2007) Low cost removal of reactive dyes using wheat bran. J Hazard Mat 146:408–416

Claus H, Faber G, Koenig H (2002) Redox-mediated decolorization of synthetic dyes by fungal laccases. Appl Microb Biotechnol 59:672–678

Cohen R, Persky L, Hazan-Eitan Z, Yarden O, Hadar Y (2002) Mn^{2+} alters peroxidase profiles and lignin degradation by the white-rot fungus *Pleurotus ostreatus* under different nutritional and growth conditions. Appl Biochem Biotechnol 102–103:415–429

Conesa A, Punt PJ, van den Hondel CA (2002) Fungal peroxidases: molecular aspects and applications. J Biotechnol 93:143–158

Conneely A, Smyth WF, McMullan G (2002) Study of the white-rot fungal degradation of selected phthalocyanine dyes by capillary electrophoresis and liquid chromatography. Anal Chim Acta 451:259–270

Couto RS, Sanroman, M., Gubitz, G.M. (2005b) Influence of redox mediators and metal ions on synthetic acid dye decolourization by crude laccase from *Trametes hirsute*. Chemosphere 58:417–422

Couto SR, Rivela I, Munoz MR, Sanroman A (2000a) Stimulation of lignolytic enzyme production and the ability to decolourise Poly R-448 in semisolid-state cultures of *Phanerochaete chrysosporium*. Biores Technol 74:159–164

Couto SR, Rivela I, Sanroman A (2000b) *In vivo* decolourization of the polymeric dye poly R-478 by corncob cultures of *Phanerochaete chrysosporium*. Acta Biotechnol 20:31–38

Couto SR, Rosales E, Sanroman MA (2005a) Decolourization of synthetic dyes by *Trametes hirsute* in expanded-bed reactors. Chemosphere 62:1558–1563

Crestini C, Jurasek L, Argyropoulos DS (2003) On the mechanism of the laccase-mediator system in the oxidation of lignin. Chemistry 9:5371–5378

d'Acunzo F, Galli C, Gentili P, Sergi F (2006) Mechanistic and steric issues in the oxidation of phenolic and non-phenolic compounds by laccase or laccase-mediator systems. The case of bifunctional substrates. New J Chem 30:583–591

Daneshvar N, Ayazloo M, Khatae AR, Pourhassan M (2007) Biological decolourization of dye solution containing Malachite Green by microalgae *Cosmarium* sp. Biores Technol 98:1176–1182

Dos Santos. AB, Bisschops, LAE, Cervantes FJ, Van Lier JB (2004) Effect of different redox mediators during thermophilic azo dye reduction by anaerobic granular sludge and comparative study between mesophilic (30°C) and thermophilic (55°C) treatments for decolourisation of textile wastewaters. Chemosphere 55:1149–1157

Duarte-Vazquez MA, Ortega-Tovar MA, Garcia-Almendarez BE, Regalado C (2002) Removal of aqueous phenolic compounds from a model system by oxidative polymerization with turnip *(Brassica napus* L. var. purple top white globe) peroxidase. J Chem Technol Biotechnol 78:42–47

Duarte-Vazquez MA, Whitaker JR, Rojo-Dominguez A, Garcia-Almendarez BE, Regalado C (2003) Isolation and thermal characterization of an acidic isoperoxidase from turnip roots. J Agric Food Chem 51:5096–5102

Duran N, Esposito E (2000) Potential applications of oxidative enzymes and phenoloxidase-like compounds in wastewater and soil treatment: a review. Appl Catal B Environ 28:83–99

Duran N, Rosa MA, D'Annibale A, Gianfreda L (2002) Applications of laccase and tyrosinases (phenoloxidases) immobilized on different supports: a review. Enzyme Microb Technol 31:907–931

Eichlerova I, Homolka L, Nerud F (2006) Ability of industrial dyes decolorization and ligninolytic enzymes production by different Pleurotus species with special attention on *Pleurotus calyptratus*, strain CCBAS 461. Process Biochem 41:941–946

Elisangela F, Andrea Z, Fabio DG, Cristiano RM, Regina DL, Artur CP (2009) Biodegradation of textile azo dyes by a facultative *Staphylococcus arlettae* strain VN-11 using a sequential microaerophilic/aerobic process. Int Biodet Biodeg 63:280–288

Enayatzamir, K., Alikhani HA., Yakhchali, B., Tabandeh, F., Rodriguez-Couto, S. (2009) Decoloration of azo dyes by *Phanerochaete chrysosporium* immobilized into alginate beads. Environ Sci Pollut Res Int (online) PMID 19259719

Fabbrini M, Galli C, Gentili P (2002). Comparing the catalytic efficiency of some mediators of laccase. J Mol Catal B Enzym 16:231–240

Fatima A, Husain Q (2007a) Polyclonal antibodies mediated immobilization of a peroxidase from ammonium sulphate fractionated bitter gourd *(Momordica charantia)* proteins. Biomol Eng 24:223–230

Ferreira-Leitao VS, Godinho da Silva J, Bon, EPS (2003) Methylene Blue and Azure B oxidation by horseradish peroxidase: a comparative evaluation of class II and class III peroxidases. Appl Catal B Environ 42:213–221

Ferreira-Leitao VS, de Carvalho, MEA, Bon, EPS (2007) Lignin peroxidase efficiency for Methylene Blue decolouration: comparison to reported methods. Dye Pigm 74:230–236

Forgacs E, Cserhati T, Oros G (2004) Removal of synthetic dyes from waste waters. Environ Int 30:953–971

Fruhwirth G, Paar A, Gudelj M, Cavaco-Paulo A, Robra KH, Gubitz G (2002) An immobilized catalase peroxidase from the alkalothermophilic *Bacillus* SF for the treatment of textile-bleaching effluent. Appl Microb Biotechnol 60:313–319

Fu Y, Viraraghavan T (2001) Fungal decolorization of dye wastewaters: a review. Biores Technol 79:251–262

Gavril M, Hodson PV, McLellan J (2007) Decoloration of Amaranth by the white-rot fungus Trametes. Can J Microbiol 53(2):313–326

Ghodake GS, Kalme SD, Jadhav JP, Govindwar SP (2009) Purification and partial characterization of lignin peroxidase from *Acinetobacter calcoaceticus* NCIM 2890 and its application in decolorization of textile dyes. Appl Biochem Biotechnol 152:6–14

Ghoreishi M, Haghighi R (2003) Chemical catalytic reaction and biological oxidation for treatment of non-biodegradable textile effluent. Chem Eng J 95:163–169

Gianfreda L, Rao MA (2004) Potential of extra cellular enzymes in remediation of polluted soils: a review. Enzyme Microb Technol 35:339–354

Giusti F, Mantovani L, Martella A, Seidenari S (2002) Hand dermatitis as an unsuspected presentation of textile dye contact sensitivity. Cont Dermatitis 47:91–95

Giusti F, Seidenari S (2003) Disperse dye dermatitis: clinical aspects and sensitizing agents. Exog Dermatol 2:6–10

Golob V Ojstrsek A (2005) Removal of vat and disperse dyes from residual pad liquors. Dye Pigm 64:57–61

Gomez JL, Bodalo A, Gomez E, Hidalgo AM, Gomez M, Murcia MD (2007) Experimental behaviour and design model of a fluidized bed reactor with immobilized peroxidase for phenol removal. Chem Eng J 127:47–57

Gomez-Toribio V, Martinez AT, Martinez MJ (2001) Oxidation of hydroquinones by the versatile ligninolytic peroxidase from *Pleurotus eryngii*. Eur J Biochem 268:4787–4793

Gottlieb A, Shaw C, Smith A, Wheatley A, Forsythe S (2003) The toxicity of textile reactive azo dyes after hydrolysis and decolourisation. J Biotechnol 101:49–56

Greene JC Baughman GI (1996) Effects of 46 dyes on population growth of freshwater green alga *Selenastrum capricornutum*. Text Chem Color 28:23–30

Guo J, Zhou J, Wang D, Yang J, Li Z (2008) The new incorporation bio-treatment technology of bromoamine acid and azo dyes wastewaters under high-salt conditions. Biodegradation 19:93–98

Hai FI, Yamamoto K, Fukushi K (2007) Hybrid treatment systems for dye wastewaters. Crit Rev Environ Sci Technol 37:315–377

Hamid M, Rehman K (2009) Potential applications of peroxidases. Food Chem 115:1177–1186

15 Peroxidases as a Potential Tool for the Decolorization and Removal of Synthetic Dyes 489

Hao JJ, Song FQ, Huang F, Yang CL, Zhang ZJ, Zheng Y, Tian XJ (2007) Production of laccase by a newly isolated deuteromycete fungus *Pestalotiopsis* sp. and its decolorization of azo dye. J Ind Microbiol Biotechnol 34:233–240

Hao OJ, Kim H, Chiang PC (2000) Decolorization of wastewater. Crit Rev Environ Sci Technol 30:449–505

Harazono K, Watanabe Y, Nakamura K. (2003) Decolorization of azo dye by the white-rot basidiomycete Phanerochaete sordida and by its manganese peroxidase. J Biosci Bioeng 95(5):455–459

Harazono K, Nakamura K (2005) Decolorization of mixtures of different reactive textile dyes by the white-rot basidiomycete *Phanerochaete sordida* and inhibitory effect of polyvinyl alcohol. Chemosphere 59:63–68

Hathway DE, Kolar GF (1980) Mechanisms of reaction between ultimate chemical carcinogens and nucleic acid. Chem Soc Rev 9:241–264

Heinfling A, Bergbauer M, Szewzyk U (1997) Biodegradation of azo and phthalocyanine dyes by *Trametes versicolor* and *Bjerkandera adusta*. Appl Environ Microb 48:261–266

Heinfling A, Martinez MJ, Martinez AT, Bergbauer M, Szewzyk U (1998a) Purification and characterization of peroxidases from the dye-decolorizing fungus *Bjerkandera adusta*. FEMS Microb Lett 165:43–50

Heinfling A, Martinez MJ, Martinez AT, Bergbauer M, Szewzyk U (1998b) Transformation of industrial dyes by manganese peroxidases from *Bjerkandera adusta* and *Pleurotus eryngii* in a manganese-independent reaction. Appl Environ Microb 64:2788–2793

Heinfling A, Ruiz-Duenas FJ, Martinez MJ, Bergbauer M, Szewzyk U, Martinez AT (1998c) A study on reducing substrates of manganese-oxidizing peroxidases from *Pleurotus eryngii* and *Bjerkandera adusta*. FEBS Lett 428:141–146

Heinfling-Weidtmann A, Reemtsma T, Storm T, Szewzyk U (2001) Sulfophtalimide as major metabolite formed from sulfonated phtalocyanine dyes by the white-rot fungus *Bjerkandera adusta*. FEMS Microbiol Lett 203:179–183

Held C, Kandelbauer A, Schroeder M, Cavaco-Paulo A, Guebitz GM (2005) Biotransformation of phenolics with laccase containing bacterial spores. Environ Chem Lett 3:74–77

Hiraga S, Sasaki K, Ito H, Ohashi Y, Matsui H (2001) A large family of Class III plant peroxidases. Plant Cell Physiol 42:462–468

Hirai H, Shibata H, Kawai S, Nishida T (2006) Role of 1-hydroxybenzotriazole in oxidation by laccase from *Trametes versicolor*: kinetic analysis of the laccase-1-hydroxybenzotriazole couple. FEMS MicrobLett 265:56–59

Hofrichter M (2002) Review: lignin conversion by manganese peroxidase (MnP). Enzyme Microb Technol 30:454–466

Hu MR, Chao YP, Zhang GQ, Xue ZQ, Qian S (2009) Laccase-mediator system in the decolorization of different types of recalcitrant dyes. J Ind Microb Biotechnol 36:45–51

Husain Q, Husain M (2008) Applications of redox mediators in the treatment of organic pollutants by using oxidoreductive enzymes: a review. Crit Rev Environ Sci Technol 38:1–42

Husain Q (2006) Potential applications of the oxidoreductive enzymes in the decolorization and detoxification of textile and other synthetic dyes from polluted water: a review. Crit Rev Biotechnol 60:201–221

Husain Q (2010) Peroxidase mediated decolorization and remediation of wastewater containing industrial dyes: a review. Rev Environ Sci Biotechnol 9 (2):117–140

Husain Q, Husain M, Kulshrestha Y (2009) Remediation and treatment of organopollutants mediated by peroxidases: a review. Crit Rev Biotehnol 24:94–119

Husain Q, Jan U (2000) Detoxification of phenols and aromatic amines from polluted wastewater by using phenol oxidases. A review. J Sci Ind Res 59:286–293

Husain Q, Karim Z, Banday ZZ (2010) Decolorization of textile effluent by soluble fenugreek (*Trigonella foenum-graecum L*) seeds peroxidase. Water Air Soil Pollut doi:1007/s11270-010-0345-9

Husain Q, Kulshrestha Y (2010) Removal of colored compounds from textile carpet industrial effluents by using immobilized turnip (*Brassica rapa*) and tomato (*Lycopersicon esculentum*) peroxidases. Water Sci Technol (in press)

Husain Q (2010) Peroxidase mediated decolorization and remediation of wastewater containing industrial dyes: a review. Rev Environ Sci Biotechnol 9:117–140

Husain Q, Ulber R (2010) Immobilized peroxidase as a valuable tool in remediation of aromatic pollutants and xenobiotic compounds: a review. Crit Rev Environ Sci Technol (in press)

Husseiny SM (2008) Biodegradation of the reactive and direct dyes using Egyptian isolates. J Appl Sci Res 4:599–606

Jadhav UU, Dawkar VV, Telke AA, Govindwar SP (2009) Decolorization of Direct Blue GLL with enhanced lignin peroxidase enzyme production in *Comamonas* sp. UVS. J ChemTechnol Biotechnol 84:126–132

Jager I, Hafner C, Schneider K (2004) Mutagenicity of different textile dye products in *Salmonella typhimurium* and mouse lymphoma cells. Mut Res 561:35–44

Jarosz-Wilkolazka A, Kochmanska-Rdest J, Malarcyk E, Wardas W, Leonowicz A (2002) Fungi and their ability to decolourize azo and anthraquinonic dyes. Enzyme Microb Technol 30:566–572

Jing W, Lihua L, Jiti Z, Hong L, Guangfei L, Ruo fei J, Fenglin Y (2009) Enhanced biodecolorization of azo dyes by electropolymerization-immobilized redox mediator. J Hazard Mat 168:1098–1104

Junnarkar N, Murty DS, Bhatt NS, Madamwar D (2006) Decolorization of diazo dye Direct Red 81 by a novel bacterial consortium. World J Microb Biotechnol 22:163–168

Kalme SD, Parshetti GK, Jadhav SU, Govindwar SP (2007) Biodegradation of benzidine based dye Direct Blue-6 by *Pseudomonas desmolyticum* NCIM 2112. Biores Technol 98:1405–1410

Kapdan IK, Kargi F (2002) Biological decolorization of textile dyestuff containing wastewater by *Coriolus versicolor* in a rotating biological contactor. Enzyme Microb Technol 30:195–199

Kariminiaae-Hamedaani HR, Sakurai A, Sakakibara M (2007) Decolorization of synthetic dyes by a new manganese peroxidase-producing white rot fungus. Dye Pigm 72:157–162

Kaushik P, Malik A (2009) Fungal dye decolourization: recent advances and future potential. Environ Int 35:127–141

Keharia H, Madamvar D (2003) Bioremediation concept for treatment of dye containing wastewater: a review. Ind J Exp Biol 41:1068–1075

Khalid A, Arshad M, Crowley DE (2008) Accelerated decolorization of structurally different azo dyes by newly isolated bacterial strains. Appl Microb Biotechnol 78:361–369

Khan AA, Akhtar S, Husain Q (2006) Direct immobilization of polyphenol oxidases on Celite 545 from ammonium sulphate fractionated proteins of potato (*Solanum tuberosum*). J Mol Catal B Enzym 40:58–63

Khelifi E, Bouallagui H, Touhami Y, Godon JJ, Hamdi M (2009) Bacterial monitoring by molecular tools of a continuous stirred tank reactor treating textile wastewater. Biores Technol 100:629–633

Khenifi A, Bouberka Z, Sekrane F, Kameche M, Derriche Z (2007) Adsorption study of an industrial dye by an organic clay. Adsorption 13:149–158

Kiernan JA (2001) Classification and naming of dyes, stains and fluorochromes. Biotech Histochem 76:261–277

Kim S J., Shoda M (1999a) Purification and characterization of a novel peroxidase from *Geotrichum candidum* Dec 1 Involved in decolorization of dyes. Appl Environ Microbiol 65:1029–1035

Kim SJ, Shoda M (1999b) Decolorization of molasses and a dye by a newly isolated strain of the fungus *Geotrichum candidum* Dec 1. Biotechnol Bioeng 62:114–119

Kim TH, Park C, Lee J, Shin EB, Kim S (2002) Pilot scale treatment of textile wastewater by combined processes (fluidized biofilm process–chemical coagulation– electrochemical oxidation). Water Res 36:3979–3988

Kirby N, Marchant R, McMullan G (2000) Decolourisation of synthetic textile dyes by *Phlebia tremellosa*. FEMS Microb Lett 188:93–96

Knapp JS, Newby PS (1999) The decolourisation of a chemical industry effluent by white rot fungi. Water Res 33:575–577

Knutson K, Kirzan S, Ragauskas A (2005) Enzymatic biobleaching of two recalcitrant paper dyes with horseradish and soybean peroxidase. Biotechnol Lett 27:753–758

15 Peroxidases as a Potential Tool for the Decolorization and Removal of Synthetic Dyes 491

Kokol V, Doliska A, Eichlerova I, Baldrain P, Nerud F (2007) Decolorization of textile dyes by whole cultures of *Ischnoderma resinosum* and by purified laccase and Mn-peroxidase. Enzym Microb Technol 40:1673–1677

Koyuncu I (2002) Reactive dye removal in dye/salt mixtures by nano filtration membranes containing vinyl sulphone dyes: effect of feed concentration and cross flow velocity. Desalination 143:243–253

Krcmar, P., Ulrich, R. (1998) Degradation of polychlorinated biphenyl mixtures by the lignin-degrading fungus *Phanerochaete chrysosporium*. Folia Microb 43:79–84

Kulshrestha Y, Husain Q (2006a) Bioaffinity-based an inexpensive and high yield procedure for the immobilization of turnip (*Brassica rapa*) peroxidase. Biomol Eng 23:291–297

Kulshrestha Y, Husain Q (2006b) Direct immobilization of peroxidase on DEAE cellulose from ammonium sulphate fractionated proteins of bitter gourd (*Momordica charantia*). Enzyme Microb Technol 38:470–477

Kulshrestha Y, Husain Q (2007) Decolorization and degradation of acid dyes mediated by salt fractionated turnip (*Brassica rapa*) peroxidases. Toxicol Environ Chem 89:255–267

Lan J, Huang X, Hu M, Li Y, Qu Y, Gao P, Wu D (2006) High efficient degradation of dyes with lignin peroxidase coupled with glucose oxidase. J Biotechnol 123:483–490

Li L, Dai W, Yu P, Zhao J,Qu Y (2009) Decolorisation of synthetic dyes by crude laccase from *Rigidoporus lignosus* W1. J Chem Technol Biotechnol 84:399–404

Lima, ROA Bazo AP, Salvadori, DMF, Rech CM, Oliveira DP, Umbuzeiro GA (2007) Mutagenic and carcinogenic potential of a textile azo dye processing plant effluent that impacts a drinking water source. Mut Res 626:53–60

Liu G, Jiti Z, Wang J, Zhou M, Lu H, Jin R (2009) Acceleration of azo dye decolorization by using quinone reductase activity of azoreductase and quinone redox mediator. Biores Technol 100:2791–2795

Liu JQ, Liu HT (1992) Degradation of azo dyes by algae. Environ Poll 75:273–278

Liu JZ, Wang TL, Ji LN (2006) Enhanced dye decolorization efficiency by citraconic anhydride-modified horseradish peroxidase. J Mol Catal B Enzym 41:81–86

Lopez C, Mielgo I, Moreira MT, Feijoo G, Lema JM (2002) Enzymatic membrane reactors for biodegradation of recalcitrant compounds: application to dye decolourisation. J Biotechnol 99:249–257

Lopez C, Moreira MT, Feijoo G, Lema JM (2004) Dye decolorization by manganese peroxidase in an enzymatic membrane bioreactor. Biotechnol Prog 20:74–81

Lopez C, Moreira MT, Feijoo G, Lema JM (2007) Dynamic modelling of an enzymatic membrane reactor for the treatment of xenobiotic compounds. Biotechnol Bioeng 97:1128–1137

Lourenco ND, Novais JM, Pinheiro HM (2000) Reactive textile dye colour removal in a sequencing batch reactor. Water Sci Technol 42:321–328

Lu L, Zhao M, Wang Y (2007) Immobilization of laccase by alginate-chitosan microcapsules and its use in dye decolorization. World J Microb Biotechnol 23:159–166

Lu R, Shen XL, Xia LM (2005) Studies on laccase production by *Coriolus versicolor* and enzymatic decoloration of dye. Chem Ind Forest Prod 25:73–76

Machado, KMG, Matheus DR, Bononi, VLR (2005) Ligninolytic enzymes production and Remazol Brilliant Blue R decolorization by tropical Brazilian basidiomycetes fungi. Braz J Microb 36:246–252

Maddhinni VL, Vurimindi HB, Yerramilli A (2006) Degradation of azo dye with horseradish peroxidase (HRP). J Ind Inst Sci 86:507–514

Magri ML, Miranda MV, Cascone O (2005) Immobilization of soybean seed coat peroxidase on polyaniline: synthesis optimization and catalytic properties. Biocatal Biotransform 23:339–346

Maki F, Yugo U, Yasushi M, Isao I (2006) Preparation of peroxidase-immobilized polymers and their application to the removal of environment-contaminating compounds. Bull Fib Tex Res Found 15:15–19

Martinez AT (2002) Molecular biology and structure function of lignin-degrading heme peroxidases. Enzyme Microb Technol 30:425–444

Mateo C, Palomo JM, Fernandez-Lorente G, Guisan JM, Fernandez-Lafuente R (2007) Improvement of enzyme activity, stability and selectivity via immobilization techniques. Enzyme Microb Technol 40:1451–1463

Matto M, Husain Q (2006) Entrapment of porous and stable concanavalin A-peroxidase complex into hybrid calcium alginate-pectin gel. J Chem Tech Biotechnol 81:1316–1323

Matto M, Husain Q (2007) Decolorization of direct dyes by salt fractionated turnip proteins enhanced in the presence of hydrogen peroxide and redox mediators. Chemosphere 69:338–345

Matto M, Husain Q (2008) Redox mediated decolorization of Direct Red 23 and Direct Blue 80 catalyzed by bioaffinity based immobilized tomato (*Lycopersicon esculentum*) peroxidase. Biotechnol J 3:1224–1231

Matto M, Husain Q (2009a) Decolorization of textile effluent by bitter gourd peroxidase immobilized on concanavalin A layered calcium alginate–starch beads. J Hazard Mat 164:1540–1546

Matto M, Husain Q (2009b) Decolorization of direct dyes by immobilized turnip peroxidase in batch and continuous processes. Ecotoxicol Environ Saf 72:965–971

Matto M, Husain Q (2009c) Calcium alginate-starch hybrid support for both surface immobilization and entrapment of bitter gourd (*Momordica charantia*) peroxidase. J Mol Catal B Enzym 57:164–170

Matto M, Satar, R, Husain, Q (2010) Application of calcium alginate-starch entrapped bitter gourd (Momordica charantia) peroxidase in the removal of colored compounds from a textile effluent in batch as well as in continuous reactor. Appl Biochem Biotechnol 158(3):512–523

McMullan G, Meehan C, Conneely A, Kirby N, Robinson T, Nigam P, Banat IM, Marchant R, Smyth WF (2001) Microbial decolourisation and degradation of textiles dyes. Appl Microb Biotechnol 56:81–87

Michniewicz A, Ledakowicz S, Ullrich R, Hofrichter M (2008) Kinetics of the enzymatic decolorization of textile dyes by laccase from *Cerrena unicolor*. Dye Pigm 77:295–302

Mielgo I, Lopez C, Moreira MT, Feijoo G, Lema JM (2003) Oxidative degradation of azo dyes by manganese peroxidase under optimized conditions. Biotechnol Prog 19:325–331

Mielgo I, Moreira MT, Feijoo G, Lema JM (2001) A packed-bed fungal bioreactor for the continuous decolorization of azo dyes (Orange II). J Biotechnol 89:99–106

Mohan SV, Prasad KK, Rao NC, Sarma PN (2005) Acid azo dye degradation by free and immobilized horseradish peroxidase catalyzed process. Chemosphere 58:1097–1105

Mohan SV, Roa CN, Prasad KK, Karthikeyan J (2002) Treatment of simulated Reactive Yellow 22 (Azo) dye effluents using *Spirogyra* species. Waste Manag 22:575–582

Mohorcic M, Teodorovic S, Golob V, Friedrich J (2006) Fungal and enzymatic decolorisation of artificial textile dye baths. Chemosphere 63:1709–1717

Moldes D, Sanroman MA (2006) Amelioration of the ability to decolorize dyes by laccase: relationship between redox mediators and laccase isoenzymes in *Trametes versicolor*. World J Microb Biotechnol 22:1197–1204

Mondal S (2008) Methods of dye removal from dye house effluent-an overview. Environ Eng Sci 25:383–396

Moreira MT, Mielgo I, Feijoo G, Lema JM (2000) Evaluation of different fungal strains in the decolourisation of synthetic dyes. Biotechnol Lett 22:1499–1503

Moreira MT, Palma C, Mielgo I, Feijoo G, Lema JM (2001) *In vitro* degradation of a polymeric dye (Poly R-478) by manganese peroxidase. Biotechnol Bioeng 75:362–368

Moreira PR, Bouillenne F, Almeida-Vara E, Malcata FX, Frere JM, Duarte JC (2006) Purification, kinetics and spectral characterisation of a new versatile peroxidase from a *Bjerkandera* sp. isolate. Enzyme Microb Technol 38:28–33

Moreira PR, Duez C, Dehareng D, Antunes A, Almeida-Vala E, Frere JM, Malcata FX, Duarte JC (2005) Molecular characterization of a versatile peroxidase from a *Bjerkandera* strain. J Biotechnol 118:339–352

Murugesan K, Dhamija A, Nam IH, Kim YM, Chang. Y.S. (2007b) Decolourization of Reactive Black 5 by laccase: optimization by response surface methodology. Dye Pigm 75:176–184

Murugesan K, Kalaichelvan PT (2003) Synthetic dye decolourization by white rot fungi Ind. J Exp Biol 41:1076–1087

15 Peroxidases as a Potential Tool for the Decolorization and Removal of Synthetic Dyes   493

Murugesan K, Nam IH, Kim YM, Chang YS (2007a) Decolorization of reactive dyes by a thermostable laccase produced by *Ganoderma lucidum* in solid state culture. Enzyme Microb Technol 40:1662–1672

Muthukumar M, Sargunamani D, Selvakumar N (2005) Statistical analysis of the effect of aromatic, azo and sulphonic acid groups on decolouration of acid dye effluents using advanced oxidation processes. Dyes Pigm 65:151–158

Nazari K, Esmaeili N, Mahmoudi A, Rahimi H, Moosavi-Movahedi AA (2007) Peroxidative phenol removal from aqueous solutions using activated peroxidase biocatalyst. Enzyme Microb Technol 41:226–233

Neamtu M, Yediler A, Siminiceanu I, Macoveanu M, Kettrup A (2004) Decolorization of Disperse Red 354 azo dye in water by several oxidation processes-A comparative study. Dyes Pigm 60:61–68

Nigam P, Armour G, Banat IM, Singh D, Marchant R (2000) Physical removal of textile dyes from effluents and solid state fermentation by dye-adsorbed agricultural residues. Biores Technol 72:219–226

Novotny C, Dias C, Kapanen A, Malachova K Vandrovcova M, Itavaara M, Lima N (2006) Comparative use of bacterial, algal and protozoan tests to study toxicity of azo and anthraquinone dyes. Chemosphere 63:1436–1442

Novotny C, Rawal B, Bhatt M, Patel M, Sasek V, Molitoris HP (2001) Capacity of *Irpex lacteus* and *Pleurotus ostreatus* for decolorization of chemically different dyes. J Biotechnol 89:113–122

Nyanhongo GS, Gomesa J, Gubitz GM, Zvauya R, Read J, Steiner W (2002) Decolorization of textile dyes by laccases from a newly isolated strain of *Trametes modesta*. Water Res 36:1449–1456

O'Neill C, Lopez A, Esteves SR, Hawkes FR, Hawkes DL, Wilcox SJ (2000) Azo-dye degradation in an anaerobic-aerobic treatment system operating on simulated textile effluent. Appl Microb Biotechnol 53:249–254

Okazaki SY, Nagasawa SI, Goto M, Furusaki S, Wariishi H,Tanaka H (2002) Decolorization of azo and anthraquinone dyes in hydrophobic organic media using microperoxidase-11 entrapped in reversed micelles. Biochem Eng J 12:237–241

Ozturk A, Abdullah MI (2006) Toxicological effect of indole and its azo dye derivatives on some microorganisms under aerobic conditions. Sci Tot Environ 358:137–142

Padmesh, TVN, Vijayaraghavan K, Sekaran G, Velan M (2005) Batch and column studies on biosorption of acid dyes on fresh water macro alga *Azolla filiculoides*. J Hazard Mat 125:121–129

Palma C Martinez AT, Lema JM, Martinez MJ (2000) Different fungal manganese-oxidizing peroxidases: a comparison between *Bjerkandera* sp. and *Phanerochaete chrysosporium*. J Biotechnol 77:235–245

Palmieri G, Giardina P, Sannia G (2005) Laccase-mediated Remazol Brilliant Blue R decolorization in a fixed-bed bioreactor. Biotechnol Prog 21:1436–1441

Pandey A, Singh P, Iyengar L (2007) Bacterial decolorization and degradation of azo dyes. Int Biodet Biodeg 59:73–84

Park C, Lee M, Lee B, Kim SW, Chase HA, Lee J, Kim S (2007) Biodegradation and biosorption for decolorization of synthetic dyes by *Funalia trogii*. Biochem Eng J 36:59–65

Passardi F, Bakalovic N, Teixeira FK, Margis-Pinheiro M, Penel C, Dunand C (2007) Prokaryotic origins of the non-animal peroxidase superfamily and organelle-mediated transmission to eukaryotes. Genomics 89:567–579

Pavanello S, Clonfero E (2000) Biological indicators of genotoxic risk and metabolic polymorphisms. Mut Res 463:285–308

Pazarlioglu NK, Urek RO, Ergun F (2005) Biodecolourization of Direct Blue 15 by immobilized *Phanerochaete chrysosporium*. Process Biochem 40:1923–1929

Pearce CI, Lloyd JR, Guthrie JT (2003) The removal of colour from textile wastewater using whole bacterial cells: a review. Dye Pigm 58:179–196

Pedroche J, del Mar Yust M, Mateo C, Fernandez-Lafuente R, Giron-Calle J, Alaiz M, Vioque J, Guisan JM, Millan F (2007) Effect of the support and experimental conditions in the intensity of the multipoint covalent attachment of proteins on glyoxyl-agarose supports: COR-

RELATION between-enzyme support linkages and thermal stability. Enzyme Microb Technol 40:1160–1166

Pereira L, Coelho AV, Viegas CA, dos Santos, MMC, Robalo MP, Martins LO (2009) Enzymatic biotransformation of the azo dye Sudan Orange G with bacterial Cot A-laccase. J Biotechnol 139:68–77

Pielesz, A., Baranowska, I., Rybakt, A., Włochowicz, A. (2002) Detection and determination of aromatic amines as products of reductive splitting from selected azo dyes. Ecotoxicol Environ Saf 53:42–47

Pisklak TJ,Macias M, Coutinho DH, Huang RS, Balkus KJ (2006) Hybrid materials for immobilization of MP-11 catalyst. Topics Catal 38:269–278

Platzek T, Lang C, Grohmann G, Gi US, Baltes W (1999) Formation of a carcinogenic aromatic amine from an azo dye by human skin bacteria *in vitro*. Hum Exp Toxicol 18:552–559

Pocedic J, Hasal P, Novotny C (2009) Decolorization of organic dyes by *Irpex lacteus* in a laboratory trickle-bed biofilter using various mycelium supports. J Chem Technol Biotechnol 84:1031–1042.

Podgornik H, Poljansek R, Perdih A (2001) Transformation of Indigo Carmine by *Phanerochaete chrysosporium* ligninolytic enzymes. Enzyme Microb Technol 29:166–172

Pogni R, Baratto C, Giansanti S, Teutloff C, Verdin J, Valderrama B, Lendzian F, Lubitz W, Vazquez-Duhalt R, Basosi R (2005) Tryptophan-based radical in the catalytic mechanism of versatile peroxidase from *Bjerkandera adusta*. Biochemistry 44:4267–4274

Pointing SB, Bucher, VVC, Vrijmoed, LLP (2000) Dye decolorization by subtropical basidiomycetous fungi and the effect of metals on decolorizing ability. World J Microb Biotechnol 16:199–205

Potthast A, Rosenau T, Fischer K (2001) Oxidation of benzyl alcohols by the laccase-mediator system (LMS)-A comprehensive kinetic description. Holzforschung 55:47–56

Prasongsuk S, Lotrakul,P., Imai T, Punnapayak H (2009) Decolourization of pulp mill wastewater using thermotolerant white rot fungi. Sci Asia 35:37–41

Pratt M, Taraska V (2000) Disperse blue dyes 106 and 124 are common causes of textile dermatitis and should serve as screening allergens for this condition. Am J Cont Dermatitis 11:30–41

Pricelius S, Held C, Sollner S, Deller S, Murkovic M, Ullrich R, Hofrichter M, Cavaco-Paulo A, Macheroux P, Guebitz GM (2007) Enzymatic reduction and oxidation of fibre-bound azo-dyes. Enzyme Microb Technol 40:1732–1738

Quintanilla-Guerrero F, Duarte-Vazquez MA, Tinoco R, Gomez-Suarez M, Garcia-Almendarez BE, Vazquez-Duhalt R., Regalado C (2008) Chemical modification of turnip peroxidase with methoxypolyethylene glycol enhances activity and stability for phenol removal using the immobilized enzyme. *J Agric Food Chem* 56:8058–8065

Regalado C, Garcia-Almendarcz BE, Duarte-Vazquez MA (2004) Biotechnological applications of peroxidases. Phytochem Rev 3:243–256

Rezaee A, Ghaneian MT, Hashemian SJ, Moussavi G Khavanin A, Ghanizadeh G (2008) Decolorization of Reactive Blue 19 dye from textile wastewater by the UV/H_2O_2 process. J Appl Sci 8:1108–1112

Robinson T, Chandran B, Nigam P (2001b) Studies on the production of enzyme by white-rot fungi for decolourisation of textile dyes. Enzyme Microb Technol 29:575–579

Robinson T, McMullan G, Marchant R, Nigam P (2001a) Remediation of dyes in textile effluent: a critical review on current treatment technologies with a proposed alternative. Biores Technol 77:247–255

Rodriguez E, Pickard MA, Vazquez-Duhalt R (1999) Industrial dye decolorization by laccases from lignolytic fungi. Curr Microb 38:27–32

Rodriguez-Couto S (2009) Enzymatic biotransformation of synthetic dyes. Current Drug Metab 10(9):1048–1054

Rothman N, Talaska G, Hayes R, Bhatnagar V, Bell D, Lakshmi V, Kashyap S, Dosemeci M, Kashyap R, Hsu F, Jaeger M, Hirvonen A, Parikh D, Davis B, Zenser T (1997) Acidic urine pH is associated with elevated levels of free urinary benzidine and N-acetylbenzidine and urothelial cell DNA adducts in exposed workers. Cancer Epidemiol Biomarkers Prev 6:1039–1042

Rotkova J, Sulakova R, Korecka L, Zdrazilova P, Jandova M, Lenfeld J, Horak D, Bilkova Z (2009) Laccase immobilized on magnetic carriers for biotechnology applications. J. Magnetism Magnetic Mat 321:1355–1340

Rott U (2003) Multiple use of water in industry-the textile industry case. J Environ Sci Health 38:1629–1639

Ruiz-Duenas FJ, Camarero S, Perez-Boada M, Martinez MJ, Martinez AT (2001) A new versatile peroxidase from *Pleurotus*. Biochem Soc Trans 29:116–122

Ryan S, Schnitzhofer W, Tzanov T, Cavaco-Paulo A, Gubitz GM (2003) An acid-stable laccase from *Sclerotium rolfsii* with potential for wool dye decolourization. Enzyme Microb Technol 33:766–774

Ryberg K, Goossens A, Isaksson M, Gruvberger B, Zimerson E, Nilsson F, Bjork J, Hindsen M, Bruze M (2009) Is contact allergy to disperse dyes and related substances associated with textile dermatitis? Brit J Derm 160:107–115

Sadhasivam S, Savitha S, Swaminathan K (2009) Redox-mediated decolorization of recalcitrant textile dyes by *Trichoderma harzianum* WL1 laccase. World J Microb Biotechnol 25:1733–1741

Saijo T, Sato T, Tanaka N, Ichiyanagi A, Sugano Y, Shoda M (2005) Precipitation diagram and optimization of crystallization conditions at low ionic strength for deglycosylated dye-decolorizing peroxidase from a basidiomycete. Acta Cryststallogr F Struct Biol Crystallz Comm 61:729–732

Satar R, Husain Q (2009a) Bitter gourd peroxidase-catalyzed degradation and decolorization of water insoluble disperse dyes mediated by 1-hydroxybenzotrialzole. Biotechnol Bioprocess Eng 14:213–219

Satar, R., Husain, Q. (2009b) Phenol-mediated decolorization and removal of disperse dyes by bitter gourd (Momordica charantia) peroxidase. Environ Technol 30(14):1519–1527

Satar R, Husain Q (2009c) Applications of Celite-adsorbed white radish (*Raphanus sativus*) peroxidase in batch process and continuous reactor for the degradation of reactive dyes. Biochem Eng J 46:96–104

Sato T, Hara S, Matsui T, Sazaki G, Saijo S, Ganbe T, Tanaka N, Sugano Y, Shoda M (2004) A unique dye-decolorizing peroxidase, DyP, from *Thanatephorus cucumeris* Dec 1: heterologous expression, crystallization and preliminary X ray analysis. Acta Cryst 60:149–152

Schlosser D, Hofer C (2002) Laccase-catalyzed oxidation of Mn^{2+} in the presence of natural Mn^{3+} chelators as a novel source of extracellular H_2O_2 production and its impact on manganese peroxidase. Appl Environ Microb 68:3514–3521

Sedighi M, Karimi A, Vahabzadeh F (2009) Involvement of ligninolytic enzymes of *Phanerochaete chrysosporium* in treating the textile effluent containing Astrazon Red FBL in a packed-bed bioreactor. J Hazard Mater 169 (1–3):88–93

Selvam K, Swaminathan K, Chae KS (2003) Decolourization of azo dyes and a dye industry effluent by a white rot fungus *Thelephora* sp. Biores Technol 88:115–119

Semple KT, Cain RB, Schmidt S (1999) Biodegradation of aromatic compounds by microalgae. FEMS Microb Lett 170:291–300

Shakeri M, Sugano Y, Shoda M (2008a) Stable repeated-batch production of recombinant dye-decolorizing peroxidase (rDyP) from *Aspergillus oryzae*. J Biosci Bioeng 105(6):683–686

Shakeri M, Shoda M (2008b) Decolorization of an anthraquinone dye by the recombinant dye-decolorizing peroxidase (rDyP) immobilized on mesoporous materials. J Mol Catal B Enzym 54:42–49

Shim J, Kim GY, Yeon KH, Cho SH, Woo JJ, Moon SH (2007) Degradation of azo dye by an electroenzymatic method using horseradish peroxidase immobilized on porous support. Kor J Chem Eng 24:72–78

Shimokawa T, Hirai M, Shoda M, Sugano Y (2008) Efficient dye decolorization and production of dye decolorizing enzymes by the basidiomycete *Thanatephorus cucumeris* Dec 1 in a liquid and solid hybrid culture. J Biosci Bioeng 106:481–487

Shin KS (2004) The role of enzymes produced by white-rot fungus *Irpex lacteus* in the decolorization of the textile industry effluent. J Microb 42:37–41

Shrivastava R, Christian V, Vyas, BRM (2005) Enzymatic decolorization of sulfonphthalein dyes. Enzyme Microb Technol 36:333–337

Soares GMB, de Amorim, MTP, Costa-Ferreira M (2001) Use of laccase together with redox mediators to decolourize Remazol Brilliant Blue R. J Biotechnol 89:123–129

Soares GM, Amorim MT, Hrdina R, Ferreire MC (2002) Studies on the biotransformation of novel diazo dyes by laccase. Process Biochem 37:581–587

Stolz A (2001) Basic and applied aspects in the microbial degradation of azo dyes. Appl Microb Biotechnol 56:69–80

Su Y, Zhang Y, Wang J, Zhou J, Lu X, Lu H (2009) Enhanced bio-decolorization of azo dyes by co-immobilized quinone-reducing consortium and anthraquinone. Biores Technol 100:2982–2987

Subramaniam S, Balasubramaniam R, Kuppuswamy K, Venkatesan G, Vasu K (2010). Bitter gourd (*Momordica charantia*) peroxidase in decoloriztion of dyes from tannery effluent. Recent Res Sci Technol 2(2):49–53

Sugano Y (2009) DyP-type peroxidases comprise a novel heme peroxidase family. Cell Mol Life Sci 66:1387–1403

Sugano Y, Matsushima Y, Shoda M (2006) Complete decolorization of the anthraquinone dye Reactive Blue 5 by the concerted action of two peroxidases from *Thanatephorus cucumeris* Dec 1. Appl Microb Biotechnol 73:862–871

Sugano Y, Nakano R, Sasaki K, Shoda M (2000) Efficient heterologous expression in *Aspergillus oryzae* of a unique dye-decolorizing peroxidase, DyP, of *Geotrichum candidum* Dec 1. Appl Environ Microb 66:1754–1758

Sugimori D, Banzawa R, Kurozumi M, Okura I (1999) Removal of disperse dyes by the fungus *Cunninghamella polymorpha*. J Biosci Bioeng 87:252–254

Sumathi S, Manju BS (2000) Uptake of reactive textile dyes by *Aspergillus foetidus.* Enzyme Microb Technol 27:347–355

Susla, M., Novotný, C., Erbanová, P., Svobodová, K. (2008) Implication of *Dichomitus squalens* manganese-dependent peroxidase in dye decolorization and cooperation of the enzyme with laccase. Folia Microbiol (Praha) 53(6):479–485

Svobodová, K., Erbanová, P., Sklenár, J., Novotný, C. (2006) The role of Mn-dependent peroxidase in dye decolorization by static and agitated cultures of *Irpex lacteus*. Folia Microbiol (Praha) 51(6):573–578

Tan L, Qu YY, Zhou JT, Li A, Gou M (2009) Identification and characteristics of a novel salt-tolerant *Exiguobacterium* sp. for azo dyes decolorization. Appl Biochem Biotechnol (online). doi:1007/s12010-009-8546-7

Tavares, APM, Cristovao RO, Gamelas, JAF, Loureiro JM, Boaventuraa, RAR, Macedo EA (2008a) Sequential decolourization of reactive textile dyes by laccase mediator system. J Chem Technol Biotechnol 84:442–446

Tavares, APM, Cristovao RO, Loureiro JM, Boaventuraa, RAR, Macedo EA (2008b) Optimisation of reactive textile dyes degradation by laccase-mediator system. J Chem. Technol Biotechnol 83:1609–1615

Tekere M, Mswaka AY, Zvauya R, Read JS (2001) Growth dye degradation and ligninolytic activity studies on Zimbabwean white rot fungi. Enzym Microb Technol 28:420–426

ten Brink HB, Dekker HL, Schoemaker HE, Wever R (2000) Oxidation reactions catalyzed by vanadium chloroperoxidase from *Curvularia inaequalis*. J Inorg Biochem 2000; 80:91–98

Tien M, Kirk TK (1983) Lignin-degrading enzyme from the hymenomycete *Phanerochaete chrysosporium* burds. Science 221:661–663

Tinoco R, Verdin J, Vazquez-Duhalt R (2007) Role of oxidizing mediators and tryptophan 172 in the decoloration of industrial dyes by the versatile peroxidase from *Bjerkandera adusta.* J Mol Catal B Enzym 46:1–7

Torres E, Bustos-Jaimes I, Le Borgne S (2003) Potential use of oxidative enzymes for the detoxification of organic pollutants. Appl Catal B Environ 46:1–15

Tullio AD, Caputi L, Malatesta F, Reale S, Angelis FD (2005) Characterization of a novel micro-peroxidase from *Marinobacter hydrocarbonoclasticus* by electrospray ionization tandem mass spectrometry. J Mass Spect 40:325–330

Ulson de Souza, SMAG., Forgiarini E, Ulson de Souza AA (2007) Toxicity of textile dyes and their degradation by the enzyme horseradish peroxidase (HRP). J Hazard Mat 147:1073–1078

Umbuzeiro GA, Freeman H, Warren SH, Kummrow F, Claxton LD (2005a) Mutagenicity evaluation of the commercial product CI Disperse Blue 291 using different protocols of the Salmonella assay. Food Chem Toxicol 43:49–56

Umbuzeiro GA, Freeman HS, Warren SH, Oliveira DP, Terao Y, Watanabe T, Claxton LD (2005b) The contribution of azo dyes to the mutagenic activity of the Cristais River. Chemosphere 60:55–64

Urek RO, Pazarlioglu NK (2005) Production and stimulation of manganese peroxidase by immobilized *Phanerochaete chrysosporium*. Process Biochem 40:83–87

Urra J, Sepulveda L, Contreras E, Palma C (2006) Screening of static culture and comparison of batch and continuous culture for the textile dye biological decolorization by *Phanerochaete chrysosporium*. Braz. J Chem Eng 23:281–290

Van Der Zee FP, Bisschops, IAE, Lettinga G, and Field JA (2003) Activated carbon as an electron acceptor and redox mediator during the anaerobic biotransformation of azo dyes. Environ Sci Technol 37:402–408

Van Der Zee FP, Cervantes FJ (2009) Impact and application of electron shuttles on the redox (bio) transformation of contaminants: a review. Biotechnol Adv 27:256–277

Verma P, Baldrian P, Nerud F (2003) Decolorization of structurally different synthetic dyes using cobalt(II)/ascorbic acid/hydrogen peroxide system. Chemosphere 50:975–979

Verma P, Madamwar D (2002a) Decolourization of synthetic dyes by a newly isolated strain of *Serratia marcescens*. World J Microb Biotechnol 19:615–618

Verma P, Madamwar D (2002b) Decolorization of synthetic textile dyes by lignin peroxidase of *Phanerochaete chrysosporium*. Folia Microb 47:283–286

Verma P, Madamwar D (2005) Decolorization of azo dyes using basidiomycete strain PV002. World J Microb Biotechnol 21:481–485

Vianello A, Zancani M, Nagy G, Macri F (1997) Guaiacol peroxidase associated to soyabean root plasma membranes oxidizes ascorbate. J Plant Physiol 150:573–577

Vineis P, Pirastu R (1997) Aromatic amines and cancer. Cancer Caus Cont 8:346–355

Vivekanand V, Dwivedi P, Sharma A, Sabharwal N Singh RP (2008) Enhanced delignification of mixed wood pulp by *Aspergillus fumigatus* laccase mediator system. World J Microb Biotechnol 24:2799–2804

Wagner M, Nicell JA (2003) Impact of the presence of solids on peroxidase-catalyzed treatment of aqueous phenol. J Chem Technol Biotechnol 78:694–702

Wang P, Dai S, Waezsada SD, Tsao AY, Davison BH (2001) Enzyme stabilization by covalent binding in nanoporous sol-gel glass for nonaqueous biocatalysis. Biotechnol Bioeng 74:249–255

Wang XY, Yu J (1998) Adsorption and degradation of synthetic dyes on the mycelium of *Trametese versicolor*. Water Sci Technol 38:233–238

Wariishi H, Kabuto M, Mikuni J, Oyadomari M, Tanaka H (2002) Degradation of water-insoluble dyes by microperoxidase 11, an effective and stable peroxidative catalyst in hydrophilic organic media. Biotechnol Prog 18:36–42

Welinder K.G. (1992) Superfamily of plant, fungal and bacterial peroxidases. Curr Opin Struct Biol 2:388–393

Wesenberg D, Kyriakides I, Agathos SN (2003) White-rot fungi and their enzymes for the treatment of industrial dye effluents. Biotechnol Adv 22:161–187

Xiao-Bin C, Jia R, Li PS, Zhu Q, Tu SQ, Tang WZ (2007) Studies on the properties and co-immobilization of manganese peroxidase. Chin J Biotechnol 23:90–95

Xu F, Deussen HJ, Lopez B, Lam L, Li K (2001) Enzymatic and electrochemical oxidation of *N*-hydroxy compounds. Redox potential, electron transfer kinetics, and radical stability. Eur J Biochem 268:4169–4176

Xu F, Kulys JJ, Duke KC, Li K, Krikstopaitis K, Deussen, HJW, Abbate E, Galinyte V, Schneider P (2000) Redox chemistry in laccase-catalyzed oxidation of N-hydroxy compounds. Appl Environ Microb 66:2052–2056

Xu F, Salmon S (2008) Potential applications of oxidoreductases for the re-oxidation of Leuco Vat or sulfur dyes in textile dyeing. Eng Life Sci 8:331–337

Yan H, Pan G (2004) Increase in biodegradation of dimethyl phthalate by *Closterium lunula* using inorganic carbon. Chemosphere 55:1281–1285

Yang QX, Yang M, Pritsch K, Yediler A, Hagn A, Schloter M, Kettrup A (2003) Decolorization of synthetic dyes and production of manganese-dependent peroxidase by new fungal isolates. Biotechnol Lett 25:709–711

Yang R, Tan H, Wei F, Wang S (2008) Peroxidase conjugate of cellulose nanocrystals for the removal of chlorinated phenolic compounds in aqueous solution. Biotechnology 7:233–241

Yu G, Wen X, Li R, Qian Y (2006) *In vitro* degradation of a reactive azo dye by crude ligninolytic enzymes from nonimmersed liquid culture of *Phanerochaete chrysosporium*. Process Biochem 41:1987–1993

Zhang F, Yu J (2000) Decolourisation of Acid Violet 7 with complex pellets of white rot fungus and activated carbon. Bioproc Biosys Eng 23:295–301

Zille A, Ramalho P, Tzanov T, Millward R, Aires V, Cardoso MH, Ramalho MT, Gubitz GM, Cavaco-Paulo A (2004) Predicting dye biodegradation from redox potentials. Biotechnol Progress 20:1588–1592

Zissi U, Lyberatos G (2001) Partial degradation of p-aminoazobenzene by a defined mixed culture of *Bacillus subtilis* and *Stenotrophomonas maltophilia*. Biotechnol Bioeng 72:49–54

Zouari-Mechichi H, Mechichi T, Dhouib A, Sayadi S, Martinez TA, Martinez JA (2006) Laccase purification and characterization from *Trametes trogii* isolated in Tunisia: decolorization of textile dyes by the purified enzyme. Enzyme Microb Technol 39:141–148

Chapter 16
Solid Waste Management Options and their Impacts on Climate Change and Human Health

Muna Albanna

Contents

16.1	Introduction	500
16.2	Definition of Waste	501
16.3	Types of Waste, Composition, and Generation Rate Increases	501
	16.3.1 Municipal Solid Waste (MSW)	503
	16.3.2 Agricultural Waste	506
	16.3.3 Industrial Waste	507
	16.3.4 Construction and Demolition Waste	508
16.4	Characteristics of Waste	509
16.5	Waste Treatment and Disposal Options	510
	16.5.1 Landfilling	510
	16.5.2 Incineration	513
	16.5.3 Composting	514
	16.5.4 Recycling	515
	16.5.5 The Cost of Waste Management Options	517
16.6	Impact of Solid Waste Management Options on Climate Change	518
16.7	Impact of Solid Waste Management Options on Human Health	520
	16.7.1 Consequent Impacts of Landfilling on Human Health	520
	16.7.2 Consequent Impacts of Incineration on Human Health	521
	16.7.3 Consequent Impacts of Composting on Human Health	522
	16.7.4 Consequent Impacts of Recycling on Human Health	522
16.8	Proper Management of Solid Waste—Preventive Measures	523
16.9	Summary	525
References		526

Abstract The recent changes in global climate are believed to be the result of growing anthropogenic greenhouse gas (GHG) emissions; mainly carbon dioxide and methane, resulting from the increased industrial activities over the years. One of the main emission sources that add to the anthropogenic greenhouse gases concentrations in the atmosphere are derived from the processes of solid waste dis-

---

M. Albanna (✉)
Water and Environmental Engineering Department, German Jordanian University,
P.O. Box: 851308, 11185 Amman, Jordan
e-mail: muna.albanna@gju.edu.jo

A. Malik, E. Grohmann (eds.), *Environmental Protection Strategies for Sustainable Development*, Strategies for Sustainability,
DOI 10.1007/978-94-007-1591-2_16, © Springer Science+Business Media B.V. 2012

499

posal. It can be shown that solid waste has adverse impacts on climate change and human health. The potential for deleterious public health and environmental effects is substantial where waste has been improperly disposed off.

Waste can be categorized predominantly into four sections as: municipal solid waste; agricultural waste; industrial waste; and hazardous waste. The economic growth and urbanization experienced over the past decades in many parts around the world have significantly escalated the quantities of the municipal solid waste. The improper disposal and the uncontrolled dumping of different types of waste have caused long term environmental and health problems, as well as degradation of land resources. Different solid waste management options have recently emerged *inter alia* including collection, processing, recycling, and disposal of the solid waste in ways that will reduce their harmful effects. This chapter will thoroughly assess the various robust and cost effective management alternatives, with the exception of hazardous waste, such as landfilling, composting, incineration, recycling, and the use of landfill gas (LFG) as a renewable source of energy. The chapter will also address the impacts of these management strategies that are reflected on the environment, the economy, and on human health.

Keywords Waste management • Climate change • Human health • Municipal solid waste • Waste Treatment

16.1 Introduction

After decades of environmental mismanagement, the challenges facing the globe are becoming numerous and convoluted; the legacy left for our current society has affected the climate, habitat, human health, water resources, and contamination levels. Many people worldwide are decrying the encroachment of industrial and development projects, which affected significantly the ecosystem in the recent decades, and many claim that their concerns have been cast aside in favor of others' financial gains. Consequently different measures should be taken universally to promote economic growth while protecting the environment, restore natural resources, and achieve human health safeguard.

The economic development of societies evidently has a major effect on the environment; since the natural resources are used, and pollution and waste are produced as results of many activities. Solid waste disposal and management is a huge challenge for all the municipalities, industries and businesses around the world, since the global society produces immense quantities of waste. Until now, the waste volumes produced are rapidly increasing due to modernized lifestyles. In addition to pollution the generated solid waste is also expensive to collect and to provide disposal for; billions of dollars are spent annually to collect, transport, dispose and treat the waste generated from municipalities, and from booming industrial and agricultural activities. Regardless of the treatment and disposal options, there will be always some negative environmental consequences: methane (CH_4) is emitted from landfills and contributes to global warming; toxic substances will be the result from incinerators operations; and, other management options may have other impacts on the environment and human health.

Finding a balance is extremely important between the two main objectives; the continuous economic development to achieve higher living standards; and, protecting the environment. Encouraging environmentally friendly economic activities such as improved energy efficiency technologies, improved farming practices, waste minimization are some of the actions towards achieving sustainable waste management.

16.2 Definition of Waste

Waste was defined as "a movable object which has no direct use, and is discarded permanently" (LaGrega et al. 2001). Though the use of the term "solid waste" suggests that the definition covers only solids, many liquids and gases may still be considered hazardous waste and as such, fall under solid waste management. In a broader definition, the Resource Conservation and Recovery Act (RCRA) which is the major federal statute that governs solid waste in the United States defines solid waste as "any garbage, refuse, sludge from a waste treatment plant, water supply treatment plant or air pollution control facility and other discarded or salvageable materials, including solid, liquid, semisolid, or contained gaseous materials resulting from industrial, commercial, mining and agricultural operations, and from community activities" (USEPA 2008a).

16.3 Types of Waste, Composition, and Generation Rate Increases

There are many waste types defined by modern systems of waste management. The Resource Conservation and Recovery Act (RCRA) delineates two categories of waste: *hazardous* and *non-hazardous* waste. In the course of this chapter, non-hazardous waste only will be discussed thoroughly.

The non-hazardous waste can consist of biodegradable and non-biodegradable materials as illustrated in Fig. 16.1. Figure 16.1 also displays the solid waste stream and their various recovery options; note that there are some components of waste that have economical value and can be recycled once correctly recovered.

The principle sources of non-hazardous solid waste that will be discussed in this chapter are: municipal solid waste (MSW) which includes household and commercial wastes; agricultural waste; non-hazardous industrial waste; and construction and demolition (C&D) waste. Types of waste are directly linked to the technological and social development of the people in their regions. Therefore, the generation

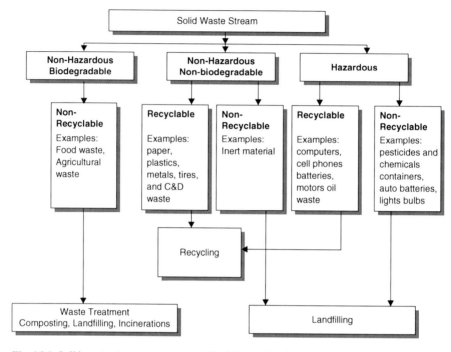

Fig. 16.1 Solid waste stream, recovery, and final disposal options

rates of different types of wastes vary between countries, and these rates have strong correlation with the levels of economical development and activities. EIONET (2009) reported that over 1.8 billion tons of waste are generated annually in Europe (estimated as 3.5 tons/person/year), and this waste is mainly made up of MSW, commercial, agriculture, non-hazardous industrial and construction and demolition. More specifically, Strange (2002) stated that the United Kingdom generates more than 400 million tons/year of non-hazardous solid waste, in the following approximate proportions, as illustrated in Fig. 16.2.

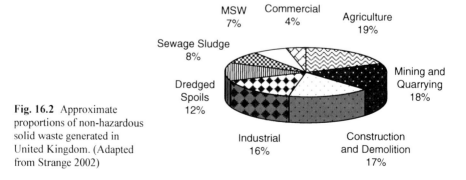

Fig. 16.2 Approximate proportions of non-hazardous solid waste generated in United Kingdom. (Adapted from Strange 2002)

It is worth noting that the generation rates of different types of wastes vary between countries around the world. In Canada, the amount of solid waste generated from all non-hazardous sources in 2002 was 30.4 million tons, of which 40% was the MSW generated in that year (Statistics Canada 2006). The generation rates of different types of non-hazardous solid waste have been increasing significantly due to several factors: population growth, modernization and technological advancements, urbanization, and the increase in industrialized manufacturing.

As for the composition of different types of waste it varies over time and location, taking into consideration the industrial development and human activities. The characteristics and quantities of the waste generated by different sectors are key elements in the development and intervention of effective and efficient solid waste management plans. Solid waste streams should be characterized into three areas: by their sources; by the types of wastes produced; and by generation rates and composition. Accurate information pertaining to these three areas is needed in order to monitor and control existing waste management systems and to be able to create regulatory, economic, and institutional decisions (World Bank 1999). The non-hazardous waste types are discussed below.

16.3.1 Municipal Solid Waste (MSW)

Municipal solid waste is the solid waste generated from residential sources, such as households, and from institutional and commercial sources such as offices, schools, hotels and other sources whose activities are similar to those of households and commercial enterprises. Although the key focus of solid waste management is on MSW, it is worth noting that MSW is only a small fraction of the total amount of waste generated as presented in the UK Solid Waste Proportions and illustrated in Fig. 16.2.

The main components of MSW are food and garden, paper and board, plastic, textile, metal and glass waste. The composition of MSW varies depending on a range of factors; the household waste reflects population density and economic prosperity, seasonality, housing standards and the presence of waste minimization initiatives (Strange 2002). Figure 16.3 shows the composition of MSW in (a) the EU countries (adapted from the European Commission report 2001), (b) New Zealand (adapted from the New Zealand Ministry for Environment 2009), (c) India (adapted from Narayana 2009), and (d) Jordan (Mrayyan 2004).

As shown in Fig. 16.3, the main components of MSW are: food and garden waste (the organic waste), paper and board waste, metals, glass, textiles, and plastics. The generation rates for the MSW differ from country to country depending on the lifestyles, population, income, and economic development. In most of the developing countries, the MSW consists mainly of organic matter, and contains less paper, plastic and metals. Moreover, all low and middle income countries have a high percentage of compostable organic matter in the urban waste stream, ranging from 40 to 85% of the total (World Bank 1999).

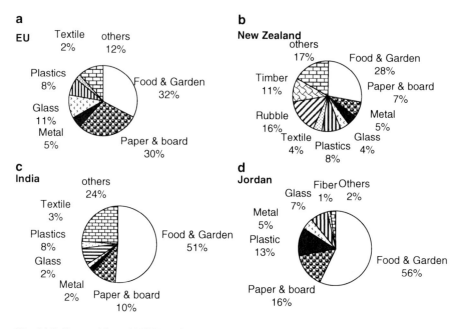

Fig. 16.3 Composition of MSW (**a**) in the European Community (adapted from European Commission Report 2001), (**b**) in New Zealand (adapted from New Zealand Ministry for Environment 2009), (**c**) in India—Thiruvananthapuram City (adapted from Narayana 2009), and (**d**) in Jordan (Adapted from Marayyan 2004)

The generation rates of MSW globally have been steadily increasing in the past two decades. Increases in municipal waste generation are related to rates of urbanization, types and patterns of consumption, household revenue, and lifestyles. The conference board of Canada (2008) has reported that Canada's municipal waste generated per capita has been steadily increasing since 1980, and this is due to the fact that Canada's per capita income and average household disposable income have also been steadily increasing, leading to increasing household consumption rates. Figure 16.4 demonstrates the changes in MSW generation in five developed countries adapted from the conference board of Canada (2008) in the period from 1985 to 2005.

The same trends were also reported by the World Bank (1999) concerning the increase in MSW generation rates in Asian countries. Figure 16.5 illustrates the MSW generated in the year 1995 compared to the expected generation in the year 2025.

From both Figs. 16.4 and 16.5, it is demonstrated that the MSW generation rates is escalating globally. The composition of the waste is expected to change as well, where packaging wastes such as paper, plastic, and glass will become more predominant in the waste stream as the economies increase and the population becomes more urbanized. EIONET (2009) reported that the packaging waste represents up to 17% of the MSW, and it arises from a wide range of sources such as: households, retail outlets, restaurants, hospitals, hotels, and manufacturing industries. The same

Fig. 16.4 Changes in the MSW generation rate in different developed countries from the year 1985 to 2005. (Adapted from the Conference Board of Canada 2008)

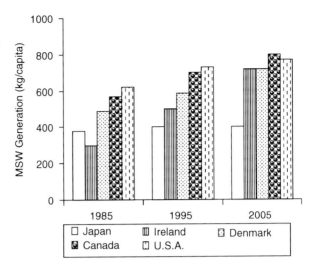

Fig. 16.5 The generated MSW in some Asian countries in the year 1995 compared to the expected generation in the year 2025. (Adapted from the World Bank 1999).

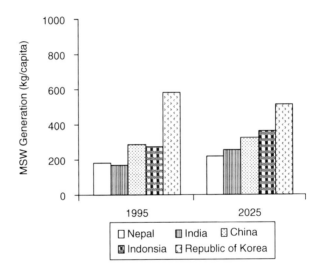

source stated that as the packaging waste has short life, it soon becomes a waste that needs to be treated or disposed off.

MSW management is expensive, and the municipal governments all over the world spend billions of dollars on waste collection, transport, and disposal. Statistics Canada (2006) reported that the MSW management services cost the Canadian Municipal Governments more than 1.5 billion-plus dollars in 2002. To achieve more sustainable MSW practices, the challenge will be to reduce the amount of MSW generated, and increase the amount of waste diverted from landfills through recycling and other initiatives in an economically feasible way. Integrated waste management systems shall be adapted while making reduced environmental impact a top priority.

16.3.2 *Agricultural Waste*

The robust economic changes and the growth of the global population have added enormous burden on the agriculture and food sector, which resulted in increased quantities of agricultural crop residues and livestock wastes. The agricultural waste produced is a result of different agricultural operations. The Environmental Management Act of British Colombia, Canada (1992) defined the agricultural operation as "any agricultural operation or activity carried out on a farm including: an operation or activity devoted to the production or keeping of livestock, poultry, farmed game, fur bearing animals, crops, grain, vegetables, milk, eggs, honey, mushrooms, horticultural products, tree fruits, berries, and; the operation of machinery and equipment for agricultural waste management or application of fertilizers and soil conditioners". Mohammadi (2006) suggested that wastes from agricultural operations are huge quantities because they can be generated throughout different operational phases, and as follows:

- Cultivation phase: this phase includes all plantation processes such as preparing land, plowing, leveling, seeding, irrigating, and fertilizing;
- Harvesting phase: this phase consists of all activities dealing with gathering products (grains, cereals, fruits, vegetables) and other by-products; and,
- Post-harvesting phase: this phase includes storing, processing, transporting and marketing.

The sources of agricultural waste are organic wastes (animal waste in the form of slurries and farmyard manures, spent mushroom compost, mucky water and silage effluent), and waste such as plastics, scrap machinery, fencing, pesticides, and waste oils, in addition to veterinary medicines (EIONET 2009).

The compositions and generation rates of agricultural waste vary between countries. The main component of the agricultural waste is the crops residues. The type and quantity of crop residue vary between countries in response to the distribution of crop. Figure 16.6 illustrates the proportionate annual production of agricultural waste: (a) in China (adapted from UNESCAP 2000), (b) in Malaysia (adapted from UNESCAP 2000), and (c) Egypt (adapted from GTZ report 2006).

Figure 16.6 shows that the composition of crop residue generated is different between countries; however, the actual generation of crop residues in any given year is a function of the crops that are grown and the size of the harvest. These factors are conditioned by changing market demand, government policy and weather (GTZ report 2006).

The generation rates of the agricultural waste differ between countries, even if they are in the same region. UNESCAP (2000) reported that the annual production of agricultural waste (manure and animal dung in addition to crop residues) in China, India, Pakistan, and the Republic of Korea is 842, 560, 84, and 25 million tons, respectively. EIONET (2009) reported that there are no overall estimates available on the quantity of agricultural waste produced in the EU, although this waste type is significant, however, as an example, Ireland has estimated that in 1998 over 80% of its national waste generated was from agricultural sources.

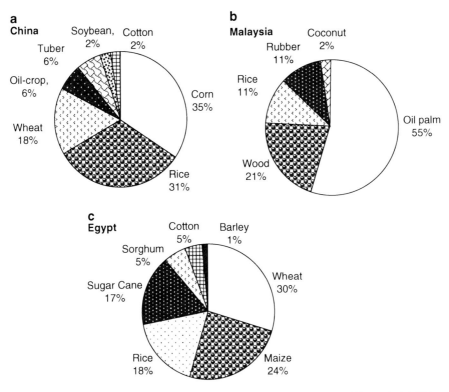

Fig. 16.6 Proportionate annual production of agricultural waste—crops residues—in (**a**) China (adapted from UNESCAP 2000), (**b**) Malaysia (adapted from UNESCAP 2000), and (**c**) Egypt (adapted from GTZ Report 2006)

There are a number of potential environmental impacts associated with agricultural waste if it is not properly managed. Nutrients run-off to surface waters can cause over enrichment of the water body which will affect the aquatic life. Improper storage of agricultural waste until disposal can pose a serious threat to the environment and surface waters—if any—due to leaching. Moreover, farming activities can increase the emissions of ammonia which will cause acidification, and CH_4 which will contribute to GHG emissions (EIONET 2009). There are several environmental regulations and codes of practices set by regulatory bodies to control the disposal and management options of agricultural waste to protect human health and the environment.

16.3.3 Industrial Waste

The non-hazardous industrial waste comprises of many different waste streams arising from a wide range of industrial processes, including but not limited to: basic metals, wood products, glass, ceramics, leather, rubber, plastics, waste from food processing, oils, waste from tobacco industry, transportation apparatus, waste from scientific research, dredging, sewage and scrap metals, paper, and many other wastes generated from process factories. Huge quantities of industrial wastes are being produced all over the globe; EIONET (2009) estimated that over 33 million tons of industrial waste was generated in Europe in 1998, while UNESCAP (2000) declared that the industrial solid waste generation in the Asian Region (including China, Japan, Republic of Korea, Hong Kong, Singapore, and others in Asian Region) is equivalent to 1,900 million tons per year. The industrial wastes are still increasing despite the national and international efforts to reduce waste from manufacturing industry. It is worth noting that the existing industrial waste management options are ubiquitously inadequate, which will cause significant challenges. Therefore, it is vital to introduce cleaner waste management technologies and to adapt waste minimization initiatives.

16.3.4 Construction and Demolition Waste

USEPA (2010) defined the construction and demolition (C&D) waste as those "materials consist of the debris generated during the construction, renovation, and demolition of buildings, roads, and bridges. The C&D materials often contain bulky, heavy materials, such as concrete, wood, metals, glass, and salvaged building components". The wastes produced from the construction and demolition (C&D) sector comprise of waste from residential, civil, and commercial construction and demolition activities (Pulikkottil and Somasundaram 2004). This type of waste includes concrete, cardboard, wood, land clearing debris, concrete masonry units, asphalt, metals, gypsum wall board, carpet, insulation, glass, and other materials used in the construction industry. Figure 16.7 illustrates the typical composition of C&D waste, as reported by EIONET (2009).

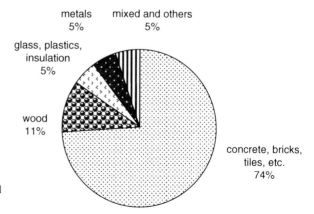

Fig. 16.7 Typical Composition of C&D Waste. (Adapted from EIONET 2009)

16 Solid Waste Management Options and their Impacts 509

In the past few decades, the growing global economy resulted in growth of the constructed building industry and as a consequent, resulted in increases in the generated C&D wastes. In order to estimate the generated waste from the C&D sector, Fatta et al. (2004) suggested the following models to calculate the C&D volume rates in building construction activities: for every 1,000 m^2 of construction, approximately 50 m^3 of waste are generated based on density value of 1.5 ton/m^3 of waste. The authors indicated that the waste quantities generated from this sector are on the rise; they estimated that the C&D waste generated in Greece in the year 2000 was 2.1 million tons with a 10% increase in the amount of waste generated from the year 1999. In Europe, the C&D generation rate is high, spawned from the demolition and renovation of old buildings. From this perspective, EIONET (2009) declared that C&D waste makes up approximately 25% of all waste generated annually in the European Union countries.

USEPA (2010) highlighted that it is imperative to take all the steps necessary to reduce and recycle the C&D materials which will result in less landfill space needed, reduce the environmental impact of producing new materials, creates jobs, and can reduce overall building project expenses through avoided purchase/disposal costs. The recycle rate from this sector can be as high as 80%.

16.4 Characteristics of Waste

Different types of waste hold substantial or potential threats to the environment and public health. These threats emerge from the characteristics of different types of wastes. There are four characteristics that may define the harmful levels of a certain type of waste (Watts 1997):

- Ignitability: The waste can be of significant threat if it can create fire under certain conditions. Examples of these phenomena include liquid wastes that contain 24% alcohol (%v/v) with a flash point of less than 60°C, and solid waste which can cause fire through spontaneous chemical changes, friction or absorption of moisture.
- *Corrosivity*: The types of waste that include acids or bases and has a pH ≤ 2.0 or ≥ 12.5, and are capable of corroding metal containers.
- *Reactivity*: Reactive waste is a type of waste that is unstable under "normal" conditions; it can form explosive mixtures with water and toxic fumes if mixed with water or heated.
- *Toxicity*: Toxic waste is one that contains concentrations of certain substances in excess of regulatory thresholds, and is expected to cause injury or illness to human health if ingested or absorbed and otherwise can cause damage to the environment.

It is worth noting that there are many types of wastes, which have one or more of the above characteristics, are discarded daily from households around the world and are disposed of in different ways. Some examples of these wastes are: paints and solvents, motors oil, light fixtures and switches, pesticides, cleaning agents, and many more items used on a daily basis. The main concern is that most of these fore-mentioned wastes are being generated from households and disposed as MSW where they are sent to landfills or incinerators, regardless of their hazardous nature and the potential harmful effect on the surrounding environment.

16.5 Waste Treatment and Disposal Options

Throughout historical developments of human civilization, the acceptable treatment and disposal options of waste created a huge challenge. As a consequence of the increasing populations of dense towns and cities, more waste was generated. The citizens of the Greek civilization in Athens, at around 500 B.C.E. were the first to suffer under the impact of the waste problem, which forced them to issue a law banning the throwing of rubbish onto the streets (Williams 2005).

The waste problem started to be recognized as a major hurdle to civilization's advancement, particularly after the industrial revolution, and the waste generation effect on societies as the result of the massive movement of population from rural areas to the denser cities. Additionally, the changed composition of the generated waste attracted vermin such as, flies, mice, rats and other pests, and posed a threat of transferring disease in the congested areas around industrial activities. In more modern history, different legislations were introduced in many countries for the proper management and disposal of waste. For example, the 1875 Public Health Act placed a duty on local authorities in the United Kingdom to arrange for the removal and disposal of wastes, while in the United States; early legislation included the 1795 law introduced in Washington, D.C. which prohibited waste disposal on streets (Williams 2005). Several local and national legislations and regulations were introduced afterwards by countries globally which incorporated proper policies and strategies for the waste disposal and treatment options.

The current practices employed in the management of solid waste vary considerably between developed and developing countries. However, the current situation across the world shows that landfilling of waste is the dominant waste disposal methodology for all categories of waste, and incineration is the main method use globally for waste treatment. According to estimates by the USEPA (2008b), 54% of MSW generate in United States is landfilled, 33.4% is recycled or composted and 12.6% is combusted with energy recovery. In Canada, the rate of waste diversion through recycle and compost in Canada is only 22% of the total waste generated (Environment Canada 2005).

The main waste disposal and treatment practices are reviewed in the subsections that follow.

16.5.1 Landfilling

A landfill is a site where waste materials are buried. Landfills are one of the oldest ways of managing waste. Landfills continue to be an important repository of solid

16 Solid Waste Management Options and their Impacts

waste for the near future since it involves the simple disposal of many types of waste, such as: MSW; industrial waste; and C&D waste. Landfills are frequently used due to the fact that there is a limit to the types of waste that can be recycled or composted.

Solid waste is delivered to landfills, and then the waste is spread out, compacted, and covered with a fresh layer of soil each day. Covering is performed in order to isolate the waste from exposure to air, and pests and as a method to control odors. When a section of the landfill is filled to capacity, it is permanently covered with a final layer of soil that will support growth of vegetation. The landfill in most cases would be lined to control and contain the leachate production.

About two-thirds of landfilled waste is biodegradable organic matter from households, businesses and industry, and other waste including inert materials, for example from construction and demolition (Surrey County Council 2010). The organic fraction of solid waste inside landfills decomposes anaerobically by methanogenic bacteria. A complex series of biological and chemical reactions begin with the burial of refuse in landfills, where an active anaerobic ecosystem generates landfill gas (LFG) as a major end product. One ton of biodegradable waste produces 200–400 m^3 of LFG (Surrey County Council 2010). The latter authors indicated that in the United Kingdom, the annual production of biodegradable waste is estimated to be 100 million tons of waste, which means that LFG is an attractive source of renewable energy, if utilized fully and properly collected and harvested.

Newly constructed landfills can be highly engineered to capture LFG emissions, with the construction of pumped systems for commercial recovery of CH_4. If collected, LFG can be used as a source of energy to generate electricity. From this view, the engineered landfills can be considered environmentally sound system for solid waste disposal. Any engineered landfill facility can be designed, constructed, operated and maintained to function effectively for as long as required for the protection of the environment. The bioreactor landfill is a new landfill technology that uses leachate recycling to accelerate the biological decomposition of food, green waste, paper and other organic wastes in a landfill, by promoting conditions necessary for the microorganisms that degrade the waste. The single most important factor in promoting waste decomposition is the moisture content of the waste. Liquids must be added to the waste mass to obtain optimal moisture content, which ranges from 35 to 45% water by weight. The bioreactor landfill technology is expected to reduce the amount of and costs associated with management of leachate (since this type is properly lined and the leachate is re-circulated within the landfill); to increase the rate of production of CH_4 for commercial purposes as a renewable source of energy (since this type of landfills is equipped with gas collection system); and to reduce the amount of land required for landfills (since this type of landfills accelerates the process of decomposition, thus the cells can be harvested). Bioreactor landfills are expected to increase this rate of decomposition and save up to 30% of space needed for landfills (Townsend et al. 2008), therefore, with the rising amounts of solid waste produced every year and scarcity of landfill spaces, bioreactor landfill is recognized as significant and environmental friendly disposal method.

The engineered landfills can be environmentally sustainable technology that is used in most of the developed countries in the recent years. However, and unfortu-

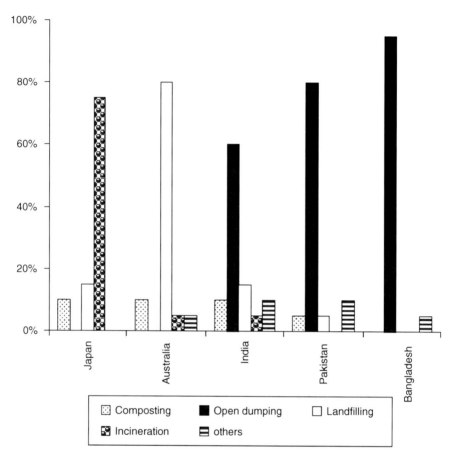

Fig. 16.8 Disposal methods for MSW in Japan, Australia, India, Pakistan and Bangladesh. (adapted from UNESCAP 2000)

nately, open dumping is still the most widespread method of solid waste disposal in many regions of the world, such as some countries in Asia and Africa. Open dumping involves the uncontrolled disposal of the waste without proper control for the leachate, dust, odor, LFG, and vermin. In some cities, open burning of wastes is still practiced at dumpsites, and in other coastal cities, waste is still simply dumped along the shoreline and into the sea (UNESCAP 2000). Figure 16.8 presents the disposal methods for MSW in five countries in Asia (the data adapted from UNESCAP 2000).

As illustrated in Fig. 16.8, the solid waste disposal methods are different amongst countries, though typically the main solid waste disposal method is either landfill or open dumps. Developed countries may rely on landfilling or incineration, depending on the available land. For example, a country with land limitation as Japan tends to use incineration as the main disposal management option, while other developed countries that do not have land limitations, such as Australia, tend to rely on landfilling as the main disposal option. However, as can be seen in Fig. 16.8, many developing countries are still disposing of their waste in open dumps, which has negative implications on the environment and on human health.

16.5.2 Incineration

Incineration is a type of thermal treatment that is recognized as one of the most widely practiced solid waste treatment, subsequent to landfilling, where the bulk of the waste is burnt with little or no pre-treatment. Incineration greatly reduces the space requirements for the same volume of waste when compared to disposal in a landfill, and can be an option for the waste disposal in countries with land limitations, since the reduction of waste volume is about a factor of 4–10 times (UNESCAP 2000). However, if not effectively managed and regulated, incinerator emissions pose environmental and health risks due to air pollution. Thompson and Anthony (2008) stated that incineration does not remove waste, it only changes it into another forms of gas, particulates and ash, which might be more hazardous while less visible than in the original form. The net climate change impacts of the incineration basically depend on how much CO_2 is released—both at the incinerator itself and in savings of fossil fuel from conventional energy sources displaced by incineration (EU 2001).

Previously, incineration processes were uncontrolled and multiple hazardous pollutants were released to the air; the current practices, with new technologies and new stringent regulations, the emission of pollutants is orders of magnitude lower than with the historic, unregulated incinerators that had little or no flue gas clean up (Rabl and Spadaro 2002). In addition, incineration plants were mainly designed purely to handle waste, but today's plants are typically designed to recover energy from the waste. The high capital and operational costs can be affected significantly by the increasingly stringent air pollution control regulations (UNESCAP 2000). Additionally, waste requires the most energy to heat and as such pre-drying the moist waste will be energy intensive, and adds up significantly to the incinerator operational costs. Due to these facts, it is estimated that sometimes the cost of waste incineration can be 10 times greater than the cost of open dumping/land filling (UNECSAP 2000). The latter authors confirmed that for these reasons, the incineration is considered inappropriate waste disposal option for the low and middle income countries. Figure 16.8 shows that 75% of the MSW generated in Japan is being treated by incineration, while 5% only of MSW waste is treated by incineration in Pakistan and India for example.

The exact composition of emissions from incinerators will vary on the type of waste being burnt at any given time, the efficiency of the installation, and the pollution control measures in place. The most common form of incineration in use at present is large scale mass burn incineration, with annual throughputs usually in excess of 100,000 tons/year; however, the mass burn incinerators require considerable upfront capital investments in addition to maintenance costs during its working life which is estimated to be around 20–30 years (EU 2001).

Unless treated properly, the emissions, the fly ash and the residues resulting from the incineration processes have serious and hazardous consequences: the emissions of acid gases, CO_2, metals and organics affect global climate; the emissions of NO_x and SO_x affect the vegetation and human health; the discharge of cooling water may have effects on water resources; and the fly ash (when landfilled) affect landfill emissions and leachate discharge (Williams 2005).

In the developed countries, there is a growing interest in alternative thermal treatment options such as gasification and pyrolysis. These technologies are based on heating the waste under carefully controlled conditions in complete absence of air, or sometimes with limited air supply, in order to breakdown the organic matter of the waste and form gaseous or liquid products that can be used as fuel for engines to generate electricity.

16.5.3 Composting

Composting process was defined by Swan et al. (2002) as the "controlled biological decomposition and stabilization of organic substances, under aerobic conditions, and allow the development of thermophilic temperatures as a result of biologically produced heat. The result is a product that has been stabilized, and is high in humic substances, and can be beneficially applied to land." The composting practices were applied by farmers for centuries, and it was mainly in the form of returning the organic residues to the land in order to maintain soil fertility and organic matter.

Waste composting has several advantages:

- The composting process can be employed as a treatment process for different types of waste, including but not limited to, MSW, sludge, biosolids, and agricultural wastes;
- The composting processes reduce the CH_4 emissions produced from the degradation of waste in landfill sites and open dumps;
- The compost improves the soil fertility and soil organic matter content. Therefore, the compost can replace soil fertilizers;
- The stabilized compost materials have been shown to degrade a wide range of organic pollutants (Albanna et al. 2010). Therefore, compost can be used for the bioremediation of contaminate sites and soils.

Composting can be carried out in many ways, ranging from small scale composting in gardens by the householders, to medium size composting by farmers or small communities, to large scale composting systems in specialized facilities. The composting processes are based on different essential activities: waste shredding where waste is shredded to reduce particle size; mixing of different feed stocks together to improve homogeneity and adjust carbon to nitrogen ratio; and adding water to increase the moisture content of dried waste.

16 Solid Waste Management Options and their Impacts

There are different composting approaches and can be summarized as follows (Swan et al. 2002):

- Windrow composting: this system is the simplest, and is based on open-air turned windrows, where the compost materials and the feed stocks are turned continuously to introduce fresh air and release heat and extra moisture.
- Aerated static piles: this approach is based on certain mechanized systems to dispense with turning and either blow or suck the air through the composting materials. The rate of aerator is linked to oxygen concentration and temperature.
- In-vessel systems: in this approach the composting feed stocks are contained in vessels that are usually enclosed, and this will result in a greater level of process and emission control. There are different types of in-vessels systems marketed according to benefits and applicability to different types of feed stocks and situations.
- Vermicomposting: This process is based on using selected species of earthworms to convert the compost organic wastes into nutrient rich compost. The process is carried out in long troughs, where the temperature is kept below 35°C. The earthworms are to fragment, and to mix and assist in the aeration of the waste.

There are three key stages of composting process: high rate composting and usually take from 4 to 40 days depending on the system type; stabilization that usually takes 20–60 days; and finally the maturation stage. Following composting main stages; grading and screening the compost are the most important steps to create products suitable for various end users.

Composting depends on the consistent activities of various range of micro-organisms to convert organic waste into stabilized compost; because of this fact, the composting processes have significant impacts on human health associated with the inhalation of organic dust that contain many of these active micro-organisms.

The commercial composting industry has grown rapidly in developed countries; Deacon et al. (2009) reported that in 2004 nearly 2 million tons were composted in United Kingdom, where 0.06 million tons were only composted in 1994. Moreover, the USEPA (1999b) report indicated that the percentage of waste composted to total waste generated has increased from 2% in 1990 to 5.6% in 1997. However, UNES-CAP (2000) reported that the composting plants in most developing Asian countries do not function at their full capacity, as well as they do not produce compost of marketable value. The high operating and maintenance costs results in compost costs that are higher than commercially available fertilizers. Also, the lack of material segregation produces compost contaminated with plastic, glass and toxic residues, consequently the produced compost will not be suitable for agriculture.

16.5.4 Recycling

Recycling is one of the modern waste reduction processes that involves the treatment or processing of used materials to make them suitable for re-use either for its original form or for other purposes (EIONET 2009). Recycling has many environmental benefits aside from reducing GHG emissions which contribute to climate change: recycling decreases the continuous consumption of raw materials and non-renewable resources, lowers the energy and other resource consumption, and reduces the air and water pollution resulting from various conventional waste disposal options. Many different materials can be recycled including various kinds of: glass, metals, paper, plastics, textiles, demolition wastes, electronics and different types of batteries. Waste materials can either be recycled for use in products similar to their original product use like paper, or can be recycled into a product which is different that the original product uses like plastics, amongst others.

Over the last few decades many incentives have increased the recycling rate in different countries. USEPA (2008b) reported that the recycling rate in the United States (U.S.) has increased from less than 10% of MSW generated in 1980 to over than 33% in 2007. The same source indicated that due to the increased recycling rate, the disposal to U.S. landfill sites has decreased from 89% of the MSW generated in 1980 to 54% in 2007. The report stated that the recycling rates of some products in 2007 were, for example, 99.2% of auto batteries, 64% of steel cans as well as for their trimmings, 54% of paper and board, and 35% for both metals and tires. The rate of recycling materials has increased also in the Asian Pacific Region as reported by UNESCAP (2000), where the recycling grew from 10% in 1990 to 22% in 1998. Most of the recyclable materials in this region were: paper and board (60% of the total recycled materials), followed by plastics, metals, and glasses.

The recycling practices will result in minimization of GHG emissions; the total recycled waste in 2007 in the U.S. eliminated GHG emissions of approximately 25 million m^3 tons of CO_2 equivalent or what is equivalent to removing more than 4.5 million cars from the road for 1 year (USEPA 2008). From the pilot recycling project in Jordan, Mrayyan (2004) reported that 20% of paper recycle in Jordan could reduce CH_4 emissions by 25% of the country's anthropogenic emissions annual production.

The ultimate benefits from recycling processes are cleaner land, air, and water, overall better community health, and a more sustainable economy for many countries who invest the time and money in recycling projects. Recycling not only reduces the volume of wastes to be disposed, but also extends material reserves of finite global resources, and saves energy needed to manufacture several products. Also, it saves different countries valuable foreign capital which would otherwise be used to import raw materials. Unfortunately, recycling projects in the developing countries face the following challenges: the lack of proper marketing for the recyclable items; lack of the technology and know how; lack of funds; and, lack of community involvements.

Recycling is a main component of modern waste reduction needed to have sustainable environmental management. It is the third component of the 3R's mantra to "Reduce, Reuse, Recycle" waste hierarchy which classify waste management strategies. Several methods for increasing the recycling rates in the future include: economic incentives for consumers to recycle spent products and their packaging; economic incentives for manufacturers of products to use recycled materials; con-

16 Solid Waste Management Options and their Impacts

sumer education to motivate people globally to separate recyclable products; incorporating recycling into initial purchase costs; and, finally enforcing mandatory recycling requirements with penalties for noncompliance.

16.5.5 The Cost of Waste Management Options

In order to have sustainable waste disposal options, it is crucial that these options reflect their full environmental and economic costs; considering that the capital and operational costs of the disposal option will be added to the external costs to the environment (Williams 2005). The costs of waste management options depend on the type of waste and the associate hazard. In many countries, landfilling of the waste is the cheapest disposal option, although landfilling is the least desirable option according to the hierarchy of waste; this indicates that the lowest cost option may not be the most environmentally acceptable option. However, landfilling costs are expected to increase due to the new regulations concerning LFG and the control of hazardous leachate.

Incineration has higher capital, and operational and maintenance costs compared to landfilling. These costs have increased lately due to more strict regulations for the operations of these facilities. As for the treatment of the industrial waste, Williams (2005) explained that incineration costs will depend on the calorific value of the waste and the amount of gas clean-up, the throughput, and the requirement for the energy generated. In all cases, the energy recovery of both landfills and incinerators reflects on the cost of disposal.

Selling the energy produced either by combustion or landfilling will result in cost reductions that can contribute to reducing or neutralizing operational costs. However, a critical point that must be taken into considerations that the energy recovery facilities (landfills or incinerators) are large facilities in size with high capital costs to construct; therefore, it is necessary to support these facilities with long term supply of waste to optimize capital investments and to guarantee sustainability.

The cost of waste disposal is affected significantly by the costs of collection and transportation. Since the municipalities are responsible for waste management options, the collection and transportation costs can be different depending on the total budget allocated for the waste management. UNESCAP (2000) reported in this context that in low income countries the collection costs represents 80–90% of the MSW budget where most of the waste is sent to open dumps, and the collection activities are carried out by labors and low levels mechanization, while the collection of the MSW in middle income countries represents 50–80% of the waste management budget. The latter source also indicated that the collection costs in high income countries represent less than 10% of the allocated waste management budget, considering that the community participation in these countries reduces costs and increases options for comprehensive management of the generated wastes, mainly by recycling and composting. Transportation costs are also critical, since it is rare to find suitable locations for landfill or incinerators sites near the points of generation which will lead to high transportation costs. The collection and transportations costs are affected by the local conditions which apply, such as: the number of collections points, the nature of containers, the size of the collecting and transporting vehicles, labor cost, frequency of collection, and the distance to disposal site (Williams 2005).

The costs of waste disposal options are variable and mainly affected by the site characteristics, cost of the land, technologies used, and the size of the planned project. Though the EU report (2001) presented a cost comparison between different waste treatment options, and declared that the average cost of landfilling is between 11 and 162 Euro/ton of treated waste, the average cost of incinerations is between 31 to 148 Euro/ton of treated waste, and the average cost of composting is 16–174 Euro/ton of treated waste. It is apparent that there is a very wide range of costs reported; it is reasonably assumed that the values towards the lower end of the price range reflect disposal charges to older or low quality facilities with low environmental protection standards. In the more advanced methodologies of engineered landfills, modern incinerators, or state-of-the-art composters are expected to see a significant increase in disposal costs.

16.6 Impact of Solid Waste Management Options on Climate Change

Methane is an important green house gas (GHG); its accumulation in the atmosphere accounts for 22.9% of total direct greenhouse effect (USEPA 2006a). The Intergovernmental Panel on Climate Change (IPCC 2001) reported that the CH_4 atmospheric life time (9–15 years) is shorter than CO_2 (200–450 years); this means that the global warming potential of CH_4 is higher for shorter time horizons. In addition, CH_4 is 25 times more effective in trapping heat in the atmosphere than CO_2 over a 100-year period (Scheutz et al. 2009).

Methane emissions arise from both natural and anthropogenic sources. The anthropogenic CH_4 emissions are rising due to increased industrial activities over the years. One of the main sources that add to the anthropogenic CH_4 in the atmosphere is solid waste disposal. In Canada, solid waste disposal is responsible for 23% of total anthropogenic CH_4 emissions (Environment Canada 2005), while in United Kingdom, landfills release 27% of the country's CH_4 emissions (Surrey Country Council 2010). In the United States, landfills are the largest human-related source of CH_4; accounting for 34% of all CH_4 emissions (USEPA 2006b).

Ninety-five percent of the solid waste sector's emissions result from landfill gas (Mohareb et al. 2004). Landfill gas (LFG) resulting from the biological degradation of organic waste inside the landfill is a flammable and potentially harmful mixture; it consists by volume of 50–65% CH_4, 35–50% CO_2 and around 1% non-methane organic compounds (NMOCs) (USEPA 2006c). The biochemical reactions that produce LFG will continue long after a landfill is capped and therefore, LFG emis-

16 Solid Waste Management Options and their Impacts 519

sions will continue even after landfill closure (Environment Canada 2006). If not collected, LFG can migrate to the surface and enter the atmosphere, consequently increasing GHG emissions.

The vast majority of existing landfills sites around the world do not have biogas capturing setups. Methane emissions from these sites are entirely released to the environment; in addition, they also emit millions of tons of volatile organic compounds. For example, the net CH_4 emissions from Canadian landfills in 2002 were estimated to be 22 million metric tonnes (Mt) of carbon dioxide equivalent (eCO_2) and only 6.6 Mt eCO_2 were captured (Environment Canada 2005). Moreover, the average emission rate of hazardous volatile substances from landfills was estimated to be 35 kg per million kg of refuse (Reinhart et al. 1992). Based on this estimation and Environment Canada (2005) statistics, the amount of NMOCs emitted from Canadian landfills is estimated to be 1,000 tonnes/year.

As for the incineration, the net climate change impacts of incineration depends on how much fossil-fuel CO_2 is released—both at the incinerator itself and in savings of fossil fuel from conventional energy sources displaced by incineration. The composition of emissions from incinerators will differ with the type of waste being burnt at any given time, and the efficiency of the pollution control measures of the incinerator. All incinerators should be routinely assessed for their effect on global warming.

There are several strategies through which the solid waste sector can reduce its GHG emissions, and can be summarized as follows:

- Source reduction: it is the most enviable option where the amounts of waste produced and sent afterwards to the disposal site are being reduced. There are significant emission reduction gains achievable through source reduction, mainly for paper and paper product wastes (Mohareb et al. 2004). The GHG emissions benefits of source reduction should lead to the encouragement of this practice since it mainly depends on the commitment of the public and the manufacturers.
- Recycling: is a significant management option that will lead to considerable reduction of GHG emissions from waste sector. Recycling will reduce the amount of waste sent to landfills and the consequent emissions due to the biodegradation of the waste. Moreover, GHG emissions will benefit from the reduction of raw materials that need to be processed, and the energy required to process these materials.
- Compsoting of the organic waste reduces GHG emissions through the conversion of rapidly decomposing matter to CO_2 opposed to CH_4 that will result from the biodegradation of these components if sent to the landfills. Also, composting diverts the waste from landfill sites and produces useful and cost effective by-products that can be used in several human activities.
- Landfill gas collection techniques: As mentioned earlier, it was estimated that 95% of GHG emissions from solid waste sector result from landfill sites, therefore, collecting and capturing LFG as a source of renewable energy is very attractive and useful techniques, and should be encouraged by all governments.

16.7 Impact of Solid Waste Management Options on Human Health

The numbers of diagnosed cancer and asthma incidents have increased persistently in recent decades and these two serious health conditions, in addition to many others, have been shown to correlate geographically with both waste treatment facilities and the presence of chemical industries, pointing to an urgent need to reduce our exposure to the health impacts of waste exposure. The following review will discuss the consequent impacts of varies solid waste management option on human health.

16.7.1 Consequent Impacts of Landfilling on Human Health

Many researchers have expressed health concerns about landfill sites and the adverse health effects of the emissions from these sites on the human health. A review of several human health cases found that there are increased risks of low birth weight, birth defects, and certain types of cancers have been reported in the residents near landfill sites, in addition to some other chronic symptoms such as fatigue, sleepiness, and headaches (Surrey County Council 2010; Redfearn and Roberts 2002). Also, a report presented by the Department of Health in the United Kingdom (1999) confirmed that there is a significant association between congenital malformations and setback distances from residences to nearby landfill site. Elliott et al. (2001) presented research that was designed to study the risks of adverse birth outcomes associated with residences within 2 km of 9,565 landfill sites operating between 1982 and 1997 in the United Kingdom, and to compare the outcomes with the results between populations living further away, as their control group. These authors reported that they found small excess risks of congenital anomalies, and low (<2.5 kg) to very low (<1.5 kg) birth weight in populations living within the 2 km boundary of the landfill sites. These authors could not clarify a mechanism to explain their finding, though stated that further research is needed to assess the adverse effect of living near landfill sites. Mrayyan (2004) reported that 30–34% of the citizens residing near the largest landfill site in Jordan experienced continuous allergy problems; 9.5–30% experienced chest inflammations and 5–13% experienced asthma attacks.

The primary exposure pathways of concern are the inhalation of the volatile organic compounds (VOCs) associated with landfill gas (LFG) which has been proven to be carcinogenic. The sources of the VOCs are the redundant substances such as household cleaning products, materials coated with or containing paints and adhesives, and other such items. The production of VOCs and their movement within LFG can be attributed to the biological decomposition of heavier organic compounds into lighter and more volatile compounds, as well as the vaporization and chemical reactions of material present in landfills (USEPA 1999a). Their rate of emission is governed by LFG production and transport mechanisms. Trace components in LFG include hydrocarbons, aromatics, halogenated hydrocarbons and inorganic compounds. The concentrations of VOCs can vary depending on several conditions such as the type of waste deposited in the landfill, the physical properties of each individual compound, and climatic conditions around the landfill site (USEPA 2003). USEPA (1999) proposed default values for VOCs concentrations for regulatory compliance purposes, but site specific information should be taken into account when determining the total VOCs concentrations for inventory purposes.

In addition to the significant impacts of LFG, the risk of pollution due to leachate seepage from landfill sites or open dumps is a major environmental concern worldwide. Leachate forms when rainwater can access the landfill site and percolates through the waste layers; as leachate passes through the landfill strata, it collects contaminants and leachate leaking from landfill sites contains harmful pollutants, such as, dissolved organic matter, in organic macrocomponents, heavy metals and xenobiotic organic compounds (Kjeldsen et al. 2002). If allowed to migrate, these contaminants, when released into the surrounding environment, will pose potential threats to soil, ground and surface water resources. Restoring, treating, and cleaning up the pollute groundwater is very difficult and expensive. The pollution of surface and ground water is of a major concern in the developing countries, since the open dumps and most of the landfills do not have leachate control or collection systems.

16.7.2 Consequent Impacts of Incineration on Human Health

Incinerators, as a result of how they are operated, discharge hundred of toxic and bio-accumulative pollutants into the atmosphere, in addition to ash. These pollutants contain toxic substances, such as dioxins, and heavy metals such as lead, mercury, and cadmium. Each one ton of waste burnt releases around 5,000 m^3 of gases containing many pollutants, and these pollutants are transported in the air and deposited in water and soil, both near and far from the incinerator (Allsopp et al. 2001).

Many researchers have expressed serious concerns about the potential health impact of waste incineration's pollutant release into atmospheric air, especially due to the fact that the older generations of waste incinerators do not operate efficiently with modern emissions control. Incinerator emissions are a major source of air pollution, emitting fine particulates, toxic metals, and more than 200 VOCs into the environment (Thompson and Anthony 2008). According to these authors, the high level of fine particulates have been associated with increased prevalence of asthma; the toxic metals accumulate in the body and have been connected to a range of emotional problems such as autism, attention-deficit hyperactivity disorder (ADHD), and learning difficulties in children, and in adults have been connected to unprovoked acts of violence, unexplained depression and Parkinson' disease. Most VOCs are carcinogens, mutagens, and hormone disrupters. Emissions from incinerators also contain other unidentified compounds which have harmful potentials such as dioxins, furans, acid gases, and the polyaromatic hydrocarbons (PAHs). It was reported that the PAHs can cause genetic changes, which will affect the human health for generations to come (Thompson and Anthony 2008).

Exposures to compounds emitted from incinerators occur by inhalation of contaminated air, or by consumption of agricultural product or soil that has been contaminated by deposition of airborne pollutants (Allsopp et al. 2001). In addition, workers at incinerator plants may also be exposed to contaminated ashes. Ash contains toxic pollutants, such as mercury, lead, chromium, and arsenic, which can contaminate vegetables if scattered in gardens. Children can also unintentionally swallow contaminated dirt on their hands if playing near discarded ash. Hu and Shy (2001) stated that several studies showed significant associations between waste incineration and lower male-to-female ratio, twinning, lung cancer, laryngeal cancer, ischemic heart disease, urinary mutagens and promutagens, and increases in blood level concentrations of certain organic compounds and heavy metals. The damage is especially problematic for children, for the elderly, and those with pre-existing respiratory conditions.

16.7.3   Consequent Impacts of Composting on Human Health

Composting, by definition, depends on the interconnected activities of a various range of micro-organisms in order to convert the organic waste substrate into a stabilized and nutrient rich material, which is the final or "stabilized" compost. In the continuously active systems, there are many health concerns that result from the composting activities regardless of the size of the facilities.

Deacon et al. (2009) reported that elevated levels of endotoxin have been measured in environments around and close to composting facilities. The authors explained that the inhalation of endotoxin at elevated concentrations have been associated with acute airway obstruction, chronic bronchitis and decreased lung function. Furthermore, Swan et al. (2002) demonstrated that varies microbial components and chemicals, such as fungi, bacteria, actinomycetes, endotoxins, mycotoxins, glucans, and volatile organic compounds, generated during the handling of the compost has potential health hazards associated with the exposure to compost bioaerosols. These adverse health effects include allergies and toxicity. These authors affirmed that to reduce the harmful effects of composting processes on human health; it is important to establish good hygiene practices at every composting facility, to understand the potential health hazard associated with the exposure to compost bioaerosols, and to examine the microbial components of bioaerosols generated during the handling of the compost to mitigate the adverse effects on the operators' and all neighbors' health.

16.7.4   Consequent Impacts of Recycling on Human Health

Gladding (2002) presented the results of a study conducted in different recycling facilities operating in the United Kingdom, where the objective of this study was

16 Solid Waste Management Options and their Impacts

to investigate the health risks of materials recycling facilities. The author demonstrated the following harmful effects on human health:

- The noise was found to exceed the recommended levels due to the operating equipments on the sites, which affected the sites operators' hearing;
- The biological agents present in the air surrounding the plants are harmful to the health of the operators working in these facilities, due to close contact with the waste materials, as well as the aerosol of organic dust from organic residues and contaminants within the materials received;
- The dust, endotoxin, and glucan were found in excess of the recommended exposure levels. As a result, the operators appeared to suffer from irritated noses, coughed with phlegm, had hoarse/parched throat, suffered from stomach problems, had headaches and influenza type symptoms, experienced nausea and were tired;
- The blood data from some operatives new to the industry showed a significant decrease in lymphocytes, monocytes, and neutrophils over a period of 10 months, which reflected inflammation in the lymph nodes.

The study concluded that several measures should be taken promptly to mitigate the harmful effects due to the exposure and handling methods of the recycled wastes, with a focus on education and training hygiene.

16.8 Proper Management of Solid Waste—Preventive Measures

The accelerated economic developments in the last two decades have led to uncontrolled population growth, rapid urbanization, and modernized life styles which affected the consumption patterns and resulted in the generation of different types of wastes that vary widely. The current solid waste disposal options contribute to several environmental problems; incineration creates toxic substances that affected the human health, while landfills emit methane and other gases that contributed to global warming. Solid waste is an inevitable product of civilization. There are critical responsibilities facing the global societies: determining how to clean up their legacy problems and restore natural resources; and implementing strategies to allow for future growth, at the same time protecting the environment and human health, maintaining biodiversity, and protecting cultural/ social values. In order to achieve more sustainable waste management practices, the challenge will be to reduce the amount of generated waste from different human activities and at the same time to increase waste diversion in an economically feasible way.

There are several actions needed to achieve sustainable waste management practices and reduce the effect of solid waste on the environment and human health, such as:

- Integrate waste management systems while making reduced environmental impacts a top priority,
- Promote the hierarchy of waste management (Reduce, Reuse, and Recycle), and encourage waste segregation. It is important to deal with the solid waste as resource and to prevent wasting potentially useful materials. The recovery, reuse and recycling is increasingly being realized as the central basis of an integrated approach to waste management,
- Source reduction: It is considered as one of the most desirable waste management options, and depends on reducing the amount of material used in a process. It is based on the reuse of the product or part of it. It optimizes the use of resources, and removes potential sources of pollution; therefore, it can provide the highest level of environmental protection. It may be possible to reduce the quantities of waste produced at each stage in the life cycle of a product. In the design phase, consideration can be given to the types of materials to be used and to the recyclability of the product at the end of its life time (EIONET 2009). It is extremely imperative to use efficient processes during manufacturing of the product mainly what concerns energy and materials. Mohareb et al. (2004) reported that the extended producer responsibility programs, applied in Germany, which place full product liability on producers can encourage effective waste reduction,
- Promote waste minimization through waste exchange. Waste minimization is the process of reducing waste generation by individuals and communities, which reduces the impact of the waste on the environment and human health. The waste exchange scheme has provided major benefits to a variety of companies, by providing savings in disposal in addition to raw materials cost (UNESCAP 2000),
- Encouraging the prevention or reduction of waste and its harmful impacts by the development of clean solid waste disposal and treatment technologies,
- Provide consumers with accurate information on the different streams of waste, the generation rates, consequences and risks, and educate them on different ways of dealing with waste, and the economic benefits of proper waste treatment and disposal. Society will consider their actions with proper regard to their environmental impacts,
- Promote private sector incentives and initiatives to move to commercializing the waste management activities in many countries. UNESCAP (2000) stated that privatization basically involves the transfer of management responsibility and/ or ownership from the public to the private sector and has proven to be a powerful means of improving the efficiency of some waste management services and disposal,
- Placing incentives for utilization of biogas energy, incentives for the installation of anti-pollution equipments, and incentives for the improvement of existing waste management schemes. These incentives can be provided in the form of tax credits or tax reduction, and will encourage the private sector to invest in waste to energy projects.

Developing robust solid waste management strategies is still considered as major challenge for different communities regardless of waste quantities or types generated.

16.9 Summary

In the last decade, signs of climate change and global warming started emerging and making changes to life and landscape all around the world. The emissions from solid waste disposal sites and their impact on the GHG global budget and human health became one of the main challenges for all communities. Moreover, the inefficient waste management has aggravated the problems and resulted in threats on natural resources. These issues clearly emphasis the importance of effective and efficient waste management techniques.

In the past few decades the generation rates of solid waste increased significantly. The generated wastes reflected losses of materials as well as energy resources, and resulted in economic and environmental cost on society due to waste collection, treatment and disposal. Integrated waste management systems while making reduced environmental impacts is one of the global top priorities. There are a number of different options that should be encouraged for the treatment and management of waste including waste prevention, minimization, re-use, recycling, and energy recovery from treatment options.

Landfilling is one of the oldest and most predominant ways of management waste. Landfills emissions if released to the atmosphere can be harmful to the environment and human health. However, properly designed and managed active landfills are capable of generating significant amounts of power. Landfill gas is recognized as an important source of renewable energy.

Incineration is another major traditional waste treatment option. Uncontrolled emissions from incinerators will affect human health in different serious manners. To address pollution from incinerators, strict standards have been set to control these harmful emissions and the pollutants they carry. The cost of incineration is greater than other traditional waste management options, and incinerators have not been able to compete financially with other waste treatment options.

Substantial environmental benefits are associated with the use of compost to improve the soil's organic status. Composting is an effective recycling strategy for turning the organic waste into a valuable recourse and useful by-products. It is estimated that 40 to 70% of the total MSW can be composted.

Recycling is one of the most important integrated solid waste management plans. Recycling reduces the amount of waste needed to be treated and consequently reduces the harmful impacts of the waste on the environment and human health. Waste materials can either be recycled for use in products similar to their original use or can be recycled into a product which is different from the original use. The benefits of recycling should further encourage recycling programs globally

Regardless of the waste treatment or disposal option, it is important to maximize the recovery of waste before looking to its disposal and residual problems. In many developing countries, waste management is often hindered by a lack of national policy directions, available funds, and allocation of responsibilities. Most of these countries lack the national planning to develop integrated waste management policies and strategies, and financing remains a critical issue in most regional waste management operations.

References

Albanna M, Fernandes L Mostafa W (2010) Kinetics of biological methane and non-methane oxidation in the presence of non-methane organic compounds in landfill bio-covers. Waste Manag 30:219–227

Allsopp M, Costner P, Johnston P (2001) Incineration and Human Health. Environ Sci Pollut Res 8(2):141–145

Conference board of Canada (2008) Environment- Municipal Waste Generation. http://www.conferenceboard.ca/hcp/details/environment/municipal-waste-generation.aspx

Deacon L, Pankhurst L, Liu J, Drew HG, Hayes ET, Jackson S, Longhurst J, Longhurst P, Pollard S, Tyrrel S (2009) Endotoxin emissions from commercial composting activities. Environ Hlth 8(1):S9

Department of Health—United Kingdom (1999) Health effects in relation to landfill sites. Environmental Chemicals Unit. http://www.advisorybodies.doh.gov.uk/land1.htm

EIONET—European Environmental Information and Observation Network (2009) What is Waste? http://scp.eionet.europa.eu/themes/waste/#10

Elliott P, Briggs D, Morris S, de Hoogh C, Hurt C, Jensen TK, Maitland I, Richerdson S, Wakefield J, Jarup L (2001) Risk of adverse birth outcomes in populations living near landfill sites. BMJ 18(7309):363–368

Environment Canada (October 2005) Landfill gas management in Canada report. http://www.ec.gc.ca/pdb/ghg/inventory_report/2005_report/tdm-toc_eng.cfm

Environment Canada (September 2006) Greenhouse gas sources and sinks. Fact sheet 8- Waste: 1990–2000. http://www.ec.gc.ca/pdb/ghg/inventory_report/1990_00_factsheet/fs8_e.cfm

Environmental Management Act of British Colombia Canada (1992) Agricultural waste control regulation. B.C. Reg. 131/92. http://www.bclaws.ca/EPLibraries/bclaws_new/document/ID/freeside/10_131_92

EU- European Commission (2001) Waste management options and climate change. Office for official publications of the European Communities Luxembourg ISBM 92-894-1733-1 http://ec.europa.eu/environment/waste/studies/pdf/climate_change.pdf

Fatta D, Papadopoulos A, Kourmoussis F, Mentzis A, Sgourou E, Moustakas K, Loizidou M Siouta N (2004) Estimation methods for the generation of construction and demolition waste in Greece. In: Limbachiya MC, Roberts JJ (eds) Construction management waste. Thomas Telford Publishing LTD, London, England. ISBN: 0727732854

Gladding TL (2002) Health risks of materials recycling facilities. In: Hester RE, Harrison RM (eds) Environmental and health impact of solid waste management activities issues in environmental science and technologies No. 18. Royal Society of Chemistry Thomas Graham House Science Park, Milton Road, Cambridge, UK pp 53–72

GTZ- International Services- Deutsche Gesellschaft für Technische Zusammenarbeit (2006) Technical policy note on agricultural waste management in Egypt. http://www.metap-solidwaste.org/fileadmin/documents/National_activities/technical_assistance/Preliminary_Report_egy.pdf

Hu SW, Shy CM (2001) Health effects of waste incineration: a review of epidemiologic studies. J Air Waste Manage Assoc 51(7):1100–1109

IPCC—Intergovernmental Panel on Climate Change (2001) IPCC working group I the scientific basis. http://www.grida.no/climate/ipcc_tar/wg1/248.htm

Kjeldsen P, Barlaz MA, Rooker AP, Baun A, Ledin A Christensen TH (2002) Present and long-term composition of MSW landfill leachate: a review. Crit Rev Environ Sci Technol 32(4):297–336

LaGrega MD, Buckingham PL, Evans JC, Environmental Resources management (2001) Hazardous waste management. McGraw-Hill Companies Inc., New York, NY, 10020. ISBN: 10:0-07-039365-6

Mrayyan B (2004) Methane gas emissions from landfill: opportunities and constraints—the case of Al-Russifa city. J Environ Assess Policy Manage 6(3):367–384

Mohammadi IM (2006) Agricultural waste management extension education (AWMEE), the ultimate need for intellectual productivity. Am J Environ Sci 2(1):10–14

16 Solid Waste Management Options and their Impacts

Mohareb A, Warith M, Narbaitz R (2004), Strategies for the municipal solid waste sector to assist Canada in meeting its Kyoto Protocol commitments. Environ Rev 12:71–95

Narayana H (2009) Municipal solid waste management in India: from waste disposal to recovery of resources? Waste Manage 29:1163–1166

New Zealand Ministry for the Environment (2009) Composition of solid waste. http://www.mfe.govt.nz/environmental-reporting/waste/solid-waste/composition/index.html

Pulikkottil JA, Somasundaram R (2004) Municipal solid waste management: an opportunity analysis and comprehensive bench marking of solid waste management practices. In: Limbachiya MC, Roberts JJ (eds) Construction management waste. Thomas Telford Publishing LTD, London, England. ISBN: 0727732854

Redfearn A, Roberts D (2002) Health effects and landfill sites. In: Hester RE, Harrison Royal RM (eds) Environmental and health impact of solid wast management activities issues in environmental science and technologies No. 18. Society of Chemistry Thomas Graham House Science Park, Milton Road, Cambridge, UK, pp 103–140

Reinhart DR, Cooper DC, Walker BL (1992) Flux chamber design and operation for the measurement of municipal solid waste landfill gas emission rates. J Air Waste Manage Assoc 42:1067–1070

Scheutz C, Kjeldsen P, Bogner J. De Visscher A, Gebert J, Hilger H, Huber-Humer M, Spokas K (2009) Microbial methane oxidation processes and technologies for mitigation of landfill gas emissions. Waste Manage Res 27:409–455

Statistics Canada (2006) Earth day by the numbers. http://www42.statcan.ca/smr08/smr08_023_e.htm

Strange K (2002) Overview of waste management options; their efficacy and acceptability. In: Hester RE, Harrison RM (eds) Environmental and health impact of solid waste management activities issues in environmental science and technologies No. 18. Royal Society of Chemistry Thomas Graham House Science Park, Milton Road, Cambridge, UK, pp 1–52

Surrey County Council (2010) Landfill sites. http://www.surreycc.gov.uk/sccwebsite/sccwspages.nsf/LookupWebPagesByTITLE_RTF/Landfill+sites?opendocument#2

Swan JRM, Crook B, Gilbert EJ (2002) Microbial emissions from composting sites. In: Hester RE, Harrison RM (eds) Environmental and health impact of solid waste management activities issues in environmental science and technologies No. 18. Royal Society of Chemistry Thomas Graham House Science Park, Milton Road, Cambridge, UK, pp 73–102

Thompson J, Anthony H (2008) "The health effects of waste incinerators" 4th Report of the British Society for Ecological Medicine. http://www.ecomed.org.uk/content/IncineratorReport_v3.doc

Townsend T, Kumar D, Ko J (2008) "Bioreactor landfill operation: a guide for development implementation and monitoring" version 1.0 (July 1 2008). Prepared for the Hinkley Center for Solid and Hazardous Waste Management Gainesville FL. http://www.bioreactor.org/BioreactorFinalReport/FinalReportVOLUME1_10/AttachmentforVOLUME8/Bioreactor_Landfill_OperationV10.pdf

UNESCAP—United Nations Economic and Social Commission for Asia and the Pacific (2000) Sustainable Asia—Waste. http://www.unescap.org/esd/environment/soe/2000/documents/CH08.PDF

USEPA—United States Environmental Protection Agency (1999a) Municipal solid waste landfills volume I: summary of the requirements for the new performance standards and emissions guidelines for municipal solid waste landfills. 453/R-96-004. Research Triangle Park, NC, USA

USEPA—United States Environmental Protection Agency (1999b) Characterization of municipal solid waste in the United States: 1998 update. http://www.epa.gov/wastes/nonhaz/municipal/pubs/98charac.pdf

USEPA—United States Environmental Protection Agency (2003) Landfill gas and how it affects public health safety and the environment. http://www.epa.gov/lmop/docs/faqs_about_LFG.pdf

USEPA—United States Environmental Protection Agency (2006a) "Global Anthropogenic Non-CO2 Greenhouse Gas Emissions: 1990–2020" U.S. Environmental Protection Agency Office of Atmospheric Programs Climate Change Division U.S. Environmental Protection Agency 1200 Pennsylvania Avenue NW Washington DC 20460

USEPA—United States Environmental Protection Agency (2006b) Methane sources and emissions October 2006. http://www.epa.gov/methane/sources.html#anthropogenic

USEPA—United States Environmental Protection Agency (2006c) State of knowledge October 2006. http://www.epa.gov/climatechange/science/stateofknowledge.html

USEPA—United States Environmental Protection Agency (2008a) Waste management resources. http://www.epa.gov/tribalcompliance/wmanagement/wmwastedrill.html

USEPA—United States Environmental Protection Agency (2008b) Municipal solid waste generation recycling and disposal in the United States: facts and figures for 2007. http://www.epa.gov/waste/nonhaz/municipal/pubs/msw07-fs.pdf

USEPA—United States Environmental Protection Agency (2010) Wastes—resource conservation—reduce reuse recycle—construction and demolition materials. http://www.epa.gov/wastes/conserve/rrr/imr/cdm/index.htm

Watts RJ (1997) Hazardous wastes: sources pathways receptors. Wiley, New York. ISBN: 0-471-00238-0

Williams P (2005) Waste treatment and disposal. Wiley, New York. West Sussex England. ISBN 0-470-84912-6.

World Bank—The International Bank for Reconstruction and Development (1999) "What a Waste: Solid Waste Management in Asia" Urban Development Sector Unit East Asia and Pacific Region 1818 H Street N.W., Washington, DC, 20433 U.S.A. http://web.mit.edu/urbanupgrading/urbanenvironment/resources/references/pdfs/WhatAWasteAsia.pdf

Chapter 17
Potential of Biopesticides in Sustainable Agriculture

M. Shafiq Ansari, Nadeem Ahmad and Fazil Hasan

Contents

17.1	Introduction	530
17.2	Biopesticide Categories	532
	17.2.1 Biochemical Pesticides	533
	17.2.2 Microbial Pesticides	548
17.3	Constraints of Biopesticides	573
17.4	Biopesticide Regulations	573
17.5	Market Prospects for Biopesticides	573
17.6	Conclusion	574
References		575

Abstract Sustainable agriculture also aims at increasing the yield of food and fiber crops and reducing the incidence of pests and diseases to such a degree that they do not cause extensive damage to crops. With the advent of chemical pesticides in 1940s, this crisis was resolved to a great extent. But the overdependence on chemical pesticides and eventual uninhibited use of them has caused serious health and environmental problems. This concern has encouraged researchers to look for better alternatives to synthetic pesticides. Biopesticides can make important contribution to sustainable agriculture and help reduce reliance on chemical pesticides. Microbial insecticide like *Bacillus thuringiensis* (*Bt*) produces a proteinic toxin which induces paralysis of the midgut and brings about cessation in feeding after being ingested by insect pests. Other promising candidates are *Beauveria bassiana* and *Metarrhizium anisopliae*. The spores penetrate the host cuticle, once inside the body, producing toxic metabolites called beauvericin (*B. bassiana*) and destruxins *(M. anisopliae)* responsible for death of the insects. Baculoviruses (Nuclear Polyhedrosis Virus and Granulosis Virus) are safe to human beings and wildlife, their specificity is very narrow. They do not infect beneficial insects and have capacity to persist in the environment, making them very suitable for use in sustainable agriculture. Semi-

---

M. S. Ansari (✉)
Department of Plant Protection, Faculty of Agricultural Sciences, Aligarh Muslim University,
Aligarh-202002, India
e-mail: mohdsansari@yahoo.com

A. Malik, E. Grohmann (eds.), *Environmental Protection Strategies for Sustainable Development,* Strategies for Sustainability,
DOI 10.1007/978-94-007-1591-2_17, © Springer Science+Business Media B.V. 2012

ochemicals: attractants and pheromones, and botanicals are important sources of agrochemicals used for the management of insect pests. They degrade rapidly and therefore, are considered safer than chemical pesticides to the environment.

The market for biopesticides is expanding rapidly: growing at some 10% per year, by the end of 2010 global sales are expected to hit the $1 billion mark and make up 4.2% of the overall pesticides market. Further research and development of biopesticides must be given high priority and public in general and agriculturists in particular will force governments to make policy decisions in reducing the use of chemical pesticides. The present paper describes the detailed discussion on the potential of biopesticides in sustainable agriculture.

Keywords Biopesticides • Sustainable agriculture • Microbial pesticides • Biochemical pesticides • Plant extracts and oils

17.1 Introduction

Sustainable agriculture involves successful management of agricultural resources that can satisfy changing human needs while maintaining or enhancing the quality of the environment and conserving natural resources at the same time. It is therefore, necessary to understand resources, human needs, environmental conditions and the quality of life in general (Pimentel et al. 2005; Dubey et al. 2010). However, sustainable agriculture also aims at reducing the incidence of pests and diseases to such a degree that they do not seriously damage crops without upsetting nature's balance (Kogan 1998). Herbivorous insects and mites, plant diseases and weeds are major impediments to crop production (Fernando et al. 2009). Pesticides are capable of rapidly killing a range of agricultural pests, over reliance on chemical pesticides has generated panoply of problems including safety risks, environmental contamination, outbreak of secondary pests normally held in check by natural enemies, decrease in biodiversity, insecticide resistance (Fig. 17.1) and resurgence of minor pests (Talekar and Shelton 1993; Perry et al. 1998; Czeher et al. 2008; Yadouleton et al. 2010).

Global pesticide consumption has been steadily on the rise since 1940's (Kogan 1998), increasing from 0.49 k ha^{-1} in 1961 to 2 kg ha^{-1} in 2004 (Table 17.1). W.H.O. estimated that about 1 million persons are affected by pesticide poisoning and 20,000 deaths occurred every year globally. Apart from these, they also cause harm to non target beneficial insects and soil micro flora (Pimentel 1992; Kumari et al. 2004). Therefore, Isman (2006) opined that botanicals have long been tested as an attractive alternative to synthetic chemical insecticides for pest management because botanicals reportedly pose little threat to the environment or to human health.

Integrated Pest Management (IPM) includes the integration of biological, cultural, physical and chemical controls, host plant resistance, and decision support tools, which may play a significant role in sustainable agriculture (Kogan 1998). Thus, it will help to minimize the negative environmental impact and other deleterious effects caused by insecticides while providing a more sustainable approach

17 Potential of Biopesticides in Sustainable Agriculture

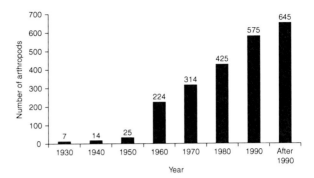

Fig. 17.1 Insecticide resistance in insects and mites. (Source: Kogan 1998)

Table 17.1 Consumption of biopesticides (Directorate of Plant Protection, Quarantine and Storage 2010)

Year	Neem	Bt	Total (MT)
1994–1995	83	40	123
1995–1996	128	47	175
1996–1997	186	33	219
1997–1998	354	41	395
1998–1999	411	71	482
1999–2000	739	135	874
2000–2001	551	132	683
2001–2002	736	166	902
2002–2003	632	143	775
2003–2004	824	157	981
2004–2005	965	139	1104
2005–2006	1717	203	1920

Bt Bacillus thuringiensis, MT million tones

to pest control, as well as to maintain crop quality, productivity and profitability (Pimentel et al. 2005).

Several entomopathogens (viruses, bacteria, fungi, and nematodes), and botanicals offer effective means of pest control when combined with other tactics such as mating disruption and the use of reduced-risk pesticides (Dev and Koul 1997; Hedin et al. 1997; Gronning et al. 2000; Pawar and Borikar 2005; Fernando et al. 2009). In addition, microbial control agents (MCAs) are safe to environment, beneficial insects, applicators, and the food supply, and can be applied just prior to harvest (El-Husseini 2006). Biopesticides can make important contributions to IPM and help reduce reliance on chemical pesticides (Shah and Devkota 2009). Hence they have a major role to play in the development of sustainable agriculture. Therefore, biopesticides often have a narrow spectrum of pest activity, which means they have a relatively low direct impact on non targets, including humans (Clemson 2007). Their use is often compatible with other control agents, and produce little or no residue (Lewis et al. 1997; Pearson et al. 2005; Clemson 2007).

In USA, there are over a thousand of biopesticide products, which are expanding rapidly: @ 4.2% of the overall pesticides market (O'Brien et al. 2009). Much of this rapid growth is due to the fact that, surprisingly, more than 80% of biopesticides are used, not by organic farmers, but by producers employing conventional farming practices. Orchard crops hold the largest share of 55% of total biopesticide use (Thakore 2006a; Fernando et al. 2009). Expert estimates, however, hold that overall pesticide use has been declining at a rate of 1.3% per year over the last decade (Thakore 2006a). This decline is attributed to increased concern about health and environment, the rise in organic agriculture or increased the demand of organic food and the emergence of alternatives, including biopesticides. The use of biopesticides is on the increase but not up to the desired level of growth (Clemson 2007). In view of this demand and the government's efforts to mitigate climate change, biopesticides are going to play an important role in future pest management programs.

This review will sieve out findings reported from field experiments and investigations on potential of biopesticides in sustainable agriculture, which strongly emphasized practical and theoretical predictions, predominantly focused on the potential of biopesticides in order to reduce pest population without upsetting the ecological balance.

17.2 Biopesticide Categories

Biopesticides may be divided into two categories: (1) naturally occurring substances which include plant extracts and semiochemicals (e.g. insect pheromones) and (2) living organisms: micro–organisms, and nematodes (Fig. 17.2).

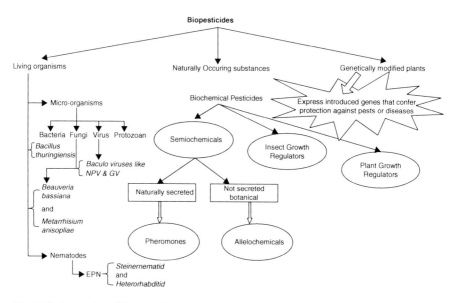

Fig. 17.2 Categories of biopesticides (*NPV* nuclear polyhedrosis virus, *GV* granulosis virus)

17.2.1 Biochemical Pesticides

Biochemical pesticides are the most closely related category to conventional chemical pesticides. Biochemical pesticides are distinguished from conventional pesticides by their non-toxic mode of action toward target organisms (usually species specific) and their natural occurrence (Steinwand 2008). The active ingredient can be a single molecule or a mixture of molecules, such as a naturally occurring mixture comprising a plant essential oil, or a mixture of structurally very similar molecules called isomers in the case of insect pheromones. While, active ingredients of biochemical pesticides occur in nature, or they may be a synthetic analog. All naturally occurring chemicals are not considered as biopesticides, because some of them are highly toxic like d-limonene which is an essential component of citrus oils and the concentrated extract of d-limonene is regulated as a conventional insecticide due to its toxic mode of action. In contrast, the oils from which d-limonene is derived normally have a non-toxic mode of action and are regulated as biopesticides. Some essential oils work as repellents, and their mode of action would be as a fragrance (Steinwand 2008). Biochemical pesticides typically fall into distinct biologically functional classes, including semiochemicals, plant extracts, natural plant growth regulators, and natural insect growth regulators. There are about 122 biochemical pesticide active ingredients registered that include 18 floral attractants, 20 plant growth regulators, 6 insect growth regulators, 19 repellents, and 36 pheromones (Mandula 2008).

17.2.1.1 Plant Extracts and Oils

Plant biodiversity provides a vast repository of biologically active compounds that find application in traditional medicines and crop protection systems (Isman 2000). Among the estimated over a half a million species of the existing plant kingdom, nearly 2,500 species belonging to 235 plant families have exhibited measurable anti-pest properties. Phytochemicals have a wide range of activity including hormonal, neurological, nutritional, and enzymatic activity, many of which still need to be explored. Plant extracts and oils are specific chemicals or mixtures of chemical components derived from a plant. This category of biopesticides is much more diverse in compositions which target the pest because of its specific mode of action. Plant extracts and oils are most often used as insecticides (Baskar et al. 2009), but can also be used as herbicides (Tworkoski 2002). The use of these plants for insecticidal purpose in storage pest control is well documented (Kostyukovsky and Trostanetsky 2006; Meena et al. 2006). The mode of action varies greatly from product to product. Whereas, sex pheromones directly interrupt the reproductive cycle of insects (Krupke et al. 2002) however, plant extracts and oils often act less directly and specifically. Some botanical extracts such as floral essence attracts insects to traps, caynene can be used as deterrents, and lemongrass oil, removes the waxy coating of leaves of weeds to cause dehydration. Extracts of *Hoodia gardoni* deterred the feeding and oviposition of *Trichoplusia ni* (Akhtar et al. 2009). Similarly, extract of *Humulus lupulus* has potential to manage the larval stage of Colorado potato beetle (Gokce et al. 2006). Plants synthesize two types of metabolites, primary and secondary. The primary plant metabolites are the starting materials for the biosynthesis of specific and genetically controlled, and enzymatically catalyzed complex compounds known as secondary metabolites. Secondary plant metabolites play an important role in the defense against insect herbivores as they act as insect repellents, feeding inhibitors, and toxins (Theis and Manuel 2003). Vacuoles and chloroplasts are the two important intracytoplasmic sites for storage of a number of hydrophilic secondary metabolites. Chloroplasts contain simple phenylpropanes and flavonoids. Lipophilic secondary metabolites (lignin) may accumulate in the cell membranes of plants (Vasconsuelo and Boland 2007) Plants also produce specialized containment structures for these compounds e.g. terpenes are generally restricted to resin ducts and glandular hairs, and accumulation of furanocoumarins is restricted to oil tubes (Combrinck et al. 2007). The largest and most structurally diversified class of naturally occurring organic products of secondary plant metabolites are the terpenes, the allied terpenoid compounds and the steroids. Recent research in plant resistance to insect pests of *Nicotiana* spp. and *Solanum* spp. have demonstrated that leaf exudates produced by them play a significant role in determining resistance to insects on such plants. Products made from plant extracts and oils can be regulated as biopesticides, e.g. pyrethrum is an extract from a species of *Chrysanthemum* and is commonly used in organic agriculture, yet it can be highly toxic to cause the paralysis and kills the insects (Casida and Quistad 1995) by altering the electrical impulses on to axons (Klaassen et al. 1996; Costa 1997). However, pyrethrum and a closely related class of synthetic insecticides called pyrethroids are regulated as conventional pesticides. Plant products that are not considered biopesticides include nicotine extracted from tobacco (highly toxic to bees), and rotenone extracted from roots of the legume family (Ware and Whitacre 2004). High concentrations of rutin and chlorogenic acids have been found in trichomes of tomato leaves. Which are toxic to corn earworm (Isman and Duffy 1982; He et al. 2008). Lignins and tanins are phenolic polymers in plants. Lignification provides mechanical strength for arial shoots and also helps to resist the attack of microorganisms and insect (Raven 1977). Tanins affect the insect digestion because of the cogulation of mucoproteins in their oral cavity (Harborne 1988). Coumarins are generally localized in roots and seed coats (Murray et al. 1982) which are ovicidal to Colorado potato beetles. Furanocoumarins are capable of photosensitizing insects. Xanthotoxin, a furanocoumarin when incorporated in an artificial diet (0.1%) with UV irradiated could cause 100% mortality to Southern army worm. Thymol is a chemical constituent of thyme essential oil obtained from *Thymus vulgaris* which is used to control the Varroa mite (*Varroa destructor*), a species that is parasitic to bees (Marco and Cecilia 2008). Thymol is volatile in nature and permeates into hives, coming in direct contact with the parasitic mites. The volatilized thymol irritates the mites, causing them to withdraw from the bees and die from starvation.

17 Potential of Biopesticides in Sustainable Agriculture

17.2.1.2 Insect Growth Regulators

Insect growth regulators are chemical compounds that alter the growth and development of insects (Siddall 1976; Wardhaugh 2005). They are three types, each with a distinct mode of action. Juvenile hormones (JH) disrupt immature development and prevent the emergence of an adult. Bowers et al. (1976) isolated and identified anti-JH substances from the plants which provide protection against insects. Later on Slama and Williams (1965) discovered the "Paper Factor" responsible for the formation of supernumerary larva in *Pyrrhocoris apterus*. Precocene I and II have been isolated from the composite plant, *Ageratum houstoniatum*. These two compounds and their analogs interfere with the JH activity that causes precocious metamorphosis and sterilization in Heteroptera, Homoptera and Orthoptera groups of insects (Bowers et al. 1976; Bowers 1991; Grafton-Cardewell et al. 2006). Third one chitin synthesis inhibitors limit the ability of the insect to produce a new exoskeleton after molting and also affecting the elasticity and firmness of endocuticle (Ishaya and Horowitz 1999; Schneider et al. 2003). Thus, chitin synthesis inhibitors leave the insect unprotected from the cuticle and from the prey, drastically reducing its chances of survival. Various chitin synthesis inhibitors including benzoylureas, buprofezin and cyromazine are regulated as conventional pesticides (Wardhaugh 2005). Azadirachtin from the neem tree, *Azadirachta indica* (Meliaceae) is possibly the most promising botanical pesticide and is effective at doses as low as 0.1 ppm and acts as a feeding deterrent to more than 100 species of herbivorous insects (Saxena 1989). Saponins have been shown to act as toxins and feeding deterrents to a number of species of mites, lepidopterans, beetles and many other insects (Ishaaya et al. 1969). Saponin toxicity is due to the binding of saponins to free sterols in the gut of insect, thereby reducing the rate of sterol uptake into the haemolymph. When sterol supply is reduced, saponin may interfere with the insect molting process (Chaieb 2010). Generally, most of the registered insect growth regulators are structurally similar to juvenile hormone, commonly known as juvenoids. Juvenoids can be naturally occurring or synthetic. They have been used predominantly indoors for both household applications such as cockroaches and mites, and greenhouse. Naturally occurring Juvenoids are known as phytojuvenoids (i.e., sesamin, sesamolin, juvacimene, juvadecene, and others), showed modest to very high JH activity have been isolated from several plant families (Bowers 1991). The discovery of juvocimenes obtained from sweet Basil, *Ocimum basilicum* (Bowers and Nishida 1980) that disrupted the insect hormonal process has led to the development of second generation of JH-active commercial products such as fenoxycarb. The effectiveness of these hormone-based insect growth regulators seems to be stable against several field crop pests (Lopez et al. 2005). There are several known insect Juvenile hormones (i.e. JH I-III, JH-0, and iso-JH-0) synthesized and secreted from the corpora allata (Miyamoto et al. 1993; Jagajathi and Martin 2010). Any disturbance in the normal hormone balance may cause a crucial disorder in the growth and development of insects. JHs control a number of processes such as embryogenesis, molting and metamorphosis, reproduction, diapause, communication, migration/dispersal, caste differentiation, pigmentation, silk production, and phase transformation. Although JHs showed insect-specific control potential, their instability and synthetic difficulties did not allow the use of JH itself for pest control. Instead, many JH analogs (or mimics) (JHAs) became attractive candidates for pest control because of the ease of synthesizing these analogs and their actions are more selective than those of other peptide and steroid hormones (Eto 1990). The first compound introduced into the market was methoprene. This is a terpenoid compound used primarily against household pests because of its low activity against agricultural pests and low residual on plants under field conditions (Smith 1995).

The 'ecdysteroides' or 'phytoecdysones' are a group of terpenoids discovered in plants (Eswaran 2009). Most steroids are nonpolar but the ecdysteroides are polar (Heftmann 1973). The term ecdysone or molting hormone is used for this group of steroids because α- and β-ecdysone were isolated from insects by Butenandt and Karlson (1954) that affect ecdysis or molting in insects. The insect molting hormones ecdysones (ecdysone) clearly resemble plant cholesterol in structure. The high concentrations of ecdysteroides in plants at strategic locations suggest a defensive function against non adapted insect herbivores. It has been shown that phytoecdysones and several synthetic ecdysone analogs severely inhibit growth, development, and reproduction when fed to several species of insects (Singh et al. 1982). As in plant growth regulators, an advantage of insect growth regulators is that they are effective when applied at a very small quantity. However, they are not species specific and impact arthropods generally including insects, and spiders.

Iridoid glucosides are present in *Plantago lanceolata* known to deter the feeding or decrease growth rate of *Melitaea cinxia* (Reudler talsma et al. 2008). Iridoid glucosides, potentially act as deterrents or toxic agent to a variety of generalist insect herbivores and have been shown to act as antifeedants to grasshoppers and lepidopteran larvae (Bowers and Puttick 1988). The feeding inhibition activity of sesqueterpenes against certain lepidopteran larvae seems to be due to the blocking of the stimulatory effects of glucose, sucrose and inositol on chemosensory receptor cells located in the insect mouth parts (Frazier 1986). Sesquiterpene lactones are poisonous to several lepidopterans, flour beetles and grasshoppers (Isman and Rodriguez 1983; Eswaran 2009). Gossypol is toxic to a number of herbivorous insects, causing an antibiosis effect on many insect pests of cotton including the tobacco bud worm, *Heliothes virescens* and cotton leaf worm, *Spodoptera littoralis* (Gang et al. 2009). The heliocides are toxic to the tobacco bud worm and to other insects (Stipanovic et al. 1986). Diterpenes are non volatile and are found in the resins of higher plants. Higher level of diterpenes resin acids such as abietic and levopimaric acids were found to increase mortality, reduce growth, and extend development time of several species of sawfly larvae (Wagner et al. 1983). Cucurbetacins, triterpenoides are feeding deterrents for a number of arthropods including cucumber leaf beetles: *Phyllotreta spp., Phaedon spp., and Ceratoma trifurcate;* stem borer, *Margonia hyalinata;* and red spider mite (Da Costa and Jones 1971). Other IGRs available for use against household and agricultural pests are fenoxycarb and pyriproxyfen (El-Shazli and Refaie 2002).

17.2.1.3 Attractants

Attractants are the important tools of insect pest management and have been described as "new, imaginative, and creative approaches to the problems of sharing our earth with other creatures" (Carson 1962). Their use in insect pest management is precise, specific, and ecologically sounds (Krupke et al. 2002).

Semiochemicals that act intraspecifically in chemical communication between individuals of the same species are called Pheromones (Karlson and Butenandt 1959). Semiochemicals that deliver behavioral message between individuals of different species are called allomones if they favour the producer and kairomones if they favor the receiver (Brown et al. 1970). The consequences of this complex biochemical and biophysical process are to produce a precise chemical communication system whereby chemically specific pheromones or kairomones produce directed insect behavior. Such responses to naturally occurring pheromones or kairomones can often be initiated with structurally optimized synthetic molecules having the requisite steriochemical structures. These are called parapheromones or parakairomones and may be cheaper and have more useful properties for insect control, such as persistence or they are less volatile, than the natural semiochemicals.

Insect Pheromones

The term 'pheromones' was first introduced by Peter Karlson and Martin Luscher in 1959. Pheromones are the secretions of exocrine glands of insects that cause a specific reaction in the perceiving individuals of the same species, for example, sexual attraction and mating behavior, alarm, aggregation, trail marking, or specific changes in physiological development, such as maturation or sexual determination (Karlson and Butenandt 1959; Krupke et al. 2002). Knowledge of specific pheromone chemicals of insects has developed very rapidly since the chemical characterization of the sex pheromones of the female silkworm moth, *Bombyx mori* as (E,Z)-10,12-hexadecadien-1-ol (Bombykol) (Butenandt et al. 1959). They obtained 5.3 mg of active ingredient by the processing of more than 300,000 adults of *B. mori.* The pheromones for more than 3000 insect species have been identified, with more than 1700 from lepidopterans alone (Klowden 2007).

Sex Pheromones

Sex pheromones are very widely, if not universally, distributed in the class insecta and have been demonstrated in at least 10 orders (Jurenka 2004; Reddy and Guerrero 2004; El-Sayed 2005; Gould et al. 2009). The sex pheromones of several hundred species of insects have been identified as specific chemical structures (Tamaki 1985; Mayer and McLaughlin 1990) (Table 17.2). Sex pheromones appear to have reached their maximum evolutionary development in the Lepidoptera where they are characteristically volatile esters, alcohols, aldehydes, and ketones produced by

538   M. S. Ansari et al.

Table 17.2 Insect sex pheromones. (Source: Mayer and McLaughlin 1990)

Insect	Compound	Structure
Sugar beet wireworm (*Limonius californicus*)	Valeric acid	
Honeybee (*Apis mellifera*)	9-Kcto-2-decenoic acid	
Cabbage looper (*Trichoplusia ni*)	*cis*-7-Dodecen-1-ol	
Fall army worm (*Laphygma frugiperda*)	*cis*-9-Tetradecen-1-ol	
Bkack carpet beetle (*Attagenus megatoma*)	*trans*-3-*cis*-5-Tetra-decadienoic acid	
Silk worm moth (*Bombyx mori*)	*trans*-10-*cis*-12-hexadecen-1-ol	
Gypsy moth (*Porthetria dispar*)	*d*-10-Acetoxy-*cis*-7-hexadecen-1-ol	
Male butterfly (*Lycorea ceres ceres*)	Cetyl acetate	
Male butterfly (*Lycorea ceres ceres*)	*cis*- Vaccenyl acetate	
Pink bollworm (*pectinophora gossypiella*)	10-Propyl- *trans*-5,9-Tridecadienyl acetate (Propylure)	
Male butterfly (*Lycorea ceres ceres*)	2,3-Dihydro-7-methyl-1 H-pyrrolizidin-1-one	

eversible glands in the last abdominal segments of the females. Sex pheromones
are blends of closely related substances that are common, as in the bollworms,
Heliothis zea and *H. virescens*. Closely related but nonsympatric species typically
have closely related sex pheromones, as demonstrated in the Sesiidae, where, of 23
species studied, 20 produce (*Z*)-3, (*Z*)-13-octadecadienol or its acetate as the major component. Sex pheromones of Coleoptera are more varied in chemical structure. For example, that produced by female Japanese beetle, *Popillia japonica*,
is (*Z*)-5-(1-decenyl-dihydro-2-(3h)-furanone. The sex pheromones of the female
Northern corn rootworm, *Diabrotica barberi* and Western corn rootworm, *D. virgifera virgifera* is (2R, 8R)-8-methyl-2-decyl propanoate and that of Southern corn
rootworm, *D. undecimpunctata Howardi* is (*R*)-10-methyl-2-tridecanone. In the
past few years a series of sex pheromones have been isolated and characterized
(Krupke et al. 2002; El-Sayed 2005; Gould et al. 2009). Several of the pheromones
that have been identified are simple straight chain aliphatic compounds. It is prob-
able that these are formed biosynthetically in the same fashion as fatty acids. For
example, it has been shown in Butenandt's laboratory that the fatty acids extracted
from *B. mori* are inactive as sex attractants, but that reduction with LiAlH, yields
a preparation with pheromone activity. These results imply that an acid with bombykol chain is present in the insect and that this acid may be converted to bom-

bykol by reduction of the carboxyl group. There have been numerous reports of males producing aphrodisiacs (Andersson 2000), but these reports have not been confirmed by the use of synthetic chemicals in biological assays. Male butterflies of the subfamily Lycorea possessed a pair of extrusible, odoriferous, brush like structure on the posterior of its abdomens. The odor-producing structure has been named "hair pencils." During mating the males extrude the "hair pencils" while in aerial pursuit of the female. The male induces the female to alight on available herbage by brushing the "hair pencils" against her antennae. "Hair penciling" the female's antennae continues until she is acquiescent, at which time copulation occurs (Brower et al. 1965). Meinwald et al. (1966) have found three compounds of the "hair pencils," a pyrrolizidine and two aliphatic esters. Chemoreceptors are present on antennae, palpi, tarsi, and ovipositor; and are incredibly sensitive to specific semiochemicals which are commonly perceived in nanogram (10^{-9} g) to pico gram (10^{-12} g) quantities, well below the level of human olfactory perception. However, the manner in which the "hair pencils" are used suggests that these compounds are used in mating. Chemoreception of sexual odorants by male insects has been demonstrated to be highly efficient and selective (Touhara 2009). Synthetic bombykol gives a positive response in 50% of the challenges at a concentration of only pg/ml of pentane. Boeckh et al. (1965) have shown that $B.$ $mori$ males will respond to air streams containing as little as 200 molecules of bombykol per cm^3. Bombykol is thus one of the most biologically active substances known to man. This remarkable level of activity may be typical of the long-range attractants, which must be highly effective because of the large volume of air through which they diffuse before reaching the olfactory receptors of the male. Other sex pheromones that have high biological activity include gyptol, and the sex substance of the fall army worm. Pheromones are used in pest management, in order to attract an insect to a trap containing a lethal pesticide or to disrupt mating. Pheromones can also be used to monitor pest populations as part of larger Integrated Pest Management (IPM) systems, particularly to determine appropriate timing and application of pesticides. Insect pheromones account for a large percentage of the biochemical pesticides on the market. In mid 2002, EPA had registered 36 pheromones, which comprised over 200 individual products (Ware and Whitacre 2004). Insect sex pheromones can be used alone to manage pest populations when pest pressure is moderate to low, such as after several years of consecutive use. Other practical uses include, in survey traps to provide information about population levels, to delineate infestations, to monitor control or eradication programs, and to warn of new pest introductions (Ware and Whitacre 2004). Advantages to the use of insect pheromones include their high species specificity and relatively low toxicity. Sex pheromones tend to be specific to a particular species or even strain of insect, making them one of the most targeted pest management strategies (Goldansaz et al. 2004). This specificity thus maintains an ecological balance by leaving undisturbed populations of other insect species and non-target organisms (Ware and Whitacre 2004). A disadvantage of insect pheromones is that they often must be used in combination with other pest management strategies to achieve the efficacy desired. This is particularly true when pest pressure is high. With high pest pressure, the male is more likely to locate a mate by simply bumping into her rather than by using pheromones to communicate over long distances (Karlson and Butenandt 1959). However, the combination of pest management strategies typically lowers pest pressure in subsequent years, creating the opportunity for the insect pheromones to be used alone. Pheromones of many species have been identified and are synthetically produced for use in insect pest management programs (Reddy and Guerrero 2004). Some pheromones attract only one type of insect, while others (such as the clearwing borer *Podosesia syringae* lure) attract several related species. A few lures (such as for Japanese beetle *P. japonica*) may include floral or food analogues that attract females as well as males. Klein et al. (1981) have shown that incorporation of the female sex pheromone of the Japanese beetle, (R,S)-5-(1-decenyl)-dihydr0-2-($3H$)-furanone together with the kairomone lure significantly improves the catch of male Japanese beetles.

Alarm Pheromones

Alarm pheromones are produced mostly by social insects to warn other colony members of danger and to recruit for colony defense (Klowden 2007). It is certainly more adaptive and effective for a species to mount a collective response to some traumatic stimulus rather than to mount an individual response (Fujiwara-Tsujii et al. 2006). In the honeybees, the release of isopentyl acetate and more than 20 other substances by alarmed workers in the act of stinging release a frenzied attack by other workers. Unlike other classes of pheromones, the alarm pheromones are produced in fairly large abundance and appear to be the least specific of all pheromones (Witte et al. 2007). The typical alarm pheromones of the formicinae ants is 4-methyl-3-heptanone, which is responsible for the fruity odors of crushed workers, and produces immediate confused and arractic behavior of all the workers in the immediate vicinity (Fig. 17.3). Of the 11,000 known ant species, all are eusocial and therefore rely heavily on pheromone communication within the colony. The glandular source of ant alarm pheromones is usually the mandibular gland but has also been found in the Dufour's gland and anal glands (Witte et al. 2007). When discharged through the mandibles onto as invader ant, the latter becomes marked as an aggressor. The green peach aphid, *Myzus persicae*, secretes the alarm pheromones (E)-β-farnesene from its cornicals when attacked by predators (Haubruge 2008; de-Vos 2010). Alarm pheromones generally consists of low-molecular-weight, highly volatile compounds that easily spread throughout a colony yet evaporate quickly to terminate the aggression when the danger no longer exists (de-Vos 2010).

Aggregation Pheromones

Unlike sex pheromones that act on only one sex, aggregation pheromones induce group formation by bringing many individuals of both sexes together (Klowden

17 Potential of Biopesticides in Sustainable Agriculture

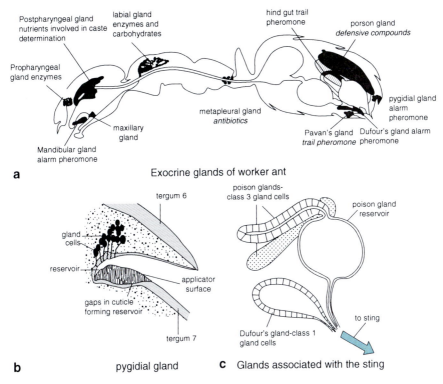

Fig. 17.3 Exocrine glands of worker ants. **a** Longitudinal section of a worker ant showing the principal glands with dark shading. All the glands shown may not be found in one species. Examples of the functions of secretions from the glands are shown in italics, but note that the secretion from a particular gland may serve different functions in different species (based on Hölldobler and Wilson 1990). **b** Pygidial gland of a worker ant, *Pachycondyla* (based on Holldobler and Traniello 1980). **c** Poison gland and Dufour's gland of a worker ant (*Myrmica*). The organization of the gland is simpler than that depicted in (**a**) (based on Billen 1986)

2007; Lacey et al. 2009). Mountain pine beetle, *Dendroctonus ponderosa*, is a major enemy of the lodgepole pine; *Pinus contorta* secretes aggregation pheromones which is a mixture of *exo*-brevicomin, trans-verbenol, and terpenoid myrcene (Lacey et al. 2009). Southern pine beetle, *D. frontalis* severely damages loblolly pine, *Pinus taeda*. The female beetle releases frontalin after the initial attack together with *trans*-verbenol, which is synergistic together with α-pinene from the tree. The male beetle finding female entrance holes releases verbenone, which inhibits male response, and subsequently *endo*-brevicomin, which acts to inhibit the response of both male and female beetle, thus shifting the attack to other trees. Norway spruce beetle, *Ips typographus* attacks the Norway spruce, *Picea abies*. Male beetles initiate the attack and release 2-methyl-3-buten-2-ol and *cis*-verbenol, which attracts both sexes (Wertheim et al. 2005). After mating

542   M. S. Ansari et al.

males release small quantities of ipsenol and ipsdienol, which are short-range inhibitors of the aggregation responses of both sexes, thus terminating the attack (Seybold 2003). Coccinelid beetles produce an aggregation pheromone that attracts large numbers of males and females to overwintering sites, and their aposematic coloration is enhanced and predators discouraged when the brightly colored insects are aggregated. As a pheromone attractant, grandlure is a mixture of cyclobutane alcohol (3 parts), a cyclohexane alcohol (4 parts), ans Z-and E-pair of cyclohexane and acetaldehydes (1.5 parts each): Grandlure is used extensively in monitoring and in removal trapping of boll weevil infestation at 1–10 traps per acre. About 7.5 million fluorescent yellow traps, each containing approximately 10 mg of grandlure, were used in the expanded boll weevil eradication program in 1988.

Trail-Marking Pheromones

They are secreted by social insects primarily by termites and ants that forage on the ground but a few non social insects also produced them. When a worker locates a resource, she lays down a trail when returning to the colony that other worker can use to find the resource (Klowden 2007). Honeybees mark the nest enterance with products from the Nasonov gland that induce workers to enter. The ant *Formica rufa* appears to use formic acid as a trail marker. Trail pheromones appear to have arisen as metabolic by-products that were eventually adapted as signals and may be exceptions to the rule that pheromones exist as specific component blends. The major trail-marking pheromones of the Texas leaf-cutting ants, *Atta texana* is methyl 4-methylpyrrole-2-carboxylate (Tumlinson et al. 1972) and that of the imported fire ant, *Solenopsis invicta*, is a mixture of the sesquiterpenes (Z,E)- and (E,E)-frarnese and (Z,E)-3,4,7,11-tetramethyl-1,3,6,10-dodecatetrene.

I. Mating Disruption Mating disruption (MD) involves the use of pheromones, i.e. the chemicals produced by an insect which evoke a specific response in the other individuals of the same species. MD is based on the principle that when a specific pheromone is released in the air in an orchard in sufficiently high quantity, the males are unable to orient to natural sources of pheromone and fail to locate the calling female and the reproduction is prevented (Krupke et al. 2002; Witzgall et al. 2010).

I.a Mechanisms Olfactory receptors in moths for the detection of pheromones are located on the antennae (Gronning et al. 2000). When exposed to a constant stimulus, e.g., pheromone, the output from sensory organs declines rapidly; this condition is known as adaptation. The sensory organs recover fairly rapidly (in about 2–3 s) once the stimulus is removed. On the other hand a high and uniform concentration of the pheromone could effectively shut down the ability of sensory organs to detect the pheromone (Witzgall et al. 2010). The exposure to high concentration of pheromone may result in the decline of behavioral response lasting several minutes or few hours. This affects the central nervous system and is referred to as habituation. In this situation nerves do not recover in the normal manner. Thus habituation caused by the exposure of moths to high concentrations of pheromones could play an important role in suppressing normal male responsiveness in the mating disruption.

I.b. False Trails Male moths of *Cydia pomonella* are attracted to false sources (Angeli et al. 2007; Witzgall et al. 2008), which critically affect the time and energy in search of females when pheromones are placed at different parts of field (Pfeiffer et al. 1993a, b). Therefore, the likelihood of a male of *C. pomonella* finding a calling female would be very low. Under these situations it may be important that all pheromone sources are releasing about the same amount of pheromone as a calling female. Male moths would cease to be attracted to a pheromone source when the concentration of pheromone gets too high i.e. above a response threshold. If this mechanism is important, it would be beneficial to have as many pheromone point sources as possible.

I.c. Masking In this mechanism the background level of the pheromone is often high enough to mask the odor trail from a calling female. In this mechanism although, the sensory system of the male moths is normally functioning but in the high background level of the pheromone the relatively low concentration trails emitted by females cannot be perceived e.g. it is effective against pink bollworm, *Pectinophora gossypeilla* by permeating the air with synthetic pheromone, gossyplure (Lykouressis et al. 2004). The natural sex attractants emanating from the female is masked causing disruption in mating and ultimate result in suppression (Witzgall et al. 2008). However, Stelinski et al. (2008) suggested that disruption of *Phyllocnistis citrella* should be effective even at high population densities given the density-independent nature of disruption for this species and the remarkably low rate of pheromone per hectare required for efficacy.

17.2.1.4 Volatile Kairomones as Lure for Insects

Semiochemicals synthesized by plants are used by insects as cues for host-plant selection, as attractants, arrestants, and feeding and oviposition stimulants. They are often 'attractive' in nanogram to microgram quantities and are used for monitoring insect infestation, for removal trapping, for timing control operations, and as a lure in poison baits (Suckling 2000; Tooker et al. 2005). A number of volatile plant kairomones have practical use as lure to attract insects to traps for monitoring populations or to bait for control. This technology began with Howlett's (1915) discovery that methyl eugenol or 3,4-dimethoxy-1allylbenzene, from lemon grass oil, *Cymbopogon nardus* was highly attractive to the male fruit flies, *Dacus diversus* and *D. zonatus* and attracts at least 60 other closely related *Dacus* spp. Methyl eugenol has both olfactory as well as phagostimulatory action and is known to attract fruit flies from a distance of 800 m (Roomi et al. 1993; Wee et al. 2007). Methyl eugenol, when used together with an insecticide impregnated into a suitable substrate, forms the basis of male annihilation technique. This technique has been successfully used for the eradication and control of several *Bacterocera* species (Cunningham 1989; Wee et al. 2002). Adult fruit flies use visual and olfactory stimuli to locate hosts and the traps that combine visual and olfactory cues proved to be most efficient for capturing fruit flies (Prokopy et al. 2003; Pinero et al. 2006). The responses of the fruit flies to visual stimuli are dependent on colour, shape and size of the stimulus (Cornelius 2000). Methyl eugenol (4-allyl-1,2-dimethoxybenzenecarboxylate) and cue-lure [4-(p acetoxyphenyl)-2- butanone] are highly attractive kairomone lures to *B. dorsalis*, and *B. cucurbitae*, respectively (Khrimian et al. 2006). Hardy (1979) estimated that at least 90% of the *Dacinae* species are strongly attracted to either methyl eugenol or to cuelure- raspberry ketone. Cue-lure has not been isolated as a natural product but is rapidly hydrolyzed to form raspberry ketone (Metcalf and Metcalf 1992). Males of at least 176 species of *Dacinae* are attracted to cue-lure- raspberry ketone, and 58 species to methyl eugenol (Metcalf 1990). Of the 46 *D.* species that are agricultural pests, 24 respond to cuelure-respberry ketone and 8 to methyl eugenol (Metcalf and Metcalf 1992). Use of methyl eugenol and cue-lure mixtures has been reported in Taiwan and Okinawa to attract *B. dorsalis* and *B. cucurbitae* (Ito et al. 1976; Liu and Lin 1993). In the event that attractiveness was not reduced, using mixtures in control programs in Hawaii may reduce the amount of pesticide placed in the environment and the cost for treatment of both *B. dorsalis* and *B. cucurbitae* (Metcalf 1990; Wee et al. 2002, 2007). Similarly, a detection trap containing mixtures of methyl eugenol and cue-lure would have many advantages.

This discovery has marked the beginning of the modern chemical ecology of insects (Metcalf 1990), and that this first chemical characterization of plant kairomone for insect control antedated the discovery of the first insect pheromone (Cornelius 2000), Bombykol ((E,Z)10,12-dodecadienol), the sex pheromone of the silkworm moth, *B. mori* (Butenandt et al. 1959) used against the trapping of *P. gossypiella* by more than 50 years. Geraniol, a volatile kairomone for the Japanese beetle, *P. japonica*, and mixtures of plant kairomones geraniol, eugenol, and phenethyl alcohol esters are widely used today to control this pest.

The use of kairomones lures for trapping Japanese beetle has had belated recognition as an important method of suppression of this pest, and was not recommended by the U.S.D.A. until Schwartz (1975) endorsed trapping as a non-insecticidal method for destroying beetles and reducing damage. The state of Maryland recommended trapping and reported that more than 369 tones of adult beetles were destroyed during a single summer. Hamilton et al. (1971) reported that mass trapping in Nantucket, Massachusetts reduced Japanese beetle populations by at least 50% over a period of 3 years. The employment of 1,400 traps at Dulles International Airport was reported to have eliminated the need to treat departing aircrafts to eliminate "hitchhiking" Japanese beetles (Klein 1981).

Pheromone traps (traps baited with these lures) are not intended for controlling pests alone, but aid in determining if a pest is present and whether a population is increasing, peaking, or decreasing. This information is essential in determining when and how often to time control actions. Different lures should never be handled at the same time since cross-contamination will affect their performance (Knight et al.

17 Potential of Biopesticides in Sustainable Agriculture

2006). Traps come in several designs, capitalizing on certain behaviors of some insects, such as a tendency to fly upward or search for protected sites. Color may also influence attractiveness (http://ccesuffolk.org).

17.2.1.5 Successful Application in Sustainable Agriculture

Application of semiochemical attractants has been most extensively practiced and has become an integral part of many IPM programs (Cornelius 2000; Suckling 2000). As of the 1990s, more than 1,000 pheromones and kairomones and their synthetic parapheromone and parakairomone analogs have been identified and are commercially available for more than 150 species of insects (Ridgway et al. 1990). In 1987, according to Cunningham et al. (Ridgway et al. 1990), 15,000 traps baited with methyl eugenol, 33,000 baited with trimedlure, and 7,000 baited with cue-lure were deployed in the Los Angeles basin. From 1980 to 1987, these traps intercepted 393 male *Ceratitis capitata,* 346 male *D. dorsalis,* 11 male *D. cucurbitae,* 8 male *D. zonatus,* 4 male *D. correctus,* and one of each of *D. scutellatus, D. bivittatus,* an *D. tryoni.* Monitoring for the appearance of male codling moth, *Cydia pomonella,* using winged traps with the female sex pheromone, provides a guide for the time of applications of cover sprays of apple orchards. Flat yellow sticky traps incorporating the female sex pheromone of the California red scale, *Aonidiella aurantii* (Maskell), are used to monitor the appearance of the male scale insects as a guide for the initiation of spray programs in orchards (Phillips 1981).

Attractants also play an important role in the removal trapping of insect pests. Classical example of using attractants for removal trapping is the use of methyl eugenol to lure male oriental fruit flies, *D. dorsalis* (Hendel), to toxic baits. This insect was eradicated from the island of Rota by dropping 2.5 in fibreboard blocks impregnated with 24 g of 97% methyl eugenol and 3% naled insecticide (3.5 g of naled acre^{-1}, 8.6 g ha^{-1}) (Steiner et al. 1965). Similarly, more than 99% reduction of populations of melon fly, *D. cucurbitae* Coquillet, was obtained over a large area of island of Hawaii, by suspending 2.5 in^2 fibreboard blocks acre^{-1}, 2–5 feet above the ground each treated with 25 g of 95% cue-lure and 5% naled. The application rate was 1.25 g of naled and 22.8 g of cue-lure acre^{-1} (Cunningham and Steiner 1972).

Pheromones play an important role in mating disruption. The first successful demonstration of mating disruption was shown for pink bollworm, *P. gossypiella,* using the parapheromone *hexalure,* or (Z)-7- hexadecenol acetate, which is about 15 times as active as the natural pheromone gossyplure (Shorey et al. 1974). The release of about 330 g of hexalure per hectare during the 16-week growing season of cotton in the Imperial Valley of California caused most female moths to remain unmated and produced a 75.5% reduction of larval population over the next generation. Commercial applications of gossyplure in the Imperial Valley in 1982, when 15,000 were treated, resulted in good commercial control with only 5% cotton crop damage, as compared to more than 30% in nearby fields treated with commercial insecticides (Justum and Gordon 1989). Wang et al. (2004) successfully captured *Plutella xylostella* from cabbage by application of sex pheromone lures. However,

Sulifoa and Ebenebe (2007) used pheromones to evaluate the larval infestation of *P. xylostella* on cabbage.

17.2.1.6 Neem in Sustainable Agriculture

Azadirachtin (Tetranortriterpenoids), a predominant active insecticidal component found in neem seeds and leaves (Butterworth and Morgan 1968; Al-Fifi 2009) and is the best known derivative (Broughton et al. 1986). It is a strong antifeedent and growth disruptor to several insect species and thus a potential candidate for use in plant protection (Ruscoe 1972; Streets and Schmutterer 1975; Ladd et al. 1978; Warthen 1979; Schmutterer and Rembold 1980) and neem formulations are environmentally safe and low toxic to non target natural enemies (Saxena 1989). Insect's growth regulations are one of the multiple functions provided by azadirachtin (Rembold et al. 1980; Mordue and Blackwell 1993). Azadirachtin does not directly kill the pest but alters the life processing behavior in such a manner that the insect can no longer feed, breed or undergo metamorphosis (Elahi 2008). More specifically, Azadirachtin disrupts molting by inhibiting biosynthesis or metabolism of ecdysone and juvenile hormone (Ware and Whitacre 2004)

Mode of Action

a. Insect Growth Regulator The insect growth regulatory effects of neem derivatives as methanolic neem leaf extracts and azadirachtin in nymphs of heteroptera (Leuschner 1972) and larvae of Lepidoptera (Ruscoe 1972) was first reported in 1972. It is a very interesting property of neem products and unique in nature, since it works on juvenile hormone, growth of larval form of an insect depends on an enzyme, ecdysone. When the neem components, especially azadirachtin enters into the body of larva, the activity of ecdysone is suppressed and the larva fails to molt, remains in the larval stage and ultimately dies (Rembold et al. 1984; Akhtar and Ismam 2004; Al-Fifi 2009). RD-9-Replin, a neem product showed at higher concentrations prolonged nymphal period of *Dysdercus koenigii* as such the number of generations were reduced to almost half of the normal population. It also affected the growth and development in the subsequent generation. Consequently, the hatching was nil in the third generation in the insects treated with RD-9Replin. If the concentration of azadirachtin was not sufficient, the larva manages to enter the pupal stage but dies at this stage and if the concentration is still less the adults emerging from the pupae are 100% malformed, absolutely sterile without any capacity for reproduction (Schmutterer 1987; Lowry and Isman 1996; Dilio et al. 1999). Extracts of *Melia azadirach* and *A. indica* caused adverse effect on survival, fecundity, development and oviposition of *P. xylostella* (Charleston et al. 2006). Azadirachtin also affected metamorphosis, longevity and reproduction of three tephritid fruit flies, where adult emergence was completely inhibited at concentrations of 14 ppm for *Ceratitis capitata* and *Bactrocera dorsa-*

17 Potential of Biopesticides in Sustainable Agriculture

lis (Stark et. al. 1990). Neem extract @ 0.25 mL/m^2 to the fifth nymphal instar of *Locusta migratoria migrotroides* led to increase mortality during molts, prolonged development and reduced fitness (Freisewinkel and Schumutterer 1991). Survivorship of nymphs and adults of *Acyrthosiphon pisum* treated with Margosan-O was reduced in concentration dependent manner (Stark and Wennergren 1995). Larval growth and development of maize stalk borer, *Chilo partellus* was totally inhibited by the application of neem bitter concentrate (NBC) @ $\geq$500 parts per billion (Saxena and Francis 1996). Sublethal dose of azadirachtin (1 µg/insect) led to inhibition of laval growth of *Helicoverpa armigera* (Gujar 1997). Six neem formulations viz. bioneem, neemazol, econeem, achook, nimata and nimbicidine @0.5 and 1% for each manifest 3 types of morphological deformities viz. larval-pupal intermediate, larva from pupa and deformed adults (Srivastava and Mathur 1998).

b. Feeding Deterrent The most important property of neem is feeding deterrence (Al-Fifi 2009). When a hungry insect larva wants to feed on the leaves, feeding is stimulated through the maxillary gland that gives a trigger, peristalysis in the alimentary canal is speeded up, the larva feels hungry and it starts feeding on the surface of the leaf (Schmutterer 1990; Sahayaraj and Paulraj 1999;). When the leaf is treated with the neem product, because of the presence of azadirachtin, salanin and melandriol thereby causing anti-peristalitic wave in the alimentary canal and this produces something similar to the vomiting sensation in the insects. Because of this sensation the insect does not feed on the neem treated surface, thus ability to swallow is also blocked (Ayyangar and Rao 1989, 1990). Diemetry et al. (1995) showed deterrent and antifeeding effect of neem azal F of neem seed kernal extract with 5% azadirachtin when offered to adults and Ist instar nymphs of cowpea aphid, *Aphis craccivora*. Neem oil is a potential antifeedant against adults of chrysomelid, *Dicladispa armigera*. Daily consumption of fresh rice leaves was 1.05 ± 0.08 mg which was reduced to 50% when leaves were treated with 6.46% NSO (Deka and Hazarika 1997). Neem azal-S and margosan –O @ 2% deter the egg laying and oviposition deterring index (ODI) was reduced by 80.7 and 52.6%, respectively (Dimetry et al. 1995). Daily spray of neem leaf extract @ 10–15% up to 10 days repelled the insect from the feeding on lemon plant (Ashok Kumar et al. 1998). Neem based insecticides i.e. ecozine, agroneem, and nemix deterred feeding by bect armyworm larvae (Greenburg et al. 2005)

c. Oviposition Deterrent The application of neem based pesticides against insects does not normally lead to obvious mortality but may result in a substantial reduction in the fecundity of insects (Al-Fifi 2009), so that the following generations may be reduced below economic threshold level. Beetles with reduced fecundity consume much less food than untreated one (Schmutterer 1987). Neem also reduces pests by not allowing the females to deposit eggs. However, the injection of azadirachtin prevented the juvenile hormone production and therefore, also prevents vitellogenin synthesis and egg production (Rembold et al. 1984). Schmutterer (1987) revealed the sterilizing effect of azadirachtin results primarily from inhibition of oogenesis and vitellogenesis. Seeds in storage are coated with neem kernal powder and neem oil prevents egg laying (Dorn et al. 1987). NSKE @ 1 L/100 L of water resulted in an effect on oviposition and mortality of beetle, *Contotracheles nenuphar* and adult fly, *Rhagoletis pomonella* adults (Prokopy and Powers 1995). Neem seed oil @ 5% deterred the oviposition by *Daccus cucurbitae* on bitter guards (Singh and Srivastav 1985). NSKE @1.25–20% and azadirachtin @ 1.25–10 ppm were evaluated as a oviposition deterrent to *B. cucurbitae* and *B. dorsalis* on pumpkin and guava, respectively (Singh et al. 2001). Neem extracts reduced the hatchability of the eggs of *D. koenigii* (Bhathal et al. 1991). Azadirachtin @ 30–60 ppm reduces the fertility and fecundity of cabbage aphid *Brevicoryne brassicae* (Koul 1998). Neem T/S (1% Azadirachtin) and neem azal-F (5% azadirachtin) @1 mL/L showed the ovicidal effect on the eggs of *Leucinodes orbonelis* (Kumar and Babu 1998). Other pesticidal activities include : (1) The formation of chitin (exoskeleton) is also inhibited (2) Mating as well as sexual communication is disrupted (3) Larvae and adults of insects are repelled (4) Adults are sterilized (5) Larvae and adults are poisoned. Females of *P. xylostella* deposited similar number of eggs when treated or untreated neemix cabbage leaves (Liu and Liu 2006).

d. Storage Insect Control Stored-product insects can cause postharvest losses, up to 9% in developed countries to 20% or more in developing countries (Phillips and Throne 2010). Throughout the tropics much of the food harvested is lost during storage. Recently attention has been given to the possible use of plants products or plant derived compounds as promising alternatives to synthetic insecticides in controlling insect pests of stored products (Khalequzzaman and Islam 1992; Liu and Ho 1999; Umoetok and Gerard 2003). The effectiveness of many plant derivatives for use against stored grain pests has been reviewed by Jacobson (1983, 1989). Neem offers the impoverished farmers and even affluent farmers wanting to replace pesticides with a natural and inexpensive alternative (Umoetok and Gerard 2003). A light coating of neem oil or kernel powder protects stored food crops for up to 20 months from all type of infestation with no deterioration or loss of palatability. Even today in rural India, people stored their grains mixed with dried neem. Neem not only protects but also prevents the further proliferation of storage pests if it has already infested the grains. The insect stops feeding on them due to anti feeding property of neem (Saxena et al. 1989). On the other hand, the oviposition deterrent of neem prevents the female insect from laying the egg during its egg laying period of its life cycle (Makanjuola 1989; Phillips and Throne 2010).

17.2.2 Microbial Pesticides

Microbial pesticides come from naturally occurring or genetically altered bacteria, fungi, algae, viruses or protozoans (Chandler et al. 2008). They suppress pests either by producing a toxin specific to the pest, causing disease, preventing establishment of other microorganisms through competition, or various other modes of action (Clemson 2007). The first commercial preparation of an entomopathogen was

17 Potential of Biopesticides in Sustainable Agriculture

sporiene developed during the mid 1940's in France (Deakon 1983). For all crop types, bacterial biopesticides claim about 74% of the market; fungal biopesticides, about 10%; viral biopesticides, 5%; predator biopesticides, 8%; and "other" biopesticides, 3% (Thakore 2006a). At present there are approximately 73 microbial active ingredients that have been registered by the US EPA. The registered microbial biopesticides include 35 bacterial products, 15 fungi, 6 non-viable (genetically engineered) microbial pesticides, 8 plant incorporated protectants, 1 protozoa, 1 yeast, and 6 viruses (Steinwand 2008). Microbial biopesticides may be delivered to crops in many forms including live organisms, dead organisms and spores. Microbial pesticides have a wide and specific range of properties that make them desirable for integrated crop management (Hajek 2004). To achieve a better system for sustainable microbial control, attention is required for the better understanding of phylogeny of microbial natural enemies (Rehner and Buckley 2005), their biogeography and factor determining bio-diversification of gene flow (Bidochka and Small 2005), and assessment of their background levels in agricultural and natural ecosystems (Mensink and Sheepmaker 2007).

17.2.2.1 Bacterial Biopesticides

In 1949, Steinhaus coined the term "microbial control" in which pathogens are employed to suppress the population of insect pests. The pathogens isolated from diseased insect were extensively studied (Steinhaus 1947, 1949, 1963; Burges and Hussay 1971). *Bacillus* was first recorded in 1901 as the cause of the damaging "Sotto" disease in silkworms in Japan (Ishikawa 1936). Berliner (1915) isolated *Bacillus thuringiensis* from Mediterranean flour moth, *Ephestia kuehniella* and it was reisolated by Mattes (1927). Both Berliner and Mattes reported a high pathogenicity of the *Bacillus* for the flour moth larvae. The bacterium occurs naturally in the soil and is distributed worldwide. Enormous biological diversity that exists within the species of *B. thuringiensis* can be illustrated by the report that 1,000 isolates of *Bt* were obtained from soil samples from five continents (Martin and Travers 1989). For the first time, *B. thuringiensis* was recognized as living insecticides by Steinhaus (1954). The biopotency has been tested against a number of insect pests belonging to order, Lepidopetra, Diptera, Hymenoptera, Orthoptera and Coleoptera (Chandler et al. 2008). It was regarded as novel insecticide with a wide host range particularly against lepidopterous pests (Estruch et al. 1996). To be effective they must come into contact with the target pest, and may require ingestion to be effective. On the basis of their pathogenecity, entomopathogenic bacteria are divided into four groups: (1) obligate pathogens (e.g. *B. popilliae*) (2) crystalliferous spore-formers (e.g. *B. thuringiensis*) (3) facultative pathogens (e.g. *B. sphaericus*) and (4) potential pathogens (e.g. *Serratia marcescens*). Of these, *B. thuringiensis* is the most widely used microbial pesticide, subspecies and strains of *B. thuringiensis* (*Bt*), accounting for approximately 90% of the biopesticide market (Chattopadhyay et al. 2004; Romeis et al. 2006). This is the most effective, with about 6,000 isolates stored in various repositories throughout the world. Each strain of this bacterium produces a different mix of proteins and specifically kills one or a few related species of insect larvae. *B. thuringiensis* produces three types of entomocidal toxins: α exotoxin, ß exotoxin, and δ endotoxin (insecticidal crystal protein, ICP) (Heimpel 1967; Chattopadhyay et al. 2004).

a. Mode of Action Insecticidal activity of *Bt* is due to the presence of crystalline protein body called parasporal body or δ -endotoxin formed during sporulation (Heimpel 1967). The molecular weight of proteins of the bipyramidal crystal (5 × 15.5 nm) is of 120,000 Dalton. The protein is degraded proteolytically into a toxin of 64–71 kDa in the gut of the host insect under high pH and also in in-vitro condition. Each bacterial cell forms a spore at one end and a crystal at the other. When the cell lyse after sporulation spore and crystal are set free into the medium and sediment together (Fast 1981). Ingested crystal toxin acts on the gut or passes through the gut into haemocoel where it exerts its effect. The toxins appeared to be confined to the gut but not in haemolymph (Fast and Videnova 1974; Fast 1975). The active toxins cause a rapid stimulation of glucose uptake by gut epithelial cells (Fast and Donaghue 1971) and also respiration results in rapid depletion of ATP (Faust and Geiser unpublished observation). Then there is break down of epithelial cells 45 min after treatment (Heimpel and Angus 1959) and the ions leak from the gut to haemolymph (Louloudes and Heimpel 1969; Fast and Donague 1971). These events within the gut cause larvae to stop feeding due to an extreme digestive discomfort (Heimpel and Angus 1959; Chattopadhyay et al. 2004). The spore on the other hand gets attached to the gut wall changing its permeability. Death of larvae may be due to septicemia, paralysis, starvation and increase in concentration of K^+ ions (Louloudes and Heimpel 1969; Chattopadhyay et al. 2004; Romeis et al. 2006). The crystal proteins may differ in biochemical and toxic properties that differ in toxicity to insect and in antigenic properties (Krywienczyk et al. 1978; Sharp and Baker 1979). However, *B. thuringiensis* spores do not usually spread to other insects or cause disease outbreaks on their own as occurs with many other pathogens (Whalon and McGaughey 1998).

Formulations of *Bt* var. *tenebrionis* and *San Diego* have been registered for use against Colorado potato beetle larvae and elm leaf beetle adults and larvae, respectively (Duan et al. 2004, Zehnder and Gelernter 1989). *Bt* is primarily used to control lepidopteran pests (moths and butterflies), which are some of the most damaging crop pests. However, *Bt* can also be used to control a broad range of other pests including specific species of mosquitoes, flies, and beetles. Approximately 525 insects belonging to various orders have been reported to be infected by *Bt* toxins (Thakore 2006b). Resistance to the crystal toxin has been reported in several insects (Gould 1988; McGhaughey 1990; Tabashnik et al. 1990; Tabashnik 1994). *Bt* endotoxins are also the most common basis for genetically modified pest resistant crops. The genetics of *B. thuringiensis* have contributed to the development of increasing numbers of novel strains. This has been accomplished by using both the natural conjugation process of *B. thuringiensis* and recombinant methodology

17 Potential of Biopesticides in Sustainable Agriculture   551

(Carlton 1988; Gonzales et al. 1982). The main objectives in genetically altering *Bt* are to increase or modify host range to increase virulence and to increase persistence. The rearrangement of genes within a strain of *Bt* may not only broaden the host range, but may also increase the virulence of that strain to a specific insect (Aronson and Wu 1989).

b. Application of *Bt* on Field Crops First commercial formulation of *B. thuringiensis* was marketed by USA in 1958. These formulations have the characteristics of being ecofriendly, safe, easy to use and cheap, as compared to chemical pesticides. The discovery of isolates which increase the capability of acting against the coleopterans (*B. thuringiensis tenebrionis*), mosquitoes *(B. sphaericus)* and dipteran *(B. moritai)* (Tanweer et al. 1998). *Bt* @ (300 g ha^{-1}) has further been used in the management of several pests, notably *P. xylostella* on cruciferous vegetables (Jayaraj et al.1992; Abdel-Razek et al. 2006). Certain adjuvants, like whole egg homogenate 1% and whole milk (0.5%) and *Catherebthus roseus* with *Bt* @ 0.5–1.0 kg formulation ha^{-1} has been found quite effective against important pests like *S. litura, E. vitella, H. armigera* and *P. xylostella*. The work on *B. popillae* is limited which causes disease in white grubs (Redmond et al. 1995; Zhang et al. 1997). In addition, *Bt* is used in agriculture as a liquid applied through overhead irrigation systems or in a granular form for control of European corn borer (Cranshaw 2008). The treatments funnel down the corn whorl to where the feeding larvae occur. This also makes *Bt* useful in applications where pesticide drift onto gardens is likely to occur, such as treating trees and shrubs. *Bt* formulations are also used against larvae of mosquitoes, black flies and fungus gnats (Cranshaw 2008). *Bt.* var. *kurstaki* was found effective against the fruit borer when mixed with botanical insecticides viz. Neemax @ 1 kg ha^{-1} and Multineem @ 2.5 l ha^{-1} and, Biotox *(Bt)* @ 1 kg ha^{-1} combined with malathion @ 0.5 kg a.i. ha^{-1} in different combinations were sprayed. Lowest fruit borer incidence was observed (8.6%) when Biotox was applied to the crop two times alternated with one malathion application followed by the treatment where malathion was applied twice alternated with one application of Biotox (10.6%) (Mishra and Mishra 2002). Spraying of *Bt* formulations viz. Delfin, Halt, Biolep and Spicturin were carried out at 15 days interval and indicated that Delfin and Halt @ 1.0 kg ha^{-1} was highly effective in reducing per cent fruit damage caused by *H. armigera* on tomato and increased the fruit yield, respectively. *B. thuringiensis (Bt)* formulations were evaluated in comparison with neem and chemical insecticides against *L. orbonalis* which showed that five sprays of Dipel 8 L @ 0.2% at 10 days interval resulted in minimum shoot (9.56%) as well as fruit (11.78%) infestation and maximum yield of marketable fruits 196.96 quintel ha^{-1} and proved to be the most effective treatment (Puranik et al. 2002). Mohan et al. (2009) found that *Bt.* Kurstaki HD-1 and Cry1 class toxins were highly toxic to *P. xylostella*. The maximum yield of healthy fruits was obtained in dipel and fenvalerate combination. Solo and joint efficacy of some biopesticides viz. azadirachtin, *B. thuringiensis* and avermectin against diamondback moth, *P. xylostella* revealed that the joint action of all biopesticides were significantly supe-

552 M. S. Ansari et al.

rior to control with recommended dose of malathion (0.05%). Highest suppression of 88.85% and 90.54% larval population was achieved from combined spraying of *Bt* (0.05%) and avermetcin (0.05%) after 3 and 14 days of spraying, respectively, followed by Joint spraying of *Bt.* (0.05%) + azadirachtin – 1,500 ppm (0.15%) recorded 87.43 and 84.85% after 3 and 14 days of spraying, respectively (Chatterjee and Senapati 2000). To control mosquito larvae, *Bt* formulations containing the israelensis strain are placed into the standing water of mosquito breeding sites. For these applications, *Bt* usually is formulated as granules or solid, slow-release rings or brickettes to increase persistency (Cranshaw 2008). Applications were made shortly after insect eggs are expected to hatch, such as after flooding due to rain or irrigation. *Bt* is longer persistent in water than on sun-exposed on leaf surfaces, but reapply if favorable mosquito breeding conditions last for several weeks. *Bt* applied for control of elm leaf beetle or Colorado potato beetle (*San Diego/tenebrionis* strain) is sprayed onto leaves in a manner similar to the formulations used for caterpillars (Lindquist 1994). *Bt* based commercial formulation are shown in Tables 17.3 and 17.4.

Table 17.3 *Bacillus* based microbial biopesticides. (Source: Pawar and Singh 1993)

Sr. No.	Species/variety/serotype	Effective against	Commercial name
1	*Bacillus thuringiensis* var. *kurstaki* sero-type H3A, 3b(HD-1)	Lepidoptera	Dipel, Delfin, Thuricide, Bactospeine
2	*Bacillus thuringiensis* var. *kurstaki* serotype H2a,3b(NRD-12)	Lepidoptera	Javeline
3	*Bacillus thuringiensis* var. *isralensis* serotype H-14	Diptera	Vactobac, Bactomos, Isoctal, Bactoulide
4	*Bacillus thuringiensis* var. *Tenebrionis*	Coleoptera	Novodor
5	*Bacillus thuringiensis* var. *Galleriae*	Lepidoptera	Spicutrin
6	*Bacillus thuringiensis* var. *aizawai* serotype H-7	Lepidoptera	Certan
7	*Bacillus sphaericus* serotype H5a,5b	Diptera	Spic, Biomass
8	*Bacillus sphaericus* serotype B 101	Diptera	Spherix
9	*Bacillus popillae*	Coleoptera	Doom
10	*Bacillus moritai*	Diptera	Lavillus

17 Potential of Biopesticides in Sustainable Agriculture

Table 17.4 *Bt.* based microbial biopesticides marketed in India. (Source: Pawar and Singh 1993)

Sr. No	Biopesticide	For controlling
1	*Bacillus thuringiensis* var. *kurstaki* serotype 3a,3b,sa-11 W.G.	*Plutella* in cabbage/cauliflower, *Heliothis* in cotton, and *Spodoptera* in cotton and castor
2	*Bacillus thuringiensis* var. *kurstaki* serotype 3a,3b,strain HD-1,8L	Bollworms in cotton, fruit borer in brinjal, okra and tomato
3	*Bacillus thuringiensis* var. *Galleriae*	*Plutella* in cabbage/cauliflower, *Heliothis* in cotton, and *Spodoptera* in cotton and castor
4	*Bacillus thuringiensis* var. *kurstaki* serotype H-14, strain 164	Mosquito larvae
5	*Bacillus sphaericus* m-1539 serotype H 5a, 5b	Mosquito larvae
6	*Bacillus thuringiensis* var. *kurstaki* serotype H-14, strain 164	Mosquito larvae

17.2.2.2 Fungal Biopesticides

There are over 750 species of fungi and 100 representing genera known to be associated with insect diseases (Benjamin et al. 2002; Zimmerman 2007). Entomopathogenic fungi are widely distributed throughout the fungal kingdom, although the majority occurs in the subdivision of Deuteromycotina (imperfect fungi) and Zygomycotina. The diseases caused by fungi are termed "mycoses". Most entomopathogenic fungi are obligate or facultative pathogens and some are symbiotic (Scholte et al. 2004). Their growth and development are limited mainly by the external environmental conditions, in particular, high humidity or moisture are adequatic with ambient temperature for sporulation and spore germination. Entomopathogenic fungi have a wide host range but, Deuteromycetes fungi attack virtually all species of insects (David 1967; Ferron 1975) and arachnids. Fungi may be associated with insects as ectoparasites and endoparasites. Insects are usually infected by spores or conidia (Zygomycotina), conidia (Deuteromycotina), Zoospores (Mastigomycotina) and ascospores (Ascomycotina). The potential of fungal pathogens in the control of insect pests has been recognized since the later part of the nineteenth century, when *M. anisopliae* was tested against the wheat cockchafer, *Anisoplia austriaca* and *Bothynoderes punctiventris*.

The immature (nymphal or larval) stages are more often infected than the mature or adult stage: in others; reverse may be the case (Shah and Pell 2003). Pupal stage is infrequently attacked and the egg stage is rarely infected by fungi (Tanada and Kaya 1993). Host specificity varies considerably; some fungi infect a broad range of hosts and others are restricted to a few or a single insect species. Those with a broad host range may consist of a variety of pathotypes (McCoy et al. 1988). *B. bassiana* and *M. anisopliea* infect over 100 different insect species in several insect orders (Keller et al. 2003), but isolates of these two fungi have a high degree of specificity (Benjamin et al. 2002). Host specificity may be associated with the physiological state of the host system i.e. insect maturation and host plant (McCoy et al. 1988), the properties of insect integument and/or with the nutritional requirements of the fungus (Kerwin and Washino 1986a, c) and the cellular defence of the host (Scholte et al. 2004; Zimmerman 2007).

a. Mode of Action The mode of action of fungi varies and depends on both the types and isolates and the target insect pests (Shah and Pell 2003). Entomopathogenic fungi invade their hosts by direct penetration of the host exoskeleton or cuticle or in some cases through natural body openings. Conidia germinate on the insect cuticle and often differentiate to form an appressorium (Brey et al. 1986) (Fig. 17.4). Infected hyphae penetrate down through the host cuticle and eventually emerge into the haemocoel and proliferate in the insect's body, producing toxins and draining nutrients to cause insect death. Generally, the insect host dies shortly after the fungus begins to develop in the haemocoel (Tanada and Kaya 1993). Death is caused by tissue destruction and, occasionally, by toxins produced by the fungus. The fungus frequently emerges from the insect's body to produce spores that, when

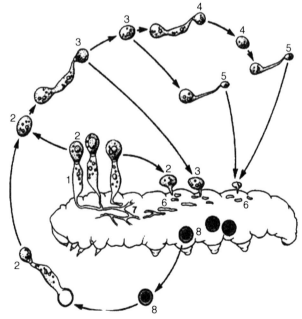

Fig. 17.4 Generalized life cycle of entomopathogenic fungus. Under favorable conditions, the endostroma in the larva produces conidiophores. (*1*) Which form conidium, (*2*) which in turn may form secondary conidium (*3*), and which in turn form tertiary conidium (*4*) in the absence of a suitable host. Capilliconidium (*5*) is formed from other conidia and may be the major infective unit. Hyphal bodies (*6*) are formed in insect's hemocoel and developed into mycelium and storma (*7*), which produces resistant or resting spore (*8*). (Adapted from Tanada and Kaya 1993)

17 Potential of Biopesticides in Sustainable Agriculture

spread by wind, rain, or contact with other insects, can spread infection. Destruxins (DTXs), a group of cyclic depsipeptides (peptides containing ester linkages) produced by *M. anisopliae*, have received most attention, and are responsible for insect mortality by the fungus (Suzuki et al. 1970; Zimmermann 2007). DTXs are insecticidal by injection (Fargues et al. 1986) and, in some cases, when ingested by mouth; toxicity is most acute to Lepidopteran larvae and adult Diptera including mosquito larvae.

B. bassiana is a fungus that grows naturally in soils throughout the world and acts as a parasite on various insect species, causing white muscardine disease (Shah and Pell 2003). The species is named after the Italian entomologist, Agostino Bassi who discovered it in 1835 as the cause of the muscardine disease of silkworms. *B. bassiana* (formerly also known as *Tritirachium shiotae*) is the anamorph (the asexually reproducing form) of the fungus species *Cordyceps bassiana*. The teleomorph (the sexually reproducing form) was discovered in 2001. *B. bassiana* is one of the most commonly occurring entomopathogenic fungal species, and shows strong pathogenicity to Lepidoptera, Hymenoptera, and Coleoptera (Harris et al. 2000; Fernandez et al. 2001; Mannion et al. 2001; Phoofolo et al. 2001). It is being used as a biological insecticide to control a number of pests such as termites, whiteflies, and beetles and it is also used in the control of the malaria transmitting mosquitos (Donald and Neil 2005). When the microscopic spores of the fungus come into contact with the body of an insect host, they germinate, penetrate the cuticle, and grow inside, killing the insect within a matter of days. Afterwards a white mold emerges from the cadaver and produces new spores. A typical isolate of *B. bassiana* can attack a broad range of insects; various isolates differ in their host range (Posada et al. 2004).

b. Metabolites Species of the genus *B.* have been reported to produce the secondary metabolites bassianin, bassiacridin, beauvericin, bassianolide, beauverolides, tenellin and oosporein (Strasser et al. 2000; Vey et al. 2001; Romeis et al. 2006). It cannot be assumed that these substances will also be produced under natural conditions in the soil or in the target host. Further, it should be kept in mind that entomopathogenic fungi naturally cause epizootics similar to those resulting from artificial inoculations. There are no reports of metabolites entering the food chain or accumulating in the environment as a result of such natural or artificial epizootics or natural metabolite secretion (Vey et al. 2001; El-Katatnya 2010).

The green muscardine fungus, *M. anisopliae* was isolated by Metchnikoff (1879) from the beetle *A. austriaca* and Steinhaus (1949) suggested its use as a microbial agent against insect pests. The colony of *M. anisopliae* appears white when young, but as the conidia mature, the colour turns to dark green (Zimmerman 2007). Culture of *M. anisopliae* contains the cyclodepsipeptides destruxins A. B, C, D and E and desmethylmestruxin B (Tamura et al. 1964; Suzuki et al. 1970). Destruxins have been considered as a new generation insecticides. They cause titanic paralysis when inoculated into larvae of *Gallerria mellonella* (Roberts 1966; Romeis et al. 2006). Destruxins are also produced in fungus infected larvae and are important in development of the symptoms (Suzuki et al. 1971; Vey et al. 1986). The rapid productions of destruxins in the larvae cause death. Destruxins are toxic to insects only by ingestion and not through the integument (Farques et al. 1986). Cytopathology occurs in the midgut cells with the change in the mitochondrial and endoplasmic reticulum.The secondary metabolites e.g. destruxin-E may act as immunosuppressant, inhibiting the cellular and/or humoral-host defense response (Vey et al. 1986). The green muscradine fungus also produces proteolytic enzyme that inhibits the protease which occur in the insect haemolymph (Zimmerman 2007).

The fungus has been used extensively in Brazil since 1972 for the control of sugarcane spittle bug, *Mahanarva postica*. The use of the pathogen increased by 67% in 1980 as compared with 1979, as a result of the increased production of the fungus in powder form for easy distribution. So, chemical treatments against this pest were confined to about 20,800 ha representing only 18.2% of the total area requiring protection (Risco 1980). In 1982, there were five registered products available for commercial production viz. Biomax, Biocontrol, Combio, Metabiol and Metapol (Soper 1982). Biomax is formulated at 2.8×10^7 viable spores per gram. It is generally remommended to be used @ 1–1.3 kg ha^{-1} in 250–300 L of water. Different strains of the fungi are available (Messias et al. 1983; Alves et al. 1984) which vary in their host specificity (Table 17.5). It is important that the mean relative humidity should be 80% during this period. Several insecticides and fungicides are found to inhibit the growth and sporulation of the fungus (Samueles et al. 1989).The fungus also occurs naturally (Barbosa et al. 1979) and during the epizootics of the disease, the amount of spores produced to kill insects is estimated to be 4.9 kg ha^{-1} or 5.9×10^{13} spores ha^{-1} (Alves et al. 1982). In Australia, the fungus was found to survive for more than 30 months (Samuels et al. 1990). Rombach et al. (1986) applied *M. anisopliae* and *B. basiana* besides three other fungi against *N. lugens* on rice in the Philippines and obtained a mortality of 63–93% 3 weeks after application in the field. The mycelium preparation sporulated on the plant and was as effective as the conidia suspension in infecting *N. lugens*. In Taiwan, this fungus is effective against *Mogania hebes* and *Alissonotum impressicolle*. Sundara Babu et al. (1983) studied the mass production and use of the fungus against coconut rhinoceros beetle, *Oryctes rhinoceros*. All the life stages of the pest are susceptible, but the management of the pest in manure pits and heaps requires very high inoculum rate in the view of the subterranean habit of the immature stages. As much as 3.4×10^{11} spores/tone of farm yard manure had to be inoculated to achieve 100% mortality (Sundara Babu (1999). In sugarcane ecosystem the fungus attacks *Pyrilla* and white grubs and causes epizootics on *Pyrilla* under favourable weather conditions (Kulshrestra and Gursahnni 1961) where the fungus caused 100% mortality of nymphs and adults. In Uttar Pradesh, it was observed that overwintering population of *Pyrilla* could be readily infected by fungus application and the infected individuals were capable of spreading the infection and induced epizootic (Varma and Singh 1987). The fungus is found safe to the principle parasite of *Pyrilla* i.e. *Epiricania melanoleuca*. Number of infested fruit by *R. cerasi* of cherry was significantly redused by 65% by application of *B. bassiana* (Claudia and Eric 2009).

Table 17.5 Application of *Beauveria bassiana* in crops. (Source: Bhattacharya et al. 2003)

Insects	Host plants	Control methods	Results	Country	References
Cydia pomonella	Apple orchards	Spraying of fungal spores @6 × 10^9 conidia/tree	50% mortality of last instars	France	Ferron et al. (1978)
Cydia pyrivora	Pear orchards	Spray of bovarine formulation (MI)	The formulation along with small amount of trichlorophon exerted 80.8% mortality while 71.8% mortality occurred by using bovarine alone	Ukraine	Drozda (1978)
Castinia licus	Sugarcane	Fungal spores sprayed @10^8 conidia/ml (MI)	Caused 27.3% mortality of larval population alone and with monochrotophos mortality increases to 45.3%	Brazil	Vilas and Alves (1988)
Ostrinia nubilalis	Maize crop	Spraying of fungal spores @ 10^{13} conidia/ha (MI)	Significant reduction in borer holes/plant	Italy	Foschi and Grassi (1985)
Cosmopolites sordidus	Banana tree	Soil treatment of *B. bassiana* (2 × 10^5 spores/mg) @4–6 g/tree (MI)	Reduction of number of survivors/tree 33 days after treatment as compared to control	Cuba	Ayala and Monzon (1997)
Leptinotarsa decemlineata	Potato plant	Spraying of bovarine on the potato plants (MI)	83% control by Bovarine within 5 days	Former USSR	Lakhidov (1979)
Leptinotarsa decemlineata	Tomato	Underground treatment of the insects with *B. bassiana* and 7.5 and 75 g/m^2	Significant reduction in adult emergence	USA	Cantwell et al. (1986)
Monochamus alternatus	Pine trees	*B. bassiana* cultured on wheat bran pellets and implanted under the bark	Average mortality of larvae were 13–81% and 43–45% in logs and treated parts of standing trees	Japan	Shimazu et al. (1992)
Myllocerimus aurolineatus	Tea plantation	Soil application @ 15–30 kg/ha (MI)	More than 80% larval mortality was recorded	China	Wu and Sun (1994)
Sphenophorus levis	Sugarcane	@ 4.9 × 10^{11} conidia/acre	92.3% mortality	Brazil	Badilla and Alves (1991)
Bemesia argentifolii	Cucumber and melon	Spore suspension @ 5 × 10^{13} conidia in 180 l of water/ ha (MI)	More than 85% mortality of 3rd and 4th instar nymphs	USA	Wright et al. (2000)
Myzus persicae	Conala plant	Application of *B. bassiana* twice @ 10^8 spores/ ml (MI)	72–86% mortality 6 days after the first spray	USA	Miranpuri and Khachatourians (1993)
Nilaparvata lugens	Rice	@ 4 × 10^{12} to 5 × 10^{12} conidia/ha (MI)	63–98% mortality after 3 weeks of application	Philippines	Rombach et al. (1986b)

ha Hectare, *MI* microbial insecticide

17.2.2.3 Viral Biopesticides

More than 20 groups of viruses are known to be insect pathogens (Martignoni and Iwai 1986). They have been placed to 12 viral families where Baculoviridae, Entomopoxviridae and Reoviridae are unique and have been seriously considered as control agents. There are over 1100 hosts, including a few mites and ticks (Martignoni and Iwai 1986).

a. Baculoviruses Viruses in the family Baculoviridae are the best known of all the insect viruses. Baculoviruses are the most common and most widely studied group of viruses pathogenic for insects (Martignoni 1984). The two most important subgroups within the family Baculoviridae are the Nuclear Polyhedrosis Viruses (NPV) and the Granulosis Viruses (GV). Although the disease symptoms of these two viral groups are different and they share the following characteristics: Both groups have an enveloped rod shaped virion (approx. 50 × 250 nm) containing a circular double stranded DNA genome with molecular weight ranging from 50 to 100 million Dalton (Bilimoria 1986). Both NPV and GV are occluded in a polyhedral proteinaceous inclusion body (IB), with diameter of 40–80 μm and length greater than 300–420 μm. Several to many virions may be contained within an IB, while only one GV virion is contained within an IB. Molecular phylogeny of occluded proteins of baculoviruses was described by Rohrmann (1986) (Fig. 17.5).

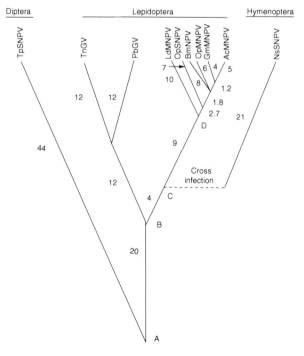

Fig. 17.5 Molecular phylogeny of occluded proteins of baculoviruses. The numbers indicate the percentage of amino acid differences between various points on the tree. The letters signify major branch points in the phylogeny and include the following: (*A*) divergence of lepidopteran and dipteran lines; (*B*) divergence of granulosis viruses (GVs) from lepidopteran singly enveloped, nuclear polyhedrosis viruses (SNPVs); (*C*) development of nuclear polyhedrosis viruses (NPVs); (*D*) development of multiply enveloped, nuclear polyhedrosis viruses (MNPVs). (Courtesy of Rohrmann 1986a and Society for General Microbiology)

17 Potential of Biopesticides in Sustainable Agriculture 559

a.1. Nuclear Polyhedrosis Virus Most NPV have been isolated from Lepidoptera (88%), Hymenoptera (6%) and Diptera (5%) (Rohrmann 1986; Saufi 2008). These viruses are easily recognized from other viruses because of the presence of unique polyhedral bodies in the cell nuclei. NPV infects the nucleus of the cell and multiplies within the nucleus. Maestri (1856) observed these bodies were called "polyedrische Koerperchen" (Polyhedral granules) (Fischer 1906) and the associated disease, Polyederkrankheit (polyhedral disease) (Wahl 1909) or polyhedrosis (Prell 1926). The nature of the polyhedra, present in the nuclei, remained unresolved for many years. The polyhedra were considered to be protozoa by Bolle (1894) who named them *Microsporidium polyhedricum*. Von Prowazek (1907, 1913) demonstrated the filterability of the infectious agent of jaundice of silkworm, but he also observed a coccus that he belived to be the etiologic agent and named it *Chlamydozoon bombyscis*. He and other workers considered the polyhedra to be a by-product of the disease (Sasaki 1910; Glaser and Chapman 1916). In the silkworm the disease called "Jaundice" in America, 'Giallume" in Italy, "Grasserie" in France, "Nobyo" in Japan and "Gelbsucht" or "Fettsucht" in Austria and Germany. The concept of viral nature was based on the wilt of gypsy moth (Glaser 1918), and the jaundice of the silkworm (Acqua 1918). Glaser and Chapman (1916) biochemically analyzed the polyhedra and concluded that they contained nucleoprotein. The viral particles were found to contain nucleic acids and proteins (Bergold 1947). This was confirmed by Wyatt (1925a, b) who established the presence of DNA in the NPVs and GVs. A NPV is normally transmitted from one insect to another by the oral ingestion of polyhedra. When ingested by a susceptible insect, the polyhedra dissolve and infectious viral particles enter midgut cell, replicate, then pass to tissues more commonly associated with NPV infections. This pathway was summarized by Granados and Williams (1986). The period of time from ingestion to polyhedra to death is from 4 days to 3 weeks and varies with different NPVs, insect hosts, the number of polyhedra ingested, the larval instar during which the polyhedra are ingested, and the environmental temperature. Many NPV infections can be transmitted from an infected female via the egg to her progeny, even though the adult moth displays no disease symptoms (Smith 1976). Whether the transmission is exclusively the result of a virus adsorbed to the egg surface or also includes transmission inside egg (transovarian) is unresolved.

a.II. Granulosis Virus Paillot (1926) was the first to describe a granulosis viral infection in the larva of the *Pieris brassicae*, he later described three similar diseases in the larvae of *Euxova segetum* (Paillot 1934, 1935, 1937). The rod shaped viral particle was detected with the electron microscope by Bergold (1948) in capsules obtained from infected larvae of the pine shoot roller, *Choristoneura murinana* and by Steinhaus et al. (1949) in capsules from the variegated cutworm, *Peridroma saucia*. The infection is called granulosis because of the presence of minute granules (Steinhaus 1949). Wyatt (1925a, b) analyzed the amino acids and nucleotides of GVs from *C. murinana* and *C. fumifera* and demonstrated that the viruses contained DNA. Others showed the virus to be double stranded, super coiled, covalently closed, circular DNA (Shvedchikova et al. 1969; Tweeten et al. 1977). Lepidop-

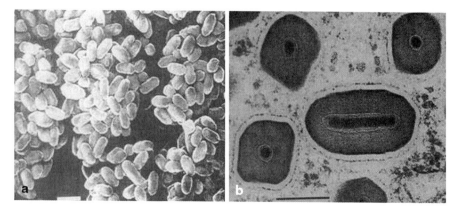

Fig. 17.6 **a** Occlusion bodies (Capsules) of a granulosis virus (GV) of fall armyworm. *Spodoptera frugiperda*. Scanning electron micrograph of capsule (bar = 0.5 µm). **b** Thin section of capsules. Central capsid sectioned longitudinally, others are transversely (bar = 0.25 µm). Note enveloped nucleocapsid proteinaceous matrix. (Adapted from Tanada and Kaya 1993)

tera are the only known hosts of granulosis viruses. GVs are generally more host specific than NPVs. Most attempts to cross transmit GVs were successful to host species of the same genus or closely related family (Groner 1986). The fat body is the primary site of infection, but the epidermis and tracheal matrix may occasionally be affected (Huger 1963). GVs are transmitted orally and via the egg. Latent infections also occur. The period between ingestion of the virus and the death of the host generally ranges between 4 and 25 days. External symptoms are not usually apparent in early stages of infection, but towards the later stages infected larvae frequently develop a lighter color. GV infection involving the epidermis cause liquefaction of the infected larvae similar to NPV infection, but when the epidermis is not involved, liquefaction does not take place (Martignoni and Iwai 1986). Occlusion bodies (Capsules) of a granulosis virus (GV) of fall armyworm *S. frugiperda* are shown in Fig. 17.6.

b. Mode of Action Viruses invade an insect's body via the gut. They multiply in tissues and can disrupt components of an insect's physiology, interfering with feeding, egg laying and movement. When ingested by the host insect, infectious virus particles are liberated internally and become active. Once in the larval gut, the virus's protein overcoat quickly disintegrates, and the viral DNA proceeds to infect digestive cells. Within a few days, the host larvae become unable to digest food, and so weaken and die (Thakore 2006b). The infected larvae hang upside down from the leaves and twigs in a characteristic way and a brownish fluid oozes from them (Fig. 17.7). This is a highly infective fluid and is readily disseminated amongst the healthy insect population. Baculoviruses are particularly attractive for use as biopesticides due to their high host specificity. Baculovirus products are sold under many trade names to control the insect pest of different crop (Table 17.6).

17 Potential of Biopesticides in Sustainable Agriculture

Fig. 17.7 Larva of armyworm, *Pseudaletia unipuncta*, dead from a NPV (Photo by Illinois Natural history Survey Extension staff)

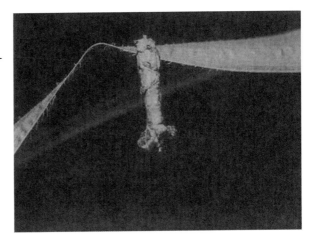

Table 17.6 Application of baculoviruses on crops. (Source: D'Amico (http://www.nysaes.cornell.edu/ent/biocontrol/pathogens/baculoviruses.html))

Commodity	Insect pest	Virus used	Virus product
Apple, pear, walnut and plum	Codling moth	Codling moth granulosis virus	Cyd-X (Granados and Federici 1986)
Cabbage, tomatoes, cotton, (and see pests in next column)	Cabbage moth, American bollworm, diamondback moth, potato tuber moth, and grape berry moth	Cabbage armyworm Nuclear Polyhedrosis Virus	Mamestrin (Jacques and Morris 1981)
Cotton, corn, tomatoes	*Spodoptera littoralis*	*Spodoptera littoralis* Nuclear Polyhedrosis Virus	Spodopterin (Jacques and Morris 1981)
Cotton and vegetables	Tobacco budworm *Helicoverpa zea*, and Cotton bollworm *Heliothis virescens*	*Helicoverpa zea* Nuclear Polyhedrosis Virus	Gemstar LC, Biotrol, Elcar (Granados and Federici 1986)
Vegetable crops, greenhouse flowers	Beet armyworm *(Spodoptera exigua)*	*Spodoptera exigua* Nuclear Polyhedrosis Virus	Spod-X (Granados and Federici 1986)
Vegetables	Celery looper (*Anagrapha falcifera*)	*Anagrapha falcifera* Nuclear Polyhedrosis Virus	none at present
Alfalfa and other crops	Alfalfa looper (*Autographa californica*)	*Autographa californica* Nuclear Polyhedrosis Virus	Gusano Biological Pesticide (Granados and Federici 1986)
Forest Habitat, Lumber	Douglas fir tussock moth *(Orgyia psuedotsugata)*	*Orgyia psuedotsugata* Nuclear Polyhedrosis Virus	TM Biocontrol (Burges 1981).
Forest Habitat, Lumber	Gypsy moth (*Lymantria dispar*)	*Lymantria dispar* Nuclear Polyhedrosis Virus	Gypchek (Insect Viruses and Pest Management)

b.1. Appearance Insects killed by baculoviruses have a characteristic shiny-oily appearance, and are often seen hanging limply from vegetation. They are extremely fragile to the touch, rupturing to release fluid filled with infective virus particles. This tendency to remain attached to foliage and then rupture is an important aspect of the virus life-cycle. As discussed above, infection of other insects will only occur if they eat foliage that has been contaminated by virus-killed larvae.

These are target specific viruses which can infect and destroy a number of important plant pests. They are particularly effective against the lepidopterous pests of vegetables. NPVs are known from a wide range of insect orders, as well as from crustacean (shrimp), but by far have been most commonly reported from lepidopterous insects from which well over 500 isolates are known (Volkmann et al. 1995).

These viruses are highly specific and do not affect beneficial insects like parasitoids and predators and are safe to fish, birds, animals and man. Considering the usefulness of NPV's there has been a growing demand amongst the farmers for these bioagents. NPV can be used in a variety of field crops, including sorghum, chickpea, cotton and maize. In sorghum, NPV is the preferred product for *Helicoverpa* management, not only because it is effective (frequently giving over 90% control) but because it preserves the full range of beneficial insects in the crop (e.g. *Microplitis* and *Trichogramma* wasps). The two products released in the USA are *Hz* NPV and its major hosts are *Heliothis/Helicoverpa* in cotton, tomatoes and legumes, and *Se* NPV against *S. exigua* in cotton, vegetables, grapes, flowers and ornamentals. The cost is $6–10/acre treatment. At this cost, virus products are competitive with standards.

Host specificity of this virus has limited its use to *H. armigera, S. litura* and *P. xylostella.* Extensive research has been conducted on the use of NPVs for tackling two major pests namely *S. litura* and *H. armigera*. Nuclear Polyhedrosis viruses like *Ha*NPV, *Sl*NPV are increasingly being used as alternatives to chemicals. Lack of commercial production has suppressed its popularity and it is used in limited areas. Applications of *Ha*NPV @ 450LE (Larval equivalent) ha^{-1} at 10 days interval were most effective on chickpea in the reduction of pod damage. Two spray of NPV @ 500 LE ha^{-1} on chickpea against *H. armigera*, resulted, the lowest pod damage of 8.6 and 7.9% when observed 7 and 14 days after treatment. The highest larval reduction of 79.9 and 81.5% was recorded at 7 and 14 days after virus application. The highest yield was recorded (17.9 q ha^{-1}) (Rai et al. 2001).

c. Interaction with other Control Agents Larval population of *H. armigera* decresed significantly by NPV @ 250 LE ha^{-1} on chick pea followed by endosulfan 350 g a.i. ha^{-1} and gave higher yield next to endosulfan (Singh and Ujagir 2001). HaNPV + endosulfan were comparable with endosulfan with increased chickpea yield from 9.1 to 12.2 q ha^{-1} in untreated control (Dahiya and Chauhan 2001). HaNPV @ 1.5×10^{12} POB ha^{-1} spray 3 times at 10 days interval with 3 sprays of endosulfan (700 g a.i. ha^{-1}) at 15 days interval were most effective in comparison to control against *H. armigera* on tomato (Singh et al. 2001). Three sprays of HaNPV @ 250 LE ha^{-1} with *T. pretiosum* 5 releases (50,000 ha^{-1} release) resulted lowest

17 Potential of Biopesticides in Sustainable Agriculture 563

larval population (1.84) and fruit damage (4.06%) and highest yield 127.83 q ha⁻¹
(Rahman et al. 2001). *Bt*-HaNPV-endosulfan was sprayed four times at 10 days
interval starting from flowering initiation stage found to be highly effective in
suppressing the larval population of *H. armigera*, reducing the pod damage and
increasing the yield in comparison to 3 sprays of endosulfan (Rao et al. 2001). Five
biorational insecticides were evaluated against the major insect pest of pigeonpea
viz; *Bt* (biolep, dipel and halt), NPV, *B. bassiana* (dispel), NSKE and Juvenile hormone (RH-2485). Out of them NPV was found best (Saxena 2001). Larval mortality
was maximum in endosulfan and monocrotophos followed by HaNPV + monocrotophos. Fruit damage also denoted the same trend (Mahalingum and Haneef 2001).
Field application and dosage of NPV are presented in Table 17.7.

Each virus only attacks particular species of insects, and they have been shown
to have no negative impacts on plants, mammals, birds, fish, or non-target insects
(D'Amico 2007a, b). This specificity is useful in preservation of natural predators when used in Integrated Pest Management systems. Baculoviruses can also
cause sudden and severe outbreaks within the host population for complete control
(Sylvar 2008). Disadvantages of baculoviruses include the need for the virus to be
ingested, resulting in lower efficacy, and their traditionally high cost of production.
The lower efficacy resulting from the need to be ingested is partially counterbal-
anced by the mode of action. When the target insect dies, the dead insect host's
body is spread on the foliage. The location and form of the infected insect carcass
increases the probability the infected carcass will be eaten by another larval host.

17.2.2.4 Entomopathogenic Nematodes

Nematodes commonly referred to as roundworms, eelworms, or threadworms, are
translucent, usually elongate, more or less cylindrical and noncellular elastic cuticular body organism. At least five orders and eleven families of nematodes contain entomopathogenic species (Tanada and Kaya 1993). Insects serve as vectors or
intermediate hosts for a number of parasitic nematodes of vertebrates (Schmidt and
Roberts 1989; Grewal et al. 2001). Close association also exists with insects and
plant nematodes (Mamiya and Enda 1972; Mamiya 1984). In 1929, R.W. Glaser
found the nematode, *Steinernema* (syn. *Neoaplectana*) *glaseri* infecting the Japanese beetle, *P. japonica*, and was the first to culture this parasitic nematode on
artificial media and use it in field tests against the beetle (Glaser and Farrell 1935;
Grewal et al. 2005). Van Zwaluwenburg (1928) listed 16 orders of insects of which
749 species had associations with nematodes. Relationships between nematodes
and insects vary from fortuitous association to obligatory parasitism. Filipjev and
Schuurmans Stekhoven (1941) placed insect nematodes, those nematodes living
within the alimentary tract of insects, and those living in the body cavity of insects,
whereas Welch (1963) classified nematodes as to whether they were external or internal parasites of insects. Nematodes that are obligate parasite of insects tend to be
monoxenous or oligoxenous. Oligoxenus species includes the mermithid *M. nigriscens*, which infects several species of grasshoppers (Christie 1937). The facultative

Table 17.7 Field application and dosages of virus formulations. (Source: Ramakrishnan 1993)

Insect species	Crop	Rate of virus application	Observation	References
Helicoverpa armigera	Chickpea	250 and 125 LE/ha	Significant reduction in larval population	Naraynan (1979)
	Redgram	Two application of 375 LE/ha 10^7 PIB/mL	Crop protected from damage by pest	Jayraj (1981)
	Lablab	125 LE/ha+0.035% endosulfan	Effective in reducing damage	Jayraj (1981)
	Sunflower	125 LE/ha	Effective control	Rabindra et al. (1986)
Spodoptera litura	Banana	1,400 LE(0.8 × 10^9)	Larval mortality 80.1–90.4%	Santharam et al. (1978)
	Groundnut	1,500 LE/ha	Yield increase 66.0%	Krishnaiah et al. (1984)
		750 LE/ha(4.6 × 10^6 PIB/pot)	As effective as chlorpyriphos 0.04% spray or carbaryl 15% dust	Sachitanandam et al. (1989)
	Tobacco nursery	93 × 10^6 PIB/mL	Reducing of damage 14.48%	Ramakrishnan et al. (1983)
	Cotton	250–500 LE/ha	Larval mortality 80.6–87.7%	Jayraj et al. (1981)
	Blackgram	1,500 LE/ha	Yield increase 16.2%	Krishnaiah et al. (1984)
		750 LE/ha	Larval mortality 87.1%	Mahadevan and
	Cauliflower	26 × 10^7 PIB/mL	Reduction of damage upto 56.4%	Chaudhri and Ramkrishnan (1980)
	Chillies	2 Applications of 250 LE/ha	Matched fenpropathrin 200 g ai/ha	Dhandapani and jayaraj(1989)
Adisura atkinsoni	Lablab	125 LE/ha + 0.035% endosulfan and 250 LE/ha + 0.035% endosulfan	Significantly reduced pod damage and increased yield	Narayanan (1987)
Mythimna seperata		2.5, 5.0, 7.5 and 10.0 × 10^6 PIB/mL	9.3% mortality in 10.0 × 10^6 PIB/ ml	Neelgund and Mathad (1974)
Spilosoma obliqua	Cowpea	10^6 PIB/mL 10^9 PIB/mL	Leaf damage reduced from 50 to 10%	Battu and Ramakrishnan (1989)

Table 17.7 (continued)

Insect species	Crop	Rate of virus application	Observation	References
Amsacta albistriga	Groundnut	2×10^5 PIB/L	Delayed larval mortality	Jayaraj et. al.(1977)
Oryctes rhinoceros	Coconut	*Baculovirus* collected in phosphate buffer, purified and concentrate was used for infection	Longevity of infected beetles reduced by 40%. Infected females were rendered sterile	Mohan et. al.(1983)
Chilo infuscatellus	Sugarcane	10^9 and 10^7 inclusin bodies, 10^5–10^9	Mortality 26–82% in virus coated eggs, 69–96% in 1st and 2nd instar, 81.4, 64.1, 54.6% in 3rd to 5th instar larvae respectively	Easwaramoorthy and Santalakshmi (1988), Easwaramoorthy and Jayaraj (1989)

LE larval equivalent, *PIB* polyhedral inclusion bodies, *ai* active ingredient, *ha* hectare parasites, *Deladenus* (syn. *Beddingea*) spp. are oligoxenic and infect siricid wood wasps, a battle associated with the wood wasps, and several species of *Rhyssa*, which are hymenopteran parasitoids of wood wasps (Bedding 1984; Grewal et al. 2004). In contrast, some Steinernematids and Heterorhabditids are polyxenic e.g. *Steinernema carpocapsae* infects more than 250 species of of insects from several orders under laboratory conditions (Poinar 1979) and studies with *Heterorhabditis bacteriophora* (syn. *Heliothidis*) indicated that it has a broad host range (Khan et al. 1976; Poinar 1979; Rishi et al. 2008).

Rhabditid nematodes have three pre-adaptations which have allowed them to evolve the life styles exhibited by the genera *Steinernema* and *Heterorhabditis* (Sudhaus 1993). First the order Rhabditida has evolved species that have a variety of associations with insects. Second, Rhabditid produce the dauer juvenile stage which confers the capacity to enter an insect and persist in the absence of food (Grewal et al. 2001, 2004, 2005). Third, Rhabditids are bacterial feeders, and hence are pre adapted to enter a mutualistic relationship with entomopathogenic bacteria inside the insect haemocoel. Microbes must be able to successfully infect their host in order to cause disease (pathogenecity), but infection is by no means sufficient to generate disease symptoms (virulence) (Finlay and Falkow 1989). Toxins are the major determinants of virulence, but many toxins play only a minor role in the colonization process and their role in microbial ecology is little understood. It is likely that many of the bacterial genes involved in the infection of insects play a role in the mutualistic interaction with the nematode, as recently highlighted by Hentschel et al. (2000).

a. Mode of infection Insect parasitic nematodes parasitize their hosts by directly penetrating through the cuticle into the hemocoel or by entering through natural openings (spiracles, mouth, and anus) (Sankar 2009). Species in the order Tylenchida possess a stylet and in the order Mermithida possess a spear, which is technically called an odontostyle. *Heterorhabditis* has an interior tooth that is used to scrape and rupture the host's cuticle (Bedding and Molyneux 1982). Nematodes infect their insect hosts passively or actively. Passive infection occurs when a mermithid deposits its eggs on the hosts food. The eggs are ingested by an insect, and the nematode, bore through the midgut and enters the hemocoel (Rishi et al. 2008). Christie (1937) has described hatching and penetration by *Mermis nigriscens* through the midgut of a grasshopper. Active infection occurs when the nematodes seek their hosts and penetrate directly through the integument into the haemocoel. Poinar and Doncaster (1965) observed the penetration process of the sphaerulariid, *Tripeus sciarae* into the dipteran host, *Bradysia paupera*. The infected adult female ensheathed in the fourth stage cuticle, produces an adhesive mass about its head. This secretion digests the anterior portion of the ensheathed cuticle and adheres the nematode to the host. The attached nematode uses its stylet and possibly some enzymes to penetrate into the host. The penetration process may take from 10 min to 2 h, and the wound is sealed by the adhesive substance after the nematode has entered the insect. Insects generally show the external evidence of a nematode infection. In mermithids, the infections usually manifest greater external morphological changes than in the other nematode groups.

Nematodes in the hemocoel of insects cause many physiological changes and these changes have been recorded from a number of insects groups (Sankar 2009). The presence of mermithids results in sterility and ultimately death of the adult insects. Adult grasshoppers infected with *Agamermis decaudata* or *Mermis nigriscens* have markedly reduced ovaries and are usually sterile (Christie 1937). Physiological basis for parasitic castration has been examined by Gordon et al. (1973) who showed that vitellogenesis proceeds normally up through the beginning of yolk deposition in female desert locust, *Schistocerca gregaria* infected with *M. nigrescens* infection (Gordon and Webster 1971). Molting of desert locust can be inhibited by *M. nigriscens* infection (Craig and Webster 1974). Black fly larvae infected with mermithids have reduced fat bodies contents (Phelps and DeFoliart 1964). Insects infected with nematodes often have abnormal behavior compared with uninfected individuals. However, abnormal behavior may be the result of other anomalies and should not be used as the sole criterion for a nematode infection. Abnormal behavior in insects is usually manifested late in nematode infection (Poinar and Gyrisco 1962).

Steinernematid nematodes carry a bacterial symbiont in their digestive system. When the infective juvenile stage, called a dauer larva, penetrates the host cuticle and enters the haemocoel, it releases the bacteria. The bacteria, *Xenorhabdus* kill the insect host within 48-hrs and the juvenile nematodes feed on the bacterial and bacterial byproducts in insect body. The nematodes develop to adult stage and lay eggs inside the host. These become infective juvenile daver larvae, which exit the dead host and infest other host insects. Steinernematid nematodes are not host specific and because they kill their host within few days, those species that can be mass produced on artificial media have been commercialized as biological insecticides.

b. Entry and Establishment An important function of the nematode is vectoring the entomopathogenic bacteria into the insect host. The symbiotic bacteria enter the insect passively, carried within the nematode gut. Following entry, the bacteria must adhere to a surface within the host to establish infection. The fimbriae of *X. nematophillus* are a potential adhesion site, which have been found to be responsible for *in vitro* agglutination by phase 1 bacteria of the haemocytes of *Galleria mellonella* (Moureaux et al. 1995). Entomopathogenic bacteria may provide nutrients for themselves and their nematode symbiont by secreting a variety of extracellular enzymes which degrade host tissue. Organisms that spread via the bloodstream usually produce siderophores or have alternative means of removing iron that is tightly complexed to host proteins, and DNA sequences which probably specify siderophore production have been found in *Pseudomonas luminescens* (french-Constant et al. 2000).

Grubs of *P. japonica* displayed grooming behavior using their legs and roster at the last segment as well as evasive reactions on sensing nematode attack (Gaugler et al. 1994). Mechanical barriers include the sieve plates which protect spiracles of many soil dwelling larvae, and entering in insect via the tracheal opening is not possible in many grubs (Forschler and Gardner 1991), leatherjackets (Peters and

Ehlers 1994) and maggots (Renn 1998); however in sawfly larvae spiracles are the most important route of entry (Georgis and Hague 1981). A second route of entry is through the mouth opening or anus. The width of both (e.g. in wireworms) may exclude infective juveniles (IJs) (Eidt and Thurston 1995), and insect mandible crush the nematode to death (Gaugler and Molloy 1981). Use of anus as entry site avoids the latter problem and represents the main route in housefly maggots and leaf miners (Renn 1998). Still, frequent defecation may expel nematodes entering the anus, and in grubs saw fly larvae invasion is more successful than via the mouth than via the anus (Georgis and Hague 1981; Cui et al. 1993). When nematode have reached the gastric caecae, malpighian tubules or the gap between the peritrophic membrane and midgut epithelium, expulsion with faeces is avoided. A general problem with parasitism via the intestinal tract is the host's gut fluid which may kill up to 40% of invading non adapted nematodes and significantly reduce penetration via the intestine (Wang et al. 1995; Grewal et al. 2005).

After successfully entering the tracheal system or the intestine, the dauer larvae still must pass through the tracheole or the gut wall, respectively. The fragile tracheole wall might be penetrated just by the mechanical pressure of the pointed nematode head. The gut wall of insect is protected by the peritrophic membrane, which can be a serious obstacle for nematodes (Forschler and Gardner 1991; Grewal et al. 2004; Shapiro-Ilan 2006). *S. glaseri* takes 4–6 h to penetrate the midgut of *P. japonica* (Cui et al. 1993). Dauer larvae of *H. bacteriophora* were observed using their proximal tooth for the rupturing of the insect body wall (Bedding and Molyneux 1982; Grewal et al. 2005) and it was long believed that only *Heterorhabditis* could penetrate the tissues like insect integument, since dauer larvae of *Steinernema* lack the tooth. There is increasing evidence that nematodes secretions are involved in penetration, at least in *Steinernema* species: protease inhibitors decrease penetration of *S. glaseri* through the gut wall of *P. japonica* (Abu Hatab et al. 1995), and the midgut epithelium cells of *G. mallonella* showed a marked histolysis in response to secretions of *S. carpocapsae* (Simoes 1998).

c. Host Resistance An insect resists nematode infection through behavioral, physical, or physiological means. Behavioral resistance occurs when the insect actively avoids or repels the nematode. Petersen (1975) reported that extremely active mosquito species had a lower prevalence of infection by the mermithid, *Romanomermis culicivorax* than less active ones. Scarab larvae may avoid infection by wiping nematodes away from the mouth (Akhurst 1986). Physical resistance occurs when the nematode cannot penetrate the integument or the cocoon of a host insect. *R. culicivorax* have difficulty in penetrating the integument of older mosquito larvae (Petersen and Wills 1970). Dauer juveniles of *S. carpocapsae* cannot penetrate the silken cocoons of hymenopteran parasitoids (Kaya and Hotchkin 1981), if a hole is made in the cocoon; infection occurs (Kaya 1978a, b). Spiracles are portals of entry for nematodes (Triggiani and Poiner 1976) but sieve plates over the spiracles, especially with scrab larvae, may deny nematode access through this entry (Akhurst 1986). Physiological resistance to infection involves the destruction of nematode by

17 Potential of Biopesticides in Sustainable Agriculture

digestive enzyme in the insect alimentary tract and the melanization and encapsulation of nematode within the haemocoel. Welch and Bronskill (1962) reported that melannotic encapsulation of *S. carpocapsae* occurred in larvae of several mosquito species. Complete encapsulation of this nematode occurs within 5 h after the penetration into the hemocoel of *Aedes aegypti* (Bronskill 1962).

d. Field Application A field study was conducted by Cheng and Boucher (1972) who showed the initial evidence that nematodes may protect plants from anthomyiid pest species. Crop yield (cabbage) was greater than the untreated group but less than diazinon- treated plants. Satisfactory increased yield was obtained against *Delia radicum* in *Brussels sprouts* (Georgis et al. 1983) after application of nematode. Entomopathogenic nematodes in the families, Steinernematidae and Heterorhabditidae have been used to suppress populations of pest insects in a variety of agro ecosystems, and in several cases their positive effects on crop yield have been shown (Lewis et al. 1998).

EPNs are the most extensively studied against the white grub (Klein 1993; Grewal et al. 2004). At least five EPN species, *S. anomaly*, *S. glaseri*, *S. kushidai*, *S. scarabei* and *H. megidis*, were originally collected and described from naturally infected white grubs and many more species have been documented to use white grub as natural hosts. *S. scapterisci* can provide long term suppression of mole cricket population after inoculative releases and may become a major means of mole cricket management in pastures in Florida (Parkman and Smart 1996) and *S. carpocapsae* for *P. xylostella* on cabbage and also *S. thermophilum* on *P. xylostella* on cabbage (Somvanshi et al. 2006).

Grewal et al. (2005) provided an extensive summary of studies on the efficacy of various species and/or strains of EPN against white grub. Nematodes that offered satisfactory field control of *P. japonica* include *S. scarabaei* (100%), *H. bacteriophora* (strain GPS 11) (34–97%), *H. bacteriophora* (strain TF) (65–92%) and *H. zealandica* (strain XI) (73–98%).

17.2.2.5 Entomopathogenic Protozoa and Microsporida

Various types of association occur between protozoa and the insects (Margulis et al. 1990; Choi et al. 2002; Jhony et al. 2006). Some of these protozoa are pathogens of both vertebrates and the insects (e.g., haemosporida causing malaria), others are commensals or weakly virulent in the insects digestive tract (e.g., gregarines); still others are highly virulent forms (e.g. microsporidia) (Weiser 1963; Margulis et al. 1990). Most of the entomopathogenic protozoa occur in the "Sporozoa" phyla: Apecomplexa and Microspora. Apicomplexa are separated into three classes: Gregarinia, Coccidia and Hematozoa (Viver and Desportes 1990). The difference between gregarine and coccida is in their gamogony; the female gamont of the gregarine gives rise to a number of gametes and that of coccidian to a single gamete (Viver and Desportes 1990). The members of the phylum Apicomplexa have typical fine structures, primarily the "apical complex", that occur in at least one stage in the life cycle (Levine 1971, 1982). Haplosporidia infect the digestive tract, fat body, oenocytoids and malpighian tubules of insects (Weiser 1963). Among the haplosporidia, *Nephridiophaga blattellae* is the best known (Woolever 1966).

The life cycle of gregarines varies greatly and is often complex. The insect ingests a mature oocyst, which is acted upon by the digestive juices in the insect's alimentary tract. Several sporozoites, eight or rarely four, emerge from the oocyst and penetrate into the midgut epithelial cells or into the haemocoel and become trophozoites. After reaching a certain stage, trophozoites emerge from the cells and destroy them in the process. In the lumen of the digestive tract, they develop subsequently into gamonts. In some gregarines the sporozoites do not penetrate the cell but remain attached as trophozoites to the cell with epimerite when it becomes a mature gamont. The gamonts undergo syzygy in pairs or in longer groups. A membrane develops around the associated gamonts to produce a cyst (gametocyst). One gamont produces microgametes and the other member produces macrogametes or the gamonts from the isogametes. The gametes fuse to form zygotes, which are the only diploid stage in the life cycle of gregarines (Grasser 1953) (Fig. 17.8). In this way gregarines cause their most severe pathology during their intracellular and intercellular developments.

The members of the class microsporea in the phylum microspora are commonly called microsporidia. The disease they cause is called microsporidiosis (Tsai and Wang 2001; Johny et al. 2006). Microsporidia rank among the smallest of eukaryotes and have an obligate intracellular habit. They possess a unicellular spore containing a uninucleate or binucleate sporoplasm and an extrusion apparatus always equipped with a polar filament and polar cap. Their life cycle stages are also ultrastructurally unique and distinct from other spore forming protozoa and play a critical role in taxonomic determinations (Keeling 2003; Tsai et al. 2003). Even though they are eukaryotic microorganisms, their ribosomal RNAs have prokaryotic properties (Ishiara and Hayashi 1968; Curgy et al. 1980). There are about 800 described species of microsporidia (Canning 1990).

As a group, the microsporidia are the most important protozoan pathogens of insects. They form the majority of the protozoa pathogenic to insects and cause economically serious diseases in beneficial and pest insects. They are the most promising protozoa for use in microbial control. All microsporidia are obligate pathogens and multiply only in living cells. Many cause severe acute infestations in insects but some produce only inapparent or chronic infections, which nevertheless may play an important role in regulating the insect populations. Microsporidia invade insects through three natural portals of entry: oral, cuticular, and ovarial pathways (Kramer 1976), venereal transmission between sexes is rare but has been reported in *Plodia interpunctella* (Kellen and Lindegren 1971) and coccinelid beetles (Toguebaye and Marchand 1984). Entrance by the oral and cuticular portals results in horizontal transmission and by the ovarial portal in vertical transmission. These types of transmissions are believed to occur commonly in the same host generation, but in the European corn borer, *Ostrinia nubilalis*, the

Fig. 17.8 Life cycle of gregarine *Stylocephalus* sp. (1–8) Developmental stages occurring in the intestine of the host (a tenebrionid beetle): (*1*) sporozoite escaping from the spore ingested by host; (*2*) free sporozoite; (*3*) differentiation of epimerite; (*4–5*) growing gamonts (syn. Trophozoites); (*6*) mature gamont detached from the intestinal epithelium; (*7*) pairing of gamonts; (*8*) encystment, (*9–12*) developmental stages occurring outside of the host; (*9*) differentiation of gametes (female gametes in the upper gamont, male ones in the lower); (*10*) fertilization; (*11*) zygote; (*12*) meiosis (*10, 11, and 12* occur under the cyst wall). (Courtesy of Vivier and Desportes 1990 and by permission of Jones and Bartlett Publishers. Drawing by S. Manion-Artz)

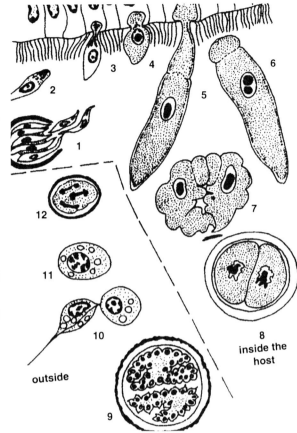

microsporidium *Nosema pyrausta* is transmitted primarily by transovarial infection (vertical) in the first generation and by vertical and horizontal transmissions in the second generation (Siegel et al. 1988). The oral portal is the most common route through which the microsporidium gains access to host tissues. *N. apis*, which infects the digestive tracts of adult bees, appears to infect only through the oral route. Transmission through faeces contaminated with microsporidium spores commonly occurs in susceptible hosts (Andreadis 1987), the larval silk glands are often infected but have not been considered as a route of horizontal transmission except in the case of a *N.* sp. of the gypsy moth, *Lymantria dispar* (Jeffords et al. 1987). Some microsporidia, like *N. kingi* infecting Drosophilids, and others infecting Lepidoptera (Tsai et al. 2003) which utilized both the oral and ovarian route. Microsporidia are intracellular pathogens, usually in the cytoplasm and infrequently in the nucleus. They produce infections that range from subacute and chronic to acute and virulent. Some microsporidia infect only a single tissue or organ, such as the midgut epithelium, muscle, or fat body. Tissue specificity, however, may vary with the host infected by the microsporidium. *N. algerae* infects the brain and nerve ganglia of *Aeder algerae*, but in *Anopheles quadrimaculutus,* it infects nearly all tissues (Hazards and Lofgren 1971). The concentration of spore dosage may also determine the types of tissue infected. A low dosage of *Vairimorpha necatrix* results in a chronic infection of mainly the fat body and some mascular tissue of *Trichopusialni,* whereas a high dosage produces an acute infection in the midgut tissue (Chu and Jaques 1979). According to Fisher and Sanborn (1964), microsporidian infection in larvae of *Tribolium castenium* caused a juvenile hormone like effect. They reported that the larvae grew faster, had supernumarary molts, and rarely, if ever pupated (Milner 1972).

An insect infected with a microsporidium generally is altered in colour, size, form, and activity, depending on the tissues or organ infected. A translucent larva turns increasingly opaque or dull, milky white as the infection progresses, the integument usually remains intact throughout infection. In pebrine of the silkworm, the dark molted areas or dark-brown spots on the integument are a pathognomic sign of this disease. In some infections, the growth of a larva is retarded or the larva is reduced in size prior to death. In others, the larva is swollen or distended in form. Microsporidium infection may alter the behavior of the honey bee. The infection by *N. apis* substantially reduces the hoarding behavior of adult bees and may affect the overall production of a colony (Rinderer and Elliott 1977a). Abnormal behavior also results through the indirect effect of the microsporidium on the hypopharyngeal glands, making them unable to secrete royal jelly (Wang and Moeller 1970).

a. Field application Host specificity testing is critical to introduction of classical biological control agents (Hokkanen et al. 2003). The field studies described, in concert with studies of natural field populations (Solter et al. 2000), provided evidence that *V. disparis*, and *N. lymantriae* by extrapolation, the more host specific *N. portugal* (Solter et al. 1997), would not endanger non-target species by host switching if introduced. Many non-target species that did not occur in high densities were collected in insufficient numbers to evaluate susceptibility to the *L. dispar* microsporidia; however, these species were recovered in roughly equivalent numbers from treated and untreated plots, suggesting that no acute mortality occurred due to spraying of the microsporidia (Leellen et al. 2010)

As predicted from earlier laboratory studies (Solter et al. 1997), *V. disparis* was less host specific than *N. lymantriae*. In general, *V.* species appear to be relatively more virulent; they attack the fat body tissues of the host, reproducing quickly and filling the target cells with spores (Solter and Maddox 1998; Vavra et al. 2006). Nevertheless, similar to findings in laboratory studies (Solter et al. 1997), several of the non-target infections produced in this field study were atypical, producing abnormal spores and/or low numbers of mature spores, or resulting in acute death of the host before optimal reproduction of the pathogen. A typical infections in non-target hosts, as well as many infections that appeared similar to those in the natural host, were noted to be most likely "dead end" infections and result in inability of the pathogen to be transmitted to nonspecific individuals (Solter et al. 1997; Solter and Maddox 1998).

17.3 Constraints of Biopesticides

Biopesticides formulations particularly of viruses and fungi, have limited spectra of activity against pest complexes as compared to chemical insecticides. Its effectiveness is a product of a dynamic interaction between the hosts, pathogens and environment. Temperature, and environmental conditions critically affect the host body temperature, has a profound effect on the efficacy of the pathogen (Blanford et al. 1998; Blanford and Thomas 2000). Biopesticide may kill a host in 4–7 days at 25–30°C. However, temperature in field conditions is not always constant and may fluctuate widely over a 24 h period and during control campaign. Furthermore, the majority of acridids appear to be active behavioral thermo regulators. The advantages of microbial insecticides over chemical insecticides in term of reduced threats to non target species and limited potential for environmental degradation are a compelling reason for increasing efforts to improve these agents and to increase their utility for IPM programs (Cuperus et al. 2004). Although, many compounds have been isolated, characterized, and evaluated as anti-insect compounds from plants. The standardization of plant based anti-insect preparations has been the biggest constraint and has subsequently hindered their potential marketability compared with conventional pesticides. However, the concept of phytosynergistic strategy and assessing the toxicity of co-occurring toxins seems to have potential and could lead to further advancement in the development of biopesticides (Koul and Dhaliwal 2001).

17.4 Biopesticide Regulations

Integrated pest management is being redefined as BIPM (Biointensive Integrated Pest Management) by the advent of biologically based alternatives, which are environment friendly and ensure food safety and security, apart from maintaining the balance of the living biota. In India, all pesticides (including biopesticides) are regulated under the Insecticid Act, 1968 and require mandatory registration from the Central Insecticide Board and Registration Committee. This body has streamlined the registration guidelines and brought the quantum of data generation to a lower scale, granting the registration of the biopesticides on a priority basis.

17.5 Market Prospects for Biopesticides

Market prospects for biopesticides are generally satisfactory, especially, to consider that various countries would focus upon increasingly improved food safety standards, a major pre-requisite for export as well as domestic markets. Improvements in primary production are a first step to achieve higher food safety in the long run. However, biocontrol products face stiff competition with low-priced generical pesticides (chemical products for which patents have phased out). Furthermore, pesticide users are often reluctant to apply biocontrol on grounds of an actual or perceived lower performance or effectiveness. Various low-quality biopesticides circulate in regional markets also contributing negative image. In cooperation with government and the private sectors the projects would promote a number of model products that will show a great promise in 'cash-crops' like oil palm and cocoa as well as staple food like rice. In these monoculture systems, resistance of pests against chemicals is an important reason to use biopesticides instead of chemicals. Another important factor in improving competitiveness is quality: in order to convince farmers and users that biopesticides are effective in agro climatic conditions.

17.6 Conclusion

The ultimate goal of a pest management program is to protect the commodity economically without disrupting the ecosystem. With this goal in mind pest management practitioners should make every attempt to prevent losses caused by pests by implementing and application of biopesticides. Insecticides should be used only when the nonchemical tactics fail to keep pests below economically damaging levels and should supplement a completely integrated program. The use of precise monitoring strategies and established economic thresholds should ensure that pest management strategies are employed appropriately. For the last few decades there has been a concern for the life-giving quality of land, the high ecological price of unsustainable conventional agriculture and the need to reorient it. Out of this concern, one thing that has emerged crystal clear is that our future lies in sustainable agriculture. The pre requisite of sustainable agriculture is conservation of natural resources, both living and non-living. But most of the resource base agriculture is under threat due to various reasons, one being chemicalisation of agriculture due to indiscriminate use of pesticides, has led to agricultural disaster and not agricultural prosperity. The challenge therefore is to make our agricultural resources more productive and economically viable, but at the same time safer to health, environment and farming communities. While present day needs do necessitate use of pesticides for increased production of food and fiber and prevention of peasts and diseases, it is being increasingly realized that we have also to minimize, if not eliminate all together, serious health hazards, and conserve and improve the environment and the local communities by taking advantage of cultural, natural and other innumerable pest control methods, and use of pesticides only to avert serious economic losses in agriculture. It is true that we cannot simply go back to an agricultural society in which there are no chemical pesticides or adopt only organic farming. Such a shift would entail unimaginable repercussions. Food production and its availability must always remain an overwhelming priority. So the best we could do is to be cautious in the use of pesticides so as to minimize the health and environmental hazards,

17 Potential of Biopesticides in Sustainable Agriculture

always weighing all risks against benefits to ensure maximum advantage and safety to man's health and his environment.

Acknowledgements We are thankful to Cambridge University Press (The Insect: Structure and Function by R.F. Chapman), John Wiley and Sons, Inc. (Introduction to Insect Pest Management, (Eds) Robert L. Metcalf and William H. Luckmann) and Academic Press (Insect Pathology by Yoshnori Tanada and Harry K. Kaya) for giving us permission to reproduce a few required figures in this manuscript. We are also thankful to one of the anonymous referee for critical comments on the earlier draft.

References

Abdel-Razek AS, Abbas MH, El-Khouly M, Abdel-Rahman A (2006) Potential of microbial control of diamondback moth, *Plutella xylostella* (Linnaeus), (Lepidoptera: Plutellidae) on two cabbage cultivars under different fertilization treatments. J Appl Sci Res 2(11):942–948

Abu Hatab M, Selvan S, Gaugler R (1995) Role of proteases in penetration of insect by the entomopathogenic nematode, *Steinernema glaseri* (Nematode: Steinernematidae). J Inverteb Pathol 66:125–130

Acqua C (1918) Ricerche sulla malattia del giallume nel baco da seta. Rend. Ist Bacologico Sc Super Agri Portici 3:245–256

Akhtar Y, Isman MB (2004) Comparative growth inhibitory and antifeedant effects of plant extracts and pure allelochemicals on four phytophagous insect species. J Appl Entomol 128:32–38

Akhtar Y, Shikano I, Isman MB (2009) Topical application of a plant extract to different life stages of *Trichoplusia ni* fails to influence feeding or oviposition behavior. Entomol Exptl Appl 132:275–282

Akhurst RJ (1986) Controlling insects in soil with entomopathogenic nematodes. Fundamental and applied aspects of invertebrate pathology. In: Samson RA, Vlak JM, Peters (eds) Fourth Int. Collq. Invertebr Pathol. Wageningen, The Netherlands, pp 265–267

Al-Fifi Zia (2009) Effect of different neem products on the mortality and fitness of adult *Schistocerca gregaria* (Forskal). JKAU 21(2):299–315

Alves BS, Risco SH, Silvernia Neta S, Machado Neta R (1984) Pathogenecity of nine isolates of *Metarhizium anisopliae* (Metsch.) Sorok to *Diatraea saccharalis* (fabr.) J Angew Entomol 97:404–406

Alves SB, Risco BSH, Almeida LC (1982) Evaluation of spore amounts of *Metarhizium anisopliae* (Metch.) Sorok. produced per hectare during the epizootic phase of the disease on sugarcane leafhopper, *Mahanarva posticata* (Stal) in Brazil. Entomol News, Internat Soc Sugarcane Technol 13:7

Andersson J, Borg-Karlson AK, Wiklund C (2000) Sexual cooperation and conflict in butterflies: a male-transferred anti-aphrodisiac reduces harassment of recently mated females. Proc Royal Soc Lond B 267:1271–1275

Andreadis T G (1987) Horizontal transmission of *Nosema pyrausta* (Microsporida: Nosematidae) in European corn borer, *Ostrinia nubilalis* (Lepidptera: Pyralidae). Environ Entomol 16:1124–1129

Angeli G, Anfora G, Baldessari M, Germinara GS, Rama F, De-Cristofaro A, Ioriatti C (2007) Mating disruption of codling moth, *Cydia pomonella* with high densities of ecodian sex pheromone dispensers. J Appl Entomol 131(5):311–318

Anonymous (1983) Proc 4th All India workshop on biocontrol, IIHR, Bangalore, 22–24 February

Armstrong E (1976) Transmission and infectivity studies on *Nosema kingi* in *Drosophila willistoni* and other drosophillids. Z Parasitenkd 50:161–165

576   M. S. Ansari et al.

Aronson A, D Wu (1989) Specificity *in vivo* and *in vitro* of *Bacillus thuringiensis* delta-endotoxin active on lepidoptera. In: Baker R, Dunn PE (eds) New directions in biological control. Liss, New York

Ashok Kumar Singh, Marya BR, Ram H (1998) Neem leaf extract controls ants on lemon plant. Neem Newsletter 15(3):37

Ayala JL, Monzon S (1997) Test on different doses of *Beauveri bassiana* for the control of the banana weevil (*Cosmopolites sordidus*) (Germar.) Centr Agric Rev Cient Facul Cien Agri 4:19–24

Ayyangar GSG, Rao PJ (1989) Azadirachtin effects on consumption and utilization of food and midgut enzymes of *Spodoptera litura* (Fabr.). Ind J Entomol 51:373–376

Ayyangar GSG, Rao PJ (1990) Changes in haemolymph constituents of *Spodoptera litura* under the influence of azadirachtin. Ind J Entomol 52:69–83

Badilla FF, Alves SB (1991) Control of sugarcane weevil *Sphenophorus levis* Vaurie (Col : Curculionidae) with *Beauveria bassiana* and *Beauveria brongniartii* under laboratory and field conditions. Turrialba 41:237–243

Barbosa JT, Riscado GM, Lima Filho M (1979) Population fluctuations of the sugarcane froghopper, *Mahanarva posticata* Stal 1855 (Hom. Cercopidae) and its natural enemies. Anais Socidade Entomol Brazil 8:39–46

Baskar K, Kingsley S, Vendan ES, Paulraj MG, Ignacimuthu S (2009) Antifeedant, larvicidal and pupicidal activities of *Atalantia monophylla* (L) Correa against *Helicoverpa armigera* Hubner (Lepidoptera: Noctuidae). Chemosphere 75:355–359

Battu GS, Ramakrishnan N (1989) Efficacy of nuclear polyhedrtosis virus for the control of *Spilosoma obliqua* (walker) on cowpea. J Biol Cont 3(2):99–102

Bedding RA (1984) Large scale production, storage and transport of insect parasitic nematodes *Neoplactana spp.* and *Heterorhabditis* spp. Ann Appl Biol 104:117–120

Bedding RA, Molyneux AS (1982) Penetration of insect cuticle by infected juveniles of *Heterorhabditis* spp (Heterorhabtidae, Nematoda) Nematologica 28:354–359

Benjamin MA, Zhioua E, Ostfeld RS (2002) Laboratory and field evaluation of the entomopathogenic fungus, *Metarhizium anisopliae* (Deuteromycetes) for controlling questing adult, *Ixodes scapularis* (Acari: Ixodidae). J Med Entomol 39(5):723–728

Berenbaum MR (1991) Coumarins. In: Rosenthal GA, Berenbaum MR (eds) Herbivores: Their Interaction with Secondary Plant Metabolites. Acad Press, New York, pp 221–249

Bergold GH (1947) Die isolierung des polyeder virus und dier nature der polyder. Z Naturforsch 2:122–143

Bergold GH (1948) Uber die kapselvirus-krankheit. Z Naturforsch 3:338–342

Berliner E (1915) Ueber die schlaffsucht der Ephestia kuhniella und Bac. thuringiensis n. sp. Z Angew Entomol 2:21–56

Bhathal SS, Singh D, Dhillon RS, Nayyar K, Singh D (1991) Ovicidal effects of neem oil and plant extract of *Agiratum conyzoides* Linn. on *Dysdercus koenigii* Fab. J Insect Sci 4(2):185–186

Bhattacharya AK, Mondal P, Ramamurthy VV, Srivastava RP (2003) *Beauveria bassiana*: A potential bioagent for innovative integrated pest management programme. In: Srivastava RP (ed) Biopesticides and Bioagents in integrated pest management of Agricultural crops. International Book Distributing Co., Lucknow, pp 381–492

Bidochka MJ, Small CL (2005) Phylogeography of *Metarhizium*, an insect pathogenic fungus. In: Vegaand Fe, Blackwell M (eds) Insect fungal associations: ecology and evolution. Oxford Univ Press, Oxford, UK, pp 28–50

Bilimoria SL (1986) Taxonomy and identification of baculoviruses. In: Granados RR, Federici BA (eds) In the biology of baculoviruses, vol 1. CRC Press, Boca Raton, Fl, pp 37–59

Billen JPJ (1986) Comparative morphology and ultrastructure of the Dufour's gland in ants (Hymenoptera: Formicidae). Entomol Gener 11(3l4):165–181

Blanford S, Thomas MB (2000) Thermoregulation in two acridid species: effects of habitat and season and implications for biological control using pathogens. Environ Entomol 29(5):1060–1069

17 Potential of Biopesticides in Sustainable Agriculture 577

Blanford S, Thomas MB, Langewald J (1998) Behavioural fever in a population of the Senegalese grasshopper, *Oedaleus senegalensis*, and its implications for biological control using pathogens. Ecol Entomol 23:9–14

Boeckh J, Kaissling KE, Schneider D (1965) Insect olfactory receptors. Cold Spring Harb Symp Quant Biol 30:263–280

Bolle G (1894) Il giallumeod il mal del grasso del baco da seta. Comunicazione preliminare. Atti Mem IP Sco Agrar Gorizia 34:133–136

Bowers MD, Puttick G M (1988) Response of generalist and specialist insects to qualitative allelochemicals variations. J Chem Ecol 14:319–334

Bowers WS (1991) Insect hormones and antihormones in plants. In: Rosenthal GA, Berenbaum MR (eds) Herbivores: their interactions with secondary plant metabolites, vol 1. Acad Press, New York, pp 431–456

Bowers WS, Nishida R (1980) Juvocimenes; potent juvenile hormone mimics from sweet Basil. Science 209:1030–1032

Bowers WS, Ohta T, Cleere JS, Marsella PA (1976) Discovery of insect anti-juvenile hormones in plants. Science 193:542–547

Brey PT, Latge JP, Prevost MC (1986) Integument penetration of the pea aphid, *Acrythosiphon pisum* by *Conidiobolus obscurus* (Entomophthoraceae). J Inverteb Pathol 48:34–41

Bronskill JF (1962) Encapsulation of rhabditoid nematode in mosquitoes. Can J Zool 1269–1275

Broughton HB, Ley SV, Slawin AMZ, Williams DJ, Morgan ED (1986) X-ray crystallographic structure determination of detigloyldihydro-azadirachtin and reassignment of the structure of limonoid insect antifeedant azadirachtin. J Chem Soc Chem Commun 46–47

Brower LP, Brower J, Cranston FP (1965) Courtship behavior of the queen butterfly, Danaus gilippus berenice (Cramer). Zoologica 50:1–39

Brown WL, Jr Eisner T, Whitlake RH (1970) Allomones and kairomones: transpacific chemical messenger. Bioscience 20:21–22

Burges HD (1981) Microbial control of pests and plant diseases. Acad Press, London UK

Burges HD, Hussey NW (1971) Microbial control of insect and mites. Acad Press, London UK

Butenandt A, Beckmann R, Stamm D, Hecker E (1959) Uber den Sexual-lockstoff des Seidenspinners *Bombyx mori*. Reindarstelling und Konstitution. Z Naturforsch 146:283–384

Butenandt A, Karlson P (1954) Uber die Isolierung cines metamorphosehormons der insektan in kristalissierter Form. Z Naturforsch 9:389–391

Butterworth JH, Morgan ED (1968) Isolation of a substance that suppresses feeding in locusts. J Chem Soc Chem Commun 1968:23–24

Byers JA (1989) Chemical ecology of bark beetles. Experientia 45:271–283

Canning EU (1990) Phyllum Microspora. In Handbook of Protoctista. Margulis L, Corliss JO, Melkonian M, Chapman DJ (eds) Jones and Bartlett, Boston, pp 53–72

Cantwell GE, Cantelo WW, Schroder RFW (1986) Effect of *Beauveria* bassiana on underground stages of Colorado beetle, *Leptinotarsa decemlineata* (Coleptera: Chrysomelidae). Gr Lak Entomol 19:81–84

Carlton BC (1988) Genetic improvements of *Bacillus thuringiensis* as a bioinsecticide. In: Robert DW, Granados RR (eds) Proc Biotechnology, Biological Pesticides and Novel Plant—Pest Resistance for Insect Pest Management, 19–20 July 1988. Cornell Univ, Ithaca, NY

Carson R (1962) Silent spring. Houghton-Mifflin, Boston

Casida JF, GB Quistad (1995) Metabolism and synergism of pyrethrins. In: Casida JE, Quistad GB (eds) Pyrethrum flowers: production, chemistry, toxicology, and uses. Oxford Univ Press, New York, pp 259–276

Chaieb I (2010) Saponins as Insecticide: a Review. Tunisian J Plant Protec 39(5):39–50

Chandlera D, Davidsona G, Grantb WP, Greavesb J, Tatchelle GM (2008) Microbial biopesticides for integrated crop management: an assessment of environmental and regulatory sustainability. Trends Food Sci Technol 19(5):275–283

Chatterjee H, Senapati SK (2000) Pestology 24(7):52

Chattopadhyay A, Bhatnagar N, Bhatnagar R (2004) Bacterial Insecticidal Toxins. Critical Rev Microbiol 30:33

Chaudhri S, Ramkrishnan N (1980) Field efficacy of baculovirus in combination with sublethal dose of DDT and endosulfan on cauliflower against tobacco caterpillar, *Spodoptera litura* (F.) Ind J Entomol 42:592–596

Cheng HH, GE Boucher (1972) Field comparison of the neoaplectanid nematode DD-136 with diazinon for control of *Hylemia spp.* on tobacco. J Econ Entomol 65:1761–1763

Choi JY, Kim JG, Choi YC, Goo TW, Chang JH, Je YH, Kim KY (2002) *Nosema* sp. isolated from cabbage white butterfly (*Pieris rapae*) collected in Korea. J Microbiol 40:199–204

Christie JR (1937) *Mermis subnigrescens,* a nematode parasite of grasshoppers. J Agric Res 55:353–364

Chu WH, Jaques RP (1979) Pathologie d'une microsporidiose de l'arpenteuse du chou, *Trichoplusia ni* (Lep: Noctuidae), par *Vairimorpha necatrix*. Entomophaga 24:229–235

Claudia D, Eric W (2009) Field applications of *Beauveria bassiana* to control the European cherry fruit fly *Rhagoletis cerasi*. J Appl Entomol (in press)

Clemson Extension [Clemson HGIC], Home and Garden Information Center (2007) Organic Pesticides and Biopesticides. URL: http://hgic.clemson.edu/factsheets/HGIC2756.htm (accessed 20 Aug 2008). Clemson (SC): Clemson University

Combrinck S, Du PlooyGW, McCrindle RI, BM Botha (2007) Morphology and Histochemistry of the glandular trichomes of *Lippia scaberrima* (Verbenaceae). Ann Botany 99(6):1111–1119

Cornelius ML, Duan JJ, Messing RH (2000) Volatile host fruit odors as attractants for the Oriental fruit fly (Diptera: Tephritidae) J Econ Entomol 93(1):93–100

Costa LG (1997) Basic Toxicology of Pesticides. In: Keifer MC (ed) Human health, effects of pesticides, occupational medicine. State of the art reviews, vol 12(2). Hanley and Belfus Inc, Philadelphia, pp 251–268

Craig SM, Webster JM (1974) Inhibition of molting of the desert locust, *Schistocerca gregaria* by the nematode parasite *Mermis nigriscens.* Can J Zool 52:1535–1539

Cranshaw WS (2008) *Bacillus thuringiensis*: Home and Garden, Insect series No 5.556

Cui L, Gaugler R, Wang Y (1993) Penetration of steinernematid nematodes (Nematoda: Steinernematidae) into Japanese beetle larvae, *Popillia japonica* (coleoptera: Scarabaeidae). J Inverteb Pathol 62:73–78

Cunningham RT (1989) Male Annihilation. In World Crop Pests. Robinson AS, Hooper G (eds) Elsevier, Amsterdam, Netherlands, pp 78–81

Cunningham RT, Steiner LF (1972) Field trial of cue-lure + naled on saturated fibreboard blocks for control of the melonfly by male- annihilation technique. J Econ Entomol 65:505–507

Cuperus GW, Berberet RC, Noyes RT (2004) The essential role of IPM in promoting sustainability of agricultural production systems for future generations. In: Koul O, Dhaliwlal GS, Cuperus G (eds) Integrated pest management: potential, constraints and challenges. CAB Internat, Wallingford, UK, pp 265–280

Curgy JJ, Vavra J, Vivaris C (1980) Presence of ribosomal RNAs with prokaryotic properties in *Microsporidia*, eukaryotic organism. Biol Cell 38:49–55

Czeher C, Labbo R, Arzika I, Duchemin JB (2008) Evidence of increasing Leu-Phe knockdown resistance mutation in *Anopheles gambiae* from Niger following a nationwide long-lasting insecticide-treated nets implementation. Malar J 7:189

D'Amico V (2007a) *Baculoviruses in Biological Control: A Guide to Natural Enemies in North America*. Ithaca (NY): Cornell Univ URL: http://www.nysaes.cornell.edu/ent/biocontrol/pathogens/baculoviruses.html (accessed 19 Aug 2008)

D'Amico V (2007b) Biological Control: A Guide to Natural Enemies in North America. Cornell University, Weeden CR, Shelton AM Hofman PM, http://www.nysaes.cornell.edu/ent/biocontrol/pathogens/baculoviruses.html

Da Costa CP, Jones CM (1971) Cucumber beetle resistance and mite susceptibility controlled by the bitter gene in *Cucumis sativus* L. Science 172:1145–1146

Dahiya B, Chauhan R (2001) Evaluation of insecticides and biopesticides on pod borer *Helicoverpa armigera* Hubn in Chickpea. 'Symp on biocontrol based pest management for quality plant protection in the current millennium. July 18–19 PAU, Ludhiana

17 Potential of Biopesticides in Sustainable Agriculture

David WAL (1967) Physiology of the insect integument in relation to the invasion of pathogens. In: Beament JWL, Treherne JE (eds) Insect physiology. Oliver and Boyd, Edinburgh, pp 17–35

De Vosa M, Chenga WY, Summersb HE, Ragusob RA, Jandera G (2010) Alarm pheromone habituation in *Myzus persicae* has fitness consequences and causes extensive gene expression changes. Proc Nat Acad Sci USA 107(33) 14673–14678

Deakon JW (1983) Microbial Control of Plant Pests and Diseases. Van Nostrand Reinbold (UK) Co Ltd. Workingham UK

Charleston D S, Rami KW, Dicke M, Vet LEM (2006) Impact of botanical extracts derived from *Melia azedarach* and *Azadirachta indica* on populations of *Plutella xylostella* and its natural enemies: A Wild test laboratory fndings. Biol Cont 39:105–114

Deka N, Hazarika LK (1997) Effect of neem seed oil on food consumption and utilization by rice hispa, Pesticide Res J 9(1):113–116

Dev S, Koul O (1997) Insecticides of natural origin. Harwood Acad, Amsterdam, p 365

Dhandapani N, Jyaraj S (1989) Efficacy of nuclear polyhedrosis virus for the control of *Spodoptera litura* on chillies. J Biol Cont 3(1):47–49

Di-illio V, Cristofaro M, Marchini D, Paola N, Dallai R (1999) Effects of a neem compound on the fecundity and longevity of *Ceratitis capitata* (diptera:Tephritidae). J Econ Entomol 92:76–82

Dimetry NA, Barkat AA, Abdalla EF, EI-Metwally HE, EL-Salam, AMEA (1995) Evaluation of two neem seed kernel extracts against *Liriomyza trifolii* (Burg) (Dipt. Agromyzidae). Anzeiger fur schadlingsk Pflanz Umweltschutz 68(2):39–41

Directorate of Plant Protection, Quarantine and Storage (2010). Promotion of Integrated Pest Management (IPM), Strengthening and Modernisation of pest management approach in India. In: Hall FR, Menn JJ (eds) Biopesticides: Use and Delivery. Humana, Totowa

Donald G, McNeil Jr (2005) Fungus Fatal to Mosquito May Aid Global War on Malaria, The New York Times, 10 June

Dorn A, Rademacher JM, Shen E. (1987) Effects of azadirachtin on reproductive organs and fertility in the large milkweed bug, *Oncopeltus fasciatus*. In: Schmutterer H, Ascher KRS (eds) Natural products from the neem tree and other tropical plants. Proc 3rd Internat neem Conf. GTZ, Eschborn, Germany, pp 273–288

Drozda VF (1978) Effectivenes of the use of Boverin against the pear tortricid. Zak Ros 24:319–335

Duan JJ, Head G, Jensen A, Reed G (2004) Effects of transgenic *Bacillus thuringiensis* potato and conventional insecticides for Colorado potato beetle (Coleoptera: Chrysomelidae) management on the abundance of fround-dwelling arthropods in Oregon potato ecosystems. Environ Entomol 33:275–281

Dubey NK, Shukla R, Kumar A, Singh P, Prakash B (2010) Prospect of botanical pesticides in sustainable Agriculture. Current Sci 98:479–480

Easwaramoorthy S, Jayaraj S (1989) Studies on the pathogeninicity of granulosis virus on the sugarcane shoot borer, *Chilo infuscatellus* Snellen. J Biol Cont 3(2):103–107

Easwaramoorthy S, Santalakshmi G (1988) Efficacy of granulosis virus in the control of sugarcane shoot borer, *Chilo infuscatellus* Snellen. J Biol Cont 2(1):26–28

Eidt DC, Thurston GS (1995) Physical deterrents to infection by entomopathogenic nematode in wire worm (Coleoptera: Elateridae) and other soil pests. Can Entomol 127:423–429

Elahi KM (2008) Social Forestry, Exotic Trees and Mango. The Daily Star Published 6 Sept 2008. URL: http://www.thedailystar.net/story.phpnid=53438 (accessed on 6 Sept 2008)

El-Husseini MM (2006) Microbial Control of insect pests: is it an effective and environmentally safe alternative? Arab J Plant Protec 24:162–169

El-Katatnya MH (2010) Virulence potential of some fungal isolates and their control-promise against the Egyptian cotton leaf worm, *Spodoptera littoralis*. Arch Phytopathol Plant protec 43(4):332–356

El-Sayed AM (2005) The pherobase: database of insect pheromones and semiochemicals. http://www.pherobase.com

El-Shazly MM, Refaie BM (2002) Larvicidal effect of the juvenile hormone mimic, pyriproxyfen on culex pipiens. J Am Mosquito Cont Assoc 18(4):321–328

Estruch J, Gregoury J, Warren W, Martha MA, Gardon JN, Joyce AC, Michael GK (1996) *Bacillus thuringiensis:* vegetative insecticidal protein with a wide spectrum of activities against lepidopterous insect. Proc Nat Acad Sci USA 93(11):5389–5394

Eswaran SV (2009) Why does the grasshopper not eat spinach. Resonance 14:978–984

Eto M (1990) Biochemical mechanism of insecticidal activities, In: Haug G, Hoffman H (eds) Chemistry of plant protection, vol 6. Springer, New York, pp 65–107

Fargues J, Robert PH, Vey A, Pais M (1986) Toxixite relative de la destruxine e pour le lepidoptera *Gelleria mellonella*. CR Acad Sci Paris 303:84–86

Fast PG (1975) Bi-mon. Res Notes Can For Serv 31:1–2

Fast PG (1981) The crystal toxin of *B. thuringiensis*. In: Burges HD (ed) Microbial control of pests and plant diseases, 1970–1980. Acad Press, London, pp 223–248

Fast PG, Donaghue TP (1971) The δ-endotoxin of *Bacillus thuringiensis*. II. On the mode of action. J Invertb Pathol 18:135–138

Fast PG, Videnova E (1974) The δ -endotoxon of Bacillus thuringiensis. V. On the occurrence of endotoxin fragments in hemolymph. J Invertb Pathol 23:280–284

Fernandez S, Groden E, Vandenberg JD, Furlong MJ (2001) The effect of mode of exposure to *Beauveria bassiana* on conidia acquisition and host mortality of Colorado potato beetle, *Leptinotarsa decemlineata*. J Inverteb Pathol 77:217–226

Fernando WGD, Ramarathnam R, Nakkeeran S (2009) Advances in Crop Protection Practices for the Environmental Sustainability of Cropping Systems. In: Peshin R, Dhawan AK (eds) Integrated Pest Management, pp 131–162

Ferron P (1975) Les champignons entomopathogenes: Evolution des recherches au cours des dix dernieres annees. Bull Sci Biol Organ Int Lutte Biol WPRS Bul

Ferron P, Vincent JJ, Dickler E (1978) Joint FAO-IAEA and IDBC-WPRS Research coordination Meeting on "The use of integrated control and the sterile insect technique for control of the moth". Hiedelberg, Nov 7–10, 1977, pp 84–87

French-Constant RH, Waterfield N, Burland V, Perna NT, Daborn PJ, Bowen D, Blattner FR (2000) A genomic sample sequence of the entomopathogenic bacterium, *Photorhabdus luminscens* w14: potential implication for virulence Appl Environ Microbiol 66:3310–3329

Filipjev I N, Schuurmans Stekhoven JH Jr (1941) A manual of agricultural helminthology E. J. Brill, Leiden, Holland

Finlay BB, Falkow S (1989) Common themes in microbial pathogenecity. Microbiol Rev 53:210–230

Fischer E (1906) Uber die Ursachen der Disposition und uber Fruhymptome der Raupenkarnkheiten. Biol Zentralbl 26:534–544

Fisher FM Jr, Sanborn RC (1964) *Nosema* as a source of juvenile hormone in parasitized insects Biol Bull 126:235–252

Fletcher BS, Prokopy RJ (1991) Host location oviposition in tephritid fruit flies. In: Reproductive behaviour of insects: individuals and populations. Chapman and Hall, London, pp 139–171

Forschler BT, Gardner WA (1991) Parasitism of *Phyllophaga hirticula* (Coleoptera: Scarabaeidae) by *Heterorhabditis heliothidis* and *Steinernema carpocapsae*. J Inverteb Pathol 58:396–407

Foschi S, Grassi S (1985) Results of treatment with *Beauvria bassiana* (Balsam.) Vull and with *Metarrhizium anisopliae* (Metsch.) Sorok on *Ostronia nubilalis* Hb Dif Piante 8:301–308

Frazier JL (1986) Perception of allelochemicals that inhibit feeding. In: Brattsten LB, Ahmad S (eds) Molecular aspects of insect-plant associations. Plenum Press, New York, pp 1–42

Freisewinkel DC, Schumutterer H (1991) Contact action of neem oil on the African migratory locust, *Locusta migratoria migratoroides*. Zeitschrif Angewandta Zool 78(2):189–203

Fujiwara-Tsujii N, Yamagata N, Takeda T, Mizunami M, Yamaoka R (2006) Behavioral responses to the alarm pheromone of the ant, *Camponotus obscuripes* (Hymenoptera: Formicidae). Zool Sci 23(4):353–358

Gang W, Harris MK, Guo J Y et al (2009) Temporal allocation of metabolic tolerance in the body of beet armyworm in response to three gossypol-cotton cultivars. Sci China Ser C-Life Sci 52(12):1140–1147

17 Potential of Biopesticides in Sustainable Agriculture 581

Gaugler R, Molloy D (1981) Instar susceptibility of *Simulium vittatum* (Diptera: Simuliidae) to the entomogenous nematode, *Neoplectana carpocapsae*. J Nematol 13:1–5

Gaugler R, Wang Y, Campbell JF (1994) Aggressive and evasive beahaviours in *Popillia japonica* (Coleoptera: Scarabeidae) larvae: defences against entomopathogenic nematode attack. J Inverteb Pathol 64:193–199

Geissman TA, Crout DHG (1969) Organic chemistry of secondary plant metabolism. Freeman Cooper and Co., San Francisco

Georgis R, Hague NGM (1981) A neoaplactenid nematode in larch sawfly *Cephalcia lariciphila* (Hymenoptera: Pamphiliidae). Ann Appl Biol 99:171–177

Georgis R, Poinar, Jr. GO, Wilson AP (1983) Practical control of the cabbage root maggot, *Hylemia brassicae* (Diptera: Anthomyiidae) by entomogenous nematodes. IRCS Medl Sci: Microbiol and Parasitol Infect Dis 11:322

Glaser RW (1915) Wilt of gypsy moth caterpillars. J Agric Res 4:101–128

Glaser RW (1918) The polyhedral virus of insects with a theoretical consideration of filterable viruses generally. Science 48:301–302

Glaser RW, Chapman JW (1916) The nature of the polyhedral bodies found in insects. Biol Bull 30:367–391

Glaser RW, Farrell CC (1935) Field experiments with the Japanese beetle and its nematode parasite. J N Y Entomol Soc 43:345–371

Gokce A, Whalon ME, Cam H, Yanar Y, Demirtas I, Goren N (2006) Plant extract contact toxicities to various developmental stages of Colorado potato beetles (Coleoptera: Chrysomelidae). Ann Appl Biol 149:197–202

Goldansaz SH, Dewhirst S, Birkett MA, Hooper AM, Smiley DWM, Pickett JA, Wadhams L, McNeil JN (2004) Identification of two sex pheromone components of the potato aphid, *Macrosiphum euphorbiae* (Thomas). J Chem Ecol 30:819–834

Gonzales JM, Jr Brown BJ, Carlton BC (1982) Transfer of *Bacillus thuringiensis* plasmids coding for d-endotoxin among strains of *B. thuringiensis* and *B. cereus*. Proc Nat Acad Sci USA 79:6951–6955

Gordon R, Webster J M, Hislop TG (1973) Mermithid parasitism, protein turnover and vitellogenesis in the desert locust, *Schistocera gregaria* Forskal. Comp Biochem Physiol 46:575–593

Gordon R, Webster JM (1971) *Mermis nigrescens:* Physiological relationship with its host, the adult desert locust *Schistocerca gregaria*. Exp Parasitol 33:66–79

Gould F (1988) Ecological genetic approaches for the design of genetically engineered crops. In: Robert DW, Granados RR (eds) Proc conf: biotechnology, biological pesticides and novel plant-pest resistance for insect pest management. Cornell Univ Ithaca, NY, pp 19–20

Gould FL, Groot AT, Vasqnez GM, Schal C (2009) Sexual communication in Lepidoptera: a need for wedding genetics, biochemistry and molecular biology. CRC Press, Boca Raton

Grafton-Cardwell EE, Leea JE, Stewartb JR, Olsenc KD (2006) Role of Two Insect Growth Regulators in Integrated Pest Management of Citrus Scales. J Econ Entomol 99(3):733–744

Granados RR, Federici BA (1986) "The Biology of Baculoviruses," vols I and II. CRC Press, Boca Raton, FL

Granados RR, Williams KA (1986) *in vivo* infection and replication of baculoviruses. In: Granados RR, Federici BA (eds) The biology of baculoviruses. CRC press, Boca Raton, Fl, pp 89–108

Grasser PP (1953) Sous embranchment des sporozoaries (Sporozoa Leuckart, 1879; Rhabdogeniae Delarge et Herouard, 1986; Telosporidia Schaudinn, 1900). Trait Zool. 1 Fasc. 2:545–690

Greenberg SM, Showler AT, Liu T (2005) Effects of neem-based insecticides on beet armyworm (Lepidoptera: Noctuidae). Insect Sci 12:17–23

Grewal P, Ehlers RU, Shapiro-Ilan DI (eds) (2005) Nematodes as Biocontrol Agents. CABI Publishing, Wallingford, UK

Grewal PS, Nardo ED, Aguillera MM (2001) Entomopathogenic nematodes: potential for exploration and use in South America. Neotrop Entomol 30(2):191–205

Grewal PS, Power KT, Grewal SK, Suggars A, Haupricht S (2004) Enhanced consistency in biological control of white grubs (Scarabaeidae) with new strains of entomopathogenic nematodes. Biol Cont 30:73–82

Groner A (1986) Specificity and safety of baculoviruses. In: Granados RR, Federici BA (eds) The biology of baculoviruses. CRC press, Boca Raton, pp 177–202

Gronning EK, Borchert DM, Pfeiffer DG, Felland CM, Walgenbach JF, Hull LA, Killian JC (2000) Effect of specific and generic pheromone blends on captures in pheromone traps by four leafroller species in mid-Atlantic apple orchards. J Econ Entomol 93:157–164

Gujar GT (1997) Biological effects of azadirachtin and plumbagin on *Helicoverpa armigera*. Ind J Entomol 59(4):415–422

Gupta PK (2004) Pesticides exposure-Indian scene. Toxicology 198:83–90

Hajek A (2004) Natural enemies: an introduction to biological control. Camb Univ Press, Camb, UK

Hamilton DW, Schwartz PH, Townshend BG, Jester CW (1971) Traps reduce an isolated infestation of Japanese beetle. J Econ Entomol 64:150

Harborne JB (1988) Introduction to ecological biochemistry, 3rd ed. Acad Press, London

Hardy DE (1979) Review of economic fruit flies of the South Pacific region. Pac Insects 20:429–432

Harris RS, Harcourt SJ, Glare TR, Rose EA, Nelson TJ (2000) Susceptibility of *Vespula vulgaris* (Hymenoptera: Vespidae) to generalist entomopathogenic fungi and their potential for wasp control. J Inverteb Pathol 75:251–258

Haubruge E (2008) Production of alarm pheromone in aphids and perception by ants and natural enemies. Essai présenté en vue de l'obtention du grade de docteur en sciences agronomiques et ingénierie biologique, François Verheggen, http://www.fsagx.ac.be/zg/personal/cv/these-verheggen.pdf

Hazard EI, Lofgren CS (1971) Tissue specificity and systematics of a Nosema in some species of *Aedes, Anopheles*, and *Culex*. J Inverteb Pathol 18:16–24

He F, Pan QH, Shi Y, Duan CQ (2008) Biosynthesis and Genetic Regulation of Proanthocyanidines in Plants. Molecules 13:2674–2703

Hedin PA, Hollingworth RM, Masler EP, Miyamoto J, Thompson DG (eds) (1997) Phytochemicals for pest control.Am Chem Soc, Washington, DC, p 372

Heftmann E (1973) Steroids. In: Miller LP (ed) Phytochemistry; organic metabolites. Van Nostrand Rienhold Co., New York, pp 171–226

Heimpel AM (1967) A critical review of Bacillus thuringiensis var. thuringiensis Berliner and other crystalliferous bacteria. Annu Rev Entomol 12:287–322

Heimpel AM, Angus TA (1959) The site of action of crystalliferous bacteria in lepidoptera larvae. J Insect Pathol 1:152–170

Hentschel U, Steinert M, Hacker J (2000) Common molecular mechanism of symbiosis and pathogenesis. Trends Microbiol News 8:226–231

Hokkanen HT, Bigler F, Burgio G, Van Lenteren JC, Thomas MB (2003) Ecological risk assessment framework for biological control agents. In: Hokkanen HMT, Hajek AE (eds) Environmental impacts of microbial insecticides: need and methods of risk assessment. Kluwer Acad Pub, The Netherlands

Holldobler B, Traniello J (1980) Tandem running pheromone in ponerine ants. Naturwissenschaften 67:360

Hölldobler B, Wilson EO (1990) The Ants, Harvard University Press ISBN 0-674-04075-9

Howlett FM (1915) Chemical reaction of fruit flies. Bull Entomol Res 6:297–305

Hudson KE (1974) Regulation of greenhouse sciarid fly populations using *Tetradonema plicans* (Nematoda: Mermithoidae). J Invertb Pathol 23:85–91

Huger A (1963) Granuloses of insect. In: Steinhaus EA (ed) Insect pathology: an advanced treatise. Acad Press, New York, pp 531–575

Ishaaya I, Horowitz AR (1999) Insecticides with novel modes of action: an overview. In: Ishaaya I, Degheele D (eds) Insecticides with novel modes of action. Springer, Berlin, pp 1–24

Ishaaya I, Birk Y, Bondi A, Tencer Y (1969) Soybean saponins IX. Studies of their effect on birds, mammals and coldblooded organisms. J Sci Food Agricul 20:433–436

Ishihara R, Hayashi Y (1968) Some properties of ribosomes from the sporoplasm of Nosema bombycis. J Invertebr Pathol 11:377–385

17 Potential of Biopesticides in Sustainable Agriculture 583

Ishikawa K (1936) Pathology of the Silkworm. Meibundo, Tokyo, p 512

Isman MB (2000) Plant essential oils for pest and disease management. Crop Protect 19:603–608

Isman MB (2006) Botanical insecticides, deterrents, and repellents in modern agriculture and an increasingly regulated world. Ann Rev Entomol 51:45–66

Isman MB, Duffy SS (1982) Toxicity of tomato phenolic compounds to the fruit worm, *Heliothis zea*. Entomol Exptl Appl 31:217–221

Isman MB, Rodriguez E (1983) larval growth inhibitors from species of *Parthenium* (Asteraceae). Phytochem 22:2709–2713

Ito T, Teruya T, Sakiyama M (1976) Attractiveness of bber- blocks containing mixture of methyl eugenol and cue-lure on *Dacus dorsalis* and *Dacus cucurbitae*. Bull Okinawa Agric Exp Stn 2:39–43

Jacobson M (1983) Phytochemicals for the control of stored product insects. In: Proc. 3rd Intnat Wkg Conf Stored Prod Entomol, Kansas State Univ, Manhattan, Kansas, USA, pp 182–195

Jacobson M (1989) Botanical pesticides: past, present and future. In: Arnason JT, Philogene BJ, Morland P (eds) Insecticides of plant origin, pp 1–10

Jacques RP, Morris ON (1981) Compatibility of pathogens with other methods of pest control and crop protection. In: Burges HD (ed) Microbial control of pests and plant diseases. Acad Press, London UK

Jagajothi A, Martin P (2010) Efficacy of andrographolide on pupal-adult transformation of *Corcyra cephalonica* Stainton. J Biopest 3(2):508–510

Jayaraj S, Santaram G, Narayanan K, Sundarrajan K, Balagurunathan R (1981) Effectiveness of nuclear polyhedral virus against field populations of tobacco caterpillar, *Spodoptera litura* on cotton. Andhra Agric J 27:26–29

Jayaraj S, Sathiah N, Sundarababu PC (1992) Biopesticides research in India: the present and future status. In: David BV (ed) Pest management and pesticides: Indian scenario. Namruth Pub, Madras, India, pp 144–157

Jayaraj S, Sundramurthy VT, Mahadevan NR, Swamyappan M (1977) Relative efficacy of nuclear Polyhedrosis virus and *Bacillus thuringiensis* Berliner in the control of groundnut red hairy caterpillar, *Amsacta albistriga* (Walker.) Madras Agric J 64:130–131

Jayraj S (1981) Control of gram catterpiller on lablab and Bengal gram with Nuclear Polyhedrosis Virus. Tamil Nadu Agric Univ Coimbatore News letter 11:1

Jefford MR, Maddox JV, O'Hayer KW (1987) Microsporidian spores in gypsy moth larval silk: A possible route of horizontal transmission. J Inverteb Pathol 49:332–333

Johny S, Kanginakudru S, Muralirangan MC, Nagaraju J (2006) Morphological and molecular characterization of a new microsporidian (Protozoa: Microsporidia) isolated from *Spodoptera litura* (Fabricius) (Lepidoptera: Noctuidae). Parasitol 1–12

Jurenka R (2004) Insect pheromone: biosynthesis, chemistry of pheromones and other semiochemical, pp 97–131

Justum AR, Gordon RFS (1989) Insect pheromones in plant protection. Wiley, Chichester, UK

Karlson P, Butenandt (1959) Pheromones (ectohormones) in Insects. Ann Rev Entomol 4:39–58

Karlson P, Luscher M (1959) Pheromones: a new term for a class of biologically active substances. Nature 183:55–56

Kaya HK (1978a) Interaction between *Neoplectana carpocapsae* (Nematoda: Steinernematidae) and *Apentalis militarisi* (Hymenoptera: Braconidae), a parasitoid of army worm, *Pseudaletia unipuncta*. J Inverteb Pathol 31:358–364

Kaya HK (1978b) Infectivity of *Neoplectana carpocapsae* and *Hetetrorhabditis heliothidis* to pupae of the parasite *Apentalis militaris*. J Nematol 19:241–244

Kaya HK, Hotchkin PG (1981) The nematode, *Neoplectana carpocapsae* Weiser and its effect on selected ichneumonid and barconid parasites. Environ Entomol 10:474–478

Keeling PJ (2003) Congruent evidence from a α- tubulin and β- tubulin gene phylogenies for a zygomycete origin of microsporidia. Fungal Genetics Biol 38:298–309

Kellen WR, Lindegren JE (1971) Modes of transmission of *Nosema plodiae* Kellen and Lindegren, a pathogen of *Plodia interpunctella* (Hubner). J Stored Prod Res 7:31–34

584 M. S. Ansari et al.

Keller S, Kessler P, Schweizer C (2003) Distribution of insect pathogenic soil fungi in Switzerland with special reference to *Beauveria brongniartii* and *Metarhizium anisopliae*. Bio Cont 48:307–319

Kerwin JL, Washino RK (1986a) Oosporogenesis by *Lagenidium giganteum*: induction and maturation are regulated by calcium and calmodulin. Can J Microbiol 32(8):663–672

Kerwin JL, Washino RK (1986c) Regulation of oosporogenesis by *Lagenidium giganteum*; promotion of sexual reproduction by unsaturated fatty acids and sterol availability. Can J Microbiol 32:294–300

Khalequzzaman M, Islam MN (1992) Pesticidal action of Dhutura, *Datura metel* Linn. Leaf extracts on *Tribolium castaneum* (Herbst). Bangladesh J Zool 20(2):223–229

Khan A, Brooks WM, Hirschman H (1976) *Chromonema helioyhidis* (Steinernematidae, Nematoda), a parasite of *Heliothis zea* (Noctuidae, Lepidoptera), and other insects. J Nematol 8:159–168

Khrimian A, Jang EB, Nagata J, Carvalho L (2006) Consumption and Metabolism of 1,2-Dimethoxy-4-(3-Fluoro-2-Propenyl)Benzene, a Fluorine Analog of Methyl Eugenol, in the Oriental Fruit Fly, *Bactrocera dorsalis* (Hendel). J Chem Ecol 32:1513–1526

Klaassen CD, Amdur MO, Doull J (eds) (1996) Casarett and Doull's toxicology. The basic science of poisons, 5th edn McGraw-Hill Co, Inc Toronto

Klein MG (1981) Mass trapping for suppression of Japanese beetles. In: Mitchell EP (ed) Management of insect pests with semiochemicals. Plenum, NY, pp 183–190

Klein MG (1993) Biological control of scarabs with entomopathogenic nematodes. In: Bedding R, Akhurst R, Kaya H (eds) Nematodes and the biological control of insect pests. CSIRO Press, East Melbourne, Aust, pp 49–58

Klein MG, Tumlinson JH, Ladd TL, Jr Doolittle (1981) Japanese beetle (Coleoptera: Scarabaeidae): response to synthetic sex attractant plus phenethyl propionate eugenol. J Chem Ecol 7:1–7

Klowden MJ (ed) (2007) Physiological Systems in Insects. 2nd ed. Acad Press, London

Knight A, Hilton R, Vanizuskrik P, Light D (2006) Using pear ester to monitor codling moth in sex pheromone treated orchards. Oregon state univ ext serv, p 8

Kogan M (1998) Integrated Pest Management: Historical Perspectives and Contemporary Developments. Ann Rev Entomol 43:243–270

Kostyukovsky M, Trostanetsky A (2006) The effect of a new chitin synthesis inhibitor, Novaluron, on various developmental stages of *Tribolium castaneum*. J Stored Prod Res 42(2):136–148

Koul O (1998) Effect of neem extract and *azadirachtin* on fertility and fecundity of cabbage aphid, *Brevicoryne brassicae* (L). Pestic Res J 10:258–261

Koul O, Dhaliwal GS (2001) Phytochemical biopesticides. Harwood Acad, Amsterdam, p 223

Koul O, Dhaliwal GS (2002) Microbial biopesticides. Taylor and Francis, London, UK

Kramer JP (1976) The extra corporial ecology of microsporidia. In: Bulla LA Jr, Cheng TC (eds) Comparative pathobiology, vol 1. Plenum Press, New York, pp 127–135

Krishnaiah K, Ramakrishnan N, Reddy PC (1984) Further trials on control of *Spodoptera litura* (Fab.) by nuclear polyhedrosis virus on black gram and groundnut. Ind J plant protect 12:81–83

Krupkea CH, Roitbergb BD, Juddc GJR (2002) Field and laboratory responses of male codling moth (Lepidoptera: Tortricidae) to a pheromone-based attract-and-kill ktrategy. Environ Entomol 31(2):189–197

Krywienczyk J, Dulmag HT, Fast PG (1978) Occurrence of two serologically distinct groups within Bacillus thuringiensis serotype 3ab variety kurstaki. J Inverteb Pathol 37:372–375

Kulshrestra RS, Gurusahni KA (1961) An observation on entomogenous fungus, *Metarhizum anisopliae* (Metsch.) Sor. Ind J Sugarcane Res Dev 5:163–164

Kumar S (2007) Insecticidal genes and their sustainable use in insect resistant transgenic crops. In: Plant Tissue culture and molecular markers: their role in improving crop productivity. IK Internat Pvt Ltd, New Delhi, pp 201–228

Kumar SP, Babu PCS (1998) Ovipositional deterrent and ovicidal effects of neem azal against brinjal shoot and fruit borer and leaf beetle. Shaspa 5(2):193–196

17 Potential of Biopesticides in Sustainable Agriculture 585

Kumari B, Madan VK, Singh J, Singh S, Kathpal TS (2004) Monitoring of pesticidal contamination of farmgate vegetables from Hisar. Environ Monit Assess 90:65–71

Lacey ES, Millar JG, Moreira JA, Hanks LM (2009) Male-Produced Aggregation Pheromones of the Cerambycid Beetles *Xylotrechus colonus* and *Sarosesthes fulminans*. J Chem Ecol 35:733–740

Ladd Tl, Jacobson M Jr, Buriff CR (1978) Japanease beetle: Extracts from neem tree seeds as feeding detterents. J Econ Entomol 71:810

Lakhidov AN (1979) Biopreprations for the control of Colorado beetle. Ilzch Rast 11:24

Lecadet MM (1970) In Microbial Toxin. In: Monite TC, Kadis S, Aul SJ (eds) *Bacillus thuringiensis* toxins the proteinaceous crystal. Acad Press, London and New York, pp 437–471

Leellen FS, Daniela K Pilarska, Michael L. McManus , Milan Zúbrik, Jan Patočka, Wei-Fone Huang, Julius Novotny (2010) Host specificity of microsporidia pathogenic to the gypsy moth, *Lymantria dispar* (L.): Field studies in Slovakia. J Inverteb Pathol (In press)

Leuschner K (1972) Effect of an unknown plant substance on a shield bug. Naturuwissenschaften 59:217–218

Levine LD (1982) Apicomplexa. In: Parker SP (ed) Synopsis and classification of living organism. McGraw Hill, New York, pp 571–587

Levine ND (1971) Uniform terminology for the protozoan subphylum Apicomplexa. J Protozool 18:352–355

Lewis EE, Campbell JF, Gaugler R (1998) A Conservation approach to using entomopathogenic nematodes in turf and landscapes. In: Conservation biological control. Acad Press, London, pp 235–254

Lewis W, van Lenteren C, Phatak S, Tumlinsen J (1997) A Total System Approach to Sustainable Pest Management. Proc Nat Acad Sci USA 94:12243–12248

Lindquist RK (1994) Integrated management of fungus gnats and shore flies. In: Robb K (ed) Proc. 10th conf. Insect and Disease management on Ornamentals. Soc Am florists, Alexandria verginia, pp 58–67

Liu TX, Liu SS (2006) Experience-altered oviposition responses to a neem-based product, Neemix, by the diamondback moth, *Plutella xylostella*. Pest Manag Sci 62:38–45

Liu YC, Lin JS (1993) The response of melon by *Dacus cucurbitae* Coquillett to the attraction of 10%MC. Plant Protec Bull 35:79–88

Liu ZL, Ho SH (1999) Bioactivity of the essential oil extracted from *Evodia rutaecarpa* Hook f. et Thomas against the grain storage insects, *Sitophilus zeamais* (Motsch) and *Tribolium castaneum* (Herbst). J Stored Prod Res 35:317–328

Lopez O, Fernandez-Bolanos JG, Gil MV (2005) New trends in pest control: the search for greener insecticides. Green Chem 7:431–442

Louloudes SJ, Heimpel AM (1969) Mode of action of *Bacillus thuringiensis* toxic crystals in larvae of silkworm, *Bombyx mori*. J Inverteb Path 14:375–380

Lowry DT, Isman MB (1996) Inhibition of aphid (Homoptera: Aphididae) reproduction by neem seed oil and azadirachtin. J Econ Entomol 89(3):602–607

Lykouressis D, Perdikis D, Michalis C, Fantinou A (2004) Mating disruption of the pink bollworm *Pectinophora gossypiella* (Saund.) (Lepidoptera: Gelechiidae) using gossyplure PB-rope dispensers in cotton fields. J Pest Sci 77:205–210

Maestri A (1856) Frammenti anatomici, fisiologici e patologici sul baco da seta (*Bombyx mori* Linn.). Fratelli fusi, Pavia

Mahadevan NR, Kumaraswamy T (1980) A note on the effectiveness of Nuclear polyhedrosis virus against *Spodoptera litura* (F.) on black gram, *Phaseolus mungo* (L.) Madras Agric J 67:138–140

Mahalingum CA, Mohammed Haneef A (2001) Effect of certain adjuvents along with HaNPV for the management of Tomato fruit borer. *Helicoverpa armigera* Hubn. Symp biocontrol based pest management for quality plant protec in the current millennium. PAU, Ludhiana, July 18–19

Makanjuola WA (1989) Evaluation of extracts of neem (*Azadirachta indica* A. Juss) for the control of some stored product pests. J stored Prod Res 25(4):231–237

Mamiya Y (1984) The pine wood nematode. In: Nickle WR (ed) LANS and insect nematodes. Marcel Dekker, New York, pp 589–626

Mamiya Y, Enda N (1972) Transmission of *Bursaphelenchus lignicolus* (Nematoda: Aphlenchoididae) by *Monochamus altenatus* (coleopteran: Cerambycidae). Nematologica 18:159–162

Mandula B (2008) Personal communication. Washington DC: USA EPA Office of Pesticide Programs, Biopesticide and Pollution Prevention Division. Retired Scientist

Mannion CM, McLane W, Klein MG, Moyseenko J, Oliver JB, Cowan D (2001) Management of early-instar Japanese beetle (Coleoptera: Searabaeidae) in field-grown nursery crops. J Econ Entomol 94:1151–1161

Marco Lodesani, Cecilia Costa (2008) Maximizing the efficacy of a thymol based product against the mite *Varroa destructor* by increasing the air space in the hive. J Apicul Res 47 (2):113–117

Margulis L, Corlis JO, Melkonian M, Chapman DJ (1990) Handbook of protoctista. Jones and Bartlett, Boston

Martignoni ME (1984) Baculovirus: An attractive biological alternative. In: Garner WY, Harvey J (eds) Chemical and biological control in forestry. ACS Symp Ser No 238, Seattle, Washington, pp 55–67

Martignoni ME, Iwai PJ (1986) A catalogue of viral disease of insects, mites, and ticks. US Dep Agric Forest Serv Pac NorthWest Res Stn Gen Tech. Rep PNW-195

Martin PAW, Travers RS (1989) World wide abundance and distribution of *Bt* isolates. Appl Environ Microbiol 55:2437–2442

Mattes O (1927) parasitare Krankheiten der Mehlomottenlarven und Versuche uber ihre Verwendbarkeit als biologisches Bekampfungsmittel (Zugleich ein Beitrag zur Zytologie der Bakterien) Ges Beford Gesamte naturwiss. Marburg 62:381–417

Mayer MS, McLaughlin JR (1990) Handbook of insect pheromones and sex attractants. CRC Press, Boca Raton, FL

McCoy CW, Samson RA, Boucias DG (1988) Entomogenous fungi. In: Ignoffo CM (ed) Handbook of natural pesticides. Microbial Insecticides, Part A. Entomogenous Protozoa and Fungi', pp 151–236

McGhaughey WH (1990) Insect resistance to *Bacillus thuringiensis*. In: Baker R, Dunn PE (eds). New directios in biological control. Liss, New York

Meena R, Su hag P, Prates HT (2006) Evaluation of ethanolic extract of *Baccharis genistelloides* against stored grain pests. J Stored Prod Res 4(4):243–249

Meinwald J, Meinwald YC, Wheeler JW, Eisner T, Brower LP (1966) Major Components in the Exocrine Secretion of a Male Butterfly (Lycorea). Science 151:583–585

Mensink BJWG, Sheepmaker JWA (2007) How to evaluate the environmental safety of microbial protection products: a proposal. Biocont Sci Technol 17:3–20

Messias CL, Roberts DW, Grefig AT (1983) Pyrolysis Gas Chromatography of the fungus *Metarhizium anisopliae*: an aid to strain identification. J Inverteb Pathol 42:393–396

Metcalf RL (1990) Chemical ecology of Dacinae fruit flies (Diptera: Tephritidae). Ann Entomol Soc Am 83:1017–1030

Metcalf RL, Metcalf ER (1992) Fruit fies of the family Tephritidae, In: Metcalf RL, Metcalf ER (eds). Plant kairomones in insect ecology and control. Routledge, Chapman and Hall, London, pp 109–152

Metchinkoff E (1879) Diseases of the larva of the grain weevil. Insects harmful to agriculture [series]; Issue III, the grain weevil. Published by the Commission attached to the Odessa Zemstvo Office for the investigation of the problem of insects harmful to agriculture. Odessa, p 32

Milner RJ (1972) *Nosema whitei,* a microsporidian pathogen of some species of *Tribolium.* III. Effect on *T. castanium.* J Inverteb Pathol 19:248–255

Miranpuri GS, Khachatourians GG (1993) Application of entomopathogenic fungus, *Beauveria bassiana* against green peach aphid, *Myzus persicae* (Sulzer) infesting canola. J Insect Sci 8:287–289

Mishra BK, Mishra PR, Pathak NC (1990) Effect of neem oil on the growth and development of Epilachna beetle, *Henosepilachna sparsa* (Hbst.). Orissa J Agric Res 2(3–4):169–172

17 Potential of Biopesticides in Sustainable Agriculture 587

Mishra NC, Mishra SN (2002) Impact of biopesticides on insect pests and defenders of okra. Ind J Plant Protec 30:99–101

Miyamoto J, Hirano M, Takimoto Y, Hatakoshi M (1993) Insect growth regulators for pest control, with emphasis on juvenile hormone analogs: Present status and future prospects. ACS Symp. Ser ACS, Washington, DC 524:144–168

Mohan KS, Jaipal SP, Pillai GB (1983) Baculovirus disease in *Oryctes rhinoceros* population in Kerela. J Plant Crops 55:154–161

Mohan M, Sushil SN, Selvakumar G, Bhatt JC, Gujar GT, Gupta HS (2009) Differential toxicity of *Bacillus thuringiensis* strains and their crystal toxins against high-altitude Himalayan populations of diamondbackmoth, *Plutella xylostella* L. Pest Manag Sci 65:27–33

Mordue-Luntz AJ, Blackwell A (1993) Azadirachtin: An Update. J Insect Physiol 39:903–924

Moureaux N, Karjanainen T, Givauden A, Borleoux P, Boemare N (1995) Biochemical characterization and agglutinating properties of of *Xenorhabdus nematophilus* F1 fimbrae. Appl Environ Microbiol 61:2707–2712

Murray RDH, Mendez J, Brown SA (1982) The natural coumarins. Wiley, Chechester, UK

Narayanan K (1979) Studies on the Nuclear Polyhedrosis Virus of Gram pod Borer, *Helionthis armigera* (Hubner) (Noctuiidae: Lepidoptera) Ph.D. thesis Tamil Nadu Agric Univ Coimbatore, India

Narayanan K (1987) Field efficacy of nuclear polyhedrosis virus of *Adisura atkinsini* Moore on field beans. J Biol Cont 1(1):41–47

Neelgund YF, Mathad SB (1974) Susceptibility of larvae of the armyworm, *Pseudaletia separata* walker (Lepedoptera: Noctuiidae) to various doses of its nuclear polyhdrosis virus. Ind J Exp Biol 12:179–181

O'Brien KP, Franjevic S, Jones J (2009) Green Chemistry and Sustainable Agriculture: The role of Biopesticides. Advancing green chemistry September 2009

Ono M, Terabe H, Hori H, Sasaki M (2003) Insect signalling: components of giant hornet alarm pheromone. Nature 424:637–638

Paillot A (1926) Su rune nouvelle maladied du noyau ou grasserie des chanilles de Pieris brassicae et un nouveau groupe de micro-organismes parasites. CR Acad Sci Paris Ser 182:180–182

Paillot A (1934) Un nouveau type de maladie a ultravirus chez les insectes. CR Acad Sci Paris Ser 198:204–205

Paillot A (1935) Nouvel ultravirus parasite d'Agrotis segetum provoquant une proliferation des tissue infectes. CR Acad Aci Paris Ser 201:1062–1064

Paillot A (1937) Nouveau type de pseudo-grasserie observe chez les chenilles d'Euxoa segetum. C R Acad Aci Paris Ser 205:1264–1266

Parkman JP, Smart Jr GC (1996) Entomopathogenic nematodes, a case study: Introduction of *S. scapterisci* in Florida. Biocont Sci Tech 6:413–419

Pawar AD, Singh B (1993) Prospects of botanicals and biopesticides. In: Parmar BS, Devakumar C (eds) Botanicals and biopesticides. SPS Publishing No.4, Soci Pesticides Sci, India and Westvill Pub House, New Delhi, pp 188–196

Pawar VM, Borikar PS (2005) Microbial options for the management of *Helicoverpa armigera* (Hubner). In: Hem Saxena AB, Rai R, Ahmad, Sanjeev Gupta (eds) Recent advances in *Helicoverpa* Management. Ind Soc Pulses Res Devel, IIPR, Kanpur, pp 193–231

Pearson DE, Callaway RM (2005) Indirect nontarget effects of host-specific biological control agents: implications for biological control. Biol Cont 35:288–298

Perry AS, Yamamoto I, Ishaaya I, Perry RY (1998) Insecticides in agriculture and environment: Retrospects and prospects.Springer, Berlin, p 261

Peters A, Ehlers RU (1994) Susceptibility of leatherjackets (*Tipula paludosa* and *Tipula oleracea; Tipulidae;* Nematocera) to the entomopathogenic nematode *steinernema feltiae.* J Inverteb Pathol 63:163–171

Petersen JJ (1975) Penetration and development of the mermithid nematode, *Reesimermis neilseni* in eighteen species of mosquitoes. J Nematol 7:207–210

588 M. S. Ansari et al.

Petersen JJ, Chapman HC, Woodard DB (1967) Prelimnery observations on the incidence and biology of mermithid nematode of *Aedes sollicitans* (Walker) in Lousiana, Mosq News 27:493–498

Petersen JJ, Wills OR (1970) Some factors affecting parasitism by mermithid nematodes in Southern house mosquito larvae. Mosq News 32:226–230

Pfeiffer DG, Kaakeh W, Killian JC, Lachance MW, Kirsch P (1993a) Mating disruption to control damage by leafrollers in Virginia apple orchards. Entomol Exptl Appl 67:47–56

Pfeiffer DG, Kaakeh W, Killian JC, Lachance MW, Kirsch P (1993b) Mating disruption for control of damage by codling moth in Virginia apple orchards. Entomol Exptl Appl 67:57–64

Phelps RJ, DeFoliart GR (1964) Nematode parasitism of Simullidae. Univ Wisconsin Res Bull 245

Phillips PA (1981) California red scale and the pheromone trap. Calif Citrogr 68(12):284–286

Phillips TW, Throne JE (2010) Biorational approaches to managing stored-product insects. Ann Rev Entomol 55:375–397

Phoofolo MW, Obrycki JJ, Lewis LC (2001) Quantitative assessment of biotic mortality factors of the European corn borer (Lepidoptera: Crambidae) in field corn. J Econ Entomol 94:617–622

Pimentel D (1992) Environment and human cost of pesticides use. Bioscience 42:740–760

Pimentel D, Zuniga R, Morrison D (2005) Update on the environmental and economic costs associated with alien-invasive species in the United States. Ecol Econ 52:273–288

Pinero JC, Jácome I, Vargas R, Prokopy RJ (2006) Response of female melon fly, *Bactrocera cucurbitae* to host-associated visual and olfactory stimuli. Entomol Exptl Appl 121:261–269

Poinar GO Jr (1979) Nematodes for biological control of insects. CRC Press, Boca Raton, Florida

Poinar GO Jr, Doncaster CC (1965) The penetration of *Typius sciarae* (Bovein) (Spherulariidae: Aphelenchoidea) into its insect host, *Bradysia paupera* Tuom. (Mycetophilidae: Diptera). Nematologica 11:73–78

Poinar GO Jr, Gyrisco GG (1962) Studies on the bionomics of *Hexamermis arvalis* Poinar and Gyrisco, a mermithid parasite of alfalfa weevil, *Hypera postica* (Gyllenhal). J Insect Pathol 4:469–483

Posada F, Vega FE, Rehner SA (2004) *Syspastospora parasitica*, a mycoparasite of the fungus *Beauveria bassiana* attacking the Colorado potato beetle *Leptinotarsa decemlineata*: a tritrophic association. J Insect Sci 4:24

Prell H (1926) Die Polyderankheiten der inseckten. In: Jordan K, Horn W (eds) In Verl. III. Int. Entomol. Kongr. Zurich 1925, vol 2. Weimar, Australia, pp 145–168

Prokopy RJ, Miller NW, Pinero JC, Barry JD, Tran LC et al (2003) Effectiveness of GF-120 fruit fly bait spray applied to border area plants for control of Melon flies (Diptera: Tephritidae). J Econ Entomol 96:1485–1493

Prokopy RJ, Powers PJ (1995) Influence of neem seed kernel extract on oviposition and mortality of *Conotrachelus nenuphear* (Col., Curculionidae) and *Rhagoletis pomonella* (Dip., Tephritidae) adults. J Appl Entomol 119(1):63–65

Puranik TR, Hadapad AB, Sulunkhe GN, Pokharakar DS (2002) Management of shoot and fruit borer, *Leucinodes orbonalis* Guenee through *Bacillus thuringiensis* formulations on brijal. J Entomol Res 26:229–232

Rahman SJ, Rao AG, Swarnarsree P, Saxena R (2001) Effectiveness of *Trichogramma pretiosum* and Ha NPV against *Helicoverpa armigera* Hubn in tomato. Symp Biocontrol Based Pest Management for Quality Plant Protection in theCcurrent Millennium. PAU, Ludhiana, July 18–19

Rai HS, Verma ML, Gupta MP (2001) Efficacy of foliar application of Nuclear Polyhedrosis Virus against Pod Borer, *Helicoverpa armigera* Hubn in Chickpea. Symp Biocontrol Based Pest Management for Quality Plant Protection in the Current Millennium. PAU, Ludhiana, July 18–19

Ramakrishnan N (1993) Baculovirus pesticides. In: Botanicals and Biopesticis.de Parmar BS, Devakumar C (eds) SPS Publ No.4, Soc Pesticides Sci India and Westvill Publ House, New Delhi, pp 178–187

Ramakrishnan N, Chaudhary S, Kumar S, Rao RSN (1981) Field efficacy of Nuclear Polyhedrosis Virus against tobacco caterpillar, *Spodoptera litura* (F.) in tobacco. Tobacco Res 7(20):129–134

17 Potential of Biopesticides in Sustainable Agriculture 589

Rao AG, Rahman SJ, Swarsree P, Saxena R (2001) Evaluation of sequential of application of biopesticides for the management of *Helicoverpa armigera* Hubn in pigeonpea. Symp Biocontrol Based Pest Management for Quality Plant Protection in the Current Millennium. PAU, Ludhiana, July 18–19

Raven JA (1977) Evolution of vascular land plants. Adv Bot Res 154–159

Ravindra RJ, Jayaraj S, Balasubramanian M (1986) Control of *Heliothis armigera* on sunflower with Nuclear Polyhedrosis Virus. J Entomol Res 9:246–248

Reddy GVP, Guerrero A (2004) Interactions of insect pheromones and plant semiochemicals. Trends Plant Sci 9(5):253–261

Redmond CT, Potter DA (1995) Lack of efficacy of *in-vivo* and putatively *in-vitro* produced *Bacillus popilliae* against field populations of Japanese beetle (Coleoptera: Scarabaeidae) grubs in Kentucky. J Econ Entomol 88:846–854

Rehner SA, Buckley E (2005) A *Beauveria* phylogeny inferred from nuclear ITS and EFI-A Sequences: evidence for cryptic diversification and links to *Cordyceps telemorphs*, Mycologia 97:84–98

Rembold H, Forster H, Czoppelt Ch, Rao JP, Sieber KP (1984) The azadirachtins: a group of insect growth regulators from the Neem Tree. In: Schmutterer H, Ascher KRS (eds) Natural pesticides from the neem tree and other tropical plants. Proc 3rd Int Neem Conf Rauischholz-Hausen, 1983, Eschborn, GTZ, pp 153–162

Rembold H, Sharma GK, Czoppelt C, Schmutterer H (1980) Evidence of growth disruption in insects without feeding inhibition by neem seed fractions, Z. Pflanzenkd. Pflanzenschutz 87:290–297

Renn N (1998) Routes of penetration of entomopathogenic nematode *steinernema feltiae* attacking larval and adult house flies (*Musca domestica*). J Inverteb Pathol 72:281–287

Reudler Talsma JH, Biere A, JA Harvey, van Nouhuys S (2008) Oviposition Cues for a Specialist Butterfly–Plant Chemistry and Size. J Chem Ecol 34:1202–1212

Ridgway RL, Silverstein RM, Inscoe MN (1990) Behavior modifying chemicals for insect management. Marcel Dekker, New York

Rinderer TE, Elliot K D (1977a) Influence nosematosis on the hoarding behavior of the honey bee. J Inverteb Pathol 30:110–111

Risco BSH (1980) Biological control of froghopper *Mahanarva postica* Stal, with the fungus *Metarhizium anisopliae* in the state of Alagoas- Brazil. Entomol Newsl Int Soc Sugarcane Tech 9:10

Rishi Pal M, Abid Hussain, Prasad CS (2008) Natural occurrence of entomopathogenic nematodes in meerut district, North India. Int J Nematol 18(2):198–202

Roberts DW (1966) Toxins from the entomogenous fungus Metarhizium anisopliae 1. Production in submerged and surface cultures, and in inorganic and organic nitrogen media. J Invertebr Pathol 8:212–221

Rohrmann GF (1986) Polyhedrin structure. J Gen Virol 67:1499–1513

Rombach MC, Aguda RM, Shephard BM, Roberts DW (1986) Infection of rice brown plant hopper, *Nilaparvatha lugens* (Homoptera: Delphacidae), by field application of entomopathogenic Hyphomycetes (Deuteromycotina). Environ Entomol 15:1070–1073

Romeis J, Meissle M, Bigler F (2006) Transgenic crops expressing *Bacillus thuringiensis* toxins and biological control. Nature Biotech 24:63–71

Roomi MW, Abbas T, Shah AH, Robina S, Qureshi AA, Sain SS, Nasir KA (1993) Control of fruit flies (Dacus sp.) by attractants of plant origin. Anzeiger fiir Schadlingskunde, Pflanzeschutz, Umwelschutz, 66:155–157

Ruscoe CNE (1972) Growth disruption effects of an insect antifeedant. Nature New Biol 236:159–160

Sachitanandam S, Rabibndra RJ, Jayaraj S (1989) Pot culture studies on the efficacy of NPV formulations against the tobacco cut worm, *Spodoptera litura* (F.) larva on groundnut. J Biol Cont 3(1):44–46

Sahayaraj K, Pulraj MG (1999) Toxicity of some plant extracts against life stages of a reduviid predator, *Rhynocoris marginatus*. Ind J Entomol 61(4):342–344

Samuels KDZ, Pinnock DE, Allosop PG (1989) The potential of *Metarhizum anisopliae* (Metschonikoff) Sorokin (Deuteromycotina: Hyphomycetes) as a biological control agent of *Inopus rubriceps* (Macquat) (Diptera: Stratiomycidae). J Australian Entomol Soc 28:69–74

Samuels KDZ, Pinnock DE, Bull RM (1990) Scarabeid larvae control in sugarcane using *Metarhizum anisopliae*. J Inverteb Pathol 55:135–137

Sankar M (2009) Investigation of indigenous entomopathogenic nematodes, *Heterorhabditis indica* as a potential biocontrol agent of insect pest in rice. PhD thesis Osmania Univ, Hyderabad, India, pp 168–179

Santharam G, Raghupati A, Easwaramoorthy S, Jayaraj S (1978) Effectiveness of Nuclear Polyhedrosis virus against field populations of *Spodoptera litura* F. on banana. Ind J Agric Sci 48:676–678

Sasaki C (1910) On the pathology of the jaundice (Gelbsucht) of the silkworm. J Tokyo College Agric 2:105–161

Saufi AEM (2008) Characterization of an Egyptian *Spodoptera littoralis* nuclear polyhedrosis virus and a possible use of a highly conserved region from polyhedron gene nucleopolyhedrovirus detection. Virol J 5:13

Saxena H (2001) Biorational management of insect pest of pigeonpea. Symp Biocontrol Based Pest Management for Quality Plant Protection in the Current Millennium. PAU, Ludhiana, July 18–19

Saxena RC (1989) Insecticides from neem. In: Arnason JT, Philogene BJR, Morand P (eds) Insecticides of plant origin. ACS Symp. Series387. Am Chem Soc, Washington, DC, pp 110–135

Saxena RC (1998) Green revolution without blues: botanicals for pest management. In: Ecological Agricultural and Sustainable Development, vol 2. Dhaliwal GS, Randhawa NS, Arora R, Dhawan AK (eds) Ind Ecol Soc Cent Res Rural and Indust Develop, Chandigarh, pp 111–127

Saxena RC, Francis OO (1996) Growth development and reproduction of *Busseola fusca (Fuller)* (Lepidoptera: Noctuidae) on semi synthetic diet containing 'neem bitter concentrate'. Neem Newletter 13(2):18

Schmidt GD, Roberts LS (1989) Foundation of parasitology 4th ed. Times Mirror Mosby College pub, Toronto, pp 243–243

Schmutterer H (1987). Fecundity-reducing and sterilizing effects of neem seed kernel extracts in the Colorado potato beetle, *Leptinotarsa decemlimeat*. In: Schmutterer H, Ascher (eds) Natural pesticides from the neem tree and other tropical plants. KRS Proc. 3rd Internt Neem Conf. Nairobi, 1986, Eschborn, GTZ

Schmutterer H (1990). Properties and potential of natural pesticides from neem tree, *Azadirachta indica*. Ann Rev Entomol 35:217–297

Schmutterer H, Rembold H (1980) Zur wirkung einiger Reinfraktianen aus samon von *Azadirachta indica* auf FraBaktivital und Metamorphose von *Epilachna varivestis* (Coleoptera: Coccinellidae). Z Ang Entomol 89:179–188

Schneider MS, Smagghe G, Gobbi A, Vinuela E (2003) Toxicity and pharmacokinetics of insect growth regulators and other novel insecticides on pupae of *Hyposoter didymator* (Hymenoptera: Ichneumonidae), a parasitoid of early larval instars of lepidopteran pests. J Econ Entomol 96(4):1054–1065

Scholte EJ, Knols BGJ, Samson RA, Takken W (2004) Entomopathogenic fungi for mosquito control: a review. J Insect Sci 4:19

Schroer S, Ehlers RU (2005) Foliar application of the entomopathogenic nematode *Steinernema carpocapsae* for biological control of diamondback moth larvae (*Plutella xylostella*). Biol Cont 33:81–86

Schwartz PH Jr. (1975) Control of insects on deciduous fruit and tree nuts in the home orchard without insecticides. USDA Home Garden Bull 211:36

Seybold SJ (2003) Biochemistry and Molecular Biology of De Novo Isoprenoid Pheromone Production in the Scolytidae. Ann Rev Entomol 48:425–453

Shah BP, Devkota B (2009) Obsolete pesticides: Their Environment and Human health hazards. J Agric Environ 10:51–56

17 Potential of Biopesticides in Sustainable Agriculture 591

Shah PA, Pell JK (2003) Entomopathogenic fungi as biological control agents. Appl Microbiol Biotechnol 413–423

Shapiro-Ilan DI, Gough Dh, Piggott SJ, Fife JP (2006) Application technology and environmental considerations for use of entomopathogenic nematodes in biological control. Biol Cont 38:124–133

Sharp ES, Baker FL (1979) Ultrastructure of the unusual crystal of HD-1 isolate of *Bacillus thuringiensis* var. *Kurstaki*. J Inverteb pathol 34:320–322

Shimazu M, Kushida T, Tsuchiya D, Mitsuhashi W (1992) Microbial control of *Monochamus alternatus* – Hope (Coleoptera: Cerambycidae) by implanting wheat-bran pellets with *Beavuveria bassiana* in infested tree trunks. J Japanese Sec 74:325–330

Shorey HH, Kaae RS and Gaston LK (1974) Sex pheromones of Lepidoptera: development of a method for pheromone control of Pectinophora gossypiella incotton. J Econ Entomol 64:347–350

Shvedchikova NG, Ulanov VP, Tarasevich LM (1969) Structure of the granulosis virus of Siberian silkworm, *Dendrolimus sibiricus* Tschetw. Mol Biol 3:283–287

Siddall JB (1976) Insect growth regulators and insect control: a critical appraisal. Environ Health Perspec 14:119–126

Siegel JP, Maddox JV, Ruesink WG (1988) Seasonal progress of *Nosema prrausta* in the European corn borer, *Ostrinia nubilalis*. J Inverteb Pathol 52:130–136

Simoes N (1998) Pathogenicity of the complex *Steinernema carpocapsae-Xenorhabdus nematophilus*: molecular aspects related with virulence. In: Simoes N, Boemare N, Ehlers RU (eds) Pathogenicity of entomopathogenic nematodes versus insect defence mechanisms: impact on selection of virulent strains. Luxembourg, Eur Com, pp 73–84

Singh H, Ujagar R (2001) Field efficacy of NPV along with adjuvants against *Helicoverpa armigera* on chickpea at Pant Nagar. Symp on Biocontrol Based Management for Quality Plant Protection in the Current Millennium. July 18–19, 2001, PAU, Ludhiana

Singh P, Russel GB, Fredericksen S (1982) The dietry effects of ecdysteroids on the development of housefly. Entomol Exptl Appl 32:7–12

Singh RP, Srivastva BG (1985). Alcohol extract of neem (*Azadirachta indica* A. Juss) seed oil as oviposition deterent for *Dacus cucurbitae* (Coq.) Ind J Entomol 45(4):497–498

Singh SP, Murphy ST, Ballal CR (2001) Augmentative Biocontrol- Proc ICAR-CABI Workshop, pp 250

Slama K, Williams CM (1965) Paper factor as inhibitor for the embryonic development of European bug, *Purrhocoris apterus*. Nature 210:329–330

Smith KM (1976) Virus-Insect Relationship. Longman, London

Smith CA (1995) Searching for safe methods of flea control. J Am Vet Med Assoc 206:1137–1143

Solter LF, Maddox JV (1998) Timing of an early sporulation sequence of microsporidia in the genus *Vairimorpha* (Microsporidia: Burenllidae). J Inverteb Pathol 72:323–329

Solter LF, Maddox JV, McManus ML (1997) Host specificity of microsporidia (Protista: Microspora) from European populations of *Lymantria dispar* (Lepidoptera: Lymantriidae) to indigenous North American Lepidoptera. J Inverteb Pathol 69:135–150

Solter LS, Pilarska DK, Vossbrinck CR (2000) Host specificity of microsporidia pathogenic to forest Lepidoptera. Biol Cont 19:48–56

Somvanshi VS, Ganguly S, Paul AVN (2006) Field efficacy of the entomopathogenic nematode *Steinernema thermophilum* Ganguly and Singh (Rhabditita: Steinernematidae) against diamondback moth (*Plutella xylostella* L.) infesting cabbage. Biol Cont 37:9–15

Soper RS (1982) Commercial mycoinsecticide. Internat. colloq invertebr pathol III. Univ Sussex, Brigton, UK, pp 98–102

Srivastava JP, Mathur YK (1998) Insect growth regulatory effect of some neem based formulations against sesamum pod borer. Abs. in ICPPMSA, 11–13 Dec, 1998, CSAU&T, Kanpur, pp 228

Stark JD, Vargas RI, Tahlman RK (1990) Azadirachtin: effects on metamorphosis, longevity, and reproduction of three tephritid fruit fly species (Diptera: Tephritidae). J Econ Entomol 83(6):2168–2174

Stark JD, Wennergren U (1995) Can population effects of pesticides be predicted from demographic toxicological studies? J Econ Entomol 88:1089–1096

Steets, R. and Schmutterer, H. (1975) Einflub van Azadirachtin auf die Lebensdauer und das Reproduktion vermogen Von *Epilachna variensis* Muls. (Coleoptera:Coccinellidae). Z Pflkrankh Pflschutz 82:176–179

Steiner LF, Mitchell WC, Harris EJ, Kozuma TT and Fujimoto MS (1965) Oriental fruit fly eradication by male annihilation. J Econ Entomol 98:961–994

Steinhaus EA (1947) Insect microbiology. Comstock, Ithaka

Steinhaus E A (1949) Principles of insect pathology. Mc Graw-Hill, New York

Steinhaus EA (1954) The effect of disease on insect population. Hilgardia 23:197–281

Steinhaus EA (1963) Insect pathology, vol. I and II. Acad Press New York

Steinhaus EA, Hughes KM, Wasser HB (1949) Demonstration of the granulosis virus of the variegated cutworm. J Bacteriol 57:219–224

Steinwand (2008). Washington DC: US Environmental Protection Agency. Biopesticide Ombudsman (Personal communication)

Stelinski LL, JR Miller, ME Rogers (2008) Mating Disruption of Citrus Leafminer Mediated by a Noncompetitive Mechanism at a Remarkably Low Pheromone Release Rate. J Chem Ecol 34:1107–1113

Stipanovic RD, Willams HJ, Smith LA (1986) Votton terpenoids inhibition of *Heliothis viriscens* development. In: Green MB, Hedin PA (eds) Natural resistance of plants to pests. ACS Symp Series 296. Am Chem Soc, Washington DC, pp 79–94

Strasser H, Vey A, Butt T (2000) Are there any Risks in Using Entomopathogenic Fungi for Pest Control, with Particular Reference to the Bioactive Metabolites of *Metarhizium*, *Tolypocladium* and *Beauveria species*? Biocont Sci Tech 10:717–735

Suckling DM (2000) Issues affecting the use of pheromones and other semiochemicals in orchards. Crop Protect 19(8–10):677–683

Sudhaus W (1993) The nematode genera *Heterorhabditis* and *Steinernema* both entomopathogenic by means of systemic bacteria, are not sister texa Verhandlungen der Deutschen Zoologischen Gesellschafi 86:146

Sulifoa JB, Ebenebe AA (2007) Evaluation of pheromone trapping of diamondback moth (*Plutella xylostella*) as a tool for monitoring larval infestations in cabbage crops in Samoa. South Pacific J Natural Sci 7:43–46

Sundara Babu PC (1999) Development of *Metarhizum anisopliae* for the management of *Oryctes rhinoceros*. In: Rabindra RJ, Santhram G, Sathiah N, Kennedy JS (eds) Emerging trends in microbial control of crop pests. TNAU, Coimbatore, pp 146–154

Sundara Babu PC, Balasubramanian M, Jayaraj S (1983) Studies on the pathogenicity of *Metarhizum anisopliae* (Metschnikoff) sorokin var. major Tulloch on *Oryctes rhinoceros* (L.) TNAU Res. Publ.I

Suzuki A, Kawakami K, Tamura S (1971) Detection of destruxins in silkworm larvae infected with *Metarrhizium anisopliae*. Agric Biol Chem, Tokyo 35:1641–1643

Suzuki A, Taguchi H, Tamura S (1970) Isolation and structure elucidation of three new insecticidal cyclodepsipeptides, destruxin C and D and dasmethyldestruxin B, produced by *Metarrhizium anisopliae*. Agric Biol Chem, Tokyo, 34:813–816

Sylvar Technologies (2008) Research. URL: http://www.sylvar.ca/content/13636 (accessed 19 Aug 2008). Fredericton (NB)

Tabashnik B E, Cushing N L, Finson N and Johnson N W (1990) Field development of resistance of *Bacillus thuringiensis* in diamondback moth (Lepidoptera: plutellidae). J Econ Entomol 83:1671–1676

Tabashnik BE (1994) Evolution of resistance to *Bacillus thuringiensis*. Ann Rev Entomol 39:47–79

Talekar NS, Shelton AM (1993) Biology, Ecology, and management of the diamond back moth. Ann Rev Entomol 38:275–301

Tamaki Y (1985) Sex pheromones. In: Kerkut GA, Gilbert LS (eds) Comprehensive insect physiology. Biochem Pharmacol 9:145–191

Tamura S, Kuyama S, Kodaria Y, Higashikawa S (1964) The structure of destruxin B, a toxic metabolite of *Oospora destructor*. Agric Biol Chem Tokyo 28:137–138

17 Potential of Biopesticides in Sustainable Agriculture 593

Tanada Y and Kaya HK (1993) Insect pathology. Acad Press, San Diego

Tanveer A, Mustafa R, Anwar Z (1998) Control of hydatidosis in rabbits through feeding local plants. Punjab Univ J Zool 13:99–113

Thakore Y (2006a) The biopesticide market for global agricultural use. Ind Biotechnol 23:192–208

Thakore Y (2006b) The new biopesticide market. Report code: CHM029B

Theis N, Manuel L (2003) The evolution of function in plant secondary metabolites. Int J Plant Sci 164:93–102

Toguebaye BS, Marchand B (1984) Etude histopathologique et cytopathologique d'une microsporidose naturelle chez la coccinelledes cucurbitacees d'Afrique, *Henoseepilachna elaterii* (*Coleoptera: Coccinellidae*). Entomophaga 29:421–429

Tomlin C (1994) A World Compendium. The Pesticide Manual. Incorporating the Agrochemicals Handbook, 10th ed. Crop Protect Publ, Bungay, Suffolk, UK,

Tooker JF, Crumrin AL, Hanks LM (2005) Plant volatiles are behavioral cues for adult females of the gall wasp, *Antistrophus rufus*. Chemoecol 15:85–88

Touhara K, Vosshall LB (2009) Sensing odorants and pheromones with chemosensory receptors. Ann Rev Physiol 71:307–332

Triggiani O, Poiner GO Jr. (1976) Infection of adult Lepidoptera by *Neoplectana carpocapsae* (Nematoda). J Inverteb Pathol 27:413–414

Tsai SJ, Lo CF, Wang CH (2003) The characterization of microsporidian isolates (Nosematidae: Nosema) from five important lepidopteran pests in Taiwan. J Inverteb Pathol 83:51–59

Tsai SJ, Wang CH (2001) Interaction of the microsporidium and nucleopolyhedrovirus in *Spodoptera litura*. Formosan Entomol 21:183–195

Tumlinson JH, Moser JC, Silverstein RM, Brownlee RG, Ruth JM (1972) A volatile trail pheromone of the leaf cutting ant, Atta texana. J Insect Physiol 18:809–814

Tweeten KA, Bulla LA Jr, Consigli RA (1977) Supercoiled circular DNA of an insect granulosis virus. Proc Nat Acad Sci USA 74:3574–3578

Tworkoski T (2002) Herbicide effect of essential oils. Weed sci 50:425–431

Umoetok SBA, Gerard MB (2003) Comparative efficacy of *Acorus calamus* powder and two synthetic insecticides for control of three major insect pests of stored cereal grain. Global J Agric Sci 2(2):94–97

Van Zwaluvenburg RH (1928) The interrelationships of insects and roundworm. Bull Exp Stn Hawai. Sugar Planrters' Assoc Entomol Ser Bull 20

Varma A, Singh K (1987) *Metarhizum anisopliae* (Metschnikoff) Sorokin in the management of *Pyrilla perpusilla* Walker, the sugarcane leafhopper. IISR Entomol Newsletter 18:12

Vasconsuelo A, Boland R (2007) Molecular aspects of the early stage of elicitation of secondary metabolites in plants. Plant Sci 172:861–875

Vavra J, Hylis M, Vossbrinck CR, Pilarska DK, Linde A, Weiser J, McManus ML, Hoch G, Solter LF (2006) *Vairimorpha disparis* n.comb. (Microsporidia: Burenellidae): a redescription of the *Lymantria dispar* (L.) (Lepidoptera: Lymantriidae) microsporidium, *Thelohania disparis* Timofejeva 1956. J Euk Microbiol 53:292–304

Vey A, Hoagland RE, Butt TM (2001) Toxic metabolites of fungal biocontrol agents. In: Butt TM, Jackson C Magan N (eds) Fungi as biocontrol agents. Progress, problems and potential. CABI Publ, Oxford, UK, pp 311–346

Vey A, Quito JM, Pais M (1986) Toxemie d'origine fongique chez les Invertebres et ses consequences cytotoxiques: Etude sur l'infection a *Metarhizium anisopliae* (Hyphomycetes, Moniliales) chez les Lepidoptera et les Coleopteres. C R Soc Biol 180:105–112

Vilas BAM, Alves SB (1988) Pathoginicity of *Beauveria spp.* and its effect associated with monochrotophos *Castina licus* (Drury, 1770). (Lepidoptera: Castiniidae) Ann Soc Entomol Brazil 17:305–332

Viver E, Desportes I (1990) Phylum epicomplexa. In: Margulis L, Corlis JO, Malkonian M, Chapma DJ (eds) Handbook of protoctista. Jones and Bartlett, Boston, pp 549–573

Volkmann LE, Blissard GW, Friesen P, Keddie BA, Possee R, Theilmann DA (1995) Family Baculoviridae, in virus taxonomy. Springer, New York, pp 104–113

Von Prowazek S (1907) chlamydozoa. II. Gelbsucht der Seidernaraupen. Arch protistenkd

Von Prowazek S (1913) Untersuchungen uber die gelbsucht der seidenraupen. Centralbe bakteriol Parasitenkd Infek 67:268–284

Wagner MR, Benjamin DM, Clancy KM, Schuh BA (1983) Influence of diterpene resin acids on feeding and growth of larch sawfly *Pristiphora erichosnii* (Hartig) J Chem Ecol 9:119–127

Wahl B (1909) Uber die polyederkrankheit der Nonne (Lymantria monacha L.) Centra lbl. Gesamte Forstwes 35:164–172

Wang Der-I, Moeller FE (1970) The division of labor and queen attendance behavior nosema-infected worker honey bees. J Econ Entomol 63:1539–1541

Wang XP, Le VT, Fang YL, Zhang ZN (2004) Trap effect on the capture of *Plutella xylostella* (Lepidoptera: Plutellidae) with sex pheromone lures in cabbage fields in Vietnam. Appl Entomol Zool 39(2):303–309

Wang Y, Campbell JF, Gaugler R (1995) Infection of entomopathogenic nematode *Steinernema glaseri* and *Heterorhabditis bacterophora* against *Poppilia japonica* (Coleoptera: Scarabaeidae) larvae. J Inverteb Pathol 66:178–184

Wardhaugh KG (2005) Insecticidal activity of synthetic pyrethroids, organophosphates, insect growth regulators, and other livestock parasiticides: An Australian perspective. Environ Toxicol Chem 24(4):789–796

Ware G, Whitacre D (2004) An Introduction to Insecticides. In: Radcliffe E, Hutchison W, Cancelado R, Radcliffe's IPM World Textbook. URL: http://ipmworld.umn.edu. St. Paul (MN): University of Minnesota

Warthen JD Jr. (1979) *Azadirachta indica*: A source of insect feeding inhibits and growth regulators. ARM-NE-4, USDA, SEA, Agricl Rev Man

Wee S, Tan KH, Nishida R (2007) Pharmacophagy of Methyl Eugenol by Males Enhances Sexual Selection of *Bactrocera carambolae*. J Chem Ecol 33:1272–1282

Wee SL, Hee AKW, Tan KH (2002) Comparative sensitivity to and consumption of methyl eugenol in three *Bactrocera dorsalis* (Diptera: Tephritidae) complex sibling species. Chemoecol 12:193–197

Weiser J (1963) Sporozoan infections. In: Steinhaus EA (ed) Insect pathology: an advanced treatised. Acad Press, New York, 2:291–334

Welch HE (1963) Nematode infection. In: Steinhaus EA (ed) In insect pathology: an advance treatise, vol 2. Academic Press, New York, pp 363–392

Welch HE, Broskill JF (1962) Parasitism of mosquito larvae by the nematodes, DD 136 (Nematoda: Neoplectanidae). Can J Zool 40:1263–1275

Wertheim B, van Baalen EJA, Dicke M, Vet L (2005) Pheromone-mediated aggregation in nonsocial arthropods: an evolutionary ecological perspective. Ann Rev Entomol 50:321–346

Whalon ME, McGaughey WH (1998) *Bacillus thuringiensis*: use and resistance management. In: Ishaaya I, Deheele D (eds) Insecticidal with novel modes of action, mechanism and application. Springer, New York, pp 106–137

Witte V, Abrell L, Attygalle AB, Wu X, Meinwald J (2007) Structure and function of Dufour gland pheromones from the crazy ant, *Paratrechina longicornis*. Chemoecol 17(1):63–69

Witzgall P, Kirsch P, Cork A (2010) Sex pheromones and their impact on pest management. J Chem Ecol 36:80–100

Witzgall P, Stelinski L, Gut L, Thomson D (2008) Codling moth management and chemical ecology. Ann Rev Entomol 53:503–522

Woolever P (1966) Life history and electron microscopy of a haplosporidian, *Nephridiophaga blattellae* (Craeley) n. comb., in the malpigihian tubules of German cockroach, *Blattella germanica* (L.) J Portozool 13:622–642

Wright MS, Henderson G, Chen J (2000) Growth response of Metarhizium anisopliae to two Formosan subterranean termite nest volatiles, naphthalene and fenchone. Mycologia 92:42–45

Wu GY, Sun JD (1994) A study of application of *Beauveria bassiana* (Bals.) Vuill. in the control of the tea brown weevil. China Tea 16:30–31

Wyatt G R (1952b) Specificity in the composition of nucleic acids. Exp Cell Res Suppl 2:201–217

Wyatt GR (1952a) The nucleic acids of some insect viruses. J Gen Physiol 36:201–205

17 Potential of Biopesticides in Sustainable Agriculture

Yadouleton AW, Padonou G, Asidi A, Moiroux N, Bio-Banganna S, Corbel V, N'guessan R, Gbenou D, Yacoubou I, Gazard K and Akogbeto MC (2010) Insecticide resistance status in *Anopheles gambiae* in southern Benin. Malar J 9:83

Zehnder GW, Gelernter WD (1989) Activity of the M-ONE formulation of a new strain of *Bacillus thuringiensis* against the Colorado potato beetle (Coleoptera: Chrysomelidae): Relationship between susceptibility and insect life stage. J Econ Entomol 82:756–761

Zhang J, Hodgman TC, Krieger L, Schnetter W, Schairer HU (1997) Cloning and analysis of the first Cry gene from *Bacillus popilliae*. J Bacteriol 179:4336–4341

Zimmermann G (2007) Review on safety of the entomopathogenic fungus, *Metarhizium anisopliae*. Biocont Sci Technol 17(9):879–920

Index

A

Absorption, 97, 98, 115, 121, 271, 299, 300, 390, 476, 509
Acanthamoeba, 281
ACC (1-aminocyclopropane-1-carboxylate) deaminase, 56, 384, 387
Acetogen, 426
Acetosyringone, 348, 483
Acetylcholinesterase inhibitor, 305
Acid black, 468, 470, 471, 473, 474, 478, 479
Acid phosphatase, 50, 201
Acinetobacter, 171, 179, 183, 366, 466, 469
Activated carbon, 5, 123, 124, 127, 128, 141, 261, 273, 275, 287, 422, 462
Adsorbable organic halide, 16, 17, 398, 414, 417–427
Adsorption, 5, 6, 14, 18–20, 55, 98, 128, 153, 198, 273, 275, 286, 299–301, 308, 309, 314, 321, 348, 422, 444, 458, 474
Aeder algerae, 148, 308, 398, 572
Aerobic degradation, 17, 19, 86, 121, 136, 138, 140, 141, 143, 144, 148, 150, 153, 299, 308, 313, 399, 423–427, 432, 459, 460, 514
Aggregation pheromone, 540–542
Agricultural waste, 4, 21, 66, 67, 69, 98, 99, 500, 501, 506, 507, 514
Agriculture, 1–4, 12, 22, 45, 54, 65–67, 69, 75, 76, 86, 167, 176, 197, 297, 305, 330, 333, 381, 389, 502, 506, 515, 532, 534, 551, 574
Agro by-product, 5, 66, 69, 70, 73–78, 80, 83, 84, 86, 87, 93–95, 98
Agrobacterium, 179, 352
Agro-industrial wastes, 4, 67, 71, 82, 84, 86, 87
Agro-processing waste, 66
Alarm pheromone, 540

Alcohol dehydrogenase, 350
Alfalfa, 1, 13, 14, 329–337, 345, 561
Alkaline DNA elution, 243
Alkaline unwinding, 10, 230, 240, 241, 243
Aminoglycoside, 7, 168, 175, 356, 357
Ampicillin, 173
Amylase, 80, 81, 83, 384
Anaerobic degradation, 5, 17, 19, 22, 87, 89, 119, 121, 140, 143, 144, 148, 205, 299, 308, 313, 316, 317, 423–427, 432, 455, 459, 460, 511
Anatoxin, 10, 264, 265, 269, 271–273, 276, 284
Anglesite (PbSO$_4$), 440
Antagonism, 287, 334, 337
Antennae, 539, 542
Anthracnosis, 332
Anthraquinone, 114–116, 154, 455, 457, 458, 461, 462, 466, 468, 470–472, 477, 480
Anthrxanhin, 95
Antibiotic, 7, 8, 163, 164, 168, 174, 175
Antibiotic resistance, 6–8, 41, 52, 164, 167–170, 173–176, 183, 184, 344, 345, 355, 357
Antibiotic resistance gene, 6–8, 164, 169
Antioxidant, 5, 16, 91, 92, 94–96, 303, 389–391
Aquificales, 360
Arbuscular mycorrhizae (AM), 4, 38, 39, 55, 211
Arthrobacter, 179
Arthrospira, 262
Arylmethane, 115, 154, 457, 458, 461
Arylsulphatase, 200, 201
Ascomycotina, 553
Ascospore, 553
Asphalt, 508
Atomic force microscopy (AFM), 15, 355

598 Index

Autoinducer, 349, 352, 359
Auto-oxidation, 68
Autosomal adenine phosphoribosyltransferase, 250
Auxin, 3, 51, 56, 384
Auxochrome, 115, 456
Azadirachtin, 535, 546–548, 551, 552
Azo, 6, 19, 115–117, 119, 120, 128, 133, 134, 140–144, 146–148, 150–154, 455–462, 465–474, 477–482
Azoreductase, 140, 142, 152, 484
Azospirillum brasilense, 347
Azotobacter chroococcum, 55, 306, 387

B

Bacillus, 14, 40, 82, 85, 93, 171, 173, 175, 277, 280, 331, 334, 335
Bacillus thuringiensis, 22, 529, 531, 549–553
Baculoviridae, 558
Baculovirus, 529, 558, 560, 562, 563, 565
Baggase, 76
Basidiomycete, 38, 82, 468, 470, 471
Bassiacridin, 555
Bassianin, 555
Bassianolide, 555
Beauveria bassiana and *Metarrhizium anisopliae*, 529, 554, 556
Beauvericin, 555
Beauverolide, 555
Benzoylurea, 535
Betaine, 384, 387
Bioaerosol, 522
Bioaugmentation, 12, 298, 312, 313
Biocontrol, 13, 14, 40, 41, 46, 273, 279, 280, 281, 282, 334, 336, 337, 386, 556, 561, 574
Biodecolourization, 19, 113, 126, 136–142, 151, 152
Biodegradation, 1, 6, 11, 12, 23, 51, 113, 119, 135–142, 148, 150, 152, 154, 298, 299, 306, 309, 316, 318, 319, 362, 364, 399, 423, 425, 454, 460–462, 468, 519
Biodiesel, 87, 89, 90, 392
Biofilm, 1, 6, 14, 15, 170, 209, 278, 285, 318, 342–367
Biofouling, 14, 365–367
Bio-fuel, 4, 67, 87–90
Biological control, 14, 40, 229, 266, 267, 281, 282, 286, 287, 333–335, 337, 572
Biological nitrogen fixation, 13, 331, 380
Biological oxygen demand (BOD), 86, 112, 120, 122, 415–419, 424, 426, 428, 429, 432, 457, 458

Biological treatment, 11, 19, 69, 136, 137, 261, 262, 268, 272, 273, 284–287, 317, 320, 423, 424, 426, 428, 429, 457, 462, 463
Biomagnification, 297
Biopesticide, 1, 22, 23, 40, 529–534, 549, 551, 553, 558, 560, 573, 574
Bioreactor, 12, 16, 19, 129, 137, 141, 250, 298, 312, 317, 318, 353, 365, 389, 511
Bioremediation, 1, 4, 12, 23, 50, 53, 113, 135–137, 142, 144, 150, 153, 177, 178, 296, 298, 299, 310, 312–321, 514
Biostimulation, 12, 312, 313, 319
Biotin, 72, 384
Bitter gourd peroxidase (BGP), 465, 475, 476, 478, 480, 484
Bleaching, 17, 120, 404, 407–409, 411–414, 416–421, 423, 424, 430, 431, 458, 468, 478, 480
Bond paper, 401
Bradyrhizobiaceae, 382
Buprofezin, 535
Burkholderia, 278, 284, 334, 381, 382, 385
Burkholderiaceae, 382

C

Calcium aluminate hydrate (CAH), 18, 437, 444, 446
Calcium hypochlorite, 412
Calcium silicate hydrate (CSH), 18, 437, 444, 446
Candida tropicalis, 316, 466
Candida utilis, 75, 77, 316
Capillary electrophoresis (CE), 14, 152, 207
Carbon dioxide, 89, 90, 137, 143, 200, 202, 298, 440, 499, 519
β-Carotene, 95, 96, 251
Carotenoid, 83, 92, 94–96, 263, 391, 458
Cation exchange capacity, 17, 55, 400
Cellulase, 81, 83, 93
Cenococcum geophilum, 44
Ceratocystis fimbriata, 84
Cerussite ($PbCO_3$), 440
Charcoal rot, 332, 333
Chemical degradation, 314
Chemical oxygen demand (COD), 112, 120, 122, 130, 131, 133, 134, 136, 143, 415, 417–419, 423, 425, 426, 428, 429, 457, 458, 466, 469
Chemical pulping, 400, 403, 405, 407, 408, 412, 416
Chemoattractant, 384
Chemoreceptor, 539

Index 599

Chipper, 403
Chitin, 71, 129, 200, 535, 548
Chitosan, 71, 129, 423, 478
Chlorinated hydrocarbon, 16, 299, 306, 313, 398, 418
Chlorinated pesticide, 316, 317
Choline, 384
Chromobacterium violaceum, 352
Chromophore, 115, 121, 148, 456, 457, 467
Chromosomal aberration, 120, 230, 232, 245–248
Cibacron red, 473, 474
Cinnabar, 164
Cis-11-methyl-dodecenoic acid, 359
Coelomycetes, 332
Colletotrichum trifolii, 332
Comet assay, 10, 230, 241, 242, 245
Commercial waste, 21, 501
Composting, 4, 12, 21, 22, 66, 69, 86, 298, 313, 318, 319, 500, 514, 515, 517–519, 522, 525
Conidia, 553, 555–557
Conjugative plasmid, 165, 175, 177, 180, 181, 183, 184
Contaminant, 8, 11, 12, 17, 18, 23, 48, 73, 117, 122, 131, 133, 142, 143, 165, 167, 178, 231, 245, 298, 299, 310, 312–314, 316, 318, 319, 321, 386, 408, 443, 444, 447, 521, 523
Copier paper, 401
Coriolus versicolor, 82, 461, 462
Corrosivity, 509
Cota-laccase, 147, 152, 459
Coulomb's law, 125
Coumarins, 391, 534
Cream wove, 401
Cryptoxanthin, 95, 96
Crystal violet, 14, 351, 478
Cue-lure, 544, 545
Cunninghamella polymorpha, 461
Cyanine, 15, 354, 458
Cyanobacteria, 1, 10, 11, 261–263, 266–269, 272, 273, 275, 277–282, 284–287
Cyanophage, 273, 277, 280, 286
Cyanophicin, 266
Cyanotoxin, 10, 261, 262, 269–273, 275–277, 280–282, 284–287
Cyclic adenosine monophosphate (cAMP), 391
Cyclooxygenase, 391
Cydia pomonella, 543, 545, 557
Cylindrospermopsin, 10, 264, 269–273, 276
Cyromazine, 535

Cystathionine, 384
Cystine, 384

D

Daidzein, 91, 383, 391
Damping-off, 13, 14, 329, 330, 332–334, 336
Debarking, 415
Decolourisation, 19, 121, 122, 125, 131–134, 137, 139–144, 147, 148, 150–153
Dedusting, 411
Dehydrogenase, 176, 199, 201, 277, 280, 350
Deinking, 413, 414
Delignification, 417, 420, 421
Denaturating gradient gel electrophoresis (DGGE), 49, 179, 181, 207, 208, 366
Denitrification, 204, 206, 424
Depithing, 411
Deuteromycotina, 432, 553
Diazotrophic bacteria, 381
Dicot, 333
Digoxigenin, 166
Diguanylate cyclase (DGC), 360
Diketopiperazine, 352
Dinitrotoluene (DNT; $C_6H_3[CH_3][NO_2]_2$), 315, 439
Diosgenin, 390
Dioxin, 16, 53, 398, 407, 414, 418, 421, 521
Direct Yellow 10, 473
DNA adduct, 233, 240, 244, 391
DNA microarray, 167
DNA strand break, 233, 239, 240, 243–245
Docosahexaenoic, 365
Dye, 1, 6, 14, 19, 20, 85, 111–154, 351, 366, 389, 404, 413, 414, 453–484
Dye decolorizing peroxidase (DYP), 468, 472, 473; *see also* Peroxidase

E

Ecdysteroide, 536
Ectomycorrhizae (EM), 38, 44, 46, 55–57
Eelworms, 563
Eicosapentaenoic acid, 365
Electroflocculation, 422
Electrokinetic coagulation, 121
Electrokinetic remediation, 443
Elemental chlorine free (ECF), 17, 408, 419
Endotoxins, 522, 550
Enterobacter, 11, 381, 426
Ester-linked fatty acid (ELFA), 211, 212
Ethyl protocatechuate (EP), 390
Ethylendiaminetetraaceticacid (EDTA), 443, 444

600   Index

Eudrilus eugeniae, 86
Eutrophication, 305, 380, 446
Excavation, 314, 318, 443
Exopolysaccharide, 346, 347, 364
Extracellular polysaccharide, 342, 344, 346–348, 355, 357, 361, 363, 367, 383

F

Fatty acid methyl ester (FAME), 205, 211
Fenugreek seed peroxidase (FSP), 478, 484
Flavonoid, 53, 92, 197, 379, 382, 383, 385, 391, 534
Flexibacter, 277, 280
Fluorescein isothiocyanate, 14, 15, 354
Fluorescence in situ hybridization (FISH), 175, 210, 246
Forage, 12, 13, 15, 194, 330–332, 334, 387, 389, 542
Formica rufa, 542
Formononetin glycoside, 383
Freudlich, 123
Fructosyl transferase, 83
Fuchsine, 466, 469
Furan, 16, 398, 407, 414, 418, 521
Furanone, 358, 359, 538, 540

G

Ganoderma lucidum, 82, 472
Gas chromatography (GC), 152, 479
Genistein, 91, 391
Genotoxic, 9, 10, 91, 230–232, 237–239, 243–246, 249–251, 456
Genotoxicity, 1, 10, 23, 229–234, 238–240, 243, 245–247, 249, 252, 253, 456
Geotrichum candidum, 461, 468, 472
Glucan, 72, 522, 523
Gluconic acid, 78, 79
β-Glucosidase, 14
Glutathione transferase, 391
Glycitein, 383
Glycoconjugate, 353, 354
Grandlure, 542
Granulosis viruse (GV), 558–560
Green fluorescent protein (GFP), 15, 53, 249, 250, 352, 354, 355
Greenhouse gas, 21, 22, 88, 90, 499, 507, 516, 518, 519, 525
Gypsum, 445, 508

H

Halorespiration, 425
Hartig net, 38

Heliocide, 536
Heliothes virescens, 536, 538, 561
Hemocoel, 566, 567, 569
Hepatototoxin, 262
2-Heptyl-3-hydroxy-4-quinolone (PQS), 352
Heterocyst, 266
High performance liquid chromatography (HPLC), 114, 152, 240
Homoserine, 384
Horseradish peroxidase (HRP), 465, 469, 473, 474, 476, 478, 479, 481, 483
Human health, 3, 11, 13, 79, 194, 231, 261–263, 297, 299, 321, 389, 445, 500, 501, 507, 509, 513–515, 520–525, 530
Humulus lupulus, 534
Hydraulic pulper, 414
Hydrocerussite ($PB_3[CO_3]_2[OH]_2$), 440
Hydrolase, 384, 385
Hydraulic drum, 409, 430
1-Hydroxybenzotriazole (HOBT), 467, 472, 474–477, 481–484
Hyphomicrobiaceae, 382
Hypoxanthine-guanine phosphoribosyl transferase (hprt), 10, 244, 250

I

Ignitability, 509
Immobilised enzyme, 317
Immobilization, 18, 20, 56, 136, 141, 281, 285, 343, 437, 442, 444–445, 477, 479–481
Immunoglobulin, 344
Incineration, 4, 21, 69, 318, 320, 500, 510, 512–514, 517–519, 521–523, 525
Incp-1β plasmid, 171, 178, 181
Index, 9, 116, 124, 201, 202, 204, 212–215, 248, 390, 420, 428, 456, 457, 547
Indigoid, 115, 154, 457, 458, 461, 462, 483
Indole-3-acetic acid (IAA), 41, 51, 387, 465
Infra-red spectroscopy (FTIR), 126
Integrated pest management, 23, 530, 539, 563, 573
Invertase, 38, 384
Iolaxanthin, 95
Iridoid glucoside, 536
Iron-manganese oxide, 18, 443, 444
Isoamyl acetate, 84
Isoamylic alcohol, 280

J

Jasmonate, 384
Juvenoid, 535

K

Kairomone, 537, 540, 543–545
Kanamycin, 7, 173, 175
Kappa number, 17, 419–422
Kloceckera apiculata, 75
Korarchaeota, 360
Kraft pulping, 405, 406, 415
Krofta saveall, 413

L

β-Lactam, 169, 172, 174, 344, 356
β-Lactamase, 169, 172–174, 344, 357
Landfill gas (LFG), 500, 511, 518–520, 525
Landfilling, 4, 21, 500, 510, 512, 513, 517, 518, 520, 525
Langmuir, 123
Laser scanning microscopy (LSM), 15, 342, 355, 366
Leachate, 18, 21, 22, 170, 236, 241, 246, 511, 512, 514, 517, 521
Leaching, 17, 18, 120, 204, 205, 296, 300, 301, 380, 507
Lead, 17, 244, 386, 437–440, 445, 521, 522
Lectin, 342, 353, 354, 384
Legume, 1, 4, 12, 13, 15, 16, 65, 70, 96, 197, 214, 330, 364, 379–381, 383–390, 392, 534, 562
Lepidopteran, 535–537, 550, 555
Lignin, 5, 16, 71, 72, 80, 82, 93, 96–98, 128, 138, 200, 399, 400, 404–408, 412, 415–420, 462, 465, 470, 472, 534
Lignin peroxidase, 82, 92, 138, 142, 144, 145, 315, 460, 461, 465, 466, 469, 470, 473, 474, 482
Lignocellulose, 71
Lindane, 305, 316
Lipopolysaccharide, 10, 249, 347
Lipoxygenase, 391
Lutein, 95, 96
Lycopene, 92, 96

M

Macrolide-lincosamide-streptogramin, 168, 173, 175
Macrophomina phaseolina, 13, 329, 332
Manganese peroxidase, 80, 82, 138, 142, 145, 146, 460, 461, 465–468, 470–473, 478, 480, 482
Magnetic resonance imaging (MRI), 15, 355
Malachite green, 462
Maplitho, 401
Mass spectrometry, 126, 152, 175, 210, 252
Massicot (PbO), 440

Mastigomycotina, 553
Mating disruption, 531, 542, 543, 545
Mayorella, 282
Mechanical pulping, 400, 405–407, 417, 418
Medicago sativa L, 13, 329, 330, 347; *see also* Alfalfa
Metagenomic, 7, 174, 209
Metal/heavy metal, 1, 4, 5, 8, 17, 18, 37, 38, 46–48, 54–57, 73, 95, 98, 114, 116, 117, 120, 127, 149, 163–167, 176–178, 180, 183, 184, 201, 212, 214, 235, 236, 239, 244, 247, 308, 310, 315, 321, 344, 349, 361, 364, 379, 380, 386, 387, 391, 392, 418, 422, 437–447, 455, 456, 463, 478, 480, 503, 508, 509, 514, 516, 521, 522
Methane, 5, 21, 87, 89, 90, 143, 426, 499, 500, 518, 519, 523
Methanogen, 87, 316, 317, 426, 511
Methyl eugenol, 543–545
Methylobacteriaceae, 382
Michaelis-menten, 150, 480
Microbial biomass, 9, 129, 136, 194–197, 201, 202, 211, 213–216
Microbial degradation, 121, 136, 202, 310, 313, 314, 425
Microbial pesticides, 548, 549
Microbial respiration, 143, 202, 203
Microcolony, 345, 346
Microcystin, 10, 11, 264, 265, 269–272, 274–276, 278, 284, 285
Microcystis, 263, 264, 266, 268, 270, 271, 280
Micronucleus, 230–232, 247, 248
Microperoxidase, 477
Microsporidiosis, 570
Mineralisation, 148, 194, 198, 200, 202–205, 214
Mixed inoculant, 331, 334
Mobile genetic element (MGE), 6, 165, 181–183
Momordica charantia, 476
Monocot, 333
Monoterpene alcohol, 84
Morphogene, 350, 351
M-toluate, 47
Mugenic acid, 384
Municipal solid waste, 1, 21, 500, 501, 503
Mutagenicity, 16, 229, 233, 234, 238, 239, 244, 247, 249, 297, 418
Mutatox, 235, 239
Mycorrhiza, 23, 35–51, 54, 55, 197; *see also* Arbuscular mycorrhizae *and* Ectomycorrhizae
Mycotoxin, 522

N

N-acyl homoserine lactone, 349, 352, 359, 364, 367

Nanofiltration, 14, 131

Nanoparticle, 286, 287, 354

Nematode, 45, 384, 531, 532, 563, 566–569

Neoxanthin, 95

Nephridiophaga blattellae, 570

Neurotoxin, 10, 261, 262, 269, 271

Niacin, 384

Nitric oxide synthase, 391

Nitrification, 130, 201, 204, 206

Nitrilotriacetate (NTA), 443

Nitrocellulose (NC; $C_6H_7[NO_2]_3O_5$), 439

Nitrogen fixation, 206, 266–268, 347; *see also* Biological nitrogen fixation

Nitrogenase, 266, 280

Nitroglycerin (NG; $C_3H_5N_3O_9$), 439

N-methionylnicotinic acid, 384

Nodularin, 269–271, 278, 284, 285

Nodulation, 4, 15, 331, 336, 383

Non-hazardous industrial waste, 21, 501, 507

Nosema pyrausta, 571

Nuclear magnetic resonance spectroscopy (NMR), 126, 152

Nuclear polyhedrosis viruse (NPV), 558–560, 562, 563

Nutrient, 3, 12, 14, 15, 17, 22, 36–40, 42–45, 47, 48, 50, 51, 54, 55, 66, 71–74, 81, 85, 86, 129, 137, 163, 194–196, 198, 200, 203, 204, 221, 263, 266, 267, 273, 278, 280, 298, 306, 309, 310, 312, 313, 315, 319, 321, 330, 341, 343, 345, 346, 349, 350, 362, 364, 367, 380, 391, 390, 424, 426, 462, 507, 515, 522, 554, 567

O

Ochromonas, 281

ONPG (o-Nitrophenyl-β-galactopyranoside), 353

Oosporein, 555

Opine, 53, 530

Organic pollutant, 11, 20, 36, 37, 47, 48, 51, 53, 54, 113, 121, 134, 144, 152, 211, 317, 321, 463, 514

Organochlorine, 235, 236, 239, 297, 305, 310, 417, 418, 424, 425

Ornithine, 384

Oscillatoria, 263, 265, 266, 270, 271, 280, 282, 462

Osmolality, 388

Osmoprotectant, 384

Osmostressed, 388

Ovipositor, 539

P

Paenibacillus, 334, 381

Palpi, 539

p-Aminobenzoic acid, 169, 384

Panthothenate, 384

Paper, 1, 6, 16, 17, 69, 73, 80–82, 93, 94, 112, 117, 118, 120, 230, 351, 389, 398–404, 406, 408–415, 418–421, 423–426, 428, 429, 431, 432, 438, 454, 456, 458, 503, 504, 508, 511, 516, 519, 530, 535

Pb immobilization, 437, 445, 446

PCR-single-strand conformation polymorphism (PCR-SSCP), 49

Pectinase, 80, 81, 83, 93

Pentachlorophenol (PCP), 299, 315, 399, 419

Peroxidase, 1, 20, 83, 92, 144–147, 316, 384, 385, 454, 460, 464, 465, 470, 472, 473, 476–481, 483–485

Persister, 345, 366

Pesticide, 12, 22, 296, 297, 299–310, 312, 314–317, 530, 532, 533, 539, 544, 549, 551, 561, 574

Phanerochaete chrysosporium, 77, 80, 82, 136, 138, 146, 315, 316, 461, 462, 465, 466, 469–471, 473, 478, 480, 482

Phenolase, 384

Phenylbenzotriazole, 234

Phenylpropanoid, 53

Pheromone, 22, 530, 532, 533, 537–546

Phosphatase, 43, 50, 200, 201, 285, 384

Phosphoglycerate mutase, 350

Phospholipid, 92, 200, 211, 215

Phospho-lipid (PLFA), 211, 212, 215, 216

Phosphomonoesterase, 200

Photodegradation, 134, 314

Phthalocyanine, 115, 154, 457, 458

Phthalocyanine dye, 116, 458, 461, 465, 482

Phycocyanin, 263, 266

Phyllobacteriaceae, 382

Physico-chemical treatment, 112, 261, 272–274, 284–286

Phytoalexin, 383, 384

Phytoanticipin, 383

Phytochemical, 16, 66, 389–391, 533

Phytoecdysone, 536

Phytophagous, 45

Phytophthora medicaginis, 331

Phytophthora sojae, 331

Phytoremediation, 36, 47, 48, 55, 56, 314, 315, 392, 443

Phytosiderophore, 384

Phytostabilization, 18, 443

Phytosterol, 390

Pimephales promelas, 120, 241

Index

Plant growth promoting rhizobacteria (PGPR), 3, 13, 15, 40, 41, 55, 331, 336, 379, 381, 382–385, 387
Plantago lanceolata, 536
Pollutants, 1, 3, 5, 8, 10, 11, 16, 17, 19–21, 35–37, 42, 48, 51, 53–57, 90, 98, 111–113, 121, 122, 126, 128, 133, 134, 137, 138, 142, 144, 152, 153, 185, 198, 211, 234, 243, 245, 247, 249, 298, 310, 312, 314, 315, 317, 320, 321, 364, 386, 392, 399, 414, 415, 419, 425, 426, 453, 454, 457, 458, 463, 464, 513, 514, 521, 522, 525
Poly-acrylamide (PAM), 141
Polychlorinated biphenyl (PCB), 53, 235, 243, 315
Polycyclic aromatic hydrocarbons (PAH), 47, 234, 235, 238, 319, 465, 521, 522
Polygalacturonase, 82, 384
Polyol, 41, 52
Polyphenol oxidase, 20, 315, 464
Polyvinyl alcohol (PVA), 141
Popillia japonica, 538, 540, 544, 563, 567–569
^{32}P-postlabeling, 240
Potcher, 409, 412, 429
Pozzolanic, 18, 444, 446
Promazine, 483
Prophage, 239
Protease, 20, 200, 239, 243, 285, 384, 477, 556, 568
Proteorhodopsin, 209
Pseudomonas, 11, 14, 40, 53, 152, 171, 177, 179, 285, 330, 331, 334–337, 342, 348, 352, 354, 363, 364, 366, 381, 385, 386, 424, 426, 459, 466, 469, 567
Pulp, 1, 16, 17, 70, 73, 74, 77, 78, 80–83, 86, 93, 112, 118, 120, 230, 398–426, 428, 429, 431, 432
Putrescine, 384
Pycnidia, 333
Pyridoxine, 384

Q

Quantum dot, 354
Quorum sensing, 342, 346, 347, 352, 366, 367

R

Raman microscopy (RM), 15, 355
Random amplified polymorphic DNA (RAPD), 244
Reactivity, 151, 509
Recombinant dye decolorizing peroxidase (rDYP), 468, 472, 478, 480

Recycling, 4, 21, 22, 67, 69, 86, 90, 112, 121, 138, 194, 297, 331, 413, 414, 422, 424, 428, 432, 441, 447, 500, 502, 511, 515–517, 519, 522–525
Red fluorescent protein, 354
Remazol blue, 473, 474
Repellent, 533, 534
Reverse osmosis, 14, 121, 131, 422, 423, 458
Rhizobacteria, 1, 15, 23, 41, 47, 48, 54, 329, 331, 334, 337, 379, 380, 383, 386, 388
Rhizobiaceae, 381, 382
Rhizodegradation, 53
Rhizodeposition, 15, 55, 197, 202, 382, 392
Rhizomorph, 38, 55
Rhizopus arrhizus, 461
Rhizosphere, 3, 4, 13, 15, 35–47, 50, 51, 53, 54, 56, 180, 197, 329, 334, 367, 379, 381–383, 385–388, 392
Rhodamine B, 6, 466, 469
Riboflavin, 384
Root apex necrosis, 336
Roundworms, 563
Runoff, 299–301, 317, 380, 441

S

Saccharomyces cerevisiae, 77, 120, 249, 316
Salicylic acid, 384
Salix, 39, 44, 45, 50
Salmonella typhimurium, 10, 120, 234, 237
Saponin, 384, 390, 391, 535
Saxitoxin, 10, 269, 271, 272, 275, 284
Scanning transmission X-ray microscopy (STXM), 15, 355
Sclerotia, 329, 332, 333, 337
Sclerotinia, 332
Sclerotinia trifoliorum, 332
Sclerotium rolfsii, 332
Selenastrum capricornutum, 120
Semiochemical, 530, 532, 533, 537, 539, 543
Serratia liquefaciens, 352
Shooting range, 1, 17, 18, 437–447
Siderophore, 3, 40, 41, 46, 50, 51, 56, 384, 567
Single cell protein, 4, 75–77, 94
Sinorhizobium meliloti, 329, 331, 334, 347, 364, 385
Sister chromatid exchange (SCE), 230, 241, 242, 246, 247
Sodification, 380
Sodium channel modulator, 305
Soil salinization, 380
Soil washing, 17, 443
SOS chromotest, 10, 238, 239

604 Index

Southern blight, 332
Southern hybridization, 166, 181
Soybean peroxidase (SBP), 465, 483
Sphingomonas, 278, 284, 285, 363
Sphingopyxis, 285
Spirulina, 76, 262
Spodoptera littoralis, 536, 561
Sporiene, 549
Stabilization/solidification, 18, 444
Stachydrine, 384
Staphylococcus, 171, 172, 342, 350, 459
Stenotrophomonas, 171, 366
Stilbenoid, 391
Streptomyces, 14, 77, 82, 279, 281, 331
Sulphachloropyridazine, 175
Sulphite pulping, 405, 406
Surface-enhanced Raman scattering (SERS), 15, 355
Susceptibility, 20, 41, 151, 205, 297, 356, 426, 477, 572
Sustainability, 1, 2, 8, 9, 13, 49, 54, 67, 86, 193–195, 199, 206, 209, 213, 216, 306, 318, 517
Sustainable, 2, 3, 5, 8, 21, 48, 53,
Sustainable agriculture, 1–3, 48, 194, 381, 392, 529–532, 545, 574
Swarming, 352
Synthetic dye, 6, 19, 85, 111–115, 117, 119, 121–123, 126, 134, 135, 137–139, 151, 152, 154, 454, 455, 459–461, 467, 468, 470–472, 477, 478, 483
Syringol, 16, 398, 418

T

Tannin, 117, 415
Tarsi, 539
Temperature-programmed desorption (TPD), 126
Tenellin, 555
Terminal restriction fragment length polymorphism (T-RFLP), 49, 207
Tetramethyl rhodamine isothiocyanate (TRITC), 15, 354
Tetranortriterpenoid, 546; *see also* Azadirachtin
Textile industry, 6, 112, 117, 120, 148, 470
Thiamine, 384
Threadworm, 563
Thymidine kinase (tk), 250
Tomato peroxidase (TMP), 476, 480, 484
Total chlorine free (TCF), 17, 408, 419
Total organic carbon (TOC), 195, 197, 214, 457, 480

Toxicity, 10, 12, 16, 18, 44, 86, 118–120, 122, 129, 137, 148, 150, 198, 209, 212, 231–233, 238, 239, 248, 264, 265, 269, 278, 282, 284, 285, 297, 304, 313, 344, 358, 414, 418, 423, 426, 441, 445, 454–457, 474, 509, 522, 535, 539, 550, 555, 573
Tradescantia, 241, 248
Trail-marking pheromone, 542
Trametes hirsuta, 461
Trametes versicolor, 82, 461, 462, 466, 469, 482, 483
Trimethoprim, 169
Trinitrobenzene (TNB; $C_6H_3N_3O_6$), 439
Trinitrotoluene (TNT; $C_6H_2[NO_2]_3CH_3$), 120, 315, 439
Triosephosphate isomerase, 350
Tritium-labelled thymidine, 243
Turnip peroxidase (TP), 465, 474, 476, 478, 480, 484

U

Ultrafiltration, 131, 272, 275, 423
umuC-assay, 237
Unscheduled DNA synthesis, 243
Up-flow anaerobic sludge blanket (UASB), 17, 426, 427
Urease, 200

V

Vacuum drum, 409, 412, 430
Vanillin, 16, 84, 398, 418, 484
Veratryl alcohol (VA), 461, 467, 469, 481–484
Vermicomposting, 5, 86, 515
Vibrio fischeri, 235, 239
Violuric acid (VLA), 481–484
VirB11-type ATPase, 180
VirB4-like ATPase, 180, 181
VirD4-type ATPase, 180
Volatile organic compound (VOC), 519–522
Volatilization, 300, 319

W

Waste minimization, 67, 297, 501, 503, 508, 524
Waste treatment, 69, 87, 171, 501, 510, 513, 518, 520, 524, 525
Wastewater treatment, 17, 19, 112, 113, 117, 118, 121, 129, 131, 148, 170–172, 181–183, 185, 298, 312, 363, 364, 414, 424, 428, 429, 445, 455, 459, 460, 463
Weathering, 38, 165, 380, 440–442, 445
White radish peroxidase (WRP), 478, 481
White rot fungi, 23, 138, 147, 315, 316, 460

Index

X

Xanthene, 115, 154, 457, 458
Xanthophyll, 95
Xanthotoxin, 534
Xenobiotic, 12, 113, 135–138, 143, 249, 297, 315, 317, 384, 459, 461, 465, 468, 472, 521

X-ray photoelectron spectroscopy (XPS), 126
Xylene cyanol, 466, 469

Z

Zeaxanthin, 95, 96
Zoospore, 553
Zygomycotina, 553